The Biochemical Genetics of Man

Second Edition

Edited by

D. J. H. BROCK
Department of Human Genetics,
University of Edinburgh, Scotland

O. MAYO
Waite Agricultural Research Institute,
University of Adelaide, Australia

1978

ACADEMIC PRESS

LONDON NEW YORK SAN FRANCISCO
A Subsidiary of Harcourt Brace Jovanovich, Publishers

ACADEMIC PRESS INC. (LONDON) LTD.
24/28 Oval Road,
London NW1

United States Edition published by
ACADEMIC PRESS INC.
111 Fifth Avenue
New York, New York 10003

Library of Congress Catalog Card Number: 77–93210
ISBN: 0–12–134760–5

Printed in Great Britain by
Cox & Wyman Ltd, London, Fakenham and Reading

Contributors

G. BECKMAN *Department of Medical Genetics, University of Umeå, S–901 85 Umeå, Sweden*

B. BOETTCHER *Department of Biological Sciences, University of Newcastle, Newcastle, New South Wales 2308, Australia*

D. J. H. BROCK *Department of Human Genetics, University of Edinburgh, Western General Hospital, Edinburgh, EH4 2HU, Scotland*

B. CARRITT *Institute of Genetics, University of Glasgow, Church Street, Glasgow G11 5JS, Scotland*

D. W. COOPER *School of Biological Sciences, Macquarie University, North Ryde, New South Wales 2113, Australia*

K. W. E. DENSON *Maternity Unit, John Radcliffe Hospital, Headington, Oxford OX3 9DU, England*

G. FRASER *Room 12A, LCDC Building, Department of Health and Welfare, Tunney's Pasture, Ottawa, Ontario K1A 012, Canada*

W. KRONE *Department of Human Genetics University of Ulm, D–7900 Ulm, West Germany*

O. MAYO *Biometry Section, Waite Agricultural Research Institute, Glen Osmond, South Australia 5064*

W. R. MAYR *Institute of Blood Group Serology, University of Vienna, 1090 Vienna, Austria*

R. G. SUTCLIFFE *Institute of Genetics, University of Glasgow, Church Street, Glasgow G11 5JS, Scotland*

M. W. TURNER *Department of Immunology, Institute of Child Health, University of London, 30 Guildford Street, London WC1N 1EH, England*

J. M. WHITE *Department of Haematology, King's College Hospital Medical School, University of London, Denmark Hill, London SE 5, England*

R. H. WILSON *Institute of Genetics, University of Glasgow, Church Street, Glasgow G11 5JS, Scotland*

U. WOLF *Institute of Human Genetics and Anthropology, University of Freiburg, 7800 Freiburg, West Germany*

Preface

In the six years since the publication of our first edition, human biochemical genetics has become much more widely accepted as a discipline in its own right. Without claiming that our book has made more than a miniscule contribution to this desirable end, we would like to note a few of the possible reasons. First of all, there is the normal process of acceptance, with its four stages (J. B. S. Haldane, 1963, *Journal of Genetics* **58**, 464):
  (i)   this is worthless nonsense;
  (ii) this is an interesting, but perverse, point of view;
  (iii) this is true, but quite unimportant;
  (iv) I always said so.
More importantly, we may point to recent successes in cell biology and in the understanding of the immune response, to the prevention of genetical disease by antenatal diagnosis and therapeutic abortion, to the use of new techniques for the precise investigation of human chromosomal aberrations and variants, and to the success of somatic cell genetics, combined with other techniques, in mapping the human genome.

However, apart from these important but isolated practical and theoretical advances, what the last six years has seen has been a massive outpouring of new facts, with little theory accompanying them. This has been particularly true as far as the detection of patterns of variation is concerned, whether the variation is normal or deleterious. Perhaps the most formidable task of the human biochemical geneticist at the moment is the encompassing of the great mass of information on normal variation within some kind of logical framework.

In this second edition, therefore, what we have attempted to do wherever possible is, without failing to cover new developments in all their important aspects, to demonstrate or at least hint at the existence of pattern wherever we have been able to find it. To this end, we have, with our authors, virtually rewritten the entire book, without, we trust, sacrificing any of the features of the first edition which were well-received.

D. J. H. BROCK and O. MAYO
April, 1978

Contents

Genetic Basis of Variation

1 The Structure and Function of Proteins
D. J. H. BROCK

2 Genes, Proteins and the Control of Gene Expression
R. G. SUTCLIFFE, B. CARRITT and R. H. WILSON

Pathological Variation

14 Unsolved Mendelian Diseases
G. R. FRASER

Genetic Basis of Variation

1 The Structure and Function of Proteins

D. J. H. BROCK

Department of Human Genetics, University of Edinburgh, Scotland

I. INTRODUCTION

The importance of proteins in general biochemistry is indicated in their naming (Greek protos or first). If anything their importance in biochemical

genetics is even greater. Ever since Beadle (1945) formulated the relationship "one gene—one enzyme", laboratory investigators have concentrated
on proteins as the biochemical units through which genetic variation may
be monitored. The discovery by Pauling *et al.* (1949) that the haemoglobin
of patients with sickle cell anaemia is electrophoretically different to the
haemoglobin of normal controls, extended these investigations to man. In
the chapters that follow, it will become clear that with few exceptions (such
as the red cell and leucocyte antigens) progress in the study of human biochemical genetics has depended on increasingly sophisticated methods of
analysing protein structure and quantity. There are as yet disappointingly
few signs that analytical methods will become sufficiently sensitive and
sophisticated to tackle variation in DNA itself. For the forseeable future
most analyses in biochemical genetics will take place at the level of protein.

   There is a second sense in which proteins are important in molecular
genetics. Analysis of eukaryotic chromatin has shown that in many, if not
all nuclei, protein is a major chemical component. There is no doubt that
DNA is the hereditary material and that RNA is the messenger material;
what is surprising is that there is so much protein in the nucleus. Histone:
DNA weight ratios are approximately $1 \cdot 0$ over a wide range of cell types
while the non-histone protein: DNA ratios vary from $0 \cdot 1$ to $1 \cdot 0$ (Bonner
et al., 1968). The functions of nuclear protein are at present poorly understood; histones, in view of their remarkable evolutionary conservatism, are
probably structurally important for chromatin as well as having a possible
regulatory role in DNA transcription. Non-histone nuclear protein is
likely to be more heterogeneous in function, though presumably much of it
will be involved in DNA and RNA regulation. Until more precise functions
for the nuclear proteins are available, biochemical geneticists will continue
to concentrate on proteins as consequences of genetic variation rather than
on proteins as agents of gene function.

II. CLASSIFICATION OF PROTEINS

There is no really satisfactory way of classifying proteins. Physico–
chemical classifications, which use solubility, shape and chemical composition as criteria, are often used (Table 1). The main objections to this
system are (1) the degree of overlap between the groups, (2) the extraordinary heterogeneity of the globular proteins, ranging from albumins, through
most enzymes to the strongly basic, low molecular weight protamines of
sperm nuclei and (3) the fact that conjugated proteins may have properties
largely determined by the prosthetic group (as in the mucoproteins) or
almost independent of it (as in the metalloproteins).

Table 1

Physico–chemical classification of proteins[a]

Type and subtype	Main properties
Fibrous (scleroproteins) collagens elastins keratins	usually insoluble, linear or elongated in shape, resistant to proteolytic digestion, characteristic or higher animals, found mainly in ectodermal and mesodermal tissues
Globular albumins globulins histones protamines prolamins glutelins	usually soluble, spheroid or ellipsoid in shape, histones and protamines found in the nucleus, albumins in the circulation, prolamines and glutelins are plant-specific, globulins the most ubiquitous proteins
Conjugated nucleoproteins mucoproteins, glycoproteins lipoproteins metalloproteins phosphoproteins chromoproteins	properties may depend on prosthetic group; thus mucoproteins (large amounts of carbohydrate) behave like carbohydrates, whereas glycoproteins (smaller amounts of carbohydrate) behave like other globular proteins

[a] Based on Young (1963) and White *et al.* (1973)

Table 2

Proteins classified by function[a]

Type	Examples
Structural	collagen, keratin, elastin, histones (?)
Storage	casein
Defence	immunoglobulins, complement factors, blood-clotting factors
Catalysis	enzymes
Transport	albumin, haemoglobin, lipoprotein
Mechanical transduction	myosin, actin
Electrical transduction	acetylcholinesterase, rhodopsin
Information carrier	hormones, acidic nuclear proteins (?), repressor substances
Information receiver	cell receptor proteins

[a] Adapted from Knowles and Gutfreund (1974)

Classifications according to function also have disadvantages (Table 2). The functions of many proteins are not clearly understood. An example is the foeto-specific a_1-globulin, alpha foetoprotein. It has been suggested that this is an embryonic form of albumin, in which case its function is transport. It is equally possible that its function is to bind and carry oestrogens to the foetal tissues, which suggests a transport protein of a rather different kind or perhaps even an information-carrying protein. If, as current experiments suggest, alpha foetoprotein functions as an inhibitor of the immune reaction, preventing rejection of the genetically incompatible foetus by its mother, it could be classified as a defence protein (Yachnin, 1976). Similarly the plasma protein caeruloplasmin may be an iron-oxidizing enzyme, a copper transporting protein or a regulator of hepatic copper homeostasis (Poulik and Weiss, 1975). The ubiquitous and important histones may be involved in structural support of chromatin or else the regulation of gene transcription through either the receipt or transmission of molecular messages.

III. GENERAL STRUCTURE OF PROTEINS

A. Amino Acids

Hydrolysis of proteins derived from a wide variety of different species—microorganisms, plants, animals or man—yields the same set of about 23 different amino acids. Twenty of these are primary in the sense that they are coded for by the genetic code (Table 3), while others like hydroxyproline, hydroxylysine and cystine are secondary modifications. The amino acids may be divided into those which are acidic together with their amides (Asp, Asn, Glu, Gln), those which are basic (Arg, His, Lys) and those which are neutral. The latter group can be further subdivided into aliphatic (Ala, Gly, Ile, Leu, Val), aromatic (Phe, Try, Tyr), hydroxyl-containing (Ser, Thr), sulphur-containing (Cys, Met) and imino amino acids (Pro).

All amino acids have a least one carboxyl and one amino group involved in proton-exchange reactions in aqueous solution. The a-carboxyl group is characterized by a dissociation constant (pKa) in the range 2 to 3, while the a-amino group, as NH_3^+, is characterized by a pKa near 10. Thus at pH values between 4 and 9 neutral amino acids exist in dipolar or ampholytic form with the carboxyl group dissociated and the amino group protonated, i.e. $H_3^+N–RCH–COO^-$. Over this intermediate range amino acids bear little net charge, and at one pH (the isoelectric point, pI) bear no net charge. The

Table 3

The amino acids, their dissociation constants and isoelectric points[a]

Amino acid	Abbreviation	pK$_1$(COOH)	pK$_2$	pK$_3$	pI
Alanine	Ala	2·34	9·69	—	6·00
Arginine	Arg	2·17	9·04 (NH$_3^+$)	12·48 (guanidinium)	10·76
Asparagine	Asn	2·02	8·80	—	5·41
Aspartic acid	Asp	1·88	3·65(COOH)	9·60(HN$_3^+$)	2·77
Cystine	Cys	1·96	8·18(SH)	10·28(NH$_3^+$)	5·07
Glutamic acid	Glu	2·19	4·25(COOH)	9·67(NH$_3^+$)	3·22
Glutamine	Gln	2·17	9·13	—	5·65
Glycine	Gly	2·34	9·60	—	5·97
Histidine	His	1·82	6·00 (imidazolium)	9·17(NH$_3^+$)	7·59
Isoleucine	Ile	2·36	9·68	—	6·02
Leucine	Leu	2·36	9·60	—	5·98
Lysine	Lys	2·18	8·95 (α–NH$_3^+$)	10·53 (ϵ–NH$_3^+$)	9·74
Methionine	Met	2·28	9·21	—	5·74
Phenylalanine	Phe	1·83	9·13	—	5·48
Proline	Pro	1·99	10·60	—	6·30
Serine	Ser	2·21	9·15	—	5·68
Threonine	Thr	2·63	10·43	—	6·53
Tryptophan	Trp	2·38	9·39	—	5·89
Tyrosine	Tyr	2·20	9·11(NH$_3^+$)	10·07(OH)	5·06
Valine	Val	2·32	9·62	—	5·96

[a] Cohn and Edsall (1942)

isoelectric point may be calculated by averaging the two dissociation constants (Table 3). If the amino acid has another acidic group, as in Asp or Glu, the pI is given by the average of the two dissociation constants for the acid groups. If it has another basic group, as in His or Lys, the pI is given by the average of the conjugate acids of the two basic groups.

B. The Peptide Bond

The fundamental chemical bond of all proteins is the peptide bond, in which the α-carboxyl group of an amino acid residue is bound to the α-amino group of another residue, to give the repeating unit shown in Fig. 1. Since the bonding electrons of the C=O group are also distributed along the C—N bond this confers partial double bond character on the N—C—O unit. In turn this restricts free rotation around the carbon and nitrogen atoms and makes the peptide backbone a planar one. In fact the bond angles and bond distances of a peptide chain are always the same whatever the nature of the constituent amino acids. The individuality of the chain is determined by the side groups (designated R in Fig. 1) and by free rotation of the C—R bond.

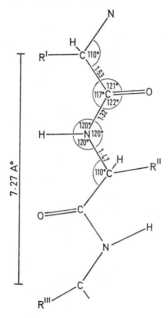

Fig. 1 The peptide bond. Fundamental dimensions as derived from X-ray crystallography

Since the peptide bond is an acidic linkage it may be hydrolysed by strong acids or strong bases. Hydrolysis with 6 mol litre^{-1} hydrochloric acid in sealed tubes for several days will quantitatively cleave the peptide bonds of most proteins, though it will also completely destroy tryptophan residues. Alkaline hydrolysis is even more destructive of the common amino acids, and only used to estimate the tryptophan content of a protein. A variety of proteolytic enzymes, such as trypsin, chymotrypsin, pepsin, papain, elastase, carboxypeptidase A and B and subtilisin will attack peptide bonds, and unlike chemical hydrolysis, may attack selectively. Thus trypsin catalyses hydrolysis of arginyl and lysyl bonds and chymotrypsin tryptophanyl, phenlalalinyl and tryrosinyl bonds. In general acid and alkaline hydrolysis is used in estimating the amino acid composition of proteins, and enzymic hydrolysis in working out amino acid sequence.

C. Primary Structure

The primary structure of a protein is defined by the sequence of amino acids within the backbone and also by the number of polypeptide chains linked by disulphide bonds. Since a polypeptide chain may contain more than 100 amino acids and since there are 20 common amino acids found in proteins the number of possible combinations of primary structures is 20^{100} or greater. Elucidation of the amino acid sequence of a protein consists of a series of steps: (1) determination of the number of independent disulphide-linked polypeptide chains, (2) cleavage of the interpeptide bonds by oxidation or reduction, (3) specific cleavage of each polypeptide chain into smaller chains of manageable size, (4) sequence determination of each of the cleavage products, (5) fitting the individual segments into a unique sequence for each of the peptide chains and (6) identifying the sites of linkage holding individual peptide chains together. This complicated process was first achieved for beef insulin (Sanger, 1956).

Amino acid sequences have now been determined for over 600 different proteins and over 200 abnormal human haemoglobins (Croft, 1976). Among human proteins where primary sequences are known are the various immunoglobulin chains, a number of polypeptide hormones, cytochromes b and c, myoglobin, fibrinopeptides and several plasma globulins. Inspection of sequences shows that the ordering of amino acids within polypeptide chains is such that virtually all possible dipeptide sequences have been found. Repeating sequences within the same molecule are rarely observed. For any given named protein the primary structure is unique both in terms of amino acid sequence and the number of inter-chain and intra-chain disulphide bonds. Differences do occur between the amino acid sequences of proteins from different species, from different tissues of the same species and as a consequence of mutational events. The subtle species variation in the primary structures of insulin is shown in Table 4.

Table 4

Species variation in the primary structures of insulin

Source	A-chain			B-chain
Position	8	9	10	30
Man	Thr	Ser	Ile	Thr
Dog	Thr	Ser	Ile	Ala
Rabbit	Thr	Ser	Ile	Ser
Cattle	Ala	Ser	Val	Ala
Pig	Thr	Ser	Ile	Ala
Sheep	Ala	Gly	Val	Ala
Horse	Thr	Gly	Ile	Ala

Though demonstration of differences in the primary structures of closely related proteins is best achieved through amino acid sequencing, a short-cut method known as finger-printing has often been used. This involves partial digestion of the protein by acid hydrolysis or proteolytic enzymes followed by determination of the product peptides by electrophoresis or chromatography. Superimposing the peptide maps on one another may reveal a single difference attributable to a particular sequence of the protein. If the primary structure of the normal protein is known it may be possible in this way to identify the position of an amino acid substitution in a mutant analogue without recourse to sequencing. Many of the sites of the amino acid changes in the mutant human haemoglobins have been characterized in this way.

D. Protein Conformation

In theory a long polypeptide chain could be looped, coiled or twisted into a large number of different shapes or conformations, the only restriction being the requirement of planarity for the peptide bond. In practice the number of shapes is limited due to the operation of other types of bonding force within and between molecules. These forces include covalent bonding, ionic bonding, hydrogen bonding, electrostatic attraction and repulsion and van der Waal's attraction and repulsion. Ordering of polypeptide chains resulting from the formation of hydrogen bonds between the carbonyl oxygen and the amide nitrogen determines the secondary structure. Further ordering of the secondary structure due to interactions between the side chains of the various amino acid residues—and involving hydrogen bonding, van der Waal's interactions and charge transfer forces—determines the tertiary structure.

The importance of hydrogen bonding in controlling protein conformation was first recognized by Pauling and Corey (1951). They pointed out that bonding between amide–NH and carbonyl C=O could be most easily achieved in two different ways; intra-chain binding in coil-type structures and inter-chain binding in sheet-like structures. For the intra-chain hydrogen bonded structure the most stable conformation was the α-helix in which the hydrogen bonds form rings of atoms as shown in Fig. 2. There are 3·6 amino acids per turn and 1·5 Å along the axis of the helix between one residue and the next. Thus the α-helix may be regarded as a spiral staircase in which the amino acid residues form the steps, with the height of the steps being 1·5 Å and the height of each complete turn 5·4 Å, making 3·6 steps per turn. For the inter-chain hydrogen bonded structure two types of sheet fill the spatial requirements, in one of which the chains are parallel and in the other anti-parallel. Because the size of side-chain groups may make it difficult to pack fully extended polypeptide chains side-by-side, both parallel and anti-parallel sheets are 'pleated', meaning that the side chain groups project above and below the plane of the sheet.

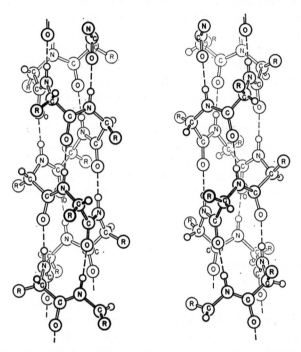

Fig. 2   The α-helix. The drawing on the left shows a left-handed helix and on the right a right-handed helix. Carbon (C), nitrogen (N) and oxygen (O) atoms are marked while hydrogen atoms are unmarked. R indicates side-groups

Proteins of sheet structure are not particularly common. They are generally fibrous and insoluble, such as silk fibroin, swan quill and kangaroo tail tendon. The α-helical structure occurs more frequently; again it tends to belong to insoluble proteins which may be reversibly stretched when wet, like keratin of hair, epidermin of skin and myosin of muscle. But most proteins in their native state exist as tight compact structures and can only be properly described in terms of tertiary structure. As has already been pointed out this involves a number of different types of bonding force, some or all of which may be acting at different points of the molecule. A proper description of the tertiary structure of a protein is only possible after the solution of its conformation by X-ray diffraction analysis of the purified crystalline material. This was first achieved for sperm whale myoglobin (Kendrew *et al.*, 1960), and has now been extended to a number of other proteins, including some of human origin like haemoglobin.

The question arises as to whether secondary and tertiary structures are entirely determined by the primary structure or whether a protein can assume that conformation which is thermodynamically most stable under the existing conditions. Evidence from experiments in which mercaptoethanol and urea have been used to break the intramolecular disulphide bonds of compact single chain globular proteins such as ribonuclease and lysozyme shows that the loss of native structure can be completely restored by reversing the experimental conditions (Sela *et al.*, 1957). If reforming of disulphide bonds were completely random, only a small proportion of the protein would have been restored to its native conformation. Similar experiments using guanidinium chloride to destroy the conformation of antibody fragments show that reversal of conditions will lead to a refolded protein with full ability to recognize antigen (Buckley *et al.*, 1963). It thus seems that the amino acid sequence or primary structure of a protein is the exclusive factor in determining its conformation (Wetlaufer and Ristow, 1973).

E. Quaternary Structure

It has been apparent for a number of years that describing proteins in terms of primary, secondary and tertiary structure is often inadequate and that an aggregation of several folded or coiled units may be necessary to confer biological activity on the assembly. The separate units need not be similar. The association of units within the final aggregate determines the so-called quaternary (or subunit) structure. Forces holding the subunits together are usually non-covalent and the ease of subunit disruption (often in strongly ionized media) has contributed to the growing knowledge of the many proteins with quaternary structure. It has often been suggested that any protein with a molecular weight greater than 60 000 daltons must consist of subunits (Frieden, 1971), though Hopkinson *et al.* (1976) have surveyed a series of human enzymes and shown that the molecular weight of subunits ranges from 13 000 to 116 000 daltons.

Aggregate structures are a feature of both fibrous and globular proteins. Fibrous proteins, by their nature, tend to form highly extended three-dimensional arrays, often without definite subunit number. In α-keratin, the basic α-helical units associate by longitudinal twining to give a coiled coil which is then embedded in matrix protein. In silk fibroin the basic pleated sheets associate through van der Waal's forces to form three-dimensional stacked arrays. In collagen three polypeptide chains are woven together in a left-handed sense to accommodate the awkard amino acid proline, and then further strengthened by inter-molecular ester cross-links.

Amongst the globular proteins subunits may be identical or different. The classical example is haemoglobin. The quaternary structure of haemoglobin is a tetramer with two α- and two β-chains, coded for by different genetic loci. Haemoglobins from a variety of different species and indeed the different normal haemoglobins (A, A_2 and F) of man all have the same tetrameric structure. Among the large number of variant human haemoglobins involving mutations in the α- and β-chain controlling loci, none has been found in which the different chains cannot aggregate. The tetrameric quaternary structure of haemoglobin is thus important for its biological function, but unlike the secondary and tertiary structures, comparatively independent of primary amino acid sequence. Mutations involving amino acids at the sensitive contact points between α- and β-chains may disturb the stability of haemoglobin, but will not destroy the structure (Chapter 11).

F. Isozymes

Until about 30 years ago it was generally accepted that proteins, like intermediary metabolites, had the same chemical structure whatever the species or organ of origin. It is now accepted that many different enzymes, defined according to the reaction they catalyse, exist in multiple forms even within the same compartment of the same tissue of the same species. Nomenclature in this area is confusing. The Standing Committee of Enzymes of the International Union of Biochemistry defines proteins with different properties but similar catalytic activity as isozymes or isoenzymes. The degree of similarity of catalytic action may be open to different interpretation. It is generally agreed that the acid and alkaline phosphatases are separate enzymes and not isozymes, as are acetylcholinesterase and pseudocholinesterase. The esterases, however (Chapter 6), are more difficult to define since they have broad and overlapping substrate specificities when tested *in vitro*. Some workers distinguish between isozymes and heterozymes (Wieland and Pfleiderer, 1962), with the former term being reserved for different forms of an enzyme within the same tissue of a single species and the latter for similar enzymes in different tissues or different species. When genetic variation within a species occurs a further complexity arises, for the enzyme products of allelic genes are usually discovered by electrophoretic techniques used in searching for isozymes and conform to the

Standing Committee definition of isozymes. Enzyme variant is a clearer term though it must be recognized that many variants will themselves consist of isozymes.

1. The origin of isozymes

Multiple loci. If an enzyme or protein is coded for by more than one genetic locus it will consist of at least two different types of polypeptide chain. This heterogeneity may not be apparent on electrophoresis, if as in the case of haemoglobin, one particular association of subunits is inherently more stable than other permutations. But in other situations a variety of other combinations of chains are both possible and observable. The prototype for this kind of isozyme pattern is lactate dehydrogenase (LDH), discussed in Chapter 6.

Genetic variation is obviously important in resolving the number of loci controlling a set of isozymes. Tissue distributions can also give useful information. Thus amongst the isozymes of aldolase the A form occurs in most tissues and is unique to muscle, the B form is predominant in liver and kidney and the C form in brain and heart (Penhoet *et al.*, 1966). With pyruvate kinase one set of isozymes occurs in red cells and another set in white blood cells. Both sets occur in liver and kidney (Bigley *et al.*, 1968). With malate dehydrogenase, malic enzyme, isocitrate dehydrogenase, glutamate-oxalate transaminase and superoxide dismutase different isozymes sets are expressed in the cytosol and mitochondrial compartments of the same tissues (Chapter 6).

Information on enzyme substructure may also be obtained from interspecific somatic cell hybridization or by hybrids generated by *in vitro* mixing of enzyme extracts from different tissues or species. Provided that the enzymes from the different species have distinctive electrophoretic mobilities, hybrid isozyme formation is usually observed in the fused or mixed cell products (Harris, 1975). If the enzyme is a monomer no hybrid formation occurs. If it is a dimer such as alkaline phosphatase, or a trimer such as nucleoside phosphorylase, three and four bands respectively are found in the hybrids (Figs 3A and B). This of course assumes that the dimers and trimers from each species contain identical subunits. When the enzyme is under multi-locus control the hybrid patterns become more complex depending on the specificity of the interactions. A dimeric structure in which each of the isozymes must contain both polypeptide chains generates a four-banded pattern (Fig. 3C); if this restriction is dropped the number increases to 10 (Fig. 3D). In the human–mouse hybrid of LDH, where each parental enzyme is a tetramer under the control of two separate genes, up to 35 electrophoretic bands may be seen. At this level of complexity electrophoretic resolution begins to fail.

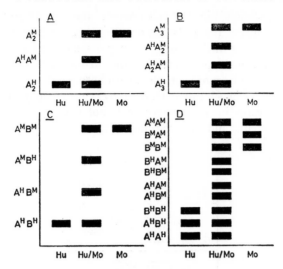

Fig. 3 Isozyme patterns in human–mouse (Hu/Mo) hybrids compared to those in parental cells. In A and B, the parental enzymes are dimers and trimers, respectively, but with identical subunits. In C and D the parental enzymes are dimers with non-identical subunits. D differs from C in that there is now no restriction on the permutations of polypeptide within the dimer

Other techniques for resolving gene control of enzymes include electrophoresis, gel-filtration and ultracentrifugation under strongly dissociating conditions. These procedures allow individual polypeptide chains to be recognized but require a start from highly purified material, a disadvantage not shared by conventional electrophoretic techniques. Hopkinson *et al.* (1976) have gathered data by these methods on 66 human enzymes and found that in 24 of them there is good evidence for more than one locus. The similarities between the properties of the isozymes coded for by the different loci controlling an enzyme are sufficient to suggest that in most cases multiple loci are the consequence of gene duplications in the course of evolution. The only exception found by Hopkinson *et al.* (1976) was for superoxide dismutase, where the cytoplasmic form of the enzyme is dimeric and the mitochondrial form tetrameric.

Multiple alleles at a single locus. Much of biochemical genetics is concerned with the effects of variation at a given locus on the gene products of that locus. Unravelling the possible isozyme patterns which may result when an individual is heterozygous or homozygous for a variant allele is thus important in understanding the complexities of protein structure. A great deal of useful and illuminating information has been derived from the study of the polymorphic red cell enzyme and serum proteins. Harris (1975) has been instructive in laying out these basic principles. Thus if the enzyme is a monomer the heterozygote may display two bands corresponding to the homozygous types. If it is a dimer, trimer or tetramer the heterozygous patterns will be more complex, with both parent-type isozymes and hybrid structures (Fig. 4). But as Harris (1975) has pointed out, the absence of hybrid structures in a heterozygote is not necessarily evidence for a monomeric protein, and the various tetrameric mutant haemoglobins

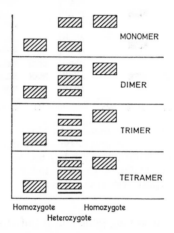

Fig. 4   Diagram of characteristic isozyme patterns expected in heterozygotes when enzymes are monomers, dimers, trimers or tetramers. Harris (1975) with permission

which give monomer-type electrophoretic patterns are testimony to this point. The mutant polypeptide chain may have structural features limiting its combination with wild type chain into a hybrid molecule, so that only self-association is possible. Polypeptides may aggregate immediately on release from the cellular protein synthesizing machinery and then be resistant to dissociation and reassociation with the products of the other allele. This was originally thought to be the mechanism whereby heterozygotes for haemoglobins A and S show only two electrophoretic bands and never the hybrid third band corresponding to $\alpha_2\beta\beta^s$. Though it is now known that the absence of haemoglobin hybrids on electrophoretograms is due to rapid dissociation into dimers and removal of more stable tetramers during the separation process, it is possible that resistance of homopolymeric units to dissociation may in other situations restrict hybrid formation. Another special situation, unique to products of X-linked genes, follows from the X-inactivation hypothesis (p. 428). Heterozygotes for G6PD variants show no hybrid formation, even though these structures may be formed *in vitro*, presumably because each isozyme is synthesized in a different cell.

The heterozygote patterns of polymeric enzymes (Fig. 4) often show characteristic symmetries, with the middle bands staining more intensely than the outer bands. For a dimer the middle band may contribute 50% of the total activity and the outer bands 25% each. Asymmetric patterns in which one outer band is much weaker than the other may be taken as evidence that one allele contributes significantly less enzyme activity than the other (Harris, 1975). Thus heterozygotes for the dimeric enzyme placental alkaline phosphatase, who have the phenotypes 3–1 and 3–2, have asymmetric electrophoretic patterns with weak bands corresponding to the homozygous 3 phenotype. This suggests that the allele PL^3 contributes less to the hybrid than the alleles PL^1 and PL^2, a conclusion borne out of the results of quantitative assays of enzyme from different phenotypes (Beckman, 1970).

Aggregation–disaggregation. The self-association of polypeptide chains into a series of aggregates of regularly increasing molecular weight has been described in a number of systems (Frieden, 1971). One of the earliest prototypes was beef liver glutamate dehydrogenase, an enzyme which undergoes a concentration-dependent polymerization influenced by the presence of coenzymes. As isolated, glutamate dehydrogenase has a molecular weight of about $1 \cdot 3 \times 10^6$ by sedimentation velocity and diffusion. The coenzyme NADPH (or simple dilution) promotes the dissociation of the polymer into subunits of molecular weight about 350 000. This disaggregation of a presumed tetramer into monomers is fully reversible. Since monomers have much greater enzymatic activity than polymers and since concentration of the enzyme is very high within the liver cell, it would seem that the polymerization reaction is important in regulating the kinetic characteristics of the enzyme *in vivo*. If the 350 000 molecular weight subunits are treated with detergent or urea a further and irreversible dissociation into subunits of molecular weight 40 000 occurs. These probably represent the basic polypeptide chains of the enzyme but are catalytically inert. Thus in this case an aggregation of the primitive subunit into a polymer of eight chains is necessary for the establishment of enzymatic activity, and then further concentration-dependent polymerization is conferred on the system to provide fine metabolic tuning.

Another system in which aggregation is of importance in controlling enzyme activity is glycogen phosphorylase, best studied in rabbit skeletal muscle. Two forms are known, in one of which, phosphorylase a, a reactive serine residue has been phosphorylated by a separate protein kinase called

phosphorylase kinase (Fig. 5). This phosphoserine group may be hydrolysed by a specific phosphatase to give phosphorylase b. Both the "a" and
"b" forms are based on aggregates of a basic polypeptide chain of molecular
weight 90 000 which has no enzymatic activity. The enzymatic activity of the
phosphorylases is mediated through AMP. With the "b" form AMP promotes association from an active dimer to a less active tetramer, an aggregation which is inhibited by the preferential binding of the substrate glycogen.

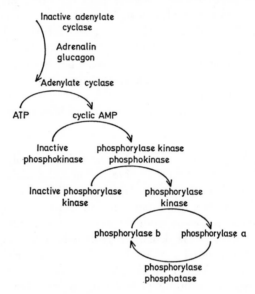

Fig. 5   Factors influencing phosphorylase a and b interconversions

With the "a" form AMP enhances the rate of the glycogen-induced
dissociation from a tetramer to a dimer. Activity is therefore uniquely
sensitive both to the levels of glycogen and of AMP, as well as to the
factors involved in the regulation of the conversion of the "a" and "b"
forms (Fig. 5).

A well-known example of protein polymerization is given by the haptoglobin phenotypes (Chapter 6). In Hp 1-1 types a single electrophoretic
band is seen, whereas Hp 2-1 and Hp 2-2 phenotypes have a series of
electrophoretic bands. The characteristic spacing of the bands suggests a
polymeric series. This has now been demonstrated by Fuller *et al.* (1973),
who isolated six discrete Hp 2-2 polymers by polyacrylamide gel electrophoresis. Quantitative amino terminal analysis and amino acid analysis
showed that each polymer consisted of α^2 and β polypeptide chains in a 1:1
ratio. Molecular weight determinations by SDS electrophoresis and ultra-

centrifugal analysis showed that each polymer differed from the next member in the series by an average increment of 54 000. This is the approximate size of a subunit consisting of an α^2 chain (17 300) and a β chain (40 000). Differential reduction with mercaptoethanol proved that the α^2 and β subunits are joined together through disulphide bonds to form the polymers. The six polymers isolated conformed to the series $\alpha_3^2\beta_3$, $\alpha_4^2\beta_4$, $\alpha_5^2\beta_5$ ... and ranged in theoretical molecular weights from 171 000 to 465 000. Thus the molecular chain formula for Hp 2-2 is $(\alpha^2\beta)_n$ where n is a series; and that for Hp 2-1 would include hybrid polymers of the type $(\alpha^1\beta)_m$ and $(\alpha^2\beta)_n$.

Secondary modifications. Secondary modifications or post-translational changes refer to the situation where the primary sequence of the protein or enzyme is modified after polypeptide synthesis. There are a number of ways in which this can happen. Acetylation of free amino groups, oxidation of sulphydryl or hydroxyl groups, the addition of phosphate groups, the removal of neuraminic acid residues and the removal of amide groups by non-specific chemical reactions could all change the electrophoretic mobility of the resulting protein. Many examples are known where storage of proteins in the frozen state without the addition of reducing agents such as mercaptoethanol or dithiothreitol will allow oxidation of a reactive cysteine residue and change both the activity and mobility of the protein. Many of the complex isozyme patterns described in Chapter 5 are thought to be a consequence of such secondary modifications.

2. The frequency of subunits

It would appear that the majority of proteins are composed of subunits (Frieden, 1971). Klotz et al. (1970) listed about 100 enzymes in which there was evidence that the native active unit was composed of several polypeptide chains. SDS gel electrophoresis, gel filtration and ultracentrifugal studies have been useful in gaining information on both subunit number and subunit molecular weight. Amongst human enzymes the most complete data come from Hopkinson et al. (1976). They examined data from the literature on the enzyme products of 100 different gene loci and came to the conclusion that 72 were polymeric (i.e. consisted of subunits) and 28 monomeric (i.e. consisted of a single polypeptide chain). Polymeric enzymes were evenly distributed across the entire range of catalytic activities: from oxidoreductases through transferases, hydrolases, lyases, isomerases to ligases. Dimers (43% of the total) were the most common type of subunit composition, though tetramers (24%) were also well-represented. Only 4% of the total were trimers and only one enzyme was found which had more than four subunits in its basic composition.

G. Multi-enzyme Complexes

Multi-enzyme complexes are systems in which proteins having different enzymatic activities are bound together in tight association. The best known examples are the fatty acid synthetases and the α-ketoglutarate dehydrogenases. In man the latter group are represented by the enzyme pyruvate dehydrogenase, which controls the intersection of the Embden-Meyerhof pathway and the citric acid cycle. The complex can be purified and separated into three components which may then be reconstituted. The three enzymes comprising the complex are pyruvate decarboxylase, dihydrolipoyl transacetylase and dihydrolipoyl dehydrogenase, assembled in a ratio of 12:1:6 (Linn et al., 1972). The presumed advantage of such systems is that different enzymes clearly obtain benefit of proximity with respect to the products of other sequentially related enzymes. If metabolic intermediates are unstable they may be stabilized by continuous association with a protein milieu.

One apparent multi-enzyme complex implicated in a human inborn error of metabolism is the α-ketodecarboxylase system whose deficiency causes maple syrup urine disease. There is evidence to suggest that three enzymes with different specificities for substrates are associated in a weak interactive way to produce the decarboxylase complex. A mutation in any one of these three specificities could then lead to a disruption of the entire complex and a loss of ability to metabolize valine, leucine and isoleucine, characteristic of maple syrup urine disease (Dancis and Levitz, 1972). A similar situation occurs with the pyruvate dehydrogenase complex where a mutation in any one of the three subunits can destroy cellular ability to convert pyruvate to acetyl-CoA (Farrell et al., 1975).

IV. GENERAL PROPERTIES OF PROTEINS

A. Proteins as Ampholytes

Like amino acids proteins have both acidic and basic behaviour. Their acidic properties are largely determined by the strongly ionized γ and β-carboxylic groups of glutamic and aspartic acids, respectively, and to a lesser extent by the weekly ionized sulphydryl group of cysteine and the phenolic hydroxyl group of tyrosine. Basic properties of proteins are conferred by the ϵ-amino group of lysine, the guanidino group of arginine and the imidazolium group of histidine. Characteristic dissociation constants for the various groups in a protein milieu are shown in Table 5. Since most

Table 5

Characteristic pK_a values for acidic and basic groups in proteins[a]

Group	pK_a at 25°C
α-Carboxyl (terminal)	3·0–3·2
β-Carboxyl (aspartic)	3·0–4·7
γ-Carboxyl (glutamic)	c. 4·4
Imidazolium (histidine)	5·6–7·0
α-Amino (terminal)	7·6–8·4
ε-Amino (lysine)	9·4–10·6
Guanidinium (arginine)	11·6–12·6
Phenolic hydroxyl (tyrosine)	9·8–10·4
Sulphydryl (cysteine)	c. 8–9

[a] Cohn and Edsall (1942)

proteins contain nearly all the ionic groups listed they are effectively buffered over most of the pH range. However, the only amino acid residue that has significant buffering capacity at pH values near neutrality is the imidazolium group of histidine.

As electrolytes, proteins migrate in an electric field, with the direction of migration being determined by the net charge of the molecule. The net charge depends on the relative proportion of acidic and basic residues in the protein and the pH of the medium. For each protein (as for each amino acid) there is a pH value at which it will not move in an electric field, the isoelectric point (pI). At pH values below the isoelectric point, the protein has a net positive charge and moves towards the cathode; at pH values above the isoelectric point it has a net negative charge and moves towards the anode. Isoelectric points for given proteins are constants and can aid in their characterization. As shown in Table 6 many proteins have two iso-

Table 6

Isoelectric points of some proteins[a]

Protein	pI	Protein	pI
Pepsin	c.1	fibrinogen	5·5, 5·8
Urinary gonadotropin	3·2, 3·3	prolactin	5·73
Keratin	3·7, 5·0	IgG₁	5·8
Albumin	4·7, 4·9	collagen	6·6, 6·8
β-Lactoglobulin	5·1, 5·3	myoglobin	6·99
Myosin	5·2, 5·5	haemoglobin	7·07
Insulin	5·35	IgG₂	7·3
β₁-Lipoprotein	5·4	cytochrome c	9·8, 10·1
α₁-Lipoprotein	5·5	lysozyme	11·0, 11·2

[a] Mahler and Cordes (1966)

electric points, indicating that more than one molecular species is present. This "microheterogeneity" is usually different to the heterogeneity responsible for isozyme patterns on electrophoresis and is often only discernible through the technique of isoelectric focusing. Isolectric focusing is dependent on the construction of smooth and stable pH gradients, a problem made much easier recently by the introduction of synthetic ampholytes covering the entire pH range. The use of acrylamide gels as supporting matrices together with synthetic ampholytes to produce a smooth continuous pH range gives isoelectric focusing enormous resolution. In fact the resolution has been so great, and the resulting patterns for many proteins so complex, that many workers have preferred electrophoresis in screening crude extracts for possible polymorphisms. However, where isoelectric focusing may make enormous contributions in the future is in the area of antibody fine-structure diversity (Williamson, 1975).

B. Molecular Weights

The molecular weights of proteins range from a lower limit of 5000 (below which they are called polypeptides) up to several millions. A variety of methods for determining molecular weight are in use and include electron microscopy, determination of osmotic pressure, sedimentation velocity and sedimentation equilibrium, gel filtration and polyacrylamide gel electrophoresis. None of these methods is more than approximate, and molecular weights determined in this way must usually be corrected when the amino acid sequence of the protein is worked out and an accurate molecular weight calculated. The value of a protein molecular weight to the biochemical geneticist is two-fold; it gives information on the size of thee structural gene controlling the protein and it can also in certain instances suggest a protein controlled by more than one genetic locus.

Hopkinson et al. (1976) have analysed the subunit molecular weights of a series of enzymes which have been studied by electrophoretic techniques. They found a continuous distribution of subunit sizes from 13 000 to 160 000 (Fig. 6) with a mean of 45 800. Further analysis of subunit molecular weight of proteins known to be monomers, dimers, or higher polymers showed that there was no significant difference between the mean subunit size of these different groupings. If the enzymes were sub-divided according to functional type (e.g. oxidoreductases, transferases, etc.) there was again no difference in the mean subunit size. When the protein was controlled by more than one locus there was a remarkable correlation between the subunit size of the product of the first locus when compared to that of the second locus. Though these data are restricted to soluble enzymes of sufficient stability to undergo electrophoresis, and cannot therefore be regarded as representative of all proteins, it does suggest that a large number of enzymes have molecular weights between 20 000 and 70 000, which in turn suggests that the size of the structural genes lie between 600 and 2100 base pairs.

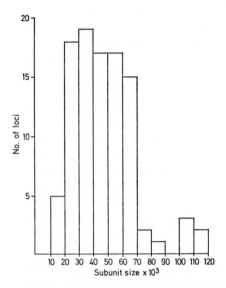

Fig. 6   Distribution of subunit sizes among the products of 99 different loci coding for enzyme structure. Hopkinson *et al.* (1976), with permission

C. Proteins as Antigens

The increasing use of immunological tests for mutant proteins has began to focus attention on the properties of proteins as antigens. Two facets of this must be considered: the capacity of the protein to provoke an immune response and to produce antibodies in the experimental animal, usually termed the immunogenicity, and the specificity of the resulting antibody combining sites, the antigenic specificity. Both are important in considering the usefulness of immunological methods in probing for the nature of a particular mutation. However, since virtually all polypeptides with molecular weights greater than 5000 are immunogenic there seems little reason why antibodies against most of the normal proteins of interest in human genetics should not be raised. What concerns us more is whether such antibodies will be capable of recognizing all, some, or hardly any of the altered protein structures that they will be used to probe for.

Antigenic specificity depends on the number and topology of the antigenic determinants. A determinant is usually a site on a protein which reacts in one to one ratio with the corresponding specific antibody population. This does not mean that the determinant is an absolutely unique structure or a single amino acid residue or that the corresponding antibody population is of unique amino acid sequence, though in some cases one or both of these statements may be true. It does mean that only one molecule of antibody can bind to the determinant at one time. In turn this means that a single determinant in a monomeric protein reacting with a truly monospecific serum cannot be detected by precipitin reactions, which require multivalent antigens (Reichlin, 1975). This is worth bearing in mind since many of the conventionally used immunological tests for proteins, such as Ouchterlony, Mancini and immunoelectrophoresis, depend on the existence of precipitating antibodies, i.e. antibodies which on forming complexes with their specific antigens will precipate from solution. Where the assumption is that precipating antibodies will not be formed, quantitative complement fixation tests are probably better employed.

There seems little doubt that antigen specificity resides in the three-dimensional structure of the protein, and that primary, secondary, tertiary and quaternary structures are all involved. Antibodies to globular proteins cross-react very poorly with the corresponding denatured protein, while immunization of animals with denatured proteins leads to antisera which have poor specificity for the native protein (Celada and Strom, 1972). Furthermore it is the surface residues which comprise most of the antigenic determinants in proteins, a conclusion which has been reached from studies on the species specific antisera for the haemoglobins and cytochrome c (Reichlin, 1975).

Rabbit antiserum to haemoglobin A reacts with both isolated α- and β-chains (Reichlin, 1970). A systematic attempt has been made by Reichlin to map the antigenic determinants in human haemoglobin by comparing single amino acid mutants and other defined variants with normal HbA using rabbit antiserum and a complement fixation test (Reichlin, 1972). He suggests that haemoglobin can be divided into a "variable" and "conservative" region, with the distinguishable mutants being located in the variable regions and the indistinguishable mutants being localized in the conservative regions (Fig. 7). However, there are positions in the variable regions that are not involved in antigenic determinants, and these presumably demarcate what are likely to be molecular limits of the determinant. For example, in the β-chain variable region 1 → 21 the distinguishable mutants

are S (β6 Glu→Val) and C (β6 Glu→Lys) at the amino terminus end.
Positions 9, 12, 22, 50, 86 and 87 in the β-chain are not involved in an
antigenic determinant since the haemoglobins Lepore (which differ from A
at these positions) are indistinguishable from A with rabbit anti-A-serum.
As pointed out earlier all the mutants that are thought to be involved in
antigenic determinants are surface residues. Furthermore, the five putative
sites (Fig. 7) are all closely associated with α-helical regions of haemoglobin.

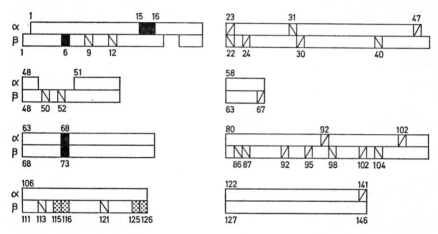

Fig. 7 Mutant haemoglobins antigenically distinguishable (filled box) and indistinguishable (diagonal box) from HbA. Residue numbers in α- and β-chains respectively are shown. Dots within a box mean that one or more of the indicated
sequence positions is involved in the same determinant. Variable regions on the left,
conservative on the right. Reichlin (1975) with permission

Direct binding measurements by fluorescence quenching with rabbit anti-
A-sera yielded a stoichiometry of 6 mol Fab per $\alpha\beta$ unit (Noble *et al.*,
1969), which is in satisfactory agreement with the five determinants
localized by mutant analysis.

When rabbits are immunized with mutant haemoglobins they produce
antisera directed against the mutant. Quantitative complement fixation tests
suggest that these anti-mutant sera are more sensitive than anti-normal
sera. This is thought to follow from the larger sequence difference between
the mutant and rabbit haemoglobin compared to the number of sequence
differences between HbA and rabbit haemoglobin. If, however, the mutant
haemoglobin is one in which the change occurred in the conservative
region the resulting antiserum normally fails to discriminate the mutant
from HbA. Thus antisera to two internal substitutions (β 67 Val→Glu and
β 24 Gly→Arg) were immunogenically silent (Reichlin, 1975).

The extensive studies on antigenic specificity of the mutant haemoglobins by Reichlin (1972, 1974, 1975) is of value in considering the more general usefulness of immunological tests. Most proteins will be multivalent in their reaction with antibodies. Thus a point mutation is unlikely to destroy the antigenicity of the protein, though it may reduce it. One would therefore expect that immunological tests for protein would be more powerful than biological ones, where a point mutation might easily abolish activity. Provided that the technology of the antibody–antigen test is appropriately carried out (which means in practice that workers do not rely on precipitin reactions but are prepared to test by complement fixation as well) one would expect most mutations to respond to normal antibody. This point is further considered in section V.

V. EFFECT OF MUTATION

A. Protein Structure

1. Amino acid substitutions

Knowledge of the genetic code together with advances in methods of working out the primary structures of protein molecules has meant that the precise nature of the amino acid sequence change in a mutant protein can often be identified. Most of this information comes from the study of the mutant haemoglobins, since starting tissue is abundant and also because the primary sequences of the α, β, γ and δ chains are well established. Some mutant haemoglobins may now be sequenced with little more effort that it takes to separate the α- and β-chains by column chromatography, to carry out a partial hydrolysis of one or both of the chains and to examine the resulting mixture of peptides against a peptide map of wild type haemoglobin. In many situations both the nature and position of the amino acid change may be identified in this way (see Chapter 11).

The majority of the mutant haemoglobins arise from a change in a single base of the triplet codon, thus giving a new codon and a substituted amino acid. These point mutations (replacement mutations) can be divided into two groups according to whether there has been a major or minor change in the structure of the affected base. Purine to purine and pyrimidine to pyrimidine changes are known as transitions, while purine to pyrimidine and pyrimidine to purine changes are known as transversions. On theoretical grounds one might expect twice as many transversions to be found as transitions. However, analysis of the β-chain variants of haemoglobin suggest that approximately as many transitions as transversions have been found. Presumably this reflects the more difficult chemical task of altering a pyrimidine to a purine and vice versa than modifying one purine to give another or one pyrimidine to give another. The mutation in haemoglobin C, involving a change from glutamic to lysine in position 6 of the β-chain, is a transition since there has been a change from guanine to adenine in the first member of the triplet codon. The mutation giving rise to haemoglobin S, on the other hand, is a transversion involving a change from adenine to thymine in the second member of the triplet codon for the same position 6 amino acid.

It is worth considering how many point mutations are possible in a protein of given size. Consider a polypeptide chain consisting of 100 amino acids. It will be coded for by 300 base pairs (ignoring initiator and terminator codons). This means that it is possible to generate 900 separate mutants, assuming that each base may be replaced by any one of the other three common bases. Of the 900 new mutant genes approximately 23% will give rise to no change in the resulting amino acids because of degeneracy of the genetic code. Another 4% will be chain terminator codons and would presumably not lead to any recognizable protein product. This leaves about 73% or 660 new alleles which should give rise to a distinctive new protein product. When one considers that few polypeptide chains contain as few as 100 amino acids, it is possible to see that very large numbers of mutant possibilities exist for even the simple one polypeptide chain proteins. The fact that multiple allelism, a feature of haemoglobin and G6DP biochemistry, is still considered unusual rather than the rule is probably a product of the inadequate methods currently available for detecting protein variants. Thus electrophoresis, the most widely used tool, will usually not detect a protein variant unless there has been a change in charge in at least one of the amino acids. This effectively eliminates about two-thirds of all possible amino acids substitutions. Thus it is of some interest that of the 287 possible point mutations of the β-chain of haemoglobin leading to a charge change (Harris 1975), between 20 and 30% have already been detected. With G6PD, a much larger protein, the 100 known variants probably represent considerably less than 10% of the theoretical total.

2. Deletion mutations

A relatively common mutational event is the deletion of part of a chromosome region with subsequent transmission of the abnormality to subsequent generations. Such mutations may be the consequence of two or more simultaneous chromosome or chromatid breaks with aberrant reunions, or they may arise because of unequal crossing-over. Frequently several genes are involved but sometimes only a sequence of bases within a gene is lost.

Because of the triplet nature of the genetic code, most deletions will result in "frameshifts" in which all the codons beyond the mutation are altered. Since this also means that all the amino acids beyond the mutation are altered it is unlikely that in most cases stable, functional polypeptides will be formed. However, losses of three (or multiples of three) bases results in deletions of one or more amino acids from sequences otherwise the same as those in the original polypeptide chain.

Unlikely as a "clean" deletion may seem (i.e. removal of some multiple of three bases so that there is no frameshift), several haemoglobin variants have been identified which are best explained in terms of such an event (Chapter 11). The first of these was haemoglobin Freiburg, in which a single amino acid residue has been deleted from position 23 of the β-chain. In haemoglobins Leiden and Tours single amino acids have also been deleted from the β-chain. In haemoglobins Gun Hill, Tochigi and St. Antoine, multiples of amino acids have been removed from the middle of the polypeptide chain leaving the remainder of it intact. Precisely how such deletions occur without destroying the polypeptide chain, remains mysterious. However, it must be assumed that these few recognized examples of clean deletions represent a tiny fraction of deletion mutations, the majority of which will not lead to a viable or identifiable product.

3. Rearrangement mutations

The normal process of crossing-over between chromatids of homologous chromosomes during meiosis is an exact process presumably dependent on gene for gene matching of whole lengths of synapsing chromosomes. In regions where DNA sequences have been duplicated, misalignment or mispairing at synapsis is more likely to happen, giving rise to rearrangements of DNA sequences and consequent alterations in the structure of the polypeptides controlled by the genes involved. This phenomenon is usually referred to as "unequal crossing-over". It was first described for a human protein in the Lepore haemoglobins, in which a misplaced synapsis between the β-chain controlling gene on one chromosome and the δ-chain controlling gene on another chromosome has lead to a fusion (δβ) product, containing the amino terminal portion of the δ-chain and the carboxyl terminal portion of the β-chain (Chapter 11).

A number of other fusion polypeptide chains, products of presumed unequal crossing-over events, have now been described. They include the Lepore haemogolobins (δβ-chain fusion), the anti-Lepore haemoglobins (βδ-chain fusion), haemoglobin Kenya (γβ-chain fusion), haemoglobins Cranston and Tak (β–β-chain fusions) and the product of the Hp^2 allele (Chapter 6). Furthermore the evidence that a single polypeptide chain of an immunoglobulin is transcribed from a single strand of RNA suggests that fusion of the V and C regions must be brought about at the level of DNA and not at the level of protein (Chapter 2). This suggests that at least in this limited context the rearrangement of genes on chromosomes leading to a fused polypeptide product is a normal rather than an exceptional process.

4. Terminator mutations

In addition to coding for the 20 common amino acids found in proteins the genetic code contains three stopwords or terminator codons. These are UAA, UAG and UGA (in the RNA nomenclature). It must be presumed that most polypeptide sequences end because the codon beyond that specifying the terminal amino acid is one of these three. It can also be presumed that if there is a mutation in a terminator codon so that it now specifies an amino acid, sequences of DNA which might exist beyond the stop point will be read off and incorporated into the polypeptide chain.

An example of this phenomenon amongst the variant haemoglobins is given by haemoglobin Constant Spring. This is an α-chain variant with 31 additional amino acids extending beyond the arginine terminus of the normal 141 sequence α-chain (Clegg et al., 1971). The amino acid in position 142 of haemoglobin Constant Spring is glutamine, for which the normal RNA codon is CAA. Other very similar α-chain variants are haemoglobins Icaria, Koya Dora and Seal Rock which likewise have 172 amino acids in the α-chain. They differ from Constant Spring in only one, highly significant, residue (Clegg and Weatherall, 1976). Icaria has lysine in position 142, Koya Dora has serine and Seal Rock has glutamic acid. The RNA codons for lysine, serine and glutamic acid are AAA, UCA and GAA. Thus if one makes the assumption that there is a sequence of DNA beyond the normal terminator codon for the α-chain (UAA), it is possible to see that there has been a mutation of the first base in the terminator codon from U to C in Constant Spring, from U to A in Icaria and from U to G in Seal Rock. In Koya Dora the mutation A→C has been in the second base of the triplet codon. Further evidence for this hypothesis has come from the discovery of another α-chain variant, heamoglobin Wayne, which has 146 residues in the α-chain (Clegg and Weatherall, 1976). These differ from position 139 onwards (Fig. 8). If one assumes a frameshift mutation in the codon for position 139 of the α-chain it is possible to explain the new amino acids in haemoglobin Wayne appearing in positions 139 through to 146, and also to explain the termination of the chain at this point.

RESIDUE NO.	138	139	140	141	142	143	144	145	146	147	148.....
NORMAL α CHAIN	Ser	Lys	Tyr	Arg							
α CONSTANT SPRING	Ser	Lys	Tyr	Arg	Gln	Ala	Gly	Ala	Ser	Val	Ala.....
α ICARIA	Ser	Lys	Tyr	Arg	Lys	Ala	Gly	Ala	Ser	Val	Ala.....
α WAYNE	Ser	Asn	Thr	Val	Lys	Leu	Gln	Pro	Arg		

	138	139	140	141	142	143	144	145	146	147	148
NORMAL α CODONS	UCU	AAA	UAC	CGU	<u>UAA</u>						
ICARIA	UCU	AAA	UAC	CGU	<u>AAA</u>	GCU	GGA	GCC	UCG	GUA	GCU
CONSTANT SPRING	UC(U)	AAA	UAC	CGU	<u>CAA</u>	GCU	GGA	GCC	UCG	GUA	GCU
WAYNE	UCA	AAU	ACC	GUU	AAG	CUG	GAG	CCU	CGG	UAG	
	Ser	Asn	Thr	Val	Lys	Leu	Glu	Pro	Arg		

Fig. 8 Amino acid residues and codons involved in terminator mutations

B. Protein Properties

1. Loss of function

If a mutation occurs at a sufficiently sensitive site within the amino acid sequence of a protein it may well abolish function completely. Such mutant proteins will then have to be recognized by methods which are independent of their biological function, i.e. enzymes can no longer be identified by the reactions they catalyse. For most proteins such recognition implies immunological identity or its function as an antigen. It is obviously possible for a protein to suffer a mutation so destructive that both biological function and antigenic function are abolished. Such situations are in general not distinguishable from gene deletions, though in specialized cases like that of β-thalassaemia it is possible to probe for the precise site of the molecular defect by messenger RNA analysis and *in situ* hybridization.

Mutations in which both protein function and antigenicity are abolished are usually referred to as CRM negative, because they give rise to no cross-reactive material in standard precipitin tests. Mutants in which biological activity has been abolished but antigenicity retained are referred to as CRM positive. Recognition of the status of a mutant protein is dependent on the availability of a specific antiserum which will usually have been raised against the purified protein itself. For human genetic diseases purified human proteins are best used in raising antisera because of the specificity of immunological tests. This very largely restricts the wide-scale testing of genetic disorders for CRM status except where cross-reactivity with heterologous antisera can be invoked.

Boyer *et al.* (1973) have attempted to assess the relative incidence of CRM positive and CRM negative mutations amongst inborn errors of metabolism. They point out that since each type can arise through a diversity of mechanisms theoretical estimations of relative incidence is impossible and that an observational one must be used. Collecting data from 260 kindreds in 24 distinct genetic disorders, they found a ratio of CRM positive to CRM negative of about 3:2. The estimate has obvious limitations and can only be at best a crude approximation. Since a number of disorders, now typed as CRM negative, will be reclassified as CRM positive as more sensitive and specific immunological tests are brought to bear on them, the 3:2 ratio is probably a minimal figure. Some of the technical problems in establishing CRM status in haemophilia A are outlined in Chapter 13.

The presence of a functionless protein or enzyme in the tissues of a person with a genetic disorder has one immediate and practical application.

As Boyer *et al.* (1973) have pointed out the concept of protein replacement therapy for genetic disease assumes that the recipient will not recognize the replaced protein as foreign. In terms of current immunological theory this also assumes that such a patient has been producing a mutant protein not too dissimilar from the one which is to be infused and who will therefore continue to see replacement protein as "self". One would expect this to be the case for most point mutations where only a single amino acid in a large polypeptide has been changed. One would be less sure about more devastating mutations, such as frame-shifts, where a large portion of the polypeptide chain has been deleted. The residual protein may provide a model of "self" which is quite misleading when the body has to cope with a rather differently structured replacement protein. It has been suggested that this type of phenomenon may underly the development of circulating antibodies in a proportion of haemophiliacs who have been typed as CRM positive (Chapter 13).

On the other hand studies by Changas and Milman (1975) and Upchurch *et al.* (1975) on patients with the Lesch-Nyhan syndrome show that even when red cell haemolysates have residual HGPRT activity the samples type as CRM negative. The existence of some residual enzyme activity proves the presence of a modicum of mutant protein still capable of acting on substrate, and throws doubts on existing immunological methodologies which often employ inadequate tests. Another puzzling situation has been reported by Gray *et al.* (1976) in their investigations of sucrase–isomaltase deficiency. They were unable to identify any enyzme activity in intestinal biopsies of seven patients with this disorder and on the basis of immunological tests typed them as CRM negative. However, heterozygous carriers of the gene who had approximately 50% of normal enzymatic activity had twice as much immunoreactive protein as catalytically active protein. This suggested that the mutant allele in the carriers was producing CRM positive protein but in the affected homozygotes CRM negative protein. It is clear that the establishment of CRM status is still in its infancy and that many investigators have not followed the results of Reichlin's investigations on the mutant haemoglobins (Section IV.C).

2. Impairmant of function

Most mutations which can be recognized at the molecular level involve some impairment of protein function. Among the enzymes of intermediary metabolism, where genetic defects have been most extensively studied, mutation usually results in a large reduction but not complete abolition of enzymatic activity. It is of course theoretically possible that impairment of metabolic homeostasis might result from an enzyme with an increased activity. This has been suggested for one clinical form of gouty arthritis in which increased activity of the rate-limiting enzyme in purine biosynthesis, phosphoribosylpyrophosphate synthetase (PRPP synthetase), has been observed. Measurement of enzyme quantity with a specific antiserum showed that the mutant PRPP synthetase had 2·5 to 3·0—fold higher specific activity than normal (Becker *et al.* 1973), and it is presumed that this accounted for the oversynthesis of purines and consequent hyperuricaemia. Curiously enough a different mutant PRPP synthetase has also been reported, again with hyperactivity but this time with normal specific activity. In this case the increased activity is thought to be due to an impaired sensitivity to the normal feedback inhibitors ADP and GDP. It thus represents the first example of a mutation in the regulator site of an allosteric enzyme (Sperling *et al.*, 1973, Zoref *et al.*, 1975). On the other hand increase in enzyme activity with normal specific activity and no regulation defects has been reported for the mutants G6PD Hektoen (Yoshida, 1970) and pseudocholinesterase Cynthiana (Yoshida and Motulsky, 1969). Neither appear to cause metabolic disadvantage.

In general, however, mutations tend to lead to reduced enzyme activity with impairment of cellular function. This may result from a altered enzyme ability to bind substrate (K_m mutants), impaired co-factor binding (as in some of the vitamin-dependent aminoacidopathies) or reduced stability (as in some of the G6PD variants). If the enzyme shares a subunit with another enzyme a mutation may abolish either one activity (hexosaminidase A in Tay Sachs disease) or both activities (hexosaminidase A and B in Sandhoff's disease). If the enzyme is part of a multi-enzyme complex impairment of function in one enzyme can appear to affect all the components of the complex (Section III.G).

Amongst the haemoglobin mutants a variety of different categories of impaired function have now been described. The best known of these is the rather special situation of haemoglobin S, where a β-chain mutation has lead to an alteration in the solubility of the deoxygenated form of the protein, without apparently affecting any of its other properties. In the different types of haemoglobin M the ferric form of haem is stabilized by mutation in an amino acid involved in haem bonding, and thus prevents haemoglobin from functioning in its true role of an efficient and flexible oxygen gatherer and distributor. This is also seen in the altered affinity haemoglobins, most of which bind oxygen so inadequately that the patient must overproduce red cells in order to provide himself with enough haemoglobin to supply peripheral tissues. Amongst the unstable haemoglobins various forms of stress, including infections, drugs and perhaps even components of the diet, can throw the protein out of solution and cause it to weaken and perhaps even ultimately destroy its surrounding red cell. These examples are discussed in Chapter 11.

As pointed out in Section II the functions of many proteins are not clearly understood and this makes the elucidation of the effects of mutation difficult. Membrane proteins, in particular, are difficult to study by conventional biochemical techniques because of their insolubility. However, recent advances have been made in probing some of the mechanisms of protein-mediated active transport. This is particularly the case for transport phenomena involving the amino acids and is discussed in Chapter 10. Another more recently investigated transport protein is the plasma membrane receptor for low density lipoprotein (LDL). As pointed out in Chapter 10 Section II.B a deficiency of this protein is responsible for both the dominantly and recessively inherited forms of familial hypercholesterolaemia. By use of radio-labelled LDL it has been possible to demonstrate that the receptor sites are localized to short segments of the plasma membrane which are indented and coated on both sides with a fuzzy material (Anderson *et al.*, 1976). These coated regions have been known for some time as specialized sites on the cell surface where adsorptive endocytosis of proteins appears to be initiated. It is possible from studies on familial hypercholesterolaemia that the mechanism of protein adsorption is similar to that of the active transport of amino acids, and involves highly specific gene-controlled receptor proteins. This discovery has already had one immediate practical consequence. Arguing that if LDL could be disguised from the receptor sites directed against it on the plasma membrane it might be able to enter the cell by an alternative, non-specific route, Basu *et al.*, (1976) have prepared a "cationized" form of LDL by chemical modification. They found that not only was it able to penetrate the fibroblasts of patients with familial hypercholesterolaemia but that its ability to perform its normal function of suppressing cholesterol synthesis and activating cholesterol ester formation was unimpaired. Since the surface of mammalian cells usually bears a negative charge (Weiss, 1973) the cationization of proteins may have a general usefulness as a means for enhancing the cellular uptake of proteins in situations when mutation has lead to a specific impairment of receptor protein function.

REFERENCES

Anderson, R. G. W., Goldstein, J. L. and Brown, M. S. (1976). *Proc. Natn. Acad. Sci. USA* **73**, 2434.

Basu, S. K., Goldstein, J. L., Anderson, R. G. W. and Brown, M. S. (1976). *Proc. Natn. Acad. Sci. USA* **73**, 3178.

Beadle, G. W. (1945). *Chem Rev.* **37**, 15.

Becker, M. A., Kostel, P. J., Meyer, L. J. and Seegmiller, J. E. (1973). *Proc. Natn. Acad. Sci. USA* **70**, 2749.

Beckman, G. (1970). *Hum. Hered.* **20**, 74.

Bigley, R. H., Stenzel, P., Jones, R. T., Campos, J. O. and Koler, R. D. (1968). *Enzymol. Biol. Chem.* **9**, 10.

Bonner, J., Dahmus, M. R., Fambrough, D., Huang, R. C., Marushige, K. and Tuan, D. Y. H. (1968). *Science* **159**, 47.

Boyer, S. H., Siggers, D. C. and Krueger, L. J. (1973). *Lancet* **ii**, 654.

Buckley, C. E., Whitney, P. L. and Tanford, C. (1963). *Proc. Natn. Acad. Sci. USA* **50**, 827.

Celada, F. and Strom, R. (1972). *Quart. Rev. Biophys.* **5**, 395.

Changas, G. S. and Milman, G. (1975). *Proc. Natn. Acad. Sci. USA* **72**, 4147.

Clegg, J. B. and Weatherall, D. J. (1976). *Br. Med. Bull.* **32**, 262.

Clegg, J. B., Weatherall, D. J. and Milner, P. F. (1971). *Nature, Lond.* **234**, 337.

Cohn, E. J. and Edsall, T. J. (1942). "Proteins, Amino Acids and Peptides as Ions and Dipolar Ions". Reinhold, New York.

Croft, L. R. (1976). "Handbook of Protein Sequences, Supplement B". Joynson–Bruvvers Ltd., Oxford.

Dancis, J. and Levitz, M. (1972). *In* "The Metabolic Basis of Inherited Disease" (J. B. Stanbury, J. B. Wyngaarden and D. S. Fredrickson, Eds.), 3rd ed. 426–439. McGraw-Hill, New York.

Farrell, D. F., Clark, A. F., Scott, C. R. and Wennberg, R. P. (1975). *Science* **187**, 1082.

Frieden, C. (1971). *Ann. Rev. Biochem.* **40**, 653.

Fuller, G. M., Rasco, M. A., McCombs, M. L., Barnett, D. R. and Bowman, B. H. (1973). *Biochemistry* **12**, 253.

Gray, G. M., Conklin, K. A. and Townley, R. R. W. (1976). *New Eng. J. Med.* **294**, 750.

Harris, H. (1975). "The Principles of Human Biochemical Genetics", 2nd ed. North-Holland, Amsterdam.

Hopkinson, D. A., Edwards, Y. H. and Harris, H. (1976). *Ann. Hum. Genet.* **39**, 383.

Kendrew, J. C., Dickerson, R. E., Strandberg, B. E., Hart, R. G., Danes, D. R., Phillips, D. C. and Shore, V. C. (1960). *Nature, Lond.* **185**, 422.

Klotz, I. M., Langerman, N. R. and Darnell, D. W. (1970). *Ann. Rev. Biochem.* **39**, 25.

Knowles, J. R. and Gutfreund, H. (1974). *In* "Biochemistry Series One; Vol. I. Chemistry of Macromolecules" (H. Gutfreund, Ed.) Butterworths, London.

Linn, T. C., Pelley, J. W., Pettit, F. H., Hucho, F., Randall, D. D. and Reed, L. J. (1972). *Arch. Biochem. Biophys.* **148**, 327.

Mahler, H. R. and Cordes, E. H. (1966). "Biological Chemistry". Harper and Row, New York.

Noble, R. W., Reichlin, M. and Gibson, Q. H. (1969). *J. Biol. Chem.* **244**, 2403.

Pauling, L. and Corey, R. B. (1951). *Proc. Natn. Acad. Sci. USA* **37**, 251.

Pauling, L., Itano, H. A., Singer, S. J. and Wells, I. C. (1949). *Science* **110**, 543.

Penhoet, E. E., Rajumar, T. V. and Ruther, W. J. (1966). *Proc. Natn. Acad. Sci. USA* **56**, 1275.

Poulik, M. D. and Weiss, M. L. (1975). "The Plasma Proteins" (F. W. Putnam, Ed.), Vol. 2, 51–109. Academic Press, New York and London.

Reichlin, M. (1970). *Immunochemistry* **7**, 15.

Reichlin, M. (1972). *J. Mol. Biol.* **64**, 485.

Reichlin, M. (1974). *Immunochemistry* **11**, 21.

Reichlin, M. (1975). *Adv. Immunol.* **20**, 71.

Sanger, F. (1956). "Currents in Biochemical Research". (D. E. Smith, Ed.). Interscience, New York.

Sela, M., White, F. H. and Anfinsen, C. B. (1957). *Science* **125**, 691.

Sperling, O., Persky-Brosh, S., Boer, P. and de Vries, A. (1973). *Biochem. Med.* **7**, 389.

Upchurch, K. S., Leyva, A., Arnold, W. J., Holmes, E. W. and Kelley, W. N. (1975). *Proc. Natn. Acad. Sci. USA* **72**, 4142.

Weiss, L. (1973). *J. Natn. Cancer. Inst.* **50**, 3.

Wetlaufer, D. B. and Ristow, S. (1973). *Ann. Rev. Biochem.* **42**, 125.

White, A., Handler, P. and Smith, E. L. (1973). "Principles of Biochemistry", 5th ed. McGraw-Hill, New York.

Wieland, T. and Pfleideuer, G. (1962). *Angew. Chem. Internat. Edn.* **1**, 169.

Williamson, A. R. (1975). *In* "Isoelectric focusing" (J. P. Arbuthnot and J. A. Beeley, Eds), 291–305. Butterworths, London.

Yachnin, S. (1976). *Proc. Natn. Acad. Sci. USA* **73**, 2857.

Yoshida, A. (1970). *J. Mol. Biol.* **52**, 483.

Yoshida, A. and Motulsky, A. G. (1969). *Am. J. Hum. Genet.* **21**, 486.

Young, E. G. (1963). *In* "Comprehensive Biochemistry" (M. Florkin and E. H. Stotz, Eds), Vol. 7, pp. 1055. Elsevier, Amsterdam.

Zoref, E., de Vries, A. and Sperling, O. (1975). *J. Clin. Invest.* **56**, 1093.

2 Genes, Proteins and the Control of Gene Expression

R. G. SUTCLIFFE, B. CARRITT and
R. H. WILSON

Institute of Genetics, University of Glasgow, Scotland

I. INTRODUCTION

A fundamental conceptual advance stemming from the work of Gregor Mendel was the distinction between the hereditary determinants (the genotype) and the characters they influence (the phenotype). In this chapter we discuss the expression of biochemical aspects of phenotype, where possible in terms of mammalian and human genetics. We establish the relationship between structural gene mutation and the resultant biochemical phenotype. We show that while this relation is in principle a simple one, the mode of expression of a structural gene mutation can take on many guises, some of which can be interpreted by the unwary as regulatory gene mutations. In prokaryotes, various forms of genetic regulation have been discovered by means of genetic analyses which cannot at present be applied to higher eukaryotes. Nevertheless, the phenomenon of differentiation in higher eukaryotes represents the most obvious and tantalizing example of the regulation of gene expression and the basis of this is explored in the latter half of the chapter.

A. Genotype and Phenotype

One gene may affect many aspects of the phenotype by pleiotropic effects. In classical phenylketonuria, a deficiency of the enzyme phenylalanine hydroxylase leads to a variety of physiological and developmental alterations in the phenotype, from changed levels of serum phenylalanine and tyrosine to abnormal brain development and lowered IQ. No gene operates on its own and the phenotype develops by a complex matrix of largely undefined interactions at all levels of gene expression. Thus, we cannot yet explain why, for example, a lack of the enzyme hypoxanthine-guanine phosphoribosyl transferase should lead to the bizarre tendency to auto-mutilation seen in the Lesch-Nyhan syndrome.

As well as single genes having multiple, pleiotropic effects on the pheno-type, many genes often interact to control apparently single aspects of structure or function. In his classic studies, Waddington showed by mutation and selection experiments in Drosophila that the products of a number of autosomal genes could interact to create the wing phenotype cross-veinless which was identical to the phenotype of different flies which were homozygous for a particular X-linked variant. This led Waddington (1957) to the concept that many genes interact to stabilize the process of development and that some are individually of minor, additive effect, while others, such as the X-linked gene cited above, are of major effect. A similar example in human genetics is found in the congenital malformation hydrocephalus. This can be inherited as either an X-linked condition, an autosomal recessive condition or more commonly as the consequence of an unknown number of genes interacting with environmental factors (McKusick, 1975).

B. The Gene

After the prophetic voices of Garrod (1909) and Bateson (1909), the major advances in elucidating gene-protein relationships came with the use of microorganisms thirty or so years later. In 1909 Bateson was to state of Mendelian unitary factors (genes) that

> the consequence of their presence is in so many instances comparable with the effects produced by ferments, that with some confidence we suspect that the operations of some units are in an essential way carried out by the formation of definite substances acting as ferments.

As a result of the study of nutritional mutants of Neurospora, and the characterization of specific enzyme deficiencies in these strains by Beadle and Tatum, Beadle (1945) was able to postulate the "one gene – one enzyme" hypothesis, that the function of genes was to direct the synthesis of specific enzymes. Subsequently, the hypothesis has been extended to proteins irrespective of function and has been modified in various ways. First, some proteins were found to be composed of two or more dissimilar subunits, as in the cases of haemoglobin and lactate dehydrogenase. This led to the hypothesis "one gene–one polypeptide chain". Secondly, many polypeptides are covalently modified after translation, either by the addition of non-protein groups such as carbohydrate or phosphate, or by polypeptide cleavage (Neurath and Walsh, 1976), as in the case of zymogen activation, prohormone cleavage, or in the activation of blood clotting factors (Chapter 13). Indeed, in the case of vitellogenin (Section III.C) and some other proteins discussed in Section VII, cleavage of a single polypeptide chain sometimes results in the formation of two or more functional products. Further, there is some evidence (Section VI.A) that in the special case of antibodies, the dictum should read "two genes–one polypeptide chain". Next, there is the recent example where one might say "one DNA sequence—two co-linear polypeptide chains". This extraordinary finding has been made in the tiny DNA phage ϕX174, which only codes for nine separate polypeptide chains. Its genome is so small that in two places pairs of genes share the same DNA sequence, but are transcribed in two different frames (Barrell et al., 1976; Lewin, 1976). Finally, it appears that certain prokaryotic and eukaryotic genes contain internal sequences which are not found in their respective mRNAs and which cannot therefore code for part of the gene product as detected in the cytoplasm. This very recent finding is discussed in Section VIII.

C. Mutation

It is now accepted that DNA is the hereditary material in almost all organisms. The sequence of deoxyribonucleotide base pairs in the double stranded DNA molecule determines, by a base-pairing mechanism, the sequence of ribonucleotide bases in a single stranded RNA transcript. The sequence of bases in the RNA is translated into a sequence of amino acids which are enzymatically linked into a growing polypeptide chain on the ribosomes of the cell. The genetic information originally in the nucleic acid four base code is translated into amino acids using adaptor molecules of tRNA, with a triplet of bases in the mRNA (a codon) coding for one amino acid in the nascent polypeptide (see Table 1). The amino acid sequence of a protein determines its structure and biological activity, so that mutations in a structural gene will to some variable extent change the biological activity of the protein molecule it codes for.

Alterations to the DNA base sequence of an organism (mutations) arise spontaneously at a low rate. It is estimated that misreplication of DNA occurs in vivo at rates in the range of 10^{-7} to 10^{-9} incorrectly incorporated bases per base replicated in higher organisms. Agents which increase this spontaneous mutation rate include ionizing radiations, ultraviolet light, many chemical agents which alkylate or otherwise modify DNA bases, chemicals that intercalate themselves into the DNA helix producing

Table 1

The genetic code[a]

First base	Second base U	C	A	G	Third base
U	UUU Phe UUC Phe UUA Leu UUG Leu	UCU UCC UCA Ser UCG	UAU Tyr UAC Tyr UAA term. UAG term.	UGU Cys UGC Cys UGA term. UGG Trp	U C A G
C	CUU CUC CUA Leu CUG	CCU CCC CCA Pro CCG	CAU His CAC His CAA Gln CAG Gln	CGU CGC CGA Arg CGG	U C A G
A	AUU AUC Ile AUA AUG Met(I)	ACU ACC ACA Thr ACG	AAU Asn AAC Asn AAA Lys AAG Lys	AGU Ser AGC Ser AGA Arg AGG Arg	U C A G
G	GUU GUC Val GUA GUG (I)	GCU GCC GCA Ala GGG	GAU Asp GAC Asp GAA Glu GAG Glu	GGU GGC Gly GGA GGG	U C A G

[a] For detailed references to derivation of the code see *Cold Spring Harb. Symp. Quant. Biol.* **31** (1966), Watson (1976)

Bases—U: uracil, C: cytosine, A: adenine, G: guanine

Amino acids—Ala: alanine, Arg: arginine, Asn: asparagine, Asp: aspartic acid, Cys: cysteine, Gln: glutamine, Glu: glutamic acid, Gly: glycine, His: histidine, Ile: isoleucine, Leu: leucine, Lys: lysine, Met: methionine (Initiation), Phe: phenylalanine, Pro: proline, Ser: serine, Thr: threonine, Trp: tryptophan, Tyr: tyrosine, Val: valine

term. = "nonsense" triplet (chain termination)

distortions, and finally DNA base analogues that mimic natural DNA bases.

A classification of the molecular consequences of mutation leads to the four classes listed below, examples of which have been identified by sequencing human haemoglobin variants (Chapter 11).

1. Missense mutations

In these, the alteration of a single base pair (point mutation) in the DNA leads to the substitution of an incorrect amino acid in an otherwise normal polypeptide.

2. Nonsense mutations

These point mutations produce one of three codons in mRNA which are not recognized by tRNAs as coding for amino acids (hence the somewhat inappropriate term "nonsense"). Instead, these codons (UAG-amber,

UAA-ochre and UGA-opal) are recognized by the ribosome as signals for the termination of protein synthesis, so that nonsense mutations lead to the premature release of the growing polypeptide chain.

3. Frameshift mutations

These mutations add or delete one or two bases. The consequences of such alterations are first, to alter the amino acid coded for at the site of addition or deletion, and secondly, to shift the reading frame of the mRNA triplet codons. As the ribosome has no mechanism for determining the correct reading phase of the mRNA, this misreading produces a string of incorrect amino acids (gibberish) until a termination signal is reached. Polar effects on the translation of genes "downstream" from the frameshift may occur with mRNAs coding for more than one polypeptide.

4. Deletion/insertion mutations

These mutations may have large portions of DNA either inserted into, or deleted from, genes. If genes are closely linked, deletions may extend over several genes. If a single mRNA molecule codes for several proteins, some deletions and insertions may show polar effects similar to nonsense and frameshift mutations.

The molecular events leading to mutational base changes are complex and may involve the systems which repair lesions in DNA, or affect the precision of crossing-over. Different classes of mutation are induced by different agents. Single base changes (missense or nonsense) are produced by chemical mutagens, ultraviolet irradiation or base analogues and can be back-mutated to wild type by these agents. In contrast the DNA-intercalating agents can induce and revert frameshift mutations, where the previously mentioned mutagens prove ineffective. Deletions and insertions may be induced by X-irradiation which produces some double strand breaks in the DNA. An insertion may give rise to wild type by loss of the inserted material, although deletions should never regain the lost DNA sequences, and thus are stable non-reverting mutations.

D. Complementation

The ability to distinguish allelic mutations from mutations in different structural genes is of use in the accurate diagnosis and treatment of closely related genetic diseases. In the laboratory, the problem of allelism is largely solved by test-breeding. If variants are recessive and non-allelic, the phenotype of the F_1 will usually be normal so that intergenic complementation is observed. Because of the rarity of genetic disease, natural complementation is very infrequent in outbred human populations though marriage has occurred between clinically indistinguishable albinos to produce non-albino children (Trevor-Roper, 1952). In many related diseases, direct tests for allelism are unnecessary as distinctions can be made simply on the basis of the mode of inheritance, on the identification of altered levels of metabolites, or on the deficiency of a specific enzyme. However, conditions due to the same enzyme deficiency may result from mutations in different structural genes and this has been investigated by various methods, including somatic cell genetics (Section V.A.).

Inherited methylmalonicacidaemia (MMA) may be caused by a deficiency of methylmalonyl-CoA mutase ("mutase" for short) which is an enzyme composed of apoenzyme (or "core" enzyme) and the co-factor 5–deoxyadenosylcobalamin which is obtained from the metabolism of cobalamin (vit. B_{12}) by three successive enzyme-driven steps. Thus, MMA can be due to a defect either of the apoenzyme or of any one of the three enzymes which generate the co-enzyme from cobalamin (Mahoney et al., 1975). Cells grown from patients with different types of MMA were fused together and the resulting heterokaryons were tested for complementation by observing the incorporation of ^{14}C-labelled propionate which occurred if functional mutase was present (Gravel et al., 1975). In these experiments complementation was observed when fusions were carried out between MMA cells of differing type and not when both parent cells bore mutations in the same gene. Similar tests have been used to probe for suspected genetic heterogeneity in diseases where there was a clear method for scoring the cellular phenotype but where the underlying biochemistry was less well-understood. These are included in Table 2. Intergenic complementation can also occur when two genes contribute to the primary structure of a multimeric enzyme. An example is the fully functional heterokaryon formed between cells from patients with Tay-Sachs and Sandhoff's disease.

Occasionally, positive complementation results are obtained even when the test involves alleles of the same gene (cistron) and this is referred to as interallelic (or intracistronic) complementation (de Weerd-Kastelein et al., 1972). It occurs in homopolymeric proteins where different variant polypeptide chains aggregate to produce a "hybrid" enzyme with partial activity, which is usually only a few percent of the activity of the normal enzyme. Because of the specificity of the aggregation involved, interallelic complementation is a relatively unusual event. It has been claimed to occur in vitro for the structural gene of galactose-1-phosphate uridylyl transferase, the enzyme deficient in galactosaemia (Nadler et al., 1970), though the authors could not rule out intergenic complementation (p. 434). Finally, complementation tests should be used with caution when dealing with

Table 2

Diseases in which *in vitro* complementation
has shown genetic heterogeneity

Disease (deficient enzyme)	Number of complementation groups	Reference
GM1 gangliosidosis (β-galactosidase)	2[a]	Galjaard et al. (1975)
GM2 gangliosidosis (hexosaminidase)	2 (Tay Sachs and Sandhoffs)	Thomas et al. (1974)
Methylmalonic acidaemia (methylmalonyl CoA mutase)	4	Gravel et al. (1975)
Maple syrup urine disease (branched chain keto acid decarboxylase)	2[a]	Lyons et al. (1973)
Xeroderma pigmentosum	4[a] +1 (De Sanctis-Cacchione)	Kraemar et al. (1971); de Weerd-Kastelein et al. (1972); Cleaver and Bootsma (1975)

[a] Diseases in which there was no clear genetic evidence for heterogeneity prior to *in vitro* complementation

regulatory mutations since regulatory DNA sites (operators and promotors, see Section II.A) are *cis*-acting and do not specify independent products.

Before the heterokaryon test had been applied to Man a simpler method of mixed cell culture had been used with some success to probe the genetic heterogeneity of the mucopolysaccharidoses. Co-culture of cells from patients with different types of mucopolysaccharidoses resulted in some cases in the reduction of intracellular mucopolysaccharides as measured by biochemical or cytological methods (Neufeld and Cantz, 1971). This type of test is of very limited application because it depends on the free diffusion of either protein or metabolites between cells, so that conclusions must be drawn cautiously when co-culture has no effect on cellular phenotype, as for example, when fibroblasts from patients with Hurler and Scheie syndromes are grown together. However, a positive effect of co-culture (the mutual correction of Hurler and Hunter cells) indicates the equivalent of complementation.

The culture experiments discussed above have not been used to study genetic heterogeneity in differentiated tissues since the latter cannot be cultivated without loss of their specialized phenotype and may in fact not grow at all (Section V). Here it may be that co-homogenization of tissue might be informative if a suitable metabolic pathway assay is available. A different approach has been used to study the genetics of differentiated functions in mouse lymphocytes. In various strains of mice there are immune response (*Ir*) genes which affect the ability of the mice to make antibody against various types of antigen, non-responsiveness usually being recessive. When crosses were made between particular strains which failed to respond to a particular hapten the resulting F_1 progeny were found to be responders (McDevitt, 1973). This suggested intergenic complementation. When B cells taken from some of these strains were incubated *in vitro* with antigen, their supernatant was found to contain protein factors which, when injected together with hapten into a complementary non-responder, enabled that animal to make antibody against the hapten. It emerged that the complementing strains differed in that the mice in one could not synthesize the appropriate factor, whereas members of the complementary strain could not respond to it (Taussig and Munro, 1976; Isac *et al.*, 1976). This type of test may well feature strongly in dissecting the genetics of the HLA region in Man.

II. SOME ESTABLISHED ASPECTS OF GENETIC REGULATION

A. The Operon

In bacteria and their viruses the structural genes coding for related or sequential functions are often found clustered together in the genome. Only where this gene cluster is transcribed into a single (polycistronic) RNA message, may it be termed an operon. An important consequence of this co-transcription of distinct genes is that they are subject to coordinate regulation of expression when the rate of RNA transcription or its initiation in this region are altered.

The best studied example of coordinate regulation in a bacterial operon is of that concerned with lactose metabolism in *Escherichia coli* (the *lac* operon; Jacob and Monod, 1961). This consists of three structural genes which code for enzymes responsible for the hydrolysis (β-galactosidase; *z* gene), uptake (β-galactoside permease; *y* gene) and modification (thiogalactoside transacetylase; *a* gene) of lactose and related compounds. Since these genes are co-transcribed, the levels of their protein products in the cell can be coordinately controlled in response to cellular requirements by an integrated system of signals operating at the level of gene transcription.

In addition to the three structural genes, the *lac* operon also consists of two types of control region, the operator (*o*) and the promoter (*p*) (see Fig. 1). The promoter region is the site at which RNA polymerase binds prior to transcribing the structural genes. However, this transcription cannot occur if the operator site, situated between the promoter and the first

R. G. SUTCLIFFE, B. CARRITT AND R. H. WILSON

structural gene is occupied by a repressor molecule. This molecule is the protein product of the regulator (i) gene, which is not strictly speaking a part of the operon, and which, in other operons, may be spatially separated from the structural genes and their control regions.

Repressor molecules continue to occupy the operator site as long as lactose is absent, thus ensuring that the lactose metabolizing facility is not present when not required. When lactose (or any of a number of artificial inducers) is present, it interacts with the repressor in such a way as to prevent the latter's attachment to the operator site. The bound RNA polymerase is then free to proceed with uncontrolled reading of the structural genes.

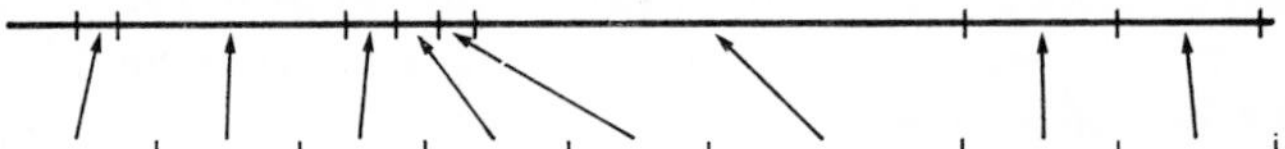

Gene	i^P	i	(spacer)	p	o	z	y	a
No. of base pairs	?	1040	—	84	21	3510	780	825
Protein product	—	repressor	—	—	—	β-galactosidase	permease	acetylase
Controlling site	i-promoter	—	—	promoter	operator	—	—	—
Mutations	i^q, i^{sq}	i^-,i^{-d},i^t,i^s	—	p^{uv5},p^{LB}	o^c	z^-, z^{-polar}	y^-	a^-

Fig. 1 Simplified structure of the *lac* region of *E. coli* showing some of the classes of mutants isolated

Since glucose is the preferred carbon source for *E. coli*, a mechanism exists whereby induction of the lactose metabolizing enzymes does not occur when glucose and lactose are present together. Thus, the presence of glucose in the cell depresses the synthesis of intracellular cyclic AMP (cAMP) which is required, together with a specific binding protein (CRP) for the attachment of RNA polymerase to the promoter region. Hence, it is only in the absence of glucose that RNA polymerase can bind to the promoter region, and only in the presence of lactose that it can transcribe the *lac* structural genes.

A number of mutations in regulatory functions are known for the *lac* operon (Table 3). These include mutations in the genes coding for the repressor and cAMP-binding proteins, and others which affect the binding properties of the control (operator and promoter) regions of the operon. A class of structural gene mutations which is unique to co-transcribed genes is that of polar mutations. These arise by insertion of inappropriate translational "stop" codons (nonsense mutations), either by point mutation or promoter-proximal frameshift, the effect of which is to prevent or depress the expression of the normal structural genes distal to the mutation.

Table 3

Regulatory mutations in the *E. coli lac* system

Mutation	Phenotype	Dominance to wild type allele
i^q i^{sq}	overproduction of repressor (mutation to more efficient i gene promoter)	*cis*-acting (dominant)[a]
i^-	constitutive expression of *lac* operon	recessive
i^{-d}	constitutive expression of *lac* operon (repressor allosterically locked into low operator-affinity conformation)	semi-dominant
i^t	induces at lower inducer concentration	semi-dominant
i^s	induces at higher inducer concentration if at all	*trans*-dominant
p^{uv5}	insensitive to cAMP effects	*cis*-acting (dominant)
p^{L8}	low levels of *lac* enzymes due to inefficient initiation of mRNA synthesis, regulation normal	*cis*-acting (recessive)
o^c	constitutive expression of *lac* genes due to impaired recognition by repressors	*cis*-acting (dominant)
z^{-polar}	no β-galactosidase, decreased levels of permease and acetylase due to polar effects	*cis*-acting (recessive)
crp^-	no expression of any genes requiring cAMP for promotion—no cAMP receptor protein	recessive
cap^-	as crp^-, very low levels of cAMP due to impaired cAMP synthesis	recessive

[a] Two mutations are said to be in *cis* in a diploid heterozygote when they occur together on only one of a homologous pair of chromosomes, and in *trans* when on different homologues. A pair of recessive mutations which produce the wild type phenotype when in *trans* and also in *cis* are defined as being in separate cistrons (genes). Conversely, two recessive mutations in the same cistron produce a wild type heterozygote in *cis* but not in *trans* (but see interallelic complementation Section I.D)

The types of negative (repressor/operator binding) or positive (CRP effects on polymerase binding) control seen in the *lac* operon are common in bacteria and their viruses. However, other forms of regulation are known, and, as examples, we cite some of the features of the regulation of gene expression in phages.

B. Bacteriophage Lambda

The phage genome can exist free in the host cytoplasm where it undergoes rapid replication leading to the production and release of progeny phages and thus to host cell lysis (the lytic cycle). Alternatively, a state of lysogeny can exist when the phage DNA is incorporated into the host chromosome and is replicated with it. The lysogenic phase is terminated in response to various signals, such as damage to the host chromosome and at this point the phage DNA excises itself from the host chromosome, and the lytic cycle begins.

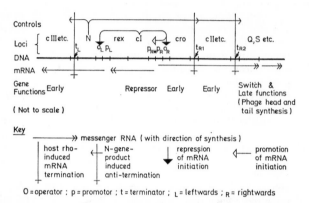

Fig. 2   A simplified diagram of the early genes in phage λ showing control being exerted in three ways: repression, promotion and anti-termination. The rightward and leftward mRNAs are drawn at different levels as, owing to the polarity of DNA, they are transcribed from different strands

In the lysogenic state, only two lambda genes are expressed, the lambda repressor (the cI gene product) and the *rex* gene, whose product protects the host cell against the lytic development of certain other unrelated phages. The lambda repressor acts negatively to prevent the expression of all other lambda genes, by binding to two operator sites (O_L and O_R) thereby preventing transcription from their promoters (P_L and P_R) of the leftward and rightward gene clusters which contain the genes required to trigger off the lytic cycle and ultimately to direct the synthesis of the viral head and tail proteins (Fig. 2). The presence of significant quantities of lambda repressor in the cell helps to maintain the synthesis of further repressor mRNA through a feed-back loop which acts positively on the P_{RM} site (an example of autogenous regulation, see Section II.D).

Each operator site in lambda contains several nearly identical sites which bind repressor (Ptashne, 1975; Maniatis and Ptashne, 1976). When fully repressed each operator binds three molecules of repressor in a cooperative manner such that the presence of a repressor molecule at one site increases the chance of repressor binding at adjacent sites. This produces a threshold concentration of repressor below which the operator is no longer repressed and provides a sensitive switch from the lysogenic to the lytic state. It is not clear how the concentration of repressor is lowered enough to effect the switch, though the process seems to be associated with damage to DNA and can be caused experimentally by ultraviolet light, X-irradiation or treatment with DNA-damaging chemicals like mitomycin-C.

Once repression is lifted from the operator sites transcription continues leftwards through the N gene and rightwards through the cro gene until transcription is terminated at leftward (t_L) and rightward (t_R) terminators. This transcription is due to host RNA polymerase which is an aggregate of various protein subunits including a host rho factor which is responsible for the termination. However, the phage N gene product is capable of overcoming this rho factor-induced termination and transcription can be extended leftward through the DNA repair, recombination and excision genes and rightward through the genes for DNA synthesis. The N gene product is, therefore, a positive regulator of these groups of genes and is essential for their expression. It has a similar effect at this second rightward terminator (t_{R2}) allowing transcription through the Q gene and beyond.

Controls also exist at the level of translation in the lytic cycle so that the mRNAs of different genes can be translated into protein at different rates (Ray and Pearson, 1975). Two factors can operate here; the efficiency with which the ribosomes can engage with specific sites on the mRNA, and the rapidity with which the ribosome can translate the mRNA into protein. For example, the genes which are grouped in the late transcription unit are co-transcribed and all the mRNAs should be present in equal amounts. However, some of the genes involved code for structural proteins of the phage head and tail which are required in large amounts whilst other genes code for enzymes which are required in far smaller amounts. The relative levels of different gene products coded for by the same mRNA molecule show an almost 1000:1 ratio between extremes. Different genes are thus translated at different rates from this messenger. It is not clear whether this regulation is achieved by the rate of ribosome engagement or by the rate of translation. (For further accessible reviews, see Davison, 1973 and Watson, 1976.)

C. Translational Control in RNA Bacteriophage MS2

Translational control is also found in the RNA phage MS2. Here the phage genome functions as a mRNA and codes for three proteins: a replicase (REP) to produce further copies of the RNA genome, a coat protein (CP) to package this RNA into a sheath of 180 protein subunits, and an attachment protein (A) to fit in the protein shell of the mature phage particle to enable it to attach to and infect further host $E.\ coli$ (Fiers $et\ al.$, 1976). The ratios of the three gene products are approximately 150 CP: 1A: 1REP protein. The three proteins are synthesized at different times in the infective cycle; coat protein is made throughout the cycle, while REP appears at an intermediate time and A protein is made later. It has been found that the coat protein translation initiation site is the only one which is accessible at the beginning of the active cycle, the other two sites being concealed in the folded parts of the molecule. Translation along the *CP* gene dissociates base pairs involved in the maintenance of some of the folded regions and discloses the initiation site for the RNA replicase (*REP*) gene, which is then translated. However, coat protein subunits bind to and mask the active replicase gene, leading to the early cessation of replicase synthesis. The replicase enzyme starts to replicate the RNA strand. During this process the ribosome binding site for the *A* gene is exposed, allowing *A* protein synthesis, until folding of the virtually complete RNA molecule again masks the *A* gene initiation site.

D. Autogenous Regulation

Regulatory gene products, such as the *lac* repressor, can act as sensors of key metabolites. In some prokaryotic and fungal systems evidence suggests that enzymes can themselves act as sensors and so regulate their own synthesis, and that of related proteins. This type of control has been termed autogenous regulation (Goldberger, 1974; Calhoun and Hatfield, 1975).

Evidence for autogenous regulation starts with the isolation of a structural gene mutation which is associated with unusual changes in the rate of synthesis of its product, or of the products of related genes. In *Salmonella typhimurium* histidine is the end product of an unbranched metabolic pathway involving nine separate enzymes, the genes for which form an operon. Although histidine represses the operon in the wild type, mutants of *S. typhimurium* exist which are resistant to histidine repression. Somewhat surprisingly, some mutations mapped in the first structural (*hisG*) gene of the operon, which specifies the enzyme PR-ATP synthetase. As *cis*-dominance was not observed in artificial diploids, it was concluded that the mutant enzyme was itself in some way responsible for the non-repressible phenotype. Details of the process are still under investigation. The enzyme appears to inhibit *his* operon transcription *in vitro*, and it can bind histidine-charged histidyl-tRNA. The charging of histidyl-tRNA could be a means by which the system detects the concentration of histidine.

No clear examples of autogenous regulation are established in higher eukaryotes. Yoshida (1970) has drawn attention to the structural gene mutation in G6PD Hektoen, which results in an approximately four-fold elevation in enzyme activity. Enzyme purification revealed that the mutant enzyme had similar specific activity and stability to the normal enzyme. It is not clear what type of regulation accounts for the increased quantity of gene product. Autogenous regulation is one possibility, and is a "short loop" system of regulation, in which the concentrations of some proteins directly regulate their own rate of synthesis. This is a different system from metabolic regulation, in which many more intermediate signals are involved before cytoplasmic events alter genetic activity. The possibility that cytoplasmic enzymes can directly affect transcription emphasizes the difficulty in distinguishing between structural and regulatory genes at the level of enzyme phenotype.

III. GENETIC REGULATION OF MAMMALIAN PROTEIN SYNTHESIS

The forms of genetic regulation known in prokaryotes provide models for some of the types of regulation which might occur during human development and which might be affected by mutation. Yet, as we have seen, control acts at a variety of levels in these simple organisms without the additional complications found in eukaryotic cells. These complications include the division of the cell into nucleus and cytoplasm, the physical separation of the processes of transcription and translation, and the modification of the primary transcript to form mRNA. In addition, the genome responsible for these phenomena is more complex than that of prokaryotes and is associated with a variety of chromosomal proteins. Finally, the process of differentiation results not only in multiple cellular phenotypes, but also in cells which are metabolically interdependent, with relatively short life spans, and which cannot at present be subjected to the type of genetic investigations used in the study of prokaryotes.

A. Regulatory Aspects of Inborn Errors

Structural gene mutation can lead to a variety of protein products, some with apparently normal function and others which are seriously defective or undetectable. The discovery of a variant protein with altered primary structure usually indicates the presence of a structural gene mutation. However, for the majority of proteins, structural alterations must be inferred from changes in function or physical properties. In the case of enzymes, these changes would include altered net charge, thermolability, allosteric properties, K_m and the serological detection of cross-reacting material (CRM). When no gene product can be detected (the null pheno-type) there may be grounds for postulating mutation in a regulatory func-tion. However, null phenotypes are often accounted for by a failure to detect a structurally altered gene product, either because it is too small (premature chain termination or large deletion), unstable, or antigenically distinct (due perhaps to a frameshift mutation). It is also worth noting that, in many cases, enzyme activity may be abolished by mutation in non-allelic structural genes which contribute to enzyme function, as in co-factor deficiency in methylmalonicacidaemia (Section I.D), or subunit deficiency in Sandhoff's disease (Chapter 10).

Null phenotypes can only be unequivocally attributed to mutations in regulatory genes or control regions by the demonstration that they map outside all of the structural genes contributing to the normal phenotype. For example, all the induced mutations producing the xanthine dehydro-genase (XDH) null phenotype in Drosophila (the so-called rosy mutants) are found to map between electrophoretic markers in the *XDH* structural gene (Gelbart *et al.*, 1974, 1976), indicating that they too are structural gene mutations. The only case of a mutation affecting XDH levels which maps outside the structural gene was found in a wild population of Droso-phila (Chovnick *et al.*, 1976). This mutation, which causes an increase in the levels of XDH, maps outside the left-hand end of the *XDH* structural gene and behaves as a *cis*-acting regulatory site.

For mutations in regulatory functions which map outside the relevant structural genes, the observation that they act only in *cis* is usually inter-preted to mean that the regulatory function is in some way concerned with the binding of RNA polymerase prior to gene transcription, or its initiation (Section II.A and Table 3). Where mutations act either in *cis* or in *trans*, a diffusible product of a regulatory gene is thought to exist.

From the wealth of evidence in this volume, it is clear that an inborn error of human metabolism must be regarded as a structural gene mutation unless there are strong indications to the contrary. Molecular analysis of the β^0-thalassaemia syndromes has suggested that the failure to synthesize β-globin mRNA is not due to extensive deletion of the β-globin structural gene (Chapter 11; Ottolenghi *et al.*, 1975). Similarly, β^+-thalassaemia appears to be due to reduced amounts of normal β-globin mRNA, rather than to normal amounts of defective message (Housman *et al.*, 1973). Although other interpretations are possible, the β-thalassaemias may repre-sent human conditions which have arisen from defects in the control of gene transcription. An interesting exception has been found in a type of β^0-thalassaemia from Ferrara, in Italy. Here it appears that normal β-globin mRNA is present which can direct the synthesis of β-globin only in a heterologous cell-free system (Conconi *et al.*, 1972; Ottolenghi *et al.*, 1977).

This implies that the condition may arise from a regulatory defect in the translation of β-globin mRNA.

There are other aspects of "structural gene" inborn errors which raise interesting questions of regulation. There appears to be a general lack of dosage compensation in heterozygotes for autosomal recessive inborn errors of metabolism. In those diseases where the protein defect is known, the apparent concentration of protein in heterozygous cells often approaches the arithmetic mean of the two homozygotes (thus providing a means of carrier detection, Chapter 10). This finding can be interpreted in a variety of ways. First, it may indicate that sufficient enzyme is present in the cells of heterozygotes to prevent significant disturbances of metabolism, and that consequently no signals are needed to increase the production of enzyme from the normal allele (metabolic "slack", Harris, 1975). This would be consistent with the recessive nature of the disease. Secondly, at some loci, both alleles may always be expressed at maximum rates, so that dosage compensation would be impossible. Thirdly, the gene is subject to a form of control insensitive to the concentration of the resulting metabolic products. Finally, it is possible that dosage compensation does occur but that in multimeric proteins it is concealed by allelic subunit interactions. If an enzyme has reduced activity because of the presence of variant as well as normal subunits, then a compensatory increase in the expression of both structural alleles in a heterozygote will not necessarily be detectable. This possibility may be important, since polymeric enzymes are more common than monomeric ones (at least among human polymorphic enzymes, Hopkinson et al., 1976) and because the same effect could result from the interaction of polypeptides specified by different loci. It may be possible to test this explanation by using dissociation techniques similar to those which revealed the subunit structure of enzymes such as LDH and hexosaminidase (Markert, 1963; Beutler and Kuhl, 1975). Dissociation and reassociation of normal enzyme in the presence of enzyme obtained from the recessive homozygote may result in a variety of heteropolymeric enzymes which have reduced enzyme activity.

One of the few well-documented examples of dosage compensation concerns the X-linked genes in Drosophila (in which gross X-inactivation does not occur). Compensation can be detected at the level of protein activity and also at the level of transcription by means of autoradiography of polytene chromosomes. The results cannot be accounted for by differential polytenization in either sex and compensation seems to depend on the presence of a Y-chromosome as well as a single dose of X-chromosome (Lucchesi, 1973).

Another regulatory aspect of "structural gene" inborn errors lies in the pleiotropic effects of a structural gene mutation on the concentrations of other proteins. This is observed in many different inborn errors. For example, defects of individual lysosomal enzymes often result in abnormally elevated levels of other lysosomal enzymes in the same individual. In familial hypercholesterolaemia the absent or defective low density lipo-protein (LDL) receptor results in reduced transfer of LDL into the cell, which in turn affects the cellular activity of enzymes associated with cholesterol metabolism. In normal cells the transfer of LDL into the cell causes an increased rate of esterification of cholesterol, presumably by enzyme induction or activation, as well as the suppression of activity of hydroxymethylglutaryl-CoA reductase which controls the rate of *de novo* cholesterol biosynthesis (Goldstein *et al.*, 1975, 1976).

Finally, we should ask whether any of the inborn errors which affect mammalian differentiation and development might be due to abnormal genetic regulation. This question arises because the changes in protein synthesis which underly differentiation represent *a priori* evidence of genetic regulation in mammalian cells (an area to be discussed in the following sections). It is probable that quite a variety of genetic abnormali-ties of differentiation and development result in developmental arrest, spontaneous abortion, or resorption, as in the case of homozygotes for the lethal or semi-lethal alleles of the *t* locus in the mouse (Bennett, 1975). An understanding of the genetic aspects of differentiation is not simply of academic interest to the human biochemical geneticist. Non-lethal develop-mental variants may be found, for example, in the locomotor and immune systems, since these processes are not essential until after birth. Although some syndromes have bizarre and biochemically uninformative phenotypes (such as the lobster-claw syndrome), others may be more amenable to investigation. Thus, in murine testicular feminization the primary defect appears to be a lack of target tissue receptors for the hormone testosterone (Bardin *et al.*, 1973; Shire, 1976; Lyon *et al.*, 1975). "Developmental" variants are also known in the immune system; for example the lack of adenosine deaminase (ADA) leads to a form of combined immunodeficiency (Giblett *et al.*, 1972) and a variety of other deficiencies exist which are inherited in a simple manner and reflect a failure of development of either the cellular or the humoral arm of the immune system. Here we may be observing variants in the regulatory processes which underly development, although, as the example of ADA deficiency shows, this does not imply that such mutations have occurred in regulatory genes in the prokaryotic sense, or even in genes whose products are expressed only in a single or restricted type of differentiated cell. Although this point applies at present to the majority of developmental mutants throughout eukaryotes, the reader is referred to the homeotic mutants of Drosophila as ones which appear to be closely involved with the stabilization, if not the genesis, of differentiation.

Homeotic mutations change the fate of individual groups of imaginal ("embryonic") cells which develop during pupation into many of the individual surface and internal structures of the adult fly. Thus a fly which is heterozygous for antennapedia develops small legs in the place of antennae (Prostlethwait and Schneiderman, 1973).

B. Expression of Specialized Proteins in Differentiated Cells

The presence of tissue-specific proteins provides some of the clearest evidence for the regulation of gene expression in eukaryotes. Although a long list can be compiled of proteins which are synthesized in restricted types of tissues or at specific stages of development, it is not clear what proportion of the structural genome is expressed in a tissue-specific manner. The technical difficulties of resolving the proteins of eukaryotic cells into single quantifiable fractions ensures that the question will remain unanswered for some time to come. Yet, even if such resolution were possible, there are obvious interpretational difficulties, the most serious of which is a definition of tissue-specificity. Specialized products, such as IgG and growth hormone, are easily categorized. So also are those proteins which appear to be ubiquitous until closer structural studies reveal that different but related gene products are performing the same functions in different tissues (see the instance of actin, Storti *et al.*, 1976). However, many ubiquitous enzymes exhibit different patterns of activity in various tissues or at certain stages during development (Chapters 5 and 9).

The tissue-specificity of gene expression can also be studied by the techniques of nucleic acid hybridization (Lewin, 1974). Axel *et al.* (1976) have studied the complexity of mRNA from chick oviduct and liver cells. The method of isolation was such that they were examining a major but incomplete sample of mRNA. Of this, about 15% was judged to be tissue-specific. Axel *et al.* also found that a small proportion of mRNA sequences was present in many (700 to 100 000) copies per cell whereas most mRNA sequences were found in 5 to 10 copies per cell. Similar results have been reported by Bishop *et al.* (1974) in human (HeLa) cells and by Galau *et al.* (1974) in the sea urchin.

Somewhat surprisingly, it has been claimed that small amounts of mRNA for erythroid α and β globins are present in heterogeneous nuclear (HnRNA) of non-erythroid cells (Humphries *et al.*, 1976). Such sequences were not detectable in the cytoplasm, and suggest that some "inappropriate" gene transcription occurs to give RNA which may then be degraded in the nucleus together with the majority of non-messenger HnRNA sequences. These preliminary data emphasize the difficulties in distinguishing at exactly what level the regulation of gene expression occurs in the sequence DNA→HnRNA→mRNA→protein→active protein.

C. The Induction of Protein Synthesis by Hormones

One of the most widely studied systems of protein regulation in higher organisms is that of protein induction by hormones. Often the effects on protein biosynthesis are rapidly reversed when hormone is removed, suggesting a form of modulative control analogous to those found in lower organisms.

Tomkins *et al.* (1969) were among the first to study the effect of steroid hormones on protein biosynthesis in cultured cells. They observed that dexamethasone or hydrocortisone increased the levels of the enzyme tyrosine aminotransferase (TAT) by 5- to 15-fold in cultures of rat minimal deviation hepatoma cells. The level of regulation involved was tested using actinomycin-D (AM-D), which is a potent inhibitor of transcription (among other effects). Surprisingly, AM-D was found to increase the activity of TAT in cells exposed to hormone by a further 10 to 30%. From this Tomkins *et al.* (1969) concluded that hormone induction did not act at the level of transcription. They deduced that a regulatory gene coded for a labile repressor which either inactivated the mRNA for TAT or increased its rate of degradation. Inhibition of transcription by AM-D would result in a rapid loss of repressor, leading to an increased stabilization of the mRNA for TAT. Since the concentration of TAT in the cell reflects its relative rates of synthesis and degradation, an increase in mRNA stability would result in a rise in cellular TAT levels. Tomkins *et al.* (1969) concluded that the superinduction hypothesis was reasonable since AM-D had no effect on the stability of TAT in their system. However, Kenny *et al.* (1974) have used more sensitive methods and have shown that the superinduction of TAT is due to a side effect of AM-D which leads to the stabilization of the TAT enzyme. A detailed but essentially indirect analysis has led these workers to conclude that steroids induce TAT by increasing the rate of synthesis of TAT mRNA, and more recent work by Steinberg *et al.* (1975) supports this view. There is some evidence that insulin may induce TAT levels by a post-transcriptional mechanism. Hepatoma cells which have been pretreated with hydrocortisone respond with greater TAT activity on subsequent exposure to insulin, whereas pretreatment with insulin did not potentiate the inductive effect of the steroid (Reel *et al.*, 1970).

The differences between inbred strains of laboratory mouse provide virtually the only source of mammalian variants which can be investigated both genetically and biochemically. In the mouse there are structural variants of β-glucurondase which are detected on the basis of either heat-liability or electrophoretic mobility and which permit the structural gene (*Gus*) to be mapped near the end of chromosome 5 (Swank *et al.*, 1973). Although β-glucuronidase is expressed in most tissues, its activity in kidney is increased 20- to 50-fold by injection of testosterone. Different strains of mice vary with respect to inducibility. In some (e.g. C57 black) β-glucuronidase levels rise from a basal level of 20 units to an induced level of 300 units after a lag period of 45 hours. In other strains (e.g. strain A/J) the increase is two to three times as great and the lag period is shorter. Crosses between the two strains give rise to a pattern of inducibility intermediate between the parental types, suggesting that the inducibility difference is due to a single gene, *Gur*, which is linked to *Gus*. Injection of testosterone into animals heterozygous at the *Gus* and *Gur* loci revealed that the "high inducer" *Gur* allele exerted most of its effect on the *cis Gus* allele. This may constitute evidence of regulation at the level of the gene, although at present the physical distinction between the *Gur* and *Gus* "loci" have not been clarified by appropriate studies.

Ganschow and Paigen (1967) have described two loci which affect the expression of β-glucuronidase post-translationally. The *Egasyn* locus codes for a protein which is located in the endoplasmic reticulum and which binds β-glucuronidase on to the reticulum. Although *Egasyn* mutants have a lysosomal β-glucuronidase, no enzyme is bound to the reticular fraction. The *beige* locus controls the excretion of β-glucuronidase from the kidney, possibly by specifying the structure of part of the lysosome membrane. Unlike the wild type mice, homozygous recessives at the *beige* locus do not secrete β-glucuronidase into their urine when injected with testosterone. The beige and egasyn mutants illustrate the role of post-translational events in the regulation of gene expression.

Work by Tata (1976) on the synthesis of vitellogenin in Xenopus liver stress the possible complexities of steroid induction. Vitellogenin is a protein which is synthesized by the female Xenopus under the influence of oestrogen. The protein is secreted into the blood and is taken up by the oocyte where it is cleaved into two yolk proteins: phosvitin and lipovitellin. When oestrogen is injected into male toads, vitellogenin is synthesized after a lag time of 2 to 3 days; synthesis reaches a peak after 8 to 15 days and falls to zero by the 30th day. If a second oestrogen injection is given 30 days after the first injection, vitellogenin synthesis starts without a lag and increases more rapidly than after only one injection. This secondary response can also be obtained in organ culture (Tata, 1976; Green and Tata, 1976). These results suggest that oestrogen causes moderately long-term changes in Xenopus liver which potentiate the induction of vitellogenin. It is not yet clear whether the changes occur at the level of the chromosome or more peripherally.

In some cases there is evidence that steroid induction acts directly at the level of the chromosome. Radiolabelling experiments show that, unlike most polypeptide hormones, steroids are bound to cytoplasmic and nuclear carrier proteins and are carried into the nucleus where the hormone binds to chromatin (Yamamoto and Alberts, 1976; Gorski and Gannon, 1976). In the oviduct of the immature chick, primitive mucosal cells differentiate under the influence of oestrogen into mitotically active populations of goblet, ciliated and tubular gland cells respectively (Oka and Schimke, 1969; O'Malley and Means, 1974), and study of this process may provide considerable insights into the mode of action of steroid hormones. Oestrogen induces the tubular gland cells to synthesize and secrete large quantities of ovalbumin, and progesterone subsequently induces the synthesis of small quantities of avidin (Woo and O'Malley, 1975). Cell free translation studies showed that the rate of synthesis of ovalbumin and avidin was directly proportional to the quantity of specific mRNA available for translation (Chan et al., 1973). This suggested that the rate of synthesis of the two proteins was controlled by the availability of their respective mRNAs. The subsequent purification of ovalbumin mRNA made it possible to develop a cDNA probe which assayed the quantity of mRNA present in the cell and showed that the number of ovalbumin mRNA sequences per cell closely correlated with the translational activity of mRNA measured in the cell free translation experiments. This indicated that the induction of ovalbumin was not subject to significant translational control. The results also argued against the existence of a post-transcriptional level of regulation whereby mRNA sequences are continually synthesized and then selectively degraded; this is because the cDNA probe used to count the number of mRNA sequences per cell would hybridize specifically to fragments of mRNA as short as 20 to 30 nucleotides (Woo and O'Malley, 1975). Finally, the cDNA probe showed that there was no amplification of the structural gene for ovalbumin during oestrogen induction (Harris et al., 1973).

A variety of enzymes and other proteins are induced by steroid hormones (Palmiter, 1976). Some, like ovalbumin and avidin, are induced within six hours of exposure to hormones (Chan et al., 1973). Others, such as murine kidney β-glucuronidase, are induced within 36 to 48 hours of the application of androgen (Swank et al., 1973). Palmiter (1976) argues that the delay in expression of the induced protein may simply represent the differing time courses for the initiation of transcription of differing genes, though it must not be assumed that all steroid inductions occur through regulatory changes at the level of gene transcription.

Unlike the membrane-soluble steroid hormones, polypeptide hormones such as adrenocorticotrophic hormone, insulin, adrenalin and glucagon bind to cell membrane receptors and their intracellular effects are mediated by the generation of the "second messenger" cyclic AMP (cAMP), through the action of adenyl cyclase (Southerland and Rull, 1960). Many of the affects of cAMP appear to be mediated directly through the cytoplasm by the activation of cAMP-dependent protein kinases (Krebs, 1972) on enzymes (e.g. phosphorylase b and glycogen synthetase), on membrane proteins, or on nuclear proteins (Kleinsmith *et al.*, 1966; Costa *et al.*, 1976). Cyclic GMP (cGMP) is antagonistic to the effects of cAMP, so that interactions between cyclic nucleotides may mediate the effect of polypeptide hormones at the level of cytoplasmic metabolism and possibly also at a transcriptional level (Goldberg, 1975).

IV. GENETIC REGULATION AND DEVELOPMENT

A. Hierarchies of Control

In prokaryotes we have seen that there are two types of regulatory elements. In the first, which can be referred to as regulatory sites, the DNA base sequence does not code for protein but is involved in the initiation and termination of transcription, as in the case of the operator or promoter sites and stop codons. The second type of regulatory element codes for a diffusible product which in prokaryotic systems is a protein which binds to the genome and acts either as repressor or activator. Thus the gene which specifies an activator or repressor protein only differs formally from a gene coding for say, β-galactosidase, in that the former produces a substance which acts upon the genome to regulate the pattern of gene expression. In mammalian cells, where much regulation could occur outside the nucleus, we should recognize that other categories of interaction exist. The point can be illustrated by returning to the effect of hormones: these regulate the pattern of gene expression, by first interacting with hormone receptors in the cytoplasm. In murine testicular feminization, the lack of receptors for testosterone leads to the failure of the hormone to induce β-glucuronidase in mouse kidney cells (see Shire, 1976). In one sense the presence of the receptor protein is as important for determining the cells' responsiveness to hormone as is the presence in the nucleus of the structural gene for β-glucuronidase. This point applies especially during early embryogenesis where subsequent differentiation may be determined by the incidence of environmental signals on undifferentiated embryonic cells.

Developmental studies in amphibia, Drosophila and other organisms have shown that positional differences between groups of cells lead to successive restrictions in their developmental fate or potential. Cells of restricted developmental fate are regarded as being determined. The morphological and biochemical changes characteristic of differentiation follow sooner or later from the cell's state of determination. Thus, cellular differentiation occurs at the end of a complex series of interactions between the nucleus, cytoplasm and environment. The extracellular signals which mediate these processes may be specialized molecules or gradients of ubiquitous proteins or metabolites; in either case the cell must be able to detect the level of such signals and eventually respond with changes in gene expression. The link between environmental signal and cellular response is probably mediated in a variety of ways. In the case of steroids, the cytoplasmic receptor protein carries the hormone into the nucleus and binds to specific regions of the genome. In the case of less specialized "metabolic" signals (assuming they exist) the identity of the receptor protein poses more of a problem. It may be useful to consider that metabolic enzymes themselves act as sensors of metabolites and bring about changes in nuclear programming. It is in this context that autogenous regulation is of interest (Section II.D), since it raises the possibility that a structural gene may be regulated by the level of its own protein product. In the case of an enzyme this could lead to a mechanism by which levels of substrate regulate synthesis of enzyme messenger as well as perhaps the expression of other genes.

There are a number of general ways by which signals received in the nucleus could alter the pattern of gene expression at the level of transcription. In some organisms there is excision of somatic DNA. Many whole chromosomes are lost during early embryogenesis in diptera such as *Miastor, Wachtiella* and *Myetiola*, although the entire chromosome sets are required for the development of germline cells (Brown and Dawid, 1969). In the female bandicoot, one of the X-chromosomes in each somatic cell is completely lost, rather than being inactivated as in other mammals (Hayman and Martin, 1965). Other types of chromosomal alterations have been observed especially in diptera and amphibia, involving changes in the primary structure of DNA. These findings were made in organisms which were particularly suitable for biochemical and cytogenetic studies; it is not known how widespread such phenomena might be or whether they play an important role during mammalian development. For example, the polytene chromosomes of diptera are found in tissues which grow by increase in cell size rather than by cell division and it has been suggested that similar chromosomes exist in the polyploid giant cells of the mouse trophoblast (M. L. H. Snow, personal communication). Amplification of the genes coding for the 18S and 28S rRNAs occurs in the primary oocytes of *Xenopus laevis* and a variety of other organisms. Here chromosomal rDNA is copied during pachytene to form a few thousand extrachromosomal rDNA sequences which synthesize 18S and 28S rRNA and so form the multiplicity of nucleoli characteristic of the diplotene oocyte. Other more complex changes in somatic DNA have been described and concern transposable controlling elements in the genome of *Zea mays* (McClintock, 1965, 1968; Fincham and Sastry, 1974), and paramutation (Brink, 1973). An arguably similar phenomenon to paramutation has been observed at the bobbed (*bb*) locus of Drosophila which codes for 18S and 28S rRNA (Ritossa, 1968; Shermoen and Kiefer, 1975; Tartof, 1975). Such phenomena have been discussed in the context of antibody diversity (Section VI) and X-inactivation (Chapter 9).

Although the evidence is fragmentary, there is a general consensus that gene modification, excision and amplification do not underly most processes of differentiation in higher eukaryotes. The nuclear transplantation experiments of King and Briggs (1952), Gurdon (reviewed 1970, 1974), McKinnell *et al.* (1969) and others provide experimental support in amphibia. In Gurdon's experiments the nucleus of an intestinal cell was transferred by micropipette into the cytoplasm of an enucleated oocyte. This rendered that egg diploid, and suitable physical shocks stimulated parthenogenetic development of the egg into an embryo. In a substantial proportion of experiments, the egg developed into a swimming tadpole, and in occasional animals, metamorphosis occurred and an adult frog developed. This provides evidence that the genome of the original somatic intestinal cell contained a full complement of genetic material and demonstrates the power of the egg cytoplasm in reprogramming the nucleus. These findings suggest that genetic regulation during development is not primarily controlled by the modification or excision of genes.

In some cases there is direct evidence that transcriptional changes occur during development. There is a relative absence of transcription in heterochromatin and this is particularly clear in the case of the human X-chromosome in somatic female cells (Chapters 3 and 9). In the case of the banded polytene chromosomes of Drosophila, where one band is generally regarded as containing one cistron (Judd *et al.*, 1972), heterochromatin can be directly correlated with low levels of RNA synthesis by the use of autoradiography to detect *in situ* incorporation of tritiated uridine. Gene transcription in Drosophila often results in expansion or decondensation of individual chromosome bands at different stages of development and in different tissues. This provides one of the few direct examples in which differentiation is accompanied by observed alterations in gene activity (Ashburner, 1970, 1972; Ashburner and Richards, 1976). Such changes in the physical structure of the genome during development are probably due to the association of chromosomal protein with the DNA and hence constitute a phenomenon mainly limited to eukaryotes. The tight super-coiling of condensed chromatin effectively prevents transcription both of the genes involved and also to some extent of nearby genes. This latter phenomenon, of which there are many examples collectively referred to as position effects (Baker, 1968), is readily exemplified by the inactivation of associated autosomal material which is translocated on to an inactivated mammalian female X-chromosome (Russell and Montgomery, 1969; Cattenach, 1975). Thus, the activity or inactivation of a gene can influence the activity of closely linked genes through the physical state of the chromatin.

B. Chromosomal Proteins

Detailed studies on the structure and properties of chromosomal proteins have contributed much to our understanding of the ways in which gene expression may be regulated at the level of the chromosome (Elgin and Weintraub, 1975). There are two major classes of chromosomal proteins: the histones and the non-histone proteins. In most organisms there are five classes of histone, which show relatively little structural heterogeneity between different tissues and considerable evolutionary conservation. These proteins are all highly basic, being particularly rich in arginine or lysine. Recent evidence indicates that histone classes F2a, 2b, 3 and 4 associate into octomers around which the DNA is wound (Olins and Olins, 1974; Thomas and Kornberg, 1975; Noll, 1976), bearing out the long-held belief that one of the functions of histone is to maintain the structure of chromatin. The limited heterogeneity of histones argues strongly against a primary role for them in specific gene inactivation, as this would require the recognition of gene sequences and a corresponding heterogeneity in primary structure. In contrast, the non-histone proteins are very hetero-geneous in nature. Some probably contribute to the structural organization of the chromosome, whilst others have enzymatic functions, for example polymerases, acetylases and protein kinases. Such enzymes may help to determine which genes are made available for transcription. Increases in transcription are often accompanied by varied protein kinase activity in the chromatin, and protein phosphorylation and other modifications may well affect the affinity of protein for the sugar–phosphate backbone of DNA. It is also probable that the specificity of DNA-dependent RNA polymerase can be altered so that selected subsets of genes are transcribed as discussed by Travers (1976). In prokaryotic systems the core polymerase protein is bound by different protein subunits which direct the specificity of the enzyme, as in the phage lambda N protein (Section II.B) and the bacterial rho, sigma and delta factors (Fukuda and Doi, 1977).

Over the past 12 years evidence has accumulated from chromatin reconstitution experiments, suggesting that non-histone proteins can control the pattern of gene transcription. In these highly artificial experiments chromatin is separated into its component parts usually by high concentrations of urea and NaCl. The chromatin is then reconstituted and is tested for its capacity to support transcription in the test tube, using prokaryotic DNA-dependent RNA polymerase. Although the resulting RNA transcripts are of much lower molecular weight than HnRNA *in vivo*, they can be analysed by nucleic acid hybridization. When chromatins were reconstituted using the non-histone proteins from different tissues of the same animal, a small proportion of different RNA sequences were transcribed. This led to the concept that certain of the non-histone proteins are responsible for the fine control of gene transcription. Since the pioneering work of Paul and Gilmour (1968), many technical advances have been made in transcribing chromatin and in analysing the resulting RNA. Using the chick oviduct system (p.58), O'Malley and colleagues have purified a nuclear receptor protein for progesterone which they have named "progestophilin" (Kuhn *et al.*, 1977). The protein appears to be a dimer, subunit A having affinity for pure DNA and subunit B having affinity for chromatin. The mode of action of progestophilin is at present unknown. However, Schwartz *et al.* (1976) have suggested that it may be possible to investigate its function *in vitro* using purified preparations of chromatin. They found that the addition of purified oviduct receptor: progesterone complexes to isolated chromatin in the presence of *E. coli* RNA polymerase lead to a pattern of RNA synthesis which was similar in time course and kinetics to that observed in oviduct cells *in vivo*.

Chromatin reconstitution experiments have also shown that non-histone proteins can specifically regulate the transcription of RNA which codes for the globin protein of chicken reticulocytes (Barrett *et al.*, 1974; Chiu, 1975). Indeed, Chiu *et al.* (1975) have obtained a 10% fraction of chromatin protein which contains the material which appears to regulate the expression of globin sequences *in vitro* and they point out that this fraction also contains proteins which have been shown to be tissue-specific by immunological methods, and which undergo some qualitative changes during differentiation and carcinogensis. Yet, exciting as these finds are, chromatin technology is still at an early stage. The concentration of reticulocyte chromatin used in Chiu's transcription experiments was perhaps one thousandth of that found *in vivo* (approximately 100 mg DNA/ml, Lin and Riggs, 1975); and attempts to study the kinetics and the location of "regulatory" proteins in chromatin can be complicated by non-specific binding of protein to DNA and other nuclear structures (as reviewed by Lin and Riggs, 1975 and Gorski and Gannon, 1976). It is likely that a minority of chromosomal proteins are involved in specific gene recognition, and at present we do not know how or to what extent such proteins control the pattern of gene transcription.

C. Models of Genetic Regulation

As the biochemical characterization of the nucleus proceeds, the results can be used to devise models of genetic regulation in eukaryotes (see Lichtenstein and Shapot (1976) for a review). The most prominent of these models have been those of Britten and Davidson (1969), Davidson and Britten (1973), Davidson et al. (1977) and Georgiev (1969), which were based on current knowledge of DNA complexity and of HnRNA size and turnover. These models are important and have provoked much discussion. However, they concentrate on the integration of regulatory signals at the level of the DNA rather than, for example, through the specificity of RNA polymerase (Travers, 1976), or through the processing and transport of mRNA from nucleus to cytoplasm (Harris, 1968; Lichtenstein and Shapot, 1976). Still less do such models have regard for the physical structure of native chromatin (Crick, 1971; Paul, 1972) and the existence of position effects. Further, we should not assume that all structural genes are regulated in the same way, or that the processes which initiate differentiation are the same as those that subsequently stabilize the differentiated phenotype.

V. EXPRESSION OF THE DIFFERENTIATED PHENOTYPE *IN VITRO*

The current understanding of genetic regulation in prokaryotes has followed from the ability to culture microorganisms under defined conditions. Thus it was to be hoped that culture of mammalian cells would provide a way of investigating the regulatory phenomena which underly the initiation and stabilization of differentiation. Over the past two decades, this hope has been realized in only the most limited way; first, because it is difficult to culture differentiated mammalian cells and secondly, because current methods of somatic cell genetics do not provide genetic analyses at a resolution which is comparable with genetic methods in prokaryotes, or with parasexual techniques in fungi. The following sections show that, despite these problems, some useful advances have been made and it seems probable that the increasing use of teratoma cells (Section V.D) will lead to more significant advances *in vitro* in the immediate future.

Cells which have been explanted from normal differentiated tissues, particularly those of epithelial origin, rarely give rise to long-term cultures which retain the specialized functions and morphologies of the original tissue. This may be because the specialized cells are overgrown by more rapidly dividing, less specialized cell types present in the tissue (Wigley, 1975). Alternatively, it might be that in adapting to the culture environment, the more specialized cell types lose their differentiated functions as a result of deficiencies in the culture medium or the need for rapid cell division.

The majority of specialized cells grow slowly *in vivo* or not at all, even where differentiation is not a terminal event, and it may be that the differentiated state of a normal cell is not compatible with its rapid and extended proliferation in culture. For example, short-term cultures of chick embryo chondrocytes (Holtzer *et al.*, 1960) and of chick embryo myoblasts (Stockdale and Holtzer, 1961) only synthesize chondroitin sulphate or myosin, respectively, in conditions of restricted growth. Similarly, pigment formation by chick iris epithelium in culture only occurs during the stationary phase of growth (Ephrussi and Temin, 1960). However, this may not be a general feature of short-term cultures of differentiated cells, as there appears to be no incompatibility between growth and the expression of specialized functions in some other primary cultures of cartilage (Cahn and Lasher, 1967).

Differentiated epithelial cells are particularly difficult to maintain in culture (Wigley, 1975). In some cases the loss of differentiated functions by such cultures may be due to the lack of an extrinsic factor; for example, the survival of foetal mouse hepatocytes *in vitro* has been shown to be dependent upon the presence in the culture medium of corticosteroid hormones (Waymouth *et al.*, 1971), in the presence of which the cells express liver-specific functions for extended periods. Similarly, lactose synthesis by cultures of bovine mammary tissue is prolonged by insulin (Ebner *et al.*, 1961).

In vivo, quite abrupt changes in differentiation occur when cell division ceases. This is obvious in the case of the nervous system and in the development of the pancreas (Wessells and Rutter, 1969). The regeneration of vertebrate skin and intestine stems from the presence of relatively undifferentiated cells which maintain a high rate of cell division. Conversely, the expansion of B immunocyte clones (Section VI) clearly depends on the presence of surface immunoglobulin of specificity appropriate to the triggering antigen. Whilst there may generally be an inverse relationship between the rate of cell division and the expression of differentiated functions, this does not argue that tissue stem cells *in vivo* or "dedifferentiated" cells *in vitro* do not have a restricted state of determination. It is equally likely that such cells still possess the regulatory processes responsible for initiating differentiation, but that overt differentiation is prevented either by the lack of suitable environmental signals or the overriding demands of cell division.

A. Analysis of Differentiation in Cultured Tumour Cells by Cell Fusion

Unlike their normal counterparts, some classes of tumour cells maintain a differentiated phenotype even in long-term culture (Table 4). The existence of such cells has permitted the use of somatic cell hybridization to probe the

Table 4

Tissue-specific products synthesized by
serially cultured mammalian cells

Product	Origin of cells	Species
5-Hydroxytryptamine Histamine	} mast cell tumour	mouse
Immunoglobulins	myeloma, osteogenic sarcoma	mouse, human
Albumin Hepatic alcohol dehydrogenase Aldolase B Induced tyrosine aminotransferase Induced alanine aminotransferase Induced tryptophan pyrrolase	} hepatoma	rat
Adrenocorticotrophic hormone Growth hormone Melanocyte stimulating hormone	} pituitary tumour	rat rat mouse
Corticosteroids	adrenal tumour	mouse
Ketosteroids	testicular Leydig cell tumour	mouse
Melanin (tyrosinase)	melanoma	hamster, mouse
Ectopic ACTH	melanoma	human
S-100 protein Acid mucins	} glioma	rat
Catechol-O-methyltransferase Tyrosine hydroxylase Acetylcholinesterase Cholinacetylase	} neuroblastoma	human mouse mouse mouse
Collagen Mucopolysaccharides	} skin, connective tissue	mouse, human
Myofibrils Creatine kinase Myosin	} skeletal muscle	chick, rat

possible genetic mechanisms which underly the differentiated state. However, when discussing the results of such studies, we must bear in mind that the starting materials are tumour cells, and that these may be abnormal, not only in their capacity for proliferation, but also with respect to the stability of their differentiated functions.

The regulation of the differentiated phenotype in cultured cells has been studied extensively using the technique of cell fusion (Davidson, 1974; Ephrussi, 1972). In the presence of inactivated Sendai virus (HVJ) (Harris, 1970; Ephrussi, 1972; Poste, 1972), the outer membranes of two mammalian cells will coalesce and the cytoplasms amalgamate. This gives rise to a binucleate (or sometimes multinucleate) cell which is called a heterokaryon if the original cells are of different origins. The heterokaryon, although metabolically active, is not capable of cell division or growth, but in rare cases (approximately one heterokaryon in 10^3 to 10^6) the two nuclei enter mitosis and the chromosomes of both cells are combined within a single nucleus. This hybrid cell is capable of extended proliferation, forming a clone in which genetic elements of both parents are present. Hybrids formed between cells derived from the same species retain all or virtually all of the genetic information of both parental cells. But, for unknown reasons, hybrids formed between certain pairs of species preferentially lose the chromosomes of one parent whilst retaining those of the other. This rapid and apparently unordered loss of chromosomes occurs most notably in the case of hybrids formed between rodent (mouse or hamster) and human cells, and provides a somatic approach to human gene mapping (Ruddle and Creagan, 1975; p. 111).

As a rule the genotypes of both parental cells contributing to a hybrid cell are co-expressed for a wide range of ubiquitous or household cellular functions, exemplified by the enzymes of glycolysis, the citric cycle and purine metabolism. Conversely, differentiated or luxury functions (Ephrussi, 1972), which were present in only one of the parents, behave in a less predictable fashion and are frequently not expressed in the hybrid cell. Sometimes this loss of expression has been shown not to be due to the loss of the structural gene for the luxury function and, in such cases, the failure to express a luxury function is referred to as extinction.

Hybrids formed between pigmented Syrian hamster melanoma cells and unpigmented mouse fibroblasts are unpigmented and do not express the enzyme dihydroxyphenylalanine ("dopa") oxidase which is required for melanin synthesis (Davidson et al., 1968). If it is assumed that the Syrian hamster gene for melanin production (that is, the structural gene or genes for dopa oxidase) is present in the unpigmented hybrid, then the experiment suggests that the presence of the mouse fibroblast genome in the same nucleus effectively prevents its expression. This in turn implies negative control of melanin production by the fibroblast genome (but not necessarily at the level of gene transcription). The further observation that hybrids containing hamster melanoma and mouse fibroblast genomes in the ratio 2:1 sometimes do synthesize melanin (Davidson, 1974) is interpretable in the same way if it is maintained that the single mouse genome is capable of suppressing dopa oxidase expression by only one of the hamster genomes. However, these experiments do not fully exclude the possibility that the failure to synthesize melanin is due to the loss from the hybrid of the hamster chromosome(s) carrying the relevant structural genes.

A proof that extinction of a differentiated function in a hybrid is due to suppression by the genome of the non-differentiated parent can only be obtained by the demonstration that this function reappears when chromosomes of the undifferentiated parent are lost from the hybrid. Such proof exists in the cases of a specific kidney esterase (Klebe et al., 1970) and in some liver-specific functions which are expressed in hepatoma cells. Rat hepatoma cells express albumin, aldolase B and hepatic alcohol dehydrogenase (ADH) and can be induced by steroids to express elevated levels of tyrosine transaminase and alanine transaminase. These liver-specific functions are not expressed in cells cultured from normal rat liver or in mouse fibroblasts. Fusions between the different cell types were used to investigate the regulation of liver-specific functions. The syntheses of aldolase B (Bertolotti and Weiss, 1972a) and hepatic ADH (Bertolotti and Weiss, 1972b) were extinguished in the hybrids of rat hepatoma × mouse fibroblast and rat hepatoma × normal rat liver cells, even though little or no chromosome loss had occurred after fusion. After a further period of time in culture, some of the hepatoma × normal liver cell hybrids were found to re-express aldolase B and ADH and also to have a reduced number of chromosomes compared with the original hybrids in which synthesis of both enzymes was extinguished. Although it was not possible to identify the parental type of the lost chromosomes, it was suggested that the re-expression of the extinguished function was brought about by the loss from the hybrids of a negative control element originally contributed by the undifferentiated parent.

When albumin synthesis was analysed serologically in the same rat hepatoma × mouse fibroblast hybrids (Peterson and Weiss, 1972) it was found that rat albumin production was partially extinguished and that no mouse albumin synthesis occurred. In these experiments, the hybrids were made by fusing rat hepatoma and mouse fibroblast cells in the ratio 1:1. In some 2:1 rat hepatoma × mouse fibroblast hybrids only mouse albumin was produced; in one 2:1 hybrid both rat and mouse albumin were produced, whilst in other such hybrids neither types were synthesized. The mouse fibroblast used in these fusions was approximately tetraploid, and so the effect of a single mouse genome on albumin production in hybrids was studied by fusing a near diploid mouse lymphoblast with the same rat hepatoma cell. Here, mouse albumin was usually produced by hybrids, irrespective of whether the parental input ratio was 1 rat:1 mouse or 2:1; rat albumin was also produced in all hybrids, but less often where the parental genome ratio was 1:1. These results constitute a proof that the genes for albumin production persist in a potentially active form in cells in which they have never been expressed. It appears that the expression of liver-specific functions is not under a rigorous form of coordinate regulation, since one function can be expressed by a hybrid in the absence of another. Aldolase B and hepatic ADH seem to be under some form of negative control, similar to that discussed for melanin, and it has been suggested that the synthesis of albumin is under some form of positive control in the hepatoma cells and that it is subject to the same sort of gene dosage effects seen in the negative control of melanin synthesis. However, in the case of albumin it seems equally possible that we are seeing competition for limiting amounts of "repressor" substance contributed by the undifferentiated parent. On either hypothesis, those hybrids which produce mouse, but not rat albumin must be assumed to have lost the relevant rat chromosome(s).

A more complex situation appears to exist in the case of hepatic tyrosine transaminase (TAT). This enzyme, although probably widely distributed throughout the body, is present in liver in high concentration, where it is also subject to induction by corticosteroid hormones. Normal liver *in vivo* and some hepatoma cells in culture show vastly elevated (10 to 20 fold) levels of TAT activity after the administration of hydrocortisone or its analogue dexamethasone. The high basal level of TAT activity and its inducibility by dexamethasone are both extinguished in hepatoma × fibroblast heterokaryons (Thompson and Gelehrter, 1971) and in the resultant hybrids (Benedict *et al.*, 1972; Schneider and Weiss, 1971). In one experiment, however, the further loss of chromosomes from a hybrid was accompanied by the reappearance of the hormone-inducible phenotype, but not of the high basal level of activity characteristic of hepatic cells (Weiss and Chaplain, 1971). These results suggest that the high basal level of hepatic TAT and its inducibility by dexamethasone are subject to separate forms of negative control, possibly emanating from different chromosomes. In fact, using hybrids formed between rat hepatoma and human fibroblast cells, it has been shown that the human fibroblast negative control "gene" for TAT inducibility is on the X-chromosome (Croce *et al.*, 1973).

The specificity of extinction is suggested by experiments in which immunoglobulin-producing cells are used as one parent of hybrids. The usual result of fusions of immunoglobulin-secreting myeloma cells with non-secreting fibroblasts is extinction of immunoglobulin synthesis (Periman, 1970; Coffino *et al.*, 1971; Klein and Wiener, 1971). But when the non-secreting parent is a thymocyte, immunoglobulin production has been observed (Parkman *et al.*, 1971) even though the thymocyte itself does not secrete immunoglobulin. When two immunoglobulin-producing cells are fused, then both parental types of molecule are secreted together with various presumed hybrid molecules; this is true even when the parents were of different species (rat and mouse) (Cotton and Milstein, 1973) and also when one of the parents was a spleen cell with no proliferative ability in culture (Köhler and Milstein, 1975).

These types of experiment provide a means of exploring the mechanisms by which differential gene expression is controlled. However, there are some difficulties in using these results as a basis for molecular models. Apart from the uncertainties introduced by the random loss of chromosomes, even from intraspecific hybrids, we are faced with problems such as the species specificity of putative "repressor" or "activator" molecules and the unknown stoichiometric relationships between such molecules and their targets. In addition to this, the cells used as the differentiated parent in most of the experiments of this sort have been tumour-derived cells. The majority of cells which retain the ability to express differentiated functions in culture are those derived from tumours (Table 4) and it might be argued that they are thus in an abnormal state with regard not only to the control of proliferation *in vivo* but also to the control of differentiated functions in culture. This consideration may make it difficult to extrapolate from these experiments to the normal situation *in vivo*.

B. Mechanisms of Extinction and Re-expression of Luxury Functions in Hybrids

The experiments discussed above are essentially simple ones and cannot be expected to provide precise genetic and molecular explanations for differentiation. The behaviour of liver aldolase B, TAT inducibility and kidney esterase indicates that these differentiated functions are susceptible to "switching" by gene products of undifferentiated cells, and that these products act over relatively long intranuclear ranges (i.e. they act in *trans*). These results also allow us to define the control of these functions by the non-differentiated cells as negative, but only if by this we mean the term as a description of the phenomenon; that is, we cannot conclude from them that the control is exerted by negatively acting molecules at the level of gene transcription, as for example the *lac* repressor. Studies in which mouse cells which can be induced to synthesize haemoglobin (Friend cells) were fused with those which cannot (mouse lymphoma or non-inducible Friend cell variants) suggest that extinction of the inducible phenotype may operate at the level of globin mRNA availability in some cases, and in others at the level of translation (Harrison *et al.*, 1976).

Molecular mechanisms cannot at present be formulated in the case of positive control of gene expression in hybrids as exemplified by the cross activation of albumin genes in mouse fibroblasts. It may be that a positively acting gene product (activator) is synthesized in the albumin-producing cell. Such an activator might be tissue specific and would therefore also be regarded as a differentiated product. To some extent this explanation is simply a redefinition of the problem since the phenomenon of tissue specificity is being invoked at the epigenetic level to account for synthesis of differentiated cytoplasmic products.

C. Analysis of Malignancy by Somatic Cell Hybridization

In an extended series of experiments, Harris, Klein and others have shown that the tumorigenicity of many cell lines is suppressed or extinguished by fusion with non- or weakly-tumorigenic cells (Klein *et al.*, 1971; Bregula *et al.*, 1971; Wiener *et al.*, 1971, 1973). They found that, although the fusion of two highly malignant mouse cell lines results in a malignant hybrid, hybrids formed between mouse cells with different degrees of malignancy generally display the level of malignancy characteristic of the less tumorigenic parent. The ability of a human cell line (HeLa) to form tumours in immuno-deficient mice has also been shown to be suppressed by fusion with a diploid human fibroblast (Stanbridge, 1976). Hybrids formed between Ehrlich ascites tumour cells and an established line of mouse cells or diploid mouse fibroblasts occasionally give rise to tumours when inoculated into irradiated hosts. However, the cells in these tumours had a reduced number of chromosomes as compared to those in the inoculum, and it was suggested that this re-expression of malignancy was due to the loss from the hybrids of specific chromosomes of the non-malignant parent (Klein *et al.*, 1971) in the same way that chromosome loss leads to the re-expression of extinguished luxury functions (p. 68).

D. Teratomas

A unique opportunity for studying differentiation under controlled conditions exists in teratomas, malignant gonadal tumours of mammals and birds. These tumours consist of a malignant embryonal carcinoma stem cell, and a wide range of cell types which can be histologically recognized as tissues such as bone, cartilage, nerve and muscle (Damjanov and Solter,

1974). Embryonal carcinoma cells are not differentiated and are the only malignant cells in the tumour. It appears that as the embryonal stem cells proliferate, a proportion develop into differentiated cell types which are not themselves malignant.

Using standard tissue culture techniques, mouse embryonal carcinoma cells can be propagated *in vitro* without loss of their malignancy or their capacity to form differentiated tumours for extended periods. Furthermore, it has recently been shown that some mouse teratoma cell lines can be induced under appropriate conditions to differentiate *in vitro* (Martin, 1975). Differentiation *in vitro* appears to give rise to cells resembling a range of somatic tissues.

Although there are some tissues which have never been found in differentiated teratomas (e.g. liver and lung), mouse teratoma cells can give rise, upon injection into suitable hosts, to tumours containing tissues representative of all three embryonal germ layers (Damjanov and Solter, 1974). Their similarity to early embryonal cells is also suggested by the fact that teratomas can be experimentally induced in mice by extrauterine implantation of early embryos. Thus embryonal carcinoma cells and most teratoma cell lines are generally considered to be of relatively undetermined developmental potential and are normally described as being pluripotent. In contrast, teratoma cell lines are known which have no potential to differentiate; these are described as nullipotent.

The capacity for normal differentiation of cultured teratoma cells is strikingly shown by experiments in which these cells are introduced into a normal mouse blastocyst (Papaioannou *et al.*, 1975; Mintz and Illmensee, 1975). The presence of genetic markers in the teratoma cell and in the host embryo made possible the demonstration that the introduced cells contributed to a wide range of tissues in the chimera which subsequently developed. In one case, colonization of the germ cells occurred, such that progeny were obtained from the chimera which had the phenotype of the mouse from which the teratoma cells was originally obtained, and this result suggests the exciting possibility that mutations obtained in somatic cells *in vitro* may be analysed genetically and phenotypically.

Early experiments had suggested that the capacity of mouse teratoma cells to form differentiated tumours was extinguished by fusion with a tumorigenic mouse fibroblast (Finch and Ephrussi, 1967). However, it has recently been shown by Miller and Ruddle (1976) that hybrids between a mouse teratoma cell line and normal mouse thymocytes not only retained the tumorigenicity of the teratoma parent, but also that the tumours induced by inoculation of hybrid cells into immunodeficient mice were differentiated.

On this evidence, it appears that the capacity of the cultured teratoma cells to differentiate *in vivo* is not subject to extinction as are many differentiated functions in other tumour cells (p. 68). The reason for this is not clear, but may conceivably be related to the unique development status of the teratoma cell which makes it insensitive to the types of control which modulate the expression of differentiated functions in developmentally committed cells.

VI. ANTIBODY BIOSYNTHESIS AND THE GENERATION OF ANTIBODY DIVERSITY

Studies on inherited protein variants indicate that most proteins are coded for by genes which are present in single copies per haploid genome. Proteins which are coded for by more than one gene usually show heterogeneity in primary structure (e.g. haemoglobin, Chapter 11, lactate dehydrogenase and many other isozymes, Chapter 5). This is also true for antibodies, which fall into about ten basic classes and sub-classes. However, within antibodies we find an additional and far greater diversification of primary structure which underlies the wide diversity of possible antibody specificities. We therefore ask how many genes code for antibodies and how such a wide variety of specificities can be expressed in association with a limited number of antibody classes.

When an animal has responded to an antigen by synthesizing an antibody of appropriate specificity it has, in a sense, changed its pattern of gene expression in response to that antigenic signal. The complex processes which underly this response are thought to be divided into two parts and this becomes clear when one considers the processes by which immunocytes become committed to the synthesis of a particular antibody specificity (Hobart and McConnell, 1975). Throughout the organism's life B stem cells are thought to develop into committed B cells which express their surface IgM or IgD of randomly selected specificity with one type of specificity per cell. In the absence of recognizable antigen these cells are transient and are continually replaced by others with different specificities. If antigen is present and binds to an appropriate immunocyte, that cell divides to form a clone of cells of virtually identical specificity (Burnet, 1959). Some of the cells develop into plasma cells and secrete antibody, others remain as B memory cells, to be triggered into further rounds of division on subsequent exposure to antigen. Thus a clone is expanded (selected) by being exposed to antigen. Similar theories apply to the lymphocytes which mediate cellular immunity, though it is still unclear whether the two arms of the immune response share the same system of antigen recognition (Munro and Bright, 1976).

A. Antibody Biosynthesis

The basic unit of an immunoglobulin consists of two heavy (H) chains and two light (L) chains connected by disulphide bridges (Chapter 12, Fig. 2). The antibody combining sites are found at the amino terminal ends of the molecule and each consists of one heavy chain sequence together with one light chain sequence (Steward, 1974; Hobart and McConnell, 1976). Thus, an IgG molecule has two antibody combining sites, one for each H–L chain pair. Studies on the individual H and L polypeptides revealed that about 110 amino acid residues at the amino terminal end of the molecule are highly variable in sequence (variable or V region) and the remaining carboxy terminal sequence (about 110 residues for L chains and 330 for H chains) are relatively constant (constant or C region). This led to the concept that separate genes code for the constant and variable regions of H and L chains. This concept has been supported by the finding that different constant sequences can be found in association with the same variable polypeptide sequence, and that recombination occurs at low frequency between genetic markers of the constant and variable regions of heavy chains (Eichmann, 1975).

Antibody specificity depends upon the primary structure of the variable polypeptides, V_H and V_L. This variability occurs in discrete hypervariable regions, each consisting of between 6 and 18 consecutive amino acid residues, which are located in the antibody combining site on the surface of the intact molecule. The specificity of the combining site is determined by the heavy and light chains' hypervariable residues, which are precisely orientated by the molecular framework provided by the highly conserved structure of the inter-hypervariable (framework) regions of the variable polypeptides (Capra et al., 1973; Wasserman et al., 1974).

If separate genes code for the variable and constant regions of an antibody molecule, we must consider how the separate sets of genetic information result in a single polypeptide chain. There are various possibilities; joining can occur between the genes, the primary RNA transcript, the mRNAs, or the proteins. Partial sequencing of the mRNA for murine κ chains has revealed a section of polynucleotide which overlaps the C_κ-V_κ junction (Milstein et al., 1974). This suggests C–V joining occurs before any protein is synthesized. Other evidence is consistent with this conclusion and indicates joining at the level of DNA. Heavy chain mutants have been found in which there is a deletion of amino acid residues across the V–C junction (Frangione et al., 1973). Further, somatic cell fusion has been carried out between mouse myeloma cells which produce isotypically and idiotypically distinct antibodies (i.e. antibodies with mutually distinct C and

V regions, respectively). The fusion hybrids also expressed both parental cell antibody types, and no antibodies of hybrid type in which the V polypeptide expressed by one myeloma cell was linked to the C polypeptide synthesized by the other parental cell type (Milstein and Köhler, 1975).

Interest in mechanisms of V–C joining has been deepened by evidence that isotypic switches occur during antibody synthesis. Initially, a clone of antibody producing cells (B cells) develops and synthesizes IgM in response to an antigenic stimulus. However, it is thought that after a period of time, often involving re-exposure to antigen, an IgG antibody is synthesized in the same clone and that IgM synthesis ceases. During this process the antibody specificity of the clone remains unchanged. This suggests that more than one class of C gene (Cμ, Cγ etc.) can be associated with the same V gene sequence, or copies of it (Imanishi et al., 1975). Similarly, human patients are known who have biclonal myelomas, in which one clone synthesizes an IgM (or IgA) antibody, and the other synthesizes IgG; yet both antibody classes have identical variable regions and light chains. This suggests that a switch in C_H gene expression occurs during the development of such myelomas, so that two distinct clones develop.

Models which are advanced to account for the union of V and C genes must draw on the genetic evidence for the number of genes involved and their linkage relation. Most of this genetic information comes from the study of immunoglobulin allotypes (Mage, 1971, 1974; Mage et al., 1973). When data from Man, mouse, rabbit and guinea pig are pooled, it is clear that allotypic variation is commonly found in the constant regions of all the major immunoglobulins. In many cases (e.g. the Gm markers of human IgG) the amino acid substitutions are simple and involve a small number of residues. Since these variants show Mendelian inheritance, it follows that there is only one gene for each respective immunoglobulin constant region. This suggests that there are 10 C_H genes in humans specifying, respectively, the four sub-classes of IgG, the two sub-classes each of IgA and IgM and the immunoglobulins D and E (see Chapter 12). In addition, there are thought to be three C_L genes for Cλ (which exists in three sub-classes) and one C_L gene for Cκ. Nucleic acid hybridization experiments support the notion that the individual heavy chain types are specified by few rather than multiple genes (Williamson, 1976).

Linkage studies in Man, mouse and rabbit have shown that there are three major clusters of immunoglobulin genes (Hood, 1973). One linkage group consists of C_H genes, the second of Cλ and the third of C_K genes. The three groups are not linked to each other. In the case of the rabbit H chain, evidence suggested that the C_H and V_H genes are closely linked. This resulted from the discovery of genetic markers (a locus allotypes) in the rabbit heavy chain V region, which were in close linkage with C_H markers. Although the physical basis of rabbit a allotypes has been reinterpreted on the basis of new evidence (see Section VI.C), the conclusion that V_H and C_H genes are linked has been confirmed by studies on inherited V-region differences in inbred strains of mice.

Individuals who are heterozygous for a simple allotypic marker express both alleles in their serum immunoglobulin. However, each individual antibody molecule is formed of only one or other allelic polypeptide, rather than of both, and immunofluorescence studies on the B cells of heterozygotes show that individual cells also express only one allele at each immunoglobulin locus. This phenomenon is known as allelic exclusion or hemizygous expression (Hood et al., 1975). Genetic studies on animals heterozygous for both V and C gene allotypes reveal that the V and C genes expressed in any immunocyte are in the cis-configuration (Kindt et al., 1970). Antibody from a trans-related $V–C$ gene pair is only detected in about 1% of circulating antibodies. This minor component may result from somatic cross-overs between the C and V genes on homologous chromosomes. Cis-limited expression of linked loci is well-known in mammalian X-chromosome inactivation (p. 428); however, the immunoglobulin loci provide the only known example for mammalian autosomal genes.

The close linkage of the C_H genes to the V_H gene pool provokes speculation about the sort of chromosomal events which might underly $V–C$ joining. The phenomena of allelic exclusion and cis-transcription indicate that it is restricted either to one homologous chromosome or to a pair of sister chromatids. If the C_H and C_L genes are only present in single copies per haploid genome, we must reject the simple theory that there are C genes linked to each and every V gene. (Such theories would also demand that each somatic V gene existed as a number of identical copies, each linked to a different C gene.) As there are few C genes, it has been suggested that some sort of transfer of genetic information occurs either by gene excision and insertion (transposition) or by the insertion of copies of V or C genes. Some attractive models are based on precedents in maize genetics (Fincham and Sastry, 1974) and the lysogenic cycle in phages (Section II.B). The reader is referred to Williamson and Fitzmaurice (1975) and Williamson (1976) for further discussion.

B. The Problem of Antibody Diversity

We do not know how antibody diversity is generated and the question is under intense investigation. Although antisera react highly specifically with antigen, animals are capable of responding to an enormous range of natural and synthetic antigens. The specificity of an antibody lies in the primary structure of its V_L and V_H polypeptide sequences, which are presumed to be coded for by DNA (since individual cells of a rapidly dividing B cell clone synthesize a virtually identical antibody or act as memory cells for that antibody). It follows that the wide range of antibody specificities must be reflected by a similar range of V genes in the animal's population of B cells. At the present time, the major question concerns these V genes. In particular, are all the somatic V genes represented in the germ line DNA, or are they generated from relatively few germ line genes by somatic recombination or mutation? The concept that antibodies have multiple specificities, and thus reduce the requirement for V genes on either a germ line or somatic hypothesis is based on firm evidence. This concept, termed "antibody polyfunctionality" by Richards et al. (1975), does not deny the specificity of antisera observed experimentally, since the serum antibodies concerned would have overlapping rather than identical ranges of specificities. If an antibody is bound to 100 different determinants this would still represent an insignificant degree of cross-reactivity if there are say, 10^6 possible antigenic determinants. Although antibody polyfunctionality reduces the total number of antibody combining sites which need to be invoked to account for the recognition of all possible natural and artificial antigens, the scale of this reduction is not known.

Theoretically, there is a second way in which the immune system may economize on V genes. If an antibody combining site is formed by one V_H and one V_L sequence, it follows that the maximum possible number of combining sites coded for by hV_H genes and lV_L genes is hl. Although not every V_H–V_L combination may be compatible with a functional antibody, it is clear that the use of separate V gene pools for H and L chains reduces the total number of V genes required. Thus 10^6 antibody combining sites might only require 10^3 V_H and 10^3 V_L genes. If the polyfunctionality factor reduced the necessary number of combining sites from 10^6 to 10^4, the number of V_H and V_L genes needed would be about 100 each. Again, the scale of these reductions can only be guessed at present.

C. V_H Markers in the Rabbit and Mouse

The discovery of three apparently allelic V_H allotypes (α allotypes) in the rabbit was initially taken as evidence for the presence of a single, or few germ line V_H genes. These markers are expressed in 70 to 90% of immuno-globulin molecules and amino acid sequencing has shown that they differ at multiple amino acid sites and are located in the V_H chain. However, the recent work of Loo *et al.* (1977) suggests that the α allotypes are not allelic. Strosberg *et al.* (1974) described a rabbit which responded to *Micrococcus lysodeikticus* with antisera containing all three α variants. Mudgett *et al.* (1975) have found that a third ("latent") α variant is synthesized at low levels in about half of the rabbits immunized with streptococcal extracts. This suggests that the α markers are isotypic rather than allotypic variants and that their pseudo-Mendelian expression is due to an allelic regulator gene or possibly to a chromosomal rearrangement within the α loci, leading to the preferential expression of a particular variant after V–C joining. The amino acid sequences which underly the group α markers are located in the highly conserved framework region of the V_H chain and do not contribute to the antibody combining site.

The fairly recent discovery of inherited V_H idiotype markers in mice represents an important advance in our ability to count V genes in the germ line. Idiotypes (Id) are antigenic determinants located on the V domain of antibody. Injection of homogeneous antibody into animals of the same or differing species results in the formation of antibody which binds to idiotype determinants. Suitable adsorption provides a specific anti-idiotype serum which can be used to detect the presence of the idiotype in sera of different animals in breeding experiments. Although the relationships between antibody structure, specificity and idiotype are highly complex, the existence of six simply inherited idiotypes provides evidence of multiple V_H genes in the germ line (see Eichmann, 1975; Weigert *et al.*, 1975, and Williamson, 1976). The idiotypes are all closely linked to the murine C_H markers and are expressed codominantly (Eichmann and Berek, 1973). Two other V markers have been described in the mouse on the basis of fine specificity analysis, both being linked to the C_H markers (Imanishi and Mäkelä, 1974, 1975). Rapid progress is now being made in this field.

D. A Gene Stitching Model

Capra and Kindt (1975) have pointed out the contrast between the numbers of V_H markers in rabbit and mouse and they have suggested that three classes of genes are stitched together to code for a particular H or L chain. The association of particular inherited murine idiotypes with different V_H subgroups (Barstad *et al.*, 1974) argues against the Capra and Kindt model (Eichmann, 1975).

E. Somatic Mutation

If there are few germ line V genes for the heavy and light chains, then somatic mutation (Cohn, 1973) or hyper-recombination (Whitehouse, 1967) would be the most probable ways of generating diversity. Though there is no experimental evidence to support any particular somatic mutation theory, a variety of immunogenetic considerations show that only certain types of theory can be reasonably entertained.

The low rates of mutation which probably occur in eukaryotic cells (10^{-8} to 10^{-10} per locus per cell division) make a simple mutation theory difficult to accept. Even if there were, for example, 10^8 mouse B stem cells generating V region diversity by such a mechanism and the cells had a life span of about two days, the rate of stem cell division would be about 0.5×10^8 per day. Given a rate of somatic mutation of 10^{-8}/locus/division this would result in one variant immunoglobulin V gene being generated per day. Although these are crude estimates it is difficult to conceive of more than 100 variants being generated per day and being available for selection by antigen-driven clonal expansion. The difficulty with this scheme is that variable polypeptides are highly diverse and cannot all be related to a single archetypal primary structure by single point mutations. Multiple base changes are required to achieve multiple amino acid substitutions and even to bring about the codon changes required to specify, for example, a glycine residue in place of a lysine residue (Table 1). Thus simple somatic mutation theories require the presence in the germ line of multiple V gene sequences which will then be modified by a more limited number of point mutations.

Higher rates of somatic mutation acting along the entire V gene will not reduce the requirement for a large number of germ line genes. This is because high rates of mutation will generate coincidental mutations in the framework regions. Cohn's model (1973) attempts to take account of the problem by allowing for a form of somatic selection which acts during the period of mutation. On his model, mutation would be followed by limited synthesis of antibody which contain variant V sequences. This hypothetical synthesis of antibody is pictured as allowing selection of those cells which produce structurally normal antibody in terms of domain folding, etc. The problem with this theory is largely statistical. Since only about 30% of the V gene codes for hypervariable amino acids, the chance of repeated mutations occurring only in the hypervariable region soon falls to a very low level as the number of rounds of mutation increase. In view of these difficulties, and if somatic mechanisms contribute heavily towards the generation of antibody diversity, we have to entertain the possibility that mutation is limited to the hypervariable sites within the V gene. This would not involve the genetic load of the framework sequence, so that higher rates of somatic mutation could be entertained, although there is no evidence that such rates occur. An alternative or possibly complementary model would involve high rates of intracistronic V gene recombination. Here, a fairly limited number of germ line V genes would undergo somatic recombination, resulting in the exchange of DNA coding for a complete or partial hypervariable sequence between one germ line V gene and another. It is doubtful whether there is a genetic precedent for such a mechanism on the scale required if recombination is to account for a major part of antibody diversity.

Increasingly, the complexity of the genome is being probed by the use of nucleic acid hybridization, which will doubtless produce much valuable data on antibody genes in the future. However, it is not yet clear that hybridization is the method of choice for counting V genes, since the latter are highly diverse in primary sequence and should therefore be analysed using a nucleic acid probe of comparable diversity, rather than one obtained from a myeloma cell line which synthesizes a homogeneous antibody. The problems associated with this approach are discussed by Williamson (1976), Williamson and Fitzmaurice (1975) and Tonegawa and Steinberg (1975).

VII. GENE CLUSTERS IN EUKARYOTES

In prokaryotes, structural genes which code for proteins with related functions are often grouped together and are subject to coordinate control of transcription. In some cases (e.g. the *lac* operon) regulation is affected by co-transcription of the structural genes from a single promoter; in other cases transcription proceeds from more than one promoter as for the early genes of phage (Section II.B). In a few cases, clusters of apparently related genes have also been described in eukaryotes and it is reasonable to ask whether this has any significance from the standpoint of regulation. However, it should be pointed out that the clustering of related genes, whether they are under regulatory control or not, may have causes and consequences which have little to do with gene regulation (p. 82).

Gene clusters have been identified in Neurospora, which are concerned with the biosynthesis of pyrimidines (*pyr-3*), aromatic amino acids (*arom*) and histidine (*his-3*) (Whitehouse, 1973; Berlyn, 1967; Pateman and King-horn, 1976). Polar mutants are known for all three systems, suggesting co-transcription of each cluster of genes. However this does not provide evidence of operons, since in the case of *pyr-3* and *his-3*, the various enzymatic functions are carried out by single polypeptides with two and three distinct catalytic activities, respectively. In the case of the *arom* cluster, the gene product is a multienzyme complex with five distinct enzyme activities (Ahmed and Giles, 1969). However, the *arom* gene product has recently been shown to be synthesized as a single polypeptide with five associated enzyme activities, which is subsequently cleaved by intracellular proteases, probably during biochemical purification (Lumsden and Coggins, 1977).

Genetic analysis of rudimentary larvae of Drosophila, which require exogenous pyrimidines for survival (Norby, 1970), has revealed tight linkage between the genes for three of the enzymes of the pyrimidine biosynthetic pathway (Soderholm *et al.*, 1975; Rawls and Fristrom, 1975). In mammals, these enzymes (carbamyl phosphate synthetase, aspartate transcarbamylase and dihydro-orotase) co-purify in biochemical separations (Shoaf and Jones, 1973; Mori and Tatibana, 1975). This may explain why mammalian cells in culture which are resistant to aspartate analogues (Kempe *et al.*, 1976) show a simultaneous elevation of these three enzyme activities.

To date, no eukaryote gene cluster has been shown to be organized like a prokaryote operon, in which distinct structural genes are transcribed to a single mRNA which is itself translated into several polypeptides (Section II.A). In yeast there is close linkage of the structural genes for three of the enzymes* of galactose metabolism which are under coordinate control (Douglas and Hawthorne, 1964, 1966; Kew and Douglas, 1976). However, this cluster does not qualify as an operon as there is no evidence that the structural genes are co-transcribed. Regulatory loci have been identified (*i*, *c*, *GAL-3*, *GAL-4*) and these are not linked to the structural gene cluster. The *i* gene product negatively controls (Section II.A) the expression of the unlinked gene *GAL-4*, through the *cis*-acting site, *c*. In turn, *GAL-4* acts positively on the expression of the structural gene cluster. Although *c* may be analogous to an operator, the *c*, *GAL-4* combination cannot be called an operon. Similar situations exist in other fungi, where linked structural genes specify proteins which are under coordinate control and which are not part of a multi-enzyme complex. However, in no case is there firm evidence of an operon, even though *cis*-acting loci have been identified (see review by Pateman and Kinghorn, 1976). It is not clear whether this reflects the biological situation or problems of ascertainment.

Although the types of genetic analysis which led to the identification of gene clusters in lower eukaryotes are not available for mammals, extensive

---

* Galactokinase, galactose-1-phosphate uridylyl transferase and uridine diphospho-galactose-4-epimerase.

polymorphisms have allowed the assignment of immunoglobulin structural genes to three groups (p. 75), and the demonstration that three linked genes contribute to the rhesus phenotype (Chapter 7). Studies on human haemoglobin variants have indicated that there is close linkage between the genes for the γ, δ and β globin chains. The fused δ/β and γ/β chains of the Lepore type haemoglobins suggest that the γ, δ and β genes are contiguous, as does the evidence from those $\delta\beta$ thalassaemias in which deletion of the δ and β genes extends into one of the γ chain genes (Chapter 11).

Other examples of close linkage of related genes in man are to be found in the structural genes coding for the non-allelic isozymes of salivary and pancreatic analyses (Merritt et al., 1972, 1973), and in the components of the histocompatibility complex where an impressive degree of natural variation has allowed fine structure mapping (Shreffler and David, 1975; Munroe and Bright, 1976). The complex contains genes which specify the histocompatibility, the mixed lymphocyte culture and a variety of immune-associated (Ia) antigens, as well as certain red cell antigens, complement factors and some cell membrane receptors for complement (see Chapter 7).

In the examples quoted above, there is little or no evidence to suggest the existence of the type of coordinate regulation of linked structural genes which is found in the lac operon (Section II.A). However, in some cases, there seem to be forms of reciprocal control; for example, during the switch from foetal (HbF) to adult (HbA) haemoglobin, an increase in the synthesis of β-chains occurs concomitantly with a corresponding decrease in γ-chain synthesis (Weatherall and Clegg, 1976). As no γ-chain mRNA is detectable in cells synthesizing HbA, this suggests a reciprocal control of transcription of the closely linked γ- and β-chain genes (although instabilty of the γ-chain mRNA cannot be excluded). Similarly, in the mouse MHC, there appears to be a reciprocal relationship between the expression of the t locus and H-2K and between tla and H-2D (Bennett, 1975; Boyse and Old, 1969). In Man, expression of only one of the two closely linked amylase loci occurs in pancreatic and salivary tissue, respectively. Whilst these observations undoubtedly raise interesting questions as to control mechanisms, it is not clear to what extent these may depend on the clustering of the affected genes.

In some cases, such as Rhesus and HLA, it is probable that linkage between the different genes involves genetic co-adaptation. Thus, close linkage would ensure that individuals receive a combination of alleles at the different loci in the cluster which confer an advantage which would be lost were recombination to occur freely between the loci. In other instances, clustering may simply be the result of gene duplication (presumably by non-homologous crossing-over) and limited divergence. The two a-globin chain genes in Caucasians probably result from a-gene duplication without divergence, whereas the δ- and β-chain genes, which differ by only 10 amino acid codons, may have arisen by duplication with some divergence. Similar considerations apply to the two γ-chain genes, which differ by only one amino acid codon.

Recent advances in molecular and biochemical techniques may soon allow the identification and analysis of mammalian gene clusters, even where genetic evidence is lacking. The use of nucleic acid hybridization, restriction endonucleases and plasmid cloning has revealed clustering of the genes for rRNA and of those for histones, and has also shown that these clusters occur in multiple copies in the mammalian genome. The genes for 28S and 18S rRNA are co-transcribed and the RNA transcript subsequently cleaved to yield 28S and 18S rRNA in the equimolar quantities required for ribosome assembly. In amphibians and mammals, the rRNA genes are tandemly repeated several hundred times; in Man, repeats of rRNA genes map to at least five distinct chromosomal sites (Ruddle and Creagan, 1975). Although coordinate regulation of 28S and 18S gene transcription occurs, each gene pair is separated from the neighbouring repeat pair by a non-transcribed spacer region. Re-initiation of transcription must therefore occur for each gene pair. It is now clear that all the major RNAs of the translational apparatus (i.e. 4S, 5S, 18S and 28S RNAs) are coded for by repeated structural genes, and it seems reasonable to assume that this reflects their fundamental role in the expression of other structural genes.

The histone genes of Man, Drosophila and the sea urchin are similarly arranged in tandemly repeated clusters. The structural genes for the five classes of histone each lie between spacer sequences; Kedes, 1976; Gross et al., 1976a; Schaffner et al., 1976; Gross et al., 1976b). At present it is not known whether the histone cluster is co-transcribed, although the mRNAs for the five histones are distinct in the cytoplasm of the cell. As in the case of the rRNAs, the histones have an important structural role, in which, with the exception of histone Hl, they are required in equimolar amounts (Section IV.B). Hl is present in chromatin in about half-molar quantities relative to the other four, and thus coordinate regulation of transcription of the entire cluster either does not occur, or there is also some form of post-transcriptional control which results in differential rates of protein synthesis from equal amounts of mRNA. Some evidence for post-transcriptional control exists (Kedes and Gross, 1969), and is also suggested by the fact that histone mRNA synthesis occurs during the G2 stage of the cell cycle in human HeLa cells, when no histone protein is synthesized.

The significance of the clustering of rRNA and histone genes is not clear, although it may be related to the fact that both clusters are highly repeated. This is not to deny that this clustering may have a regulatory significance, although it should be emphasized that the rRNA genes are not translated and cannot be regarded as typical of mammalian structural genes.

VIII. INSERTED SEQUENCES IN STRUCTURAL GENES

Since the elucidation of the genetic code, it has generally been assumed that the base sequence of a gene will correspond to the amino acid sequence of its protein product. These assumptions drew some support from genetic mapping studies which showed that there was a co-linear relationship between the sites of structural gene mutations and the position of consequent changes in amino acid sequence (Yanofsky et al., 1964; Yanofsky and Horn, 1972; Sarabhai et al., 1964; Cellis et al., 1973). The later discovery that many RNA molecules contain terminal sequences in excess of those required to code for their proteins was simply explained by the concepts of post-transcriptional modification (i.e. addition of polyadenylate "tails" to the 3′ end of the mRNA molecule) and of transcribed but non-translated "leader" and other sequences at the 5′ and 3′ ends of the gene which may be involved in message function.

However, recent results using the latest techniques of cDNA synthesis, plasmid cloning, restriction endonuclease digestion and nucleic acid hybridization have revealed puzzling features in the structure of the structural genes for rabbit and mouse β-globin (see Jeffreys and Flavell, 1977) mouse immunoglobulin γ_{II}-chains (see Nature **269**, 648 (1977)), chick ovalbumin (Breathnach et al., 1977), Drosophila rRNA (Glover and Hogness, 1977; White and Hogness, 1977) and a coat protein of the DNA virus SV40 (Aloni et al., 1977). It appears that these genes contain internal sequences which are not found in their respective mRNA transcripts and which cannot therefore code for a part of the gene product as detected in the cytoplasm. (In the case of Drosophila 28S rRNA, of course, the transcript is itself the gene product.) In the rabbit and mouse β-globin genes, the non-coding insert is approximately equal in size to the whole of the coding sequence which is split into two parts. In the case of the chick ovalbumin gene, the experiments leave open the possibility that the three separate coding regions may be very remote from each other, conceivably even on different chromosomes.

These results raise some very basic issues for genetics and for this reason deserve the closest possible scrutiny. An obvious suggestion might be that these additional sequences are found only in aberrant copies of reiterated genes and that the functional copies will show the expected co-linearity with their product. This may be the case for the Drosophila 28S ribosomal gene, as "normal" copies of the gene are found in the same cell. In the other cases, however, there is no evidence for multiple genes, and in any case the experiments would be expected to assay for that gene *most nearly* co-linear to the corresponding mRNA. It is not our wish to enter into a detailed critique of these exciting techniques; suffice it to say that, in the case of the rabbit β-globin gene, at least, there seems at present little reason to doubt that the structure of the functional gene is as described.

Whether these DNA inserts play a role in the control of gene expression is a matter of conjecture and first we must wait to see how inserts are omitted from mRNAs. There would seem to be three classes of possible mechanism. The RNA polymerase could fail to transcribe the insert and move over or past it to the separated fraction(s) of coding sequence, resulting in a covalently linked transcript of the coding sequences. Secondly, the inserted sequence could be transcribed with the coding sequences but be removed from a precursor of the mRNA by specific enzymes. Finally, the coding sequences might be transcribed separately and the RNAs then be joined to form a single mRNA. Different "joining" mechanisms may have predictably different consequences for the expression of normal or mutant structural genes which contain inserts (and may be of relevance in human β-thalassaemia). A genetic error in joining could map within the outer limits of a single structural gene to yield a defective protein. Alternatively, abnormal protein from defectively joined messengers could be due to defects in the enzymatic machinery for joining, the genes for which would map elsewhere in the genome. More generally, we are left wondering how widespread such inserted sequences are in higher organisms and whether they are limited to the genes for differentiated functions such as globin, ovalbumin and immunoglobulin. Finally we must again beware ignoring alternatives to the concept that gene transcription is regulated through *cis*-acting DNA sites which map *outside* the 5' end of a structural gene.

IX. CONCLUSIONS

Little is known of the mechanism of genetic regulation in higher eukaryotes. Most of our thinking is based upon mechanisms which have been identified in prokaryotes in which thorough genetic studies have been possible and it appears that even in these simple organisms, regulatory mechanisms exist at most conceivable levels of cellular organization. The mammalian cell is more complex and the differentiation of cells into somatic and germ line renders it intrinsically difficult to carry out further genetic studies on the mechanism of differentiation of somatic cells. Up to now, attempts to carry out such studies have been confounded by the difficulties of culturing differentiated cells *in vitro* and in generating and selecting suitable genetic mutants in somatic cells. For these reasons, developmental geneticists have concentrated on the analysis of lower eukaryotes such as fungi and Drosophila.

The biochemical genetics of Man and other higher organisms has revealed a wide variety of inherited changes in the protein phenotype. However, detailed protein analyses show that most of these mutations are located in structural genes rather than in regulatory genes or in regulatory DNA sites. In some respects, this does not significantly affect the work of the human biochemical geneticist, who is primarily concerned with the analysis of normal and pathological variation at the protein level, rather than with mechanisms of genetic regulation *per se*. Nevertheless, some of the pleiotropic effects he studies at the protein level may represent interesting regulatory phenomena, some of which may act at the level of gene transcription either quantitatively or even qualitatively. Similarly, when detecting heterozygotes by means of "half-normal" levels of the protein in question, it should be recognized that this method depends upon the absence of dosage compensation which itself raises an interesting question of regulation.

At present, the general lack of suitable genetic methods for searching for regulatory genes in mammals has led to a large number of studies on the protein phenotype of the nucleus and the cytoplasm of differentiated cells, in the hope that the structure of the former will provide an understanding of the genesis of differentiation in the latter. Although important and exciting results have stemmed from biochemical studies on steroid hormone induction and from reconstitution experiments on chromosomal proteins, the lack of suitable mutant cells prevents an independent assessment of the types of genetic regulation which may exist in higher eukaryotes.

It is to be hoped that the culture of teratoma cells may provide a system in which aspects of the genetics of mammalian development can be studied, ultimately at the molecular level. This is because teratoma cells can differentiate either *in vivo* or *in vitro* and because it may be possible for cells which have been mutagenized *in vitro* to develop into germ cells *in vivo*, where their formal genetics and developmental effects can be studied with more confidence.

REFERENCES

Ahmed, S. I. and Giles, N. H. (1969). *J. Bacteriol.* **99**, 231.
Aloni, Y., Dhar, R., Laub, O., Horowitz, M. and Khoury, G. (1977). *Proc. Natn. Acad. Sci. USA* **74**, 3686.
Ashburner, M. (1970). *Adv. Insect Physiol.* **7**, 1.
Ashburner, M. (1972). *Exp. Cell Res.* **71**, 433.
Ashburner, M. and Richards, G. (1976). *Symp. R. Ento. Soc. Lond.* **8**, 203.
Axel, R., Feigelson, P. and Schutz, G. (1976). *Cell* **7**, 247.
Baker, W. K. (1968). *Adv. Genet.* **14**, 133.
Bardin, C. W., Bullock, L. P., Sherins, R. J., Mowszowicz, I. and Blackburn, W. R. (1973). *Rec. Prog. Hormone Res.* **29**, 65.
Barrell, B. G., Air, G. M. and Hutchison, C. A. (1976). *Nature, Lond.* **264**, 34.
Barret, T., Maryanka, D., Hamlyn, P. H. and Gould, H. J. (1974). *Proc. Natn. Acad. Sci. USA* **71**, 5057.
Barstad, P., Weigert, M., Cohn, M. and Hood, L. (1974). *Proc. Natn. Acad. Sci. USA* **10**, 4096.
Bateson, W. (1909). "Mendel's Principles of Heredity". Cambridge University Press.
Beadle, G. W. (1945). *Physiol. Rev.* **25**, 643.
Benedict, W. F., Nebert, D. W. and Thompson, E. B. (1972). *Proc. Natn. Acad. Sci. USA* **69**, 2179.
Bennett, D. (1975). *Cell* **6**, 441.
Berlyn, M. S. (1967). *Genetics* **57**, 561.
Bertolotti, R. and Weiss, M. C. (1972a). *J. Cell Physiol.* **79**, 211.
Bertolotti, R. and Weiss, M. C. (1972b). *Biochimie* **54**, 195.
Beutler, E. and Kuhl, W. (1975). *Nature, Lond.* **258**, 262.
Bishop, J. O., Morton, J. G., Rosbash, M. and Richardson, M. (1974). *Nature, Lond.* **250**, 199.
Boyse, E. A. and Old, L. J. (1969). *Ann. Rev. Genet.* **3**, 269.
Bregula, U., Klein, G. and Harris, H. (1971). *J. Cell Sci.* **8**, 673.
Breathnach, R., Mandel, J. L. and Chambon, P. (1977). *Nature, Lond.* **270**, 314.
Brink, R. A. (1973). *Ann. Rev. Genet.* **7**, 129.
Britten, R. J. and Davidson, E. H. (1969). *Science* **165**, 149.
Brown, D. D. and Dawid, I. B. (1969). *Ann. Rev. Genet.* **3**, 127.
Burnet, F. M. (1959). "The Clonal Selection Theory of Acquired Immunity". Cambridge University Press.
Cahn, R. D. and Lasher, R. (1967). *Proc. Natn. Acad. Sci. USA* **58**, 1131.
Calhoun, D. N. and Hatfield, G. W. (1975). *Ann. Rev. Microbiol.* **29**, 275.
Capra, J. D. and Kindt, T. J. (1975). *Immunogenetics* **1**, 417.
Capra, J. D., Wasserman, R. L. and Kehoe, J. M. (1973). *J. Exp. Med.* **138**, 410.
Cattenach, B. M. (1975). *Ann. Rev. Genet.* **9**, 1.
Cellis, J. E., Smith, J. D. and Brenner, S. (1973). *Nature New Biol.* **241**, 130.
Chan, L., Means, A. R. and O'Malley, B. W. (1973). *Proc. Natn. Acad. Sci. USA* **70**, 1870.
Chiu, J-F., Tsai, Y-H., Sakuma, K. and Hnilica, L. S. (1975). *J. Biol. Chem.* **250**, 9431.
Chovnick, A., Gelbart, W. M., McCarron, M., Osmond, B., Candido, P. M. and Baillie, D. L. (1976). *Genetics* **84**, 233.

Cleaver, J. E. and Bootsma, D. (1975). *Ann. Rev. Genet.* **9**, 19.
Coffino, P., Knowles, B., Nathenson, S. G. and Scharff, M.D. (1971). *Nature New Biol.* **231**, 87.
Cohn, M. (1973). *In* "The Biochemistry of Gene Expression in Higher Organisms" (J. K. Pollack and J. W. Lee Eds) 574. D. Reidel Publishing Co.
Conconi, F., Rowley, P. T., del Senno, L., Pontremoli, S. and Volpato, S. (1972). *Nature New Biol.* **238**, 83.
Costa, M., Gerner, E. W. and Russell, D. H. (1976). *J. Biol. Chem.* **251**, 3313.
Cotton, R. G. H. and Milstein, C. P. (1973). *Nature, Lond.* **244**, 42.
Crick, F. H. C. (1971). *Nature, Lond.* **234**, 25.
Croce, C. M., Litwack, G. and Koprowski, H. (1973). *Proc. Natn. Acad. Sci. USA* **70**, 1268.
Damjanov, I. and Solter, D. (1974). *Curr. Topics Pathol.* **59**, 69.
Davidson, E. H. and Britten, R. J. (1973). *Q. Rev. Biol.* **48**, 565.
Davidson, E. H., Klein, W. H. and Britten, R. J. (1977). *Dev. Biol.* **55**, 69.
Davidson, R. L. (1974). *Ann. Rev. Genet.* **8**, 195.
Davidson, R. L., Ephrussi, B. and Yamomoto, K. R. (1968). *J. Cell Physiol.* **72**, 115.
Davison, J. (1973). *Br. Med. Bull.* **29**, 208.
de Weerd-Kastelein, E. A., Keijzer, W. and Bootsma, D. (1972). *Nature New Biol.* **238**, 80.
Douglas, H. C. and Hawthorne, D. C. (1964). *Genetics* **49**, 837.
Douglas, H. C. and Hawthorne, D. C. (1966). *Genetics* **54**, 911.
Ebner, K. E., Hageman, E. C. and Larson, B. L. (1961). *Exp. Cell. Res.* **25**, 373.
Eichmann, K. (1975). *Immunogenetics* **2**, 491.
Eichmann, K. and Berek, C. (1973). *Eur. J. Immurol*, **3**, 599.
Elgin, S. C. R. and Weintraub, H. (1975). *Ann. Rev. Biochem.* **44**, 725.
Ephrussi, B. (1972). "Hybridization of Somatic Cells". Princeton University Press.
Ephrussi, B. and Temin, H. (1960). *Virology* **11**, 547.
Fiers, W., Contreras, R., Duerinck, F., Haegeman, G., Iserentant, D., Merregaert, J., MinJou, W., Molemans, F., Raeymaekers, A., Van den Berghe, A., Volckaert, G. and Ysebaert, M. (1976). *Nature, Lond.* **260**, 500.
Finch, B. W. and Ephrussi, B. (1967). *Proc. Natn. Acad. Sci. USA* **57**, 615.
Fincham, J. R. S. and Sastry, G. R. K. (1974). *Ann. Rev. Genet.* **8**, 15.
Frangione, E., Lee, L., Haber, E. and Bloch, K. J. (1973). *Proc. Natn. Acad. Sci. USA* **70**, 1073.
Fukuda, R. and Doi, R. H. (1977). *J. Bacteriol.* **129**, 422.
Galau, G. A., Britten, R. J. and Davidson, E. H. (1974). *Cell* **2**, 9.
Galjaard, H., Hoogeveen, A., Keijzer, W., Wit-Verbeek, H. A. deW. and Reuser, A. J. J. (1975). *Nature, Lond.* **257**, 60.
Ganschow, R. and Paigen, K. (1967). *Proc. Natn. Acad. Sci. USA* **58**, 938.
Garrod, A. E. (1909). "Inborn Errors of Metabolism", Oxford University Press, Oxford.
Gelbart, W. M., McCarron, M., Pandey, J. and Chovnick, A. (1974). *Genetics* **78**, 869.
Gelbart, W. M., McCarron, M. and Chovnick, A. (1976). *Genetics* **84**, 211.
Georgiev, G. P. (1969). *J. Theor. Biol.* **25**, 473.
Giblett, E. R., Anderson, J. E., Cohen, F., Pollara, B. and Meuwissen, H. J. (1972). *Lancet ii*, 1067.

Glover, D. M. and Hogness, D. S. (1977). *Cell* **10**, 167.
Goldberg, M. L. (1975). *Life Sci.* **17**, 1747.
Goldberger, R. F. (1974). *Science* **183**, 810.
Goldstein, J. L., Dana, S. E., Brunschede, G. Y. and Brown, M. S. (1975). *Proc. Natn. Acad. Sci. USA* **72**, 1092.
Goldstein, J. L., Sothani, M. K., Faust, J. R. and Brown, M. S. (1976). *Cell* **9**, 195.
Gorski, J. and Gannon, F. (1976). *Ann. Rev. Physiol.* **38**, 425.
Gravel, R. A., Mahoney, M. J., Ruddle, F. H. and Rosenberg, L. E. (1975). *Proc. Natn. Acad. Sci. USA* **72**, 3181.
Green, C. D. and Tata, J. R. (1976). *Cell* **7**, 131.
Gross, K., Probst, E., Schaffner, W. and Birnstiel, M. L. (1976a). *Cell* **8**, 455.
Gross, K., Schaffner, W., Telford, J. and Birnstiel, M. L. (1976b). *Cell* **8**, 479.
Gurdon, J. B. (1970). *Proc. R. Soc. B.* **176**, 303.
Gurdon, J. B. (1974). *In* "The Control of Gene Expression in Animal Development". Clarendon Press, Oxford.
Harris, H. (1968). "Nucleus and Cytoplasm" 2nd ed. Clarendon Press, Oxford.
Harris, H. (1970). "Cell Fusion". Clarendon Press, Oxford.
Harris, H. (1975). *In* "The Principles of Human Biochemical Genetics" 2nd ed., 232–242. North Holland Publishing Co., Oxford and Amsterdam.
Harris, S. E., Means, A. R., Mitchell, W. M. and O'Malley, B. W. (1973). *Proc. Natn. Acad. Sci. USA* **70**, 3776.
Harrison, P. R., Affara, N., Conkie, D., Rutherford, T., Somerville, J. and Paul, J. (1976). *In* "Progress in Differentiation Research" (N. Müller-Bérat, Ed.) North Holland Publishing Co., Amsterdam.
Hayman, D. L. and Martin, P. G. (1965). *Genetics* **52**, 1201.
Hobart, M. J. and McConnell, I. (1975). "The Immune System" Blackwell, London.
Holtzer, H., Abbott, J. and Lash, J. (1960). *Proc. Natn. Acad. Sci. USA* **46**, 1533.
Hood, L. (1973). *Stadler Genet. Symp.* **5**, 73.
Hood, L., Campbell, J. H. and Elgin, S. C. R. (1975). *Ann. Rev. Genet.* **9**, 305.
Hopkinson, D. A., Edwards, Y. H. and Harris, H. (1976). *Ann. Hum. Genet.* **39**, 383.
Housman, D., Forget, B. G., Skoultchi, A. and Benz, E. J. (1973). *Proc. Natn. Acad. Sci. USA* **70**, 1809.
Humphries, S., Windass, J. and Williamson, R. (1976). *Cell* **7**, 267.
Imanishi, T. and Mäkelä, O. (1974). *J. Exp. Med.* **140**, 1498.
Imanishi, T. and Mäkelä, O. (1975). *J. Exp. Med.* **141**, 840.
Imanishi, T., Hurme, M., Sarvas, H. and Mäkelä, O. (1975). *Eur. J. Immunol.* **5**, 198.
Isac, R., Mozes, E. and Taussig, M. J. (1976). *Immunogenetics* **3**, 409.
Jacob, F. and Monod, J. (1961). *J. Mol. Biol.* **3**, 348.
Jeffreys, A. J. and Flavell, R. A. (1977) *Cell* **12**, 1097.
Judd, B. H., Shen, M. W. and Kaufman, T. C. (1972). *Genetics* **71**, 139.
Kedes, L. H. (1976). *Cell* **8**, 321.
Kedes, L. H. and Gross, P. (1969). *Nature, Lond.* **223**, 1335.
Kempe, T. D., Swyryd, E. A., Bruist, M. and Stark, G. R. (1976). *Cell* **9**, 541.
Kenny, F. T., Lee, K.—L. and Ihle, J. N. (1974). *In* "Control Processes in Neoplasia" (M. A. Mehlman and B. W. Hanson, Eds). Academic Press, New York and London.
Kew, O. M. and Douglas, H. C. (1976). *J. Bacteriol.* **125**, 33.

90   R. G. SUTCLIFFE, B. CARRITT AND R. H. WILSON

King, T. J. and Briggs, R. (1952). *Proc. Natn. Acad. Sci. USA* **38**, 455.
Kindt, T. J., Mandy, W. J. and Todd, C. W. (1970). *Biochemistry* **9**, 2028.
Klebe, R. J., Chen, T. and Ruddle, F. H. (1970). *Proc. Natn. Acad. Sci. USA* **66**, 1220.
Klein, E. and Wiener, F. (1971). *Exp. Cell Res.* **67**, 251.
Klein, G., Bregula, U., Wiener, F. and Harris, H. (1971). *J. Cell Sci.* **8**, 659.
Kleinsmith, L. J., Allfrey, V. G. and Mirsky, A. E. (1966). *Science* **154**, 781.
Köhler, G. and Milstein, C. (1975). *Nature, Lond.* **256**, 495.
Kraemer, K. H., Coon, H. G., Petinga, R. A., Barrett, S. F., Rahe, A. E. and Robbins, J. H. (1972). *Proc. Natn. Acad. Sci. USA* **72**, 59.
Krebs, E. G. (1972). *Curr. Topics Cell Reg.* **5**, 99.
Kuhn, R. W., Schrader, W. T., Coty, W. A., Conn, P. M. and O'Malley, B. W. (1977). *J. Biol. Chem.* **252**, 308.
Lewin, B. (1974). *In* "Gene Expression" Vol. 2. Wiley, London.
Lewin, R. (1976). *New Sci.* **72**, 148.
Lichtenstein, A. V. and Shapot, V. V. (1976). *Biochem. J.* **159**, 783.
Lin, S-Y, and Riggs, A. D. (1975). *Cell* **4**, 107.
Loo, W. van der, Baetzelier, P. de, Hamers-Casterman, C. and Hamers, R. (1977). *Eur. J. Immunol.* **7**, 15.
Lucchesi, J. C. (1973). *Ann. Rev. Genet.* **7**, 225.
Lumsden, J. and Coggins, J. R. (1977). *Biochem. J.* **161**, 599.
Lyon, M. F., Glenister, P. H. and Lamoreux, M. L. (1975). *Nature, Lond.* **258**, 620.
Lyons, L. B., Cox, R. P. and Dancis, J. (1973). *Nature, Lond.* **243**, 533.
Mage, R. G. (1971). "Prog. Immunol. 1st Int. Congr. Immunol" (B. Amos, Ed.) 47. Academic Press, New York and London.
Mage, R. G. (1974). *Curr. Topics Microbiol. Immunol.* **63**, 131.
Mage, R. G., Lieberman, R., Potter, M. and Terry, W. D. (1973). *In* "The Antigens" (M. Sela, Ed.), 299. Academic Press, London and New York.
Mahoney, M. J., Hart, A. C., Steen, V. D. and Rosenberg, L. E. (1975). *Proc. Natn. Acad. Sci. USA* **72**, 2799.
Maniatis, T. and Ptashne, M. (1976). *Sci. Am.* **234**, 64.
Markert, C. L. (1963). *Science* **140**, 1329.
Martin, G. R. (1975). *Cell* **5**, 229.
McClintock, B. (1965). *Brookhaven Symp. Biol.* **18**, 162.
McClintock, B. (1968). *Symp. Soc. Dev. Biol.* **26**, 84.
McDevitt, H. O. (1973). *In* "Medical Genetics" (V. A. McKusick and R. Claiborne, Eds). H.P. Publishing Co., New York.
McKinnell, R. G., Deggins, B. A. and Labat, D. D. (1969). *Science* **165**, 394.
McKusick, V. A. (1975). "Mendelian Inheritance in Man", 4th ed. Johns Hopkins University Press, Baltimore.
Merritt, A. D., Rivas, M. L. and Ward, J. C. (1972). *Nature, Lond.* **239**, 243.
Merritt, A. D., Rivas, M. L., Bixler, D. and Newell, R. (1973). *Am. J. Hum. Genet.* **25**, 510.
Miller, R. A. and Ruddle, F. H. (1976). *Cell* **9**, 45.
Milstein, C. and Köhler, G. (1975). *Nature, Lond.* **256**, 495.
Milstein, C., Brownlee, G. G., Cartwright, E. M., Jarvis, J. M. and Proudfoot, N. J. (1974). *Nature, Lond.* **252**, 354.
Mintz, B. and Illmensee, K. (1975). *Proc. Natn. Acad. Sci. USA* **72**, 3585.
Mori, M. and Tatibana, M. (1975). *J. Biochem.* **78**, 239.

Mudgett, M., Fraser, B. A. and Kindt, T. J. (1975). *J. Exp. Med.* **141**, 1448.
Munro, A. and Bright, S. (1976). *Nature, Lond.* **264**, 145.
Nadler, H. L., Chacko, C. M. and Rachmeler, M. (1970). *Proc. Natn. Acad. Sci. USA* **67**, 976.
Neufeld, E. F. and Cantz, M. J. (1971). *Ann. New York Acad. Sci.* **179**, 580.
Neurath, H. and Walsh, K. A. (1976). *Proc. Natn. Acad. Sci. USA* **73**, 3825.
Noll, M. (1976). *Cell* **8**, 349.
Norby, S. (1970). *Hereditas* **73**, 11.
Oka, T. and Schimke, R. T. (1969). *J. Cell. Biol.* **41**, 816.
Olins, A. L. and Olins, D. E. (1974). *Science* **183**, 330.
O'Malley, B. W. and Means, A. R. (1974). *Science* **183**, 610.
Ottolenghi, S., Lanyon, W. G., Williamson, R., Weatherall, D. J., Clegg, J. B. and Pitcher, C. S. (1975). *Proc. Natn. Acad. Sci. USA* **72**, 2294.
Ottolenghi, S., Comi, P., Giglioni, B., Williamson, R., Vullo, G. and Conconi, F. (1977). *Nature, Lond.* **266**, 231.
Palmiter, R. D., Moore, P. B., Mulvihill, E. R. and Emtage, S. (1976). *Cell* **8**, 557.
Papaioannou, V. E., McBurney, M. W., Gardner, R. L. and Evans, M. J. (1975). *Nature, Lond.* **258**, 70.
Parkman, R., Hagemeijer, A. and Merler, E. (1971). *Fed. Proc.* **30**, 530.
Pastan, I. and Adhya, S. (1976). *Bacteriol. Rev.* **40**, 527.
Pateman, J. A. and Kinghorn, J. (1976). *In* "The Filamentous Fungi" (J. D. Smith and D. R. Berry, Eds). 159. Edward Arnold, London.
Paul, J. (1972). *Nature, Lond.* **238**, 444.
Paul, J. and Gilmour, R. S. (1968). *J. Mol. Biol.* **34**, 305.
Periman, P. (1970). *Nature, Lond.* **228**, 1086.
Peterson, J. A. and Weiss, M. C. (1972). *Proc. Natn. Acad. Sci. USA* **69**, 571.
Poste, G. (1972). *Int. Rev. Cytol.* **32**, 157.
Prostlethwait, J. H. and Schneiderman, H. A. (1973). *Ann. Rev. Genet.* **7**, 381.
Ptashne, M. (1975). *Harvey Lect.* **69**, 143.
Rawls, J. M. and Fristrom, J. W. (1975). *Nature, Lond.* **255**, 738.
Ray, P. N. and Pearson, M. L. (1975). *Nature, Lond.* **253**, 647.
Reel, J. R., Lee, K.-L. and Kenny, F. T. (1970). *J. Biol. Chem.* **245**, 5800.
Richards, F. F., Konigsberg, W. H. and Rosenstein, R. H. (1975). *Science* **187**, 130.
Ritossa, F. M. (1968). *Proc. Natn. Acad. Sci. USA* **59**, 1124.
Ruddle, F. H. and Creagan, R. P. (1975). *Ann. Rev. Genet.* **9**, 407.
Russell, L. B. and Montgomery, C. L. (1969). *Genetics* **63**, 103.
Sarabhai, A., Stretton, A. O. W. and Brenner, S. (1964). *Nature, Lond.* **201**, 13.
Schaffner, W., Gross, K., Telford, J. and Birnstiel, M. (1976). *Cell* **8**, 471.
Schneider, J. A. and Weiss, M. C. (1971). *Proc. Natn. Acad. Sci. USA* **68**, 127.
Schwartz, R. J., Kuhn, R. W., Buller, R. E., Schrader, W. T. and O'Malley, B. W. (1976). *J. Biol. Chem.* **251**, 5166.
Shermoen, A. W. and Kiefer, B. I. (1975). *Cell* **4**, 275.
Shire, J. G. M. (1976). *Biol. Rev.* **51**, 105.
Shoaf, W. T. and Jones, M. E. (1973). *Biochemistry* **12**, 4039.
Shreffler, D. C. and David, C. S. (1975). *Adv. Immunol.* **20**, 125.
Soderholm, G., Schwartz, M. and Norby, S. (1975). *FEBS Lett.* **53**, 148.
Southerland, E. W. and Rull, T. W. (1960). *Pharmacol. Rev.* **12**, 265.
Stanbridge, E. J. (1976). *Nature, Lond.* **260**, 17.
Steinberg, R. A., Levinson, B. B. and Tomkins, G. M. (1975). *Cell* **5**, 29.
Steward, M. W. (1974). "Immunochemistry". Chapman and Hall, London.

Stockdale, F. E. and Holtzer, H. (1961). *Exp. Cell Res.* **24**, 508.
Storti, R. V., Coen, D. M. and Rich, A. (1976). *Cell* **8**, 521.
Strosberg, A. D., Hamers-Casterman, C., van der Loo, W. and Hamers, R. (1974). *J. Immunol.* **113**, 1313.
Swank, R. T., Paigen, K. and Ganschow, R. E. (1973). *J. Mol. Biol.* **81**, 225.
Tartof, K. D. (1975). *Ann. Rev. Genet.* **9**, 355.
Tata, J. R. (1976). *Cell* **9**, 1.
Taussig, M. J. and Munroe, A. J. (1976). *Fed. Proc.* **35**, 2061.
Thomas, G. H., Taylor, H. A., Miller, C. S., Axelman, J. and Migeon, B. R. (1974). *Nature, Lond.* **250**, 580.
Thomas, J. O. and Kornberg, R. D. (1975). *Proc. Natn. Acad. Sci. USA* **72**, 2626.
Thompson, E. B. and Gelehrter, T. D. (1971). *Proc. Natn. Acad. Sci. USA* **68**, 2589.
Tomkins, G. M., Gelehrter, T. D., Granner, D., Martin, D., Samuels, H. H. and Thompson, E. B. (1969). *Science* **166**, 1475.
Tonegawa, S. and Steinberg, C. (1975). *In* "The Generation of Antibody Diversity; A New Look" (A. J. Cunningham, Ed.). 175. Academic Press, London and New York.
Travers, A. (1976). *Nature, Lond.* **263**, 641.
Trevor-Roper, P. D. (1952). *Br. J. Ophthal.* **36**, 107.
Waddington, C. H. (1957). "Strategy of the Genes". Allen and Unwin, London.
Wasserman, R. L., Kehoe, J. M. and Capra, J. D. (1974). *J. Immunol.* **113**, 954.
Watson, J. D. (1976). "Molecular Biology of the Gene" (3rd ed.) W. A. Benjamin, California.
Waymouth, C., Chen, H. W. and Wood, B. G. (1971). *In Vitro* **6**, 371.
Weatherall, D. J. and Clegg, J. B. (1976). *Ann. Rev. Genet.* **10**, 157.
Weigert, M., Potter, M. and Sachs, D. H. (1975). *Immunogenetics* **1**, 511.
Weiss, M. C. and Chaplain, M. (1971). *Proc. Natn. Acad. Aci. USA* **68**, 3026.
Wessells, N. K. and Rutter, W. J. (1969). *Sci. Am.* **220**, 36.
White, R. L. and Hogness, D. S. (1977). *Cell* **10**, 177.
Whitehouse, H. L. K. (1967). *Nature, Lond.* **215**, 371.
Whitehouse, H. L. K. (1973). "Towards an Understanding of the Mechanism of Heredity" (3rd ed.). Edward Arnold, London.
Wiener, F., Klein, G. and Harris, H. (1971). *J. Cell Sci.* **8**, 681.
Wiener, F., Klein, G. and Harris, H. (1973). *J. Cell Sci.* **12**, 253.
Wigley, C. B. (1975). *Differentiation* **4**, 25.
Williamson, A. R. (1976). *Ann. Rev. Biochem.* **45**, 467.
Williamson, A. R. and Fitzmaurice, L. C. (1975). *In* "The Generation of Antibody Diversity: A New Look" (A. J. Cunningham, Ed.). Academic Press, London and New York.
Woo, S. L. C. and O'Malley, B. W. (1975). *Life Sci.* **17**, 1039.
Yamamoto, K. and Alberts, B. M. (1976). *Ann. Rev. Biochem.* **45**, 721.
Yanofsky, C., Carlton, B. C., Guest, J. R., Helsinki, D. R. and Henning, U. (1964). *Proc. Natn. Acad. Sci. USA* **51**, 266.
Yanofsky, C. and Horn, V. (1972). *J. Biol. Chem.* **247**, 4494.
Yoshida, A. (1970). *J. Mol. Biol.* **52**, 483.

3 Chromosomes and Protein Variation

W. KRONE and U. WOLF

*Department of Human Genetics, University of Ulm, West Germany
and Institute of Human Genetics and Anthropology, University
of Freiburg, West Germany*

I. INTRODUCTION

The biochemical consequences of numerical and structural anomalies of the human chromosome complement are investigated for two main reasons: to contribute to our knowledge of the location of genes on human chromosomes, and to further our understanding of the complex syndromes caused by chromosomal aberrations.

Gene assignments to human chromosomes and their more precise localization on chromosome segments are primarily achieved by the methods of somatic cell genetics, notably by the combined biochemical and cytogenetic analysis of the segregation of marker genes and chromosomes in interspecific hybrid cells. Structural aberrations like deficiencies and rearrangements have proven to be valuable tools in these studies. In this way and via the analysis of instructive pedigrees, studies of these minor chromosomal aberrations have contributed much more to our knowledge of the chromosomal map than those of the major trisomies.

With regard to our understanding of these syndromes it has become commonplace to attribute them to genetic imbalance. Were it known how genetic balance is achieved in normal chromosome complements, this term could be used with much more justification. But it is precisely our ignorance in this respect which has stimulated research on the consequences of chromosomal aberrations and — in confirmation of Sir William Harvey's famous recommendation (see Chapter 14 Section VII) — much has in fact been learned this way, especially with the classical objects of genetic research like Drosophila, *Zea mays*, and, more recently, fungi.

In mammals the investigation of the intricate mechanisms which bring about the arrest of development of a chromosomally "unbalanced" zygote or the development of a grossly malformed organism is still in a very preliminary state. Progress along these lines will depend on an interdisciplinary combination of embryology, cell physiology, molecular biology and biochemistry, to name only the most important fields involved. Theoretical considerations of the possible effects of chromosomal aberrations on gene expression, differentiation and proliferation may yield testable working hypotheses.

In accordance with the subject matter of this chapter, the discussion is largely restricted to the biochemical approach and preference is given to those studies dealing with the effects of chromosomal aberrations on defined species of proteins, mostly enzymes.

II. THE HUMAN CHROMOSOMES

A. Identification and Linear Differentiation

1. Chromosome banding patterns by differential staining
The possibility of correlating chromosome and protein variation has greatly improved in recent years. This progress traces back to the introduction of a

number of differential staining procedures at the beginning of this decade, which are still being added to and refined. In this chapter, the characteristics of the individual chromosomes after differential staining will not be discussed in detail, but can be referred to in a number of reviews and monographs such as Caspersson and Zech (1973), Miller *et al.* (1973), Dutrillaux

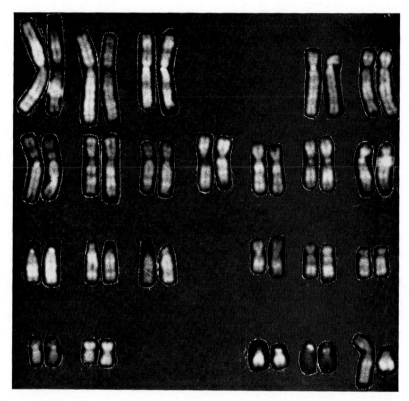

Fig. 1   Male karyotype showing Q-banding arranged according to the Paris Conference (1972) (Courtesy of Dr. W. Vogel, Freiburg)

and Lejeune (1975), Schwarzacher (1976). Directions on nomenclature and documentation of cytogenetic data were agreed upon at the Paris Conference (1971) and a Supplement (1975) to it. A karyotype representative of the current resolving power of the chromosome banding pattern after differential staining is depicted in Fig. 1.

The banding patterns allow for a sub-division of the karyotype in units of 0·5 to 1 μ. Thus, it has become possible to notice structural deviations within the size range of 0·5 μ. This, however, does not mean that each

96 W. KRONE AND U. WOLF

recognizable aberration of this size can be identified in terms of a particular chromosome segment, since some segments of non-homologous chromosomes show a very similar sequence of banding. Estimations of the number of bands distinguishable in the haploid mitotic karyotype vary from 500 in the light microscope (Dutrillaux, 1975) to nearly 700 in the electron microscope (Bahr et al., 1973), and they may even approach the number of genes, if estimated from electron micrographs of prematurely condensed G1-chromosomes, lying in the order of 10^4 to 10^5 (Schwarzacher, 1976). A recent investigation of early mitotic stages by Yunis (1976) yielded between 843 and 1256 bands in late prophase chromosomes.

While there are various techniques of differential staining which make possible the identification of all individual chromosomes, certain chromosome regions can be characterized by only one staining procedure.

In the Q (quinacrine)-banding technique, introduced by Caspersson et al. (1970) (the first method to allow for identification of all human chromosomes), the distal ends are not stained except for the long arm of the Y-chromosome and the satellites of the acrocentrics. In contrast, R (reverse)-banding (Dutrillaux and Lejeune, 1971) demonstrates the segments complementary to the Q-bands and consequently stains the distal segments (as does a variant, T (telomere)-banding (Dutrillaux, 1973) which represents the remainder of R-banding after the application of strong denaturation conditions).

Both Q- and R-banding do not stain the centromeres or the secondary constrictions. The G (Giemsa)-bands (Schnedl, 1971; Sumner et al., 1971) coincide with the Q-bands, but with some C-positive segments being additionally stained. The C (constitutive heterochromatin)-bands (Arrighi and Hsu, 1971; Yunis et al., 1971) are, by definition, the areas of constitutive heterochromatin which are essentially the centromeric regions, secondary constrictions and the distal long arm of the Y-chromosome.

Among a large number of modifications producing banding patterns, is the method applying proteolytic enzymes resulting in G-bands (Dutrillaux et al., 1971; Seabright, 1972), and an acridine orange fluorescence technique resulting in R-bands (Dutrillaux et al., 1973; de la Chapelle et al., 1973).

2. Chromosome replication patterns

The conventional method of demonstrating DNA replication patterns by autoradiography after ^{3}H-thymidine incorporation has largely been replaced by techniques using BUdR, resulting in a much higher resolution of the patterns.

Zakharov and Egolina (1972) have shown that those chromosome regions replicating under exposure to BUdR are delayed in mitotic spiralization.

Proceeding from the finding of Galley and Purkey (1972) which showed that the introduction of heavy atoms into biopolymers perturbs fluorescence after staining with an acridine dye, Latt (1973) succeeded in demonstrating chromosome replication patterns. He observed that chromosome segments substituted with BUdR are less fluorescent with the benzimidazol derivative "Hoechst 33258" (Hilwig and Gropp, 1972) than non-substituted segments. In a more refined version of the method, Giemsa staining was introduced to differentiate between substituted and unsubstituted chromatin (Perry and Wolff, 1974). The complete sequence of the late replication patterns of human chromosomes was subsequently studied (Epplen *et al.*, 1975; Epplen and Vogel, 1975; Grzeschik *et al.*, 1975). By and large there is a good correlation between the BUdR patterns and the G-bands, some minor differences including one or other late replicating region which is only faintly stained by the G-banding technique. However, in contrast to all the banding techniques mentioned, the late replicating X-chromosome is, of course, clearly visible (Fig. 2). The early replicating patterns, in turn, correspond to the R-bands as is to be expected (Kim *et al.*, 1975; Epplen *et al.*, 1976).

B. Variability

Critical examination of chromosome variability as the possible basis of altered phenotypic expression encounters two problems, both of which lie in the nature of the chromosomes themselves. On the one hand, there exist chromosome aberrations without any detectable consequence on the phenotype, so that these aberrations belong in the range of normal variability and may give rise to mere chromosomal polymorphism. On the other hand, chromosome aberrations may affect the phenotype, i.e. lead to clinical manifestations, but the aberrations are not easily detected or may even escape detection at all, due to the smallness of the affected chromosome segments or to hidden translocations.

The cause of these difficulties is the functional heterogeneity of chromatin, as discussed in Section II.D. Therefore, certain chromosomal regions may be duplicated or deleted without any effect (functionally inert material, non-informative DNA), while minimal deviations of other chromosomal segments may result in profound physical or mental disturbances. In addition, the cytogeneticist is confronted with the problem of whether a chromosome rearrangement is balanced or unbalanced.

This question cannot be solved by chromosome analysis alone. While the reciprocal exchange of chromosome segments without loss is readily compatible with normal development in most cases, there may exist

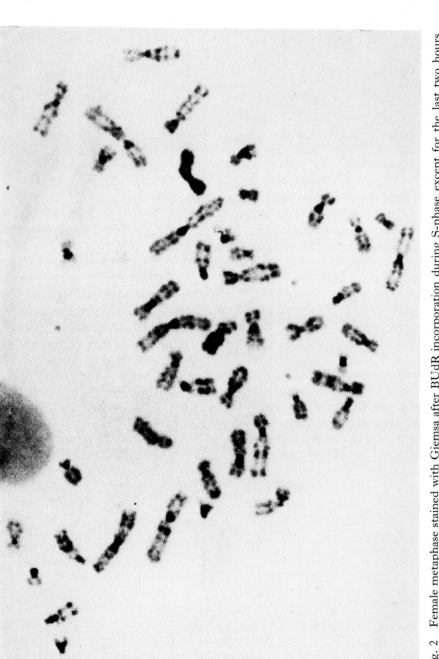

Fig. 2 Female metaphase stained with Giemsa after BUdR incorporation during S-phase except for the last two hours. Note the prominent staining of the late replicating X (Courtesy of Dr. W. Vogel, Freiburg)

situations of structural rearrangement exhibiting clinical anomalies without identifiable loss or gain of chromosome material (Wilson et al., 1973). Since all such cases are sporadic, the congenital defects may have an independent origin, or may be due to position effects (Engel et al., 1971).

1. Normal variation

There is extensive, and presumably normal variation in certain chromosome regions giving rise to chromosomal structural polymorphism. In particular the short arms and satellites of the acrocentrics (group D- and G-chromosomes), the secondary constrictions of chromosomes 1, 9, and 16 (and the satellites of 17), and the long arm of the Y-chromosome are prone to variation. In a synopsis of the results of three newborn series comprising 27 587 individuals, Nielsen and Sillesen (1975) report on an average frequency of such variants of about 16 per 1000. Looking more closely into the variation of constitutive heterochromatin, Craig-Holmes et al. (1973) determined a higher number of C-band variants than the number of individuals studied. The C-band heterochromatin was shown to vary in pedigrees as well as within individuals, apparently due to unequal exchanges of repetitive DNA during meiosis and also in somatic cells (Craig-Holmes et al., 1975). Applying other chromosome banding methods, additional variable regions appeared, e.g. the proximal long arm region of chromosome 3 fluorescing after Q-staining in 50% of the individuals studied (Schnedl, 1973). The simultaneous employment of G-, Q- and C-banding techniques on the same preparation revealed an average frequency of variants of $5 \cdot 08 \pm 0 \cdot 23$ per individual (McKenzie and Lubs, 1975). The frequencies vary in different populations, and for the Q-band intensity, even an age dependence may exist (Buckton et al., 1976). The lateral asymmetry of constitutive heterochromatin (see Section II.D), originally detected in the mouse (Lin et al., 1974), was found to be the basis of a new type of polymorphism. Inherited variation of both kind and extent of the asymmetric patterns was found in the heterochromatin of chromosomes 1, 15, 16, and Y (Angell and Jacobs, 1975) and of chromosome 9 (Galloway and Evans, 1975). If all possible normal variants as detectable by the various banding techniques are taken into consideration, every person will have a unique karyotype (Dutrillaux, 1975).

2. Pathological variation

Since there exists a quasi-continuous transition from normal to pathological variation, it is nearly impossible to predict whether or not a chromosomal structural deviation results in pathological effects. Each new case, if not identical with a well established chromosome syndrome, has to be considered individually. In view of this problem, K. Pätau designed a comprehensive study on a sample of patients with congenital anomalies and a control population as early as 1963, the "Madison Blind Study". The findings were published in a series of articles (for the most recent one, see Doyle, 1976), in which the chromosomal variants could be attributed to three classes of karyotype–phenotype relationship:

> Those which amount to gross genetic imbalance and obviously are the cause of congenital anomalies; those which are equally frequent in normal persons and therefore presumably do not cause the syndrome; and those in which the relationship is as yet unclear (Magnelli, 1976).

Some examples may be given. The Robertsonian translocation, of two acrocentric chromosomes with loss of all or most of the short arms, is a deficiency without phenotypic consequences; there is, however, the predisposition to trisomic offspring. On the other hand, the full clinical picture of Down's syndrome is expressed in patients with apparently normal chromosomes, while only the barely identifiable distal segment of chromosome 21 may be hidden somewhere in the karyotype (Williams *et al.*, 1975). In this connection, Niebuhr (1974) introduced the term "pathogenetic band" which refers here to the single band 21q22. There are an increasing number of clinical conditions which can be ascribed to the duplication or deletion of a minor chromosome segment (Schinzel, 1976). Among the carriers of a rather conspicuous chromosome anomaly (an additional small centric fragment), some exhibit clinical manifestations, while others appear normal, depending on the nature of that fragment (Ayraud *et al.*, 1976).

C. Variability and Linear Differentiation

Normal and pathological variation are not independent of the linear differentiation of the chromosomes. This interdependence points to functional implications of the banding patterns. As already mentioned, there is a striking concordance of the late replication patterns and the G(or Q)-bands (Epplen *et al.*, 1975; Epplen and Vogel, 1975). It is assumed as a general phenomenon that the late replicating segments are heterochromatin (Lima-de-Faria and Jaworska, 1968), and that consequently, the G-bands by and large will represent heterochromatin (the same, of course, is true for the C-bands), while the R-bands are composed of euchromatin. This assumption is supported by the studies of Sanchez and Yunis (1974), revealing that repetitious DNA is localized in the Q- and C-positive regions (and, in addition, in some of the telomeres) (Section II.D).

Indeed, normal chromosome variation extends mainly, if not exclusively, to Q- and C-bands (McKenzie and Lubs, 1975), while there seems to be no variation in the R-bands (except for some heterochromatic regions also fluorescing with R-banding). Pathological chromosome variation, likewise, is not distributed at random within the human genome. A synoptic consideration by Hoehn (1975) led to the conclusion that a negative correlation exists between occurrence and frequency of pathological chromosome aberrations and the portion of R-positive material involved (Fig. 3).

In a cautious appreciation of the present state of knowledge in the banding age of cytogenetic analysis, it must be admitted that we are still far from relevant, unequivocal correlations between chromosome variation and phenotype in many cases, even if the chromosome identification problems have largely been solved. This means that we cannot always predict what the consequences of a particular chromosome aberration may be, and to a minor extent also that we may still miss minor deviations which may be of significance for the carrier or his offspring. With an increasing knowledge of where particular genetic information is localized on chromosome segments, these uncertainties will be reduced, and we are approaching this aim rather rapidly.

D. Human DNA and the Number of Genes in Man

The mammalian genome is characterized by a remarkably uniform DNA content. While chromosome number varies between 2n = 6 (female Indian muntjak; Wurster and Benirschke, 1970) and 2n = 92 (*Anatomys leander*; Gardner, 1971), the amount of DNA lies within narrower limits, the lowest and the highest values observed differing by a factor of less than 10 (Bachmann, 1972). The diploid human cell nucleus in the G1 phase of the cell cycle contains, according to well controlled cytophotometric measurements (Bachmann, 1972), $7 \cdot 3 \times 10^{-12}$g of DNA (range: $6 \cdot 6$ to $8 \cdot 0$). Taking the G + C content of mammalian DNA (41%) into account, the weight of an average nucleotide pair in the human genome is $1 \cdot 0259 \times 10^{-21}$g. Dividing the DNA content by this figure yields the number of nucleotide pairs as $7 \cdot 1 \times 10^9$. Any reasonable assumption on the average number of nucleotides per gene will yield gene numbers in the range of 2×10^6 to 6×10^6 per haploid human genome (Vogel, 1964). Such a high number of genes cannot be reconciled with the rate with which each of these genes tends to undergo mutation. If, for instance, the mutation rate for dominant lethals is assumed to be 5×10^{-6} per gamete per generation, and there are 2×10^6 genes, there should be on the average 10 such mutations per individual. The proportion of unaffected zygotes would follow from the zero term of the Poisson distribution to be e^{-10}, which shows that no species

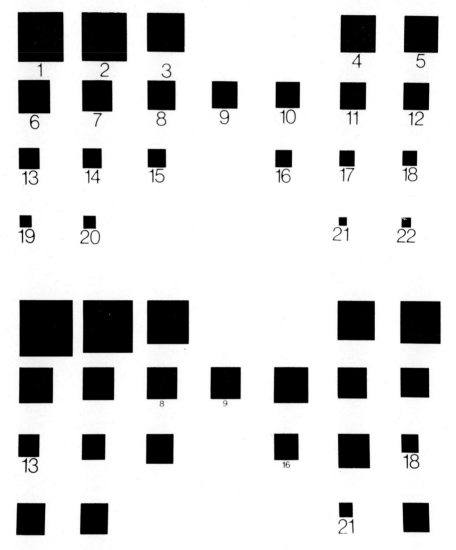

Fig. 3 Representation of the human autosomes by size and banding pattern according to Hoehn (1975)
Upper panel: DNA equivalents taken from Mendelsohn *et al.* (1973); Lower panel: arithmetical summation of size and R band values, which were taken from Holmberg and Jonasson (1973) and express the proportion of dark to light bands for a given chromosome after application of the R banding procedure (Courtesy of Dr Hoehn, Seattle and the University of Chicago Press)

with the assumed gene number and a small litter size would ever be able to propagate.

It is very difficult to estimate the rate of foetal loss due to dominant lethal gene mutations. For X-linked recessive lethals Krehbiehl (1966) calculated a rate of $0·0036 \pm 0·0007$ per generation per X-chromosome. About 6% of the haploid human complement is contained in the X-chromosome. Hence, the total rate of recessive lethal mutations would be $0·0036/0·06 = 0·06$ per gamete and generation. The majority of X-linked recessive genes of man show mutation rates below 10^{-6} (Stevenson and Kerr, 1967). If, for the purpose of this discussion, we make the disputable assumption of an average mutation rate of 5×10^{-7} we obtain a gene number of $1·2 \times 10^5$ per gamete. This is an order of magnitude less than the figures derived from the amount of DNA and the size of an average structural gene. About 1200 nucleotide pairs are required to code for an average polypeptide chain whose molecular weight, roughly 46 000 daltons was computed from the subunit sizes of 99 human enzymes by Hopkinson *et al.* (1976). For $1·2 \times 10^5$ genes of this size a total number of $1·44 \times 10^8$ nucleotide pairs would provide the genetic information, which is only 4% of the number of nucleotide pairs of the haploid human genome!

A similar discrepancy between the amount of DNA and the number of mutable gene loci exists in *Drosophila melanogaster*. The haploid amount of DNA in this organism corresponds to $1·8 \times 10^8$ nucleotide pairs, about a twentieth of that of the human complement. The DNA of Drosophila could thus code for $1·5 \times 10^5$ polypeptide chains of the above mentioned average size. Yet studies with spontaneous and induced mutations, occurring in well defined regions of the Drosophila genome, establish essentially a one–to–one correspondence between the number of identifiable genetic functions (complementation groups) and the number of bands of the respective segment in the polytene chromosomes (Lefevre, 1974; Judd and Young, 1974). There are about 5000 bands in the polytene chromosomes of *D. melanogaster*, and hence, only a thirtieth of the number of genes for which Drosophila DNA could potentially code. Since the mutation studies identify complementation groups rather than structural genes proper, the problem narrows down to the question of the structural organization of these functional units, whose cytological equivalents are the chromomeric bands. These problems were more thoroughly discussed by Muller (1965) who came to the conclusion that there are 30 000 genes in the human complement, an estimate identical with that of Vogel (1964) which was derived by similar reasoning.

The analysis of the reassociation kinetics of denatured eukaryotic DNA by various techniques has afforded considerable progress in the elucidation of the sequence organization of eukaryotic genomes (Davidson *et al.*, 1973).

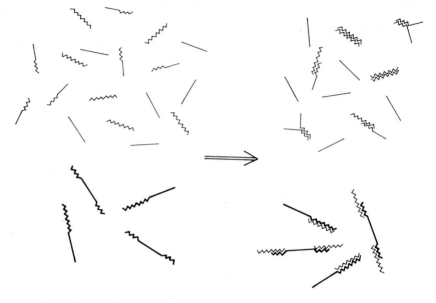

Fig. 4   ∧∧∧∧∧∧∧∧∧repetitive sequences
         ─────────────unique sequences
Upper part: short single stranded fragments are reassociated under conditions
which allow only repetitive sequences to reanneal; most of the unique sequences
remain single stranded.
Lower part: a limiting amount of radioactively labelled (heavy line) long single
stranded fragments is added; due to interspersed repetitive sequences the unique
sequences become trapped—in this simplified instance completely—within the
reassociated fraction
Oversimplification of the types of possible fragments and of the range of fragment
sizes

Studies of the dependence of the reannealing fraction on the length of the
DNA-molecules when challenged with sheared DNA of uniform fragment
size were especially revealing. The longer the DNA molecules tested, the
larger is the amount of DNA recovered in the rapidly renaturing fraction.
The principle is illustrated in Fig. 4. A general pattern of the sequence
organization in eukaryotic DNA emerges from these investigations: segments of single-copy DNA (so-called unique sequences), alternate with
repetitive sequences of variable, intermediate redundancy (Davidson *et al.*,
1975). In the euchromatin of the sea urchin and of vertebrates there are
usually two different patterns of interspersion constituting about 70 % of the
DNA: one comprising short repetitive segments, 400 to 600 nucleotide pairs
in length, with unique segments of about twice that length, and the other

one, which consists of very long unique sequences interspersed with short repetitive ones. Some of the middle repetitive sequences also occur in clusters forming long stretches of DNA, partly containing tandem repeats of shorter segments. The highly repetitive material, often seen as satellite DNA upon density gradient centrifugation, is concentrated largely within the constitutive heterochromatin. Satellite DNA forms clusters without or with little interspersion of sequences of lower redundancy. Although conforming qualitatively with this general pattern, the quantitative relationships in the Drosophila genome are quite different (Manning *et al.*, 1975). Stretches of the middle repetitive sequences with an average length of 5600 nucleotide pairs are interspersed with even longer segments of unique sequences of 13 000 nucleotide pairs on average. The range of the DNA contents of the chromomeric bands of *D. melanogaster* extends from 5000 to 100 000 nucleotide pairs, with an average of about 30 000. This corresponds well with the amount of DNA contained in pairs of unique and middle repetitive sequences. In their theory on the mechanisms of gene regulation in eukaryotes, Davidson and Britten (1973) offer an interpretation of the functional unit which is compatible with this general scheme. According to their ideas, the chromomere in its simplest form consists of one structural gene linked to a proximal cluster of more or less repetitive sequences which serve as regulatory elements. Mutations affecting the regulatory sequences upset the function of the structural gene, for instance by abolishing its transcription or by interference with the correct processing of the giant transcript of the functional unit. Hence, enough DNA serving a single genetic function is accounted for by this model to be in keeping with the estimates of gene numbers based upon mutation studies. The general interspersion patterns detected in eukaryotic genomes will necessitate modification of the Britten and Davidson theory. There is also in this model a certain preoccupation with the repetitiveness of regulatory sequences, a considerable part of which might well be as unique as the classical structural genes (as suggested by the enormous size of some of the interspersed unique sequences in vertebrates and of those in *D. melanogaster*). There are, on the other hand, families of redundant structural genes (like the ribosomal cistrons, 5S RNA, tRNA, and histone genes), all containing spacer sequences part of which may have a regulatory significance similar to the repetitive sequences interspersed between single-copy DNA.

Thus, the results of the molecular analysis of sequence organization are compatible with the estimates of gene number made on the basis of the mutational load. Both approaches lead to the conclusion that the functional unit of the eukaryotic genome contains more (often much more) DNA than is needed for the sequence of the respective classical gene. The implications of this notion for the nature of the numerous pleiotropic mutants in Man and other higher organisms are obvious. It finds additional support from the correlation between the induced mutation rate per locus and the genome size. While an increase of the overall mutation rate with increasing genome size is to be expected, a proportionality of the latter to the mutation rate (per unit mutagenic action) per locus is rather surprising (Abrahamson et al., 1973; Heddle and Athanasiou, 1975). It can best be interpreted in terms of a dependence of the target size of the mutable loci on the genome size.

In the human genome 42% of the DNA consists of repetitive sequences (Jones, 1974). The bulk of this highly heterogeneous fraction belongs to the species of molecules having an intermediate redundancy and is interspersed with single-copy DNA. About 80% of human DNA is involved in the alternating pattern (Schmid and Deininger, 1975) of unique and repetitive sequences already mentioned as the characteristic sequence organization of eukaryotic genomes. According to Schmid and Deininger (1975), 52% of human DNA consists of a pattern of relatively short unique segments of about 2000 nucleotide pairs, while the remaining part contains much longer segments. The length of the repetitive sequences has an average of about 400 nucleotide pairs. If we disregard the very rapidly reassociating fraction of the "inverted repeats" which seem to be randomly distributed in the human genome, the small fraction of highly repetitive satellite DNA remains to be considered.

There are at least eight species of satellite DNA in the human genome. They comprise between 6 and 10% of the human DNA (Jones, 1973; Saunders et al., 1975). Their isolation was accomplished mainly by caesium salt density-gradient centrifugation in the presence of heavy metal ions (Ag, Hg), and by MAK-column chromatography. Satellites I, II and III have a lower $G+C$ content than main band DNA, while satellite IV is undistinguishable from the bulk DNA in this respect. The minor species of satellite DNA, designated A, B, C and D contain more than 41% $G+C$. There are pronounced differences between the various human satellite DNAs in the extent of mismatching upon reassociation after denaturation. This is based on differing degrees of sequence divergence and internal heterogeneity, properties which bear an interesting relationship to the respective evolutionary age of the satellite DNA (Jones, 1974).

The chromosomal localization of the satellite DNAs was investigated by in situ hybridization with heavily labelled complementary RNA transcribed in vitro from the purified satellite DNAs. As in other species human satellite DNA is confined to constitutive heterochromatin (Gosden and Mitchell, 1975). Although the centric regions of most of the human chromosomes contain satellite DNA of one or several kinds, there are preferential sites for satellites I, II and to some extent III (Gosden et al.,

1975). The main site of the satellite DNA with the highest $A+T$ content is the brightly fluorescent distal part of the long arm of the Y-chromosome (Jones et al., 1974). About 70% of the DNA of the human Y-chromosome is recovered upon digestion with the restriction endonuclease Hae III in the form of two fragments occurring 6200 and 4400 times, respectively per Y-chromosome (Cooke, 1976). Part of this male specific DNA has also been isolated by competing out of male DNA all sequences common to both sexes with an excess of female DNA (Kunkel et al., 1976). It is the first fragment of the human genome isolated, which is specific for a single defined chromosome. Its functional significance and its relationship to satellite I are not clear at present. The approach serves as a valuable model for further attempts to isolate specific human DNA fragments from restriction endonuclease digests, for instance of the DNA of man–rodent hybrid cells containing only one human chromosome.

Other major locations of human satellite DNAs are the secondary constriction of chromosomes 1, 9 and 16 for satellite II, and the seondary constriction of chromosome 9 for satellites III and IV (Gosden et al., 1975). All four major satellite DNAs seem also to be present on the distal long arm of the Y-chromosome, as is satellite I in the secondary constriction of chromosome 9. It is interesting to note that these chromosome regions show a wide range of polymorphic variability. They also exhibit the phenomenon of lateral asymmetry (Angell and Jacobs, 1975; Galloway and Evans, 1975). When there are significant differences in the thymine content between the two strands of DNA and one cycle of replication is completed in the presence of BUdR, the quenching effect of the base analogue on fluorescence with "Hoechst 33258" will affect one chromatid more strongly than the other. The occurrence of these laterally asymmetric BUdR-patterns in the paracentric heterochromatin of chromosomes 1, 9, 15, 16 and Y suggests that there are pronounced inter-strand differences with respect to the thymine content in some of the human satellite DNAs (see also Section II.B).

Since satellite DNA does not contain meaningful genetic information and represents only a small fraction of the human genome, its presence does not add any new aspects to the problem of gene number in Man. However, its special properties seem to predispose it to interindividual variability and rapid evolutionary change (Jones, 1974).

It has become clear from this discussion that estimates of the number of genes in eukaryotic genomes depend strongly on the definition of the gene. If this term is considered as synonymous with "functional genetic unit" or with complementation group, the number of genes in the human complement will amount to between 10^4 and 10^5. If the "gene" is understood in the broader sense of any sequence which is transcribed at least once during the lifespan of an organism, gene number may be five to ten times higher. This, however, is again not equivalent to the total amount of genetic information which includes those sequences serving as recognition sites for molecules involved in the regulation and control of gene function. From this point of view, the enormous amount of DNA in the genomes of higher animals shows the immense efforts that are required to further our understanding of their highly integrated functions.

III. MAPPING

The remarkable extension of the spectrum of genetic markers expressed in cell cultures and the ingenious exploitation of somatic cell hybrids have brought about the impressive recent progress in human gene mapping. New methods of chromosome identification have greatly facilitated the utilization of interspecific hybrid cells. On this basis it also became possible to use chromosome rearrangements for the regional assignment of gene loci. About 120 genes have been assigned to human chromosomes and none of the autosomes is left without its marker gene. Half of these were localized by the methods of somatic cell genetics.

But the classical method of correlation of gene expression with the segregation of marker chromosomes has also been applied to the problem of human gene mapping. The approach of deletion mapping is based on the detection of hemizygous expression in an obligate heterozygote for an autosomal gene. One important source of error is the existence of a silent allele in one of the parents. While this method had resulted earlier in the exclusion of numerous marker genes from chromosome segments (Bender *et al.*, 1967; Bender and Burckhardt, 1970; Aitken *et al.*, 1976), it was successfully applied for the first time in the assignment of the red cell acid phosphatase locus to chromosome 2 (see Section IV.A.4). The localization of the MN blood group locus to band 2q14 was possible on the basis of its hemizygous expression in a patient who had lost this segment due to a complex rearrangement (German and Chaganti, 1973). The study of the segregation of polymorphic protein or blood group markers in families carrying a deletion or a more complex rearrangement will continue to provide valuable information about the human gene map.

Normal chromosome variants can, of course, also be utilized for this purpose. The highly variable segment of constitutive heterochromatin on chromosome 1, formerly known as a secondary constriction and, in its exaggerated form as an "uncoiler element", allowed the Duffy blood group locus to be localized on this chromosome (Donahue *et al.*, 1968; Ying and

Ives, 1968). Similarly, segregation analysis of the haptoglobin variants in a family with a marker chromosome 16 has made possible the assignment of the gene for the haptoglobin-α-chain to the long arm of this chromosome (Magenis *et al.*, 1970; and cf. p. 446). Utilization for this kind of study of the more frequent chromosomal polymorphic markers (see Section II.C) will almost certainly yield further information on gene assignments and linkage relationships.

The most important breakthrough in human gene mapping was made by the introduction of interspecific cell hybridization. The special usefulness of man–rodent hybrid cells is based on the gradual loss of human chromosomes from the hybrid clones as proliferation proceeds. Thus, the expression of human markers can be correlated with the presence of human chromosomes or chromosome segments. The broad spectrum of markers used ranges from isozyme systems to complex cell surface receptors and other properties of the cell membrane. A comprehensive review of the methods, results, and many relevant aspects of somatic cell genetics was given by Ruddle and Creagan (1975), Grzeschik (1973) and Davidson and de la Cruz (1974).

Let us illustrate some of the basic principles by the example of one of the more recent assignments: that of tryptophanyl-tRNA-synthetase (EC 6.1.1.2) to chromosome 14 (Denney and Craig, 1976). This is the first enzymatic component of the protein synthesizing machinery whose structural gene has been localized in a mammalian genome. The authors first developed a method of detecting the activity of tryptophanyl-tRNA synthetase on electrophoretograms. This was done by the autoradiographic localization of the tryptophanyl-tRNA formed by the enzyme after incubation with ^{14}C-tryptophan, ATP, and the mixture of yeast tRNAs. It was then ascertained that the human and the mouse enzyme could be separated by electrophoresis. Some of the cell hybrids used were prepared from an 8-azaguanine resistant heteronuclear mouse cell line (lacking HGPRT) and normal human lymphocytes. Selection under the conditions of artificial auxotrophy for purine bases (HAT medium) eliminated the mouse parent cells while the lymphocytes do not proliferate without mitogenic stimulation. The combined karyotypic and electrophoretic analysis of a series of hybrid clones and subclones showed a perfect correlation between the presence of human chromosome 14 and the human form of try-tRNA synthetase. Synteny with the gene for nucleoside phosphorylase which had been previously assigned to chromosome 14 was also demonstrated by the same type of correlation. If the method of detection of aminoacyl-tRNA synthetase activity proves to be generally applicable, man will be the first eukaryotic species in which the genes coding for this most interesting group of enzymes have been localized.

The elucidation of the synteny of genes and the localization of syntenic groups of genes on defined chromosomes can be achieved by similar strategies to those described above. A variety of selection methods were developed including drug sensitivity, nutritional auxotrophy and sensitivity against molecules interacting with the cell surface. The next step, the analysis of gene order, that is mapping in the strict sense, requires the study of chromosome segments.

Chromosome rearrangements in the human parent cell have repeatedly been used to establish gene order by segregation studies with hybrid clones. One of the best examples is the investigation of the arrangement of four X-linked genes. Various X-autosome translocations with well-defined break points were used to establish the sequence on the long arm of the X-chromosome: centromere—PGK : a-GAL : $HGPRT$: $G6PD$ (Pearson *et al.*, 1975). While these translocations were derived from human subjects, intraspecific and interspecific translocation occurring spontaneously within the composite genome of the hybrid cell were also used for the regional assignment of human genes (Hamerton *et al.*, 1975). New staining methods help to identify the chromosome segments involved in the interspecific rearrangements (Friend *et al.*, 1976).

The frequency of chromosomal rearrangements in hybrid cells can be greatly increased by exposure of one of the parent cells to ionizing radiation prior to cell fusion. In the course of the numerous breakage and reunion events thereby elicited linkage groups are disrupted. The separation of syntenic genes will occur with a frequency directly related to their distance. This recent refinement of the mapping analysis with somatic cell hybrids was introduced by Goss and Harris (1975). They determined the frequency of co-expression of unselected X-linked markers in cell hybrids prepared from irradiated normal human lymphocytes and HGPRT-deficient hamster cells. The $HGPRT$ locus on the human X-chromosome thus provided the selective marker. Analysing independently arising clones in selective medium (HAT) for the co-expression of human PGK, a-GAL and G6PD, the authors confirmed the order of these genes, previously established by several laboratories using various X-autosome translocations, as outlined above. Needless to say, this method becomes a particularly valuable tool when one gene can be "fixed" by the selective advantage which it conveys to the hybrid clones. Chromosomal analysis of the clones is not necessary for conclusions to be drawn; knowledge of the chromosome which carries the selected marker gene suffices for the assignment of the respective linkage group to that particular chromosome.

Going to still smaller chromosome fragments the attempts to achieve gene transfer to recipient cells with chromosome suspensions or with DNA preparations should be briefly mentioned. The first successful co-transfer of two human genes previously shown to be closely linked was reported by Willecke *et al.* (1976). These authors used chromosome suspensions prepared from a human lymphoblastoid cell line and were able to demonstrate the simultaneous transfer of the genes coding for cytosol thymidine kinase (TK) and for galactokinase (GALK) to TK-deficient mouse L cells. The stability of the incorporation of the human chromosome fragment, which could not be detected microscopically, depended on the duration of the exposure of the "transformed" clones to the selective medium (HAT). Hence, a loosely associated "transgenote" can undergo stable integration into the recipient's genome. If the efficiency of this method can be improved it can provide a valuable means of high resolution mapping. Whether the same can be expected from the incorporation of free DNA molecules into the genome of human cells remains to be shown. The successful transfer of bacterial genes via the DNA of transducing phages is encouraging. In analogy to bacterial transformation, Merril *et al.* (1971) were able to transfer the *E.coli* gene of galactose-1-phosphate-uridylyltransferase into the cultured fibroblasts from a patient with galactosaemia. Similarly, Horst *et al.* (1975) used cells from a patient afflicted with GM1-gangliosidosis to show the transfer of the *E.coli* structural gene for β-galactosidase.

A rather surprising new approach to human gene mapping, unrelated to somatic cell genetics, was recently discovered by Linder *et al.*, 1975. Several patients whose somatic karyotype exhibited heterozygosity for various normal markers of the centric regions, carried ovarian teratomas which were homozygous for the same markers. The conclusion can hardly be escaped that these tumours have a parthenogenetic origin and arise from one of the products of the first meiotic division. Cross-overs will separate proximal genes less frequently from the centric region than distal ones. In fact, isozyme markers, for which the patients were also heterozygous, remained so in the tumours in some instances and became homozygous in others (p. 426).

IV. PROTEIN STUDIES IN CHROMOSOMAL DISORDERS

On the basis of the recent gene assignments to human chromosomes, many of the early studies on enzyme activities in trisomy syndromes can now be viewed as having dealt with secondary, if not unspecific effects. This is certainly true for the elevation of the activities of galactokinase and galactose-1-phosphate-uridylyltransferase in white blood cells of children with trisomy 21, reported by several authors during the years 1964 to 1970. This effect was not directly related to an increased gene dosage, since the structural genes for these enzymes have since been assigned to 17q21–17q22 for the kinase and to the region distal from 3q21 for the transferase, respectively. Also, the localization of the structural gene for 6-phosphogluconate dehydrogenase (PGD) to a segment distal from 1q32 has invalidated the earlier assumption that this gene belongs to the chromosome suffering a deletion in chronic myelogenous leukaemia (now known to be chromosome 22). While the earlier studies, often conducted according to the chance availability of aberrations, aimed at the localization of genes on the basis of gene dosage relationships, such quantitative relationships are now being searched for, firstly to confirm well-known gene assignments, and secondly to investigate the consequences of well-defined genetic imbalances. In a discussion of biochemical investigations of aneuploidy it seems therefore appropriate to deal first with those studies using the products of genes which are already assigned to a definite chromosome or chromosome segment.

A. Studies of the Products of Localized Genes

1. Anomalies of chromosome 21

There are two well-documented gene dosage effects in trisomy 21 involving the function of genes which had been previously assigned to this chromosome by clonal segregation studies with man–mouse hybrid cells (Tan et al., 1973). One of these genes codes for an interferon-induced factor responsible for the antiviral state (AVS); the product of the other one is superoxide dismutase-S (SOD_S or SOD_1 also known as indophenoloxidase-A (IPO-A); see Chapter 5). Upon infection with virtually any kind of virus a defensive response is elicited in animal cells via the production of interferon which induces a transient resistance to subsequent virus infection, the AVS. The establishment of the AVS depends on the induction of the antiviral protein (AVP) by interferon. In the reovirus system this protein was shown to be a component of the ribosomes of interferon-treated cells and to inhibit specifically the translation of viral messages essential for virus replication (Samuel and Joklik, 1974; Wiebe and Joklik, 1975). Quantitative evaluation of the cellular capacity to attain an AVS in response to interferon can be made by measuring the amount of interferon which yields a certain degree of protection of cell cultures against a virus supplied at a given multiplicity. This approach was used by Tan et al. (1974a) in a comparison of the interferon response of fibroblast cultures derived from patients having trisomy 13, 18, or 21, respectively, with those from chromosomally normal persons. A three- to seven-fold reduction of interferon was needed by the 21-trisomic cells to elicit the same protective response as in the normal cells and in those with the other trisomic conditions. This disproportionate increase of interferon sensitivity in trisomy 21 confirms the assignment of a gene coding for an antiviral factor to chromosome 21, as does the likewise disproportionate decrease of interferon sensitivity of cells with monosomy 21. Extending the comparison to this lowered dose of chromosome 21, Tan (1975) found the following relationship between interferon sensitivity (Y, that is the reciprocal amount of interferon needed to produce a given protective response) and number of chromosomes 21 present (X):

$$\log Y = 0 \cdot 61 \ X.$$

The question arises whether the genetic information present on chromosome 21 is the structural gene of AVP or whether it codes for another factor involved in the establishment of the AVS, e.g. an interferon-specific cell surface receptor. The latter hypothesis is supported by the finding of Revel et al. (1976) that the protective effect of interferon on human cells can be prevented by mouse antibodies directed against the surface of man–mouse hybrid cells containing chromosome 21. The extent of this inhibition decreases with increasing number of chromosomes 21. Furthermore, it is not the intensity of the AVS achieved at saturating interferon concentrations that depends on the number of chromosomes 21, but rather the amount of interferon by which a certain antiviral response is elicited. Interestingly enough, the growth inhibitory effect of interferon on human cells also depends in an analogous way on the number of chromosomes 21 (Tan, 1976); a given amount of interferon inhibits the growth of 21-trisomic cells more strongly than that of normal cells whose growth in turn is more strongly inhibited than that of 21-monosomic cells.

That the interferon-receptor rather than the AVP is encoded on chromosome 21 was also proposed by Chany et al. (1975), who not only confirmed the exaggerated interferon response of 21-trisomic cells, but also found a normal interferon response in 11 triploid fibroblast strains. This restoration of the normal response in triploid cells provides strong evidence for regulator genes controlling the interferon-induced AVS but not being syntenic with the antiviral gene on chromosome 21. Genetic balance in this case obviously means that, if sex chromosomes can be excluded, the normal function requires a one-to-one relationship between the structural and the regulator genes involved. In man–mouse cell hybrids the production of human interferon depends on the presence of chromosomes 2 and 5 (Tan et al., 1974b). If the preliminary evidence is confirmed suggesting the presence on chromosome 16 of a regulatory element which controls the synthesis of antiviral factors (Chany et al., 1975), the interferon system of virus defence would require the cooperative function of genes on at least four different autosomes (2, 5, 16, 21). It thus provides an unique opportunity for the investigation of genetic balance and its disturbance by various aneuploid conditions.

The second gene dosage effect observed in trisomy 21 concerns an enzymatic defence mechanism. In all oxygen metabolizing cells appreciable amounts of intermediates of oxygen reduction are produced, among which the superoxide radical O_2^- contributes significantly to oxygen toxicity (Fridovich, 1974). The ubiquitous superoxide dismutases act as scavengers of the O_2^- radicals by shifting an electron between two such ions, thereby creating an O_2 molecule and, with consumption of two protons, hydrogen peroxide which in turn is the substrate of catalases and peroxidases. There are two forms of superoxide dismutases in human cells; a tetrameric, mitochondrial enzyme called superoxide dismutase-2 (SOD_M) or indophenol oxidase B (IPO-B), and the dimeric enzyme of the cytosol, SOD_S or IPO-A. Since the reduction of tetrazolium compounds by O_2^- is prevented by superoxide dismutases their position on electrophoretic media can be detected by achromatic zones on the light-induced deeply stained background of the formazans (Brewer, 1967). Erythrocytes contain only the supernatant enzyme. Using carefully standardized measurements of the size of these achromatic zones in electrophoretograms of haemolysates Sichitiu et al. (1974) demonstrated significantly increased activity of SOD_S in trisomy 21; the ratio of activities for trisomy 21 to normal was 1·495. The application of a more direct method of determination of enzyme activity by the same group of investigators (Sinet et al., 1974) yielded a similar result. The most precise measurements of SOD_S activity were made by Crosti et al. (1976) using a polarographic method. The ratio of average activities obtained in this study was again very close to the expected value of 1·5 and firmly establishes the gene dosage effect for erythrocytes. It would also be of interest to perform such comparative measurements of SOD_S activity with cells actively engaged in protein synthesis like cultured fibroblasts and lymphoblasts.

This, however, would demand the separation of SOD_S from its mitochondrial isozyme, or selective inhibition of one of the isoenzymes, as done by Sinet et al. (1975a) in confirming the SOD_S dosage effect in blood platelets.

Comparative measurements of SOD_S activity in several partial monosomies and partial trisomies of chromosome 21 enabled Sinet et al. (1976) to assign the structural gene of this enzyme to the sub-band 21q22.1. This sub-band belongs to the segment of chromosome 21 whose triplication is responsible for most of the symptoms of Down's syndrome (Aula et al., 1973).

2. Adenine phosphoribosyltransferase (APRT; EC 2.4.2.7) and chromosome 16

The assignment of the structural gene for APRT to chromosome 16 was made possible by the invention of a selective system by Kusano *et al.* (1971): exposure to the antibiotic alanosine which inhibits the synthesis of adenylic acid (AMP) from inosinic acid (IMP) makes the cells dependent on an exogenous supply of adenine. This can only be converted to AMP by APRT. The correction of an APRT-deficiency in appropriately selected mouse cells by the human *APRT* gene in man–mouse hybrid cells was used by Tischfeld and Ruddle (1974) to localize the *APRT* gene on chromosome 16. The analogy of this selective system to the HAT-system used for the selection of cells with hypoxanthine-guanine-phosphoriboyl-transferase (HGPRT) is obvious.

Marimo and Giannelli (1975) measured the APRT activity in amnion cell cultures of nine abortuses with trisomy 16 and of a number of control strains. They found the trisomic cells to have an average 60% increased APRT activity as compared with the controls. There were no differences in the activities of G6PD, HGPRT and adenosine kinase.

3. C-Group chromosomes

Quantitative investigations of the expression of cell surface antigens in heterozygotes and homozygotes for a given allele often correspond to the expected 1:2 dosage relationship. This applies to the thymocyte leukaemia antigen (T1) system of the mouse and to its interaction with the H-2 system (Boyse *et al.*, 1968) and is also observed in the HLA system of Man (Svejgaard, 1969; White *et al.*, 1973). Also the male-specific HY antigen is proportionately expressed according to the number of Y chromosomes present (as discussed in Section IV.D.2). Chromosome 6 which carries the major histocompatibility complex (MHC) of man is rarely involved in chromosomal anomalies. Trisomy 6 was observed once in a series of 152 spontaneous abortions (Kajii *et al.*, 1973). The serological findings mentioned above justify the expectations that at least some of the cell surface antigens specified by the component loci of the MHC (see Chapter 8) will be proportionately expressed in numerical anomalies affecting chromosome 6.

Among the three gene loci coding for adenylate kinases, the AK_1 gene specifies the primary structure of myokinase, the muscle isozyme which is strongly expressed in erythrocytes. The polymorphism existing at this locus (see Chapter 5) enabled Schleuterman *et al.* (1973) to detect its linkage to the *ABO* locus, which in turn was already known to be linked to the "nail patella gene" (*Np*). Thus, the autosomal linkage group AK_1; *Np*; *ABO* was established. Ferguson-Smith *et al.* (1976) investigated a group of

116   W. KRONE AND U. WOLF

patients with anomalies of chromosome 9. Among these patients all segments of chromosome 9 occur at least once as duplication or deficiency. A
43 % increase of adenylate kinase activity above the normal level was found
in the haemolysates of one patient who carried a duplication of the segment
distal from 9q33. This is another example of a gene dosage effect allowing
the regional localization of a gene already assigned to a definite chromosome.

The loci for the subunits A and B of isozymes of lactate dehydrogenase
(LDH) have been assigned to the short arms of chromosomes 11 and 12,
respectively. The similarity of the banding patterns of these two chromosomes and the obvious homology of the two polypeptide chains have
elicited speculations about the evolutionary origin of this topographic
relationship. The preservation of ancient homology of chromosomes 11 and
12 as a result of tetraploidization during vertebrate evolution is a reasonable but unproven hypothesis. A semi-quantitative analysis of the LDH isozymes of haemolysates from a patient with partial monosomy of the short
arm of chromosome 12 was performed by Mayeda *et al.* (1974). The presence of a more than three-fold elevated amount of LDH 4 (A_3B) and the
reduction in the amount of LDH 1 (B_4) indicate a decrease in the amount of
B-subunit. Overall activity of LDH was measured in haemolysates of a
patient with partial trisomy of the short arm of chromosome 12 by Rethoré
et al. (1975). The published values show an increase of 75 % above the
mean activity of a control population. White blood cells and fibroblast
cultures derived from the patient have normal overall LDH activity.
Unfortunately the contribution of the B-subunit to total activity was not
determined, so that the significance of these findings is not clear. It is
remarkable that the genes coding for two other enzymes of the glycolytic
pathway were assigned to segments of the short arm of chromosome 12 on
the basis of dosage effects: glyceraldehyde-3-phosphate dehydrogenase
(Rethoré *et al.*, 1976) and triosephosphate isomerase (Rethoré *et al.*, 1977).
Together with enolase (the gene for one of whose isozymes is also localized
on chromosome 12) these enzymes belong to a group of five exhibiting
constant proportions of their specific activities in many systems (Pette,
1965). Despite the non-synteny of the locus coding for another member of
this group (phosphoglycerate kinase) the suggestion of coordinate expression of these functionally related, syntenic genes is tempting (but cf. p.
445).

4. Anomalies of chromosomes of the A-group
Among the genes localized on the long arm of chromosome 1 is the locus
which codes for fumarate hydratase (FH) (EC 4.2.1.2). Like many other
enzymes of energy metabolism FH seemed to exist in two molecular

species, a cytoplasmic and a mitochondrial form. The kinetic and immuno-logical identity of these enzymes was proven by Tolley and Craig (1975) who also assigned the gene to chromosome 1. Braunger *et al.* (1977) measured FH activity in fibroblast cultures established by Norwood and Hoehn (1974) from a newborn child affected with a trisomy of most of the long arm of chromosome 1 due to an unbalanced translocation. In com-parison to the cells of the balanced father and to several controls these cells exhibited a 1·5-fold increase in specific activity. Several reference enzymes were used to eliminate the influence of external parameters. The gene dos-age effect was ascertained immunologically with a rabbit antiserum against pig heart FH, this being immunologically identical to human FH (Tolley and Craig, 1975).

The first example of successful deletion mapping in Man was the assign-ment of the locus of the acid phosphatase (ACP_1) of erythrocytes (EC 3.1.3.2) to a segment of the short arm of chromosome 2 (Ferguson-Smith *et al.*, 1973). The well-known polymorphism existing at this locus (Chapter 5) aided in the detection of hemizygous expression in a child monosomic for this particular segment. Since each of the various alleles occurring at the ACP_1 locus contributes a specific portion of activity the segregation of a silent allele in this family could be excluded by quantitative measurements.

In the case of a partial trisomy of the terminal segment of 2p Magenis *et al.* (1975) found elevated activity of ACP_1 with triallelic expression. Thus, single, double, and triple doses of this gene locus are proportionately expressed in erythrocytes.

This gene locus is also active in other tissues and in cultured fibroblasts (Swallow *et al.*, 1973). It would be of interest to investigate the dosage relationships in cells actively engaged in protein synthesis. The assign-ment of the ACP_1 gene to the short arm of chromosome 2 was confirmed by Hamerton *et al.* (1975) by segregation studies with human–hamster somatic cell hybrids.

In a woman heterozygous for a pericentric inversion of chromosome 3 a recombinant chromosome 3 arose which suffered a deletion in the short arm and carried a duplication of a substantial part of the long arm (Allderdice *et al.*, 1975). Children receiving this chromosome from their mother will be partially trisomic and partially monosomic for the respective segments of chromosomes 3. Since the structural gene(s) for galactose-1-phosphate-uridylyltransferase (GALT) (EC 2.7.7.10) was localized on chromosome 3 (Tedesco *et al.*, 1974), this anomaly provided the opportunity of a regional assignment on the basis of a gene dosage effect. The 1·44-fold increased activity of GALT found in the erythrocytes of a patient who carried the unbalanced chromosome (Allderdice and Tedesco, 1975) clearly assigns this gene to the duplicated segment. This interpretation is corroborated by

Table 1

Autosomal gene dosage effects

Gene product name	code[a]	Chromosome or chromosome segment	Kind of anomaly	Kind of cells studied[b]	Quantitative effect[c]	Reference
Fumarate hydratase, FH	(13685/86)	1q2 or 3——1qter	partial trisomy	F	1·57 1·6[a]	Braunger et al. (1977)
Acid phosphatase-1, ACP₁	(17150)	2q23——2pter	partial monosomy	E	0·58	Ferguson-Smith et al. (1973)
Acid phosphatase-1, ACP₁	(17150)	2q23——2pter	partial trisomy	E	1·39[e]	Magenis et al. (1975)
Galactose-1-phosphate-uridylyltransferase, GALT	(23040)	3q21——3qter	partial trisomy	E	1·44	Allderdice and Tedesco (1975)
Adenylate kinase-1, AK₁	(10300)	9q33——9qter	partial trisomy	E	1·43	Ferguson-Smith et al. (1976)
Glyceraldehyde-3-phosphate dehydrogenase, GAPD	(13840)	12p12.2——12pter	partial trisomy	E	1·47 1·37	Rethoré et al. (1976)

Enzyme (McKusick no.)[a]	Location	Type	Cell[b]	Ratio[c]	Reference
Triosephosphate isomerase, TPI (19045)	12p12.2——12pter	partial trisomy	E	1·86 2·20	Rethoré et al. (1977)
Nucleoside phosphorylase, NP (16405)	14q11——14q21	partial trisomies	E	1·49-1·73	George and Francke (1976)
Adenine-phosphoribosyl-transferase, APRT (10260)	16	trisomy	A	1·69	Marimo and Giannelli (1975)
Superoxide-dismutase-1, SOD-1 (14745)	21	trisomy	E	1·45	Crosti et al. (1976)
Superoxide-dismutase-1, SOD-1 (14745)	21 21q22.1	trisomy partial trisomy	E E	1·56 1·75	Sinet et al. (1976)
Superoxide-dismutase-1, SOD-1 (14745)	21	trisomy	P	1·56	Sinet et al. (1975a)

[a] According to McKusick's catalogues (1975)
[b] A: amnion cells; E: erythrocytes; F: fibroblasts; P: blood platelets
[c] Ratio of specific enzyme activities: abnormal/normal
[d] Immunologically ascertained
[e] The normal value was taken to be equal to two-thirds of the sum of the specific activities contributed by two B-alleles and one A-allele

the well-known dosage effect in heterozygotes for the galactosaemia allele of GALT. Besides this deficiency allele there is an electrophoretic variant occurring with a considerable frequency in human populations (see Chapter 5). This polymorphism and the fluorescence polymorphism at the centric region of chromosome 3 could be used to measure the frequency of crossing over between the centromere of chromosome 3 and the *GALT* locus.

A summary of the well ascertained autosomal gene dosage effects on identified gene products is presented in Table 1.

B. Further Biochemical Studies in Autosomal Anomalies

Early biochemical investigations of the malformation syndromes caused by chromosomal aberrations were often initiated without a reasonable working hypothesis. The search for gene dosage effects has been a poor excuse for measuring everything measurable, notably the enzyme activities routinely determined in the clinical laboratory. The results of these studies were discussed at some length in the first edition of this book (1972). Some of the more pertinent observations are summarized in Table 2. Most of these examples (and all of those omitted) represent rather indirect relationships between the chromosomal anomaly and the effect observed. A number of intriguing speculations can be made about these findings. In the context of this chapter, however, only those effects which are better defined in molecular and biochemical terms deserve a more detailed discussion.

1. Persistence of foetal and embryonic haemoglobins in trisomy 13

The discovery of the haemoglobin anomalies in trisomy 13 was not made by chance. After the characterization of the two early embryonic haemoglobins, Gower I and Gower II, by Huehns *et al.* (1961) the developmental pattern of the human haemoglobins could be depicted in terms of the differential rates of synthesis of the various polypeptide chains which, according to their relative abundance, become incorporated into the different species of haemoglobin.

In order to find out whether or not the genetic control of the development of the human haemoglobins is affected by one of the autosomal trisomies, Huehns *et al.* (1964a, b) studied the haemoglobins present in children with trisomy 13, trisomy 18 and trisomy 21. The following observations were made in children afflicted with trisomy 13:

(a) The level of Hb F was elevated and it decreased at a slower rate with increasing age
(b) Small amounts of Hb γ_4 (Barts) were regularly present

Table 2

Some of the less well defined consequences of human chromosomal aberrations

Condition and effected molecule	Effect	Reference
Trisomy 21		
leucocyte alkaline phosphatase (LAP)	elevated in granulocytes but not in thrombocytes possessing identical enzyme	Tangheroni *et al.* (1971)
phosphofructokinase (PFK; E C 3.7.1.11)	elevated in erythrocytes but not in white bloods cells, thrombocytes and fibroblasts	Layzer and Epstein (1972)
total acetylesterase activity in parotid saliva	increased; other enzymes, like amylase and acid phosphatase show normal levels	Winer and Chauncey (1975)
unknown component(s) of DNA repair system	reduced DNA repair synthesis after u.v.; increased yield of dicentrics after X-rays	Lambert *et al.* (1976)
serotonin (5-hydroxytryptamine)	decreased content in platelets, possibly due to decreased ATPase and defective ion transport	McCoy *et al.* (1974)
mitochondrial superoxide dismutase(SOD_M)	decreased activity in blood platelets	Sinet *et al.* (1975a)
glutathione peroxidase	increased activity in red blood cells	Sinet *et al.* (1975b)
18p-Syndrome	hypothyroidism of various kinds	Hansen and Bartalos (1975)
Trisomy 13		
haemoglobin	persistence of foetal and embryonic haemoglobins during postnatal life	Huehns *et al.* (1964a, b)
Trisomy 9p		
somatotropin (GH)	abnormally low secretion after stimulation	Fujita *et al.* (1976)
Trisomy 8		
blood coagulation factor VII	reduced concentration in serum; other components of haemostatic system show normal levels	Grouchy *et al.* (1974)

(c) During the neonatal period small amounts of Hb Gower II could be detected

None of these anomalies was found in trisomy 18 and in mongoloid children. An abnormally high level of Hb F in an older child with trisomy 13 was independently found by Powars *et al.* (1964). These results were subsequently confirmed by numerous studies of D-trisomic children of various ages (references cited by Kazazian, 1974).

The haemoglobin anomalies characteristic of trisomy 13 were not only observed in patients with an extra chromosome 13 but also in cases of trisomy due to the presence of Robertsonian translocations involving chromosome 13 (Huehns *et al.*, 1964a; Walzer *et al.*, 1966; Wilson *et al.*, 1967; Pinkerton and Cohen, 1967; Yu *et al.*, 1970) and in cases with partial trisomy of chromosome 13 (Bloom and Gerald, 1968; Yunis and Hook, 1966; Hoehn *et al.*, 1971). Investigation of a number of partial trisomies of chromosome 13 by Noel *et al.* (1976) allowed the conclusion that the haemoglobin anomalies are associated with trisomy of a segment distal from 13q21. In a case of partial deletion of chromosome 13 comprising this median segment an increased amount of Hb A_2 (11%) in the absence of a thalassaemic trait was observed by these authors. This cannot be interpreted as a clear case of an "anti-symptom" to the delay of the γ to $\beta\delta$ switch in trisomy 13, since an acceleration of the switch rather than an exaggeration of its result would be expected. Nevertheless, these data provide additional proof of the location of genetic factors involved in the regulation of erythroid differentiation on a median segment of the long arm of chromosome 13.

In analogy to Jacob–Monod type regulatory circuits Huehns *et al.* (1964a, b) proposed molecular models to explain the delayed switches from embryonic to foetal and from foetal to adult haemoglobin synthesis. These models are based on the imbalance between structural and regulatory genes: with an increased dosage of the structural genes the concentration of negative regulators ("repressors") could become insufficient to shut down ϵ- and γ-chain synthesis; an excess of positive regulators ("activators") could, on the other hand, provoke the continued synthesis of these chains. More elaborate models can be constructed which postulate a causal relationship between activation of the synthesis of later chains and the repression of the synthesis of the earlier ones (cf. Chapter 11). Aneuploidy of chromosome 13 may, however, affect a much more comprehensive programme of differentiation, since Lee *et al.* (1966) have shown that the development of red cell carbonic anhydrase is also retarded in trisomy 13, like that of the synthesis of the β- and the δ-chain. Since these anomalies of haemoglobin development are specifically associated with trisomy 13, they cannot simply be the consequence of a general developmental retardation.

Our understanding of these phenomena will ultimately depend on the insight into the switch mechanisms controlling the transition from the synthesis of the ϵ-chain to the γ-chain and the subsequent transition from the synthesis of the latter to that of the β- and δ-chains. Neither of the switch mechanisms seems to be coupled with the transitions of the histological location of haemoglobin synthesis.

Not only has the presence of ϵ-chains been detected in embryonic normocytes by Kleihauer *et al.* (1967), but traces of Hb Gower II were also found in the cord blood of all newborn babies tested so far (Kohne and Kleihauer, personal communication). In these normal newborns and in children with trisomy 13 no megaloblasts have been observed in blood smears. The persistence of the ϵ-chain in trisomy 13 therefore can not be due to an abnormal prolongation of the mesoblastic phase of haematopoiesis.

In view of the significance that "aneuploid" states (merogenotes and deletion mutants) have had and still have for the elucidation of regulatory mechanisms in microorganisms it seems mandatory that every case of anomaly affecting chromosome 13 be thoroughly examined with respect to the quantitative and qualitative aspects of its haemoglobins.

2. Trisomy 21 and DNA repair

The stability of all prokaryotic and eukaryotic genomes depends among other factors on the action of various DNA repair mechanisms. Gene mutations affecting one or other component of these protective mechanisms have helped to unravel the sequences of steps required from the recognition of a DNA lesion to its ultimate repair. The first chromosomal disorder discovered which seems to interfere with repair processes, is trisomy 21. Sasaki and Tonomura (1969) were the first to show that lymphocytes from patients with trisomy 21 irradiated with X-rays in the G0-phase of the cell cycle yield a higher rate of chromosomal aberrations than normal lymphocytes. Such investigations were recently extended to u.v.-irradiation and to combined treatment with X-rays and u.v. light (Holmberg, 1974; Lambert *et al.*, 1976). Besides confirming the results of the earlier investigations these authors found a 25 to 30% reduction in the u.v.-induced unscheduled DNA synthesis in white blood cells from patients with trisomy 21. This comparatively small but reproducible defect in excision repair of u.v.-induced lesions is also reflected by the low synergistic effect of combined X-ray and u.v. irradiation in white blood cells with trisomy 21: while normal cells show a doubling of the number of dicentric chromosomes per cell upon this combined treatment as compared with the effect of X-rays alone, there is only a small elevation in the trisomic cells. Apparently, the repair process elicited by the u.v.-irradiation normally enhances the rate of successful rearrangements of X-ray induced breaks, but the reduction of the activity of the u.v.-repair system in trisomy 21 eliminates most of this synergistic action.

These studies were performed with the heterogeneous populations of white blood cells. Interestingly enough, the much more uniform cell populations of cultured fibroblasts with trisomy 21 also show a 25 to 30% reduction of the rate and capacity of repair synthesis after u.v. irradiation (Rebhorn and Bayreuther, 1977). This correlates well with the more pronounced decrease of cloning efficiency of fibroblasts with trisomy 21 after u.v. exposure also observed by these authors.

The specificity of these effects can not yet be evaluated because no other chromosomal aberration has been investigated in a similar way. Perhaps the most exciting aspect of these findings is that they establish a relationship between Down's syndrome and diseases caused by defects in DNA repair mechanisms, like xeroderma pigmentosum, ataxia telangiectasia, Fanconi anaemia, and progeria, all of these having in common an increased risk of developing various kinds of malignancies. It is interesting to note in this context that chromosome 21 belongs to the eight chromosomes found most frequently to be involved in the aberrations which occur in neoplasms (Mittelman and Levan, 1976). The causal relationship between trisomy 21 and the anomalous response to X-rays and u.v.-irradiation is at present unknown. The conclusion can, however, hardly be escaped, that these defects are of major importance in the high rate of chronic lymphocytic and acute leukaemias occurring in patients with trisomy 21, and likewise, in the precocious ageing which is reflected in the reduced lifespan of fibroblasts *in vitro* (Bayreuther, personal communication).

C. Discussion

None of the biochemical findings listed in Tables 1 and 2 provides a basis for an understanding at the molecular level of the complex syndromes caused by chromosomal aberrations. The gene dosage effects listed in Table 1 indicate that in the cells studied transcription seems to be the rate-limiting step for the expression of these genes. This does not, however, necessarily imply that the genes are constitutively expressed. The rate of transcription could be controlled by closely linked regulatory elements included in either the supernumerary or the missing chromosome segment respectively.

With respect to the amount of genetic information involved in the causation of chromosomal syndromes there are two concepts which need not necessarily be mutually exclusive: the effects of aneuploidy may be due to the additive action of the many genes localized on the chromosome segment or chromosome involved in the anomaly; alternatively, these effects may depend on the presence of single dosage-sensitive loci whose duplication or deficiency is of highly pleiotropic consequence. This problem has been thoroughly investigated in *D. melanogaster* (Sandler and Hecht, 1973). Utilizing crosses between X-ray induced Y-autosomal translocation heterozygotes Lindsley *et al.* (1972) were able to produce regional duplications and deficiencies covering about 85 % of the autosomal complement of

Drosophila. The results show that, as in Man, duplication is more easily tolerated than deficiency. There are very few dosage-sensitive loci; in fact only one wild type gene causes lethality in its duplicated as well as in its hemizygous state. It was concluded that in Drosophila the effects on viability and on the phenotype are due to the additive action of most or all of the genes involved in the aneuploidy. The relevance of these interesting findings for the interpretation of the effects of human aneuploidy should be evaluated cautiously. There are fundamental differences between mammals and Drosophila in genetic regulatory mechanisms like dosage compensation and sex determination and also with respect to the effect of triploidy, which is viable and phenotypically normal in Drosophila. Thirty times more DNA is required to control the normal development of a human being as compared with the fruit fly. The number of dosage-sensitive loci in Man could thus considerably exceed the small number found in Drosophila. Hence, a combination of both possibilities mentioned above, would probably be encountered if a similar approach were feasible in Man.

Alterations of the dosage of a classical structural gene itself is not the only way in which aneuploidy can influence its activity. Hyperploid chromosome segments not containing the respective structural genes for the enzymes studied were shown to influence the activities of α-glycerophosphate dehydrogenase and NADP-dependent isocitrate dehydrogenase in *D. melanogaster* (Rawls and Lucchesi, 1974). Some of these chromosome segments were located on the same chromosome as the structural gene, some were not. It may be recalled in this context that relationships of a possible regulatory nature also exist between unlinked genes in the human genome, as exemplified by the X-linked agammaglobulinaemias. These observations caution against the assignment of gene loci to chromosomes on the basis of an apparently proportionate gene-expression alone! In each of nine trisomies of the Jimson weed (*Datura stramonium*) Carlson (1972) found an elevation of a specific enzyme activity, but in only two of these was there independent proof for the localization of the gene on the respective trisomic chromosome. Such independent evidence is provided for instance by the quantitative ratios with which different alleles are expressed in heterozygous trisomics. The recently discovered gene dosage effect for isocitrate dehydrogenase in whole 12 to 13 day mouse embryos with trisomy 1 is conclusive with regard to gene localization on the basis of this kind of evidence (Epstein *et al.*, 1977). In view of the many steps between transcription and the assembly of the final gene product, these dosage effects are astonishing examples of the precision with which these processes can occur in the eukaryotic cell. The third possibility should, nevertheless, be taken into account, namely that the majority of the effects of aneuploidy are due to alterations of the amount of regulatory information, rather than to changes in the dosage of classical structural genes. In an explicit discussion of this hypothesis, Vogel (1973) stresses the relative phenotypic neutrality of the dosage effects observed in the heterozygotes of many recessively inherited metabolic disorders.

Prokaryotic models of imbalance between regulatory and structural information (in the classical sense) were discussed on the basis of the operon concept of Jacob and Monod in the first edition of this book (1972). They are of limited value for the interpretation of the aneuploid eukaryotic systems, but useful information would certainly be obtained by investigations directed at the functioning of the well elaborated prokaryotic regulatory systems in merogenotes and other "aneuploid" states. For instance the phenomenon called escape synthesis is well understood in molecular terms. It occurs when the amount of repressor molecules becomes limiting because of an increased number of copies of the operator. Bacterial operons may thus become constitutively expressed when an induced prophage carrying the respective operator begins to replicate. If we apply this classical model of escape synthesis (Krell *et al.*, 1972) to a eukaryotic trisomic condition with three copies of an acceptor site for a repressor protein whose genetic information is present only twice, this mechanism could account for the presence or the persistence of gene functions normally repressed in comparable metabolic or developmental situations. Numerous model situations can be constructed with the appropriate permutations of the elements of the Jacob–Monod regulatory scheme, particularly if various mutants at the regulatory loci and the possibility of positive regulation are taken into account.

Although some superficial analogies exist there is no example of a regulatory circuit in mammals which can be interpreted satisfactorily in terms of the prokaryotic models. The highly integrated networks of the Britten–Davidson theory of gene regulation in eukaryotes (Britten and Davidson, 1969; Davidson and Britten, 1973), on the other hand, provide a more productive basis for speculation about the consequences of aneuploidy. The concept of producer-gene batteries functionally integrated via receptor-genes by the products of integrator-genes which in turn respond to sensor-sequences offers a rationale for the pleiotropy of chromosomal aberrations: any kind of disturbance of the relative amounts of these elements must have far-reaching consequences. Duplication of a sensor-gene with its adjacent set of integrator-genes would be expected to result in an exaggerated response of many producer genes to an external stimulating signal, like a hormone, inorganic ions, or a developmental inducer. Disruption of the linkage between a sensor-gene and the appropriate integrator-genes, or a break within an integrator-gene cluster would abolish the coordinated response to the external stimulus. Similar inferences can be made concerning the linkage relationship between the receptor-genes and their adjacent producer-genes. The model implies primarily unspecific repression and sequence-specific activation of genes and gene batteries. Sequence-specific inactivation in response to external stimuli could be easily adopted. Further modifications will become necessary when more information is available on the arrangement of genes and about the nature of the intergenic sequences in eukaryotic genomes.

Another kind of pleiotropic effect of aneuploidy can result from influences on the various components of the complex machinery of gene expression. The regulatory mechanisms discussed above are primarily concerned with transcriptional control, i.e. with control of the availability of the template, mediated mainly by DNA–protein interaction. Whether or not the relative abundance of the various molecular species of RNA polymerase and their subunits is of any regulatory significance is at present unknown (Chambon, 1975). The only proteins known to be involved in the selectivity of transcription are the steroid hormone receptors (p. 56). The kinetics of hormone binding to nuclear sites of various degrees of specificity does not suggest a pronounced gene dosage sensitivity, as far as the amount of the receptor is concerned (Yamamoto and Alberts, 1976). Increased dosage of the targets on the other hand would be expected to bring about enhanced primary and secondary responses to the hormonal stimulus. Modification of chromatin proteins, such as acetylation and phosphorylation, is an early event in many gene activation systems. We do not know whether rate limiting steps are involved at this level, but a dosage sensitivity of these reactions would probably exert a pronounced influence on the extent and duration of the response elicited by the activating stimulus.

There is increasing evidence that gene expression can also be regulated at one or other of the post-transcriptional reactions. The steps potentially involved in a eukaryotic cell are:

(a) Modification of the primary transcript by nucleolytic cleavage, "capping" at the 5′-end, and addition of a polyadenylic acid sequence at the 3′-end
(b) Association of mRNA with proteins and its transport into the cytoplasm
(c) Storage of the mRNA as a ribonucleoprotein particle in the cytoplasm and its decoating, which makes it available for translation
(d) Formation of the initiation complex by association with the ribosomal subparticles, the initiator tRNA, GTP, and initiation factors
(e) Initiation, elongation, and termination
(f) Availability of the various species of tRNA and of the isoaccepting subspecies; modification of tRNA; aminoacylation of tRNA

(g) Synthesis of the components of ribosomes
(h) Post-translational modification of the polypeptide chain

A detailed discussion of the various ways in which aneuploidy can influence the functioning and the rate of these reactions is beyond the scope of this chapter. As an example that can be made let us discuss one point: the possible significance of the availability of tRNA species and subspecies. One of the remarkable facts emerging from recent analyses of the composition of the tRNA pattern in eukaryotic cells is that differentiated cells contain specific sets of tRNA molecules. In particular, those cells synthesizing one major kind of protein contain a spectrum of tRNAs functionally adapted to the optimal translation of the small number of mRNA species present (Garel, 1974). This not only pertains to the distribution of the various amino acid specificities but also to the composition of the individual sets of isoaccepting tRNAs (Hilse and Rudloff, 1975). Furthermore, the patterns of isoacceptors undergo characteristic alterations during embryonic development, as shown in Drosophila by Grigliatti et al. (1973). The codon specificity of isoaccepting tRNAs in eukaryotic systems probably restricts the validity of the wobble-hypothesis to prokaryotic organisms. This codon specificity of isoacceptors is also the basis for the structure-rate hypothesis of Itano (1965, 1966) which states that the availability of isoaccepting tRNAs is of regulatory significance for the rate of protein synthesis. Whether this is true in normal cells remains to be demonstrated. The structure-rate hypothesis was used to explain the relatively low levels of abnormal haemoglobins in heterozygotes for various types of β-chain variant. On this theoretical basis essential distortions of the normal pattern of proteins synthesized can be expected in cells trisomic or deficient for a chromosome segment carrying one or other gene-cluster of tRNA. Altered rates of translation would result for those mRNAs which contain the codon specifically accommodated by this particular species of tRNA. The comparative analysis of the tRNA patterns in model organisms which offer a broad spectrum of chromosomal aberrations is a worthwhile task.

Similar reasoning can be made about other components of the system of protein synthesis, especially about the possible significance of the amounts of the various kinds of initiation factors, since initiation seems to be the step primarily subject to regulation (Lodish, 1976).

D. Expression of Gonosomal Genes

1. The X-chromosome

The existence of numerous genetic variants of glucose-6-phosphate dehydrogenase (G6PD) has rendered this enzyme one of the most favoured subjects of human biochemical genetics, pharmacogenetics, somatic cell genetics and population genetics. The enzyme has been purified from human erythrocytes by Yoshida (1966) and by Cohen and Rosemeyer (1969), who were able to demonstrate its tetrameric structure. Some of the alleles at the X-linked *G6PD* locus have been used to confirm the Lyon hypothesis by demonstration of the existence of two cell populations in tissues and in cell cultures of heterozygotes (Lyon, 1974). The absence of hybrid molecules of G6PD in the cells of heterozygotes is additional evidence in favour of the Lyon hypothesis (Yoshida, 1967). Heteropolymeric G6PD molecules are formed when two different alleles are active within the same cell. This was observed in cell cultures derived from a sex chromatin negative triploid foetus (69, XXY), being A/B heterozygous at the *G6PD* locus (Weaver and Gartler, 1975). The A/B heteropolymer was also found in a hexaploid clone of hybrid cells prepared from G6PD-A fibroblasts and G6PD-B leucocytes (Migeon *et al.*, 1974).

Early investigations of G6PD activity in patients with X-aneusomies (XXX; XXXX; XXY) demonstrated complete inactivation of all but one of the X-chromosomes present, at least for this gene locus (Grumbach *et al.*, 1962; Therkelsen and Petersen, 1967). These results were questioned by Steele (1970), who found significant differences between the specific G6PD activities of female and male embryonic lung fibroblasts and of female and male newborn erythrocytes, respectively. Lactate dehydrogenase and hypoxanthine-guanine phosphoribosyl transferase (HGPRT) did not exhibit any difference in this study. The conclusion seemed unavoidable, that, while dosage compensation is complete at the *HGPRT* locus, it is incomplete at the *G6PD* locus at the ages of the cell donors examined. This interpretation could, however, be abandoned on the basis of a more meticulous investigation of these relationships. Cloned embryonic lung fibroblasts from an A/B-heterozygote show either the A- or the B-phenotype in accordance with the Lyon hypothesis. Thus, hemizygous expression of the *G6PD* locus in female embryonic fibroblasts is compatible with a sex difference in specific G6PD activity (Steele and Owens, 1973; Steele and Migeon, 1973). Since skin fibroblasts of the same foetuses did not exhibit a difference of G6PD activity a tissue specific regulatory mechanism seems to adjust G6PD activity to its male and female levels, respectively. In fact, the sex difference of G6PD activity prevailing in embryonic lung fibroblasts is compensated post-natally by an increase of the G6PD activity in the male. Autosomal regulatory genes which influence the level of G6PD in erythrocytes of mice were postulated by Hutton (1971) to explain the differences observed between various inbred strains.

At some stages during the differentiation of the mammalian germ line both X-chromosomes are active. The presence of the A/B-heteropolymeric form of G6PD in mature oocytes of adult females and in those of a foetus at 16 weeks gestation clearly demonstrates the activity of both *G6PD* alleles (Gartler *et al.*, 1972; 1973). In the germ cells of a 12 week ovary, however, the G6PD heteropolymer could no longer be detected (Gartler *et al.*, 1975). Thus, X-chromosome differentiation also occurs in primordial female germ cells and is apparently maintained until the oogonia pass into meiotic prophase. Thereafter the inactive X-chromosome undergoes re-activation and X-linked genes are proportionately expressed. Comparisons of the activities of enzymes encoded by X-linked genes were performed with oocytes and one-to-two cell embryos of mice with the maternal sex chromosome constitution XO and XX, respectively. A striking 1:2 relationship was demonstrated for G6PD and HGPRT by Epstein (1969, 1972) and for phosphoglycerate kinase (PGK) by Kozak *et al.* (1974). There is, however, no difference between the G6PD activities of early male and female preimplantation embryos, both being derived, of course, from the meiotic product of XX oogonia (Brinster, 1970). Hence, in early pre-implantation embryos, G6PD activity and apparently also that of the products of other X-linked genes, reflect the synthetic activity in the pre-ovulatory phase. Protein synthesis during the maturation of the oocyte provides the enzyme inventory of the ovum and the earliest embryonal stages. These findings may well apply also to Man.

In the male germline on the other hand the X-chromosome undergoes inactivation prior to or during the differentiation of the primary sperma-tocyte. According to Lifschytz and Lindsley (1972) this is true for all animals which have developed male heterogamety. This X-inactivation process seems to be a prerequisite of a successful completion of spermato-genesis. Influences interfering with this mechanism cause partial or com-plete male sterility. This is considered the basis of the male-limited sterility in X-autosome translocation heterozygotes in mice (Russell and Mont-gomery, 1969). If X-inactivation at spermatogenesis affects only one X-chromosome the continued activity of supernumerary X-chromosomes would be a plausible cause of the arrest of germ cell differentiation and sterility in 46, XX males and in the various forms of Klinefelter's syndrome (Lyon, 1974).

X-chromosome differentiation during early embryogenesis is generally considered a random event. Variation of the composition of the resulting mosaic can be described in good approximation by the binomial distribu-tion (Nesbitt and Gartler, 1971). Significant deviations from expectation can arise in several ways.

It is generally held that X-inactivation precedes the processes of cell differentiation. The precursor cell pool from which a given organ is ulti-mately built up may therefore be very small at the time of X-inactivation.

In a proportion of heterozygotes given by the binomial rule the entire organ may be formed from cells containing the same X-chromosome in the activated or in the inactivated state, respectively. These two types of heterozygotes, exhibiting in a given tissue one or other homozygous phenotype, will occur with equal frequencies, and the number of heterozygotes detected by the analysis of that particular tissue will be smaller than expected. A frequency distribution of this kind was in fact observed by Hitzeroth *et al.* (1977), who analysed the G6PD phenotypes in erythrocytes of a large group of South African Negro women. From their data it can be concluded that the number of precursor cells of the erythropoietic system is not larger than three or four. Hence, sampling from a very small precursor cell pool is one of the reasons for the occurrence of heterozygotes for X-linked genes who express the homozygous phenotype in one or another of their organs. Some of the uniform HGPRT-positive red cell populations found in obligate heterozygous women may therefore not necessarily result from selection against the HGPRT-deficient clones (de Bruyn, 1976). In the presence of X-chromosomal anomalies, however, somatic selection against the genetically unfavourable cells is of major importance. X-chromosomes suffering any kind of a deletion are found inactivated in all cells of their carriers. Since preferential inactivation of the abnormal X-chromosome is very unlikely (for reasons which cannot be discussed in this context) selection against the cells containing the normal X in the inactive state is the most plausible explanation. Accordingly, the fate of an X-chromosome involved in a translocation with an autosome depends on the balanced or unbalanced state of the rearrangement. With some notable exceptions the normal X-chromosome is found inactivated in balanced X-autosomal translocations. Thus, genetic balance with regard to the translocated segments of the autosome and of the X-chromosome is maintained. Likewise, with one exception (Mikkelsen and Dahl, 1973), the abnormal X-chromosome is found uniformly inactivated in unbalanced X-autosomal translocations. The patterns of inactivation are further complicated by the fact that the inactivation may or may not spread to the autosomal segment, depending on the structure of the translocation chromosome. An interpretation of these complex inactivation patterns of abnormal X-chromosomes was given by Therman and Pätau (1974) and Therman *et al.* (1974). These authors argue that the condensation of X-chromatin starts from, and therefore requires the presence of a proximal segment on Xq. X-chromosomes lacking this segment cannot become inactivated and render the cell functionally unbalanced. This attractive hypothesis also explains the absence of any well documented case of i(Xp) and of telocentric Xp.

X-chromosome differentiation in early embryonal development is itself genetically controlled. One of the elements involved seems to be somehow

132   W. KRONE AND U. WOLF

related to the proximal segment on the long arm of the X-chromosome just mentioned. Other factors were postulated on the basis of non-random inactivation patterns which could not be explained by selection or by structural anomalies of the X-chromosome. Thus, we are dealing with another possible cause of a non-random inactivation pattern: genetically controlled preferential inactivation or activation of one X-chromosome.

This inherited non-randomness of X-chromosome differentiation was discovered in mice by Cattanach and Isaacson (1965, 1967). Initially detected by its influence on the activity of autosomal genes inserted into the X-chromosome (i.e. on the variegation type position effect in Cattanach's translocation), this X-chromosomal element was later shown to control the activity of other X-linked genes (Cattanach, 1975). The factor was designated as "X-chromosome controlling element" (Xce). Its functional significance is revealed especially by its allelic "high variegation" mutant ("O^{hv}"; Drews et al., 1974; Ohno et al., 1974), which exerts a pronounced cis effect on the expression of closely linked genes, like Blotchy, Tabby and Tfm. The X-chromosome carrying this allele of the controlling site remains in the active state. This is so far the clearest line of evidence for the genetic control of X-chromosome differentiation, and any theory of this process will have to take these findings into account.

In view of this situation in the mouse it can be assumed that X-differentiation is also under genetic control in Man. The existence of genetic factors causing biased, non-random inactivation patterns is suggested by studies of manifesting carriers of sex-linked recessive diseases. The occurrence of affected heterozygotes for recessive X-linked genes can be expected: they represent one of the ends of the binomial distribution of all possible inactivation patterns. An unexpected finding is, however, that they tend to occur more frequently among first degree relatives of manifesting carriers than in a random sample of definite heterozygotes. This was observed in Duchenne muscular dystrophy by Moser and Emery (1974), and a rather conspicuous family with Fabry's disease was described by Ropers et al. (1977). In this latter family, the eight daughters of a patient clearly formed two classes: a "high group" of four with the expected 50% α-galactosidase A-activity, and a "low group" having an average of 20% activity in their white blood cells. Since selection was unlikely because of high inter-tissue correlation of enzyme activity within the individual patients, the authors favour the attractive hypothesis that in their family (as well as in those described by Moser and Emery) a gene is segregating which brings about the preferential inactivation of the X-chromosome carrying the normal allele. Although the X-chromosomal or autosomal location of this factor is not established, the analogy with the mouse studies is obvious. Systematic investigations of the inheritance of inactivation patterns in heterozygotes for other X-linked genes will help to elucidate the nature of the factor involved. Since this genetic element should also influence the choice of the active X-chromosome in X-aneusomies, investigation of 47,XXX women in populations polymorphic for *G6PD* alleles should also be worthwhile.

The completeness of X-inactivation is challenged by the accumulating evidence that the *Xg* blood group locus is not subject to inactivation (see Chapter 7). This seems to be true, however, only for normal X-chromosomes. Pedigree analysis and the frequency distribution of Xg phenotypes in cases of structurally abnormal X-chromosomes (found in the inactivated state due to selection, see above) show that the *Xg* locus is inactivated (Sanger *et al.*, 1971). This apparent dependence of the non-completeness of inactivation on the structural integrity of the X-chromosome contributes to the complexity of the X-differentiation mechanism.

These matters were discussed in the context of this chapter because they provide the background for future biochemical investigations. Sequence specific interaction of DNA with proteins and DNA modification will most likely by involved in this exciting field of endeavour (Riggs, 1975).

2. The Y-chromosome

Holandric inheritance seems to be confined so far to sexual differentiation and to the morphological peculiarities of the Y-chromosome itself. Male sexual differentiation requires at least two steps: the development of the undifferentiated embryonic gonad into a testis and the maturation of the testicular tissues into a functional organ. A critical analysis of the effects of Y-chromosomal anomalies on sexual development led Siebers *et al.* (1973) to postulate separate locations of the genetic information governing these two processes. According to these authors a proximal segment of the long arm carries information responsible for testis determination, and the short arm contains information required for testicular maturation. In a distal segment of the non-fluorescent part of the long arm there seems to be additional information which is required for normal spermatogenesis, since its deletion causes azoospermy (Tiepolo and Zuffardi, 1976).

A new era in the investigation of the function of the Y-chromosome was initiated by the discovery of the male-specific transplantation antigen HY (Eichwald and Silmser, 1955). Upon male to female transplantation within highly inbred strains of mice this cell surface antigen elicits the production of antibodies which cross-react with male cells of all mammalian species tested so far (Wachtel *et al.*, 1974). The fundamental importance of this antigen became apparent when it was also detected in non-mammalian vertebrates like amphibia and birds as a specific attribute of the heterogametic sex (Wachtel *et al.*, 1975a). In view of its ubiquitous occurrence in the heterogametic sex and its remarkable conservation during evolution, Ohno (1976) postulated that HY (or HW) antigen represents the product of a Y- (or W-) chromosomal gene, determining the gonad of the heterogametic sex. In this hypothesis testicular differentiation is interpreted in analogy to other types of organogenesis which require specific cell–cell recognition mediated by surface antigens (Bennett, 1975). HY antigen resides on the surface of all XY-cells. Gonadal cells as opposed to somatic cells possess an HY-receptor which mediates testicular organogenesis by specific cell aggregation. Anchorage sites for HY antigen are also present on female cells as are the specific HY-receptors on female gonadal cells. Hence, a few XY- or XXY-cells would suffice to tag an excess of surrounding XX-cells with HY antigen, thereby imposing on them active participation in the differentiation of a male gonad. This, however, requires that the HY antigen is a diffusible agent at least across short distances. Such a hormone-like action of HY appeared likely on the basis of an investigation of the HY status of the gonads of the virilized female co-twins (freemartins) of cattle twins (Ohno et al., 1976). Dissemination of HY antigen and its uptake by the surface of XX-cells would also explain the fact that XX-males are HY positive, if all of these cases can be assumed to arise from an XXY zygote or from XY/XX mosaics and to have lost their Y-chromosome from all of their somatic and most of their gonadal cells. The presence of a Y-chromosome has not, however, been detected in all instances of XX-males. Some of them may owe their aberrant sexual differentiation to an autosomal dominant mutation analogous to the sex reversed "Sxr" mutation in mice (Cattanach et al., 1971). This mutation causes a masculinization of chromosomally female mice which have subsequently been shown to be HY positive (Bennett et al., 1975). It is interpreted by Wolf (1976) as a mutation of a regulatory element closely linked to the autosomal gene which codes for HY antigen. In the Jacob–Monod terminology this element would be equivalent to an operator. In its normal state the HY locus is irreversibly induced—probably during a sensitive phase of development—by the product of a Y-chromosomal regulatory gene. The Sxr mutant represents— according to this hypothesis—the constitutive allele of this operator sequence. Hence, HY is produced in the absence of a Y-chromosome. The presence of Y-autosomal translocation in the Sxr-mice and the existence of Y-containing cells in XX-males is not required. (The assumption of a Y-autosome translocation, in any case, is not compatible with the transmittance of a functionally intact Y-chromosome in the pertinent families.) Even in those XX-males who arise from an XXY zygote by the loss of the Y-chromosome the continued production of HY antigen by all cells would be the result of the irreversible induction of the autosomal HY locus prior to the loss of the Y-chromosome. Further predictions of Wolf's hypothesis are the familial occurrence of XX-males, which indeed was reported recently (de la Chapelle, 1976), and the existence of an autosomal recessive XY-gonadal dysgenesis due to deleterious mutation or loss of the HY locus. This clinical condition may be heterogeneous. While the majority of cases show X-chromosomal recessive transmission, the occurrence of sporadic cases prevents the exclusion of autosomal recessive inheritance.

The proportionate expression of the HY antigen in XYY-males (Wachtel et al., 1975b) is generally considered as evidence in favour of the hypothesis of Y-linkage of the HY locus. These quantitative relationships can, however, also be interpreted in accordance with the regulatory model described above.

The isolation and partial characterization of repetitive DNA specifically localized in the heterochromatic segment of the Y-chromosome was already discussed in Section II.D. This DNA fraction and its quantitative and presumably also its qualitative variability represent the first known molecular aspects of holandric inheritance.

V. ANEUPLOIDY AND THE CELL CYCLE

Embryogenesis can be envisaged as an intricate process of differential production and proliferation of various interacting cell populations. If the cell population kinetics are to result in a normally developing organism, the timing of this process must be of fundamental importance. The temporal order depends on the rate with which the various kinds of cells enter into and pass through the stages of the cell cycle. The systematic investigation of the genetic control of the cell cycle has only recently become possible by the isolation of cell cycle mutants (literature cited by Liskay, 1974). Nevertheless, it has been argued for a long time that aneuploidy affects the timing of the cell cycle. The general growth retardation observed in most autosomal trisomy syndromes prompted the early hypothesis that the mere presence of a supernumerary chromosome as such means for the cell an unspecific burden which causes a prolongation of the cell cycle. The reduction of cell numbers in many body organs of newborns affected by one of the three major autosomal trisomies (Naeye, 1967) would thus be unrelated to the genetic information carried by the respective chromosome. Specific effects on the growth of certain cell populations, on the other hand, are interpreted as arising from an enhanced activity of the particular chromosome in these cells during a distinct developmental stage (Mittwoch, 1971). The view that the amount of excess chromatin rather than the genetic information it carries exerts the major influence on the cell cycle does not, however, provide a satisfactory explanation. Malformation syndromes caused by autosomal deficiencies used also to be explained by intrauterine growth retardation. Birth weight and body height in trisomy 8, on the other hand, lie in the normal range (Kakati *et al.*, 1973). Furthermore, there are at least 32 gene loci involved in the control of the cell cycle in *Saccharomyces cerevisiae* (Hartwell *et al.*, 1973). Many more such genes might be necessary to steer a mammalian cell through the stages of the cell cycle and to keep this process appropriately coupled with cellular differentiation. The concerted action of the products of these genes might well be balanced so delicately that any disturbance of the rate of synthesis and turnover is more likely to delay rather than to accelerate the cell cycle. The involvement of specific information in the effects of aneuploidy on the timing of the cell cycle can also be inferred from the role of cell–cell recognition in the regulation of the kinetics of proliferation of the various cell populations during embryogenesis. The specificity of the cell to cell interaction is mediated by the antigenic pattern on the cell surface which depends on information distributed over many, if not all chromosomes of the complement (Weiss and Green, 1967).

Despite the formidable difficulties in standardizing the growth conditions of human homonuclear fibroblasts derived from different donors, comparative studies of cell cycle parameters utilizing this *in vitro* system are being published in increasing numbers. Barlow (1972) compared the performance of components of 45,XO/46,XY-, 45,XO/46,XXqi-, and of 45,XO/47,XXX-mosaics respectively. Irrespective of the initial proportions of the two cell populations the 45,XO cells completed the cell cycle faster than any one of the other components. Similar comparisons of 45,XO/46,XX- and 46,XX/47,XXX-mosaics would be required to find out whether X-chromosomes influence the duration of the cell cycle in proportion to their number. It would also be of interest to examine *in vitro* the effect of the Y-chromosome which exerts a well-known positive effect on the rate of growth *in vivo* (Borgaonkar and Shah, 1974). In the comparison of normal female and male cells, however, the isogenic background is not as easily achieved as it is in the mosaic studies.

It is obvious from the data summarized in Table 3 that a supernumerary chromosome 21 causes a prolongation of the cell cycle, possibly by increasing the duration of G_2 and S. Conversely, Kukharenko *et al.* (1974) found a shortened S-phase in a strain derived from a foetus with monosomy 21. The possible connection of these findings with the decreased lifespan *in vitro* of cells with trisomy 21 and with their reduced DNA repair capacity (see Section IV.B.2) is intriguing.

The effects seem not, however, to be specific for chromosome 21. Paton *et al.* (1974) observed a prolonged G2 phase also in trisomy 18, and a shortened S-phase, as in their 21-monosomic strain was also found by

Kukharenko *et al.* (1974) in cells with various other chromosomal aberrations. It is interesting to note that the phases of the cell cycle were indistinguishable from those of normal cells in a triploid strain investigated by Kuliev *et al.* (1975).

Table 3

Studies on growth parameters in trisomy 21

Increased proportion of cells with intermediate DNA content in fibroblasts; suggestion of prolonged S-phase	Mittwoch (1967)
Retarded rate of DNA synthesis in fibroblasts; measured by the rate of dilution of labelled DNA during proliferation.	Kaback and Bernstein (1970)
Significant increase of the population doubling time in fibroblasts; decrease of *in vitro* lifespan	Schneider and Epstein (1972)
Increased population doubling time in fibroblasts	Segal and McCoy (1974)
Increased duration of G_2-and possibly of S-phase in fibroblasts (LM-method[a])	Paton *et al.* (1974)
Prolonged G_2-phase in fibroblasts (LM–method[a])	Kukharenko *et al.* (1974)
Decreased duration of cell cycle time in lymphocytes (BUdR-method)	Dutrillaux and Fosse (1976)

[a] Percentage of labelled mitoses as a function of time after ^{3}H-thymidine pulse

The effects of aneuploidy on the timing of the cell cycle should be most easily detectable in rapidly proliferating tissues, like the haematopoietic system. There are in fact many haematological abnormalities connected with the human chromosomal aberrations. The best known examples are in trisomy 21 (Sigler and Zinkham, 1972): the frequent neonatal bone marrow dysfunction, the increased risk of childhood leukaemia, the decreased lobe count in neutrophils, the relatively low amount of HbF as opposed to an increased amount of Hb A_2 and a pronounced polycythaemia (Wilson *et al.*, 1968; Kohne and Kleihauer, 1975); and in trisomy 13: hypersegmentation of the granulocytes and the persistence of foetal and embryonic haemoglobins, discussed in Section IV.B.

The hypothesis seems justified that many or all of these abnormalities are causally related to the effects of the respective chromosomal aberration on the timing of the cell cycle.

Studies on the differential proliferation of the components of chromosomal mosaics did not contribute much to the elucidation of the influence of aneuploidy on the schedule of the cell cycle. The behaviour of normal/21-trisomic mosaics is heterogeneous. Follow-up studies with blood cultures showed cell selection during early childhood in favour of normal cells in some and abnormal cells in other patients. The proportions of the two cell populations in adult normal/21-trisomic mosaics do not seem to be subject to further change (Taylor, 1970). Similar variability was observed by Nielsen (1976) in various gonosomal mosaics.

Variable development of the proportions of the two cell populations was also observed in fibroblast cultures derived from one patient with a normal/ trisomy 21 mosaic (Taylor, 1970). In the cultures established from a 46,XX/47,XX,+16 mosaic abortus the normal cells outgrew the trisomic cells after 16 passages *in vitro* (Ikeuchi and Sasaki, 1975). Fibroblast cultures derived from a 46,XY/47,XY,+18 mosaic, on the other hand, maintained their initial proportion of the component cell populations through many generations (Beratis *et al.*, 1972).

Comparative determinations of cell cycle attributes will have to be performed on a much larger scale in order to further our understanding of the influence of aneuploidy on this process. The utilization of BUdR incorporation and differential staining of substituted DNA will bring about rapid progress in this field.

VI. NUCLEOLUS ORGANIZATION

The human chromosome complement is no exception to the general rule that nucleolus organizing regions (NOR) tend to become recognizable at metaphase as poorly stained segments called secondary constrictions. The classical way of assigning nucleolus organizers to secondary constrictions has been to detect their association with nucleoli during mitotic and meiotic prophase. By this approach the human NORs were localized at the secondary constrictions of the acrocentric chromosomes (mitosis: Ohno *et al.*, 1961; Ferguson-Smith and Handmaker, 1961; male meiosis: Ferguson-Smith, 1964). More detailed analyses of the association of nucleoli with the bivalents of the acrocentric chromosomes were performed by Hungerford (1971) for the pachytene of spermatogenesis and by Stahl and Luciani (1972) for the much less readily accessible early stages of oogenesis.

The essential genetical information stored within the DNA of the NORs codes for the ribosomal precursor RNA. After processing which takes place largely within the nucleoli, about half of this precursor is preserved in the form of the two cytoplasmic, high molecular weight species of RNA, 28S and 18S RNA. Hence, these molecules are complementary to large parts of the DNA of the NORs ("rDNA") and, under appropriate conditions, can form hybrid molecules with it. The experimental conditions, under which such RNA:DNA hybrids are formed can be achieved *in situ* and *in vitro*.

To determine the number of copies of rDNA contained in the human genome Bross and Krone (1972) applied the method of RNA:DNA hybridization *in vitro* and found (using a DNA content per G1-cell-nucleus of 7×10^{-12}g) 500 copies per diploid complement. This redundancy allows the application of the *in situ* hybridization method. Chromosomes on microscopic slides resist careful denaturation of their DNA by heat, acid or alkali. The binding of heavily labelled complementary RNA to the single strands of chromosomal DNA thus exposed can be detected autoradiographically. The first application of this method to the human chromosome complement (Henderson *et al.*, 1972) revealed the presence of rDNA on the short arms of all five pairs of acrocentric chromosomes (13, 14, 15, 21 and 22). In fact, no other site acquired radioactive label after hybridization *in situ* with (28S + 18S) rRNA. This was confirmed by Evans *et al.* (1974), who were also able to localize the ribosomal genes more precisely to the secondary constrictions, the so-called "satellite stalks", rather than the satellites. Interindividual variability of the amount of rDNA was predicted by these authors on the basis of their finding of an increased level of labelling of an elongated constriction on chromosome 15 which represents one of the numerous morphological variants observed in the human population. This variability between individuals was demonstrated by the quantitative filter method of RNA:DNA hybridization. Using DNA from white blood cells and from cultured fibroblasts Bross *et al.* (1973a, b) found vastly differing amounts of rDNA not only among normal persons but also within a group of patients with trisomy 21 and among carriers of a D/G-translocation. The rDNA content of the normal human genome is about $0 \cdot 03 \pm 0 \cdot 0075 \%$. The short arms of the acrocentric chromosomes belong to the most variable regions of the human genome. The suspicion that part of this morphological variability might correspond with the variability of the amount of rDNA was verified by Dittes *et al.* (1975). They performed a combined biochemical and cytogenetical study of a family with a 15/21-translocation. A detailed cytogenetic analysis of the short arms of the acrocentric chromosomes and of the centric region of the translocation chromosome allowed the interpretation of each individual's amount of rDNA in terms of number and size of his or her nucleolar constrictions. Hence, the accessory NORs resulting from trisomy 21 and trisomy 13, respectively, will only result in the expected gene dosage effect of $1 \cdot 0 : 1 \cdot 1$, if their karyotypes are similar to those of the controls with respect to all the other nucleolar constrictions. The same is true for the expected gene dosage effect of $1 \cdot 0 : 0 \cdot 9$ in comparisons of normal individuals with translocation-trisomics of the type 46, $-$ D, $+$ t(DqGq); and for the expected ratio of $1 \cdot 0 : 0 \cdot 8$, if balanced translocation carriers are compared with normal persons. Only incidentally (Bross *et al.*, 1973) or after careful selection of the probands (Guanti and Petrinelli, 1974) could these dosage effects be demonstrated. Hence, the size of each of the human NORs varies within a similar range.

The variable positions of the break points giving rise to Robertsonian translocations between acrocentric chromosomes contribute to the variability of rDNA amounts among balanced translocation carriers and among translocation trisomics. *In situ* hybridization is especially suitable in detecting the presence of a NOR in the centric region of a translocation chromosome. So far, this has been done only in very special situations (Johnson *et al.*, 1974; Warburton *et al.*, 1973). Until recently *in situ* hybridization and the analysis of the participation of the translocation chromosome in satellite associations were the only criteria by which the presence of NOR material in its centric region could be evaluated. The introduction of the highly specific silver staining for NORs (Goodpasture and Bloom, 1975) has opened a convenient way for a systematic investigation of this and many other problems of NOR research. Inactivated NORs, however, do not accept the silver stain (Miller *et al.*, 1976). This makes the silver staining method an ideal means of studying the activation of NORs during embryogenesis (Eugel *et al.*, 1977) and their inactivation during male and female meiosis. To prove unequivocally the absence of multiple ribosomal cistrons from a suspected site will still require *in situ* hybridization.

Does the interindividual variability of the amount of rDNA have any functional significance? The situation in other organisms might help to answer this question. Similar variability occurs in natural populations of plants and animals, notably in amphibia. Various mechanisms exist which compensate for the loss of substantial numbers of ribosomal cistrons. Perhaps the simplest of these mechanisms is an enhanced rate of transcription, as acting in the heterozygote for the anucleolate mutant of *Xenopus laevis* (Brown and Gurdon, 1964). The most intricate mechanism on the other hand seems to be the "magnification" of bobbed mutants in Drosophila (Ritossa *et al.*, 1974).

A mechanism of more general significance regulating the amount of rRNA per cell was disclosed by comparative studies with species which are in a diploid–tetraploid relationship to each other. Specimens from diploid and from tetraploid populations of the amphibian *Odontophrynus americanus* show the expected 1:2 relationship of their amounts of rDNA (Schmidtke *et al.*, 1976a). Nevertheless, the RNA content in kidney cells (Becak and Goissis, 1971) and a variety of other parameters are in the range typical of diploid species of Odontophrynus (Becak and Pueyo, 1970). Ancestral tetraploidization has created a similar diploid–tetraploid relationship between various species of Cyprinid fish. While the tetraploid species contain twice the number of ribosomal cistrons of the diploid species (Schmidtke *et al.*, 1975), there is no difference in cellular RNA content, nuceolar area and other parameters which are more indirectly related to the transcription of the ribosomal genes (Schmidtke *et al.*, 1976b). These findings suggest that the adjustment of the amount of rRNA per cell to the level of the diploid cell is of major importance in the process of functional diploidization of tetraploid genomes. Tetraploid species within other families of fish (Isospondyli) not far enough advanced in this process in fact show proportional expression of the increased amount of rDNA (Schmidtke *et al.*, 1976b).

There is evidence that compensatory mechanisms which counterbalance the loss or gain of ribosomal cistrons, also operate in human cells. On the cytological level of analysis alterations of the number of acrocentric chromosomes per cell nucleus could be expressed in interphase as changes in nucleolar mass. This is usually assessed in terms of the number of nucleoli or, more precisely, by the total nucleolar volume per nucleus. Particularly suitable as a system for the investigation of this problem are meningiomas, generally benign brain tumours which are characterized cytogenetically by a preferential loss of chromosome 22 and of additional acrocentric chromosomes (Zang and Singer, 1967; Singer and Zang, 1970; Mark, 1970). Meningiomas that have lost two or more acrocentric chromosomes show a significantly decreased average number of nucleoli as compared with meningiomas having a normal karyotype (Zankl and Zang, 1972). The average nucleolar volume, however, remains constant and independent of the number of acrocentric chromosomes (Zankl *et al.*, 1973). The same is true for meningioma cells having an excess of two to six acrocentric chromosomes. Hence, whatever kind of regulatory mechanism is operating, it seems to act in both directions.

Another indication of the existence of a regulatory mechanism controlling the NORs comes from recent studies of satellite associations. The frequency of participation of an acrocentric chromosome in satellite associations depends primarily on the size of its nucleolar constriction (Zankl and Zang, 1974; Schmid *et al.*, 1974; Phillips, 1975): the longer the secondary constriction, the higher the frequency of participation in satellite associations. Warburton *et al.* (1976) investigated the correlation between the tendency of acrocentrics to associate and the amount of rDNA present in their NORs, as determined by quantitative *in situ* hybridization. While the relationship between these parameters confirmed the general rule in most instances, there were a few remarkable exceptions suggesting that other influences, besides the amount of rDNA, may also determine the frequency with which certain acrocentrics engage in satellite associations.

The association behaviour seems to be a rather stable character of each individual acrocentric chromosome. Often there are reproducible quantitative differences between the association tendency of homologues. This attribute of one homologue apparently depends on the presence of a NOR on the other. Hansson (1975) investigated the behaviour of individual acrocentrics whose homologues were involved in Robertsonian translocations. These single free homologues (like the free chromosome 21 in a translocation-trisomic 46, XX, t(21q21q) or the free chromosomes 13 and 14 in a translocation carrier, 45,XY, t(13q14q)) showed the highest association indices in 11 of the 12 individuals studied. Their association with each other (when possible) was the predominant type of association found. The high tendency to engage in satellite associations was not caused by the presence of particularly conspicuous nucleolar constrictions, on these chromosomes. The translocation chromosomes were not associated with acrocentric chromosomes.

Robertsonian translocation is not the only way by which NORs can get lost. There are variants among the human acrocentric chromosomes which lack a secondary constriction, like the short arm deletions (p —) or the heterochromatic elongated short arms (p+). Both types of variants show a very low tendency to engage in satellite associations. De Capoa et al. (1976) studied an impressive example of a nonassociating 15p+ chromosome whose homologue exhibited an association index significantly higher than that of the other D-chromosomes.

These findings are not only highly suggestive of quantitative compensation of the loss or gain of NORs in man. They also seem to demonstrate a possible uniqueness of the individual human NORs. The increased association tendency of chromosomes 14 and 21 in lymphocytes from hyperthyroid patients and in those cultured in the presence of thyroid hormones points in the same direction (Nilsson et al., 1975). A direct relationship between nucleolar activity (synthesis and maturation of rRNA) and the quantitative aspects of satellite association remains to be demonstrated.

Theoretically, uniqueness of each of the five pairs of NORs could result from the sequences of the rDNA, or the regulatory circuits into which the individual NORs are integrated. The latter possibility would suggest cell type specific or stage specific expression of nucleolar functions. In either case, individuals with homozygous deficiencies of any one of the human NORs would not be expected to be viable. Uniqueness of any kind would explain the maintenance of the high number of five NORs in Man and the most closely related primates.

During the diplotene stage of oogenesis there are one to three main nucleoli associated with the bivalents of the acrocentric chromosomes. The emergence of between 15 and 40 micronucleoli during the same stage has recently raised the question of accessory NORs in the human genome (Stahl et al., 1975). These micronucleoli are frequently found associated with paracentric heterochromatin, notably of chromosomes 1, 9 and 16.

In interphase of somatic cells the C-band of chromosome 1 is also closely associated with the nucleoli (Hartung *et al.*, 1975). The sum of the volumes of the micronucleoli of diplotene oocytes hardly amounts to the equivalent of a main nucleolus. *In situ* hybridization studies with 28S and 18S rRNA do not confirm the presence of accessory NORs. A low degree of amplification or the dispersal of material derived from the main nucleoli seem, at present, good alternatives to the hypothesis of a cell-specific activation of minor sites of rDNA.

Alterations of the amount of rDNA are not the only way in which chromosomal anomalies can interfere with the ribosomal system. Besides 28S RNA the large ribosomal subunit contains 5S RNA. A major cluster of genes coding for 5S RNA is located on a distal segment of chromosome 1. During an early stage of maturation of the large ribosomal subunit, 5S RNA associates with a precursor ribonucleoprotein particle within the nucleolus. It is not known how dosage effects involving the 5S RNA gene cluster would influence the maturation of ribosomes.

There are furthermore some 70 structural genes coding for the various ribosomal proteins. In Drosophila some of these genes are very closely linked to the NOR on the X-chromosome (Steffensen, 1973). Some ribosomal proteins exert important functions in protein synthesis and ribosomal stability. Severe quantitative disturbances of their supply could adversely affect the metabolic basis of growth and regeneration. Beyond these more specialized functions the nucleolus might be involved in the coordinate control of maturation and release of mRNA and tRNA (Sidebottom and Deak, 1976). This would render the nucleolar system one of the most vulnerable parts of the machinery maintaining genetic balance.

VII. NEW TRENDS IN THE ANALYSIS OF THE HUMAN GENOME

It has become obvious from the data summarized in this chapter that biochemical investigations of the chromosomal aberration syndromes have contributed a remarkable share of information to our knowledge of the human genome. This does not only pertain to the dosage effects of already localized genes but also to the observations which still await interpretation in terms of molecular biology. A situation has in fact emerged which is highly reminiscent of the state of Drosophila genetics, with the important difference that mutants cannot be "constructed" but are provided, albeit rarely, by nature.

On the other hand, the contribution of these studies to our understanding of the causal relationships between the complex malformation syndromes and their underlying chromosomal aberrations is still disappointing. The most severe effects of genomic imbalance occur during intrauterine life. Thus, inaccessibility in a literal sense is one of the main obstacles for this kind of investigation. The mouse has become a very suitable model organism, however, since all autosomal trisomics are available through the work of Gropp and his co-workers (1975). These authors were able to produce various kinds of trisomic offspring from crosses of "all acrocentric" mice with those heterozygous for two metacentric chromosomes which are homologous for one of their arms (isochromosomes). Besides providing the most valuable information about the morphological and histological effects of aneuploidy, this system lends itself to biochemical exploitation.

The inventory of human chromosomal anomalies covers an increasing spectrum of partial monosomies and partial trisomies. In addition, an increasing variety of balanced translocations will become available for studies of position effects influencing the expression of genes already assigned to the chromosome segments involved. Artificial production of aneuploid cells by chemical induction of non-disjunction, by laser micro-beam-irradiation, or by the fusion of euploid cells with colchicine-induced nucleated cell fragments (Cremer et al., 1976) offers promising new approaches. One limitation of in vitro studies with chromosomally aberrant cells is the fact that many specimens from early abortions do not yield viable cell cultures. In a recent study of 1655 abortuses only 53% could be propagated as cell cultures (Creasy et al., 1976). Improvements in the recovery of the aborted material and of the cell culture procedures should help to overcome this difficulty. It cannot, however, be expected, that cell culture studies will provide a delineation of a complete set of "triplo-lethal" and/or "haplo-lethal" gene loci or chromosome segments in analogy to those studied with inbred strains of D. melanogaster (Lindsley et al., 1972). In the highly outbred human population multiple allelism at the loci involved in conjunction with environmental factors bring about the broad range of variation observed in the phenotypic expression and via-bility of aneuploids. Furthermore, the failure of aneuploid homonuclear cells to grow in culture may bear no relationship at all to the prenatal death of the aneuploid conceptus. As demonstrated by the existence of hetero-nuclear cell lines a broad range of karytypic variability is in fact compatible with growth in vitro. Hence, the expression of specific differentiated functions by cell cultures should be a more suitable parameter for the investigation of the effects of aneuploidy at this level. Prenatal patho-histology of the aneuploid conceptus, a newly emerging field of interest, could certainly provide the information necessary to select the more promising biochemical approaches in this respect.

At the molecular level a new phase in the analysis of the eukaryotic genome was initiated by the introduction of the class II restriction endonucleases (Nathans and Smith, 1975). These enzymes belong to the pro-karyotic restriction–modification systems which protect the cells against foreign DNA. They cleave both strands of DNA at specific sequences, called palindromes, consisting of four to six nucleotide pairs arranged in such a way that the two halves of the sequence are complementary (e.g. GAATTC in the case of the *E. coli* restriction endonuclease RI). The first area of the application of restriction endonucleases was the analysis of viral genomes. Because of the rare occurrence of palindromes in these relatively small DNA molecules a few large fragments are formed after exhaustive digestion whose arrangement within the viral DNA is analysed in formal analogy to the methods of peptide sequencing of proteins. In the genomes of higher vertebrates consisting of some 10^9 nucleotide pairs palindromes are frequent and hence, a continuum of size classes of fragments is formed after digestion with a restriction enzyme (Botchan *et al.*, 1973; Southern and Roizes, 1973). Satellite DNA may yield restriction fragments of a more uniform size class depending on the regularity of the occurrence of palin-dromes in its repeating unit. Much has been learned by restriction analysis about the structure and the sequence heterogeneity of satellite DNA (e.g. Botchan, 1974). The detection and isolation of the human Y-chromosome specific satellite DNA belongs to the most exciting advances achieved by this technique. Its application to the DNA of fractionated chromosomes should ultimately make it possible to produce a complete chart of restriction fragments of the eukaryotic genome. The information content of fragments could then be tested with *in vitro* systems of transcription and translation. The precise pattern of interspersion of DNAs of various degrees of redun-dancy with DNA of structural genes could be analysed. Integration sites of tumour virus DNA could be identified. The binding of specific regulatory proteins to specific chromosomal sites could more easily be studied. The deepest insights into the functional significance of sequence arrangements would be gained, including the identification of the transcription units which correspond to the various classes of heterogeneous nuclear RNA. Although the human genome will almost certainly not be the first one to be analysed in this way, the knowledge derived from the analysis of the genomes of lower eukaryotes (Saccharomyces, Drosophila) will facilitate this most ambitious enterprise.

Isolation of the DNA of single genes has become possible in those cases where the mRNA can be obtained in a highly purified form. If this RNA is used as a template for the synthesis of complementary DNA with a primer sequence immobilized on a cellulose support, immobilized cDNA is obtained which can be used for affinity chromatography. The first eukary-otic gene purified by this method is that of chick ovalbumin (Anderson and Schimke, 1976). This more subtle type of analysis is also suitable for the isolation of DNA molecules comprising the sequences adjacent to the structural gene. The elucidation of the functional significance of the immediate neighbourhood of a gene and of the clustering of structural genes have thus become definitely possible.

In addition to these few examples of approaches at the molecular level there are in the area of somatic cell genetics many new ways of making accessible new marker systems and increasing the resolution of mapping procedures (for review, see Ruddle and Creagan, 1975). Furthermore, the recent advances in the identification and classification of banding patterns in chromosomes of prometaphase and late prophase (Yunis and Sanchez, 1975) allow the topographical relationships to be analysed with much higher precision. Obviously, our ultimate understanding of the human genome as an integrated structural and functional unit depends on the cooperative effort of many disciplines. To cope with the complexity which must be expected on the basis of present knowledge is probably the greatest obstacle to a genuine synthesis. After the impressive progress made by the analytical approach, such a synthesis might require major advances in theoretical biology.

REFERENCES

Abrahamson, S., Bender, M. A., Conger, A. D. and Wolff, S. (1973). *Nature, Lond.* **245**, 461.
Aitken, D. A., Ferguson-Smith, M. A. and Dick, H. M. (1976). *Cytogenet. Cell Genet.* **16**, 256.
Allderdice, P. W. and Tedesco, T. A. (1975). *Lancet ii*, 39.
Allderdice, P. W., Browne, N. and Murphy, D. P. (1975). *Am. J. Hum. Genet.* **27**, 699.
Anderson, J. N. and Schimke, R. T. (1976). *Cell* **7**, 331.
Angell, R. R. and Jacobs, P. A. (1975). *Chromosoma* **51**, 301.
Arrighi, F. E. and Hsu, T. C. (1971). *Cytogenetics* **10**, 81.
Aula, P., Leisti, J. and Von Koskull, H. (1973). *Clin. Genet.* **4**, 241.
Ayraud, N., Noel, B., Lloyd, M., Letourneau, J. and Martinon, J. (1976). *J. Génét. Hum.* **24**, 81.
Bachmann, K. (1972). *Chromosoma* **37**, 85.
Bahr, G. F., Mikel, U. and Engler, W. F. (1973). *In* "Nobel Symposium 23" (T. Caspersson, L. Zech, Eds), 280. Academic Press, New York and London.
Barlow, P. W. (1972). *Humangenetik* **14**, 122.
Becak, W. and Goissis, G. (1971). *Experientia* **27**, 345.
Becak, W. and Pueyo, M. T. (1970). *Exp. Cell Res.* **63**, 448.
Bender, K. and Burckhardt, K. (1970). *Humangenetik* **9**, 95.
Bender, K., Ritter, H. and Wolf, U. (1967). *Humangenetik* **4**, 85.

Bennett, D. (1975). *Cell* **6**, 441.
Bennett, D., Boyse, E. A., Lyon, M. F., Mathieson, B. J., Scheid, M. and Yanagisawa, K. (1975). *Nature, Lond.* **257**, 236.
Beratis, N. G., Hsu, L. Y. F., Kutinsky, E. and Hirschhorn, K. (1972). *Can. J. Genet. Cytol.* **14**, 869.
Bloom, G. E. and Gerald, P. S. (1968). *Am. J. Hum. Genet.* **20**, 495.
Borgaonkar, D. S. and Shah, S. A. (1974). *Prog. Med. Genet.* **10**, 135.
Botchan, M. (1974). *Nature, Lond.* **251**, 288.
Botchan, M., McKenna, G. and Sharp, P. A. (1973). *Cold Spring Harbor Symp. Quant. Biol.* **38**, 383.
Boyse, E. A., Stockert, E. and Old, L. J. (1968). *J. Exp. Med.* **128**, 85.
Braunger, R., Kling, H., Krone, W., Schmid, M. and Olert, J. (1977). *Hum. Genet.* **38**, 65.
Brewer, G. J. (1967). *Am. J. Hum. Genet.* **9**, 674.
Brinster, R. L. (1970). *Biochem. Genet.* **4**, 669.
Britten, R. J. and Davidson, E. H. (1969). *Science* **165**, 349.
Bross, K. and Krone, W. (1972). *Humangenetik* **18**, 137.
Bross, K. and Krone, W. (1973). *Humangenektik* **18**, 71.
Bross, K., Dittes, H., Krone, W., Schmid, M. and Vogel, W. (1973). *Humangenetik* **20**, 223.
Brown, D. D. and Gurdon, J. B. (1964). *Proc. Natn. Acad. Sci. USA* **51**, 139.
de Bruyn, C. H. M. M. (1976). *Hum. Genet.* **31**, 127.
Buckton, K. E., O'Riordan, M. L., Jacobs, P. A., Robinson, J. A., Hill, R. and Evans, H. J. (1976). *Ann. Hum. Genet.* **40**, 99.
Carlson, P. S. (1972). *Mol. Gen. Genet.* **114**, 273.
Caspersson, T. and Zech, L. (Eds) (1973). "Nobel Symposium 23". Academic Press, New York and London.
Caspersson, T., Zech, L., Johansson, C. and Modest, E. J. (1970). *Chromosoma* **30**, 215.
Cattanach, B. M. (1975). *Ann. Rev. Genet.* **9**, 1.
Cattanach, B. M. and Isaacson, J. H. (1965). *Z. Vererbungsl.* **96**, 313.
Cattanach, B. M. and Isaacson, J. H. (1967). *Genetics* **57**, 331.
Cattanach, B. M., Pollard, C. E. and Hawkes, S. G. (1971). *Cytogenetics* **10**, 318.
Chambon, P. (1975). *Ann. Rev. Biochem.* **44**, 613.
Chany, C., Vignal, M., Couillin, P., Cong, N. V., Boué, A. J. and Boué, A. (1975). *Proc. Natn. Acad. Sci. USA* **72**, 3129.
Chapelle, A. de la, Schröder, J., Selander, R. K. and Stenstrand, K. (1973). *Chromosoma* **42**, 365.
Cohen, P. and Rosemeyer, M. A. (1969). *Eur. J. Biochem.* **8**, 8.
Comings, D. E. (1974). *In* "Birth Defects" (Motulsky, A. G., Lenz, W., Eds). Proceedings of the 4th International Conference, Vienna. Excerpta Medica.
Cooke, H. (1976). *Nature, Lond.* **262**, 182.
Craig-Holmes, A. P., Moore, F. B. and Shaw, M. W. (1973). *Am. J. Hum. Genet.* **25**, 181.
Craig-Holmes, A. P., Moore, F. B. and Shaw, M. W. (1975). *Am. J. Hum. Genet.* **27**, 178.
Creasy, M. R., Crolla, J. A. and Alberman, E. D. (1976). *Hum. Genet.* **31**, 177.
Cremer, T., Zorn, C., Cremer, C. and Zimmer, J. (1976). *Exp. Cell Res.* **100**, 345.
Crosti, N., Serra, A., Rigo, A. and Viglino, P. (1976). *Hum. Genet.* **31**, 197.
Davidson, E. and Britten, R. J. (1973). *Quart. Rev. Biol.* **48**, 565.

148 W. KRONE AND U. WOLF

Davidson, E., Hough, B., Amenson, C. S. and Britten, R. J. (1973). *J. Mol. Biol.* **77**, 1.

Davidson, E., Galau, G. A., Angerer, R. C. and Britten, R. J. (1975). *Chromosoma* **51**, 253.

Davidson, R. L. and de la Cruz, F. F. (Eds) (1974). "Somatic cell hybridization". Raven Press, New York and North Holland Publishing Co., Amsterdam.

De Capoa, A. Ferraro, M., Archidiacono, N., Pelliccia, F., Rocchi, M. and Rocchi A. (1976). *Hum. Genet.* **34**, 13.

De la Chapelle, A. (1976) 5th Int. Congr. Hum. Genet. Exc. Med. Amsterdam, I.C.S. No 397.

Denny, R. M. and Craig, J. W. (1976). *Biochem. Genet.* **14**, 99.

Dittes, H., Krone, W., Bross, K., Schmid, M. and Vogel, W. (1975). *Hum. Genet.* **26**, 47.

Donahue, R. P., Bias, W. B., Renwick, J. H. and McKusick, V. A. (1968). *Proc. Natn. Acad. Sci. USA* **61**, 949.

Doyle, C. T. (1976). *Hum. Genet.* **33**, 131.

Drews, U., Blecher, S. R., Owen, D. A. and Ohno, S. (1974). *Cell* **1**, 3.

Dutrillaux, B. (1973). *Chromosoma* **41**, 395.

Dutrillaux, B. (1975). "Sur la Nature et l'Origine des Chromosomes Humains". Monogr. Ann. Génét. L'Expansion Scientifique, Paris.

Dutrillaux, B. and Fosse, A.-M. (1976). *Ann. Genet.* **19**, 95.

Dutrillaux, B. and Lejeune, J. (1971). *C.R. Acad. Sci., Paris* **272**, 2638.

Dutrillaux, B. and Lejeune, J. (1975). In "Advances in Human Genetics 5" (H. Harris, K. Hirschhorn, Eds) 119. Plenum Press, New York and London.

Dutrillaux, B., Grouchy, J. de, Finaz, C. and Lejeune, J. (1971). *C.R. Acad. Sci., Paris* **273**, 587.

Dutrillaux, B., Laurent, C., Couturier, J. and Lejeune, J. (1973). *C.R. Acad. Sci., Paris* **276**, 3179.

Eichwald, E. J. and Silmser, C. R. (1955). *Transplant. Bull.* **2**, 154.

Engel, W., Vogel, W. and Reinwein, H. (1971). *Cytogenetics* **10**, 87.

Engel, W., Zenzes, M. T. and Schmid, M. (1977). *Hum. Genet.* **38**, 57.

Epplen, J. T. and Vogel, W. (1975). *Humangenetik* **30**, 337.

Epplen, J. T., Siebers, J.-W. and Vogel, W. (1975). *Cytogenet. Cell Genet.* **15**, 177.

Epplen, J. T., Bauknecht, T. and Vogel, W. (1976). *Hum. Genet.* **31**, 117.

Epstein, C. J. (1969). *Science* **163**, 1078.

Epstein, C. J. (1972). *Science* **175**, 1467.

Epstein, C. J., Tucker, G., Travis, B. and Gropp, A. (1977). *Nature, Lond.* **267**, 615.

Evans, H. J., Buckland, R. A. and Pardue, M. L. (1974). *Chromosoma* **48**, 405.

Ferguson-Smith, M. A. (1964). *Cytogenetics* **3**, 124.

Ferguson-Smith, M. A. and Handmaker, S. D. (1961). *Lancet i*, 638.

Ferguson-Smith, M. A., Newman, B. F., Ellis, P. M. and Thomson, D. M. G. (1973). *Nature New Biol.* **243**, 271.

Ferguson-Smith, M. A., Aitken, D. A., Turleau, C. and de Grouchy, J. (1976). *Hum. Genet.* **34**, 35.

Fialkow, P. J. (1974). *New Engl. J. Med.* **291**, 26.

Fridovich, J. (1974). *Adv. Enzymol.* **41**, 35.

Friend, K. K., Dorman, B. P., Kucherlapati, R. S. and Ruddle, F. H. (1976). *Exp. Cell Res.* **99**, 31.

Fujita, H., Shimazaki, M., Takeuchi, T., Hayakawa, Y. and Oura, T. (1976). *Hum. Genet.* **31**, 271.

Galley, W. C. and Purkey, R. M. (1972). *Proc. Natn. Acad. Sci. USA* **69**, 2198.
Galloway, S. M. and Evans, H. J. (1975). *Exp. Cell Res.* **94**, 454.
Gardner, A. L. (1971). *Experientia* **26**, 1088.
Garel, J. P. (1974). *J. Theor. Biol.* **43**, 211.
Gartler, S. M., Liskay, R. M., Campbell, B. K., Sparkes, R. and Gant, N. (1972). *Cell Differ.* **1**, 215.
Gartler, S. M., Liskay, R. M. and Gant, N. (1973). *Exp. Cell Res.* **82**, 464.
Gartler, S. M., Andina, R. and Gant, N. (1975). *Exp. Cell Res.* **91**, 454.
George, D. L. and Francke, U. (1976). *Science* **194**, 851.
German, J. and Chaganti, R. S. K. (1973). *Science* **182**, 1261.
Goodpasture, C. and Bloom, S. E. (1975). *Chromosoma* **53**, 37.
Gosden, J. R. and Mitchell, A. R. (1975). *Exp. Cell Res.* **92**, 131.
Gosden, J. R., Mitchell, A. R., Buckland, R. A., Clayton, R. P. and Evans, H. J. (1975). *Exp. Cell Res.* **92**, 148.
Goss, S. J. and Harris, H. (1975). *Nature, Lond.* **255**, 680.
Grigliatti, T. A., White, B. N., Teuer, G. M., Kaufman, T. C., Holden, J. J. and Suzuki, D. T. (1973). *Cold Spring Harbor Symp. Quant. Biol.* **38**, 461.
Gropp, A., Kolbus, U. and Giers, D. (1975). *Cytogenet. Cell Genet.* **14**, 42.
Grouchy, J. de, Josso, F., Beguin, S., Turleau, C., Jalberg, P. and Laurent, C. (1974). *Ann. Génét.* **17**, 105.
Grumbach, M. M., Marks, P. A. and Morishima, A. (1962). *Lancet i*, 1330.
Grzeschik, K.-H. (1973). *Humangenetik* **19**, 1.
Grzeschik, K.-H., Kim, A. M. and Johannsmann, R. (1975). *Humangenetik* **29**, 41.
Guanti, G. and Petrinelli, P. (1974). *Cell Differ.* **2**, 319.
Hamerton, J. L., Mohandas, T., McAlpine, P. J. and Douglas, G. R. (1975). *Am. J. Hum. Genet.* **27**, 595.
Hansen, J. and Bartalos, M. (1975). *Hormone Res.* **6**, 28.
Hansson, A. (1975). *Hereditas* **81**, 101.
Hartung, M., Fouet, C. and Stahl, A. (1975). *Ann. Génét.* **18**, 247.
Hartwell, L. H., Mortimer, R. K., Culotti, J. and Culotti, M. (1973). *Genetics* **74**, 267.
Heddle, J. A. and Athanasiou, K. (1975). *Nature, Lond.* **258**, 361.
Henderson, A. S., Warburton, D. and Atwood, K. C. (1972). *Proc. Natn. Acad. Sci. USA* **69**, 3394.
Hilse, K. and Rudloff, E. (1975). *FEBS Lett.* **60**, 380.
Hilwig, J. and Gropp, A. (1972). *Exp. Cell Res.* **75**, 122.
Hitzeroth, H. W., Bender, K., Ropers, H.-H. and Geerthsen, J. M. P. (1977). *Hum. Genet.* **35**, 175.
Hoehn, H. (1975). *Am. J. Hum. Genet.* **27**, 676.
Hoehn, H., Wolf, U., Schumacher, H. and Wehinger, H. (1971). *Humangenetik* **13**, 34.
Holmberg, M. (1974). *Nature, Lond.* **249**, 448.
Holmberg, M. and Jonasson, J. (1973). *Hereditas* **74**, 57.
Hopkinson, D. A., Edwards, Y. H. and Harris, H. (1976). *Ann. Hum. Genet.* **39**, 383.
Horst, J., Kluge, F., Bayreuther, K. and Gerok, W. (1975). *Proc. Natn. Acad. Sci. USA* **72**, 3531.
Huehns, E. R., Flynn, F. V., Butler, E. A. and Beaven, G. H. (1961). *Nature, Lond.* **189**, 496.
Huehns, E. R., Hecht, F., Keil, J. V. and Motulsky, A. G. (1964a). *Proc. Natn. Acad. Sci. USA* **51**, 89.

150 W. KRONE AND U. WOLF

Huehns, E. R., Dance, N., Beaven, G. H., Hecht, F. and Motulsky, A. G. (1964b). *Cold Spring Harbor Symp. Quant. Biol.* **29**, 327.

Hungerford, D. A. (1971). *Cytogenetics* **10**, 23.

Hutton, J. J. (1971). *Biochem. Genet.* **5**, 315.

Ikeuchi, T. and Sasaki, M. (1975). *Humangenetik* **30**, 167.

Itano, H. A. (1965). *In* "Abnormal Haemoglobins in Africa" (J. H. P. Jonxis, Ed) 3. Blackwell, Oxford.

Itano, H. A. (1966). *J. Cell Physiol.* Suppl. 1, **67**, 65.

Johnson, L. D., Harris, R. C. and Henderson, A. S. (1974). *Humangenetik* **21**, 217.

Jones, K. W. (1973). *J. Med. Genet.* **10**, 273.

Jones, K. W. (1974). *In* "Chromosomes Today" Vol. 5 (P. L. Pearson and K. R. Lewis, Eds) 305. John Wiley, New York.

Jones, K. W., Purdom, J. F., Prosser, J. and Corneo, G. (1974). *Chromosoma* **49**, 161.

Judd, B. H. and Young, M. W. (1974). *Cold Spring Harb. Symp. Quant. Biol.* **38**, 573.

Kaback, M. M. and Bernstein, L. H. (1970). *Ann. N.Y. Acad. Sci.* **171**, 526.

Kajii, T., Ohama, K., Niikawa, N., Ferrier, A. and Avirachan, S. (1973). *Am. J. Hum. Genet.* **25**, 539.

Kakati, S., Nihill, M. and Sinha, A. K. (1973). *Humangenetik* **19**, 293.

Kazazian, H. H. (1974). *Seminars Haematol.* **11**, 525.

Kim, M. A., Johannsmann, R. and Grzeschik, K.-H. (1975). *Cytogenet. Cell Genet.* **15**, 363.

Kleihauer, E. F., Tang, T. E. and Betke, K. (1967). *Acta Haematol.* **38**, 264.

Kohne, E. and Kleihauer, E. (1975). *Klin. Wschr.* **53**, 111.

Kozak, L. P., McLean, G. K. and Eicher, E. M. (1974). *Biochem. Genet.* **11**, 41.

Krehbiehl, E. (1966). *Am. J. Hun. Genet.* **18**, 127.

Krell, K., Gottesman, M. E. and Parks, J. S. (1972). *J. Mol. Biol.* **68**, 69.

Kukharenko, V. I., Kuliev, A. M., Grinberg, K. N. and Terskih, V. V. (1974). *Humangenetik* **24**, 285.

Kuliev, A. M., Kukharenko, V. I., Grinberg, K. N., Mikhailov, A. T. and Tamarkina, A. D. (1975). *Humangenetik* **30**, 127.

Kunkel, L. M., Smith, K. D. and Boyer, S. H. (1976). *Science* **191**, 1189.

Kusano, T., Long, C. and Green, H. (1971). *Proc. Natn. Acad. Sci. USA* **68**, 82.

Lambert, B., Hansson, K., Bui, T. H., Funes-Cravioto, F., Lindsten, J., Holmberg, M. and Strausmanis, R. (1976). *Ann. Hum. Genet.* **39**, 293.

Latt, S. A. (1973). *Proc. Natn. Acad. Sci. USA* **70**, 3395.

Layzer, R. B. and Epstein, C. J. (1972). *Am. J. Hum. Genet.* **24**, 533.

Lee, C. S. N., Boyer, S. H. and Bowen, P. (1966). *Johns Hopkins Med. J.* **118**, 374.

Lefevre, G. (1974). *Ann. Rev. Genet.* **8**, 51.

Lifschytz, E. and Lindsley, D. L. (1972). *Proc. Natn. Acad. Sci. USA* **69**, 182.

Lima-de-Faria, A. and Jaworska, H. (1968). *Nature, Lond.* **217**, 138.

Lin, M. S., Latt, S. A. and Davidson, R. L. (1974). *Exp. Cell Res.* **86**, 392.

Linder, D., McCaw, B. K. and Hecht, F. (1975). *New Engl. J. Med.* **292**, 13.

Lindsley, D. L. and Sandler, L. (1972). *Genetics* **71**, 157.

Liskay, R. M. (1974). *J. Cell. Physiol.* **84**, 49.

Lodish, H. F. (1976). *Ann. Rev. Biochem.* **45**, 39.

Long, C., Chan, T., Levytska, V., Kusano, T. and Green, H. (1973). *Biochem. Genet.* **9**, 283.

Lyon, M. F. (1974). *Proc. R. Soc. Lond. B.* **187**, 243.

Magenis, R. E., Hecht, F. and Lovrien, E. W. (1970). *Science* **170**, 85.

Magenis, R. E., Koler, R. D., Lovrien, E., Bigley, R. H., Duval, M. C. and Overton, K. M. (1975). *Proc. Natn. Acad. Sci. USA* **72**, 4526.

Magnelli, N. C. (1976). *Clin. Genet.* **9**, 169.

Manning, J. E., Schmid, C. W. and Davidson, N. (1975). *Cell* **4**, 141.

Marimo, B. and Giannelli, F. (1975). *Nature Lond.* **256**, 204.

Mark, J. (1970). *Eur. J. Cancer* **6**, 489.

Mayeda, K., Weiss, L., Lindahl, R. and Dully, M. (1974). *Am. J. Hum. Genet.* **26**, 59.

McCoy, E. E., Segal, D. J., Bayer, S. M. and Strynadka, K. D. (1974). *New Engl. J. Med.* **291**, 950.

McKenzie, W. H. and Lubs, H. A. (1975). *Cytogenet. Cell Genet.* **14**, 97.

Mendelsohn, M. L., Mayall, B. H., Bogart, E., Moore, D. H. and Perry, B. H. (1973). *Science* **179**, 1126.

Merril, C. R., Geier, M. R. and Petricciani, J. C. (1971). *Nature, Lond.* **233**, 398.

Migeon, B. R., Norum, R. A. and Corsaro, C. M. (1974). *Proc. Natn. Acad. Sci. USA* **71**, 937.

Mikkelsen, M. and Dahl, G. (1973). *Cytogenet. Cell Genet.* **12**, 357.

Miller, D. A., Dev, V. G., Tantravahi, R. and Miller, O. J. (1976). *Exp. Cell Res.* **101**, 235.

Miller, O. J., Miller, D. A. and Warburton, D. (1973). *In* "Progress in Medical Genetics 9" (A. G. Steinberg, A. G. Bearn Eds). Grune and Stratton, New York and London.

Mittelman, F. and Levan, G. (1976). *Hereditas* **82**, 167.

Mittwoch, U. (1967). *In* "Mongolism". Ciba Fdn. Study Grps. **25**, 51.

Mittwoch, U. (1971). *J. Med. Genet.* **9**, 92.

Moser, H. and Emery, A. E. H. (1974). *Clin. Genet.* **5**, 271.

Muller, H. J. (1965). *In* "Heritage from Mendel" (R. A. Brink and E. D. Styles, Eds). 419. University of Wisconsin Press, Madison.

Naeye, R. L. (1967). *Biol. Neonat.* **11**, 248.

Nathans, D. and Smith, H. O. (1975). *Ann. Rev. Biochem.* **44**, 273.

Nesbitt, M. N. and Gartler, S. M. (1971). *Ann. Rev. Genet.* **5**, 143.

Niebuhr, E. (1974). *Humangenetik* **21**, 99.

Nielsen, J. (1976). *Hum. Genet.* **32**, 203.

Nielsen, J. and Sillesen, J. (1975). *Humangenetik* **30**, 1.

Nilsson, C., Hansson, A. and Nilsson, G. (1975). *Hereditas* **80**, 157.

Noel, B., Quack, B. and Rethore, M. O. (1976). *Clin. Genet.* **9**, 593.

Norwood, T. H., Hoehn, H. (1974). *Humangenetik* **25**, 79.

Nowell, P. C., Jensen, J., Gardner, F. (1975). *Humangenetik* **30**, 13.

Ohno, S. (1976). *Cell* **7**, 315.

Ohno, S., Trujillo, J. M., Kaplan, W. D. and Kinosita, R. (1961). *Lancet ii*, 123.

Ohno, S., Geller, L. N. and Kan, J. (1974). *Cell* **1**, 175.

Ohno, S., Christian, L. C., Wachtel, S. S., Koo, G. C. (1976). *Nature, Lond.* **261**, 597.

Paris, Conference (1971). "Standardization in Human Cytogenetics. Birth Defects". Original Article Series VIII, 7, 1972. The National Foundation, New York.

Paris Conference (1971). (Suppl. 1975). "Standardization in Human Cytogenetics. Birth Defects". Original Article Series XI, 9, 1975. The National Foundation, New York.

Paton, G. R., Silver, M. F. and Allison, A. C. (1974). *Humangenetik* **23**, 173.

Pearson, P. L., Sanger, R. and Brown, J. A. (1975). *Cytogenet. Cell Genet.* **14**, 190.
Perry, P. and Wolff, S. (1974). *Nature, Lond.* **251**, 156.
Pette, D. (1965). *Naturwissenschaften* **52**, 597.
Phillips, R. B. (1975). *Humangenetik* **29**, 309.
Pinkerton, P. H. and Cohen, M. M. (1967). *J. Am. Med. Assoc.* **200**, 547.
Powars, D., Rohde, R. and Graves, D. (1964). *Lancet i*, 1363.
Rawls, J. M. and Lucchesi, J. C. (1974). *Genet Res.* **24**, 59.
Rebhorn, H. and Bayreuther, K. (1977). Personal communication.
Rethoré, M.-O., Kaplan, J.-C., Junien, C., Cruveiller, J., Dutrillaux, B., Aurias, A., Carpentier, S., Lafourcade, J. and Lejeune, J. (1975). *Ann. Génét.* **18**, 81.
Rethoré, M.-O., Junien, C., Malpuech, G., Baccichetti, C., Tenconi, R., Kaplan, J.-Cl., de Romeuf, J. and Lejeune, J. (1976). *Ann. Génét.,* **19**, 140.
Rethore, M.-O., Kaplan, J. C., Junien, C. and Lejeune, J. (1977). *Hum. Genet.* **36**, 235.
Revel, M., Bash, D. and Ruddle, F. H. (1976). *Nature, Lond.* **260**, 139.
Riggs, A. D. (1975). *Cytogenet. Cell Genet.* **14**, 9.
Ritossa, F., Scalenghe, F., Di Turi, N. and Conti, A. M. (1974). *Cold Spring Harb. Symp. Quant. Biol.* **38**, 483.
Ropers, H. H., Wienker, T. F., Grimm, T., Schroetter, K. and Bender, K. (1977). *Am. J. Hum. Genet.* **29**, 361.
Ruddle, F. H. and Creagan, R. P. (1975). *Ann. Rev. Genet.* **9**, 407.
Russell, L. B. and Montyomery, C. S. (1969). *Genetics* **63**, 103.
Samuel, C. E. and Joklik, W. K. (1974). *Virology* **58**, 476.
Sanchez, O. and Yunis, J. J. (1974). *Chromosoma* **48**, 191.
Sandler, L. and Hecht, F. (1973). *Am. J. Hum. Genet.* **25**, 332.
Sanger, R., Tippett, P. and Gavin, J. (1971). *J. Med. Genet.* **8**, 417.
Sasaki, M. S., and Tonomura, A. (1969). *Jap. J. Hum. Genet.* **14**, 81.
Saunders, G. F., Chuang, C. R. and Sawada, H. (1975). *Acta Haematol.* **54**, 227.
Schinzel, A. (1976). *Adv. Int. Med. Ped.* **38**, 37.
Schleuterman, D. A., Bias, W. B., Murdoch, J. L. and McKusick, V. A. (1969). *Ann. J. Hum. Genet.* **21**, 606.
Schmid, C. and Deininger, P. L. (1975). *Cell* **6**, 345.
Schmid, M. Krone, W. and Vogel, W. (1974). *Humangenetik* **23**, 267.
Schmidtke, J., Zenzes, M. T., Dittes, H. and Engel, W. (1975). *Nature, Lond.* **254**, 426.
Schmidtke, J., Becak, W. and Engel, W. (1976a). *Experientia* **32**, 27.
Schmidtke, J., Schulte, B., Kuhl, P. and Engel, W. (1976b). *Biochem. Genet.* **14**, 975.
Schnedl, A. (1971). *Chromosoma* **34**, 448.
Schnedl, W. (1973). *Arch. Genet.* **46**, 65.
Schneider, E. L. and Epstein, C. J. (1972). *Proc. Soc. Exp. Biol. Med.* **141**, 1092.
Schwarzacher, H. G. (1976). "Chromosomes in Mitosis and Interphase", 54. Springer-Verlag, Berlin, Heidelberg, New York.
Seabright, M. (1972). *Chromosoma* **36**, 204.
Segal, D. J. and McCoy, E. E. (1974). *J. Cell Physiol.* **83**, 85.
Sichitiu, S., Sinet, P. M., Lejeune, J. and Frezal, J. (1974). *Humangenetik* **23**, 65.
Sidebottom, E. and Deak, I. I. (1976). *Int. Rev. Cytol.* **44**, 29.
Siebers, W., Vogel, W., Hepp, H., Bolze, H., Dittrich, A. (1973). *Humangenetik* **19**, 57.
Sigler, A. T. and Zinkham, W. H. (1972). *Birth Defects. Orig. Art. Ser.* **8**, 50.

Sinet, P. M., Allard, D., Lejeune, J. and Jerome, H. (1974). *C. R. Acad. Sci. (Paris)* D **278**, 3267.

Sinet, P. M., Couturier, J., Dutrillaux, B., Poisonnier, M. Raoul, O., Rethoré, M.-O., Allard, D., Lejeune, J. and Jerome, H. (1976). *Exp. Cell Res.* **97**, 47.

Sinet, P. M., Lavelle, F., Michelson, A. M. and Jerome, H. (1975a). *Biochem. Biophys. Res. Comm.* **67**, 904.

Sinet, P. M., Michelson, A. M., Bazin, A., Lejeune, J. and Jerome, H. (1975b). *Biochem. Biphys. Res. Comm.* **67**, 910.

Singer, H. and Zang, K. D. (1970). *Humangenetik* **9**, 172.

Southern, E. M. and Roizes, G. (1973). *Cold Spring Harbor Symp.* **38**, 429.

Stahl, A. and Luciani, J. M. (1972). *Humangenetik* **14**, 269.

Stahl, A., Luciani, J. M., Devictor, M. Capodano, A. M., and Gagné, R. (1975). *Humangenetik* **26**, 315.

Steele, M. W. (1970). *Nature, Lond.* **227**, 496.

Steele, M. W. and Migeon, B. R. (1973). *Biochem. Genet.* **9**, 163.

Steele, M. W. and Owens, K. E. (1973). *Biochem. Genet.* **9**, 147.

Steffenson, D. M. (1973). *Nature New Biol.* **244**, 231.

Stevenson, A. C. and Kerr, C. B. (1967). *Mutat. Res.* **4**, 339.

Sumner, A. T., Evans, H. J. and Buckland, R. A. (1971). *Nature New Biol.* **232**, 31.

Svejgaard, A. (1969). *Vox Sang.* **17**, 112.

Swallow, D. M., Povey, S. and Harris, H. (1973). *Ann. Hum. Genet.* **37**, 31.

Tan, Y. H. (1975). *Nature, Lond.* **253**, 280.

Tan, Y.-H. (1976). *Nature, Lond.* **260**, 141.

Tan, Y.-H., Tischfield, J. and Ruddle, F. H. (1973). *J. Exp. Med.* **137**, 317.

Tan, Y.-H., Schneider, E. L., Tischfeld, J., Epstein, C. J. and Ruddle, F. H. (1974a). *Science* **186**, 61.

Tan, Y.-H., Creagan, R. P. and Ruddle, F. H. (1974b). *Proc. Natn. Acad. Sci. USA* **71**, 2251.

Tangheroni, W., Cao, A., Lungarotti, S., Coppa, G., De Virgiliis, S. and Furbetta, M. (1971). *Clin. Chim. Acta* **35**, 165.

Taylor, A. J. (1970). *Nature, Lond.* **227**, 163.

Tedesco, T. A., Diamond, R., Orwiszewski, K. G., Boedecker, H. J. and Croce, C. M. (1974). *Proc. Natn. Acad. Sci. USA* **71**, 3483.

Therkelsen, A. J. and Petersen, G. B. (1967). *Exp. Cell Res.* **48**, 681.

Therman, E. and Pätau, K. (1974). *Humangenetik* **25**, 1.

Therman, E., Sarto, G. and Pätau, K. (1974). *Chromosoma* **44**, 361.

Tiepolo, L. and Zuaffardi, O. (1976). *Hum. Genet.* **34**, 119.

Tischfeld, J. A. and Ruddle, F. H. (1974). *Proc. Natn. Acad. Sci. USA* **71**, 45.

Tolley, E. and Craig, I. (1975). *Biochem. Genet.* **13**, 867.

Vogel, F. (1964). *Z. Mensch. Vererb. Konstit. Lehre* **37**, 291.

Vogel, F. (1973). *Humangenetik* **19**, 41.

Wachtel, S. S., Koo, G. C., Zuckerman, E. E., Hammerling, U., Scheid, M. P. and Boyse, E. A. (1974). *Proc. Natn. Acad. Sci. USA* **71**, 1215.

Wachtel, S. S., Koo, G. C. and Boyse, E. A. (1975a). *Nature, Lond.* **254**, 270.

Wachtel, S. S., Koo, G. C., Breg, W. R., Elias, S., Boyse, E. A. and Miller, O. J. (1975b). *New Engl. J. Med.* **293**, 1070.

Walzer, S., Gerald, P. S., Breau, G., O'Neill, D. and Diamond, L. K. (1966). *Pediatrics* **38**, 419.

Warburton, D., Henderson, A. S., Shapiro, L. R. and Hsu, L. J. F. (1973). *Am. J. Hum. Genet.* **25**, 439.

Warburton, D., Atwood, K. C. and Henderson, A. S. (1976). *Cytogenet. Cell Genet.* **17**, 221.

Weaver, D. D. and Gartler, S. M. (1975). *Humangenetik* **28**, 39.

Weiss, M. and Green, H. (1967). *Proc. Natn. Acad. Sci. USA* **58**, 1104.

White, A. G., da Costa, A. J., Darg, C. (1973). *Tissue Antig.* **3**, 123.

Wiebe, M. E. and Joklik, W. K. (1975). *Virology* **66**, 229.

Willecke, K., Lange, R., Krüger, A. and Reber, T. (1976). *Proc. Natn. Acad. Sci. USA* **73**, 1274.

Williams, J. D., Summitt, R. L., Martens, P. R. and Kimbrell, R. A. (1975). *Am. J. Hum. Genet.* **27**, 478.

Wilson, M. C., Melli, M. and Birnstiel, M. L. (1973). *Biochem. Biophys. Res. Comm.* **61**, 354.

Wilson, M. G., Schroeder, W. A., Graves, D. A. and Krach, V. D. (1967). *New Engl. J. Med.* **277**, 953.

Wilson, M. G., Schroeder, W. A. and Graves, D. A. (1968). *Pediatrics* **42**, 349.

Wilson, M. G., Fujimoto, A., Shinno, N. W. and Towner, J. W. (1973). *Cytogenet. Cell Genet.* **12**, 209.

Winer, R. A. and Chauncey, H. H. (1975). *J. Dent. Res.* **54**, 62.

Wolf, U. (1976). Presented at the 5th Meeting of the Cytogenetics Section of the "Gesellschaft für Anthropologie und Humangenetik". Basle, Switzerland.

Wurster, D. H. and Benirschke, K. (1970). *Science* **168**, 1364.

Yamamoto, K. R. and Alberts, B. M. (1976). *Ann. Rev. Biochem.* **45**, 721.

Ying, K.-L. and Ives, E. (1968). *Can. J. Genet. Cytol.* **10**, 575.

Yoshida, A. (1966). *J. Biol. Chem.* **241**, 4966.

Yoshida, A. (1967). *Proc. Natn. Acad. Sci. USA* **57**, 835.

Yu, F.-C., Gutman, L. T., Huang, S.-W., Fresh, J. W. and Emanuel, I. (1970). *J. Med. Genet.* **7**, 132.

Yunis, J. J. (1976). *Science* **191**, 1268.

Yunis, J. J. and Hook, E. B. (1966). *Am. J. Dis. Child.* **111**, 83.

Yunis, J. J. and Sanchez, O. (1975). *Humangenetik* **27**, 167.

Yunis, J. J., Roldan, L., Yasmineh, W. G. and Lee, J. C. (1971). *Nature, Lond.* **231**, 532.

Zakharov, A. F. and Egolina, N. A. (1972). *Chromosoma* **38**, 341.

Zang, K. D. and Singer, H. (1967). *Nature, Lond.* **216**, 84.

Zankl, H. and Zang, K. D. (1972). *Virchows Arch. Abt. B. Zellpath.* **11**, 251.

Zankl, H. and Zang, K. D. (1974). *Humangenetik* **23**, 259.

Zankl, H., Stengel-Rutkowski, S. and Zang, K. D. (1973). *Virchows Arch. Abt. B. Zellpath.* **13**, 113.

Normal Variation

4 Polymorphism, Selection and Evolution

O. MAYO

Biometry Section, Waite Agricultural Research Institute, Glen Osmond, South Australia

I. INTRODUCTION

The most remarkable finding of the biochemical investigation of genetical differences in populations has been the extreme degree of variability which exists. One of the central problems of population genetics is to explain the origin and persistence of this variability. In this chapter, the various possible reasons are explored, while in the next four chapters, the extent of the variability is considered.

In 1922, Fisher pointed out that gene frequencies at any locus could be held constant by selection which favoured heterozygotes as against homozygotes. This mechanism which maintains genetical variation is now known as balanced polymorphism (Ford, 1940), by contrast with transient polymorphism, where one allele is replacing another on account of the advantage which it confers in the homozygous state. Ford defined polymorphism in general as "the occurrence together in the same locality of two or more discontinuous forms of a species in such proportions that the rarest of them cannot be maintained by recurrent mutation". (Only genetically determined polymorphism is considered here, though other kinds exist.) To some extent, this definition depends on acceptance of the view that no genotype is selectively neutral, a reasonable view from the work of Fisher (1930). However, in using the concept of effective population size originally due to Sewall Wright, Kimura and Crow (1964) proposed an alternative. The effective size of a population, roughly speaking, is the size of that randomly mating population with equal numbers of the two sexes which will yield the same amount of genetical variation as the population in question; thus, it is a relative definition. Kimura and Crow showed that in a population with effective size N_e with mutation rate μ to new alleles at a given locus, at least $1 + 4N_e\mu$ neutral alleles can be maintained (for example, if $\mu = 10^{-5}$, mutation can maintain at least five neutral alleles in a population of 100 000). More recently, Kimura and Ohta (1975) have suggested that for one particular model for the mutational basis of isozyme polymorphism (Chapter 5), the appropriate value will be $\sqrt{1+8N_e\mu}$, giving a value of three for the same numerical example. It is not clear which is the more appropriate expression, and indeed other variations will arise for X-linked loci (Mayo, 1976) or for populations where there is migration (Avery, 1975), but nonetheless it does seem that if truly neutral alleles can exist, a substantial degree of biochemical variability will be observed. Whatever the cause of the variability, a good working definition of polymorphism is to say that it exists when two or more forms are present each with a frequency of more than 1%. If the most common allele has a frequency of more than 99%, random effects (sampling) and mutation become more important, as discussed in Section II.A, and in addition the locus is unlikely to be useful as a marker (Chapter 9).

At the moment, about 100 polymorphisms are known in man; most of these are discussed in the next five chapters, and it has become clear that each methodological advance, such as gel electrophoresis, leads to the discovery of further variation. Harris (1966) and Lewontin (1967) have suggested that man may be polymorphic at as many as one-third of all structural loci, and Motulsky (1970) has speculated that we may therefore have up to 20 000 polymorphisms still to discover. It seems hardly conceivable that all of these can currently still be undergoing natural selection, though given the immense time available many may in the past have been important in evolution. Faced with this dilemma, one can either resort to the theory of neutral mutation which arises from the work of Kimura and Crow (see Kimura and Ohta, 1971 for discussion), or attack particular polymorphisms to elucidate their special selective effects, if any.

In this chapter, the theory of balanced polymorphism is discussed in some detail, to explain why it should be important. Then the evidence for selection on known polymorphisms is discussed, together with that on certain anthropological differences. Finally, a consideration of polymorphism in evolution is presented.

II. SELECTION

A. Theoretical Considerations

1. Balanced polymorphism

The argument for the stabilization of gene frequencies by natural selection is very simple. Consider a locus with two alleles, A_1 and A_2, with fitnesses (less than or equal to unity) as shown:

genotypes	A_1A_1	A_1A_2	A_2A_2
fitnesses	α	β	γ
frequences at fertilization	p^2	$2pq$	q^2
frequencies after selection	$\dfrac{p^2\alpha}{T}$	$\dfrac{2pq\beta}{T}$	$\dfrac{q^2\gamma}{T}$

Suppose that mating in the (very large) population is at random with no selective differences between sexes. Then if the frequency of A_1 is p and of A_2 is q, frequencies of genotypes at birth and after selection are as shown above, where T is a constant of proportionality, i.e.

$$T = p^2\alpha + 2pq\beta + q^2\gamma$$

The frequency of the allele A_1 is now

$$p' = (p^2\alpha + pq\beta)/T$$

i.e.
$$\Delta p = p' - p$$

i.e.
$$\Delta p = \frac{(p^2\alpha + pq\beta) - pT}{T}$$

If selection is not to change the gene frequencies, $\Delta p = 0$,

i.e.
$$p' = p$$

i.e.
$$p^2\alpha + pq\beta = pT$$

whence
$$p = \frac{\gamma - \beta}{\alpha - \beta + \gamma - \beta} \text{ and } q = \frac{\alpha - \beta}{\alpha - \beta + \gamma - \beta}$$

These are real non-trivial solutions if and only if $\alpha < \beta$ and $\gamma < \beta$, that is if the heterozygote is fitter than both homozygotes. (Conventionally, therefore, one writes $\alpha = 1-s$, $\beta = 1$, $\gamma = 1-t$.) This argument has been expanded to many alleles at a locus by a number of workers (Owen, 1953; Mandel, 1959) and an equivalent result holds for a sex-linked locus (Haldane, 1926; Bennett, 1958; Haldane and Jayakar, 1964).

By assuming that populations are indefinitely large and ignoring mutation, one treats the problem deterministically; population size complicates the argument considerably, but it must be considered especially in man, where populations have increased dramatically in size in recent generations.

The reason for the importance of population size in determining the stability of a polymorphism can be seen from the following simple argument. In a population of size N, the variance of gene frequency will be

$$\frac{q(1-q)}{2N}$$

This can never be larger than $1/8N$, but for a small population $1/8N$ is substantial by comparison with Δp, which from the calculations above must be

$$\frac{pq\,(tq - sp)}{1 -(sp^2 + tq^2)}$$

when s and t are small. In other words, for small s and t, the amount of variation in p which random sampling can produce can be larger than the change Δp due to natural selection.

It has long been known (Fisher, 1930) that most newly arisen mutants will be rapidly lost from a population unless they have a very large selective advantage, as will all disadvantageous or neutral alleles in the absence of mutation (with one allele remaining, of course). Thus, since it is unlikely *a priori* that the two different homozygotes will have the same fitness, one might expect that chance fixation would occur even with heterozygous advantage, probably with loss of the allele giving the less fit homozygote. It has further been shown by Ewens and Thomson (1970), following Robertson (1962), that polymorphisms with extreme fitness values ($s > 5t$ or $t > 5s$, and $2N_e(s+t) \leqslant 8$) are unlikely to become established following the occurrence of a new mutation in a population consisting solely of the more fit homozygotes. (On the other hand, if $s \simeq t$, the polymorphism is relatively likely to become established.) Should a polymorphism with extreme fitness values become established, Ewens and Thomson (1970) have further shown that the resultant situation is by no means stable, since fixation through drift may occur more rapidly than would be expected for a locus with two neutral alleles starting at the same frequencies in the same sized population.

This peculiar property of selection against homozygotes means that one cannot regard, for example, the gene frequencies at the locus determining fibrocystic disease as equilibrial, even though the frequency of the disease (if it is indeed determined by a single locus) is such that it could not be maintained by mutation alone (Mayo, 1970). The same reservation applies to phenylketonuria and galactosaemia, in which conditions Manwell and Baker (1970) asserted that the heterozygotes must be advantageous. Rao and Morton (1973) have presented evidence that these high frequencies of deleterious traits in certain populations (Table 1, Chapter 10) can in fact arise by chance alone.

In general, reasoning from observed genotypic frequencies to selective intensities should be avoided, for reasons both methodological (Cannings and Edwards, 1969) and practical (Reed, 1968b). It is only when a polymorphic locus has homozygotes with clearly defined differences and has been established for more than a few generations that evidence on equilibrium is likely to be unequivocal (Allison, 1964).

About 100 biochemically ascertained polymorphisms are now known in Man. If all of these are maintained by selection, one has to account first for differences in gene frequencies between populations and secondly for the joint mode of action of selection on more than one polymorphism. The first point is considered in Section III.A.

The problem of selection acting at more than one locus has long been known to be difficult to approach theoretically (Bodmer and Felsenstein, 1967) particularly when it is not known whether loci are selected on independently (i.e. the fitnesses are multiplicative) or in some other way. Different loci need not be independently affected. For example, a primitive man suffering from dominant optic atrophy would be at no greater disadvantage in life if he also carried the gene determining protanopia on his X-chromosome. However, if different polymorphic loci affected fitness independently, one would have relative fitnesses for two such loci like this:

	A_1A_1	A_1A_2	A_2A_2
B_1B_1	$(1-s_1)(1-s_2)$	$(1-s_2)$	$(1-t_1)(1-s_2)$
B_1B_2	$(1-s_1)$	1	$(1-t_1)$
B_2B_2	$(1-s_1)(1-t_2)$	$(1-t_2)$	$(1-t_1)(1-t_2)$

Here s_1 represents the relative disadvantage of A_1A_1 compared with A_1A_2, regardless of the genotype at the other locus, t_1 represents the comparable relative disadvantage of A_2A_2; similarly for s_2 and t_2. If the selection against any homozygote in the absence of other selective pressures were of the order of 1%, the polymorphism would not be very stable in a small population, yet the fitnesses of the double homozygotes would be only 98%

of the best genotype. Extrapolating to 50 polymorphisms, the lowest fitnesses will be 60% of the highest, though of course such extremely homozygous individuals will be very rare and the variance of the number of homozygous loci is quite small (Sved *et al.*, 1967). With such a system the cost of maintaining so many polymorphisms will be prodigious, since with random mating the proportion of heterozygotes at birth at any locus will be no more than a half, so that the frequency of genotypes with fitnesses close to one will be very low. To resolve this dilemma, one can postulate that selection is dependent upon population size or density and occurs when some threshold number is attained (Sved, 1968a). The relative fitnesses of different genotypes are thus all-important, whereas on the simple multiplicative model absolute fitnesses are decisive. This model is in accord with known data on the numbers of polymorphisms, but its predictions for the depression of viability on inbreeding do not seem to be close to those observed. Instead, one can postulate that many or most polymorphisms are neutral (Crow, 1970; Kimura, 1968a; Kimura and Ohta, 1969). This hypothesis can readily be tested experimentally on plants and animals (Ford, 1971; Clarke, 1975), but in Section II.B we shall discuss some of the vast quantity of direct and indirect information from man. A third possibility is linkage disequilibrium, which warrants separate consideration.

2. Linkage disequilibrium

It has been known for over 60 years that, in the absence of migration, mutation and selection, gene and genotypic frequencies for a single autosomal locus in a large population attain a stable equilibrium after one generation of random mating. However, for two or more loci this result does not hold. For two loci, gametic frequency equilibrium does not obtain in the first generation, but is approached at the rate of $1-\theta$, where θ is the frequency of recombination between the two loci (Robbins, 1918). Thus, even if the loci are unlinked, equilibrium frequencies for the gametes are only approached asymptotically. Considering two diallelic loci having alleles A_1, A_2 and B_1, B_2 with frequencies p_1, q_1 and p_2, q_2 respectively, the coefficient of linkage disequilibrium (or gametic disequilibrium) is given by $D = g_1 g_4 - g_2 g_3$ where g_1, g_2, g_3 and g_4 are the frequencies of the gametes A_1B_1, A_2B_1, A_1B_2 and A_2B_2 respectively. Now if $D \neq 0$, this may be the result of linkage between the loci, of small population size, of stratification within the population, of migration, or of selection. Lewontin (1974) has been perhaps the most forceful advocate of the view that linkage is an absolutely critical determinant of the degree of disequilibrium. For example, Franklin and Lewontin (1970) concluded that in many cases, a chromosome containing many closely linked loci would virtually behave as a unit under natural selection, yielding extremely high disequilibrium for unfixed loci. This is perhaps the extreme case, but the effect of selection at one or more loci on other loci may well be important. The disturbance to genotypic frequencies at loci linked to a locus undergoing directional selection has been termed the hitchhiker effect, and is currently under close investigation (Ohta and Kimura, 1975; Wagener and Cavalli-Sforza, 1975); it is probably more important than Ohta and Kimura suggest. (See also Thomson (1977).) The effect of population size on disequilibrium has not yet been properly elucidated, though Sved (1971) has produced an approximate relationship, as follows:

$$D^2 = \frac{p_1\, q_1\, p_2\, q_2}{1+4N_e\theta \left[\frac{(1-\theta/2)}{(1-\theta)^2}\right]}$$

Sinnock and Sing (1972b) have attempted to apply Sved's methods in an investigation of multilocus genetical systems in the defined population of Tecumseh, Michigan, but without reaching any definite conclusions. This study of Sinnock and Sing (1972a, b) is as yet the most extensive investigation of disequilibrium in human populations, and they were unable to identify any new two locus associations which would indicate the existence of selection for multilocus genotypes. However, they only used about a dozen of the polymorphisms available, so that this is perhaps not surprising.

Apart from the well-known interactions between blood groups in the determination of the consequences of maternal foetal incompatibility, discussed in Section II.B.3 below, and the special case of the HLA loci (Chapter 8), several other interactions have been reported. For example, it has been suggested that there is an excess of the haptoglobin allele Hp^1 among the offspring of ABO incompatible matings (Kirk, 1971). Vana and Steinberg (1975), in confirming this by a study in the Hutterite religious communities, have proposed that it arises as a secondary effect from the variation in frequency of Hp^1 between ABO genotypes, which in increasing order of Hp^1 frequency run *OO, AO, BO, AB, AA, BB*. It is not known why this should be so; and of course the two loci are on different chromosomes (p. 436). Turner (1969) has claimed that there is an epistatic interaction between the closely linked *MN* and *S* loci of the MNSs system, such that double homozygotes are at a disadvantage relative to single homozygotes. If true, this is just the kind of situation which could arise from the following simple argument, unrelated to selection at any particular locus. As noted above, up to one-third of all structural loci may be polymorphic. Now the human genotype contains about 3.10^9 base pairs of DNA (Kimura, 1968a). If 1% of this is in the form of structural loci coding for polypeptides of about 100 amino acids, there will be about 10^5 loci, so there could well be about 3.10^4 polymorphisms. If the average human chromosome is about 2·5 Morgans long, there will be about five polymorphisms per unit recombination, which must lead to linkage disequilibrium (Sved, 1968b; Lewontin, 1970) which could swamp the effects of individual polymorphisms.

B. Selection in Human Populations

The maintenance of most polymorphisms is not well understood, for few of the well established systems present any obvious defects in their common homozygotes. Morton *et al.* (1966) list seven identifiers of the action of natural selection on a polymorphism: relevant environmental differences between populations with different gene frequencies; systematic departures from Hardy-Weinberg gene frequency equilibrium; associations with disease; age or generation changes in genotype frequencies; differences between sexes; differences between genotypes in fertility and mortality; and systematic departures from Mendelian ratios. The haemoglobin A/S polymorphism provides positive evidence on almost all counts except, of course, differences between sexes. This is unusual; in most cases, only some of the possibilities can be tested and not all have been treated in the same detail, since it has until recently only rarely been anyone's prime aim to investigate selective balance in Man. A final possible method for detection of selective forces not discussed by Morton is linkage disequilibrium, which might display the advantage or disadvantage possessed by certain gene combinations.

1. Direct evidence from polymorphisms

The polymorphism most generally regarded as definitely explained in selective terms is that involving the locus determining the β chains of haemoglobin. Here, the substitution β^6 Glu→Val gives haemoglobin S, so called because the erythrocytes containing it take on sickle shapes when oxygen tension is low, which does not occur with the normal haemoglobin A. The difference is clearly mediated by a single change in a single codon, so there are two alleles, $Hb\beta^A$ and $Hb\beta^S$. The similarity of the world distribution of $Hb\beta^S$ and of malaria caused by *Plasmodium falciparum* led to the suggestion that the heterozygotes (A/S) were to some extent protected against malaria (see Allison (1955) for a review of the early work). The apparent decrease in the frequency of $Hb\beta^S$ with age (Table 2) and the analyses of incidence of parasitism, parasite density and malarial mortality all support the hypothesis that A/S children are less likely to die from malignant tertian malaria. Thus, the postulated balance is as follows, with relative fitnesses from Allison (1955):

genotype	$Hb\beta^A\ Hb\beta^A$	$Hb\beta^A\ Hb\beta^S$	$Hb\beta^S\ Hb\beta^S$
selective agent	malaria	—	haemolytic anaemia
relative fitness	0·80	1	0·25

The exact nature of the protection in A/S individuals has been postulated by Luzzato et al. (1970) to be the rapid removal from the circulation of parasitized A/S red cells by phagocytosis. This would happen, they suggested, because parasitized cells sickle more readily than do non-parasitized cells and they were able to demonstrate this. For a recent review, see Ringelhann et al. (1976).

Associated geographically with malaria are two other polymorphic genetical defects, thalassaemia and glucose-6-phosphate dehydrogenase deficiency. It is probable that similar mechanisms of heterozygous advantage to the malarial challenge have played a major role in their high frequencies in endemically malarial areas (Luzzato et al., 1969). Glucose-6-phosphate dehydrogenase deficiency leads in some cases to acute haemolytic crises on ingestion of or other exposure to fava beans (favism) and there is evidence that deficient individuals carrying the erythrocytic acid phosphatase allele ACP^b are less likely to suffer such haemolysis (Bottini et al., 1971). While this evidence certainly requires confirmation, it is apparently the first detected association between a selectively engendered polymorphism and another polymorphism of unknown maintenance.

Weatherall et al. (1976) have also found an apparent case of linkage disequilibrium in the loci associated with malarial tolerance, affecting alleles at the locus for the African form of β-thalassaemia and the β- and δ-chain haemoglobin structural loci.

2. Associations with disease

Because of the rapid development of knowledge of red cell antigens through recognition of maternal-foetal incompatibility and through increasing use of blood transfusion, large bodies of data were quickly built up which recorded various blood groups for people with given diseases. Such data could not long remain unanalysed and by 1953 a relationship had been established between cancer and blood group A (Aird et al., 1953); many others have followed, particularly with the availability of simple and elegant statistical techniques for demonstrating the significance of any apparent association. As an example, the widely used statistical procedure of Woolf (1955) is outlined in Chapter 9.

In Table 1, a number of associations are shown. Those marked with an *a*

Table 1

Some significant associations between polymorphisms and disease

Polymorphism	Disease	Phenotypes compared	Relative risks	Reference
[a]ABO blood groups	duodenal ulcer	O:A	1·90	McConnell (1969)
[a]ABO blood groups	gastric ulcer	O:A	1·19	McConnell (1969)
[a]ABO blood groups	carcinoma of colon and rectum	O:A	0·90	Vogel (1970)
[a]ABO blood groups	breast cancer	O:A	0·92	Vogel (1970)
[a]ABO blood groups	rheumatic heart disease	O : not−O	0·90	Haverkorn and Goslings (1969)
[a]ABO blood groups	pernicious anaemia	O:A	0·80	Vogel (1970)
[a]ABO blood groups	diabetes mellitus	O:A	0·93	Vogel (1970)
[a]ABO blood groups	thromboembolism	A:A_1	0·55	Allan (1970)
[a]ABO blood groups	myocardial infarction	O : not−O	0·68	Allan and Dawson (1968)
[a]ABO blood groups	atherosclerosis	O:A	0·69	Kingsbury (1971)
ABO blood groups	multiple sclerosis	O : not−O	1·36	MacDonald et al. (1976)
[a]Secretor system	duodenal ulcer	se:Se	1·46	McConnell (1969)
[a]Secretor system	gastric ulcer	se:Se	1·19	McConnell (1969)
[a]Secretor system	rheumatic heart disease	se:Se	1·28	McConnell (1969)
Secretor system	distant metastases	se:Se	0·46	Chakravartti (1967)
Secretor system	lung cancer	se:Se	0·41	Chakravartti (1967)
Rhesus blood groups	gastric ulcer	D+:D−	0·77	Chakravartti (1967)
Rhesus blood groups	oesophageal cancer	D+:D−	0·64	Chakravartti (1967)
Rhesus blood groups	multiple cancer	D+:D−	0·55	Chakravartti (1967)
Rhesus blood groups	pancreatic cancer	D+:D−	0·72	Chakravartti (1967)
Rhesus blood groups	breast cancer	D+:D−	1·26	Chakravartti (1967)
Rhesus blood groups	ulcerative colitis	C−:cc	1·82	Chakravartti (1967)
Rhesus blood groups	multiple sclerosis	cde/cde : all others	2·21	MacDonald et al. (1976)

System	Disease		Value	Reference
MNSs system	hypertonia	MM:MN	1·87	Chakravartti (1967)
MNSs system	ulcerative colitis	S+:S−	2·86	Chakravartti (1967)
MNSs system	breast cancer	S−:S+	1·88	Boston Collaborative Drug Surveillance Programme (1971)
Lewis	Rheumatic disease	Le(+):Le(−)	1·34	Chakravartti (1967)
Lewis	diabetes mellitus	Le(+):Le(−)	1·41	Chakravartti (1967)
Haptoglobin	leukaemia	Hp(1−1):Hp(2−2)	3·23	Wendt et al. (1968)
Gc	psoriasis	Gc(1−1):Gc(2−)	0·80	Wendt et al. (1968)
Gc	kuru	Gc(Ab−Ab): Gc(1−1)	3·03	Wiesenfeld and Gajdusek (1975)
Gm	psoriasis	Gm(2+): Gm(2−1)	1·76	Wendt et al. (1968)
Gm	acute leukaemia	Gm(b+):(b−)	0·36	Wendt et al. (1968)
Leucocyte group 5 system	acute lymphoblastic leukaemia	5a:not−5a	13·16	Warren et al. (1977)
Lp	myocardial infarction	Lp(a+): (a−)	1·85	Wendt et al. (1968)
Lp	malignant tumour	Lp(a+): (a−)	1·90	Wendt et al. (1968)
[a]PTC taste sensitivity	nodular goitre	T−:tt	0·47	Azevedo et al. (1965)
PTC taste sensitivity	dental caries	T−:tt	0·72	Chung et al. (1961)
PTC taste sensitivity	poliomyelitis	T−:tt	0·53	Brand (1964)
[a]α_1–antitrypsin deficiency	emphysema	normal: deficient	27·67	Tarkoff et al. (1968)
[a]HL–A systems (B locus)	ankylosing spondylitis	B27: not−B27	120·90	Svejgaard et al. (1975)
[a]HL–A systems (A locus)	psoriasis vulgaris	A13: not −A13	4·30	Svejgaard et al. (1975)
Aryl hydrocarbon hydroxylase inducibility	bronchogenic carcinoma	high inducibility: low inducibility	4·13	Kellerman et al. (1973)

[a] Consistent and well-established

168   O. MAYO

appear to be well established and consistent. This list is not intended to be exhaustive and no mention is made of definite absences of association, since many diseases, from loiasis (parasitism by *Microfilaria loa*; Ogunba, 1970) to schizophrenia (Chakravartti, 1967, and Chapter 9, Section IV.A) have been investigated for association with one or more polymorphisms. In particular, the association between HLA-B27 and ankylosing spondylitis is only the most striking of the established associations between this polymorphism and diseases. The leucocyte antigen loci are discussed in detail in Chapter 8 Section IV.

The problem is to interpret and to make use of the available data. Wiener (1965) has written

> . . . studies on blood groups and disease have led to no useful results to date. No one has used blood grouping in differential diagnosis, i.e. the fact of a patient's being group O would not affect one's decision as to whether he had a peptic ulcer. Moreover, no one has offered any satisfactory explanation of how one's blood group could directly affect one's susceptibility to duodenal ulcer or to carcinoma . . .

However, this point of view is not constructive, as Clarke (1961) indicated earlier, since only a few of the many possible polymorphisms have yet been investigated and the *a priori* chance of detecting a clinically significant association is very low. More constructive criticisms of these investigations emphasize that association may be real but reflect not pleiotropy or genetical linkage but environmental or historical factors. For example, Reed (1969) discussed the fact that cirrhosis of the liver is associated with the Duffy phenotype Fy(a+b+) in American Negroes but not in Caucasians and indicates that the simplest explanation is that the disease is more frequent in Caucasians and so Negroes with some Caucasian ancestry, as indicated by their Duffy phenotype, are more likely than other Negroes to develop the disease. It is of interest that the Fy(a−b−) phenotype found in Negroes but not Caucasians may be another malaria-resistant phenotype (Gelpi and King, 1976).

Associations between thromboembolism and other cardiovascular diseases such as coronary artery disease and particular phenotypes in the ABO and Lewis systems are consistent with the finding that cholesterol level in A individuals is raised relative to that in O individuals (Mayo *et al.*, 1969; Oliver *et al.*, 1969; Langman *et al.*, 1969), since cholesterol is regarded as playing an important role in the causation of such disease (Keys, 1956, 1975). Beckman and Olivecrona (1970) have provided evidence that the association between blood groups and serum-cholesterol is not mediated through the other well-established associations between blood groups and intestinal alkaline phosphatase, but it seems probable that a biochemical explanation will shortly be achieved (McConnell, 1969; Oliver *et al.*,

1969). Similarly, the association between absence of secretion of water-soluble ABH antigens and gastric or duodenal ulcer seems likely to be explained in biochemical terms (McConnell, 1966), as does that between goitre and the inability to taste PTC, since PTC is related chemically to several goitrogens from crucifers (Azevêdo et al., 1965; Milunicová et al., 1969; Fraser, 1961).

The case of aryl hydrocarbon hydroxylase (AHH) inducibility is a remarkable instance where investigation has travelled in the opposite direction from known biochemistry through a disease association to, apparently, a new polymorphism. It has been observed that polycyclic hydrocarbons, which are also carcinogenic, have this effect through their metabolites. AHH is one of the enzymes which participate in the generation of carcinogens from aryl hydrocarbons, and it has been shown to be genetically heterogeneous in its inducibility (Kellerman et al., 1973a). There now appear to be three phenotypes, showing high, medium and low inducibility, the first of which is at a frequency of about 0·10 in North American whites. In persons with bronchogenic carcinoma, however, the proportion is 0·30, giving a relative risk ratio of over four (Kellerman et al., 1973b; Knudson, 1975). Since AHH is inducible by tobacco smoke in cultivated fibroblasts, this seems a clear case where a genetical polymorphism of unknown function contributes strongly to a recently discovered epidemic disease generally regarded as being of environmental origin (e.g. U.S. Public Health Service, 1964). This work needs confirmation and elaboration.

In another case where a polymorphism appeared likely to be implicated causally with a particular disease (that of Lp(a) lipoprotein and coronary heart disease), what originally appeared to be a polymorphism determined at a single locus appears now to be far more complex, with quantitative variation between individuals superimposed on the presence of the lipoproteins which is detectable in 75% of a British sample (Walton et al., 1974; see also p. 300). Thus, the association between Lp(a) lipoprotein and coronary heart disease and hyperlipidaemia is of a different kind from the others discussed here, though it may in fact give more of an indication of the mechanism involved, viz. that the relevant disease states may be related to overall increases in various serum constituents to be found in atheromatous plaques, but not to any one minor constituent. Perhaps this example serves as a final warning against interpreting associations causally.

3. Prezygotic and early zygotic selection
Direct evidence of selection of gametes is difficult to obtain in Man. The best established case appears to be the very high sex ratio of zygotes.

Even here, however, it is not perfectly certain as yet that this is in fact due to some advantage of Y-bearing sperm over X-bearing sperm, though this is the simplest explanation.

If the finding can be confirmed that the HLA-D phenotype of each spermatozoon is the product of the haploid genotype of that spermatozoon (Halim and Festenstein, 1975; Arnaiz-Villena and Festenstein, 1976), then this may open the way to further investigation of gametic selection. So may the discovery of a polymorphism in a seemingly sperm-specific diaphorase (Caldwell et al., 1976), though in this case the phenotype is diploid, i.e. is determined by the individual's genotype, not that of his gametes.

Selection of early zygotes is more readily studied. This can be done by examining the incidence of sterility, abortions, stillbirths and neonatal deaths from different mating types or by looking for distorted segregation ratios. However, the history of the excess of MN children from heterozygote by heterozygote (MN × MN) matings does not inspire confidence in the latter method. This excess was observed (Taylor and Prior, 1939), accounted for satisfactorily as an artefact of the test system used without recourse to natural selection (Wiener, 1951), observed again (Chung et al., 1961) and disposed of again (Morton et al., 1966). (Race and Sanger, 1975, have suggested that the topic should be given a seemly burial.) Similarly, the associations found by Morton et al. (1966) between O × O matings and increased post-natal deaths and decreased live children were not supported by the findings of Reed (1968a, b), who, comparing this study with earlier work of his own, found the two inconsistent. While other workers (e.g. Kircher, 1968) have found varying effects of ABO blood groups on neonatal survival, the results have been conflicting. However, with the publication of the survey of Plank and Buncher (1975) it seems clear that maternal ABO phenotype affects sex ratio, the proportion of males being highest from AB mothers, with Rhesus groups irrelevant. The mechanism remains unknown.

Good evidence of powerful selective forces at the perinatal stage comes essentially from studies of maternal–foetal incompatibility, which may of course be relevant in the sex ratio disturbance just mentioned. Here the presence of a foetus whose red cells bear an antigen not present on the mother's cells may induce the production of antibody in the mother if foetal red cells enter the maternal circulation. This happens perhaps most frequently with the Rhesus system, where in about 10 % of all pregnancies (in Anglo-Saxons), the mother will be D negative and the foetus D positive. This incompatibility has resulted in much foetal death, though its extent varies from population to population, since Rhesus gene frequencies are quite variable. For example, all Australian Aborigines are D positive (Abbie, 1966), compared with only 80 % of British people (Race and Sanger, 1968).

Maternal–foetal incompatibility from the naturally occurring ABO anti-bodies of a mother can also cause foetal death (Cohen and Sayre, 1968). In 1943, Levine noticed that mothers of children with haemolytic disease of the newborn caused by anti-D were more often compatibly mated with respect to the ABO system than would be expected by chance. Thus, it seemed that the more rapid elimination from the maternal circulation of ABO incompatible cells prevented the production of anti-D antibodies. This has led various workers to treat D negative mothers with anti-D antibodies after the birth of a D positive child, thus preventing the natural formation of anti-D (Finn, 1970).

From the point of view of natural selection, the most remarkable features of this haemolytic disease of the newborn are first that the selection (foetal loss) will all be of heterozygotes at one locus or the other, so that, in the absence of other selective forces, the polymorphism must be unstable (Haldane, 1942) and secondly that the two polymorphisms interact so that doubly heterozygous foetuses are at an advantage in an incompatible uterus. Much effort has been expended on attempting to ascertain the point of balance of the two polymorphisms but this must be different in different populations, if one assumes that the populations are actually at their equili-brial gene frequencies, so that the selective forces must be, or have been, different (Brues, 1963; Levin, 1967; Workman, 1968). This hypothesis is now untestable but certainly, as has been pointed out in the previous section and elaborated further in this section, the forces maintaining even one polymorphism (ABO) are extremely complicated at the phenotypic level (perhaps because of pleiotropy). For example, ABO incompatibility does not seem to contribute to infertility *per se* (Solish and Gershowitz, 1969).

However, the evidence is good that ABO or Rh incompatibility alone results in increased foetal death compared with compatible matings and that joint incompatibility has a lower risk than Rh incompatibility alone (Race and Sanger, 1968; Cohen and Sayre, 1968; Cohen, 1970). Other blood group polymorphisms, such as the Kell and Kidd systems, can also produce haemolytic disease of the newborn and this evidence of selective loss of heterozygotes is thus well established.

4. Viability and mortality

Examination of different age strata of populations can yield evidence of reduced gene or phenotype frequency. Reed (1968a) and Morton *et al.* (1966) found in Oakland, California and North Eastern Brazil, respectively, no evidence of association between age and phenotype for the ABO, Rhesus, P, MNSs, Kell or Duffy blood groups. However, van Houte and Kesteloot (1972), surveying 42 804 members of the Belgian Armed Forces,

172   O. MAYO

found a significant decline in the frequency of blood group A, especially
marked in those over 55. As this confirmed earlier reports by Jörgensen and
Schwarz (1968) and Shreffler et al. (1971), it appears to be a real phenomenon, though once again without explanation.

Looking at polymorphisms known to involve powerful selective forces,
Petrakis et al. (1970) found that there was, in a sample of American
Negroes, a decline with age in the frequency of glucose-6-phosphate dehydrogenase (G6PDH) deficiency in males, but this was not noticed for the
sickle-cell trait. The results of several surveys are shown in Table 2. The
positive findings of Rucknagel and Neel (1961) and Petrakis et al. (1970)
were not repeated by Fraser (1966) nor by Lisker et al. (1965). These latter
surveys were of course carried out in regions where malaria was endemic
until very recently.

Thus, surveys of this kind may well not produce convincing evidence of
natural selection. Reed (1968b) in another context pointed out that many
thousands of pregnancies would need to be observed to detect a difference
between two mating types in reproductive performance as high as 5% and
the same stricture applies here; as the environments where these surveys
were made have differed considerably, one would not expect their findings to
be consistent unless selective forces were very great and uniform.

These conclusions are reinforced by those of Hiorns and Harrison (1970).
They have considered the theoretical problem of comparisons of groups
before and after the postulated process of selection, and present figures
illustrating the sample size necessary for the detection of selection maintaining a balanced polymorphism at equilibrium and of selection against a
recessive homozygote. They conclude that

> with realistic selection coefficients, it is often not possible (by such compari-
> sons) to detect selection ... especially when gene frequencies deviate from
> mid range, since the sample sizes required are impractically large.

III. EVOLUTION

A. Gene Flow and Anthropology

1. Genetics of isolates

Clear differences exist between certain populations of humans and these
populations are often termed races; at the biochemical, rather than the
anthropometric level, other differences exist. For example, Race and
Sanger (1968) list the antigen V of the Rh blood group system, the phenotype Fy(a-b-) of the Duffy system and the antigen Js^a of the Kell system as

Table 2

Phenotypic frequencies in different age groups for sickle cell trait and glucose-6-phosphate dehydrogenase deficiency

Trait	5–20	Age group 21–50	51 +	Overall	Reference
Hb S	0·1221	0·1111	0·0922	0·1123	Rucknagel and Neel (1961)[a]
Hb S	0·0764	0·1072	0·0997	0·0985	McCormick and Kashgarian (1965)[b]
Hb S	0·0989	0·1237	0·0789	0·1069	Lisker et al. (1965)[c]
Hb S	0·0408	0·0533	0·0667	0·0453	Lisker et al. (1965)[d]
Hb S	0·1031	0·0874	0·0793	0·0878	Petrakis et al. (1970)
gd	0·0586	0·0876	0·0263	0·0673	Lisker et al. (1965)[c]
gd	0·0146	0·0133	0·0000	0·0138	Lisker et al. (1965)[d]
gd	0·1212	0·0557	0·0387	0·0573	Petrakis et al. (1970)[a]

[a] Significant differences between age groups (at 5 % level)
[b] Ages at death
[c] Cujinicuilapa (Mexico)
[d] Omotopec (Mexico)

being "almost diagnostic of the Negro origin of a sample of blood". Similarly, the locus determining red cell catalase is polymorphic in Japan and China but apparently not in Europe (Ohkura, 1968). Again, the allele $Gm^{2,17,21}$ in the Gm serum protein polymorphism has only been detected in the Ainu of Japan (Steinberg and Kageyama, 1970). (See Section IV.E.5 of Chapter 8 for a detailed discussion of differences in leucocyte antigen frequencies in different populations.)

However, closer examination of racial differences than this must be fraught with difficulty: first because human populations have mostly increased greatly in size in recent generations, so that they are unlikely to be in genetical equilibrium; secondly because one is not certain of the degree of isolation between populations in the past, so that one cannot make exact inferences about the meaning of observed differences, and thirdly, one is not certain of the nature or extent of the selective forces acting on the polymorphisms used as markers. For example, Shokeir and Shreffler (1970) report the following frequencies of the caeruloplasmin allele Cp^A: in Caucasians, 0·0028; in American Negroes, 0·05; and in Nigerians, 0·15. They point out that the proportion of Caucasian ancestry in American Negroes is probably about 0·2 to 0·3 (Reed, 1969), whereas the frequency of Cp^A indicates that this proportion is 0·66. This would be an exciting finding, perhaps indicating selection against Cp^A in America, were it not for the fact that Bajatzadeh and Walter (1969) have found a frequency of Cp^A of 0·013 in Germany, suggesting that frequencies among Caucasians in the United States could themselves very well have been subject to sampling or selective effects. Buettner-Janusch (1969) has indeed questioned the general validity of using biochemical population genetics for any classification below the specific level; this is discussed further in Section III.B below.

It might be thought that examination of populations known to have been isolated for a considerable time could shed light on the selective forces which tend to split species or to maintain them as recognizable entities. However, such isolates are hard to find (cf. Roberts, 1975). For example, Pygmies in the Congo might be expected to be racially distinct, yet Pygmy women were often taken as wives by other people of the Congo. Even so, these Pygmies, quite apart from their distinctive size, seem to have maintained some isolation, as shown by much lower frequencies of V and Js^a in the Pygmies compared with their neighbours (Fraser et al., 1966). Frequencies of clearly deleterious traits may also rise considerably in partial isolates. For example, in one of the communities on Krk in Yugoslavia, Fraser (1964) found a raised frequency of dwarfism, in another a raised frequency of spastic quadriplegia with cataract and mental deficiency.

Morton (1968) has emphasized the problems of studying the genetics of primitive isolates, pointing out that data obtained from them are by their nature unrepeatable. Against this, it must be observed that isolates should be studied before they cease to exist (Neel, 1969) and that, as more is learnt about selective forces in larger populations, these lessons can then be applied to stored data concerning isolates.

Among the conclusions from Neel's surveys of the Xavante Indians, perhaps the most important is that population structure appears to be similar to Wright's (1943, 1966) model for rapid change, viz. small, partly isolated breeding groups related by persistent migration (Neel and Salzano, 1967). Remarkably, Nei and Imaizumi (1966b) have come to the same conclusions about the population of Japan. Neel has also concluded that as much genetical variation is to be found in a Xavante village as in Hamburg (Niswander *et al.*, 1967). This finding, if correct, emphasizes the need for a proper understanding of the mechanisms which maintain polymorphism. Another finding of note is the conclusion (Neel and Salzano, 1967; Neel, 1969) that if the Xavante population structure is typical of primitive groups, past bottlenecks are unlikely to account for different responses to inbreeding in different populations, since the tribes seem to maintain consistently high levels of inbreeding ($F=0\cdot1$ or more). A bottleneck is, in evolutionary terms, a drastic diminution in population size with results which include chance fixation at some loci and increase in frequency of deleterious alleles at others.

The objections of Morton to the use of information from isolates do not apply to the extensive studies of religious isolates in advanced societies. The most notable of these is that of McKusick *et al.* (1964 and many later papers) on the Old Order Amish in the United States of America. These people originated as a splinter group of Mennonites in Emmental, Switzerland and, after migration to North America between 1690 and 1770, have largely remained isolated, having a limited degree of emigration and virtually no immigration. As a well-documented and cohesive group, to some extent of the investigator's society, they offer immense advantages over the fragile, vanishing cultures of, for example, the Xavante Indians. Furthermore, they are philoprogenitive and have good genealogical records, high consanguinity and little illegitimacy. With all these advantages, it is not surprising that apparently new genetical diseases, such as cartilage-hair hypoplasia, a form of dwarfism (McKusick *et al.*, 1965), have been recognized and that other genetical diseases such as Ellis-van Creveld syndrome have been most extensively studied in the Amish because of their remarkably elevated gene frequencies. While studies of religious isolates cannot tell one anything about human evolution before civilization, in the hands of workers such as McKusick they can certainly lead to advances in clinical genetics.

2. Genetical distance

An essential problem in the population-genetical approach to race is that of interpretation; if the results of comparisons of polymorphisms in two ethnic groups agree with the previous conclusions about the groups, this is confirmation, but if the new findings conflict, they may be explained by sampling, selection or isolation. Despite all these objections, however, there exist extensive comparisons of races using the techniques of population genetics. In general, the aim is to introduce, to validate and to apply to known populations a measure of "genetic distance" designed to give an objective comparison of the amount of evolutionary divergence and genetical difference between populations. The word "distance" was introduced because the measures used are usually analogous to geometrical distance in multi-dimensional space.

Cavalli-Sforza and Edwards, in a series of papers (1965, 1967, 1969), have made the most ambitious attack on the problem of genetical differentiation. Starting with the proposition that the forces causing differentiation between populations can only be of two kinds, local differences in selection and random genetical drift (Cavalli-Sforza, 1969) they have attempted to derive a metric, f, which will incorporate all the available information about simple genetical differences between populations. Techniques for the measurement of genetical distance are described in Chapter 9, Section IV.B. Values of f found by Cavalli-Sforza and Edwards's general method are shown in Table 3.

They have applied their methods more extensively than other workers in this area, but a variety of methods exist; Nei and Imaizumi (1966a, b) have used another, the results of which are shown in Table 3. This method is adequate if differentiation occurs at random, and the evidence from the ABO locus in Japan is in agreement with this (Nei and Imaizumi, 1966a). This suggests that selection has been less important than drift, in the development of Japanese subgroups.

Still another method has been developed by Morton and his colleagues (Morton et al., 1968a, b) using the population-structure theory of Malecot (1948). Values obtained by this method are also shown in Table 3.

Disagreements between distances based on gene frequency data and those based on anthropometry exist, but in the absence of fossil and historical evidence, as Cavalli-Sforza and Piazza (1975) have noted, gene frequency data on surviving populations will be the least unsatisfactory data for phylogenetical analysis, at least at the level of comparisons of different populations. Analyses of cultural differences are not concerned with the same processes, so that it is hardly surprising that, for example, linguistic and genetical distances differ.

Table 3

Estimates of genetical distances, f, in different populations

Differentiation between	N_e	Time available for differentiation (in years)	f	Reference
New Guinea tribes	3000	5000	0·045	Cavalli-Sforza (1969)
Australian tribes	200		0·40	Cavalli-Sforza (1969)
South American Indian tribes	1000	15000–30000	0·082	Cavalli-Sforza (1969)
Africa main groups	10000		0·042	Cavalli-Sforza (1969)
Major human races			0·38	Cavalli-Sforza (1969)
Northern Italian villages			0·004–0·04	Cavalli-Sforza (1969)
Japanese prefectures			0·001	Nei and Imaizumi (1966a)
Isolated Japanese populations			0·002	Nei and Imaizumi (1966b)
Swiss Alpine isolate			0·005[a]	Morton et al. (1968)
Swiss random individuals			0·0004	Morton et al. (1968)

[a] Mean coefficient of kinship

B. General Considerations and Conclusions

Evidence has been presented which points strongly to the existence of substantial selective forces acting on polymorphisms. In few cases is the evidence absolutely unequivocal, but this is to be expected for two reasons. First, as already emphasized, vast homogeneous samples are needed to demonstrate the existence of selective differences of the order of 1% and differences much larger cannot *a priori* be expected to affect many polymorphisms. Secondly, linkage and genotypic equilibrium can be expected to be present in few populations, so that consistent selective effects will not be observed.

The facts that most human populations have recently expanded considerably, that few populations have remained completely isolated in recent centuries and that infectious disease is gradually ceasing to be an important factor in mortality (though Neel (1969) has suggested that it may not have been selective in the Darwinian sense) also mean that local differences in gene frequencies need mean nothing in evolutionary terms.

For all these reasons, one should, as Buettner-Janusch (1969) has pointed out, be wary of intraspecific classification of people. Since it seems that Japan and the Xavante Indian villages may have similar breeding structures (though on a different scale), it may well be asked what use, apart from its clinical significance, information about natural selection in human populations may be. The answer is that the role of polymorphism is central to one's understanding of intra-specific evolution, but the nature of the determination of polymorphism has not yet been agreed upon as the evidence is inadequate and conflicting.

Observing the enormous amount of genetical change between species over the course of mammalian evolution (up to two base pair replacements per generation, by some estimates), Kimura (1968a, b) and King and Jukes (1969) have been led to the conclusion that most of this change has been random, with most of the variants which have arisen having been neutral. Fisher (1930) showed how small a selective difference had to be for a gene to be neutral in a large population, but if $N_e s \ll 1$, it will be effectively neutral (Kimura, 1968b). Kimura felt that the variants must be neutral because otherwise any population would have to suffer too great a cost in "selective deaths" (Haldane, 1957). Haldane calculated that only one gene replacement per 300 generations would normally be possible by natural selection and this implied that replacements much in excess of this must be neutral. It has been shown (Sved, 1968a; Moran, 1970) that Haldane's arguments need not be correct, but nevertheless it does seem that selection for heterozygotes, to be effective, must involve a large loss of defective individuals. The truncation selection model of Sved (1968a) is difficult to anchor in solid physiology, but seems ecologically sound. The more direct evidence of selection which is obtained, of course, the more likely is an understanding of genetical interaction. What little specifically evolutionary evidence is available (e.g. Blundell and Wood, 1975, on the evolution of insulin and cf. the general work of Holmquist and Moise, 1975) appears to indicate that changes in amino acid constitution have, for the most part, not been random.

Unravelling of the complex ancestry of the modern races of Man seems unlikely to come from investigations of differences at single loci, nor from investigations of genetical distances between populations, because we do not know in any particular case whether selection or drift has been important. Cavalli-Sforza's (1969) conclusion that drift is important in geographical microdifferentiation, e.g. between tribes or villages but less important in larger scale groups, is clearly reasonable, but not unexpected. The important effects of isolation, known from other species (Mayr, 1963), are becoming more clear in Man, but it is still hard to explain, for example the high frequency of Tay-Sachs disease in Ashkenazi Jews (Myrianthopoulos and Aronson, 1966; Mayo, 1970; Wagener and Cavalli-Sforza, 1975). If cases like these are polymorphisms, they certainly involve substantial mortality, yet they are clearly not typical of polymorphisms. On the other hand, selection for the different ABO phenotypes seems to occur either after reproduction (ulcers, cancer etc.) or against heterozygotes (incompatibility). With the elucidation of the selective forces affecting more than a few polymorphisms still hopefully awaited, the evolutionary role of polymorphism in man is certainly not clear.

REFERENCES

Abbie, A. A. (1966). *Homo* **17**, 73.
Aird, I., Bentall, H. H. and Roberts, J. A. F. (1953). *Br. Med. J.* **1**, 799.
Allan, T. M. (1970). *Lancet i*, 303.
Allan, T. M. and Dawson, A. A. (1968). *Br. Heart J.* **30**, 377.
Allison, A. C. (1955). *Cold Spring Harb. Symp. Quant. Biol.* **20**, 239.
Allison, A. C. (1964). *Cold Spring Harb. Symp. Quant. Biol.* **29**, 137.
Arnaiz-Villena, A. and Festenstein, H. (1976). *Lancet ii*, 707.
Avery, P. J. (1975). *Genet. Res.* **25**, 145.
Azevêdo, E., Krieger, H., Mi, M. P. and Morton, N. E. (1965). *Am. J. Hum. Genet.* **17**, 87.
Bajatzadeh, M. and Walter, H. (1969). *Humangenetik* **8**, 134.
Beckman, L. and Olivecrona, T. (1970). *Lancet* i, 1000.

180   O. MAYO

Bennett, J. H. (1958). *Aust J. Biol. Sci.* **11**, 598.
Blundell, T. L. and Wood, S. P. (1975). *Nature, Lond.* **257**, 197.
Bodmer, W. F. and Felsenstein, J. (1967). *Genetics* **57**, 237.
Boston Collaborative Drug Screening Programme (1971). *Lancet i*, 301.
Bottini, E., Lucarelli, P., Agostino, R., Palmarino, R., Businco, L. and Antognoni, G. (1971). *Science* **171**, 409.
Brand, N. (1964). *Ann. Hum. Genet.* **27**, 233.
Brues, A. M. (1963). *Am. J. Phys. Anthrop.* **21**, 287.
Buettner-Janusch, J. (1969). *Trans. N.Y. Acad. Sci.* **31**, 128.
Caldwell, K., Blake, E. T. and Sensabaugh, G. F. (1976). *Science* **191**, 1185.
Cannings, C. and Edwards, A. W. F. (1969). *Am. J. Hum. Genet.* **21**, 245.
Cavalli-Sforza, L. L. (1965). *In* "Heritage from Mendel" (R. A. Brink and E. D. Styles, Eds). University of Wisconsin Press, Madison.
Cavalli-Sforza, L. L. (1969). *Proc. 12th Intern. Congr. Genet.* **3**, 405.
Cavalli-Sforza, L. L. and Edwards, A. W. F. (1967). *Am. J. Hum. Genet.* **19**, 233.
Cavalli-Sforza, L. L. and Piazza, A. (1975). *Theor. Popul. Biol.* **8**, 127.
Chakravartti, M. R. (1967). *Humangenetik* **5**, 1.
Chung, C. S., Matsunaga, E. and Morton, N. E. (1961). *Jap. J. Genet.* **6**, 1.
Clarke, B. C. (1975). *Genetics* **79**, 101.
Clarke, C. A. (1961). *Prog. Med. Genet.* **1**, 81.
Cohen, B. H. (1970). *Am. J. Hum. Genet.* **22**, 441.
Cohen, B. H. and Sayre, J. E. (1968). *Am. J. Hum. Genet.* **20**, 310.
Crow, J. F. (1970). *In* "Biomathematics" (K. Kojima, Ed.) Vol. 1, 128. Springer-Verlag, Berlin.
Ewens, W. J. and Thomson, G. (1970). *Ann. Hum. Genet.* **33**, 365.
Finn, R. (1970). *In* "Modern Trends in Human Genetics" (A. E. H. Emery, Ed.) Vol. 1. Butterworths, London.
Fisher, R. A. (1922). *Proc. R. Soc. Edinb.* **42**, 321.
Fisher, R. A. (1930). "The Genetical Theory of Natural Selection". Clarendon Press, Oxford.
Ford, E. B. (1940). *In* "The New Systematics" (J. Huxley, Ed.) 493. Clarendon Press, Oxford.
Ford, E. B. (1971). "Ecological Genetics" (3rd ed.) Chapman and Hall, London.
Franklin, I. R. and Lewontin, R. C. (1970). *Genetics* **65**, 707.
Fraser, G. R. (1961). *Lancet i*, 964.
Fraser, G. R. (1964). *J. Génét. Hum.* **13**, 32.
Fraser, G. R. (1966). *Am. J. Hum. Genet.* **18**, 538.
Fraser, G. R., Giblett, E. R. and Motulsky, A. G. (1966). *Am. J. Hum. Genet.* **18**, 546.
Gelpi, A. P. and King, M. (1976). *Science* **191**, 1254.
Haldane, J. B. S. (1926). *Proc. Camb. Phil. Soc. Math. Phys. Sci.* **23**, 838.
Haldane, J. B. S. (1942). *Ann. Eugen.* **11**, 333.
Haldane, J. B. S. (1957). *J. Genet.* **55**, 511.
Haldane, J. B. S. and Jayakar, S. D. (1964). *J. Genet.* **59**, 29.
Halim, K. and Festenstein, H. (1975). *Lancet ii*, 1255.
Harris, H. (1966). *Proc. R. Soc. B.* **164**, 298.
Haverkorn, M. J. and Goslings, W. R. O. (1969). *Am. J. Hum. Genet.* **21**, 360.
Hiorns, R. W. and Harrison, G. A. (1970). *Hum. Biol.* **42**, 53.
Holmquist, R. and Moise, H. (1975). *J. Mol. Evol.* **6**, 1.
Jörgensen, G. and Schwarz, G. (1968). *Humangenetik* **5**, 254.

Kellerman, G., Luyten-Kellerman, M. and Shaw, C. R. (1973a). *Am. J. Hum. Genet.* **25**, 327.
Kellerman, G., Shaw, C. R. and Luyten-Kellerman, M. (1973b). *N. Eng. J. Med.* **289**, 934.
Keys, A. (1956). *In* "Cardiovascular Epidemiology" (A. Keys and P. D. White, Eds). Hoeben-Harper, New York.
Keys, A. (1975). *Atherosclerosis* **22**, 149.
Kimura, M. (1968a). *Nature, Lond.* **217**, 624.
Kimura, M. (1968b). *Genet Res.* **11**, 247.
Kimura, M. and Crow, J. F. (1964). *Genetics* **49**, 725.
Kimura, M. and Ohta, T. (1969). *Genetics* **61**, 763.
Kimura, M. and Ohta, T. (1971). "Theoretical Aspects of Population Genetics", Princeton University Press, Princeton.
Kimura, M. and Ohta, T. (1975). *Proc. Natn. Acad. Sci. USA* **72**, 2761.
King, J. L. and Jukes, T. H. (1969). *Science* **164**, 788.
Kingsbury, K. J. (1971). *Lancet i*, 199.
Kircher, W. (1968). *Humangenetik* **6**, 171.
Kirk, R. L. (1971). *Ann. Hum. Genet.* **34**, 329.
Knudson, A. G. (1975). *Genetics* **79**, 305.
Langman, M. J. S., Elwood, P. C., Foote, J. and Ryrie, D. R. (1969). *Lancet ii*, 607.
Levin, B. R. (1967). *Am. J. Hum. Genet.* **19**, 288.
Levine, P. (1943). *J. Hered.* **34**, 71.
Lewontin, R. C. (1967). *Am. J. Hum. Genet.* **19**, 681.
Lewontin, R. C. (1970). *In* "Towards a Theoretical Biology" (C. H. Waddington, Ed.) Vol. 3. Aldine Publishing Co., Chicago.
Lewontin, R. C. (1974). "The Genetic Basis of Evolutionary Change". Columbia University Press, New York.
Lisker, R., Loria, A. and Cordova, M. S. (1965). *Am. J. Hum. Genet.* **17**, 179.
Luzzato, L., Usanga, E. A. and Reddy, S. (1969). *Science* **164**, 839.
Luzzato, L., Nwachuku-Jarrett, E. S. and Reddy, S. (1970). *Lancet i*, 319.
MacDonald, J. L., Roberts, D. F., Shaw, D. A. and Saunders, M. (1976). *J. Med. Genet.* **13**, 30.
Malecot, G. (1948). "Les Mathematiques de l'Heredité". Masson et Cie, Paris.
Mandel, S. P. H. (1959). *Heredity* **13**, 289.
Manwell, C. and Baker, C. M. A. (1970). "Molecular Biology and the Origin of Species: Heterosis, Protein Polymorphism and Animal Breeding". University of Washington Press, Seattle.
Mayo, O. (1970). *Ann. Hum. Genet.* **33**, 307.
Mayo, O. (1976). *Hum. Hered.* **26**, 263.
Mayo, O., Fraser, G. R. and Stamatoyannopoulos, G. (1969). *Hum. Hered.* **19**, 86.
Mayr, E. (1963). "Animal Species and Evolution". Harvard University Press, Cambridge.
McConnell, R. B. (1966). "The Genetics of Gastro-Intestinal Disorders". Oxford University Press, Oxford.
McConnell, R. B. (1969). *In* "Progress in Medical Genetics". A. G. Steinberg and A. G. Bearn, Eds, Vol. 6, 63. Heinemann, London.
McCormick, W. F. and Kashgarian, M. (1965). *Am. J. Hum. Genet.* **17**, 101.
McKusick, V. A., Hostetler, J. A. and Egeland, J. A. (1964). *Bull. Johns Hopkins Hosp.* **115**, 203.

McKusick, V. A., Eldridge, R., Hostetler, J. A., Ruangvit, U. and Egeland, J. A. (1965). *Bull. Johns Hopkins Hosp.* **116**, 285.

Milunicová, A., Jandová, A. and Skoda, V. (1969). *Hum. Hered.* **19**, 398.

Moran, P. A. P. (1970). *Ann. Hum. Genet.* **33**, 245.

Morton, N. E. (1968). *Am. J. Phys. Anthrop.* **28**, 191.

Morton, N. E., Krieger, H. and Mi, M.P. (1966). *Am. J. Hum. Genet.* **18**, 153.

Morton, N. E., Miki, C. and Yee, S. (1968a). *Am. J. Genet.* **20**, 411.

Morton, N. E., Yasuda, N., Miki, C. and Yee, S. (1968b). *Am. J. Hum. Genet.* **20**, 420.

Motulsky, A. G. (1970). *Humangenetik* **9**, 246.

Myriathopoulos, N. C. and Aronson, S. M. (1966). *Am. J. Hum. Genet.* **18**, 313.

Neel, J. V. (1969). *Proc. 12th Intern. Congr. Genet.* **3**, 389.

Neel, J. V. and Salzano, F. M. (1967). *Am. J. Hum. Genet.* **19**, 554.

Nei, M. and Imaizumi, Y. (1966a). *Heredity* **21**, 9.

Nei, M. and Imaizumi, Y. (1966b). *Heredity* **21**, 183.

Niswander, J. D., Keiter, F. and Neel, J. V. (1967). *Am. J. Hum. Genet.* **19**, 490.

Ogunba, E. O. (1970). *J. Med. Genet.* **7**, 56.

Ohkura, K. (1968). *In* "Genetics in Asian Countries" *12th Intern. Congr. Genet., Tokyo.*

Ohta, T. and Kimura, M. (1975). *Genet. Res.* **25**, 313.

Oliver, M. F., Geigerova, H., Cumming, R. A. and Heady, J. A. (1969). *Lancet ii*, 605.

Owen, A. R. G. (1953). *Proc. 8th Intern. Congr. Genet.* **2**, 1240.

Petrakis, N. L., Wiesenfeld, S. L., Sams, B. J., Collen, M. F., Cutler, J. L. and Siegelaub, A. B. (1970). *N. Engl. J. Med.* **282**, 767.

Plank, S. J. and Buncher, C. R. (1975). *Hum. Hered.* **25**, 226.

Rao, D. C. and Morton, N. E. (1973). *Am. J. Hum. Genet.* **25**, 594.

Race, R. R. and Sanger, R. (1968, 1975). "Blood Groups in Man" (5th and 6th ed.) Blackwell, Oxford.

Reed, T. E. (1968a). *Am. J. Hum. Genet.* **20**, 119.

Reed, T. E. (1968b). *Am. J. Hum. Genet.* **20**, 129.

Reed, T. E. (1969). *Science* **165**, 762.

Ringelhann, B., Hathorn, M. K. S., Jilly, P., Grant, F. and Parniczky, G. (1976). *Am. J. Hum. Genet.* **28**, 270.

Robbins, R. B. (1918). *Genetics* **3**, 375.

Roberts, D. F. (1975). *In* "Modern Trends in Human Genetics" Vol. 2 (A. E. H. Emery, Ed.) 221. Butterworths, London.

Robertson, A. (1962). *Genetics* **47**, 1291.

Rucknagel, D. L. and Neel, J. V. (1961). *Prog. Med. Genet.* **1**, 158.

Shokeir, M. H. K. and Shreffler, D. C. (1970). *Biochem. Genet.* **4**, 517.

Shreffler, D. C., Sing, C. F., Neel, J. V., Gershowitz, H. and Napier, J. (1971). *Am. J. Hum. Genet.* **23**, 150.

Sinnock, P. and Sing, C. F. (1972a). *Am. J. Hum. Genet.* **24**, 381.

Sinnock, P. and Sing, C. F. (1972b). *Am. J. Hum. Genet.* **24**, 393.

Solish, G. I. and Gershowitz, H. (1969). *Am. J. Hum. Genet.* **21**, 23.

Steinberg, A. G. and Kageyama, S. (1970). *Am. J. Hum. Genet.* **22**, 319.

Sved, J. A. (1968a). *Am. Nat.* **102**, 283.

Sved, J. A. (1968b). *Genetics* **59**, 543.

Sved, J. A. (1971). *Theor. Popul. Biol.* **2**, 125.

Sved, J. A., Reed, T. E. and Bodmer, W. F. (1967). *Genetics* **55**, 469.

Svejgaard, A., Platz, P. J., Ryder, L. P., Staub Nielsen, L. and Thomsen, M. (1975). *Transplant. Rev.* **22**, 3.
Tarkoff, M. P., Kueppers, F. and Miller, W. F. (1968). *Am. J. Med.* **45**, 220.
Taylor, G. L. and Prior, A. M. (1939). *Ann. Eugen.* **9**, 18.
Thomson, G. (1977). *Genetics* **85**, 753.
Turner, J. R. G. (1969). *Ann. Hum. Genet.* **33**, 197.
U.S. Public Health Service (1964). "The Health Consequences of Smoking". The Public Health Service Review, U.S. Department of Health, Education and Welfare. U.S. Government Printing Office, Washington, D.C.
Vana, L. R. and Steinberg, A. G. (1975). *Am. J. Hum. Genet.* **27**, 224.
van Houte, O. and Kesteloot, H. (1972). *Acta Cardiol.* **27**, 527.
Vogel, F. (1970). *Am. J. Hum. Genet.* **22**, 464.
Wagener, D. K. and Cavalli-Sforza, L. L. (1975). *Am. J. Hum. Genet.* **27**, 348.
Walton, K. W., Hitchens, J., Magnini, H. N. and Khan, M. (1974). *Atherosclerosis* **20**, 323.
Warren, R. P., Storb, R., Nguyen, D. D. and Thomas, E. D. (1977). *Lancet ii*, 509.
Weatherall, D. J., Clegg, J. B., Milner, P. F., Marsh, G. W., Bolton, F. G. and Serjeant, G. R. (1976). *J. Med. Genet.* **13**, 20.
Wendt, G. G., Kruger, J. and Kinderman, I. (1968). *Humangenetik* **6**, 281.
Wiener, A. S. (1951). *Am. J. Hum. Genet.* **3**, 179.
Wiener, A. S. (1965). *Am. J. Hum. Genet.* **17**, 369.
Wiesenfeld, S. L. and Gajdusek, D. C. (1975). *Am. J. Hum. Genet.* **27**, 498.
Woolf, B. (1955). *Ann. Hum. Genet.* **19**, 251.
Workman, P. L. (1968). *Hum. Biol.* **40**, 260.
Wright, S. (1943). *Genetics* **28**, 114.
Wright, S. (1966). *Proc. Natn. Acad. Sci. USA* **55**, 1074.

5 Enzyme Polymorphism

G. BECKMAN

Department of Medical Genetics. University of Umeå, Sweden

I. INTRODUCTION

Enzyme Variability

During the last 15 years a great amount of work has been undertaken in many laboratories in order to try to disclose the extent and character of genetically controlled enzyme variations. Markert and Møller (1959) introduced the term, isozyme (or isoenzyme) to describe the existence of different molecular forms of an enzyme which catalysed the same biochemical reaction but differed in their electrophoretic properties. Isozymes were later found to differ quite often not only electrophoretically but also in a number of other physiochemical properties (cf. Vesell, 1968). There has been some confusion, however, as to the exact meaning of the term isozyme from a genetical point of view, as some molecular forms of an enzyme are controlled by genes at separate loci while others are controlled by genes at the same locus (alleles). Some workers reserve the term isozyme for molecular forms of an enzyme controlled by genes at separate loci (Shaw, 1969) while molecular forms of an enzyme controlled by genes at the same locus are called enzyme variants or isozyme variants. The situation is further complicated by the fact that an isozyme or isozyme variant can also be controlled by genes at two separate loci. Enzyme variants will hereforth mean allelic gene products if not otherwise stated.

The information available today on genetically controlled enzyme variation in Man as well as in other species is quite extensive (cf. Markert, 1975). Extensive surveys of the population data available on enzyme variants in different populations have among others been published by Giblett (1969), Beckman (1972), Bhasin and Fuhrmann (1972) and Mourant et al. (1976).

Origin and Detection of Enzyme Variants

Alleles coding for enzyme variants may like isozymes originate from duplications and deletions of the genetic material. Enzyme variants are most probably, however, the result of point mutations at a structural gene locus leading to amino acid substitutions in the enzyme protein. The number of different alleles, which theoretically can be generated through mutations at any given structural gene locus, is of course enormous. Most enzyme variants differ presumably only in one amino acid. Depending on the type of amino acid that has been exchanged and where it is located in the molecule, the substitution may cause alterations in properties like the net charge, kinetic properties, rate of synthesis and stability of the enzyme protein.

The technique commonly used to disclose enzyme variants is electrophoresis in starch gels, but polyacrylamide and agarose are also suitable gel media. After electrophoretic separation, histochemical staining procedures are used to visualize various enzymes. The combination of these two methods is referred to as the zymogram technique. Various quantitative assay methods may also be applied to disclose enzyme variants associated with altered enzyme activity.

The electrophoretic technique is so far unsurpassed in tracing enzyme variants. Only small amounts of material are required and the technique is convenient for screening a large number of samples in a relatively short time. It should be remembered, however, that only about one-third of the enzyme variants (or mutants) can be discovered using the zymogram technique, those variants which carry an amino acid substitution involving a charge change, or an amino acid substitution not involving a charge change but where there has been a major change in the conformation of the resulting protein. See Johnson (1976) for a general discussion of the detection problem.

The material most suitable for population, family and twin investigations of enzyme variants is red blood cells. Other materials also suitable are placental tissue, white blood cells, cell cultures but rarely serum.

This chapter will not describe in detail all the different methods commonly used in enzyme genetics as well as specific staining procedures for the individual enzymes. The reader is instead asked to consult the original papers cited in the text and existing reviews on the subjects: Giblett (1969), Yunis (1969), Brewer (1970), Beutler (1971), van Someren et al. (1974), Harris and Hopkinson (1976).

Gel electrophoresis methods are discussed in Chapter 6.

The Concept of Polymorphism

The concept of polymorphism, as discussed in Chapter 4, was introduced by Ford (1940, 1965), as a description of an observed level of variability which requires selection for its maintenance. In view of the current controversy regarding the maintenance of such variability, in this chapter the working rule mentioned on page 158 is used as a definition of polymorphism. This rule was introduced by Morton (1967), and according to it, an enzyme is polymorphic if it shows variants controlled by more than one allele with frequencies exceeding 1% in at least one population. The alleles with frequencies between 0·01 and 0·99 are referred to as polymorphs, while alleles with frequencies below 0·01 are called idiomorphs. The term monomorphs is used for alleles with frequencies higher than 0·99. While a wide variety of opinions exist regarding the proper definition of enzyme polymorphism (cf. Motulsky, 1975) for the purposes of the present chapter Morton's definition is most appropriate.

Enzymes which have been found to show polymorphism in at least one population are listed in Table 1. A division of enzymes into those that show polymorphism and those that do not must of course be regarded as preliminary especially concerning enzymes for which only limited population data are available. The fact that no variants are found or only rare variants are discovered in studies of several thousand samples from one population

Table 1

Enzymes which have been found to be polymorphic in at least one population

Acetylcholinesterase	AChE	Glutathione peroxidase	GP
Acid α-glucosidase	α-GLU	Glyoxalase I	GLO
Acid phosphatase	ACP_1	Hexokinase	HK_3
Aconitase	$ACON_s$	Lactate dehydrogenase	LDH
Adenosine deaminase	ADA	Liver acetyl transferase	Ac
Adenylate kinase	AK	Malate dehydrogenase (NAD	
		dependent)	S-MDH
Alcohol dehydrogenase	ADH_2, ADH_3	Malate dehydrogenase (NADP	
		dependent)	ME_M
α-Amylase	Amy_1, Amy_2	Pepsinogen	Pg
Aryl hydrocarbon hydroxylase	AHH	Peptidase	Pep A, B, C and D
Carbonic anhydrase	CA_{II}	Phosphoglucomutase	PGM_1, PGM_3
Cytidine deaminase	CDA	6-Phosphogluconate dehydrogenase	
			6-PGD
α-Fucosidase	α-FUC	Phosphoglycerate kinase	PGK
Galactokinase	GALK	Placental alkaline phosphatase	PL
Galactose-1-phosphate uridylyl			
transferase	GALT	Pyridoxine kinase	PNK
Glucose-6-phosphate dehydrogenase		Serum cholinesterase	E_1, E_2
	G6PD		
Glutamic oxaloacetic transaminase		Superoxide dismutase	SOD_R
	GOT_S, GOT_M		
Glutamic pyruvic transaminase	GPT	Unspecific esterases	EsD
		Uridine monophosphate kinase	UMPK

does not exclude the possibility that the enzyme might be polymorphic in another population.

Superoxide dismutase (SOD) represents an example of an enzyme found to be polymorphic only in some rather limited areas. Other examples are phosphoglycerate kinase (PGK) and peptidase C (PEPC). Glucosephosphate isomerase (GPI) represents an example of an enzyme for which so far only rare variants have been discovered despite large population studies. One of the phosphoglucomutase loci, PGM_1, illustrates the occurrence of two alleles in high and rather similar frequencies in all populations studied. In contrast to PGM_1 placental alkaline phosphatase (PL) shows three common alleles but with distinct and typical frequency differences between the three major racial groups. Some variants of the enzyme glucose-6-phosphate dehydrogenase (G6PD) occur in certain populations in frequencies high enough to fit the criteria of polymorphism. Restriction of certain polymorphs to certain populations may imply the existence of local selective agents, e.g. in the case of G6PD and resistance to malaria (p. 164).

Nomenclature

The nomenclature used in enzyme genetics has quite often been rather confusing. Different research groups may for example use different names for the same enzyme and different abbreviations to designate locus, alleles and phenotypes. Some workers prefer number to designate genes, enzyme variants and phenotypes, others prefer letters, place names or family names. The use of place names has become quite cumbersome on account of the great number of enzyme variants known today for many enzymes. The most awkward custom is undoubtedly the use of place names or family names to designate heterozygous phenotypes. The nomenclature proposed by Harris and co-workers has been accepted by the Committee on Nomenclature of the International Congress on Gene Mapping (Giblett, 1976). This nomenclature allows a clear distinction between concepts such as variant, gene (allele), phenotype and genotype, and the alleles are referred to by numbers in order of discovery. It has been used here, as far as possible.

II. POLYMORPHIC ENZYME SYSTEMS

Acetylcholinesterase (EC 3.1.1.7)

A genetic variation of erythrocyte acetylcholinesterase AChE has been described by Coates and Simpson (1972). Three phenotypes AChE1, AChE2–1 and AChE2 were observed. Family studies indicated that these three phenotypes were controlled by two autosomal alleles $AChE^1$ and $AChE^2$. The enzyme is strongly bound to the erythrocyte membrane and must first be solubilized from the membrane before electrophoretic separation is possible, which makes the AChE system less suitable for large-scale population studies. It is claimed that this enzyme is the first example of a polymorphic variation in a stromal enzyme.

Acid α-glucosidase (EC 3.2.1.20)

A deficiency in the enzyme acid α-glucosidase occurs in the Type II glycogen storage disease known as Pompe's disease (cf. Howell, 1972; and see p. 522). Recently a polymorphism in acid α-glucosidase, αGLU, was reported by Swallow et al. (1975). The authors refer to the method of demonstrating the αGLU isozyme patterns and the polymorphism as "affinity" electrophoresis. In this technique starch serves both as support medium and enzyme substrate. The three phenotypes which can be distinguished are referred to as αGLU 1, αGLU 2–1 and αGLU 2 and family studies indicate that the three types are controlled by two autosomal alleles αGLU^1 and αGLU^2. In a population survey using placental extracts Swallow et al. (1975) found the following gene frequencies among Caucasians, $\alpha GLU^1 = 0.97$ and $\alpha GLU^2 = 0.03$.

Acid Phosphatase (EC 3.1.3.2.)

Acid phosphatase has been regarded as occurring in a red cell as well as in a tissue specific form based on differences in substrate specificity (Hopkinson et al., 1963; Hopkinson et al., 1964; Hopkinson and Harris, 1969). The substrate originally recommended for the detection of red cell acid phosphatase isozymes was phenolphthalein diphosphate (Hopkinson et al., 1963), in contrast to α- and β-naphthylphosphate for acid phosphatases present in other tissues. Later, however, Sørensen (1970) verified the results of Babson et al. (1959) that red cell acid phosphatase can hydrolyse β-naphthylphosphate but not α-naphthylphosphate. Other substrates to which red cell acid phosphatase has been found to be active are p-nitrophenylphosphate (Hopkinson et al., 1963) and 4-methyl umbelliferyl dihydrogen phosphate (White and Butterworth, 1971). Swallow et al. (1973) have shown, using methyl umbelliferyl phosphate as substrate, that the locus for red cell acid phosphatase is active in other tissues, thereby disproving one of the original findings.

Three gene loci for acid phosphatase have so far been identified and are referred to as ACP_1, ACP_2 and ACP_3 (cf. Swallow et al., 1973). A genetic polymorphism at the ACP_1 locus has been demonstrated (Hopkinson et al., 1963). Five acid phosphatase patterns could be observed after separation of haemolysates on starch gel electrophoresis and subsequent staining of the gel with phenolphthalein diphosphate as substrate. The five phenotypes of the enzyme originally found by Hopkinson et al. (1963) suggested the existence of a sixth type. The five phenotypes were referred to as A, BA, B, CA and CB. The sixth phenotype C was later reported by Lai et al. (1964). Family studies (Hopkinson et al., 1963) showed that the different acid phosphatase types were controlled by three autosomal allelic genes. The alleles were originally referred to as P^a, P^b and P^c but should according to the new terminology be referred to as ACP_1^A, ACP_1^B and ACP_1^C (cf. Swallow et al., 1973). The genetic hypothesis has been verified through family studies by other investigators (Giblett and Scott, 1965; Fuhrmann and Lichte, 1966; Karp and Sutton, 1967; Speiser and Pausch, 1967).

192 G. BECKMAN

The six common phenotypes of red cell acid phosphatase are shown in Fig. 1. The homozygous types show two zones with enzyme activity while the heterozygotes show four zones, viz. a mixture of the two homozygous patterns. The molecular relationship between the two zones in the homozygous types is not quite clear, but there is some indication that they represent conformational isomers, viz. the zones are interconvertible (Fisher and Harris, 1971a, b; Harris, 1975) and show some physico-chemical differences (Fisher and Harris, 1969, 1971b).

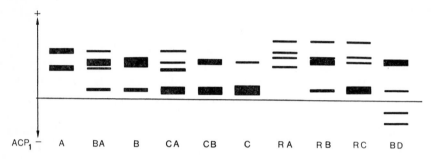

Fig. 1   Schematic drawing showing the major components of the six common and some rare phenotypes of acid phosphatase. Electrophoretic separation performed in starch gel using a citrate–phosphate buffer system pH 5·6

The structural differences in acid phosphatase, controlled by the three alleles ACP_1^A, ACP_1^B and ACP_1^C, result in a difference in the net ionic charge which accounts for the different electrophoretic mobilities between phenotypes (Luffman and Harris, 1967). These structural differences are also associated with quantitative differences in enzyme activity. The acid phosphatase activities attributable to the three alleles ACP_1^A, ACP_1^B and ACP_1^C are close to a ratio of 2:3:4 (Spencer et al., 1964a; Modiano et al., 1967; Shinoda, 1967; Jenkins and Corfield, 1972; Eze et al., 1974). In a random population made up of different phenotypes, enzyme activity shows a unimodal distribution curve (Fig 2), which, consists of a summation of a series of separate but overlapping distributions corresponding to each of the different phenotypes (Harris, 1966). Thermostability also varies between phenotypes. Type C is the most heat stable followed by B and then by A. The heterozygotes are intermediate in heat stability to the respective homozygote types (Luffman and Harris, 1967). No difference in substrate specificity has been observed between the different phenotypes (Scott, 1966; Luffman and Harris, 1967).

Four additional variants controlled by four rare alleles (ACP_1^R, ACP_1^D, ACP_1^O and ACP_1^E) have been described. One unusual phenotype referred to as RA was described by Giblett (1967) and by Karp and Sutton (1967).

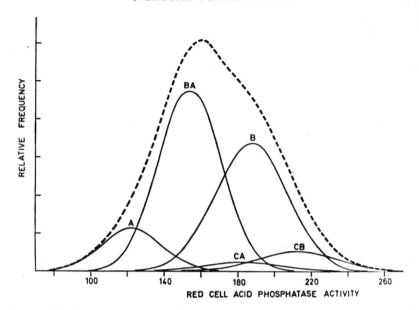

Fig. 2   Distribution of red cell acid phosphatase activity controlled by genes at the ACP₁ locus. The curves are constructed from data based on the English population. The broken line represents the activity in the total population studied and the solid lines the enzyme activity attributable to the different phenotypes. (Reprinted from Spencer *et al.*, 1964a, with permission)

The level of activity attributable to the ACP_1^R allele is similar to the activity of the ACP_1^A allele (Jenkins and Corfield, 1972). The D variant has been found in a new phenotype BD (Karp and Sutton, 1967; Giblett, 1969). A family study showing the segregation of the ACP_1^D allele was reported by Lamm (1970). Segregation data for red cell acid phosphatase in a large family published by Herbich *et al.* (1970) suggest the occurrence of a silent allele ACP_1^O. Individuals heterozygous for the ACP_1^O allele were reported to have only half the expected activity of normal red cell acid phosphatase. A second family with a segregating ACP_1^O allele has been reported (Herbich and Meinhart, 1972). The ACP_1^E allele was recently described by Sørensen (1975).

For summaries of the distribution of the three alleles ACP_1^A, ACP_1^B and ACP_1^C in different populations see Beckman (1972), Jenkins and Corfield (1972) and Mourant *et al.* (1976). The ACP_1^B allele is the most frequent one in all populations studied so far except Eskimos, Athabascan Indians and Swedish Lapps where the ACP_1^A allele is somewhat more frequent. The frequency of ACP_1^C allele does not exceed 10% in any population studied so far and is mostly below 1% except in Caucasian populations. The ACP_1^R allele occurs only in Negro populations and exceeds 1% in certain areas in Africa (cf. Jenkins and Corfield, 1972).

At the other two acid phosphatase loci ACP_2 and ACP_3 only rare variants have been encountered (Beckman and Beckman, 1967; Beckman 1970b; Beckman *et al.*, 1970; Swallow and Harris, 1972).

Aconitase (EC 4.2.1.3)

The enzyme catalyses the interconversion of citric acid, *cis*-aconitic acid and isocitric acid. The enzyme occurs both in a cytoplasmic and a mitochondrial form (Dickman and Speyer, 1954).

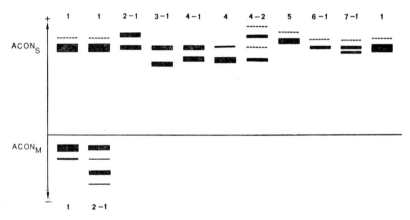

Fig. 3   Phenotypes of soluble and mitochondrial aconitase controlled by genes at the $ACON_S$ and $ACON_M$ loci respectively. (After Slaughter *et al.*, 1975)

Genetical variation in the cytoplasmic form of aconitase was first described by Schmitt and Ritter (1974). Two variants of the soluble form were found in a population study made on placental material from southern Germany. The phenotypes were referred to as s-AH 1, s-AH 2-1 and s-AH 3-1 (Fig. 3). The heterozygous types can be explained by assuming two alleles $s\text{-}AH^2$ ($ACON_S^2$) and $s\text{-}AH^3$ ($ACON_S^3$) in combination with the common allele $s\text{-}AH^1$ ($ACON_S^1$). The population data were as follows: $ACON_S^1 = 0.991$, $ACON_S^2 = 0.005$ and $ACON_S^3 = 0.004$.

Another six phenotypes of cytoplasmic aconitase have been reported by Slaughter *et al.* (1975). These are referred to as 4-1, 4-2, 5, 6-1 and 7-1 (Fig. 3). Slaughter *et al.* (1975) refer to the genes as $ACON_S^4$, $ACON_S^5$ ... $ACON_S^7$. Family studies support the notion that the different variants of aconitase are controlled by allelic genes at an autosomal locus.

A rare variant of mitochondrial aconitase designated $ACON_M2\text{-}1$ was also observed in the same material. This rare phenotype is most probably controlled by a common allele $ACON_M^1$ together with a rare allele $ACON_M^2$. Aconitase is polymorphic in the Nigerian population studied by Slaughter *et al.* (1975) but not in the European population. The gene frequencies are as follows: $ACON_S^1$ 0·99 in Europeans compared to 0·85 in the Nigerian population. The $ACON_S^2$, $ACON_S^3$ and $ACON_S^6$ genes are very rare in the European population. The $ACON_S^2$ and $ACON_S^4$ genes occur, however, in 3% and 12% of the Nigerian population respectively. The rare $ACON_M^2$ gene was only observed in the Nigerian population. The existence of a silent allele of the cytoplasmic form of aconitase seems quite possible.

Adenosine Deaminase (EC 3.5.4.4.)

Adenosine deaminase (ADA) is an aminohydrolase which catalyses the deamination of adenosine to inosine and is hence an important enzyme in the purine salvage pathway. Isozymes of adenosine deaminase in a number of different tissues and organs have been studied by Edwards *et al.* (1971a).

Edwards *et al.* (1971a) proposed that as many as four loci may be involved in the genetical control of the isozymes of adenosine deaminase activity observed in different tissues. Beckman *et al.* (1973a) in a study of isozymes in different blood cells proposed two gene loci to explain the adenosine deaminase pattern observed in various blood cells. This hypothesis was based on the observation that the genetical variation observed in red cell adenosine deaminase was not observed in another adenosine deaminase component. Purification of adenosine deaminase from a number of different tissues has revealed two major molecular species of adenosine deaminase with quite different molecular weight (Akedo *et al.*, 1970, 1972; Nishihara *et al.*, 1973). The form with low molecular weight of the enzyme is believed to be the primary gene product. Nishihara *et al.* (1973) have isolated a "tissue conversion factor" which is supposed to act as an aggregator of the low molecular weight isozyme to the high form. Evidence that the adenosine deaminase isozymes in erythrocytes as well as in tissues are coded by alleles at the same gene locus has been presented by Hirschhorn *et al.* (1973) and Hirschhorn (1975).

Figure 4 shows the three different electrophoretic patterns of adenosine deaminase called ADA 1, ADA 2-1 and ADA 2. Family studies have shown that these three phenotypes are determined by two common alleles, ADA^1 and ADA^2, at an autosomal locus (Spencer *et al.*, 1968; Hopkinson *et al.*, 1969; Tariverdian and Ritter, 1969; Dissing and Knudsen, 1970; Renninger and Bimboese, 1970). Rare variants designated ADA 3-1 (Hopkinson *et al.*, 1969), ADA 4-1 (Dissing and Knudsen, 1969), ADA 5-1 (Detter *et al.*, 1970) and ADA 6-1 (Radam *et al.*, 1974, 1975) have been found (Fig. 4).

The occurrence of a silent allele ADA^O has been suggested by Giblett *et al.* (1972) and demonstrated by Brinkmann *et al.* (1973) and Chen *et al.* (1974a) in family studies.

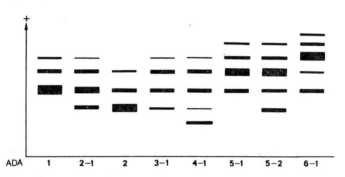

Fig. 4   Phenotypes of adenosine deaminase. (After Radam *et al.*, 1974)

A total lack of adenosine deaminase activity has been observed in association with severe combined immunodeficiency (SCID). Family studies confirmed the possibility of a silent allele, two sick children being apparently homozygous for the ADA^O gene. Scott *et al.* (1974) found that heterozygotes for the ADA^O gene could be detected with an accuracy of 90% by quantitation of the adenosine deaminase activity in erythrocyte haemolysates. Since the original report by Giblett *et al.* (1972) on the association between adenosine deaminase deficiency and SCID, about a dozen similar cases have been described (cf. Meuwissen *et al.*, 1975; and p. 674).

Further studies have revealed however that there is not always a total lack of adenosine deaminase activity in SCID patients. Residual activity of adenosine deaminase has been demonstrated in a number of cases (Van der Weyden *et al.*, 1974; Chen *et al.*, 1975; Hirschhorn *et al.*, 1976). Hirschhorn *et al.* (1976) proposed that the adenosine deaminase activity observed in fibroblasts from SCID patients represents a mutant enzyme coded for by a defective gene at a single ADA locus. Recently Trotta *et al.* (1976) have shown that adenosine deaminase activity can be restored in red cells from SCID patients as well as heterozygotes if certain measures are taken when assaying the enzyme. Evidence was presented for the production of an adenosine deaminase inhibitor in individuals with SCID and the authors proposed that the production of this inhibitor was under genetical control.

Hence the questions concerning the number of gene loci involved and the existence of a "null allele" remain open. The possibility of prenatal diagnosis of SCID based on the absence of adenosine deaminase activity in cultured amniotic cells seems feasible (Chen and Scott, 1973; Hirschhorn and Beratis, 1973; Chen et al., 1975; Hirschhorn et al., 1975) if precautions are taken when assaying the enzyme.

Adenylate Kinase (EC 2.7.4.3)

A number of different adenylate kinase isozymes have been found and the tissue distribution of them varies. Three gene loci AK_1, AK_2 and AK_3 are involved in the control of these isozymes. Wilson et al. (1976) and Khoo and Russell (1972) have shown that AK isozymes respond differently to inhibition by $AgNO_3$ and that two groups of isozymes could be identified in this way. Beckman et al. (1973a) in a study of isozymes of various blood cells presented evidence that AK isozymes are controlled by two gene loci. Wilson et al. (1976) have, on the basis of sensitivity to inhibition by $AgNO_3$ and substrate specificity, presented evidence for three gene loci. The AK_1 and AK_2 loci control isozymes sensitive to inhibition by $AgNO_3$ and the AK_3 locus those resistant to $AgNO_3$ inhibition. (AK_2 isozymes are more resistant to inhibition than AK_1.) The isozymes controlled by the AK_1 and AK_2 loci catalyse reversibly the following reaction:

$$2 \text{ ADP} = \text{ATP} + \text{AMP}$$

Another set of isozymes controlled by the AK_3 loci shows activity with either $GTP + AMP$ or $ITP + AMP$ as substrates but no activity with ATP and AMP as substrates (cf. Wilson et al., 1976). AK_1 and AK_2 isozymes have different mobilities on electrophoresis.

Three different electrophoretic patterns were described after separation of haemolysates in starch gel media by Fildes and Harris (1966). The different patterns, referred to as AK_1 1, AK_1 2-1 and AK_1 2 (Fig. 5), are controlled by two alleles AK_1^1 and AK_1^2 at the AK_1 locus. Both AK_1 1 and AK_1 2 show at least three zones with enzyme activity. The slowest migrating zone shows the strongest staining intensity. The heterozygous type AK_1 2-1 appears to represent a mixture of the components present in types AK_1 1 and AK_1 2 in roughly equal amounts. Skude and Jakobsson (1970) have presented a simpler, quicker technique using an agar gel instead of a starch gel.

Family data (Fildes and Harris, 1966) are consistent with the segregation of two autosomal allelic genes, AK_1^1 and AK_1^2. These results have been confirmed by Harris (1966), Harris et al. (1968), Rapley et al. (1967) and Bowman et al. (1967). Three rare phenotypes called AK_1 3-1 (Bowman et al., 1967), AK_1 4-1 (Rapley et al., 1967) and AK_1 5-1 (Santachiara-Benerecetti et al., 1972b), have been reported (Fig. 5).

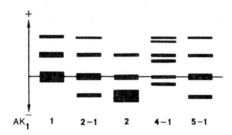

Fig. 5   Three common and two rare phenotypes of adenylate kinase controlled by genes at the AK_1 locus

In adults it has been shown that the red cell enzyme activity in individuals with the AK_1 1 phenotype is higher than in individuals with the AK_1 2-1 phenotype (Rapley and Harris, 1970; Modiano et al., 1970b). In red cells from cord blood, no such difference between phenotypes was found but the activity was lower than in adults. A silent allele AK^O has been observed in three generations of an English family, leading to half-normal enzyme activities in heterozygotes AK_1 1-0 (Singer and Brock, 1971).

Population data on the adenylate kinase polymorphism have been summarized by Beckman (1972) and Mourant et al. (1976). The frequency of the AK_1^2 allele is low in all populations tested so far; in Europeans 1 to 5 %, in Negroes below 1 % and in American Indians this allele has not been observed. The highest frequency (up to 15 %) has been found among some Asian population groups.

Alcohol Dehydrogenase (EC 1.1.1.1)

Alcohol dehydrogenase (ADH) catalyses the oxidation of ethanol to acetaldehyde and is found principally in the liver and other visceral organs.

The subunit structure of human alcohol dehydrogenase is determined by genes at three separate loci, designated ADH_1, ADH_2 and ADH_3 (Smith et al., 1971). The subunits controlled by these three gene loci are referred to as α, β and γ respectively. They are of similar molecular weight, 40 000, (Smith et al., 1973b) and combine to dimeric functional forms by free recombination. The activity of the three gene loci varies from early foetal to adult life (cf. Smith *et al.*, 1971) and free recombination between the subunits α, β and γ explains the differences in electrophoretic patterns observed during development (cf. Smith *et al.*, 1971, 1972 and 1973a). The *ADH₁* locus is primarily active during early foetal life (to about 20 weeks of gestation) and only one isozyme with the subunit structure $\alpha\alpha$ is present. A progressive increase in the synthesis of β-subunits whose structure is controlled by the *ADH₂* locus occurs somewhat later in foetal life. Free

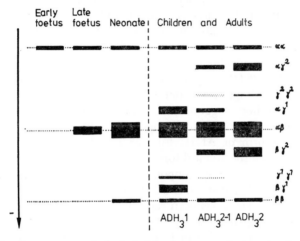

Fig. 6 Subunit structure of alcohol dehydrogenase during development. The polymorphism controlled by the ADH₃ locus is illustrated. (Reprinted from Harris, 1975, with permission)

recombination lead to the synthesis of three dimeric forms $\alpha\alpha$, $\alpha\beta$ and $\beta\beta$ which are all present at birth. During the first years of postnatal life the γ-subunit is synthesized and another three dimeric forms of ADH are hereby formed, $\alpha\gamma$, $\beta\gamma$ and $\gamma\gamma$. Hence the six multiple molecular forms or isozymes present in the adult liver are controlled by genes at three different loci (Fig. 6).

A genetic polymorphism occurs at two loci *ADH₂* and *ADH₃* but so far no genetic variation has been observed at the *ADH₁* locus.

An atypical form of alcohol dehydrogenase associated with an increased level of enzyme activity was reported by von Wartburg *et al.* (1965). Other differences between the typical and atypical forms have also been described (Von Wartburg and Schürch, 1968; Smith *et al.*, 1973a). The atypical form was found in 12 out of 59 individuals tested in a study from London (Von Wartburg and Schürch, 1968). Smith *et al.* (1972) used the pH activity ratio (activity at pH 11·0/activity at pH 8·8), a method described by von

Wartburg *et al.* (1965), to screen for the atypical form of ADH in a sample of 166 livers from individuals varying in age from 90 years to 28 weeks. Sixteen samples had the atypical form while 150 had the usual pH ratio phenotype. The atypical form appears to occur in about 6% of the European population. The atypical form of ADH was not observed in 56 samples from young foetuses (9 to 22 weeks gestation) which is to be expected as the ADH_2 locus is not active at that stage of development. No differences between sexes have been observed. Slight differences in the electrophoretic pattern have been observed between the usual and unusual phenotypes (Smith *et al.*, 1971). The atypical form found in livers also occurs in the lungs of the same individuals but not in kidney or intestine. It is interesting to note that the so-called atypical form which occurs in 6% of European populations occurs in about 90% of a Japanese population studied by Fukui and Wakasugi (1972), yet another example of poor nomenclature. As liver alcohol dehydrogenase is one of the important enzymes in ethanol oxidation, the authors suggested that the reported difference in alcohol tolerance between Japanese and European, at least in part, could be related to the atypical form of ADH. Further population data supporting an association between the atypical form of ADH and alcohol sensitivity have been presented by Stamatoyannopoulos *et al.* (1975). The atypical variant was found in 2·8% by Azêvedo *et al.* (1976) in a study from Brazil. A new allele at the ADH_2 locus has been reported by Azêvedo *et al.* (1976). The allele has been designated ADH_2^{Bahia}.

The ADH_3 locus is active in foetal and adult intestine and in kidney (Smith *et al.*, 1971) but also in stomach and liver (Smith *et al.*, 1972). Three phenotypes have been observed, ADH₃ 1, 2-1 and 2 (Fig. 6), controlled by two autosomal alleles ADH_3^1 and ADH_3^2. The frequencies of the two alleles are about 0·6 and 0·4 among Caucasians. A somewhat lower frequency for the ADH_3^2 allele, 0·14, has been reported by Azêvedo *et al.* (1976) in a Bahian population which is a triracial mixture of native Indians, African Negroes and Caucasians. There is some indication in their data that the frequency of the ADH_3^2 allele is low in Negroes.

α-Amylase (EC 3.2.1.1)

Human α-amylase is controlled by genes at two separate but closely linked loci, AMY_1 and AMY_2 (Merritt *et al.*, 1972), on chromosome 1 (Kamarýt *et al.*, 1971; Merritt *et al.*, 1973a). The genes on the AMY_1 locus control the amylase produced by the salivary glands while the AMY_2 locus controls the synthesis of the pancreatic amylase. Merritt *et al.* (1973b), have shown that salivary and pancreatic amylases can be separated into two fractions by means of electrophoresis. They state, however, that serum and urine should only be used for the study of pancreatic amylase to avoid misclassification and confusion.

Genetical heterogeneity in amylase was originally described by Kamarýt and Laxová (1965, 1966). Genetic variants of both salivary and pancreatic amylases have been observed. The genetics of salivary amylase has been reviewed by Ward et al. (1971) and further data have been presented by Merritt et al. (1973b) and Karn et al. (1975). According to Merritt et al. (1973b) the variant phenotypes are named with respect to the variant isozyme present (Fig. 7). Seven different salivary amylase phenotypes have been observed and designated, AMY_1A to AMY_1G, with AMY_1A being the common phenotype (cf. Merritt et al., 1973b). Boettcher and de la Lande (1969) have shown that the phenotypic frequencies of salivary amylase among Australian Aborigines differed from individuals with European descent and Ward et al. (1971) reported that AMY_1 variants are relatively rare in Caucasians. Merritt et al. (1973b) found the combined frequencies of the AMY_1 variants in American Caucasians to be 0·0073 compared to 0·0769 for Afro-Americans.

The three phenotypes controlled by genes at the AMY_2 locus are referred to as AMY_2 A, B and C. AMY_2A represents the common phenotype. The frequency of the AMY_2B phenotype is 0·105 in Caucasian Americans and 0·029 in Afro-Americans according to Merritt et al. (1973b). The AMY_2C phenotype is apparently restricted to Negro populations where the frequency of the phenotype is quite high, 4·7% in Afro-Americans and 33·3% in Native Africans (Merritt et al., 1973b).

Family data indicate an autosomal dominant mode of inheritance for both salivary and pancreatic amylase variants (Kamarýt and Laxová, 1965, 1966; Merritt et al., 1973b).

Aryl Hydrocarbon Hydroxylase (1.14.1.1)

Aryl hydrocarbon hydroxylase, AHH, catalyses the oxidation of a number of aryl hydrocarbons. A polymorphism, reflecting the capacity of induction of aryl hydrocarbon hydroxylase in human leucocytes in response to polycyclic hydrocarbons, was reported by Kellerman et al. (1973a). Two alleles, one for low induction AHH^a and one for high induction, AHH^b, were proposed based on the observation that individuals could be grouped into three categories, namely high (BB) intermediate (AB) and low (AA) inducers. The frequencies of the three phenotypes were 9, 46 and 34% respectively in the U.S. Caucasian population. Population and family studies indicated autosomal co-dominant inheritance.

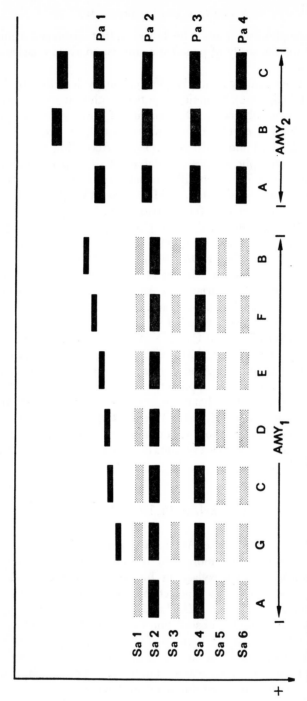

Fig. 7   Migration rates of α-amylase isozymes controlled by genes at the AMY_1 and AMY_2 loci. (After Merritt *et al.*, 1973)

The capacity of induction of AHH is studied in cultured leucocytes (Busbee et al., 1972) but can also be studied in cultured fibroblasts and epithelial cells. Cantrell et al. (1973) measured the level of AHH in human pulmonary alveolar macrophages and found that the enzyme might be induced by cigarette smoke in vivo. Kellerman et al. (1973b) reported that an association apparently existed between the ability to respond to arylhydrocarbons by induction of AHH and lung cancer. However, it may be that the situation is not as clearcut as the reported relative risk (cf. p. 169) would indicate, as AHH variability may not be a simple Mendelian trait. Possible localization of a structural gene for AHH on chromosome 2 (Brown et al., 1976) may make the analysis simpler.

Carbonic Anhydrase (EC 4.2.1.1)

Carbonic anhydrase, CA, catalyses reversibly the conversion of carbon dioxide and water to carbonic acid and is a major protein of the red blood cell. The enzyme occurs in two forms referred to as CA I and CA II (Tashian and Shaw, 1962; Tashian, 1965, 1969) determined by genes on two autosomal gene loci, CA_I and CA_{II} (Hopkinson et al., 1974a). Abbreviations synonymously used for the two forms of the enzyme are CA B and CA C respectively (Nyman and Lindskog, 1964). Inherited electrophoretic variants of CA I occur in most populations but in low frequencies (Carter et al., 1972). The common form is referred to as CA Ia and the rare variants as CA 1b Michigan (Shaw et al., 1962) (= CAP Mut described by Funakoshi and Deutsch, 1970) CA Ic Guam (Tashian et al., 1963, 1966), CA Ic Filipino (Lie-Injo, 1967), CA Id Michigan (Shows, 1967), CA Ie Michigan (Tashian et al., 1968) CAIe Portsmouth, CAIe Hall (Carter et al., 1972), CAIf (Carter et al., 1973). Autosomal co-dominant inheritance has been suggested for the rare variants.

A polymorphism of CA II was described by Moore et al. (1971). They used immunoelectrophoresis to differentiate between CA I and C II and were at the same time also able to discern a polymorphism at the CA_{II} locus. Family studies indicated co-dominant autosomal inheritance which has been verified (Lin and Deutsch, 1972; Moore et al., 1973). The polymorphism was observed in American Negroes and 18% of the individuals tested appeared to be heterozygotes and 1% homozygotes. Hopkinson et al. (1974a) have developed a new method which makes it possible to study carbonic anhydrase after separation on starch gel electrophoresis using fluorogenic substrates to visualize enzyme activity. With this method they were able to distinguish three phenotypes $CA_{II}1$, CA_{II} 2-1 and $CA_{II}2$

(Fig. 8). These phenotypes were found to correspond to those described by Moore *et al.* (1971). CA II appears to be polymorphic in Negroes but not in Caucasian and Asian populations (Hopkinson *et al.*, 1974a, Welch, 1975). The frequency of the CA_2^2 allele varies between 5 and 20%. (The new nomenclature recommended for the two gene loci is CA_1 for CA_I and CA_2 for CA_{II}.)

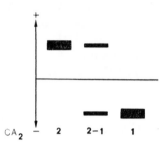

Fig. 8   Three phenotypes of carbonic anhydrase controlled by the CA_2 locus. (After Hopkinson *et al.*, 1974)

Cytidine Deaminase (EC 3.5.4.5.)

Cytidine deaminase (CDA) catalyses the deamination of cytidine to uridine and ammonia as well as the deamination of some other nucleoside analogues (Chabner *et al.*, 1974). Cytidine deaminase activity is high in leucocytes, particularly in polymorphonuclear leucocytes, while in erythrocytes there is little or no activity (Chabner *et al.*, 1974; Abell and Marchand, 1973).

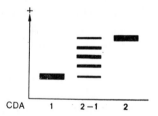

Fig. 9   Three phenotypes of cytidine deaminase. (After Teng *et al.*, 1975)

A genetical polymorphism of cytidine deaminase has been discovered (Teng *et al.*, 1975). The different enzyme patterns referred to as CDA 1, CDA 2-1 and CDA 2 (Fig. 9) can be identified by means of starch gel electrophoresis of extracts of polymorphonuclear leucocytes followed by enzymatic staining.

Family studies indicate that the enzyme patterns are controlled by two common allelic genes CDA^1 and CDA^2 at an autosomal locus. Teng *et al.* (1975) found the frequencies of the CDA^1 and CDA^2 genes to be 0·65 and 0·35 respectively in a study of Caucasian blood donors living in Seattle, U.S.A.

Diaphorase (Methaemoglobin Reductase, EC 1.6.2.2)

Two methaemoglobin reductases or diaphorases, one dependent on NADH as co-enzyme and the other on NADPH, catalyse the reduction of methaemoglobin, and are important in protecting haemoglobin from oxidizing agents.

Quite a number of rare variants of NADH-diaphorase have been discovered, all presumably controlled by allelic genes at one locus (Williams and Hopkinson, 1975). Most of the variants have a normal enzyme activity but differ in electrophoretic charge and are discovered by electrophoresis of haemolysates in population screening studies. Others, in contrast, have been discovered through their associated decreased enzyme activity. Low activity of the enzyme results in methaemoglobinaemia which can be associated with a progressive neurological disorder (Kaplan *et al.*, 1974). It should be noted that the activity of NADH-diaphorase is in general significantly lower in newborns than in adults (Ross, 1963; Eng *et al.*, 1972).

The most common phenotype of NADH-diaphorase has been designated DIA 1 and the rare phenotypes observed in various population studies as DIA 2-1, 3-2, 4-1, 5-1, 6-1 (Hopkinson *et al.*, 1970) and DIA 7-1 (Williams and Hopkinson, 1975). The different phenotypes occur in people heterozygous for the common allele DIA^1 in combination with one of the rare alleles DIA^2, DIA^7 etc. Family data are consistent with co-dominant autosomal inheritance. No case of methaemoglobinaemia has been reported to be associated with any of these variants.

The population data on NADH-diaphorase have recently been compiled by Williams and Hopkinson (1975). So far the enzyme has not been found to be polymorphic in any of the different populations studied. According to the authors the incidence of variant diaphorase phenotypes is of the order of about 2/1000.

Leroux *et al.* (1975) have proposed that NADH diaphorase is identical to or shares a common subunit with cytochrome b_5 reductase (EC 1.6.2.2). They further propose that there are two different forms of NADH diaphorase/cytochrome b_5 reductase, one soluble, the other microsomal. Deficiency of the soluble form alone causes methaemoglobinaemia and of both forms, methaemoglobinaemia with mental retardation.

Enolase (EC 4.2.1.11)

Enolase catalyses the interconversion of 2-phosphoglycerate and phosphoenol pyruvate in the glycolytic pathway. The isozyme variations observed in different tissues are controlled by three different loci (Pearce et al., 1976; Chen and Giblett, 1976). These three loci are referred to as ENO_1, ENO_2 and ENO_3. A genetical variant of erythrocyte enolase has been described and referred to as $ENO_1 2-1$ (Giblett et al., 1974b). Family studies indicated autosomal co-dominant inheritance. The common gene is designated ENO_1^1 and the rare gene ENO_1^2.

Esterase (Non-specific, EC 3.1.1.1)

Four non-specific human red cell esterases, A, B, C (Tashian, 1969) and D (Hopkinson et al., 1973) have been identified by means of electrophoresis of red cell haemolysates.

Esterase A can be separated into three zones A_1, A_2 and A_3 by electrophoresis. Genetical variants are very rare, but two different phenotypes have been observed (Tashian and Shaw, 1962; Tashian, 1965, 1969).

No variants of esterases B (butyryl esterase) and C (acetyl esterase) have been reported yet. However, most of the esterase activity of human tissues can be attributed to the butyryl esterase isozymes, and Coates et al. (1975) have suggested that these are coded for by at least three separate structural gene loci.

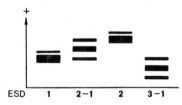

Fig. 10   Three common and one rare phenotype of esterase D

A polymorphism of esterase D (ESD) has been described by Hopkinson et al. (1963). Esters of 4-methyl-umbelliferone are used as substrates to detect esterase D. Three common phenotypes EsD 1, 2-1, and 2 have been found (Fig. 10). Family data show that the three phenotypes are controlled by two autosomal alleles $ES\ D^1$ and $ES\ D^2$ (Hopkinson et al., 1973; Benkmann and Goedde, 1974). A third phenotype ES D 3-1 has been reported (Bender and Frank, 1974; Bargagna *et al.*, 1975). Gene frequency data published on *ES D* in different populations have been compiled by Welch and Lee (1974) and by Koster *et al.* (1975). The differences between various European populations are small. The frequency of the *ES D²* gene is about 10% in European populations, somewhat lower (6 to 10%) in African populations and definitely higher (23 to 35%) in Asian populations. Welch and Lee (1974) reported a *ES D²* frequency of 24% among Lapps, but Beckman and Beckman (1976b) found a frequency of 6% among Swedish Lapps, which is somewhat lower than the Swedish population in general (8 to 9%). Various methods for the separation and identification of the different phenotypes of esterase D have been described by Hopkinson *et al.* (1973), Kuhnl *et al.* (1974), Bender and Frank (1974) and Benkmann and Goedde (1975).

α-Fucosidase (EC 3.2.1.51)

The enzyme α-fucosidase, αFUC, apparently plays an important function in the metabolism of glycolipids and glycoproteins containing fucose. Multiple zones with enzyme activity can be detected after separation of leucocyte and other tissue extracts by starch gel electrophoresis (Turner *et al.*, 1975).

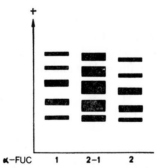

Fig. 11   Three phenotypes of α-L-fucosidase. (After Turner *et al.*, 1975)

Isoelectric focusing in thin layer acrylamide improves the separation and makes it possible to distinguish three phenotypes of α-fucosidase from leucocyte extracts. These have been designated αFUC 1, αFUC 2-1 and αFUC 2 (Fig. 11), and family studies show that the three phenotypes are controlled by two autosomal co-dominant alleles *αFUC¹* and *αFUC²* (Turner *et al.*, 1975). Treatment of the tissue extracts before separation with neuraminidase improves the enzyme patterns and makes classification easier (Turner *et al.*, 1975). The frequency data published by Turner *et al.* (1975) are as follows: $aFUC^2$, 25% among white Americans compared to 5% among black Americans. The author raised the question whether the $aFUC^2$ gene is absent among African Negroes. The specific activity of the three common phenotypes indicates that the $aFUC^1$ gene is associated with lower enzyme activity even though the variability within each phenotypic class is quite high (Turner *et al.*, 1975).

Inherited deficiency of the enzyme is associated with a progressive accumulation of glycolipids and glycoproteins containing fucose during early infancy. The disease, which is known as α-fucosidosis, leads to a progressive physical and mental deterioration at an early age (Dorfman and Matalon, 1972). Autosomal recessive inheritance is most probable, which means that individuals with α-fucosidosis are homozygous for a silent or defective gene. This gene is most probably allelic to the $aFUC^1$ and $aFUC^2$ genes. Even though α-fucosidosis is rare the frequency of the defective allele in heterozygous combination in the population cannot be neglected. The presence of a silent allele might for example explain some of the variability in enzyme activity found in the 1-1 and 2-2 phenotypes.

Ng *et al.* (1976) have found that the level of serum α-fucosidase activity is quite variable among adults and very low activity levels can be obtained for serum even though the activity in leucocytes is normal. Family studies indicate that two different types of individuals occur; those with "low" activity and others with "normal" activity of α-fucosidase in their sera. It is suggested that the release of α-fucosidase to serum is under genetical control. It is also suggested that the release process leads to a modification of the tissue enzyme as a slight difference in electrophoretic mobility exists between the tissue α-fucosidase and the enzyme found in serum.

Galactokinase (EC 2.7.1.6)

Galactokinase (GALK) catalyses the first step in the galactose–glucose conversion pathway. A deficiency of the enzyme is associated with juvenile cataract, an autosomal recessive disorder. Three alleles have been identified so far at the $GALK$ locus. These are referred to as $GALK^A$ (the common allele), $GALK^G$ and $GALK^P$ (Tedesco *et al.*, 1975). Homozygosity for the $GALK^G$ allele leads to a deficiency of the enzyme and to juvenile cataract. The $GALK^G$ allele appears to be very rare but so far little information is available on its frequency (Shih *et al.*, 1971; Mayes and Guthrie, 1968). According to Mellman *et al.* (1975) there is some indication that deficiency of the enzyme has a more complex background. The frequency data in Negro populations so far published may according to Tedesco *et al.* (1975)

be less accurate than originally thought due to the discovery of a third allele $GALK^P$. The $GALK^P$ allele is polymorphic among Negroes. Individuals homozygous for the $GALK^P$ gene have about 30% of the activity observed in individuals who are homozygous for the $GALK^A$ allele (Tedesco et al., 1972).

Galactose-1-Phosphate Uridylyl Transferase (EC 2.7.7.12)

Galactose (or hexose)-1-phosphate uridylyl transferase (GALT) catalyses the second step in the galactose–glucose interconversion. The enzyme, which is important in the metabolism of galactose, occurs in most mammalian tissues (Bertoli and Segal, 1966). The subunit structure of the enzyme has been studied (Nadler et al., 1970; Tedesco, 1972; Hammersen et al., 1975). Hammersen et al. (1975) found that the subunit composition varies, which can explain the heterogeneous picture of the enzyme which has been found between different tissues.

A deficiency of galactose-1-phosphate uridylyl transferase leads to what is commonly referred to as classical galactosaemia with an autosomal recessive inheritance. Individuals who lack the enzyme are homozygotes for a rare allele originally designated gt (Beutler et al., 1965, 1966) while the common allele was referred to as Gt^+. According to the new nomenclature (Tedesco et al., 1975) the gt allele should be referred to as $GALT^G$ and the common allele Gt^+ as $GALT^A$. The deficiency observed in galactosaemic patients is apparently not due to impaired synthesis of the enzyme protein but to the synthesis of an enzyme protein with defective catalytic function (Tedesco, 1972).

Data on the frequency of the rare $GALT^G$ allele in a number of different populations has been gathered by Tedesco et al. (1975). The frequency of the $GALT^G$ allele appears to vary quite remarkably between different populations. However, when estimating the frequency of the $GALT^G$ allele, the occurrence of another and rather common allele must be considered. Tedesco et al. (1975) found that the mean GALT activity is somewhat lower in Caucasians than in Negroes. The explanation for this difference is according to these authors and others to be found in the occurrence of the Duarte variant (Beutler et al., 1966) in a rather high frequency among Caucasians (Beutler, 1973; Ng et al., 1969). Enzyme activity is reduced to about 50% in the red blood cells of individuals who are homozygotes for the $GALT^D$ (previously Gt^D) allele while those who are heterozygotes have 75% of what is regarded as the normal value (Beutler et al., 1965). If only quantitative measurements for the detection of galactosaemia heterozygotes are applied the frequency of the $GALT^G$

allele will be overestimated. Caution must be taken to exclude individuals homozygous or heterozygous for the $GALT^D$ allele (Beutler, 1973). Avoiding this pitfall, Tedesco *et al.* (1975) estimated the frequency of the $GALT^G$ allele to be 0·0024.

The frequency of the $GALT^D$ allele was found to be about 0·07 among Caucasians, 0·03 among Negroes and very rare among Orientals (Ng *et al.*, 1973). No clinical disorder has been reported to be associated with the $GALT^D$ variant (Beutler *et al.*, 1966). The enzyme protein from an individual homozygous for the Duarte variant appears to have a slightly faster

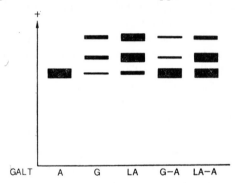

Fig. 12   Phenotypes of galactose-1-phosphate uridylyl transferase. (After Ng *et al.*, 1973)

electrophoretic mobility compared to that of the normal type (Mathai and Beutler, 1966), but is otherwise kinetically indistinguishable (Beutler and Baluda, 1966).

Three other variants of GALT have been discovered. Two of these, "Rennes" (Schapira and Kaplan, 1969) and "Indiana", are rare and have only been found in galactosaemic patients. These two variants appear to represent unstable forms of the enzyme. No clinical disorder was found to be associated with the third variant "Los Angeles" (Fig. 12). This variant is associated with a slightly elevated enzyme activity in the red cells. The "Los Angeles" allele is polymorphic in Caucasian and Negro populations, but has so far not been found among Orientals (Ng *et al.*, 1973).

Glucosephosphate Isomerase (EC 5.3.1.9)

Glucosephosphate isomerase (GPI) also known as phosphohexose isomerase (PHI) and phosphoglucose isomerase, catalyses the reversible conversion of glucose-6-phosphate to fructose-6-phosphate. No tissue specific isozymes of phosphohexose isomerase appear to occur (Detter *et al.*,

1968; Payne *et al.*, 1972) and occurrence of different alleles at the *GPI* locus can explain both the quantitative and the qualitative genetical variation of the enzyme which has been observed.

A common type and a number of rare phenotypes of GPI have been observed in electrophoretic screening of haemolysates from a great number of unrelated individuals from various ethnic groups (Detter *et al.*, 1968; Fitch *et al.*, 1968; Omoto and Blake, 1972). The common type is referred to as GPI 1 and the rare types as GPI 2-1 to GPI 10-1 (Fig. 13). Family studies indicate autosomal co-dominant inheritance (Detter *et al.*, 1968; Wendt and Kirchberg, 1971).

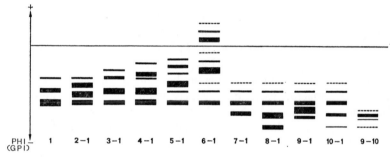

Fig. 13   The common phenotype GPI 1 and a number of rare phenotypes of glucosephosphate isomerase. (After Detter *et al.*, 1968)

The common phenotype GPI 1 consists of three components, of which the one with the most cathodic mobility at an alkaline pH shows the strongest staining intensity. All the rare phenotypes mentioned consist of three major components, one of which corresponds to the major zone in the common phenotype. The major zones in type GPI 6-1 migrate towards the anode, all other components in all types so far described migrate towards the cathode at pH 8·0. The triple-banded heterozygous pattern indicates that the enzyme is a dimer (Detter *et al.*, 1968).

Detter *et al.* (1968) described one individual with an electrophoretic pattern of GPI which suggested that the individual was heterozygous for two rare alleles (GPI^9 and GPI^{10}). The GPI 9-1 phenotype was carried by the father. The staining intensity of the rare heterozygous phenotype GPI 9-10 was strongly reduced indicating a decreased enzyme activity. Two cases of homozygosity for the GPI^9 allele in two children in one family were reported by Tariverdian *et al.* (1970a). Both children suffered from non-spherocytic haemolytic anaemia and deficiency of enzyme activity. Reduced enzyme activity was also observed in fibroblasts cultured from the two affected children (Krone *et al.*, 1970).

A third GPI variant associated with severe haemolytic anaemia was reported by Paglia *et al.* (1969). The family members affected by the anaemia were regarded as homozygotes for a rare variant. The electrophoretic mobility of the enzyme variant was only slightly slower than that of GPI 1 and heterozygotes between the rare variant and the common variant could not be distinguished from the normal homozygous pattern (GPI 1) by means of electrophoresis. The staining intensity of the rare heterozygous phenotype was, however, reduced compared to that of the common phenotype GPI 1.

Baughan *et al.* 1968 described a case with non-spherocytic haemolytic anaemia and GPI deficiency. Family studies indicated an autosomal recessive inheritance.

Today over 20 individual cases of haemolytic anaemia associated with complete or partial GPI deficiency are known (Paglia and Valentine, 1974; Schröter *et al.*, 1974; Van Biervliet *et al.*, 1975; Vives-Corrons *et al.*, 1975; Paglia *et al.*, 1975). A defect in PHI appears therefore to be one of the more common reasons for non-spherocytic haemolytic disease. The clinical severity of the anaemia varies quite widely, from well-compensated chronic haemolysis to anaemias which require transfusions. This difference can be explained by the fact that some of the alleles code for the production of a catalytically inactive or quantitatively deficient enzyme while others govern structural alterations that may adversely affect enzyme stability (Paglia and Valentine, 1974). Heterozygosity for two defective alleles appears to be the most common reason behind a deficiency. No special clinical finding other than anaemia appears to be associated with the chronic haemolytic disease, which points to a defect in GPI itself. Most cases associated with haemolytic anaemia have been found among Europeans but a few have also been observed among Japanese (Miwa *et al.*, 1973a, b) and Mexicans (Paglia *et al.*, 1975). The variant phenotypes of GPI, viz. those with retained catalytic activity and identified by means of electrophoresis of vast population samples, all appear to be very rare among Caucasians. Surveys have been published by Detter *et al.* (1968), Omoto and Blake (1972) and Mourant *et al.* (1976).

Glucose-6-Phosphate Dehydrogenase (EC 1.1.1.49)

Glucose-6-phosphate dehydrogenase, G6PD, catalyses the oxidation of glucose-6-phosphate to 6-phosphogluconate with a simultaneous reduction of NADPH. The enzyme is thereby of great importance in regulating the intracellular concentration of NADPH, which is of prime importance in red cell function.

The literature on G6DP is enormous. Only the main genetical aspects of the enzyme will be considered here. A number of comprehensive review articles have been written on different aspects of the enzyme (Beutler, 1969a, 1970; Giblett, 1969; Kirkman, 1971; Motulsky, 1972, 1975; Luzatto, 1973).

The great genetical variability of G6PD known today began to be uncovered when Hockwald *et al.* (1952) found that certain individuals of the American Negro population developed acute haemolysis when given the antimalarial drug primaquine, an early example of a pharmacogenetical problem (p. 493). Dern *et al.* (1954) presented evidence that the haemolysis was due to an intrinsic factor of the red cells. Later it was shown (Carson *et al.*, 1956), that individuals developing haemolysis were deficient in the enzyme glucose-6-phosphate dehydrogenase in their red blood cells. A great number of drugs are now known to have the same effect as primaquine. A list of such drugs was published by WHO in 1967 (also Beutler, 1969a; Giblett, 1969; Motulsky, 1972; Oski and Stockman, 1973). The haemolytic crises associated with these drugs were soon found to be only indirectly associated with the deficiency of G6PD, but directly with a failure in the reduction of NADP to NADPH (Beutler, 1969a; Giblett, 1969).

The disease favism was also found to be due to a deficiency of glucose-6-phosphate dehydrogenase (Zinkham *et al.*, 1958; Larizza *et al.*, 1958; Szeinberg *et al.*, 1958). This disease which manifests itself as an acute haemolytic anaemia after the ingestion of fava beans, has long been known and feared in certain population groups in the Middle East and in some Mediterranean countries.

Quantitative determinations of G6PD activity in red blood cells from a Negro population revealed that male Negroes could be divided into two distinct groups: deficient or non-deficient. The deficient group had only 15% of the enzyme activity found in the non-deficient group. A more nearly continuous distribution from the highest to the lowest enzyme values was observed among females. This became explicable when family studies showed that the locus for G6PD was located on the X-chromosome (Childs *et al.*, 1958).

It seemed reasonable to assume that the deficiencies observed in the Negro and Mediterranean populations were due to two different gene mutations at the G6PD locus. This assumption was based on certain characteristic differences; for example the enzyme activity was reduced to 3 to 4% in the "Mediterranean" type compared to 15% in the "Negro" type. The "Mediterranean" type of deficiency was also associated with a reduction of enzyme activity in white cells and other tissues (Ramot *et al.*, 1959) while the reduction was quite small or absent in the "Negro" type (Marks *et al.*, 1959). Hence two different variants of G6PD associated with reduced enzyme activity appeared to occur.

The fact that G6PD can occur in different molecular forms was proposed by Kirkman (1959), Marks et al. (1961) and verified by Boyer et al. (1962) and Kirkman and Hendrickson (1963). Two common types of the enzyme could be demonstrated after separation of haemolysates from Negro males by means of starch gel electrophoresis and subsequent staining of the gel for enzyme activity. The two molecular forms were referred to as A and B. Men have either the A or the B component but never both, which is obvious as the locus for G6PD is on the X-chromosome. Females on the other hand can be either homozygotes for A or B or AB heterozygotes (but see p. 512). The G6PD variant found in Negroes with G6PD deficiency showed the same electrophoretic mobility as the A type but the Mediterranean variant showed the same electrophoretic mobility as the B type.

Four different alleles are assumed to control the different types of G6PD described. These alleles are referred to as Gd^B, Gd^A, Gd^{B-} (or $Gd^{Mediterranean}$) and Gd^{A-}. Gd^B is the most frequent allele in all populations so far studied and is referred to as the normal type. Individuals with the type A show on the average 10% less enzyme activity compared to type B individuals. The difference in electrophoretic mobility between B and A is due to a single amino acid substitution where asparagine in B is replaced by aspartic acid in A (Yoshida, 1967). No other differences have been reported between the B and A variants. Increased instability of the variant enzyme proteins seems to be the main reason for the low enzyme activity of the A- and B-variants. This is based on the observation that the activity of G6PD in individuals with G6PD A- is almost as high as the activity in individuals with the A and B types when younger cells are examined, viz. cell preparations containing a large proportion of reticulocytes (Yoshida et al., 1967; Piomelli et al., 1968). Hence the deficiency associated with the A-variant depends on an increased breakdown of the enzyme in vivo and not an alteration of the catalytic efficiency of the enzyme nor a reduced rate of synthesis of the enzyme as a result of an altered protein structure. This breakdown is even greater in red cells from individuals carrying the B-variant.

Today about 100 variants of G6PD have been described. Some variants can be considered as common at least in some populations while most of them are quite rare. The variants can be divided into three main groups: variants associated with (a) normal enzyme activity, (b) enzyme deficiency, where exogenous agents are required to induce the haemolysis and (c) enzyme deficiency resulting in chronic haemolytic diseases. No detailed account will be given of the characteristics associated with each of the different variants. Information can instead be obtained from the data compiled by Yoshida et al. (1971), Beutler and Yoshida (1973) and Mourant et al. (1976), while certain aspects will be discussed on p. 493. Standardized methods for the characterization of G6PD variants have been published by the WHO Scientific Group (1967).

G6PD variants occur all over the world but most variants have been discovered in populations living in tropic or sub-tropic areas. The most common variant not associated with any haemolytic disease is Gd A. The *Gd^A* allele appears to be restricted to Negro populations (Boyer *et al.*, 1962). Both the *Gd^A* and the *Gd^A-* alleles occur at a frequency of about 20% in most of the African male populations. Examples of other G6PD variants which can be regarded as common at least in some populations are Gd B-, Gd Canton (McCurdy *et al.*, 1966), Gd Athens (Stamatoyannopoulos *et al.*, 1967), Gd Constantine (Kissen and Cotte, 1970), Gd Mahidol (Panich *et al.*, 1972; Panich and Sungnate, 1973) and Gd Mali (Kahn *et al.*, 1973).

G6PD variants associated with various degrees of deficiencies are in general found in Africa and in an area extending from the Mediterranean throughout South-West Asia, the Philippines and Indonesia. This is about the same distribution as the disease malaria. Motulsky (1960) proposed that individuals with G6PD deficiency benefit from a selective advantage towards malaria in the same way as individuals heterozygous for the sickle cell gene *Hb^S*. A limitation to the proliferation of the malaria parasite in red cells of individuals with an abnormal metabolic activity due to G6PD deficiency was supposed to constitute the selective mechanism. Evidence both for and against the hypothesis of selective advantage has been published throughout the years. However, the evidence for a protective effect has predominated and a selective advantage appears to be a fact (Luzzatto, 1973; Motulsky, 1975; see also p. 165).

From the fact that the *G6PD* locus was on the X-chromosome, it had to be diploid in the female (XX) but haploid in the male (XY). However, no marked differences in the average level of enzyme activity were found between males and females (Marks, 1958). The same level of enzyme activity was also observed in individuals with extra X-chromosomes (Grumbach *et al.*, 1962; Davidson *et al.*, 1963a; Harris *et al.*, 1963b). The explanation for this phenomenon is based on or supports the hypothesis proposed by Lyon (1961). According to this hypothesis, discussed on pp. 428 and 511, a female can be regarded as a mosaic with respect to traits controlled by genes on the X-chromosome and a heterozygous female should have two different red cell populations. This has been demonstrated by Beutler and Baluda (1964), Sansone *et al.* (1963), Tönz and Rossi (1964) and Stamatoyannopoulos *et al.* (1966, 1967).

Davidson *et al.* (1963b) cultured and cloned skin fibroblasts from a female heterozygous for G6PD deficiency. They found that the different clones had either the normal or the deficient level of enzyme activity. Cell cultures of certain tumours (leiomyomas) obtained from single cells as well as from small skin biopsies gave similar results (Gartler and Linder, 1964; Linder and Gartler, 1965; Gartler et al., 1966). The electrophoretic pattern of red blood cells of females who are AB heterozygotes shows no hybrid enzyme, which could be expected as the enzyme is a hexamer of identical subunits (Yoshida et al., 1967). Hybridization of the two G6PD variants A and B in vitro leads, however, to the formation of a hybrid enzyme (Yoshida et al., 1967). This observation supports the hypothesis that only one of two G6PD alleles is functioning at the same time in a single female cell. One exception is however represented by the oocytes, where both X-chromosomes are active (Gartler et al., 1972).

Glutamic Oxaloacetic Transaminase (EC 2.6.1.1)

Glutamic oxaloacetic transaminase (GOT) catalyses the reversible conversion of aspartate and α-ketoglutarate to oxaloacetate and glutamate. The enzyme occurs in two distinct molecular forms (Fleischer et al., 1960). One form is found in the soluble fraction of the cell, GOT_S, while the other is associated with the mitochondria GOT_M.

A polymorphism was originally described for the mitochondrial form of GOT by Davidson et al. (1970). Electrophoretic separation of placental extracts from different individuals made it possible to identify three deviating isozyme patterns. The deviant types were referred to as variants I, II and III. Two of the variants, I and II, were observed in a population study by Hackel et al. (1972). They introduced a new nomenclature and referred to the common phenotype as GOT_M 1, the variant I as GOT_M 2-1 and the variant II as GOT_M 3-1 (Fig. 14). The variant III was not observed in their study. Two additional phenotypes were observed by Hackel et al. (1972) GOT_M 3 and GOT_M 3-2. The three alleles at the GOT locus are referred to as GOT_M^1, GOT_M^2 and GOT_M^3.

The compiled population data of Davidson et al. (1970), Hackel et al. (1972) and Ananthakrishnan et al. (1972) show that the GOT_M^2 allele is polymorphic. The gene frequency was found to be 1·7% in a European population as well as in a Negro population. It was 1% in Nigerians but missing in a Negro population living in the U.K. The GOT_M^3 allele was found in 4 to 7% in Negro populations but was not observed in Europeans.

One genetic variant of GOT_S (Fig. 14) was described by Davidson et al. (1970) in a series of 860 placental extracts from Caucasians, Negroes and Puerto Ricans. Hackel et al. (1972) found no variants in 614 placental samples of Caucasian, Negro and Indian–Pakistani origin. Two electrophoretic variants of GOT_S have been described by Chen and Giblett

(1971b). The common phenotype is designated GOT_S 1 and the new phenotypes observed as GOT_S 2-1 and GOT_S 3-1 and respective alleles as GOT_S^1, GOT_S^2 and GOT_S^3. The GOT_S^2 gene is polymorphic in Japanese and American Indians. The frequency of GOT_S^3 in Yakima and Peruvian Indians was 0·018 and 0·028 respectively.

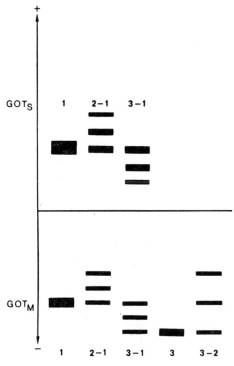

Fig. 14   Phenotypes of soluble and mitochrondrial glutamic-oxaloacetic transaminase controlled by two separate gene loci GOT_S and GOT_M. (After Chen and Giblett, 1971 and Hackel *et al.*, 1972)

Glutamic Pyruvic Transaminase (EC 2.6.1.2)

Glutamic pyruvic transaminase (GPT), also known as alanine aminotransferase, catalyses the reversible conversion of L-alanine and α-ketoglutarate to L-glutamate and pyruvate. One cytoplasmic and one mitochondrial form of the enzyme have been characterized (Boyd, 1961, 1966, Bodansky *et al.*, 1966). Genetic variation has so far only been observed in the cytoplasmic form of GPT. Three common phenotypes, GPT 1, GPT 2-1, and GPT 2 (Fig. 15) occur which are controlled by two autosomal alleles GPT^1 and GPT^2 (Chen and Giblett, 1971a; Kömpf 1971; Chen *et al.*,

1972a). Seven additional rare alleles, GPT^3 to GPT^8 (Fig. 15) including one GPT^0 allele (Chen *et al.*, 1972a; Gussman and Schwartzfischer, 1972; Olaisen, 1973b; Spielmann *et al.*, 1973), have been discovered. Of the rare alleles, the GPT^3 gene has been observed among Caucasians, GPT^4 in Africans, GPT^5 among Eskimos (Chen *et al.*, 1972a), the GPT^6 gene in people from New Guinea, the Philippines and Japan (Ishimoto and Kawata, 1974), GPT^7 in the Norwegian population (Olaisen 1973a) and the GPT^8 gene in the Italian population (Santachiara-Benerecetti *et al.*, 1975). Compilations of gene frequencies (cf. Olaisen and Teisberg, 1972; Lahav

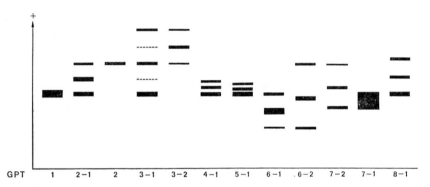

Fig. 15 Phenotypes of glutamic-pyruvic transaminase. (After Chen *et al.*, 1972a; Olaisen, 1973 and Santachiara—Benerecetti *et al.*, 1975)

and Szeinberg, 1972; Seth, 1974; Scozzari *et al.*, 1975; Mourant *et al.*, 1976) show that the enzyme is polymorphic in all populations studied so far. The highest frequency of the GPT^1 gene has been found in Zambia (92%) (Welch, 1972) and other African populations. The lowest frequency of the GPT^1 gene (31%) is reported from the Philippines (Chen *et al.*, 1972a). The frequencies of the five rare alleles have been below 1% in all populations so far studied. The occurrence of a silent allele has not been considered in most population studies. The frequency of the silent allele is probably around 0·015 to 0·020 (cf. Spielmann *et al.*, 1973).

The mean activities for the three common phenotypes varies (Chen *et al.*, 1972a; Welch 1972). The GPT^1 gene product in red cells has about three times the catalytic activity of the GPT^2 allele. The enzyme level appears to respond to physiological changes which accompany starvation, pregnancy, exercise and the administration of certain drugs (Chen *et al.*, 1972a). One can therefore speculate that differences in geographic distribution of the GPT genes might be associated with environmental factors. Individuals heterozygous for the GPT^0 allele have all been found to be healthy (Olaisen, 1973b; Spielmann *et al.*, 1973).

Glutathione Peroxidase (EC 1.11.1.9)

Glutathione peroxidase (GPX) catalyses the oxidation of glutathione by hydrogen peroxide and other organic peroxides. It is the only enzyme known to contain selenium as an integral part of its structure. Selenium deficiency in Man leads to a decrease in enzyme activity; this is analogous to "white muscle" disease in lambs (Goodwin, 1974).

GPX deficiency has been reported in a number of patients with mild haemolytic anaemias (Necheles et al., 1970a, b; Nishimura et al., 1972), but a causal link between the enzyme lesion and the anaemia has not been established with certainty. The role of selenium in this disorder may be compared with that of riboflavin in glutathione reductase deficiency (see p. 486).

Two separate polymorphic systems involving this enzyme have been reported by Beutler and his colleagues. One is revealed by mobility differences on electrophoresis of red blood cells (Beutler and West, 1974). The so-called Thomas variant having increased mobility was found in 25 of 392 samples from a Negro population as against three of 388 Caucasians (Beutler et al., 1974). Enzyme activity did not appear to differ markedly between phenotypes.

In contrast to the Thomas variant, extensive quantitative variability in enzyme activity in smaller samples from Jews and Gentiles was reported by Beutler and Matsumoto (1975) to be mainly the result of additive gene action at a single autosomal locus, the two alleles of which appeared to be in Hardy-Weinberg equilibrium. Should further investigation confirm the statistical resolution of this spectrum of variation into three separate distributions, one for each genotype, and should the Thomas variant be determined by the same locus, a situation analogous to red cell acid phosphatase (p. 192) would have been detected.

Glutathione Reductase (EC 1.6.4.2)

Glutathione reductase (GSR) catalyses the reduction of oxidized glutathione by NADPH to reduced glutathione. The enzyme is important in maintaining the stability of the structure of red cells and enzymes containing sulphydryl groups.

There are a number of reports in the literature on glutathione reductase deficiency in red blood cells associated with haematological disorders of varying severity (Carson et al., 1961; Löhr and Waller, 1962; Carson and Frischer, 1966; Prins et al., 1966). Certain drugs also seem to influence the

220   G. BECKMAN

activity of glutathione reductase and hence to induce haemolytic anaemia. Many of the drugs are similar in their chemical composition to the ones known to induce haemolysis in glucose-6-phosphate dehydrogenase deficiency (Waller, 1968). Few family studies are available but the deficiency of glutathione reductase seems to show autosomal dominant inheritance (Waller, 1968; Blume *et al.*, 1968b). However, the more recent observations of Beutler and co-workers (Beutler, 1969b; Beutler and Srivastava, 1970; Paniker *et al.*, 1970), that red cell enzyme levels are influenced by a subject's dietary intake of riboflavin, have cast some doubts on the precise nature of so-called glutathione reductase deficiency.

A variant form of glutathione reductase has been identified after separation of haemolysates on starch gel electrophoresis (Long, 1966; Kaplan and Beutler, 1968). A single zone with a somewhat faster anodal mobility than that of the normal component was observed in some samples. Family studies indicated that individuals with the fast moving zone of glutathione reductase activity were homozygotes for an unusual allele. This variant was found to be polymorphic in the American Negro population but not in Americans of Caucasian origin. The frequency of the variant allele was 0·113 in the Negro population.

Glycerol 3-Phosphate Dehydrogenase (EC 1.1.1.8)

Glycerol 3-phosphate dehydrogenase (GPD) catalyses the reduction of glycerol 3-phosphate to 3-glycerol.

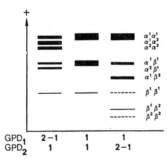

Fig. 16 Phenotypes of glycerol-3-phosphate dehydrogenase and their subunit composition controlled by genes at the *GPD₁* and *GPD₂* loci. (After Hopkinson *et al.*, 1976.) The *GPD₁* locus controls the α-chain and *GPD₂* the β-chain

Hopkinson *et al.* (1974b) found that the isozyme pattern observed in different tissues appears to be controlled by genes at two separate loci, in agreement with the proposal by Kömpf *et al.* (1971, 1972). The two loci are referred to as GPD_1 and GPD_2. The common phenotype observed in most tissue extracts is designated GPD_1 1/GPD_2 1 which implies homozygosity at both loci. Two rare phenotypes have been observed, GPD_1 2-1/ GPD_2 1 and GPD_1 1, GPD_2 2-1 (Fig. 16) in 707 skeletal muscle extracts analysed from *post mortem* specimens of North European origin (Hopkinson *et al.*, 1974b).

Glyoxalase I (EC 4.4.1.5)

The two enzymes glyoxalase I and II act together and form the glyoxalase system which converts methylglyoxal into lactate. Glyoxalase I of the red cells catalyses the irreversible conversion of glutathione and methylglyoxal to s-lactoyl-glutathione.

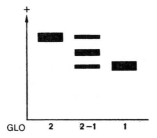

Fig. 17   Three phenotypes of glyoxalase I. (After Kömpf *et al.*, 1975a)

Red cell glyoxalase I, also known as lactoyl-glutathione-lyase is polymorphic (Kömpf *et al.*, 1975a). Three phenotypes, referred to as GLO 1, 2-1 and 2 (Fig. 17), have so far been described. Data obtained from population and family studies suggest that the three phenotypes are controlled by two autosomal alleles: GLO^1 and GLO^2. The GLO^1 gene frequency has been estimated to be 0·39 in a German population by Kömpf *et al.* (1975b). No genetical variants of glyoxalase II were observed by Charlesworth (1972) in a study of 687 North Americans, both Negro and Caucasian.

Hexokinase (Glucokinase, EC 2.7.1.1)

Hexokinase, HK, catalyses the conversion of glucose to glucose-6-phosphate. Deficiencies of the enzyme have been reported in association with chronic haemolytic anaemia (Valentine *et al.*, 1967; Keitt, 1969; Moser *et al.*, 1970; Necheles *et al.*, 1976b).

Four different forms of the enzyme HK I, HK II, HK III and HK IV have been characterized. The tissue distribution, electrophoretic mobility, substrate specificity etc. of the different hexokinases have been reviewed by Povey *et al.* (1975). A polymorphism of HK III in polymorphonuclear leucocytes has been reported (Povey *et al.*, 1975). Two phenotypes have so far been identified, HK III 1 and HK III 2-1 (Fig. 18). Family studies

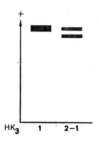

Fig. 18   Two phenotypes of hexokinase controlled by genes at the HK_3^1 locus. (After Povey *et al.*, 1975)

indicate that HK III 2-1 is heterozygous for two alleles HK_3 and HK_3^2. The frequency of the HK_3^2 allele is estimated to be 0·015. The locus for HK_3 is most probably different from the gene loci which control hexokinase in erythrocytes.

Isocitrate Dehydrogenase (EC 1.1.1.42)

The NADP-dependent isocitrate dehydrogenase (ICD) catalyses the oxidative decarboxylation of L-isocitrate to α-ketoglutarate. The mitochondrial and cytoplasmic forms of the enzyme (Henderson, 1965, 1968; Turner *et al.*, 1974a) are under separate genetical control (Chen *et al.*, 1972b).

Genetical variation of cytoplasmic or soluble isocitrate dehydrogenase, ICDs, was originally described by Chen *et al.* (1972b). Three rare variants were assumed to represent heterozygotes between a common allele ICD_S^1 and one of the rare alleles ICD_S^2, ICD_S^3 and ICD_S^3. Another two rare variant alleles, ICD^5 and ICD^6, have been reported by Turner *et al.* (1974b). Both these alleles appear to code for enzyme products which are less stable and cannot be detected in red cells.

Lactate Dehydrogenase (EC 1.1.1.27)

Lactate dehydrogenase (LDH) catalyses the reversible conversion of pyruvate to lactate and occurs in virtually all tissues. LDH can be separated into five zones upon electrophoresis (Markert and Möller, 1959). The zones are referred to as LDH 1, 2, 3, 4 and 5 in order of relative mobility towards the anode at an alkaline pH (LDH 1 being the fastest band). The enzyme is a tetramer made up of four subunits.

The subunits are of two different kinds, called A and B (or M and H), and are controlled by separate gene loci, LDH_A and LDH_B on chromosome 11 and 12 respectively (Boone et al., 1972; Chen et al., 1973). The subunit structure of the different LDH components is: LDH $1=B_4$, LDH $2=AB_3$, LDH $3=A_2B_2$, LDH $4=A_3B$ and LDH $5=A_4$ and the quantitative ratio of the five components is $1:4:6:4:1$ when the proportions of A and B are equivalent (Apella and Markert, 1961; Markert, 1963). The proportions of the subunits A and B vary considerably between different tissues (Vesell and Bearn, 1962; Starkweather et al., 1965; Fritz et al., 1969, 1971) and change during foetal development (Masters and Holmes, 1972 and West-hamer, 1973). Another type of LDH is found in extracts of spermatozoa and testes (Blanco and Zinkham, 1963; Evrev et al., 1970). This isozyme is called LDH X and is found only in postpubertal testes. This zone migrates between LDH-3 and LDH-4. Davidson et al. (1965) suggested that a third locus, LDH_C, controls the synthesis of the LDH X.

Electrophoretic variants of LDH, the result of mutations at either of the two loci controlling the A and B subunits, have been observed in erythrocytes (Boyer et al., 1963; Nance et al., 1963; Kraus and Neely, 1964; Davidson et al., 1965; Mourant et al., 1968; Blake et al., 1969; Das et al., 1970b, 1972). The following are examples representing mutations at the A locus: LDH Mem-1, LDH Mem-2, LDH Mem-4 (Kraus and Neely, 1964) and LDH Cal-1 (Das et al., 1970a), while LDH Mem-3 (Kraus and Neely, 1964) and LDH Mad-1 (Das et al., 1970b) represent mutations at the B locus.

The LDH variants are all quite rare. The frequency in Caucasian populations of heterozygotes of LDH variants appears to be of the order of one in a 1000, the LDH Mem-4 variant being the most common. Both A and B LDH variants have been found in African and American Negroes and the frequency of the variant is close to 1% (Mourant et al., 1976). The highest frequencies of LDH variants observed so far are in certain areas of New Guinea (Blake et al., 1969; Sinnett et al., 1970) and India (Das et al., 1970b, 1972). The LDH Cal-1 variant occurs in certain areas at 4% (Das et al., 1972; Saha et al., 1974; Kirk, 1975). One case of partial LDH deficiency, without any clinical stigmata, has been reported (Kitamura *et al.*, 1971). The deficiency appeared to depend on homozygosity for a deficiency mutation at the LDH_B locus.

Liver Acetyl Transferase

Liver acetyl-transferase is important for the acetylation of the drug isoniazid. This drug is commonly used in the chemotherapy of tuberculosis. Individuals can be divided into two groups: "rapid inactivators" and "slow inactivators" depending on the effectiveness of the acetylation process and hence of the inactivation of the drug (Knight *et al.*, 1959; Evans *et al.*, 1960, 1961). No significant difference in the therapeutic effect of the drug has been observed between "slow" or "rapid" inactivators (Evans, 1963) though smaller differences in response cannot be ruled out (Ellard *et al.*, 1972). Some side effects have been noticed, however. One of the complications with prolonged isoniazid therapy is peripheral neuropathy, which seems to occur more often in individuals who are "slow inactivators" (Devadetta *et al.*, 1960; Evans and Clarke, 1961). Side effects have also been observed in "slow inactivators" taking the drug β-phenylethyl-hydrazide in the treatment of depression (Evans *et al.*, 1965). Other aspects of these problems are discussed on p. 455.

It has been demonstrated in family studies that the difference between "slow" and "rapid" inactivators is genetically determined. Two common alleles are proposed, Ac^S and Ac^R. The "slow inactivators" are homozygotes ($Ac^S Ac^S$) and the "rapid inactivators" are either heterozygotes ($Ac^S Ac^R$) or homozygotes ($Ac^R Ac^R$) for the other allele (Knight *et al.*, 1959; Evans *et al.*, 1960). There are variations between different populations as to the frequencies of "slow" and "rapid" inactivators (Schloot *et al.*, 1967; Mourant *et al.*, 1976). About 50% of Caucasians and Negroes are slow inactivators, while the frequency is about 10% in the Japanese population.

Malate Dehydrogenase (NAD Dependent) (EC 1.1.1.37)

Malate dehydrogenase (MDH) catalyses the reversible conversion of malate to oxaloacetate using NAD as co-factor. One form of MDH is found in the cytoplasm and another form is bound to the mitochondria (Christie and Judah, 1953; Shrago, 1965). The cytoplasmic or soluble MDH is referred to as S-MDH while the MDH bound to the mitochondria is referred to as M-MDH Wilkinson, 1970).

Genetic variations have been observed in both S-MDH and M-MDH. Red cell lysates have been used in population studies in the search for genetical variants of S-MDH. Extracts of leucocytes and placentae have been used for the study of M-MDH, since red cells lack mitochondria and hence M-MDH. A genetical variant of S-MDH was reported by Davidson and Cortner (1967a). As the genetical variability in S-MDH was not reflected in the M-MDH, Davidson and Cortner (1967a) concluded that S-MDH and M-MDH were under independent genetical control.

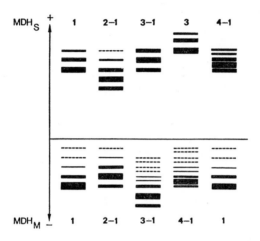

Fig. 19 Phenotypes of soluble and mitochondrial malate dehydrogenase controlled by genes at separate MDH_S and MDH_M loci

The genetical variant of S-MDH reported by Davidson and Cortner (1967a) was found in one Negro woman out of 2910 Americans tested. Of these individuals 1440 were Caucasians and 1470 Negroes. This deviating phenotype has been referred to as the "Buffalo variant". Another deviating phenotype of S-MDH but with different electrophoretic properties was reported by Blake et al. (1970a). This phenotype has been referred to as "New Guinea – 1". The occurrence of the two phenotypes in a number of different population groups has been studied and summarized by Leakey et al. (1972). They introduced a new nomenclature and referred to the deviating phenotype described by Davidson and Cortner (1967a) as S-MDH 2-1 and the "New Guinea – 1" phenotype (Blake et al., 1970a) as S-MDH 3-1 (Fig. 19). Another rare phenotype of S-MDH referred to as 4-1 has been described by Beckman and Christodoulou (1974). The S-MDH 4-1 phenotype was observed in a mixed population group from Hawaii. A phenotype with apparently the same electrophoretic mobility has been found in a study of 200 individuals from Bangladesh (Mourant *et al.*, 1976). The common and rare genes controlling the different phenotypes of S-MDH are referred to as MDH_S^1, MDH_S^2, MDH_S^3 and MDH_S^4. Of all the different population groups of the world studied so far (Beckman and Christodoulou, 1974; Mourant *et al.*, 1976) the highest frequency of the MDH_S^2 gene has been found in Ethiopians (Harrison *et al.*, 1969). The only MDH_S gene which may be polymorphic is MDH_S^3, which has been found in relatively high frequencies (2% or more) in certain tribes in New Guinea (Blake *et al.*, 1970a; Leakey *et al.*, 1972).

Deviating phenotypes of M-MDH have been reported in addition to the M-MDH 2–1 phenotype originally described by Davidson and Cortner (1967b). The two rare phenotypes (Fig. 19) have been referred to as M-MDH 3–1 and 4–1 (Beckman and Christodoulou, 1974).

The discovery of the M-MDH 2–1 type (Davidson and Cortner, 1967b) was claimed to be the first example of genetical variation in human mitochondrial enzymes, since the autosomal codominant mode of inheritance indicated that mitochondrial MDH is controlled by nuclear DNA in Man and not by mitochondrial DNA as has been proposed for some other mitochondrial enzymes (Sager, 1964).

Malate Dehydrogenase (NADP Dependent) (Malic Enzyme EC 1.1.1.40)

A polymorphism in malic enzyme (ME), viz. the mitochondrial NADPH-linked malate dehydrogenase, has been described (Cohen, 1971; Cohen and Omenn, 1972). The enzyme occurs both in a cytoplasmic (ME$_S$) and a mitochondrial form (ME$_M$). Electrophoretic separation of 154 brain

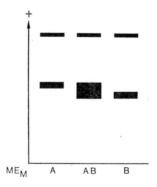

Fig. 20 Three phenotypes of mitochrondrial malic dehydrogenase ME$_M$. (After Cohen and Omenn, 1972)

extracts followed by a specific staining for the enzyme made it possible to distinguish three phenotypes. These are referred to as ME_MA, ME_MAB and ME_MB (Fig. 20). The frequencies of the two alleles ME_M^a and ME_M^b were in Caucasians 0·31 and 0·69 and in Negroes 0·18 and 0·85. One out of the 154 brain extracts examined showed a deviating isozyme pattern most probably representing a rare variant of ME_S.

Nucleoside Phosphorylase (EC 2.4.2.1)

Nucleoside phosphorylase (NP) catalyses the phosphorylytic cleavage of inosine to hypoxanthine and ribose-1-phosphate. The enzyme is widely distributed and a specially high activity of the enzyme has been found in erythrocytes (Huennekens *et al.*, 1956; Tsuboi and Hudson, 1957).

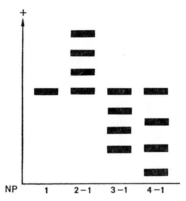

Fig. 21 The common phenotype NP 1 and three rare phenotypes of nucleoside phosphorylase. (After Edwards *et al.*, 1971)

Electrophoretic separation of haemolysates of erythrocytes reveals a rather complex system of isozymes which is further complicated by the occurrence of secondary isozymes as the result of ageing of erythrocytes (Edwards *et al.*, 1971b; Turner *et al.*, 1971). A common type NP 1 and three rare types NP 2-1, 3-1 and 4-1 (Fig. 21) have been observed by Edwards *et al.* (1971b). The rare types 3-1 and 4-1 were found in the English population and the rare 2-1 phenotype in Indian and Greek populations. As the enzyme is a trimer (Edwards *et al.*, 1971b, Edwards *et al.*, 1973) a four-banded pattern is found in heterozygotes.

The possibility of a silent allele for nucleoside phosphorylase has been proposed by Giblett *et al.* (1975). They found a young girl with absence of NP activity. This girl suffered from defective T-cell immunity and it was suggested that she was homozygous for a "silent" allele at the NP gene locus. Her parents, who are first cousins, were found to have less than half of the NP enzyme activity in their red cells. The isozyme pattern of the red cell of the parents was suggestive of "a molecular hybridization between catalytically active and inactive subunits" (Giblett *et al.*, 1975).

The enzyme NP is involved in the metabolism of purines and NP and ADA act next to each other in the sequential catabolism of purine. The presence of an inhibitory factor instead of a defect in the synthesis of ADA in SCID patients has been discussed on p. 197. A similar situation cannot be excluded in cases of defective NP activity.

Pepsinogen (EC 3.4.23)

Pepsinogen is a proenzyme or zymogen which is converted to the active enzyme pepsin by hydrochloric acid in the gut. Seven electrophoretically distinct forms of pepsinogen (Pgl-Pg7) have been demonstrated in the human gastric mucosa (Samloff, 1969; Samloff and Townes, 1969). These seven different molecular forms of pepsinogen have been divided into two groups. The five zones with the fastest migration towards the anode (Pgl-Pg5) are found in the fundus and the body of the stomach, and they are referred to as group I pepsinogens. Group II consists of Pg6 and Pg7. They occur in the pyloric antrum and the proximal duodenum.

The pepsinogens belonging to group I are always present in the urine. Group II pepsinogens are, however, rarely found in urine (Samloff and Townes, 1969). Pg5 in group I is sometimes missing in the gastric mucosa and hence also in the urine. Individuals with Pg5 present are referred to as A and those without Pg5 are classified as B.

Family studies indicate that absence of Pg5 is inherited as a simple autosomal recessive trait. The structural locus therefore has two alleles Pg^a and Pg^b, Pg^a being dominant (Samloff and Townes, 1970). Samloff and Townes (1970) proposed that another locus or other loci are involved in the determination of the other pepsinogens.

The frequency of the Pg B phenotype is 14% among American Caucasians (Samloff and Townes, 1970) and 20% among American Negroes (Samloff *et al.*, 1973; Townes and White, 1974). The Pg B phenotype has not been observed in American Chinese, Japanese and Filipino populations (Samloff *et al.*, 1973).

Bowen *et al.* (1972) have reported another phenotype with an intense staining of the Pg4 band. This new phenotype, designated Pg C may be controlled by a third allele Pg^c. Townes and White (1974) have described a variant phenotype lacking both Pg4 and Pg5, which has been called B[1], possibly determined by another allele, Pg^{b1}.

Weitkamp and Townes (1975) have reviewed the genetics of pepsinogen and claim that a total of 18 different phenotypes can be distinguished. Attempting to clarify this genetical determination, Weitkamp et al. (1975) examined the linkage relationships of Pg4 and Pg5 with other markers. They reported probably linkage of Pg5 to the HLA region, supporting the proposal mentioned above that at least some of the pepsinogens are controlled by genes at separate loci, since evidence for the linkage of Pg4 to HLA was much weaker. If only one locus is involved in determining the variation described above in Pg4 and Pg5, then Pg^a and Pg^c are codominant, and Pg^{b1} is recessive to both. However, Pc^c may not be allelic to Pg^a, Pg^b and Pg^{b1}.

Peptidase

Lewis and Harris (1967) were able to identify five different peptidases in human red blood cells by means of starch gel electrophoresis. The peptidases differed in their substrate specificity, molecular size and electrophoretic mobility and were found to be controlled by genes at separate loci (Lewis and Harris, 1967, 1969a, b). The peptidases are referred to as A, B, C, D and E. A sixth peptidase (F) was reported by Harris (1969). Another peptidase, S, was described by Rapley et al. (1971). This peptidase is not present in red cells but is found in most other tissues. Peptidase S differs from the other peptidases in its molecular size and electrophoretic mobility.

Two different methods have been used to demonstrate peptidase activity after electrophoretic separation of the different peptidases. One, described by Lewis and Harris (1967), is based on the liberation of an amino acid from the peptide. This amino acid must be susceptible to oxidative deamination by L-amino acid oxidase. This method can be used with peptides where the following amino acids are liberated: leucine, isoleucine, methionine, phenylalanine, tyrosine and tryptophan. The other method described by Rapley et al. (1971) depends on the ultraviolet absorption of liberated alanine from various peptidases. (Table 2)

Genetical variation has been observed in peptidases A and B by Lewis and Harris (1967), peptidase D (prolidase, EC 3.4.3.7) by Lewis and Harris (1969b) and peptidase C by Santachiara-Benerecetti (1970).

Peptidase A is a dipeptidase which can hydrolyse a variety of peptides. Three common peptidase A patterns, PEPA 1, PEPA 2-1 and PEPA 2 (Fig. 22), have been identified (Lewis and Harris, 1967). Peptidase A has been found to be polymorphic in a Negro population but not in Caucasians (Lewis and Harris, 1967; Lewis et al., 1968). A series of rare alleles, PEPA³,

Table 2

The relative activities of peptidases S, A, B, C, D, E and F with twenty-five dipeptides, six tripeptides, one tetrapeptide, leucyl-β-naphthylamide, leucimanide and leucyl nitroanilide. (From Rapley *et al.*, 1971 with permission)

Substrate	Peptidase						
	S	A	B	C	D	E	F
Val–Leu	++	+++	–	–	–	–	–
Val–Ala	+	+++	–	–	–	–	–
Ala–Gly	++	+++	–	–	–	–	–
Gly–Leu	–	+++	–	+	–	–	–
Leu–Gly	+	+++	–	+	–	–	–
Gly–Pho	–	++	–	+	–	–	–
Ala–Lys	+	++	–	+	–	–	–
Ala–Glu	–	++	–	+	–	–	–
Leu–Ala	++	++	–	+++	–	–	–
Gly–Ala	–	++	–	++	–	–	–
Leu–Leu	+++	++	–	++	–	+	–
Gly–Try	+	++	–	++	–	–	–
Ala–Ala	–	+	–	(±)	–	–	–
Lys–Leu	++	+	+	+++	–	+	–
Lys–Tyr	++	+	+	+++	–	+	–
Pro–Phe	(±)	++	+	+++	–	–	–
Pro–Leu	–	++	+	+++	–	–	–
Phe–Leu	+++	++	++	+++	–	+	–
Phe–Tyr	+++	++	++	+++	–	+	–
Leu–Tyr	+++	+++	++	+++	–	+	–
Ala–Tyr	+	++	+++	+	–	(±)	–
Ala–His	+	++	–	++	–	–	–
Leu–Gly–Gly	+	–	+++	–	–	(±)	–
Ala–Gly–Gly	++	–	+++	–	–	–	–
Ala–Ala–Ala	++	–	+++	–	–	+	–
Leu–Leu–Leu	+++	–	++	–	–	+	+
Leu–Gly–Phe	+	–	+	–	–	(±)	–
Tyr–Tyr–Tyr	+	–	+	–	–	(±)	+
Phe–Gly–Phe–Gly	+	–	–	–	–	++	–
Leu–Pro	–	–	–	–	++	–	–
Phe–Pro	–	–	–	–	++	–	–
Ala–Pro	–	–	–	–	++	–	–
Leu–β–Naphthylamide	(±)	–	–	–	–	+	–
Leucinamide	+	–	–	–	–	(±)	–
Leu–Nitroanilide	(±)	–	–	–	–	+	–

PEPA⁴ (Lewis and Harris, 1967) and *PEPA⁵*, *PEPA⁶* (Lewis *et al.*, 1968), have been described. The rare alleles occurred together with the common allele *PEPA¹*. The heterozygous patterns showed a hybrid enzyme since the enzyme is a dimer with free recombination between monomers (Lewis and Harris, 1969c).

Lewis *et al.* (1968) suggested that the variant polypeptide chain in the phenotype PEPA 5-1 contained a reactive sulphydryl group. Evidence was

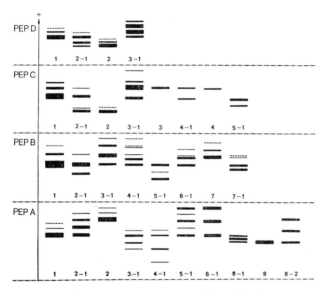

Fig. 22   Phenotypes of peptidases A, B, C and D. The drawing shows relative migration rates, the point of application is not included. (After Lewis and Harris, 1967, 1969b; Lewis *et al.*, 1968; Lewis 1973; Blake *et al.*, 1970; Santachiara-Benerecetti, 1970 and Povey *et al.*, 1972)

also presented which suggested that *PEPA⁵* coded for a peptide chain in which glutamic acid is substituted for lysine (Lewis *et al.*, 1968). The influence of seven different thiol reagents was therefore tested on the electrophoretic mobility of the PEPA 5-1 phenotype by Sinha and Hopkinson (1969). The results suggested the presence of a reactive sulphydryl group in the variant polypeptide chain coded for by the allele *PEPA⁵*.

Sinha *et al.* (1970) observed that the level of peptidase A activity varied considerably among individuals with the phenotype PEPA 1. Family studies indicated that there may exist two alleles which control the synthesis of different molecular forms of the enzyme with the same electrophoretic mobility but with different enzyme activity. Hence a *PEPA¹ strong* allele and a *PEPA¹ weak* allele could be postulated. This hypothesis is supported by the existence of variable electrophoretic patterns in the triple-banded peptidase A heterozygotes. Some heterozygotes do not show the expected 1:2:1 ratio of staining intensity (Sinha et al., 1970). A comparison of peptidase A from red blood cells with weak and strong enzyme activities respectively showed no marked difference in pH-activity curves, Michaelis constants or on CM-Sephadex columns. A slight difference in thermostability was, however, observed suggesting that peptidase A from red blood cells with weak enzyme activity was somewhat more thermostable (Sinha et al., 1971).

Later Lewis (1973) was able to demonstrate a polymorphism of peptidase A in leucocytes and other tissues. This polymorphism is associated with quantitative variation of peptidase A activity only in erythrocytes. A slightly modified method of electrophoretic separation made it possible to distinguish a new isozyme pattern in leucocyte extracts. Three common patterns, PEPA 1, PEPA 8-1 and PEPA 8 can be distinguished. Family studies indicate that they are controlled by two common alleles $PEPA^1$ and $PEPA^8$. The $PEPA^8$ allele is equivalent to the postulated $PEPA^{1\ weak}$ allele. The frequency of the $PEPA^8$ allele is 0·25 in the English population compared with 0·088 in Nigerians (Lewis, 1973).

Four different types of peptidase B were originally described by Lewis and Harris (1967) namely. PEPB 1, B 2-1, B 3-1 and B 4-1 (Fig. 22). Three additional types PEPB 5-1, B 6-1 and B 7-1 were described by Blake et al. (1970b). No hybrid enzyme has been observed in the heterozygous types. The heterozygous patterns appear as a mixture of two homozygous types (Lewis and Harris, 1967) which is to be expected if the enzyme is a monomer.

Peptidase C has been found to be polymorphic in the Babinga pygmies (Santachiara-Benerecetti, 1970). Three alleles $PEPC^1$, $PEPC^2$ and $PEPC^0$ occurred with the frequencies 0·778, 0·014 and 0·208 respectively. No clinical abnormality was observed among 11 individuals homozygous for the $PEPC^0$ allele. Additional phenotypes PEPC 3-1, PEPC 4-1, PEPC 3, PEPC 4 and PEPC 5-1 (Fig. 22) have been described by Povey et al. (1972). The PEPC 3 and 4 variants show the same electrophoretic mobility, but they can be distinguished since PEPC 4 is relatively unstable in vivo. The frequency of the $PEPC^3$ allele is about 0·0037 in Caucasians. $PEPC^4$ occurs at about 0·0075 at Caucasians and 0·08 in Negroes (Povey et al., 1972).

Three alleles have been reported at the peptidase D (prolidase) locus (Lewis and Harris, 1969b). The phenotypes observed are referred to as: PEPD 1, D 2-1, D 2 and D 3-1. The heterozygous types show a hybrid enzyme. The frequency of the $PEPD^2$ allele has been found to be 0·011 in Caucasians. 0·024 in Negroes and 0·0085 in Indians (Lewis and Harris,

1969b). A deficiency of the enzyme has been reported in a boy with iminopeptiduria (Powell *et al.*, 1974). Family studies suggest an autosomal recessive inheritance.

Peptidase S has been observed only in tissues other than red blood cells (Rapley *et al.*, 1971). The enzyme has a slower anodal mobility than the other peptidases. No individual variations have been found so far.

Population data on the peptidases A, B, C and D are scarce so far and no definite conclusions can be drawn concerning the extent of polymorphism in different populations, but it seems as if peptidases A and C are polymorphic only in Negro populations, while peptidase D is polymorphic in Caucasian, Negro and Oriental populations.

Phosphoglucomutase (EC 2.7.5.1.)

Phosphoglucomutase (PGM) is an important enzyme in glycogen mobilization and is also found in most tissues. The enzyme catalyses the interconversion of glucose-1-phosphate and glucose-6-phosphate. The enzyme is controlled by alleles at three loci, designated PGM_1, PGM_2 and PGM_3 (Hopkinson and Harris, 1965, 1966 and 1968).

The existence of a polymorphism in phosphoglucomutase was first described by Spencer *et al.* (1964b). A total of nine zones with enzyme activity can be demonstrated after separation of haemolysates on starch gel electrophoresis with subsequent staining of the gel. These zones are referred to as a, b, c, d, e, f, g, h and i in order of mobility from the cathode towards the anode. Three of the fast moving zones e, f and g are found in most individual samples tested. The zones a, b, c and d vary between individuals. Three different patterns were originally identified by Spencer *et al.* (1964b) and referred to as PGM_1 1, PGM_1 2-1 and PGM_1 2 (Fig. 23). Family studies imply that these types are controlled by two common autosomal allelic genes PGM_1^1 and PGM_1^2 at the PGM_1 locus (Spencer *et al.*, 1964b; Hopkinson and Harris, 1969; Monn, 1969). The homozygotes PGM_1 1 and PGM_1 2 show the zones a+b and c+d respectively. The heterozygotes appears as a mixture of the two homozygous types, viz. they show all four zones a, b, c and d.

In addition to the three common PGM_1 phenotypes many rare phenotypes have been described. The electrophoretic patterns indicate that these phenotypes are heterozygotes between one of the common alleles and a rare allele at the same locus. Seven rare alleles, PGM_1^3 to PGM_1^7 (Hopkinson and Harris, 1965, 1966) and PGM_1^8 (Harris *et al.*, 1968; Palmarino *et al.*, 1975) have so far been found. Additional alleles at the PGM_1 locus appear, however, to exist. These are referred to as $PGM_1^{3\ Okinawa}$ and $PGM_1^{5\ Japan}$,

234   G. BECKMAN

$PGM_1^{6\ African}$, $PGM_1^{6\ Japan}$, $PGM_1^{6\ Mal}$, and $PGM_1^{6\ Kadar}$ (Blake and Omoto, 1975). Brinkman et al. (1972) have presented evidence for the occurrence of a silent allele.

The more anodically migrating zones e, f and g were found to be controlled by genes at a separate locus called PGM_2 (Hopkinson and Harris, 1965, 1966). A number of rare variants were observed, all representing heterozygous combinations between the common allele PGM_2^1 and one of the following rare alleles: PGM_2^2 and PGM_2^3 (Hopkinson and Harris, 1966),

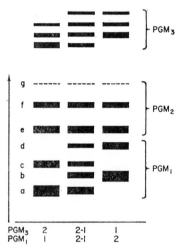

Fig. 23   Three common phenotypes of phosphoglucomutase PGM₁ and PGM₃

PGM_2^4, and PGM_2^5 (Parrington et al., 1968), $PGM_2^{5\ Trio}$ (Geerdink et al., 1974). A sixth allele PGM_2^6 or $PGM_2^{6\ Pyg}$ was reported by Santachiara-Benerecetti and Modiano (1969). Later Santachiara-Benerecetti et al. (1972a) reported a PGM_2 variant in an Indian population. This new allele is referred to as $PGM_2^{6\ Ind}$ to differentiate the allele from $PGM_2^{6\ Pyg}$ (Blake and Omoto, 1975). Further alleles at the PGM_2 locus are PGM_2^8 (Santachiara-Benerecetti et al. (1972b), PGM_2^9 (Woodfield et al., 1974) and PGM_2^{10} (Woodfield et al., 1974; Blake and Omoto, 1975).

A third locus PGM_3 has been found to control some additional zones with phosphoglucomutase activity which are present only in tissues other than red cells (Hopkinson and Harris, 1968). Two alleles at the PGM_3 locus control three different enzyme patterns PGM₃ 1, PGM₃ 2-1 and PGM₃ 2 (Fig. 23). A third phenotype PGM₃ 3-1 has been reported by van Wierst et al. (1973).

McAlpine et al. (1970a) reported that 85 to 95% of the total PGM activity was attributable to the PGM_1 locus in most tissues with the exception of fibro blasts and red cells. The PGM_2 and PGM_3 loci were responsible for only 2 to 15 % and 1 to 2 % of the activity respectively. In fibroblasts the PGM_3 locus was responsible for 7 %, while in red cells and muscle this locus contributed nothing to the activity. The PGM_1 and PGM_2 loci contributed apparently equal parts of the PGM activity in red cells. A kinetic study of the isozymes, determined by the three loci PGM_1, PGM_2 and PGM_3, has been done using glucose-1-phosphate and ribose-1-phosphate as substrates together with varying concentrations of the co-enzyme glucose-1,6-diphosphate (Quick et al., 1972, 1974). The results show that the PGM_1 isozymes are more effective in catalysing the phosphoglucomutase reaction than the isozymes controlled by the PGM_2 and PGM_3 loci. The PGM_2 isozymes functioned both as phosphoglucomutases and phosphoribomutases depending on the concentration of the co-enzyme glucose-1,6-diphosphate. The PGM_3 isozyme showed poor phosphoglucomutase activity compared to the PGM_1 and PGM_2 isozymes and the primary role of the PGM_3 isozymes remains obscure.

Modiano et al. (1970a) have observed no significant differences in the PGM activity in red cells of persons of PGM_1 1, PGM_1 2-1 and PGM_1 2 phenotypes. This finding is complicated by the expression of the PGM_2 locus in red cells and a final assessment of the activities attributable to the different alleles at the two loci must await separation of the isozymes involved.

The thermostability differs between the enzyme products of the three PGM loci in the following order $PGM_2 > PGM_1 > PGM_3$ but no inter- or intra-allelic thermostability differences were observed (McAlpine et al., 1970b). Each of the three gene loci control a set of isozymes which are all monomers. Within each set of isozymes the one with the most cathodic mobility represents the primary form synthesized. The molecular weights of the isozymes controlled by PGM_1, PGM_2 and PGM_3 loci are 51 000, 61 000 and 53 000 respectively (McAlpine et al., 1970c). Within each set of the isozymes a more anodic mobility represent secondary changes which are associated with the lifespan of the particular cell (Fisher and Harris, 1972). Various sulphydryl reagents have been shown to effect the mobility and reactivity of the different isozymes (Fisher and Harris, 1972; Greene and Dawson, 1973).

The population data gathered on phosphoglucomutase are vast and a number of review articles have been written through the years (Giblett, 1969; Hopkinson and Harris, 1969; Bhasin and Fuhrmann, 1972; Beckman, 1972; Blake and Omoto, 1975; Mourant et al., 1976). The PGM_1^1 allele occurs in 65 to 80 % of most populations. The PGM_1^2 allele occurs in relatively high frequencies (21 to 25 %) in different Lappish groups

(Beckman *et al.*, 1971). Considering the European population as a whole there appears to be an increase in the frequency of the PGM_1^2 allele from west to east (Mourant *et al.*, 1976). Bonne *et al.* (1970) found a rather high frequency of the PGM_1^2 allele in a Jewish population but this high frequency was not found in other Jewish populations (Mourant *et al.*, 1976). The lowest PGM_1^2 frequencies (10% or below) have been found in Koreans (Bajatzadeh *et al.*, 1969), North American Indians (Scott *et al.*, 1966; Alfred *et al.*, 1969, 1970), Chipaya in Bolivia (Quilici *et al.*, 1970), Yanomama in Venezuela (Arends *et al.*, 1967), Aborigines in Australia (Kirk *et al.*, 1971) and in a population of the Western Highlands of New Guinea (Sinnett *et al.*, 1970). A low frequency (10%) is also noted for the Ainu population of Northern Japan (Giblett, 1967). Otherwise the PGM_1^2 gene frequency is about 20 to 30% in Asian populations (Blake and Omoto, 1975).

Many of what can be regarded as rare alleles at the PGM_1 locus appear to show quite marked geographic or ethnic frequency variations. One example is the PGM_1^3 allele which occurs in about 10% in certain isolated areas of New Guinea and the Western Caroline Islands. Another example is the PGM_1^7 allele, also polymorphic on the Western Caroline Islands where it occurs in 4 to 8% of the population (Blake *et al.*, 1973).

The frequency of the PGM_3^2 allele ranges from 0·24 to 0·27 in European populations (Hopkinson and Harris, 1968; Lamm, 1969; Monn and Gjønnaess, 1971; Herzog and Drdová, 1971) which is low compared to 0·66 for a Nigerian population (Hopkinson and Harris, 1968). The lowest frequency of the PGM_3^2 allele, 0·19, has been found in Japanese (Ishimoto, 1969). The frequency of the PGM_3^2 allele in Australian Aborigines and natives of Papua New Guinea is 0·41 and 0·48 respectively (van Wierst *et al.*, 1973).

6-Phosphogluconate Dehydrogenase (EC 1.1.1.44)

The oxidative decarboxylation of 6-phosphogluconate to ribulose-5-phosphate in the hexosemonophosphate (HMP) shunt is catalysed by the enzyme 6-phosphogluconate dehydrogenase (6-PGD) an enzyme which occurs in most tissues.

Inherited variation in 6-PGD was first described by Fildes and Parr (1963). The nomenclature on 6-PGD is somewhat confusing. The nomenclature introduced by Bowman *et al.* (1966) was used by Beckman (1972). The nomenclature introduced by Davidson (1967) and Parr and Fitch (1967) appears, however, to be more widely accepted and will therefore be used.

Fig. 24 shows the common electrophoretic patterns of 6-PGD observed in haemolysates and white blood cells. The most common phenotype A consists of a single zone a. The phenotypes C and AC both show three zones, a, b and c, but with different intensities (Parr, 1966; Bowman *et al.*, 1966). Extracts of white cells and of other tissues show the same electrophoretic pattern as that observed in the red cells of the individual. The relative staining intensities of the different zones may, however, vary between tissues (Parr, 1966; Parr and Fitch, 1967). On prolonged incubation, additional faster-moving components appear (Fildes and Parr, 1964; Bowman *et al.*, 1966; Giblett, 1967). These additional components have not been observed in extracts from other tissues including white cells (Parr, 1966; Davidson, 1967).

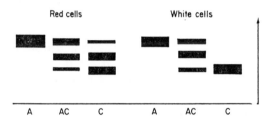

Fig. 24   Three common phenotypes of 6-phosphogluconate dehydrogenase in red and white blood cells

Family studies have shown that the phenotypes A and C represent the homozygous types controlled by two co-dominant autosomal allelic genes, PGD^A and PGD^C respectively, and that AC represents the heterozygous pattern (Parr, 1966; Bowman *et al.*, 1966; Parr and Fitch, 1967; Carter *et al.*, 1968). The enzyme is a dimer with free recombination between monomers. Hence the heterozygote shows as expected a hybrid enzyme, viz. a combination of two different polypeptide subunits (Fildes and Parr, 1964; Parr, 1966; Parr and Fitch, 1967). The explanation as to the existence of three zones in the AC phenotype is not conclusive (Giblett, 1969) but it seems from data presented by Ajmar *et al.* (1968) that a product of the reaction between NADP and stromal NADPase might be able to modify both the structure and the activity of 6-PGD. The PGD^C allele is associated with a slightly reduced enzyme activity (Parr, 1966).

A number of rare variants of 6-PGD have been described. Some of the variants are not recognizable by electrophoresis but only as a decrease in the enzyme activity.

Some of the rare electrophoretic 6-PGD variants are as follows: "Richmond" (Parr, 1966; Davidson, 1967), "Hackney" (Parr, 1966), "Friendship" (Davidson, 1967), "Elcho" (Blake and Kirk, 1969),

"Freiburg" (Tariverdian *et al.*, 1970b), "Singapore" (Blake *et al.*, 1973), "Wancoat", "Canberra", "Kadar", "Caspian", "Bombay", "Natal" (Blake *et al.*, 1974), and "Oshakati" (Jenkins and Nurse, 1974).

The following phenotypic variants have been found to be associated with altered enzyme activity: the "Ilford" variant (Parr and Fitch, 1964, 1967) which is most probably due to a "silent" allele PGD^O resulting in a phenotype with 50% of the normal activity but with the same electrophoretic mobility as the A type (supposed genotype PGD^A/PGD^O); the Newham variant with 40 to 50% of normal activity (Parr and Fitch, 1964, 1967) represents the result of a combination of the PGD^O allele with the PGD^C allele and the "Whitechapel" variant (Parr and Fitch, 1967), which is associated with a reduction of the enzyme activity to 1 to 5% of normal value. The "Whitechapel" variant represents homozygozity for the PGD^W allele, an allele which controls a product which is quite unstable in the absence of the PGD^A gene (Parr and Fitch, 1967). The heterozygote PGD^A PGD^W is referred to as the "Dalston" variant and results in a reduction of the enzyme activity to 75% (Parr and Fitch, 1967). Data presented by Dern *et al.* (1966) suggest that genes at other loci may influence the amount of the PGD gene product. Partial deficiency of 6PGD, associated with haemolytic anemia has been found in two individuals (Scialom *et al.*, 1966; Lauseckar *et al.*, 1965).

The population data so far gathered show that PGD^A is the most common allele in all populations tested so far (Tills *et al.*, 1970, 1971; Beckman, 1972 and Mourant *et al.*, 1976). The PGD^C allele occurs at between 1 to 5% in most European populations with one clear exception of the Swedish Lapps where the frequency of the gene is about 13% (Beckman *et al.*, 1971). The highest frequencies of the PGD^C allele (14 to 25%) occur in certain rather isolated population groups such as the Bhutan in India (Tills *et al.*, 1970, 1971) and various African subpopulations (Gordon *et al.*, 1967; Harrison *et al.*, 1969; Jenkins and Nurse, 1974). It has been hypothesized that the PGD^C gene frequency is higher in populations living at a high altitude (Mourant *et al.*, 1976). Quite a significant difference in the frequency of the PGD^C gene has been observed among people living at different altitudes in Ethiopia (Harrison *et al.*, 1969). In accordance with such a hypothesis is the high frequency of the PGD^C allele in the Bhutan population in India. The rather high frequency of the PGD^C allele in the Swedish Lapps does not support a hypothesis concerning selective advantage of the PGD^C allele at high altitudes (Beckman *et al.*, 1971). The lowest frequencies of the PGD^C gene occur in American Indians where the gene is lacking in most groups (Mourant *et al.*, 1976).

Three of the PGD alleles which are rare in most populations occur in frequencies high enough to consider the gene as polymorphic in certain isolated areas. The PGD^{Elcho} gene occurs at about 1·5% in certain groups of Australian Aborigines (Blake and Kirk, 1969), the PGD^{R} allele at about 1% in the Caribs of Dominica (Harvey et al., 1969) and at 1 to 4% in certain African subpopulations (Jenkins and Nurse, 1974) and finally PGD^{Kadar} at 4% in the tribal Kadar population (Blake et al., 1974).

Phosphoglycerate Kinase (EC 2.7.2.3)

Phosphoglycerate kinase (PGK) catalyses the reversible conversion of 1, 3-diphosphoglycerate to 3-phosphoglycerate.

Chen et al. (1971) were able to identify three different isozyme patterns after electrophoretic separation of haemolysates followed by staining of the gel for phosphoglycerate kinase according to the method described by Beutler (1969c). The different patterns were referred to as PGK 1, PGK 2-1 and PGK 2 (Fig. 25). The data obtained from family studies were in

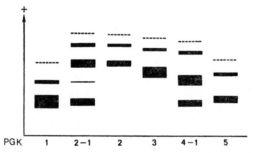

Fig. 25   Phenotypes of phosphoglycerate kinase. (After Chen et al., 1971)

accordance with the results of Valentine et al. (1969) namely that the locus for PGK is on the X-chromosome. The X-linkage has been confirmed by Khan et al. (1971). Additional variants of PGK have been described. The phenotypes observed (Fig. 25) have been referred to as PGK 3 (Chen and Giblett, 1972), PGK 4-1 (Chen and Giblett, 1972; Omoto and Blake, 1972), PGK 4 (Omoto and Blake, 1972) and PGK 5 (Chen and Giblett, 1972). Note that PGK 3, 4 and 5 are not homozygotes, but hemizygotes, viz. males.

The activity attributable to the PGK^{2} allele is not different from that of the PGK^{1} allele (Chen et al., 1971). Dosage compensation in the female appears to occur for PGK in the same way as for glucose-6-phosphate dehydrogenase (Omoto and Blake, 1972). No genetical variants were observed in population surveys of Europeans, Negroes, Filipinos, Chinese and Asiatic Indians (Beutler, 1969c; Chen et al., 1971; Blake et al., 1972). The PGK 2

240 G. BECKMAN

variant was first observed by Chen *et al.* (1971) in some areas of New Guinea and Samoa. The frequency of the PGK^2 gene was estimated to be 0·014 in certain areas of New Guinea. Additional population data have been published by Chen and Giblett (1972) showing that PGK variants are not restricted to people of Oceana but can be found also in other populations but in very low frequencies. Omoto and Blake (1972) found in a population study of certain areas in Asia and the Pacific that the variant phenotypes of PGK were restricted to certain quite well-defined areas. Thus in the Micronesian population (Ulithi Atoll) the otherwise rare PGK^2 gene occurred at a frequency of 8%. The PGK^4 gene was observed in one area in Irian Jaya (Mappi and Digul Rivers) and in two areas on New Guinea (western and eastern highlands). The frequency of the PGK^4 allele in these areas was about 4%.

Associations between haemolytic anaemia, mental disorder and hereditary deficiency of phosphoglycerate kinase have been observed in patients in several unrelated families (Kraus *et al.*, 1968; Valentine *et al.*, 1969; Hjelm and Wadam, 1970; Mazza *et al.*, 1970; Carties *et al.*, 1971; Miwa *et al.*, 1972; Konrad *et al.*, 1973; Arese *et al.*, 1973). Yoshida and Miwa (1974) studied such a patient and found a residual enzyme activity of 5%. Characterization of this enzyme revealed among other things a deviating electrophoretic mobility compared with other known variants of the enzyme. The residual enzyme activity represents most probably a variant of the enzyme controlled by an allele at the PGK locus. The enzyme variant has been designated PGK Matsue (Yoshida and Miwa, 1974). The enzyme activity decreased with the age of the erythrocytes indicating an increased rate of degradation for the variant enzyme. The variant enzyme also showed a lower affinity for the substrate.

Phosphoglycerate Mutase (EC 2.7.5.3)

Phosphoglycerate mutase, PGAM, catalyses the reversible reaction of 3-phosphoglycerate to 2-phosphoglycerate in the glycolytic pathway. The enzyme is widely distributed in tissues and the isozyme patterns have been described by Chen *et al.* (1974b), Omenn and Cheung (1974) and Kamel *et al.* (1975). Chen *et al.* (1974b) found two variants out of 3104 blood samples examined. The common type is referred to as PGAM 1 and the rare types as PGAM 2-1 and PGAM 3-1. The 2-1 phenotype was found to segregate in a Caucasian family. The 3-1 phenotype was found in a Congolese male.

Placental Alkaline Phosphatase (EC 3.1.3.1)

Alkaline phosphatase in Man occurs in a number of different molecular forms. Placental alkaline phosphatase (PL) is distinct from other human alkaline phosphatases in its electrophoretic mobility (Boyer, 1961, 1963), immunological specificity (Boyer, 1963; Sussman et al., 1968) and heat-stability (McMaster et al., 1964), among other factors. It appears to be coded for by a different gene from that for liver alkaline phosphatase (Badger and Sussman, 1976).

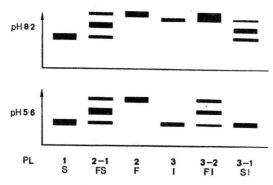

Fig. 26   Six common phenotypes of placental alkaline phosphatase examined at two different pH's. New (numbers) and old (letters) nomenclature

The existence of a polymorphism in placental alkaline phosphatase was first proposed by Boyer (1961) and verified by Robson and Harris (1965) and Beckman et al. (1966). Robson and Harris (1965, 1967) reported that most placentae could be classified into six common phenotypes when the electrophoretic separation was performed at two different pH's, 8·6 and 6·0. The common phenotypes were called S, FS, F, SI, FI and I and are according to the new nomenclature introduced by Donald and Robson (1974) called PL 1, 2-1, 2, 3-1, 2-3 and 3 respectively (Fig. 26). The study of twin placentae (Robson and Harris, 1965, 1967) shows that the six phenotypes are controlled by three common alleles PL^1, (PL^s_i), PL^2 (PL^f) and PL^3 (PL^i). The electrophoretic mobility of the PL3 variant is indistinguishable from that of the PL1 variant when examined at an acid pH. The PL 2-1, 3-1 and 2-3 phenotypes of placental alkaline phosphatase all show a triple-banded electrophoretic pattern. The zone of intermediate mobility represents a hybrid enzyme suggesting that placental alkaline phosphatase is a dimer and that the hybrid enzyme is formed through a process of random combination of monomers (Robson and Harris, 1965;

Beckman *et al.*, 1966). The fact that placental alkaline phosphatase is a dimer has been verified by Gottlieb and Sussman (1968). They were able to separate the enzyme from placental extracts into two monomers each having the same molecular weight.

The different phenotypes of alkaline phosphatase present in placental extracts can also be visualized after electrophoresis of serum from women during the last trimester of pregnancy (Boyer, 1961; Beckman *et al.*, 1966). Beckman (1970a) examined the relation between placental alkaline phosphatase phenotype and enzyme activity in pregnancy sera and placental extracts from a series of about 400 deliveries. The results from electrophoretic examinations and quantitative measurements of the enzyme in pregnancy sera and placental extracts indicated that the activity attributable to the PL^3 allele was about half that of the PL^1 and PL^2 alleles. Thomas and Harris (1971) have shown that of the six different phenotypes the PL 3 type is the least thermostable. This difference in thermostability might explain the difference in activity *in vivo* attributable to the PL^3 allele described by Beckman (1970a).

The genetical diversity of placental alkaline phosphatase is very large (Beckman, 1970b; Donald and Robson, 1974). The discrimination of some of the rare phenotypes is, however, highly dependent on the resolution of the electrophoretic techniques. As reference samples have not been exchanged for comparison of all the different phenotypes described in the literature it is not possible to give any exact figure as to the total number of different alleles involved. Donald and Robson (1974) identified 48 phenotypes of placental alkaline phosphatase in a survey of 5000 placentae of Caucasian, Negro and Asiatic Indian origin. They proposed that the phenotypes were determined by 18 different alleles at the PL locus. Hence approximately 3% of all placentae investigated show a rare phenotype.

The occurrence of a "null" allele PL^0 has been proposed by Donald and Robson (1974). The proposal is based on an analysis of collected population data from different investigators. In most population studies a small excess of homozygotes and a corresponding deficiency of heterozygotes are observed. The occurrence of a "null" allele might explain the lack of placental alkaline phosphatase observed in a dizygotic twin pair by Beckman *et al.* (1967). The children were both found to display dysostosis craniofacialis (Crouzon's syndrome).

Three common alleles PL^1, PL^2 and PL^3 have been found in various ethnic groups (Beckman, 1972; Beck and Ananthakrishnan, 1974). The PL^2 allele has its highest frequencies (24 to 34%) among Caucasians, while considerably lower frequencies are found among Asiatic mongoloid groups (2 to 8%) and African Negroes (2 to 5%). The PL^3 allele is relatively common among Asiatic mongoloid groups, especially Chinese and Japanese

(20 to 24%), but lower among Caucasians and Negroes (4 to 9%). The PL^1 allele is the most common allele in all ethnic groups so far studied. The series of rare alleles described have mostly shown frequencies well below 1%. The only exception is the PL^4 allele which occurs in 2·9% among Indians and in 1% among Chinese according to Robson and Harris (1967).

The function of placental alkaline phosphatase is unknown. However, the marked increase of alkaline phosphatase activity from the last weeks of the third trimester of the pregnancy to term and the occurrence of the enzyme in the serum of the pregnant woman (Beckman et al., 1966) suggest that the enzyme must have some important physiological role. It has been suggested that the PL^2 allele may have a protective effect in genetical incompatibility. Bottini et al. (1972a) found a significant increase in the frequency of this allele in infants incompatible with their mother in the ABO and Rh blood group systems. The relationships between the PL^2 allele and incompatibility has also been observed on the population level (Bottini et al., 1972b; Ananthakrishnan et al., 1974). Genetical incompatibility in the PL system itself does not seem to influence the maternal serum level of placental alkaline phosphatase (Beckman and Beckman, 1976a).

Possible prenatal selection against the PL^2 gene has been reported by Beckman et al. (1972). A rare type of placental alkaline phosphatase with peculiar electrophoretic properties referred to as the D-variant has been studied by Beckman and Beckman (1968) and Beckman (1970b). A significant increase of abortions in previous pregnancies was found in a series of deliveries from Sweden, where the placenta showed the D_2-variant (referred to earlier as the D-variant) (Beckman and Beckman, 1975b). Also among deliveries after various complications during pregnancy the D_2-variant was found in an increased frequency compared to a series of unselected, consecutive deliveries.

In 1973 Inglis et al. claimed that the D-variant (D_2) occurred in a frequency of 40 to 50% among cancer patients displaying ectopic heat-stable alkaline phosphatase, viz. "the Regan izozyme", a finding which is highly remarkable considering that the D-variant is controlled by a gene which in all populations studied so far is rare (0·001 to 0·007). A re-examination of their data by Beckman and Beckman (1975b) suggests that the D-variant is not associated with cancer. Furthermore the identity between "the Regan isozyme" and placental alkaline phosphatase must still be regarded as an open question. Genetical, immunological and clinical aspects of placental alkaline phosphatase have been reviewed by Beckman and Beckman (1976a).

Pyridoxine Kinase (EC 2.7.1.35)

Pyridoxine kinase (PNK) catalyses the phosphorylation of pyridoxine to pyridoxine phosphate. Chern and Beutler (1974, 1975) found a decreased activity of the enzyme in Afro-Americans. Further studies (Chern and Beutler, 1976) have shown that the mean PNK activity is 40% lower among Afro-Americans. Two alleles were postulated based on population and family studies. The alleles are referred to as PNK^H and PNK^L. No electrophoretic differences could be found between the three genotypes. The low activity was found to be associated with an increased degree of degradation of the enzyme in red cells during ageing of the cell.

Pyruvate Kinase (EC 2.7.1.40)

Pyruvate kinase catalyses the conversion of phosphoenolpyruvate to pyruvate. The enzyme is controlled by three different gene loci and the products of these loci are referred to as PK_L, PK_{M1} and PK_{M2} (Bigley et al., 1968; Blume et al., 1968a; Hopkinson et al., 1976). Studies of pyruvate kinase isozymes, especially in different neoplastic and foetal tissues, have been done by Balinsky et al. (1973) and Kamel and Schwartz-fischer (1975).

The PK_L isozyme occurs only in erythrocytes and liver. A deficiency in erythrocyte pyruvate kinase, associated with congenital non-spherocytic anaemia, was first described by Valentine et al. (1961). Since then a large number of cases have been described (Tanaka and Paglia, 1971; Kahn et al., 1975). The severity of the haemolytic disease found varies from a life threatening anaemia to a rather well-compensated condition. Tanaka and Paglia (1971) and Kahn et al. (1975) proposed that this variable severity most probably was due to the occurrence of different deficiency mutants.

The genetical heterogeneity of PK has mainly been studied through kinetic analysis. Thin layer polyacrylamide gel electrophoresis was used by Imamura et al. (1973) and Nakashima et al. (1974) to identify variants of pyruvate kinase. Miwa et al. (1975) using a combination of different methods were able to identify four different PK variants in a series of 200 individuals. Some variants did not express altered kinetic properties but could be identified by their altered electrophoretic mobility.

Pyruvate kinase deficiencies have been reported from various parts of the world but appear to be most common among North Europeans or persons of North European origin (Tanaka, 1969; Tanaka and Paglia, 1971). An exceptionally high frequency of pyruvate kinase deficiency has been reported in an Amish isolate (Bowman et al., 1965). The frequency of pyruvate kinase deficiency is difficult to estimate but might be around 1% among individuals of North European origin. The number of different alleles responsible for this deficiency and their frequencies are not known.

Serum Cholinesterase (EC 3.1.1.8)

Serum cholinesterase, also known as pseudocholinesterase, differs in its wider substrate specificity from the enzyme acetylcholinesterase which is present in nervous tissues and red blood cells. Its normal function is not precisely known. Both quantitative and qualitative variations of serum cholinesterase have been reported controlled by genes at two separate loci E_1 and E_2. A number of alleles at the E_1 locus have been detected by means of quantitative determinations in combination with different inhibitors. The alleles are referred to as E_1^a, E_1^u, E_1^s, E_1^f and $E_1^{n\text{-}butyl\ alcohol}$ according to the nomenclature by Motulsky (1964). Recently Scott (1973) has shown that the E_1^s allele is most probably two alleles, one associated with a total lack of the enzyme and the other with a very low level of enzyme activity. The enzyme variants controlled by genes at the second locus, E_2, have been demonstrated by means of starch gel electrophoresis (Harris et al., 1962).

Bourne et al. (1952) and Evans et al. (1952) showed that individuals sensitive to the muscle relaxant suxamethonium had an unusually low level of serum cholinesterase. About one out of every 2000 Europeans seemed sensitive to this drug, viz. a normal dose of the drug resulting in an extremely prolonged muscular paralysis and respiratory apnoea. Relatives of these individuals were found to have reduced levels of serum cholinesterase activity (Lehmann and Ryan, 1956). These results suggested that individuals with a low level of enzyme activity were homozygotes for a rare or "atypical" allele which was referred to as E_1^a, while individuals with a somewhat decreased level of enzyme activity were heterozygotes $E_1^u/E_2^a,E_1^u$ being the normal or "usual" allele. The difference in enzyme activity was found to be due to an altered catalytic property of the enzyme (Kalow and Genest, 1957; Kalow and Davies, 1958; Davies et al., 1960; Harris and Whittaker, 1961). The use of various inhibitors made it possible to divide individuals into different groups according to the presence of typical or atypical serum cholinesterase. Kalow and Genest (1957) developed a simple test using dibucaine as an inhibitor. The degree of inhibition in per cent was called the dibucaine number (DN). The mean DN value was 80 ± 2 for the normal or "usual" type of the enzyme and 20 $\pm$ 4 for the "atypical" enzyme. A third group was also observed with a mean DN value of 62 $\pm$ 4. Individuals with intermediate dibucaine numbers represented heterozygotes, the other two groups were homozygotes. The heterozygotes were not particularly sensitive to suxamethonium or similar drugs. The frequency of the E_1^a allele varies between 1 and 3% in most European populations but appears to be very rare or absent in Oriental, African and American Indian populations (Beckman, 1972; Whittaker and Reys, 1975; Mourant et al., 1976).

The third allele, the "silent" allele E_1^s, was postulated as a result of family studies of the inheritance of the "atypical" allele E_1^a (Kalow and Staron, 1957; Harris et al., 1960; Lidell et al., 1962; Simpson and Kalow, 1964; Dietz et al., 1965; Szeinberg et al., 1966). Individuals homozygous for the E_1^s allele appear to lack cholinesterase and were as expected extremely sensitive to suxamethonium (Lidell et al., 1962; Doenicke et al., 1963; Hodgkin et al., 1965; Goedde et al., 1965). The phenotypic expression of the E_1^s allele was soon found, however, to be quite heterogeneous as the enzyme deficiency was not complete in all individuals (Altland and Goedde, 1970). The occurrence of two allelic genes associated with cholinesterase deficiency is according to Scott (1973) the most plausible explanation for the observed heterogeneity.

The E_1^s allele is apparently very rare, about 1/100 000 (Hodgkin et al., 1965; Simpson, 1966) in the populations so far studied (Mourant et al., 1976). There is, however, one exception. In the Eskimos of Western Alaska, the frequency of the E_1^s allele is about 11% (Gutsche et al., 1967; Scott et al., 1970). Scott (1973) found the frequency of individuals heterozygous for two deficiency genes to be 1·2% in the Eskimo population of Western Alaska. The E_1^a allele has not been observed in Eskimos (Vergnes and Quilici, 1970).

A fourth "fluoride resistant" allele was discovered by Harris and Whittaker (1961). They found that serum cholinesterase was inhibited by low concentrations of sodium fluoride. Individuals could be divided into three groups: "usual", "intermediate" and "atypical" and there was a close correspondence between this classification and the one based on dibucaine numbers with some exceptions. Family studies indicated the existence of the fourth allele E_1^f (Harris and Whittaker, 1962; Lidell et al., 1963; Whittaker, 1967). A combination of the dibucaine and fluoride inhibition tests was necessary in order to distinguish the new phenotypes. The occurrence of the E_1^f allele appears to be about 1 to 2% in most populations (Mourant et al., 1976).

Another rare allele at the E_1 locus was reported by Whittaker (1968) who used n-butyl alcohol to determine the "alcohol number".

Four different zones of enzyme activity can be distinguished after separation of serum cholinesterase on starch gel electrophoresis. These zones controlled by the E_1 locus are known as C_1, C_2, C_3 and C_4 in order of mobility from the anode towards the cathode. The C_4 component represents most of the enzyme activity in serum.

A fifth component, cholinesterase C_5 (Fig. 27), controlled by the allele E_2^+ at the E_2 locus has been observed (Harris *et al.*, 1962). Autosomal dominant inheritance is suggested for the C_5 trait (Harris *et al.*, 1963a). Individuals with the C_5 component are either heterozygotes or homozygotes and have on the average 25% higher levels of enzyme activity compared

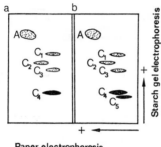

Fig. 27 Esterase patterns in two-dimensional electrophoresis. An albumin-associated esterase, C_1, C_2, C_3, C_4 and C_5 cholinesterases. (After Harris *et al.*, 1962)

to those without this component (Harris *et al.*, 1963b). The C_5-trait occurs at about 10% in most populations (Beckman, 1972; Mourant *et al.*, 1976). Rare types have been described by Ashton and Simpson (1966), Van Ros and Druet (1966), Neitlich (1966) and Gallango and Arends (1969).

The genetical control of serum cholinesterase is very complex and great caution is recommended when determining different cholinesterase phenotypes.

Superoxide Dismutase (EC 1.15.1.1)

The enzyme superoxide dismutase (SOD) is identical to the enzyme described by Brewer (1967) and referred to as indophenol oxidase (Beckman, 1973; Lippitt and Fridovich, 1973). A number of other synonymous names have also been used (Beckman and Beckman, 1975). Superoxide dismutase catalyses the dismutation of O_2 radicals to yield hydrogen peroxide and oxygen (McCord and Fridovich, 1968, 1969) according to the following reaction:

$$0^{\cdot}_2 + 0^{\cdot}_2 + 2H^+ \longrightarrow O_2 + H_2O_2$$

Two isozymes of superoxide dismutase, SOD A (cytoplasmic) and SOD B (mitochondrial) have been found in extracts of human organs, tissues and cell cultures (Beckman et al., 1973b). Isozymes A and B differ in immunological specificity (Beckman and Holm, 1975). The gene locus for SOD A is on chromosome 6 and for SOD B on chromosome 21 (Ruddle, 1973).

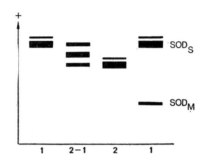

Fig. 28   Three different phenotypes of soluble superoxide dismutase (SOD$_S$). The sample to the right shows also mitochondrial superoxide dismutase (SOD$_M$) which is non-polymorphic

Three different phenotypes of the soluble form have been found (Fig. 28). These are referred to as SOD$_A$ 1, SOD$_A$ 2-1 and SOD$_A$ 2 and the two genes which control the three phenotypes are SOD_A^1 and SOD_A^2 (Beckman, 1973). Formal genetical studies verify that the three phenotypes are controlled by two autosomal alleles (Beckman et al., 1975). No variants of SOD$_B$ have so far been observed (Beckman and Beckman, 1975a). The population data available (Beckman and Beckman, 1975a) show that the frequency of the SOD_A^2 allele is low in most parts of the world with few exceptions (Beckman and Beckman, 1975b; Kirk, 1975). The frequency of SOD_A^2 is about 2·5% in the borderline area between Sweden and Finland (Beckman, 1973; Beckman and Pakarinen, 1973). The same frequency is found on the Orkney Islands which are known to have a high frequency of Scandinavian settlers (Welch and Mears, 1972; Welch et al., 1973). Founder effect together with genetic drift appear to provide the most plausible explanation for the relatively high frequency of SOD_A^2 in these rather isolated populations.

The physiological function of the enzyme, which appears to be to protect organisms metabolizing oxygen against the potentially deleterious effect of superoxide free radicals (McCord et al., 1971; Fridovich, 1972,

1974a, b and c) makes it pertinent to ask whether the rare allele SOD_A^2 is associated with any disease. As the SOD_A^2 allele is associated with a decreased stability of the enzyme (Marklund et al., 1976) a disposition for diseases which stem from an increased instability of cells and/or cell organelles could be suspected. So far, however, no association with disease has been found (Beckman et al., 1975). A protective effect of superoxide dismutase against radiation-induced chromosome damage has been observed (Nordenson et al., 1976). This is in accordance with the physiological function of superoxide dismutase, namely to dismutate O_2 radicals. These radicals may by themselves be hazardous to DNA (White et al., 1971) but to a greater extent exert their effect by reacting with H_2O_2 to produce the very reactive OH radical.

Triosephosphate Isomerase (EC 5.3.1.1)

The enzyme catalyses the interconversion of dihydroxyacetone phosphate and glyceraldehyde-3-phosphate. The enzyme is widely distributed in tissues and occurs in a number of different isozymes which are most probably controlled by alleles at two loci, TPI_A and TPI_B (Peters et al., 1973). The TPI locus has been assigned to chromosome 12.

Peters et al. (1973) examined the erythrocyte isozyme pattern in 2477 unrelated individuals of different ethnic origin. Two deviating patterns were observed in the European sample of 1703 individuals. The common phenotype is referred to as TPI 1 and the rare phenotypes as TPI 2-1 and TPI 3-1. One TPI 3-1 phenotype has also been observed in placental material collected from Negroes in Ibadan (1 out of 417). A number of cases with deficiency of triosephosphate isomerase and a moderate to severe haemolytic anaemia have been reported (Schneider et al., 1968). These have also been associated with a progressive neurological disease. (See p. 499 for a discussion of enzyme defects and tissue distribution.)

Uridine Monophosphate Kinase (EC 2.7.4.4)

The enzyme uridine monophosphate kinase (UMPK) catalyses the phosphorylation of uridine monophosphate to uridine diphosphate.

Four different phenotypes, UMPK 1, 2-1, 2 and 3-1 (Fig. 29) controlled by three alleles $UMPK^1$, $UMPK^2$ and $UMPK^3$ have been described (Giblett et al., 1974a). Formal genetical studies show a co-dominant autosomal inheritance (Giblett et al., 1974a; Kuhn et al., 1975).

Frequency data from a study of the American population by Giblett *et al.* (1974a) are as follows for the *UMPK²* gene: Caucasians 0·045, Negroes 0·011 and Orientals 0·071. The frequency of the *UMPK³* gene is very low in the Caucasian population, while a relatively high frequency 0·104 has been found in Cree Indians.

The frequency of the *UMPK* alleles in a Japanese population are as follows: *UMPK¹* 0·95 and *UMPK²* 0·05 (Haradae *et al.*, 1975) which is in quite good agreement with the data obtained by Giblett *et al.* (1974a) for American Orientals. Only the *UMPK¹* gene was observed in a study of African Negroes (Giblett *et al.*, 1974a).

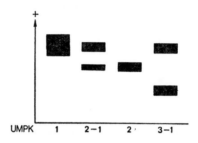

Fig. 29   Four different phenotypes of uridine monophosphate kinase. (After Giblett *et al.*, 1974)

The *UMPK²* allele is associated with lower catalytic activity. Individuals with the phenotype UMPK 2 have in their red cells about one-third of the activity found in individuals with the phenotype UMPK 1. Heterozygotes, UMPK 2-1, have an intermediate activity (Giblett *et al.*, 1974a). Giblett *et al.* (1974a) point to a possible association between a low UMPK activity and a decreased immune response.

III. AN ATTEMPT AT A SYNTHESIS

Fifteen years ago genetically controlled enzyme variations were looked upon as rare phenomena; today the occurrence of enzyme variation is considered the rule rather than the exception. Moreover, the occurrence of enzyme polymorphism has turned out to be a more common phenomenon than was anticipated at first. A search, trying to disclose the extent of polymorphism in Man, was originally initiated by Harris and his colleagues and a number of other workers have joined in. In other species population studies with the same aim have been undertaken by Hubby and Lewontin (1966), Lewontin and Hubby (1966) and Selander and Kaufman (1973).

Point mutation appears to be the most common origin of new enzyme variants. What happens once the new gene (or allele) has been introduced and why some alleles reach high frequencies remain unsolved questions. Today two schools can be identified, the "neutralists" and the "selectionists". Selection has been regarded as the important factor in maintaining enzyme variability. On the other hand Kimura (1968, 1969) and Kimura and Ohta (1971) have proposed that random genetical drift can account for polymorphic genes, which largely can be regarded as devoid of selective value. These conflicting views concerning enzyme polymorphism have recently been discussed very widely and extensively (Salzano, 1975), and have been considered in Chapter 4 in the broader context of polymorphism in general.

Here, we should merely record that Price (1976), using the possibly slightly inappropriate method of Ewens (1972), has shown that approximately half of a sample of 17 enzyme loci seem to be affected by selection. These loci are PGM_2, DIA, $PEPA$, $PEPD$, TPI, ADH and NP. More generally, we can ask, how frequent is polymorphism, or what percentage of all loci are polymorphic? Harris and Hopkinson (1972) found that 25 to 30% of all loci apparently are polymorphic and that the average heterozygosity per locus for the polymorphic alleles was about 7%. As only about one-third of the mutations can be detected by the zymogram technique the incidence of enzyme polymorphism is presumably much higher. Similar data indicating that polymorphism is a widespread phenomenon in other species have been presented by Selander and Kaufman (1973). For vertebrates including Man they calculated the average heterozygosity per locus to be about 6% while 15% was obtained for invertebrates.

Meaningful comparisons concerning the frequency of rare variants in different enzymes are hard to make due to the great variations in the number of samples analysed for each enzyme. Moreover the technical problems involved in distinguishing rare variants might lead to underestimation. Harris et al. (1974) have studied the incidence of rare alleles at 43 enzyme loci in the English population. They regarded an allele as rare when the gene frequency was less than 0·005. The conclusion they reached was that the occurrence of rare alleles is a common phenomenon and that rare alleles most probably occur at all loci coding for enzyme structure. However, they point out further that the frequency of rare alleles at any one locus is quite variable. The average heterozygosity per locus due to rare alleles was estimated to be 1·76 per 1000 for all the loci or 1 to 2 individuals per 1000. No difference in the incidence of rare alleles could be found between polymorphic and other loci.

The red blood cell is the most accessible material for large scale population studies. Therefore most data on enzyme variability have been obtained on enzymes present in red cells. This restriction may mean that the data on controlled variability in enzymes are not representative. Data pointing to decreased variability in the glycolytic enzymes present in the brain and erythrocytes have been presented by Cohen *et al.* (1973). The authors suggested that the restricted genetical variation may be due to negative selection against variants unfavourable to metabolic function. Other attempts have been made to view variation in a still broader context. For example, Zouros (1976) has carried out an extensive survey of the degree of heterozygosity for monomeric or multimeric enzymes apparently serving similar biochemical functions, and has noted that in all cases, detected heterozygosity was higher for monomeric than for multimeric enzymes. Zouros's conclusion was that unconditional overdominance (i.e. hetero-zygous advantage as on page 164) must therefore be regarded as an unlikely generator of the degree of variability seen. However, this conclusion seems to presuppose that the subunit structure of a particular enzyme has no adaptive or evolutionary significance, which seems *a priori* most unlikely. In our present state of ignorance, broad generalizations seem most unlikely to be correct.

There have been two main approaches to the study of enzyme variability in human populations, of which one can be referred to as the clinical and the other as the population approach; in this chapter, we have of course been largely concerned with the latter. The two approaches are now per-haps beginning to converge for a common benefit, viz. a better understand-ing of the importance and consequences of genetical variability both in health and disease. The clinical line of research actually started with Garrod's proposition in the beginning of this century that diseases could be caused by inborn errors of metabolism (Chapter 10). These classical "inborn errors" have mostly been found to be due to homozygosity for an allele coding for a deficient enzyme. As an example, about one-third of all patients with chronic congenital non-spherocytic haemolytic anaemia have through the years been found to suffer from specific enzyme deficiencies. Now approximately 15 different enzymes are known for which a hereditary deficiency associated with anaemia has been observed.

Quantitative methods are generally used to assess the enzyme deficiency in clinical cases. The introduction of the zymogram technique made it feasible to study large population samples and great genetical variability due to charge differences was revealed in a number of enzymes. Most variants disclosed appeared, however, not to be associated with any disease. Extended population and family studies and a rapid increase in the number of enzymes available for study using the zymogram technique soon revealed the occurrence of rare "null" or "silent" alleles without deleterious effects in some enzyme systems. The frequency of these "null" alleles is about the same as for the alleles which cause inborn errors of metabolism. Thus alleles coding for deficient enzymes in such diseases should rightly be regarded as special cases or extremes in a broad spectrum of genetical variability. The classification of alleles into "normal" and "deficient" is therefore not always as simple as previously thought.

Enzyme deficiencies have previously in spite of a quite variable clinical picture been looked upon as homogeneous entities. Combinations of electrophoretic and quantitative methods in enzyme studies have often revealed a heterogeneous picture. Such heterogeneity has been found, e.g. in deficiencies in pyruvate kinase, phosphohexose isomerase and NADH-diaphorase associated with non-spherocytic haemolytic anaemia. In these cases different structural genes have been found to cause enzyme deficiency in different families. In perhaps the majority of enzyme systems studied so far at least one of the alleles has been found to be associated with a decrease in enzyme activity, for example: glucose-6-phosphate dehydrogenase (p. 212), red cell acid phosphatase (p. 191), 6-phosphogluconate dehydrogenase (p. 236), glutamic pyruvic transaminase (p. 217), placental alkaline phosphatase (p. 241) and adenylate kinase (p. 197). In most cases these "hypomorphs" seem to represent genes coding for an enzyme variant with decreased activity but with no known clinical effects in heterozygotes.

ACKNOWLEDGEMENTS

The author wishes to thank Professor Lars Beckman for his most helpful review of the manuscript. This work was supported by the Swedish Medical Research Council (Project No. 03x-2725).

REFERENCES

Abell, C. W. and Marchand, N. W. (1973). *Nature New Biol.* **244**, 217.
Ajmar, F., Scharrer, B., Hashimoto, F. and Carson, P. E. (1968). *Proc. Natn. Acad. Sci. USA* **59**, 538.
Akedo, H., Nishihara, H., Shinkai, K. and Komatsu, K. (1970). *Biochim. Biophys. Acta* **212**, 189.
Akedo, H., Nishihara, H., Shinkai, K., Komatsu, K. and Ishikawa, S. (1972). *Biochim. Biophys. Acta* **276**, 257.
Alfred, B. M., Stout, T. D., Birkbeck, J., Lee, M. and Petrakis, N. L. (1969). *Am. J. Phys. Anthrop.* **31**, 391.
Alfred, B. M., Stout, T. D., Lee, M., Birkbeck, J. and Petrakis, N. L. (1970). *Am. J. Phys. Anthrop.* **32**, 329.
Altland, K. and Goedde, H. W. (1970). *Biochem. Genet.* **4**, 321.
Ananthakrishnan, R., Beck, W. and Walter, H. (1972). *Humangenetik* **17**, 89.

Ananthakrishnan, R., Beck, W. and Walter, H. (1974). *Humangenetik* **23**, 31.

Apella, E. and Markert, C. L. (1961). *Biochem. Biophys. Res. Commun.* **6**, 171.

Arends, T., Brewer, G., Chagnon, N., Gallango, M. L., Gershowitz, H., Layrisse, M., Neel, J., Shreffler, D., Tashian, R. and Weitkamp, L. (1967). *Proc. Natn. Acad. Sci. USA* **57**, 1252.

Arese, P., Bosia, A., Gallo, E., Mazza, V. and Pescarmona, G. P. (1973). *Eur. J. Clin. Invest.* **3**, 86.

Ashton, G. C. and Simpson, N. E. (1966). *Am. J. Hum. Genet.* **18**, 438.

Azevêdo, E. S., Olimpio da Silva, M. C. and Neto, J. T. C. (1976). *Ann. Hum. Genet.* **39**, 321.

Babson, A. L., Read, P. A. and Philips, G. E. (1959). *Am. J. Clin. Path.* **32**, 83.

Badger, K. S. and Sussman, H. H. (1976). *Proc. Natn. Acad. Sci. USA* **73**, 2201.

Bajatzadeh, M., Walter, H. and Pálsson, J. (1969). *Humangenetik* **7**, 353.

Balinsky, D., Cayanis, E. and Bersohn, I. (1973). *Biochemistry* **12**, 863.

Bargagna, M., Domenici, R. and Morali, A. (1975). *Humangenetik* **29**, 251.

Baughan, M. A., Valentine, W. N., Paglia, M. D., Ways, P. O., Simon, E. R. and DeMarsh, Q. B. (1968). *Blood* **32**, 236.

Beck, W. and Ananthakrishnan, R. (1974). *Humangenetik* **25**, 127.

Beckman, G. (1970a). *Hum. Hered.* **20**, 74.

Beckman, G. (1970b). Thesis, University of Lund, Sweden.

Beckman, G. (1972). In "The Biochemical Genetics of Man" (D. J. H. Brock and O. Mayo, Eds) Chapter 5. Academic Press, London and New York.

Beckman, G. (1973). *Hereditas* **73**, 305.

Beckman, L. and Beckman, G. (1967). *Biochem. Genet.* **1**, 145.

Beckman, L. and Beckman, G. (1968). *Acta Genet.* **18**, 543.

Beckman, G. and Beckman, L. (1975a). *In* "Isozymes" (C. L. Market, Ed.), Vol. 4, 781–795. Academic Press, New York and London.

Beckman, G. and Beckman, L. (1975b). *Hereditas* **81**, 85.

Beckman, L. and Beckman, G. (1976a). *In* "Protides of Biological Fluids" Vol. 14 (H. Peeters, Ed.). Pergamon Press, Oxford.

Beckman, G. and Beckman, L. (1976b). *Hereditas* **82**, 403.

Beckman, G. and Christodoulou, C. (1974). *Hum. Hered.* **24**, 294.

Beckman, G. and Holm, S. (1975). *Hereditas* **80**, 1.

Beckman, G. and Pakarinen, A. (1973). *Hum. Hered.* **23**, 346.

Beckman, L., Björling, G. and Christodoulou, C. (1966). *Acta Genet.* **16**, 59.

Beckman, L., Beckman, G., Christodoulou, C. and Ifekwunigwe, A. (1967). *Acta Genet.* **17**, 406.

Beckman, G., Beckman, L. and Tärnvik, A. (1970). *Hum. Hered.* **20**, 81.

Beckman, G., Beckman, L. and Cedergren, B. (1971). *Hereditas* **69**, 243.

Beckman, G., Beckman, L. and Magnússon, S. S. (1972). *Hum. Hered.* **22**, 473.

Beckman, G., Beckman, L. and Tärnvik, A. (1973a). *Hereditas* **73**, 31.

Beckman, G., Lundgren, E. and Tärnvik, A. (1973b). *Hum. Hered.* **23**, 338.

Beckman, G., Beckman, L. and Nilsson, L.-O. (1975). *Hereditas* **79**, 43.

Bender, K. and Frank, R. (1974). *Humangenetik* **23**, 315.

Benkmann, H.-G. and Goedde, H. W. (1974). *Humangenetik* **24**, 325.

Benkmann, H.-G. and Goedde, H. W. (1975). *Humangenetik* **27**, 343.

Bertoli, D. and Segal, S. (1966). *J. Biol. Chem.* **241**, 4023.

Beutler, E. (1969a). *Pharm. Rev.* **21**, 73.

Beutler, E. (1969b). *J. Clin. Invest.* **48**, 1957.

Beutler, E. (1969c). *Biochem. Genet.* **3**, 189.

Beutler, E. (1970). *Humangenetik* **9**, 250.
Beutler, E. (1971). "Red Cell Metabolism. A Manual of Biochemical Methods". Grune and Stratton, New York and London.
Beutler, E. (1973). *Isr. J. Med. Sci.* **9**, 1323.
Beutler, E. and Baluda, M. C. (1964). *Lancet* i, 189.
Beutler, E. and Baluda, M. C. (1966). *J. Lab. Clin. Med.* **67**, 947.
Beutler, E. and Matsumato, F. (1975). *Blood* **46**, 103.
Beutler, E. and Srivastava, S. K. (1970). *Nature, Lond.* **226**, 759.
Beutler, E. and West, C. (1974). *Am. J. Hum. Genet.* **26**, 255.
Beutler, E. and Yoshida, A. (1973). *Ann. Hum. Genet.* **37**, 151.
Beutler, E., Baluda, M. C., Sturgeon, P. and Day, R. (1965). *Lancet* i, 353.
Beutler, E., Baluda, M. C., Sturgeon, P. and Day, R. (1966). *J. Lab. Clin. Med.* **68**, 646.
Beutler, E., West, C. and Beutler, B. (1974). *Ann. Hum. Genet.* **38**, 163.
Bhasin, M. K. and Fuhrmann, W. (1972). *Humangenetik* **14**, 204.
Bigley, R. H., Stenzel, P., Jones, R. T., Campos, J. O. and Koler, R. D. (1968). *Enzym. Biol. Clin.* **9**, 10.
Blake, N. M. and Kirk, R. L. (1969). *Nature, Lond.* **221**, 278.
Blake, N. M. and Omoto, K. (1975). *Ann. Hum. Genet.* **38**, 251.
Blake, N. M., Kirk, R. L., Pryke, E. and Sinnett, P. (1969). *Science* **163**, 701.
Blake, N. M., Kirk, R. L., Simons, M. J., Alpers, M. P. (1970a). *Humangenetik* **11**, 72.
Blake, N. M., Kirk, R. L., Lewis, W. H. P. and Harris, H. (1970b). *Ann. Hum. Genet.* **33**, 301.
Blake, N. M., Kirk, R. L., McDermid, E. M., Omoto, K. and Ahuja, Y. R. (1972). *Hum. Hered.* **22**, 123.
Blake, N. M., McDermid, E. M., Kirk, R. L., Ong, Y. W., Simons, M. J. (1973). *Singapore Med. J.* **14**, 2.
Blake, N. M., Saha, N., McDermid, E. M. and Kirk, R. L. (1974). *Humangenetik* **21**, 347.
Blanco, A. and Zinkham, W. H. (1963). *Science* **139**, 601.
Blume, K. G., Löhr, G. W., Rüdiger, H. W. and Schalhorn, A. (1968a). *Lancet* i, 529.
Blume, K. G., Gottwik, M., Löhr, G. and Rüdiger, H. W. (1968b). *Humangenetik* **6**, 163.
Bodansky, O., Schwartz, M. K. and Nisselbaum, J. S. (1966). *Adv. Enzyme Reg.* **4**, 299.
Boettcher, B. and de la Lande, I. A. (1969). *Aust. J. Exp. Biol. Med. Sci.* **47**, 97.
Bonné, B., Sarah, A., Modai, M., Godber, M. J., Mourant, A. E., Tills, D. and Woodhead, B. G. (1970). *Hum. Hered.* **20**, 609.
Boone, C. M., Chen, T. R. and Ruddle, F. H. (1972). *Proc. Natn. Acad. Sci. USA* **69**, 510.
Bottini, E., Lucarelli, B., Pigram, P., Palmarino, R., Spennati, G. F. and Orzalesi, M. (1972a). *Am. J. Hum. Genet.* **24**, 495.
Bottini, E., Lucarelli, P. and Gloria, F. (1972b). *Am. J. Hum. Genet.* **24**, 505.
Bourne, J. G., Collier, H. O. J. and Somers, G. F. (1952). *Lancet* i, 1225.
Bowen, P., Sissons, W., Beiner, M., Harris, H. and Hopkinson, D. A. (1972). *Clin. Res.* **20**, 929.
Bowman, H. S., McKusick, V. A. and Dronamraju, K. R. (1965). *Am. J. Hum. Genet.* **17**, 1.

Bowman, J. E., Carson, P. E., Frischer, H. and deGaray, A. L. (1966). *Nature, Lond.* **210**, 811.

Bowman, J. E., Frischer, H., Ajmar, F., Carson, P. E. and Gower, M. K. (1967). *Nature, Lond.* **214**, 1156.

Boyd, J. W. (1961). *Biochem. J.* **81**, 434.

Boyd, J. W. (1966). *Biochim. Biophys. Acta* **113**, 302.

Boyer, S. H. (1961). *Science* **134**, 1002.

Boyer, S. H. (1963). *Ann. N.Y. Acad. Sci.* **103**, 938.

Boyer, S. H., Porter, I. H. and Weilbacher, R. G. (1962). *Proc. Natn. Acad. Sci. USA* **48**, 1868.

Boyer, S. H., Fainer, D. C. and Watson-Williams, E. J. (1963). *Science* **141**, 642.

Brewer, G. J. (1967). *Am. J. Hum. Genet.* **19**, 674.

Brewer, G. J. (1970). "An Introduction to Isozyme Techniques". Academic Press, London and New York.

Brinkmann, B., Koops, E., Klopp, O., Heindl, K. and Rüdiger, H. W. (1972). *Ann. Hum. Genet.* **35**, 363.

Brinkmann, B., Brinkmann, M. and Martin, H. (1973). *Hum. Hered.* **23**, 603.

Brown, S., Wiebel, F. J., Gelboin, H. V. and Minna, J. D. (1976). *Proc. Natn. Acad. Sci. USA* **73**, 4628.

Busbee, D. L., Shaw, C. R. and Cantrell, E. T. (1972). *Science* **178**, 315.

Cantrell, E. T., Warr, G. A., Busbee, D. L. and Martin, R. R. (1973). *J. Clin. Invest.* **52**, 1881.

Carson, P. E. and Frischer, H. (1966). *Am. J. Med.* **41**, 744.

Carson, P. E., Flanagan, C. L., Ickes, C. E. and Alving, A. S. (1956). *Science* **124**, 484.

Carson, P. E., Brewer, G. J. and Ickes, C. E. (1961). *J. Lab. Clin. Med.* **58**, 804.

Carter, N. D., Fildes, R. A., Fitch, L. I. and Parr, C. W. (1968). *Acta Genet.* **18**, 109.

Carter, N. D., Tashian, R. E., Huntsman, R. G. and Sacker, L. (1972). *Am. J. Hum. Genet.* **24**, 330.

Carter, N. D., Tanis, R. J., Tashian, R. E. and Ferrell, R. E. (1973). *Biochem. Genet.* **10**, 399.

Carties, P., Habibi, B., Leroux, J. P. and Marchand, J. C. (1971). *Nouv. Rev. Franc. Hémat.* **11**, 565.

Chabner, B. A., Johns, D. G., Coleman, C. N., Drake, J. C. and Evans, W. H. (1974). *J. Clin. Invest.* **53**, 922.

Charlesworth, D. (1972). *Ann. Hum. Genet.* **35**, 477.

Chen, S.-H. and Giblett, E. R. (1971a). *Science* **173**, 148.

Chen, S.-H. and Giblett, E. R. (1971b). *Am. J. Hum. Genet.* **23**, 419.

Chen, S.-H. and Giblett, E. R. (1972). *Am. J. Hum. Genet.* **24**, 229.

Chen, S.-H. and Giblett, E. R. (1976). *Ann. Hum. Genet.* **39**, 277.

Chen, S.-H. and Scott, C. R. (1973). *Am. J. Hum. Genet.* **25**, 21A.

Chen, S.-H., Malcolm, L. A., Yoshida, A. and Giblett, E. R. (1971). *Am. J. Hum. Genet.* **23**, 87.

Chen, S.-H., Giblett, E. R., Anderson, J. E. and Fossum, B. L. G. (1972a). *Ann. Hum. Genet.* **35**, 401.

Chen, S.-H., Fossum, B. L. G. and Giblett, E. R. (1972b). *Am. J. Hum. Genet.* **24**, 325.

Chen, S.-H., Scott, C. R. and Giblett, E. R. (1974a). *Am. J. Hum. Genet.* **26**, 103.

Chen, S.-H., Anderson, J. E., Giblett, E. R. and Lewis, M. (1974b). *Am. J. Hum. Genet.* **26**, 73.

Chen, S.-H., Scott, C. R. and Swedberg, K. R. (1975). *Am. J. Hum. Genet.* **27**, 46.
Chen, T. R., McMorris, F. A., Creagan, R., Ricciuti, F., Tischfield, J. and Ruddle, F. H. (1973). *Am. J. Hum. Genet.* **25**, 200.
Chern, C. J. and Beutler, E. (1974). *Am. J. Hum. Genet.* **26**, 20A.
Chern, C. J. and Beutler, E. (1975). *Science* **187**, 1084.
Chern, C. J. and Beutler, E. (1976). *Am. J. Hum. Genet.* **28**, 9.
Childs, B., Zinkham, W., Brown, E. A., Kombro, E. L. and Torbet, J. V. (1958). *Bull. Johns Hopkins Hosp.* **102**, 21.
Christie, G. S. and Judah, J. D. (1953). *Proc. R. Soc.* B. **141**, 420.
Coates, P. M. and Simpson, N. E. (1972). *Science* **175**, 1466.
Coates, P. M., Mestriner, M. A. and Hopkinson, D. A. (1975). *Ann. Hum. Genet.* **39**, 1.
Cohen, P. T. W. (1971). *Fed. Proc.* **30**, 1207.
Cohen, P. T. W. and Omenn, G. S. (1972). *Biochem. Genet.* **7**, 303.
Cohen, P. T. W., Omenn, G. S., Motulsky, A. G., Chen, S.-H. and Giblett, E. R. (1973). *Nature New Biol.* **241**, 229.
Das, S. R., Mukherjee, S. K., Das, S. K., Ananthakrishnan, R., Blake, N. M. and Kirk, R. L. (1970a). *Humangenetik* **9**, 107.
Das, S. R., Mukherjee, B. N., Das, S. K., Blake, N. M. and Kirk, R. L. (1970b). *Indian J. Med. Res.* **58**, 866.
Das, S. R., Mukherjee, B. N. and Das, S. K. (1972). *Humangenetik* **14**, 151.
Davidson, R. G. (1967). *Ann. Hum. Genet.* **30**, 355.
Davidson, R. G. and Cortner, J. A. (1967a). *Nature, Lond.* **215**, 761.
Davidson, R. G. and Cortner, J. A. (1967b). *Science* **157**, 1569.
Davidson, R. G., Migeon, B. R., Borden, M. and Childs, B. (1963a). *Bull. Johns Hopkins Hosp.* **112**, 318.
Davidson, R. G., Nitowsky, H. M. and Childs, B. (1963b). *Proc. Natn. Acad. Sci. USA* **50**, 481.
Davidson, R. G., Fildes, R. A., Glenn-Bott, A. M., Harris, H. and Robson, E. B. (1965). *Ann. Hum. Genet.* **29**, 5.
Davidson, R. G., Cortner, J. A., Rattazzi, M. C., Ruddle, F. H. and Lubs, H. A. (1970). *Science* **169**, 391.
Davies, R. O., Marton, A. V. and Kalow, W. (1960). *Can. J. Biochem.* **38**, 545.
Dern, R. J., Beutler, E. and Alving, A. S. (1954). *J. Lab. Clin. Med.* **44**, 171.
Dern, R. J., Brewer, G. J., Tashian, R. E. and Shows, T. B. (1966). *J. Lab. Clin. Med.* **67**, 255.
Detter, J. C., Ways, P. O., Giblett, E. R., Baughan, M. A., Hopkinson, D. A., Povey, S. and Harris, H. (1968). *Ann. Hum. Genet.* **31**, 329.
Detter, J. C., Stamatoyannopoulos, G., Giblett, E. R. and Motulsky, A. G. (1970). *J. Med. Genet.* **7**, 356.
Devadetta, S., Gangaharam, P. R., Andrews, R. H., Fox, W., Raemakrishnan, C. V., Selkon, J. B. and Velu, S. (1960). *Bull. Wld. Hlth. Org.* **23**, 587.
Dickman, S. R. and Speyer, J. F. (1954). *J. Biol. Chem.* **206**, 67.
Dietz, A. A., Lubrano, T. and Rubenstein, H. M. (1965). *Acta Genet.* **15**, 208.
Dissing, J. and Knudsen, J. B. (1969). *Hum. Hered.* **19**, 375.
Dissing, J. and Knudsen, J. B. (1970). *Hum. Hered.* **20**, 178.
Doenicke, A., Gurtner, T., Kreutzberg, G., Remes, I., Spiess, W. and Steinbere-ithner, K. (1963). *Acta Anaesth. Scand.* **7**, 59.
Donald, L. J. and Robson, E. B. (1974). *Ann. Hum. Genet.* **37**, 303.
Dorfman, A. and Matalon, R. (1972). *In* "The Metabolic Basis of Inherited

Diseases" (J. B. Stanbury, J. B. Wyngaarden and D. S. Fredrickson, Eds), 1245. McGraw-Hill, New York.

Edwards, Y. H., Hopkinson, D. A. and Harris, H. (1971a). *Ann. Hum. Genet.* **35**, 207.

Edwards, Y. H., Hopkinson, D. A. and Harris, H. (1971b). *Ann. Hum. Genet.* **34**, 395.

Edwards, Y. H., Edwards, P. A. and Hopkinson, D. A. (1973). *FEBS Lett.* **32**, 235.

Ellard, G. A., Aber, V. R., Gammon, P. T., Mitchison, D. A., Lakshminarayan, S., Citron, K. M., Fox, W. and Tall, R. (1972). *Lancet* **i**, 340.

Eng, L.-I.L., Loo, M. and Fah, F. K. (1972). *Br. J. Haemat.* **23**, 419.

Evans, D. A. P. (1963). *Am. J. Med.* **34**, 639.

Evans, D. A. P. and Clarke, C. A. (1961). *Br. Med. Bull.* **17**, 234.

Evans, D. A. P., Manley, K. E. and McKusick, V. A. (1960). *Br. Med. J.* **2**, 485.

Evans, D. A. P., Storey, P. B. and McKusick, V. A. (1961). *Bull. Johns Hopkins Hosp.* **108**, 60.

Evans, D. A. P., Davidson, K. and Pratt, R. T. C. (1965). *Clin. Pharmacol. Ther.* **6**, 430.

Evans, F. T., Gray, P. W. S., Lehmann, H. and Silk, E. (1952). *Lancet* **i**, 1229.

Evrev, T., Zhivcov, S. and Russev, L. (1970). *Hum. Hered.* **20**, 70.

Ewens, W. J. (1972). *Theor. Popul. Biol.* **3**, 87.

Eze, L. C., Tweedie, M. C. K., Bullen, M. F., Wren, P. J. J. and Evans, D. A. P. (1974). *Ann. Hum. Genet.* **37**, 333.

Fildes, R. A. and Harris, H. (1966). *Nature, Lond.* **209**, 261.

Fildes, R. A. and Parr, C. W. (1963). *Nature, Lond.* **200**, 890.

Fildes, R. A. and Parr, C. W. (1964). *Proc. 6th Intern. Cong. Biochem. 229.*

Fisher R. A. and Harris, H. (1969). *Ann. N. Y. Acad. Sci.* **166**, 380.

Fisher, R. A. and Harris, H. (1971a). *Ann. Hum. Genet.* **34**, 431.

Fisher, R. A. and Harris, H. (1971b). *Ann. Hum. Genet.* **34**, 439.

Fisher, R. A. and Harris, H. (1972). *Ann. hum. Genet.* **36**, 69.

Fitch, L. I., Parr, C. W. and Welch, S. G. (1968). *Biochem. J.* **110**, 56P.

Fleischer, G. A., Potter, C. S. and Wakim, K. G. (1960). *Proc. Soc. Exp. Biol. Med.* **103**, 229.

Ford, E. B. (1940). *In* "The New Systematics" (J. S. Huxley, Ed.). Oxford University Press, Oxford.

Ford, E. B. (1965). "Genetic Polymorphism", Faber and Faber, London.

Fridovich, I. (1972). *Acc. Chem. Res.* **5**, 321.

Fridovich, I. (1974a). *In* "Advances in Enzymology" (A. Meister, Ed.), Vol. 41, 35–97. John Wiley, New York.

Fridovich, I. (1974b). *In* "Molecular Mechanisms of Oxygen Activation" (O. Hayaishi, Ed.). Academic Press, New York and London.

Fridovich, I. (1974c). *New Engl. J. Med.* **290**, 624.

Fritz, P. J., Vesell, E. S., White, E. L. and Pruitt, K. M. (1969). *Proc. Natn. Acad. Sci. USA* **62**, 558.

Fritz, P. J., White, L. E., Vesell, E. S. and Pruitt, K. M. (1971). *Nature New Biol.* **230**, 119.

Fuhrmann, W. and Lichte, K. H. (1966). *Humangenetik* **3**, 121.

Fukui, M. and Wakasugi, C. (1972). *Jap. J. Legal. Med.* **26**, 46.

Funakoshi, S. and Deutsch, H. F. (1970). *J. Biol. Chem.* **245**, 4913.

Gallango, M. L. and Arends, T. (1969). *Humangenetik* **7**, 104.

Gartler, S. M. and Linder, D. (1964). *Cold Spring Harb. Symp. Quant. Biol.* **29**, 253.

Gartler, S. M., Ziprowski, L., Krakowski, A., Ezra, R., Szeinberg, A. and Adam, A. (1966). *Am. J. Hum. Genet.* **18**, 282.

Gartler, S. M., Liskay, R. M., Campbell, B. K., Sparkes, R. and Gant, N. (1972). *Cell Differ.* **1**, 215.

Geerdink, R. A., Bartstra, H. A. and Hopkinson, D. A. (1974). *Hum. Hered.* **24**, 40.

Giblett, E. R. (1967). *In* "Advances in Immunogenetics" (T. J. Greenwalt, Ed.), 114–132. Lippincott, Philadelphia.

Giblett, E. R. (1969). "Genetic Markers in Human Blood", Blackwells, Oxford.

Giblett, E. R. (1976). *Cytogenet. Cell. Genet.* **16**, 65.

Giblett, E. R. and Scott, N. M. (1965). *Am. J. Hum. Genet.* **17**, 425.

Giblett, E. R., Anderson, J. E., Cohen, F., Pollara, B. and Meuwissen, H. J. (1972). *Lancet* **ii**, 1067.

Giblett, E. R., Anderson, J. E., Chen, S.-H., Teng, Y.-S. and Cohen, F. (1974a). *Am. J. Hum. Genet.* **26**, 627.

Giblett, E. R., Chen, S.-H., Anderson, J. E. and Lewis, M. (1974b). *Cytogenet. Cell. Genet.* **13**, 91.

Giblett, E. R., Ammann, A. J., Wara, D. W., Sandman, R. and Diamond, L. K. (1975). *Lancet* **i**, 1010.

Goedde, H. W., Gehring, D. and Hoffmann, R. A. (1965). *Humangenetik* **1**, 607.

Goodwin, K. O. (1974). *In* "Trace Elements in Soil–Plant–Animal Systems" (D. J. D. Nicholas and A. R. Egan, Eds), 259–270. Academic Press, New York and London.

Gordon, H., Keran, M. M. and Vooijs, M. (1967). *Nature, Lond.* **214**, 466.

Gottlieb, A. J. and Sussman, H. H. (1968). *Biochim. Biophys. Acta* **160**, 167.

Greene, J. M. and Dawson, D. M. (1973). *Ann. Hum. Genet.* **36**, 355.

Grumbach, M. M., Marks, P. A. and Moroshima, A. (1962). *Lancet* **i**, 1330.

Gussman, S. and Schwarzfischer, F. (1972). *Z. Rechtsmed.* **70**, 251.

Gutsche, R. B., Scott, E. M. and Wright, R. C. (1967). *Nature, Lond.* **215**, 322.

Hackel, E., Hopkinson, D. A. and Harris, H. (1972). *Ann. Hum. Genet.* **35**, 491.

Hammersen, G., Mandell, R. and Levy, H. (1975). *Ann. Hum. Genet.* **39**, 147.

Haradae, S., Itoh, M. and Misawa, S. (1975). *Humangenetik* **29**, 255.

Harris, H. (1966). *Proc. R. Soc. B.* **164**, 298.

Harris, H. (1969). *Proc. R. Soc. B.* **174**, 1.

Harris, H. (1975). "The Principals of Human Biochemical Genetics" 182. North-Holland Publishing Co., Amsterdam.

Harris, H. and Hopkinson, D. A. (1972). *Ann. Hum. Genet.* **36**, 9.

Harris, H. and Hopkinson, D. A. (1976). "Handbook of Enzyme Electrophoresis in Human Genetics". North Holland, Amsterdam.

Harris, H. and Whittaker, M. (1961). *Nature, Lond.* **191**, 496.

Harris, H. and Whittaker, M. (1962). *Ann. Hum. Genet.* **26**, 59.

Harris, H., Whittaker, M., Lehmann, H. and Silk, E. (1960). *Acta Genet.* **10**, 1.

Harris, H., Hopkinson, D. A. and Robson, E. B. (1962). *Nature, Lond.* **196**, 1296.

Harris, H., Hopkinson, D. A., Robson, E. B. and Whittaker, M. (1963a). *Ann. Hum. Genet.* **26**, 359.

Harris, H., Hopkinson, D. A., Spencer, N., Court Brown, W. M. and Mangle, D. (1963b). *Ann. Hum. Genet.* **27**, 59.

Harris, H., Hopkinson, D. A., Luffman, J. E. and Rapley, S. (1968). *In* "Hereditary Disorders of Erythrocyte Metabolism" (E. Beutler, Ed.), Vol. 1, 1–20. Grune and Stratton, New York.

Harris, H., Hopkinson, D. A. and Robson, E. (1974). *Ann. Hum. Genet.* **37**, 237.

260 G. BECKMAN

Harrison, G. A., Kuchermann, C. F., Moore, M. A. S., Boyce, A. J., Baju, T., Mourant, A. E., Godber, M. J., Glasgow, B. G., Kopeć, A. C., Tills, D. and Clegg, E. J. (1969). *Philos. Trans.* B **256**, 147.

Harvey, R. G., Godber, M. J., Kopeć, A. C., Mourant, A. E. and Tills, D. (1969). *Hum. Biol.* **41**, 342.

Henderson, N. S. (1965). *J. Exp. Zool.* **158**, 263.

Henderson, N. S. (1968). *Ann. N. Y. Acad. Sci.* **151**, 429.

Herbich, J. and Meinhart, K. (1972). *Humangenetik* **15**, 345.

Herbich, J., Fisher, R. A. and Hopkinson, D. A. (1970). *Ann. Hum. Genet.* **34**, 145.

Herzog, P. and Drdová, A. (1971). *Humangenetik* **13**, 64.

Hirschhorn, R. (1975). *J. Clin. Invest.* **55**, 661.

Hirschhorn, R. and Beratis, N. G. (1973). *Lancet* **ii**, 1217.

Hirschhorn, R., Levytska, V., Pollara, B. and Meuwissen, H. J. (1973). *Nature New Biol.* **246**, 200.

Hirschhorn, R., Beratis, N., Rosen, F. S., Parkman, R., Stern, R. and Polmar, S. (1975). *Lancet* **i**, 79.

Hirschhorn, R., Beratis, N. and Rosen, F. S. (1976). *Proc. Natn. Acad. Sci. USA* **73**, 213.

Hjelm, M. and Wadam, B. (1970). *Proc. 13th Int. Congr. Hematol., Munich*, 121 (abst.).

Hockwald, R. S., Arnold, J., Clayman, C. B. and Alving, A. S. (1952). *J. Am. Med. Assoc.* **149**, 1568.

Hodgkin, W. E., Giblett, E. R., Levine, H., Baur, W. and Motulsky, A. G. (1965). *J. Clin. Invest.* **44**, 486.

Hopkinson, D. A. and Harris, H. (1965). *Nature, Lond.* **208**, 410.

Hopkinson, D. A. and Harris, H. (1966). *Ann. Hum. Genet.* **30**, 167.

Hopkinson, D. A. and Harris, H. (1968). *Ann. Hum. Genet.* **31**, 359.

Hopkinson, D. A. and Harris, H. (1969). In "Biochemical Methods in Red Cell Genetics" (J. J. Yunis, Ed.), 337–375. Academic Press, New York and London.

Hopkinson, D. A., Spencer, N. and Harris, H. (1963). *Nature, Lond.* **199**, 969.

Hopkinson, D. A., Spencer, N. and Harris, H. (1964). *Am. J. Hum. Genet.* **16**, 141.

Hopkinson, D. A., Cook, P. J. L. and Harris, H. (1969). *Ann. Hum. Genet.* **32**, 361.

Hopkinson, D. A., Corney, G., Cook, P. J. L., Robson, E. B. and Harris, H. (1970). *Ann. Hum. Genet.* **34**, 1.

Hopkinson, D. A., Meistriner, M. A., Cortner, J. and Harris, H. (1973). *Ann. Hum. Genet.* **37**, 119.

Hopkinson, D. A., Coppock, J. S., Mühlemann, M. F. and Edwards, Y. H. (1974a). *Ann. Hum. Genet.* **38**, 155.

Hopkinson, D. A., Peters, J. and Harris, H. (1974b). *Ann. Hum. Genet.* **37**, 477.

Hopkinson, D. A., Edwards, Y. H. and Harris, H. (1976). *Ann. Hum. Genet.* **39**, 383.

Howell, R. R. (1972). In "The Metabolic Basis of Inherited Disease" (J. B. Stanbury, J. B. Wyngaarden and D. S. Fredrichson, Eds), 149–173. McGraw-Hill, New York.

Hubby, J. L. and Lewontin, R. C. (1966). *Genetics* **54**, 57.

Huennekens, F. M., Nurk, E. and Gabrio, B. W. (1956). *J. Biol. Chem.* **221**, 971.

Imamura, K., Tanaka, T., Nishina, T., Nakashima, K. and Miwa, S. (1973). *J. Biol. Chem., Japan* **74**, 1165.

Inglis, N. R., Kirley, S., Stolbach, L. L. and Fishman, W. H. (1973). *Cancer Res.* **33**, 1657.

Ishimoto, G. (1969). *Jap. J. Hum. Genet.* **14**, 183.

Ishimoto, G. and Kuwata, M. (1974). *Jap. J. Hum. Genet.* **18**, 373.
Jenkins, T. and Corfield, U. (1972). *Ann. Hum. Genet.* **35**, 379.
Jenkins, T. and Nurse, G. T. (1974). *Ann. Hum. Genet.* **38**, 19.
Johnson, G. B. (1976). *Genetics* **83**, 149.
Kahn, A., Boivin, P. and Lagneau, J. (1973). *Humangenetik* **18**, 261.
Kahn, A., Marie, J., Galand, C. and Boivin, P. (1975). *Humangenetik* **29**, 271.
Kalow, W. and Davies, R. O. (1958). *Biochem. Pharmac.* **1**, 183.
Kalow, W. and Genest, K. (1957). *Can. J. Biochem.* **35**, 339.
Kalow, W. and Staron, N. (1957). *Can. J. Biochem.* **35**, 1305.
Kámarýt, J. and Laxova, R. (1965). *Humangenetik* **1**, 579.
Kámarýt, J. and Laxova, R. (1966). *Humangenetik* **3**, 41.
Kámarýt, J., Adámek, R., and Vrba, M. (1971). *Humangenetik* **11**, 213.
Kamel, R. and Schwarzfischer, F. (1975). *Humangenetik* **28**, 65.
Kamel, R., Berg, K., Schwarzfischer, F. and Wischerath, H. (1975). *Humangenetik* **27**, 53.
Kaplan, J. C. and Beutler, E. (1968). *Nature, Lond.* **217**, 256.
Kaplan, J. C., Leroux, A., Bakouri, S., Grangaud, J. P. and Benadbadji, M. (1974). *Nouv. Rev. Franc. Hémat.* **14**, 755.
Karn, R. C., Rosenblum, B. B., Ward, J. C. and Merritt, A. D. (1975). *In* "Isozymes –IV" (C. L. Markert, Ed.). Academic Press, London and New York.
Karp, G. W. Jr. and Sutton, H. E. (1967). *Am. J. Hum. Genet.* **19**, 54.
Keitt, A. S. (1969). *J. Clin. Invest.* **48**, 1997.
Kellermann, G., Shaw, C. R. and Luyten-Kellermann, M. (1973a). *New Engl. J. Med.* **289**, 934.
Kellermann, G., Luyten-Kellermann, M. and Shaw, C. R. (1973b). *Am. J. Hum. Genet.* **24**, 327.
Khan, P. M., Westerveld, A., Grzeschik, K. H., Deys, B. P., Garson, O. M. and Siniscalco, M. (1971). *Am. J. Hum. Genet.* **23**, 614.
Khoo, J. C. and Russell, P. J. (1972). *Biochim. Biophys. Acta* **268**, 98.
Kimura, M. (1968). *Nature, Lond.* **217**, 624.
Kimura, M. (1969). *Proc. Natn. Acad. Sci. USA* **63**, 1181.
Kimura, M. and Ohta, I. (1971). *Nature, Lond.* **229**, 467.
Kirk, R. L. (1975). *In* "Isozymes" (C. L. Markert, Ed.), Vol. 4, 169–180. Academic Press, New York and London.
Kirk, R. L., Blake, N. M., Moodie, P. M. and Tibbs, G. J. (1971). *Hum. Biol.* **1**, 54.
Kirkman, H. N. (1959). *Nature, Lond.* **184**, 1291.
Kirkman, H. N. (1971). *Adv. Hum. Genet.* **2**, 1.
Kirkman, H. N. and Hendrickson, E. M. (1963). *Am. J. Hum. Genet.* **15**, 241.
Kissen, C. and Cotte, J. (1970). *Enzym. Biol. Clin.* **11**, 277.
Kitamura, A., Iijima, N., Hashimoto, F. and Hiratsuka, A. (1971). *Clin. Chim. Acta* **34**, 419.
Knight, R. A., Selin, M. J. and Harris, H. W. (1959). *Trans. Conf. Chemother. Tuberc.* **18**, 52.
Kömpf, J. (1971). *Humangenetik* **14**, 76.
Kömpf, J., Ritter, H. and Schmitt, J. (1971). *Humangenetik* **13**, 75.
Kömpf, J., Ritter, H. and Schmitt, J. (1972). *Humangenetik* **14**, 103.
Kömpf, J., Bissbort, S., Gussman, S. and Ritter, H. (1975a). *Humangenetik* **27**, 141.
Kömpf, J., Bissbort, S. and Ritter, H. (1975b). *Humangenetik* **28**, 249.
Konrad, P. N., McCarthy, D. J., Mauer, A. M., Valentine, W. N. and Paglia, D. E. (1973). *J. Pediat.* **82**, 456.

Köster, B., Leupold, H. and Mauff, G. (1975). *Humangenetik* **28**, 75.

Kraus, A. P. and Neely, C. L. (1964). *Science* **145**, 595.

Kraus, A. P., Langston, M. F. Jr. and Lynch, B. L. (1968). *Biochem. Biophys. Res. Commun.* **30**, 173.

Krone, W., Schneider, G., Schultz, D., Arnold, H. and Blume, K. G. (1970). *Humangenetik* **10**, 224.

Kuhn, B., Bissbort, S., Kömpf, J. and Ritter, H. (1975). *Humangenetik* **28**, 255.

Kuhnl, P., Nowicki, L. and Spielmann, W. (1974). *Z. Rechtmed.* **75**, 179.

Lahav, M. and Szeinberg, A. (1972). *Hum. Hered.* **22**, 533.

Lai, L. Y. C. I., Nevo, S. and Steinberg, A. G. (1964). *Science* **145**, 1187.

Lamm, L. U. (1969). *Hereditas* **61**, 282.

Lamm, L. U. (1970). *Hum. Hered.* **20**, 329.

Larizza, P., Brunetti, P., Grignani, F. and Venture, S. (1958). *Haematologica* **43**, 205.

Lausecker, C., Heidt, P., Fischer, D., Hartley, H. and Löhr, G. W. (1965). *Arch. Franc. Pédiat.* **21**, 789.

Leakey, T. E., Coward, A. R., Warlow, A. and Mourant, A. E. (1972). *Hum. Hered.* **22**, 542.

Lehmann, H. and Ryan, E. (1956). *Lancet* i, 124.

Leroux, A., Junier, C., Kaplan, J.-C. and Bamberger, J. (1975). *Nature, Lond.* **258**, 619.

Lewis, W. H. P. (1973). *Ann. Hum. Genet.* **36**, 267.

Lewis, W. H. P. and Harris, H. (1967). *Nature, Lond.* **215**, 351.

Lewis, W. H. P. and Harris, H. (1969a). *Ann. Hum. Genet.* **33**, 89.

Lewis, W. H. P. and Harris, H. (1969b). *Ann. Hum. Genet.* **32**, 317.

Lewis, W. H. P. and Harris, H. (1969c). *Ann. Hum. Genet.* **33**, 93.

Lewis, W. H. P., Corney, G. and Harris, H. (1968). *Ann. Hum. Genet.* **32**, 35.

Lewontin, R. C. and Hubby, J. L. (1966). *Genetics* **54**, 595.

Lidell, J., Lehmann, H. and Silk, E. (1962). *Nature, Lond.* **193**, 561.

Lidell, J., Lehmann, H. and Davies, D. (1963). *Acta Genet.* **13**, 95.

Lie-Injo, L. E. (1967). *Am. J. Hum. Genet.* **19**, 130.

Lin, K.-T. and Deutsch, H. F. (1972). *J. Biol. Chem.* **247**, 3761.

Linder, D. and Gartler, S. M. (1965). *Science* **150**, 67.

Lippitt, B. and Fridovich, I. (1973). *Arch. Biochem. Biophys.* **159**, 738.

Löhr, G. W. and Waller, H. D. (1962). *Med. Klin.* **57**, 1521.

Long, W. K. (1966). *Proc. 3rd. Intern. Congr. Hum. Genet., Chicago,* 59 (abst.).

Luffman, J. E. and Harris, H. (1967). *Ann. Hum. Genet.* **30**, 387.

Luzzatto, L. (1973). *Israel J. Med. Sci.* **9**, 1484.

Lyon, M. F. (1961). *Nature, Lond.* **190**, 372.

Markert, C. L. (1963). *Science* **140**, 1329.

Markert, C. L. (ed.) (1975). "Isozymes I–IV.", Academic Press, London and New York.

Markert, C. L. and Møller, F. (1959). *Proc. Natn. Acad. Sci. USA* **45**, 753.

Marklund, S., Beckman, G. and Stigbrand, T. (1976). *Eur. J. Biochem.* **65**, 415.

Marks, P. A. (1958). *Science* **127**, 1338.

Marks, P. A., Gross, R. T. and Harwitz, R. E. (1959). *Nature, Lond.* **183**, 1266.

Marks, P. A., Szeinberg, A. and Banks, J. (1961). *J. Biol. Chem.* **236**, 10.

Masters, C. J. and Holmes, R. S. (1972). *Biol. Rev.* **47**, 309.

Mathai, C. K. and Beutler, E. (1966). *Science* **154**, 1179.

Mayes, J. S. and Guthrie, R. (1968). *Biochem. Genet.* **2**, 219.

Mazza, V., Arese, P., Bosia, A., Gallo, E. and Pescarmona, G. P. (1970). *Proc. 13th Intern. Cong. Hematol.*, Munich, 121.
McAlpine, P. J., Hopkinson, D. A. and Harris, H. (1970a). *Ann. Hum. Genet.* **34**, 169.
McAlpine, P. J., Hopkinson, D. A. and Harris, H. (1970b). *Ann. Hum. Genet.* **34**, 61.
McAlpine, P. J., Hopkinson, D. A. and Harris, H. (1970c). *Ann. Hum. Genet.* **34**, 177.
McCord, J. M. and Fridovich, I. (1968). *J. Biol. Chem.* **243**, 5753.
McCord, J. M. and Fridovich, I. (1969). *J. Biol. Chem.* **244**, 6049.
McCord, J. M., Keele, B. B. Jr. and Fridovich, I. (1971). *Proc. Natn. Acad. Sci. USA* **68**, 1024.
McCurdy, P. R., Kirkman, H. N., Naiman, J. L., Jim, R. T. S. and Pickard, B. M. (1966). *J. Lab. Clin. Med.* **67**, 374.
McMaster, Y., Tennant, R., Clubb, J. S., Neale, F. C. and Posen, S. (1964). *J. Obstet. Gynaec. Br. Cwlth.* **71**, 735.
Mellman, W. J., Rawnsley, B. E., Nicholas, C. W., Needelman, M. T., Mennuti, M. T., Malone, J. and Tedesco, T. A. (1975). *Am. J. Hum. Genet.* **27**, 748.
Merritt, A. D., Rivas, M. L. and Ward, J. C. (1972). *Nature, Lond.* **239**, 243.
Merritt, A. D., Lovrien, E. W., Rivas, M. L. and Conneally, P. M. (1973a). *Am. J. Hum. Genet.* **25**, 523.
Merritt, A. D., Rivas, M. L., Bixler, D. and Newell, R. (1973b). *Am. J. Hum. Genet.* **25**, 510.
Meuwissen, H. J., Pollara, B., Pickering, R. J. and Porter, I. H., ed. (1975). "Combined Immunodeficiency Disease and Adenosine Deaminase Deficiency; A Molecular Defect". Academic Press, New York and London.
Miwa, S., Nakashima, K., Oda, S., Ogawa, H., Nagafuji, H., Arima, M., Okuna, T. and Nakashima, T. (1972). *Acta Haemat. Jap.* **35**, 57.
Miwa, S., Nakashima, K., Oda, S., Oda, E., Matsumoto, N. and Fukumoto, Y. (1973a). *Acta Haemat. Jap.* **36**, 65.
Miwa, S., Nakashima, K., Oda, S., Matsumoto, N., Ogawa, H., Kobayashi, R., Kotani, M., Hatara, A., Onaya, T. and Yamada, T. (1973b). *Acta Haemat. Jap.* **36**, 70.
Miwa, S., Nakashima, K., Ariyoshi, K., Shinohara, K., Oda, E. and Tanaka, T. (1975). *Br. J. Haemat.* **29**, 157.
Modiano, G., Filippi, G., Brunelli, F., Frattaroli, W. and Siniscalo, M. (1967). *Acta Genet.* **17**, 17.
Modiano, G., Scozzari, R., Gigliani, F., Santolamazza, C., Afeltra, P. and Frattaroli, W. (1970a). *Hum. Hered.* **20**, 86.
Modiano, G., Scozzari, R., Gigliani, F., Santolamazza, C., Spennati, G. F. and Saini, P. (1970b). *Am. J. Hum. Genet.* **22**, 292.
Monn, E. (1969). *Hum. Hered.* **19**, 1.
Monn, E. and Gjønnaess, H. (1971). *Hum. Hered.* **21**, 254.
Moore, M. J., Funakoshi, S. and Deutsch, H. F. (1971). *Biochem. Genet.* **5**, 497.
Moore, M. J., Deutsch, H. F. and Ellis, F. R. (1973). *Am. J. Hum. Genet.* **25**, 29.
Morton, N. E. (1967). *Am. J. Hum. Genet.* **19**, 23.
Moser, K., Ciresa, M. and Schwartzmeier, J. (1970). *Med. Welt.* **21**, 1977.
Motulsky, A. G. (1960). *Hum. Biol.* **32**, 28.
Motulsky, A. G. (1964). *In* "Progress in Medical Genetics" (A. G. Steinberg and A. G. Bearn, Eds), Vol. 3, 44–74. Grune and Stratton, New York.

Motulsky, A. G. (1972). *Fed. Proc.* **31**, 1286.

Motulsky, A. G. (1975). *In* "The Role of Natural Selection in Human Evolution" (F. M. Salzano, Ed.), 271–291. North-Holland Publishing Co., Amsterdam.

Mourant, A. E., Beckman, L., Beckman, G., Nilsson, L. O. and Tills, D. (1968). *Acta. Genet.* **18**, 553.

Mourant, A. E., Kopeć, A. C. and Domaniewska-Sobozak, K. (1976). "The Distribution of the Human Blood Groups and Other Polymorphisms", 2nd ed. Oxford Medical Publications, Oxford University Press.

Nadler, H. L., Chacko, C. M. and Rachmeler, M (1970). *Proc. Natn. Acad. Sci. USA* **67**, 976.

Nakashima, K., Miwa, S., Oda, S., Tanaka, T., Imamura, K. and Nishina, T. (1974). *Blood* **43**, 537.

Nance, W. E., Claflin, A. and Smithies, O. (1963). *Science* **142**, 1075.

Necheles, T. F., Rai, U. S. and Cameron, D. (1970a). *J. Lab. Clin. Med.* **76**, 593.

Necheles, T. F., Steinberg, M. H. and Cameron, D. (1970b). *Br. J. Haemat.* **19**, 605.

Neitlich, W. (1966). *J. Clin. Invest.* **45**, 380.

Ng, W. G., Bergren, W. R., Fields, M. and Donnell, G. N. (1969). *Biochem. Biophys. Res. Commun.* **37**, 354.

Ng, W. G., Bergren, W. R. and Donnell, G. N. (1973). *Ann. Hum. Genet.* **37**, 1.

Ng, W. G., Donnell, G. N., Koch, R. and Bergren, W. R. (1976). *Am. J. Hum. Genet.* **28**, 42.

Nishihara, H., Ishikawa, S., Shinkai, K., Akedo, H. (1973). *Biochim. Biophys. Acta* **302**, 429.

Nishimura, Y., Chida, N. Hayashi, T. and Asakawo, T. S. (1972). *Tohoka J. Exp. Med.* **108**, 207.

Nordenson, I., Beckman, G. and Beckman, L. (1976). *Hereditas* **82**, 125.

Nyman, P. O. and Lindskog, S. (1964). *Biochim. Biophys. Acta* **85**, 141.

Olaisen, B. (1973a). *Humangenetik* **19**, 289.

Olaisen, B. (1973b). *Hum. Hered.* **23**, 595.

Olaisen, B. and Teisberg, P. (1972). *Hum. Hered.* **22**, 380.

Omenn, G. S. and Cheung, S. C.-Y. (1974). *Am. J. Hum. Genet.* **26**, 393.

Omoto, K. and Blake, N. M. (1972). *Ann. Hum. Genet.* **36**, 61.

Oski, F. A. and Stockman, J. A. III. (1973). *In* "Current Problems in Pediatrics" (L. Gluck, Ed.). Vol. 4 Nr. 2. Year Book Med. Publ., Inc., Chicago.

Paglia, D. E. and Valentine, W. N. (1974). *Am. J. Clin. Path.* **62**, 740.

Paglia, D. E., Holland, P., Baughan, M. A. and Valentine, W. N. (1969). *New Engl. J. Med.* **280**, 66.

Paglia, D. E., Paredes, R., Valentine, W. N., Dorantes, S. and Konrad, P. N. (1975). *Am. J. Hum. Genet.* **27**, 62.

Palmarino, R., Scacchi, R., Corbo, R. M., Lucarelli, P., Salsini, G., Christofori, G., Osti, L., Menini, C. and Vullo, C. (1975). *Humangenetik* **29**, 349.

Panich, V. and Sungnate, T. (1973). *Humangenetik* **18**, 39.

Panich, V., Sungnate, T., Wasi, P. and Na-Nakorn, S. (1972). *J. Med. Assoc., Thailand* **55**, 576.

Paniker, N. V., Srivastava, S. K. and Beutler, E. (1970). *Biochim. Biophys. Acta* **215**, 456.

Parr, C. W. (1966). *Nature, Lond.* **210**, 487.

Parr, C. W. and Fitch, L. I. (1964). *Biochem. J.* **93**, 28C.

Parr, C. W. and Fitch, L. I. (1967). *Ann. Hum. Genet.* **30**, 339.

Parrington, J. M., Cruickshank, G., Hopkinson, D. A., Robson, E. B. and Harris, H. (1968). *Ann. Hum. Genet.* **32**, 27.

Payne, D. M., Porter, D. W. and Gracy, R. W. (1972). *Arch. Biochem. Biophys.* **151**, 122.

Pearce, J., Edwards, Y. H. and Harris, H. (1976). *Ann. Hum. Genet.* **39**, 263.

Peters, J., Hopkinson, D. A. and Harris, H. (1973). *Ann. Hum. Genet.* **36**, 297.

Piomelli, S., Corash, L. M., Davenport, D. D., Miraglia, J. and Ambrosi, E. L. (1968). *J. Clin. Invest.* **47**, 940.

Povey, S., Corney, G., Lewis, W. H. P., Robson, E. B., Parrington, J. M. and Harris, H. (1972). *Ann. Hum. Genet.* **35**, 455.

Povey, S., Corney, G. and Harris, H. (1975). *Ann. Hum. Genet.* **38**, 407.

Powell, G. F., Rasco, M. and Maniscalco, R. M. (1974). *Metabolism* **23**, 505.

Price, G. R. (1976). *Ann. Hum. Genet.* **39**, 471.

Prins, H. K., Oort, M., Loos, J. A., Zürcher, C. and Beckers, T. (1966). *Blood* **27**, 145.

Quick, C. B., Fisher, R. A. and Harris, H. (1972). *Ann. Hum. Genet.* **35**, 445.

Quick, C. B., Fisher, R. A. and Harris, H. (1974). *Eur. J. Biochem.* **42**, 511.

Quilici, J. C., Ruffié, J. and Marty, Y. (1970). *Nouv. Rev. Franç. Hémat.* **10**, 727.

Radam, G., Strauch, H. and Prokop, O. (1974). *Humangenetik* **25**, 247.

Radam, G., Strauch, H. and Vavrusa, B. (1975). *Humangenetik* **26**, 151.

Ramot, R., Fisher, S., Szeinberg, A., Adam, A., Sheba, C. and Gafni, D. (1959). *J. Clin. Invest.* **38**, 2234.

Rapley, S. and Harris, H. (1970). *Ann. Hum. Genet.* **33**, 361.

Rapley, S., Robson, E. B., Harris, H. and Smith, M. S. (1967). *Ann. Hum. Genet.* **31**, 237.

Rapley, S., Lewis, W. H. P. and Harris, H. (1971). *Ann. Hum. Genet.* **34**, 307.

Renninger, W. and Bimboese, Ch. (1970). *Humangenetik* **9**, 34.

Robson, E. B. and Harris, H. (1965). *Nature, Lond.* **207**, 1257.

Robson, E. B. and Harris, H. (1967). *Ann. Hum. Genet.* **30**, 219.

Ross, J. D. (1963). *Blood* **21**, 51.

Ruddle, F. H. (1973). *Nature, Lond.* **242**, 165.

Sager, R. (1964). *New Engl. J. Med.* **271**, 352.

Saha, N., Kirk, R. L. and Undevia, J. V. (1974). *Am. J. Hum. Genet.* **26**, 723.

Salzano, F. M. (Ed.) (1975). "The Role of Natural Selection in Human Evolution". North-Holland Publishing Co., Amsterdam.

Samloff, I. M. (1969). *Gastroenterology* **57**, 659.

Samloff, I. M. and Townes, P. L. (1969). *Gastroenterology* **56**, 1194.

Samloff, I. M. and Townes, P. L. (1970). *Science* **168**, 144.

Samloff, I. M., Liebman, W. M., Glober, G. A., Moore, J. O. and Indra, D. (1973). *Am. J. Hum. Genet.* **25**, 178.

Sansone, G., Rasore-Quartino, A. and Veneziano, G. (1963). *Pathologica* **55**, 371.

Santachiara-Benerecetti, S. A. (1970). *Am. J. Hum. Genet.* **22**, 228.

Santachiara-Benerecetti, S. A. and Modiano, G. (1969). *Am. J. Hum. Genet.* **21**, 315.

Santachiara-Benerecetti, S. A., Cattaneo, A. and Khan, P. M. (1972a). *Am. J. Hum. Genet.* **24**, 680.

Santachiara-Benerecetti, S. A., Cattaneo, A. and Khan, P. M. (1972b). *Hum. Hered.* **22**, 171.

Santachiara-Benerecetti, S. A., Beretta, M. and Pampiglione, S. (1975). *Hum. Hered.* **25**, 276.

Schapira, F. and Kaplan, J. C. (1969). *Biochem. Biophys. Res. Commun.* **35**, 451.

266 G. BECKMAN

Schloot, W., Blume, K. G. and Goedde, H. W. (1967). *Humangenetik* **4**, 274.
Schmitt, J. and Ritter, H. (1974). *Humangenetik* **22**, 263.
Schneider, A. S., Dunn, I., Ibsen, K. H. and Weinsten, I. M. (1968). *In* "Hereditary Disorders of Erythrocyte Metabolism" (E. Beutler, Ed.), Grune and Stratton, New York.
Schröter, W., Koch, H. J., Wonneberger, B., Kalinowsky, W., Arnold, H., Blume, K. G. and Hüther, W. (1974). *Pediat. Res.* **18**, 18.
Scialom, C., Najean, Y. and Bernard, J. (1966). *Nouv. Rev. Franç. Hémat.* **6**, 452.
Scott, C. R., Chen, S.-H., Giblett, E. R. (1974). *J. Clin. Invest.* **53**, 1194.
Scott, E. M. (1966). *J. Biol. Chem.* **241**, 3049.
Scott, E. M. (1973). *Ann. Hum. Genet.* **37**, 139.
Scott, E. M., Duncan, I. W., Ekstrand, V. and Wright, R. C. (1966). *Am. J. Hum. Genet.* **18**, 408.
Scott, E. M., Weaver, D. D. and Wright, R. C. (1970). *Am. J. Hum. Genet.* **22**, 363.
Scozzari, R., Trippa, G., Barberio, C. and Menini, C. (1975). *Humangenetik* **26**, 147.
Selander, R. K. and Kaufman, D. W. (1973). *Proc. Natn. Acad. Sci.* **70**, 1875.
Seth, S. (1974). *Humangenetik* **23**, 223.
Shaw, C. R. (1969). *Int. Rev. Cytol.* **25**, 297.
Shaw, C. R., Syner, F. N. and Tashian, R. E. (1962). *Science* **138**, 31.
Shih, V.-E., Levy, H. L., Karolkewicz, V., Houghton, S., Efron, M. L., Isselbacher, K. J., Beutler, E. and MacCready, R. A. (1971). *New Engl. J. Med.* **284**, 753.
Shinoda, T. (1967). *Jap. J. Hum. Genet.* **11**, 252.
Shows, T. B. (1967). *Biochem. Genet.* **1**, 171.
Shrago, E. (1965). *Arch. Biochem.* **109**, 57.
Simpson, N. E. (1966). *Am. J. Hum. Genet.* **18**, 243.
Simpson, N. E. and Kalow, W. (1964). *Am. J. Hum. Genet.* **16**, 180.
Singer, J. D. and Brock, D. J. H. (1971). *Am. J. Hum. Genet.* **35**, 109.
Sinha, K. P. and Hopkinson, D. A. (1969). *Ann. Hum. Genet.* **33**, 139.
Sinha, K. P., Lewis, W. H. P., Corney, G. and Harris, H. (1970). *Ann. Hum. Genet.* **34**, 153.
Sinha, K. P., Lewis, W. H. P. and Harris, H. (1971). *Ann. Hum. Genet.* **34**, 321.
Sinnett, P., Blake, N. M., Kirk, R. L., Lai, L. Y. C. and Walsh, R. J. (1970). *Arch. Phys. Anthropol. Oceania* **5**, 236.
Skude, G. and Jakobsson, A. (1970). *Hum. Hered.* **20**, 319.
Slaughter, C. A., Hopkinson, D. A. and Harris, H. (1975). *Ann. Hum. Genet.* **39**, 193.
Smith, M., Hopkinson, D. A. and Harris, H. (1971). *Ann. Hum. Genet.* **34**, 251.
Smith, M., Hopkinson, D. A. and Harris, H. (1972). *Ann. Hum. Genet.* **35**, 243.
Smith, M., Hopkinson, D. A. and Harris, H. (1973a). *Ann. Hum. Genet.* **37**, 49.
Smith, M., Hopkinson, D. A. and Harris, H. (1973b). *Ann. Hum. Genet.* **36**, 401.
Sørensen, S. A. (1970). *Clin. Genet.* **1**, 294.
Sørensen, S. A. (1975). *Am. J. Hum. Genet.* **27**, 100.
Sparkes, R. S., Carrel, R. E. and Paglia, D. E. (1969). *Nature, Lond.* **224**, 367.
Spencer, N., Hopkinson, D. A. and Harris, H. (1964a). *Nature, Lond.* **201**, 299.
Spencer, N., Hopkinson, D. A. and Harris, H. (1964b). *Nature, Lond.* **204**, 742.
Spencer, N., Hopkinson, D. A. and Harris, H. (1968). *Ann. Hum. Genet.* **32**, 9.
Spielmann, W., Kuhnl, P., Rexrodt, Ch. and Hänsel, G. (1973). *Humangenetik* **18**, 341.

Spieser, P. and Pausch, V. (1967). *Vox. Sang.* **13**, 12.

Stamatoyannopoulos, G., Papayannopoulou, T., Bacopoulos, C. and Motulsky, A. G. (1966). *Am. J. Hum. Genet.* **18**, 417.

Stamatoyannopoulos, G., Papayannopoulou, T., Bacopoulos, C. and Motulsky, A. G. (1967). *Blood* **29**, 87.

Stamatoyannopoulos, G., Chen, S.-H. and Fukui, M. (1975). *Am. J. Hum. Genet.* **27**, 789.

Starkweather, W. H., Cousineau, L., Schoch, H. K. and Zarafonetis, C. J. (1965). *Blood* **26**, 63.

Sussman, H. H., Small, P. A. and Cotlove, E. (1968). *J. Biol. Chem.* **243**, 160.

Swallow, D. M. and Harris, H. (1972). *Ann. Hum. Genet.* **36**, 141.

Swallow, D. M., Povey, S. and Harris, H. (1973). *Ann. Hum. Genet.* **37**, 31.

Swallow, D. M., Corney, G., Harris, H. and Hirschhorn, R. (1975). *Ann. Hum. Genet.* **38**, 391.

Szeinberg, A., Sheba, C. and Adam, A. (1958). *Nature, Lond.* **181**, 1256.

Szeinberg, A., Pipano, S. and Ostfeld, E. (1966). *J. Med. Genet.* **3**, 190.

Tanaka, K. R. (1969). *In* "Biochemical Methods in Red Cell Genetics" (J. J Yunis, Ed.). Academic Press, New York and London.

Tanaka, K. R. and Paglia, D. E. (1971). *Sem. Hemat.* **8**, 367.

Tariverdian, G. and Ritter, H. (1969). *Humangenetik* **7**, 176.

Tariverdian, G., Arnold, H., Blume, K. G., Lenkeit, U. and Löhr, G. W. (1970a). *Humangenetik* **10**, 218.

Tariverdian, G., Ropers, H., Op't Hof, J. and Ritter, H. (1970b). *Humangenetik* **10**, 355

Tashian, R. E. (1965). *Am. J. Hum. Genet.* **17**, 257.

Tashian, R. E. (1969). *In* "Biochemical Methods in Red Cell Genetics" (J. J. Yunis, Ed.), 307–336. Academic Press, New York and London.

Tashian, R. E. and Shaw, M. W. (1962). *Am. J. Hum. Genet.* **14**, 295.

Tashian, R. E., Plato, C. C. and Shows, T. B. (1963). *Science* **140**, 53.

Tashian, R. E., Riggs, S. K. and Yu, Y. S. L. (1966). *Arch. Biochem.* **117**, 320.

Tashian, R. E., Shreffler, D. C. and Shows, T. B. (1968). *Ann. N. Y. Acad. Sci.* **151**, 64.

Tedesco, T. A. (1972). *J. Biol. Chem.* **247**, 6631.

Tedesco, T. A., Bonow, R., Miller, K. L. and Mellman, W. J. (1972). *Science* **178**, 176.

Tedesco, T. A., Miller, K. L., Rawnsley, B. E., Mennuti, M. T., Spielman, R. S. and Mellman, W. J. (1975). *Am. J. Hum. Genet.* **27**, 737.

Teng, Y.-S., Anderson, J. E. and Giblett, E. R. (1975). *Am. J. Hum. Genet.* **27**, 492.

Thomas, D. M. and Harris, H. (1971). *Ann. Hum. Genet.* **35**, 221.

Tills, D., Van den Branden, J. L., Clements, V. R. and Mourant, A. E. (1970). *Hum. Hered.* **20**, 523.

Tills, D., Van den Branden, J. L., Clements, V. R. and Mourant, A. E. (1971). *Hum. Hered.* **21**, 305.

Tönz, O. and Rossi, E. (1964). *Nature, Lond.* **202**, 606.

Townes, P. L. and White, M. R. (1974). *Am. J. Hum. Genet.* **26**, 252.

Trotta, P. P., Smithwick, E. M. and Balis, M. E. (1976). *Proc. Natn. Acad. Sci. USA* **73**, 104.

Tsuboi, K. K. and Hudson, P. B. (1957). *J. Biol. Chem.* **224**, 879.

Turner, B. M., Fisher, R. A. and Harris, H. (1971). *Eur. J. Biochem.* **24**, 228.

Turner, B. M., Fisher, R. A. and Harris, H. (1974a). *Ann. Hum. Genet.* **37**, 455.
Turner, B. M., Fisher, R. A., Garthwaite, E., Whale, R. J. and Harris, H. (1974b). *Ann. Hum. Genet.* **37**, 469.
Turner, B. M., Turner, V. S., Beratis, N. G. and Hirschhorn, K. (1975). *Am. J. Hum. Genet.* **27**, 651.
Valentine, W. N., Tanaka, K. R. and Miwa, S. (1961). *Trans. Assoc. Am. Phys.* **74**, 100.
Valentine, W. N., Oski, F. A., Paglia, D. E., Baughan, M. A., Schneider, A. S. and Naiman, J. L. (1967). *New Engl. J. Med.* **276**, 1.
Valentine, W. N., Anderson, H. M., Paglia, D. E., Jaffé, E. R., Konrad, P. N. and Harris, S. R. (1969). *New Engl. J. Med.* **280**, 528.
Van Biervliet, J. P., Vlug, A., Bartstra, H., Rotteveel, J. J., de Vaan, G. A. M. and Staal, G. E. J. (1975). *Humangenetik* **30**, 35.
Van der Weyden, M. B., Buckley, R. H. and Kelley, W. N. (1974). *Biochem. Biophys. Res. Commun.* **57**, 590.
Van Ros, G. and Druet, R. (1966). *Nature, Lond.* **212**, 543.
Van Someren, H., van Henegouwen, H. B., Los, W., Wurzer-Figurelli, E., Doppert, B., Vervloet, M. and Khan, P. M. (1974). *Humangenetik* **25**, 189.
Van Wierst, B., Blake, N. M., Kirk, R. L., Jacobs, D. S. and Johnson, D. G. (1973). *Aust. J. Exp. Biol. Med. Sci.* **51**, 857.
Vergnes, H. and Quilici, J. C. (1970). *Ann. Génét.* **13**, 96.
Vesell, E. S. and Bearn, A. G. (1962). *Proc. Soc. Exp. Biol. Med.* **111**, 100.
Vesell, E. S. (Ed.) (1968). *Ann. N. Y. Acad. Sci.* **151**, 5.
Vives-Corrons, J. L., Rozman, C., Kahn, A., Carrera, A. and Triginer, J. (1975). *Humangenetik* **29**, 291.
Von Wartburg, J. P. and Schürch, P. M. (1968). *Ann. N. Y. Acad. Sci.* **151**, 936.
Von Wartburg, J. P., Papenberg, J. and Aebi, H. (1965). *Can. J. Biochem.* **43**, 889.
Waller, H. D. (1968). In "Hereditary Disorders of Erythrocyte Metabolism" (E. Beutler, Ed.), 185–204. Grune and Stratton, New York.
Ward, J. C., Merritt, A. D. and Bixler, D. (1971). *Am. J. Hum. Genet.* **23**, 403.
Weitkamp, L. R. and Townes, P. L. (1975). In "Iszoymes IV: Genetics and Evolution", 829–838. Proceedings 3rd International Conference on Isozymes, New Haven, Connecticut, April 1974. Academic Press, New York and London.
Weitkamp, L. R., Townes, P. L. and May, A. G. (1975). *Am. J. Hum. Genet.* **27**, 486.
Welch, S. G. (1972). *Hum. Hered.* **22**, 190.
Welch, S. G. (1975). *Humangenetik* **27**, 163.
Welch, S. G. and Lee, J. (1974). *Humangenetik* **24**, 329.
Welch, S. G. and Mears, G. W. (1972). *Hum. Hered.* **22**, 38.
Welch, S. G., Barry, J. V., Dodd, B. E., Griffiths, P. D., Huntsman, R. G., Jenkins, G. C., Lincoln, P. J., McCathie, M., Mears, G. W. and Parr, C. W. (1973). *Hum. Hered.* **23**, 230.
Wendt, G. G. and Kirchberg, G. (1971). *Humangenetik* **11**, 175.
Westhamer, S., Freiburg, A. and Amaral, L. (1973). *Clin. Chim. Acta* **45**, 5.
White, I. N. H. and Butterworth, P. J. (1971). *Biochim. Biophys. Acta* **229**, 193.
White, J. R., Vaughan, T. O. and Yeh, V.-S. (1971). *Fed. Proc.* **30**, 1145.
Whittaker, M. (1967). *Acta Genet.* **17**, 1.
Whittaker, M. (1968). *Acta Genet.* **18**, 325.
Whittaker, M. and Reys, L. (1975). *Hum. Hered.* **25**, 296.
WHO (1967). *Wld. Hlth. Org. Tech. Rep. Ser. No. 366.*

Wilkinson, J. H. (1970). "Isoenzymes". Chapman and Hall, London.
Williams, L. and Hopkinson, D. A. (1975). *Hum. Hered.* **25**, 161.
Wilson, D. E. Jr., Povey, S. and Harris, H. (1976). *Ann. Hum. Genet.* **39**, 305.
Woodfield, D. G., Scragg, R. F. R., Blake, N. M., Kirk, R. L. and McDermid, E. M. (1974). *Hum. Hered.* **24**, 507.
Yoshida, A. (1967). *Proc. Natn. Acad. Sci. USA* **57**, 835.
Yoshida, A. and Miwa, S. (1974). *Am. J. Hum. Genet.* **26**, 378.
Yoshida, A., Steinmann, L. and Harbart, P. (1967). *Nature, Lond.* **216**, 275.
Yoshida, A., Beutler, E. and Motulsky, A. G. (1971). *Bull. Wld. Hlth. Org.* **45**, 243.
Yunis, J. J. (Ed.) (1969). "Biochemical Methods in Red Cell Genetics". Academic Press, London and New York.
Zinkham, W. H., Lenhard, R. E. J. R. and Childs, B. (1958). *Bull. Johns Hopkins Hosp.* **102**, 169.
Zouros, E. (1976). *Nature, Lond.* **262**, 227.

6 Inherited Variation in Plasma Proteins

D. W. COOPER

School of Biological Sciences, Macquarie University, New South Wales, Australia

I. INTRODUCTION AND SCOPE OF CHAPTER

Since Smithies'(1955a, b) reports of the high degree of resolution of protein mixtures which may be obtained by electrophoresis in starch gels, geneticists have devoted an increasing amount of attention to the study of naturally occurring genetical variation in the electrophoretic mobility of enzymes and other proteins. Over the last 21 years the molecular sieving effects of starch and acrylamide have been used to develop a number of electrophoretic techniques with high resolving power. The development of immunological techniques for recognizing antigenic variation in proteins (Ouchterlony, 1948) and their combination with electrophoresis in the technique of immunoelectrophoresis (Grabar and Williams, 1953) have also revealed genetic variation in proteins. The ready availability of human serum and the clinical importance of plasma proteins have resulted in the application of these techniques on a very large scale.

The principal aim of this chapter is to describe the main features of the human plasma protein polymorphisms. Variation is said to be polymorphic if the least common allele determining it occurs in a frequency of 0·01 or greater. Any allele which occurs at lower frequencies is said to determine uncommon or rare variation. This somewhat arbitrary definition is now generally used by population geneticists. Two subsidiary aims are to describe something of rare variation and also quantitative variation in plasma proteins. Accounts of many aspects of work on inherited variation in plasma proteins may be found in books by Prokop and Uhlenbruck (1969), Giblett (1969), Harris (1975), Manwell and Baker (1970), Ramot *et al.* (1974), and Putnam (1975).

II. TECHNIQUES FOR RECOGNIZING INHERITED VARIATION IN PROTEINS

A. Gel Electrophoresis

Laboratory manuals for electrophoresis of proteins have been written by Nerenberg (1966) and Gordon (1975). A useful short general account has been given by Lewis (1970).

1. Starch gel electrophoresis

Starch gel electrophoresis was invented by Smithies (1955a, b). Its great resolving power over previous techniques results from the combination of electrophoretic resolution and molecular sieving which comes about because the pore sizes of the starch gel have approximately the same distribution as the molecular sizes of the proteins. Vertical electrophoresis is capable of somewhat better resolution than horizontal electrophoresis (Smithies, 1959a). The technique has been described in full by several writers but the most comprehensive review is still Smithies (1959b). In principle the technique consists of casting a starch gel in a mould. Samples are introduced either with filter paper inserts or into preformed slots. Electrophoresis is carried out over a period of two to twenty hours and the gel then sliced horizontally to expose the centre of the gel where resolution is best. Protein and enzyme stains have been successfully applied to the detection of individual components. The technique has been extraordinarily widely used and this will no doubt continue. Its principal disadvantages over acrylamide electrophoresis are the greater difficulty of casting and handling the gel and the need to slice it to obtain the plane of good resolution. However, where an opaque rather than translucent surface is required, e.g. for photography of a fluorescent reaction product, starch is to be preferred. The relationship between gel concentration, relative migration rates and molecular weights has been investigated by Smithies (1964) and Ferguson (1964). The measurement of the effect of gel concentration upon the migration of a protein of unknown molecular weight relative to the effect upon several proteins of known molecular weight can be used to estimate this parameter.

2. Polyacrylamide gel electrophoresis (PAGE)

Electrophoresis in acrylamide gels is in principle the same as electrophoresis in starch gels. The technique was developed somewhat later than starch, being first described by Raymond and Weintraub (1959). The first full accounts of the method which are generally available have been given by Ornstein (1964). A recent symposium edited by Allen (1974) surveys later developments. Gels are prepared by dissolving the required weight of acrylamide with buffer solution and polymerizing it by the addition of catalysts, the action of which is expedited by light. Raymond's apparatus uses a slab of gel to which multiple samples may be applied. The apparatus described by Ornstein and Davis is now generally referred to as disc electrophoresis. It involves separation of single samples in gels cast in lengths of small glass or perspex tubing (6 mm diameter $\times$ 60 mm length). While disc electrophoresis has great resolving power for the individual sample, comparisons between different tubes even in the same run are somewhat difficult to make. For genetic work the slabs to which multiple samples may be applied, thereby allowing the use of an internal standard, are much to be preferred. The advantage of acrylamide is that it is much easier to prepare and does not need to be sliced. It is also possible to control its pore size much more exactly than that of starch, from which a variable amount of water may be lost during the boiling and degassing procedures. However, it is not as inert a medium as starch. In particular, the persulphate often used as one of the catalysts is capable of oxidizing proteins (Fantes and Furminger, 1967). In some cases it is therefore preferable to use riboflavin rather than persulphate as the catalyst, particularly at acid pH.

3. Gradient acrylamide gel electrophoresis

The gradient gel (Gradipore) technique of acrylamide electrophoresis has been developed by Margolis (Margolis and Kenrick, 1967a, b, 1968, 1969a; Margolis, 1969). It seems currently to be the one-dimensional slab electrophoresis technique with the greatest resolving power for serum. The principle involves a gradient of acrylamide concentration from a minimum of about 4% at the sample insertion end of the gel up to a maximum of 20 to 30% at the other. The gradient is usually linear but other functions are possible. The proteins are forced to migrate through increasingly smaller pores in the gel, until ultimately they reach a pore size through which they cannot pass. Pore size is governed by acrylamide concentration and the stationary point by protein size and shape. The process is a self-regulating one and the final order of protein bands is roughly a reflection of their molecular weights. The runs are not however usually taken to the point where all proteins become stationary. The high resolution achieved is due to the concentrating effect exerted on each band by a frictional force acting upon the front part of the band (because of the higher gel concentration in front) and the electrophoretic force acting in the opposite direction. Two-dimensional runs utilizing a combination of disc electrophoresis (Ornstein, 1964; Davis, 1964) and gradient gel electrophoresis give very high resolution of human serum (Margolis and Kenrick, 1969a). By varying the shape of the gradient it should be possible to achieve good resolution of parts of an electrophoretogram which are not normally well resolved. The technique has been used to demonstrate the existence of 2–2 haptoglobin bands in very small quantities in 2–1 sera (Sheridan et al., 1969). The results of electrophoresis of human serum in a Gradipore apparatus are shown in Fig. 1.

4. Isoelectric focusing

A recently perfected technique which is beginning to be applied to proteins is isoelectric focusing. An electric current is applied to an "ampholyte" solution (Ampholine, LKB, Sweden). Ampholytes are a mixture of aliphatic aminocarboxylic acids with molecular weights between 300 and 600. The electric current creates a pH gradient between the anode and cathode. Proteins in the solution take up the position in the gradient which has the pH of their isoelectric point. Very good resolution can be achieved between proteins with similar isoelectric points. The separation can be carried out in free solution or in acrylamide (polyacrylamide gel isoelectric focusing—PAGIF). The main disadvantage of the technique is the expense of the ampholytes, a commercial preparation. The PAGIF technique is reviewed in several papers in the symposium edited by Catsimpoolas (1973) and by various writers in Allen (1974).

5. Cellulose acetate

Electrophoresis on cellulose acetate has been widely used for detecting inherited variation in enzymes (see Meera Khan, 1971) but it has not been much used for plasma polymorphisms, probably because the resolution

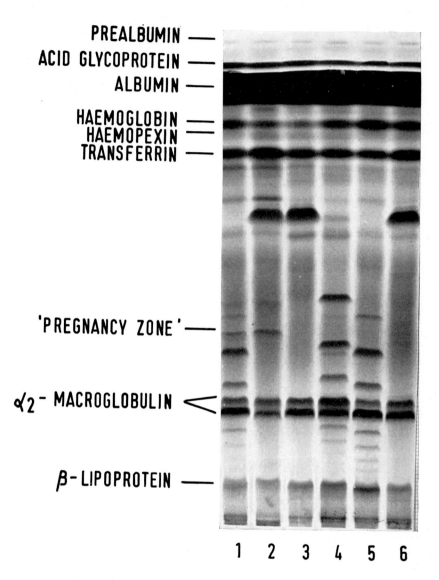

PREALBUMIN ——
ACID GLYCOPROTEIN ——
ALBUMIN ——

HAEMOGLOBIN ——
HAEMOPEXIN ——
TRANSFERRIN ——

'PREGNANCY ZONE' ——

α_2- MACROGLOBULIN <

β- LIPOPROTEIN ——

1 2 3 4 5 6

Fig. 1   Patterns of human serum proteins after 24 hours electrophoresis in 4 to 24% concave polyacrylamide gradient gel. 1. Hp2-2; 2. Hp1-1; 3. Hp1-1; 4. Hp2-1; 5. Hp2-2; 6. Hp1-1. The Hp bands lie on both sides of α_2-macroglobulin. Note the presence of faint Hp2-2 bands in the Hp2-1 types. The direction of migration is towards the top of the page, i.e. to the anode. (The pregnancy zone in slots 1 and 2 is a band seen in many pregnant women and women taking oestrogen-containing oral contraceptives, Margolis and Kenrick, 1969b). (Photograph courtesy of Dr J. Margolis)

which can be achieved is not as great as in gel electrophoresis. However, it is quicker and simpler than these techniques. Where specific means of locating proteins are available, it might well prove very useful, particularly for mass screening of large samples.

B. Immunological Techniques

Except for haemagglutination inhibition these techniques have been described by Ouchterlony and Nilsson (1973) and by Arquembourg (1975). Quantitative aspects of immunoelectrophoresis are reviewed by a number of authors in a symposium edited by Axelsen (1975).

1. Ouchterlony technique (immunodiffusion in two dimensions)

This technique involves the reaction of antigen and antibody in a gel medium, usually agar. It enables the investigator to determine the presence or absence of specific antigens in solutions such as serum. In one of the simplest forms, agar is poured at about a 1% concentration on to a smooth surface, e.g. a microscope slide. Wells are then made in the agar in one of a number of precise patterns. The fundamental principles can be exemplified by a pattern in which the centres of the three wells are equidistant from each other. Antiserum is put into one and antigen solutions into the other two. The reactants diffuse outwards and precipitates form where antibody and antigen encounter each other in equivalent proportions. The position of the line of precipitate is determined by the relative concentrations of antibody and antigen and their rates of diffusion. The rate of diffusion is determined by the molecular weight.

If the antigens in the two wells are the same, the line of precipitate will form a continuous arc between both pairs of wells (Fig. 2a). If one of the two antigen wells contains an antigen not present in the other plus the one which is present in both, a spur will form pointing towards the well with only the one antigen (Fig. 2b). This is a reaction of partial identity. If both contain an antigen not shared by the other plus one in common, a reaction of double partial identity may be seen (Fig. 2c). In reactions of partial identity, the spurs are less thick than the line of precipitate on the other side of the point of intersection. If the two antigen wells contain only unlike antigen, a reaction of non-identity may be seen (Fig. 2d). In this, the two lines of precipitate are without any tendency to coalesce and without any diminution of thickness beyond the point of intersection.

Reactions of partial identity may be viewed as a mixture of reactions of identity and non-identity. The coincidence of the lines of precipitate could be because the antigens involved are on the same molecule or because two different molecules have both precipitated at the same place. If the latter, alteration of the concentration of the reactants may shift the positions of precipitation so that the two are separated.

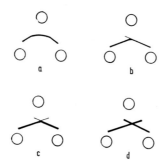

Fig. 2   The four basic types of reaction observed in the Ouchterlony or immuno-diffusion test. The antiserum well is at the top in each case and the two lower wells contain (different) antigen samples: (a) reaction of identity; (b) reaction of partial identity, with the well to the right having an extra specificity not found in the left; (c) reaction of double partial identity; (d) reaction of non-identity

The patterns of reactions shown in Fig. 2 are rather idealized and in practice other more complex patterns may be obtained, particularly if several antibody–antigen reactions are involved. For a discussion of these the reader is referred to Ouchterlony (1967), van Oss (1967), and Ouchterlony and Nilsson (1973).

2. Mancini technique (single radial immunodiffusion, SRID)
In its present form this method of quantitation is largely due to Mancini (Mancini *et al.*, 1964). Agarose containing a specific antiserum is poured on to a glass plate and allowed to set. Holes are cut into the agarose and carefully measured volumes of antigen-containing solution placed in them. A halo of antibody–antigen precipitate develops around the hole. After a time its diameter stabilizes. The area of the halo is directly proportional to the amount of antigen initially placed in the hole. As with "rocket" electrophoresis, internal standards allow the technique to be calibrated. The range of sensitivities can be varied by changing the antiserum concentration.

3. Immunoelectrophoresis
The technique of immunoelectrophoresis was first developed by Grabar and Williams (1953). Since then it has been very widely used in medicine,

genetics, immunology and biochemistry and has been extensively reviewed; see for example Crowle (1961), Ouchterlony (1967), van Oss (1967), Hirschfeld (1960), Peetom (1963), Schultze and Heremans (1966) and Grabar and Burtin (1964). In essence, it consists of a combination of electrophoresis and immunodiffusion (Ouchterlony, 1967). Electrophoresis is carried out in agar or agarose, usually in veronal buffer on a microscope slide. Then a long rectangular well is cut alongside the separated proteins, antisera inserted into it and the proteins are detected by the resulting arcs of antibody–antigen precipitates. Up to 30 human serum protein components can be detected in this way.

4. Antibody–antigen crossed electrophoresis (AACE)

The antibody–antigen crossed electrophoresis (AACE) technique of Laurell (1965, 1966) has been widely used in protein polymorphism studies in the last five years. Here the plasma to be resolved is first separated in

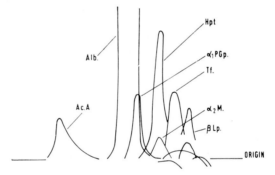

Fig. 3   Lines of antibody–antigen precipitate formed after antibody–antigen crossed electrophoresis (AACE). Electrophoresis in agarose is carried out in the horizontal plane first, the run removed and placed in another gel made of agarose and antibody. Electrophoresis is then carried out in the vertical plane, with most of the proteins migrating towards the anode (top of the page). The origin is that for the vertical dimension run. Ac.A = acetylated albumin, Alb = albumin, a_1PGp = a_1-precipitable glycoprotein, Tf = transferrin, a_2M = a_2-macroglobulin, βLp = β-lipoprotein, Hpt = haptoglobin. The antiserum is sheep anti-human serum. Redrawn from Studd et al. (1970)

conventional agarose gel electrophoresis, the agarose containing the separated proteins cut out and placed in a second agarose gel. The second gel is made up with antiserum to the plasma being resolved. Electrophoresis is then carried out at right angles to the direction of the previous run. The results are illustrated in Fig. 3. The antibody–antigen complexes form bell-shaped curves. These begin to appear as shallow bows soon after the second electrophoresis begins. They eventually stabilize at a particular height when all the antigen (serum) has been electrophoresed out of the interior part of the curve. For a given concentration of an antibody component the height of a peak is roughly proportional to the concentration of the corresponding antigen provided that the time of second electrophoresis is sufficiently long (Ouchterlony and Nilsson, 1973). When antisera to particular serum components are available, these components may be readily detected by AACE. The accuracy and reproducibility with which it may be used to quantitate serum protein levels offers the possibility that polymorphism in amounts of individual proteins may be detectable, something which is not possible with earlier electrophoresis techniques. AACE is more sensitive than conventional immunoelectrophoresis because electrophoresis rather than diffusion brings antibody and antigen together.

5. "Rocket" immunoelectrophoresis (Laurell, 1966)

This is like AACE except that the first electrophoretic run is omitted. It is very useful for measuring the concentration of specific proteins. Agarose containing a specific antiserum is poured on to a glass plate and holes for antigen solutions are cut out along a line parallel with one edge of the plate. Electrophoresis is performed so that the antigen migrates out of the holes towards the farther edge. Migrating peaks of antibody–antigen precipitates form which become stationary when antigen excess is no longer possible. The pH is chosen so that the antibody migrates in the opposite direction to the antigen; this is usually 8·6. The height of the stationary peak is proportional to antigen concentration, as in AACE. Calibration is done by including reference antigen solutions in each run.

6. Immunofixation electrophoresis

This technique was first developed by Alper and Johnson (1969). Electrophoresis is carried out in either agarose or starch gel and the position of particular components is fixed by precipitating them with a specific antiserum. The unfixed proteins are then washed out and the precipitate stained. Alper and Johnson (1969) suggest that greater sensitivity may be achieved by the radioactive labelling of a second antiserum to the immunoglobulin of the species used to make the antiserum used for immunofixation. Autoradiography would then detect the position of the antibody–antigen precipitate following application of the second antiserum. The technique of immunofixation has been successfully applied to the Gc, caeruloplasmin and C3 complement component polymorphisms. Like AACE it seems to offer the possibility of identifying polymorphisms in the low-concentration proteins of plasma.

7. Haemagglutination inhibition

The human immunoglobulin antigenic systems Km (Inv), Gm and Am are routinely detected by a haemagglutination inhibition test. The antisera for Gm typing often come from patients with rheumatoid arthritis, for Am groups from individuals who have suffered post-transfusion anaphylactic shock, and Km antisera from normal individuals who have received trans-fusions. The antisera react with immunoglobulins from individuals positive for the antigens. The Km and Gm agglutination inhibition tests are basic-ally derivatives of the Coombs test for the presence of incomplete antibodies (Steinberg, 1962; Natvig and Kunkel, 1968): O Rh(+) red blood cells aₑe coated with incomplete anti-D from an individual who possesses the Gm or Km antigen being tested for. More recently myeloma proteins with appropriate antigen have been used to coat the red cells. An incubation period is allowed to sensitize the cells. If the typing serum is added, these sensitized cells will be agglutinated. The observation which lead to the discovery of the Gm groups was that if the typing serum is first incubated with the test serum, agglutination is inhibited in some but not all cases (Grubb, 1956). Sera positive for the antigen inhibit, sera negative for it do not. Controls are used to ensure that the test serum itself is not agglutinat-ing and that the typing serum is active. The agglutination inhibition test as applied to the Am group involves coating the red cells with Am(+) immunoglobulin with the $CrCl_3$ technique of Gold and Fudenberg (1967), rather than by using incomplete anti-D. Bis-diazotized benzidine will also coat immunoglobulin on to red cells (Natvig and Kunkel, 1967).

Gm typing sera from rheumatoid arthritis patients are referred to as "Ragg" sera, for "*R*heumatoid *agg*lutinating". These are highly potent, but are often multi-specific and exhibit inconvenient prozone effects, reviewed by Steinberg (1962). Gm typing sera are also found in sera from non-rheumatoid individuals, usually from patients who have been transfused or from pregnant women. This was first demonstrated by Ropartz *et al.* (1960). Such antisera are referred to as "SNagg" sera, for "*S*erum *N*ormal *agg*lutinating". SNagg sera are less potent, are usually monospecific and do not exhibit prozone effects (Giblett, 1969). Gm typing sera may also be made by injecting rabbits (Ropartz, 1965; Litwin and Kunkel, 1966) and rhesus monkeys (Hess and Bütler, 1962; Alepa and Steinberg, 1964) with human serum or immunoglobulin. In addition to the agglutination inhibi-tion test, some Gm typing sera can be used in an Ouchterlony test (Kunkel *et al.*, 1966). The antisera are produced in primates. Reviews of the nature and origin of Gm and Km antisera are to be found in Steinberg (1967), Giblett (1969) and Grubb (1970).

III. POLYMORPHISMS

A. Established and Highly Probable Polymorphisms

1. Haptoglobin

Haptoglobin is an α_2-glycoprotein discovered by Polonovski and Jayle (1938, 1940), who observed that the addition of serum caused a marked enhancement of the peroxidase activity of haemoglobin. The physiological function of haptoglobin seems to be to combine specifically and stoichiometrically with haemoglobin. The complex is rapidly removed from circulation by the liver, thus preventing damage to the renal tubules by deposition of haemoglobin and also preventing loss of iron via the kidney (Sutton, 1970). Smithies (1955a, b) detected haptoglobin polymorphism using his starch gel electrophoresis technique. He defined three patterns (Fig. 4), Hp1-1, Hp2-1 and Hp2-2, the difference being controlled by two co-dominant alleles, Hp^1 and Hp^2 (Smithies and Walker 1955). His discovery initiated a very large series of investigations, such that the haptoglobin polymorphism is the best described serum protein polymorphism apart from the immunoglobulins. This work has been reviewed frequently (Kirk, 1968a; Giblett 1969; Harris, 1975; Buettner-Janusch 1970; Sutton 1970). The lessons learnt from haptoglobin have directed the study of other protein polymorphisms. The most distinctive feature of the polymorphism is the unusual nature of its molecular basis. The polypeptide produced by the Hp^2 gene is almost twice as large as that produced by the Hp^2 allele and is essentially a duplication of it; this enlarged polypeptide tends to form polymers, hence the complex patterns observed on the gels.

Sera are typed for Hp by adding sufficient Hb to the serum to saturate the Hp and staining for peroxidase activity after electrophoresis at a pH in the range 8 to 9. The stain is either benzidine (Smithies, 1959a, b) or o-dianisidine (Owen et al., 1960). Under these conditions, Hp1-1 is a single fast moving band, while Hp2-1 and Hp2-2 are much more complicated. Hp2-1 has the Hp1-1 band plus a series of slower bands whose intensity decreases with increasing mobility. Hp2-2 lacks the Hp1-1 band and has a similar series of slower bands, each slightly slower than the corresponding Hp2-1 band (Fig. 4). With fine resolution and sensitive staining, the stronger Hp2-2 bands can be seen in the Hp2-1 type (Sheridan et al., 1969; Sutton, 1970). Bearn and Franklin (1958), Allison (1959) and Smithies (1959b) suggested that the slower forms were a series of polymers of increasing size and that their resolution on starch gels was due to the molecular sieving effect exerted by that medium. It was also suggested that the bands peculiar to Hp2-1 are the result of random intracellular association between Hp^1 and Hp^2 gene products. Further work has established the correctness of these suggestions and given a detailed idea of the exact molecular basis of the multiple banding.

The first step in this was taken by Smithies (1960), who introduced a method for dissociating purified haptoglobin into its constituent polypeptides, which proved to be of two types, α and β. The method involves the

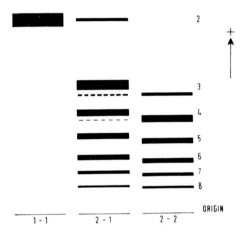

Fig. 4 The three basic haptoglobin types as defined by conventional starch gel or acrylamide electrophoresis. Note the presence of the first two HP2-2 bands in the Hp2-1 type which can be seen with very sensitive techniques (see also Fig. 1). The numbers on the right indicate the number of β (non-polymorphic) chains in the haptoglobin polymer series

reductive alkylation of disulphide bonds followed by electrophoresis in high urea (8M) starch gels at acid pH. The polymorphism resides in the α-chains. Rare β-chain variants have been found and the loci for the two chains are not closely linked (Cleve and Deicher, 1965; Javid, 1967). Using this technique Smithies et al. (1962a, b) and Connell et al. (1962) showed that the Hp[1] class was heterogeneous giving rise to either α^{1F} (fast) or α^{1S} (slow) polypeptides on urea gels. Hp1-1 could be α^{1F}, α^{1S} or both, while Hp2-1 showed either α^{1F} or α^{1S} in combination with α^2, which was slower than either (Fig. 5). This "subtyping" as Smithies's (1960) procedure is now usually called shows the existence of three polymorphic alleles, Hp^{1F}, Hp^{1S} and Hp^2. The α^{1F} and α^{1S} types could not be resolved on ordinary starch gels at alkaline pH. Fågerhol and Jacobsen (1969) have shown that starch gel electrophoresis of serum at pH 4·9 apparently resolves the Hp^{1F} product from the Hp^{1S} product.

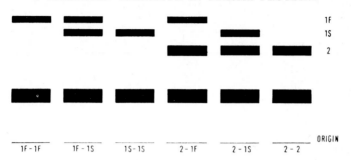

Fig. 5   The haptoglobin subtypes definable after isolation of haptoglobin and electrophoresis in urea and thioglycollate. The band common to all is attributed to the β polypeptide. IF and IS represent the two polypeptides produced by the Hp^{1F} and Hp^{1S} alleles and 2 that produced by Hp^2. Redrawn from Connell *et al.* (1966)

The molecular weights of α^{1F} and α^{1S} are about 9000, while that of α^2 is about 17 000 and that of the β chain about 42 500 (Black *et al.*, 1970). The chymotryptic peptide map of α^2 is identical with a mixture of the maps for α^{1F} and α^{1S} plus one extra peptide. This observation and the fact that α^2 was almost double the size of the other two polypeptides led Smithies *et al.* (1962b) to propose that the Hp^2 allele arose as a result of the partial duplication of the Hp^2 genes, by an unequal cross-over between the Hp^{1F} and Hp^{1S} alleles. This idea is supported by the sequence data for the three chains (Black and Dixon, 1968; Malchy and Dixon 1973). The α^{1F} and α^{1S} chains each have 83 amino acid residues and differ only by a single residue at position 54, lysine in α^{1F} and glutamic acid in α^{1S}. The N terminal sequence of α^2 from position 1 to 71 is identical with the N terminal sequence of an α^1 chain from its 1 to 71 positions. The C terminal sequence of α^2 from positions 72 to 142 corresponds to the C terminal sequence of an α^1 chain from positions 12 to 83 (Fig. 6). It has not yet been established whether α^{1F} or α^{1S} forms the N terminal sequence.

$$\overset{1}{\text{H}_2\text{N-Val}}\ldots\ldots\ldots\ldots\ldots\overset{71}{\text{Cys.Glu.Ala.Val}}\ldots\ldots\overset{83}{\text{Gln-COOH}}$$

$$\overset{1}{\text{H}_2\text{N-Val}}\ldots\ldots\ldots\ldots\ldots\text{Cys..Glu.Ala.Asp.Asp}\ldots\ldots\ldots\ldots\overset{142}{\text{Gln-COOH}}$$

$$\overset{1}{\text{H}_2\text{N-Val}}\ldots\overset{12}{\text{Ile.Ala.Asp.Asp}}\ldots\ldots\ldots\ldots\overset{83}{\text{Gln-COOH}}$$

Fig. 6   Origin of Hp^2 by non-reciprocal crossing-over in an Hp^{1F} Hp^{1S} heterozygote. The $Hp\alpha^2$ chain, thus generated, is shown in the middle, with $Hp\alpha^1$ chains on top and bottom. The exact point of the cross-over cannot be established from the structure of the hp-chains, the genetic code being compatible with crossing within the region of the underlined sequences or within the first codon on either side of the underlined sequences

284 D. W. COOPER

An inspection of the base sequences of the codons determining amino acids in positions 9 to 17 and 67 to 75 in the α^1 polypeptides shows many similarities. This internal homology could have been responsible for mispairing at meiosis with the subsequent production of the α^2-chain (Black and Dixon, 1968).

If this mechanism for the production of the α^2-chain is the correct one, it should also be possible to make a partial triplicate of the basic α^1 polypeptides by mispairing between an Hp^1 and Hp^2 allele. An unusual rare variant, Hp Johnson, has been found with an α-chain larger than the usual α^2 and this may represent such a triplication, although so far detailed molecular weight and sequencing studies do not appear to have been published.

The evolutionary relationships of both the α- and β-chains are intriguing. The α-chain of haptoglobin and the L chain of immunoglobulins have similar primary structures and may have a common evolutionary origin (Black and Dixon, 1968). Haptoglobin may be looked upon as a kind of "auto-antibody" removing haemoglobin from a part of the body where it is unwanted, i.e. from plasma. The β-chain however shows a distant relationship to the chymotrypsin family of serine proteases. In a comparison of 171 residues (approximately 60% of the chain) some 30% were found to be identical with residues occurring in either bovine trypsin, bovine chymotrypsinogen A, bovine chymotrypsin B, porcine elastase or bovine thrombin B-chain (Kurosky et al., 1974). Some internal homology also occurs in the N terminal region of the β-chain, which may have arisen by an ancient unequal cross-over event, in the same way as Hp^2 may have arisen in the recent lineage of Man.

The quaternary structure of haptoglobin is not completely understood. Hp1-1 has a molecular weight of 98 000 (Cheftel and Moretti, 1966; Moretti et al., 1966) and is almost certainly composed of two α- and two β-chains, possibly joined together in the same way as the L and H chains of immunoglobulin G. Sutton (1970) has summarized the results from a series of investigations on the structure of the slower polymer forms of Hp2-1 and Hp2-2. He suggests that the Hp2-1 series consist of a series $\alpha^1\alpha^2\beta_n$, where n varies from 3 (for the fastest non-Hp1-1 band in Hp2-1) up to 7 to 10, according to the number of multiple bands observed in the particular serum (Fig. 4). Hp1-1 is $\alpha_2^1\beta_2$. Similarly Hp2-2 consists of a series $\alpha_2^2\beta_n$ with n varying in like manner. Support for this idea comes from studies on haemoglobin binding, which occurs through the α-chain of Hp. Each β-chain appears to be able to bind one $\alpha\beta$ haemoglobin dimer (not $\alpha_2\beta_2$ tetramers). Thus Hp1-1 in the presence of less than saturating amounts of Hb shows an intermediate form of molecular weight around 135 000 (Fig. 7). The fully saturated form has a molecular weight of 167 000 (Hamaguchi, 1967; Ogawa et al., 1968). The first band of the Hp2-1 and Hp2-2 series each have three different Hp-Hb complex forms and the second has four in each case. The other bands are too close together to permit analysis.

Variation in haptoglobin level has also received attention. There is considerable variation in the number and strength of the 2-1 bands in Hp2-1 types. Moreover, the stronger the 1-1 band in Hp2-1 types, the fewer the number of 2-1 bands. It seems that the amount of α^2 substance may differ from individual to individual, and that this characteristic is inherited, although the precise mode of inheritance is not clear. Both Giblett and Steinberg (1960) and Sutton and Karp (1964) have proposed that there may be different Hp^2 alleles with respect to rate of production

Fig. 7   The effect upon haptoglobin of adding increasing quantities of haemoglobin to a Hp1-1 serum. With quantities of haemoglobin insufficient to saturate the haptoglobin, an intermediate complex is formed. It probably consists of one haptoglobin molecule and one "half" haemoglobin molecule (equivalent to an $\alpha\beta$ dimer). The fully saturated complex consists of one haptoglobin and one haemoglobin molecule

of the α^2 polypeptides. There are also many people who lack detectable haptoglobin and are referred to as HpO. The cause of a considerable proportion of these cases can be ascribed to conditions which lead to increased red cell turnover but there is also some suggestion that other cases may have a genetic basis, whose precise nature is still unclear (Sutton, 1970). Neonatal and young children also have low or undetectable levels of haptoglobin.

The population genetics of the haptoglobin types have been investigated in great detail and are summarized by Kirk (1968a) and Giblett (1969). The highest frequency of Hp^2 is found amongst Asians (approximately 0·75), while Europeans have frequencies of about 0·60, and Negroid populations in Africa frequencies of about 0·30 to 0·40. Considerable variation exists in aboriginal American and Pacific Island populations.

Investigations of selective forces which may possibly operate upon the haptoglobin polymorphism have taken two forms, analysis of segregation ratios and correlations with various diseases. While segregation ratios are usually Mendelian, an interaction with the *ABO* locus has been found (Kirk *et al.*, 1970; Kirk, 1971; Macdonald and Papiha, 1974; Vana and Steinberg, 1975) and perhaps with sex ratio (Brackenridge, 1973). The reasons for the association with the ABO phenotypes are unclear. It does not appear to be based on ABO incompatibility or population stratification, although all investigators regard the association as real. A number of investigations have been carried out on the association of Hp type and variety of diseases (Rundle *et al.*, 1975).

Teisberg and Gjone (1974) found strong evidence that the *Hp-α* locus is closely linked to that for the enzyme lecithin cholesterol acyl-transferase (LCAT). The *Hp-α* locus has been localized between the middle and terminal point of chromosome 16 (Robson *et al.*, 1969).

2. Transferrin (Tf)

Transferrin (siderophilin) is a β-globulin glycoprotein whose function is to transport iron from intestine to sites in the body such as the bone marrow, where it is utilized in the synthesis of haemoglobin, myoglobin and other iron-containing proteins. Under normal conditions serum transferrin is only about one-third saturated with iron and may therefore act as the primary defence against acute iron intoxication. It is also capable of binding other metals such as copper, manganese, cobalt, zinc and possibly chromium. Whether the capacity to bind these elements is of physiological significance is still to be established. It is also possible that transferrin may protect against infection by bacteria by successfully competing for ferric iron essential for their growth (Giblett, 1969).

Inherited variation in the electrophoretic mobility of transferrin was first discovered by Smithies (1957) using starch gel electrophoresis. He described it as β-globulin variation. Its transferrin nature was subsequently established by Allison, Poulik and Sutton (cited by Smithies and Hiller, 1959). Giblett *et al.* (1959) described the technique of ^{59}Fe autoradiography following gel electrophoresis which is now generally used for its identification. On starch gel it migrates as a distinct prominent dark band between the postalbumin region and haemoglobin-saturated Hp1-1 and this may be used to type it routinely, without recourse to autoradiography. The genetics of human transferrin has been extensively studied since the discovery of these inherited differences. Reviews of this work can be found in Bowman (1968), Kirk (1968b), Giblett (1969) and Buettner-Janusch (1970). The review of Giblett gives a comprehensive account of the physiology and chemistry of transferrin.

Human serum transferrin saturated with iron migrates as a single major band in homozygotes upon electrophoresis in starch and acrylamide at alkaline pH's. Heterozygotes possess two bands. The transferrin of the cord blood of premature infants and of spinal fluid possess multiple bands which are due to varying numbers of sialic acid residues per molecule (Parker and Bearn, 1962; Parker *et al.*, 1963b). These negatively charged molecules cause the transferrin to migrate faster towards the anode at alkaline pH's. Iron-free and iron-saturated transferrin have different mobilities (Nute, 1968, cited in Buettner-Janusch, 1970). Old and infected sera tend to show multiple transferrin bands not found in fresh sera, which in some cases are due to the splitting off of sialic acid residues by the enzym neuraminidase.

All human populations possess a normal type designated C. In addition they usually possess one or more variant types, the frequencies of whose determinant genes are usually rather low and are always less than 50%. Those faster than C are designated B, those slower, D. Particular variants are identified by a subscript. At first these were numbers which showed the relative mobility of the variant, but as more were discovered, this system broke down and geographical subscripts came into use. At present 18 variants are known, which are in order of fastest to slowest in electrophoretic mobility: B_{Lae}, B_0, B_{0-1} $B_{Atlanti}$, B_1, B_{1-2}, B_2, B_3, (C), $D_{Adelaide}$, D_0, D_{Wigan}, D_{0-1}, $D_{Montreal}$, D_{Chi}, D_1, D_{Fin}, D_2, and D_3 (Giblett, 1969). D_{Fin} is not definitely distinct from D_2. Barnett and Bowman (1968) have described another variant, $B_{Lambert}$, with mobility near that of B_1. Its relationship to other variants is not yet determined. The family data available show that the difference between C and any variant is determined by a single gene pair exhibiting codominance. Only four of these genes are found in frequencies of the order of 0·01 or greater. They are D_{Chi} amongst the Chinese and related groups, D_1 amongst Caucasians and B_{0-1} amongst Navajo Indians. Since there are no populations with two variants in appreciable frequencies, families in which more than one variant are segregating are rare. Individuals with B_2D (Beckman, 1962), B_1B_2 and B_1D_1 (Beckman and Holmgren, 1963) and $B_{1-2}B_2$ (Robinson et al., 1963) have been described. They do not possess a C band nor any "hybrid" band, which suggests that their respective genes are allelic. It is very likely that a series of multiple alleles at a single locus determines all the human variants. Transferrin polymorphism occurs extensively in other vertebrates. In cattle, sheep, horse, several primate species (Manwell and Baker (1970) and Buettner-Janusch (1970)), these are determined by a series of multiple alleles occurring in high frequency in the one population. There is no animal species in which evidence for a second locus has been found.

In further and conclusive support of the argument that only one locus is involved are the data which show that the molecule has only one long polypeptide (Greene and Feeney, 1968; Mann et al., 1970; Palmour and Sutton, 1971; MacGillivray and Brew, 1975). Partial sequencing indicates the existence of internal homology (about 40%) between the amino terminal half and the carboxyl terminal half (MacGillivray and Brew, 1975). These writers suggest that an ancestral gene duplication has occurred at some distant point in the phylogenetic development of vertebrate transferrins, which would be comparable with the origin of Hp^2 in the recent lineage of Man. They conclude from their studies and those of Williams (1974, 1975) that transferrin is composed of two homologous regions, each having a single Fe binding site and some conformational independence, including absence of disulphide bridges between the two homologous domains. The two iron-binding sites appear to be functionally equivalent (Harris and Aisen, 1975).

The fingerprinting of tryptic digests of TfC and variants suggests that the differences between them are probably due to single amino acid substitutions. This has been shown for D_1 (Wang and Sutton, 1965; Jeppson and Sjöquist, 1966; Jeppson, 1967b), B_2 (Wang et al., 1966) and D_{Chi} (Wang et al., 1967). No evidence has so far been found to suggest that differences in the sialic acid residues are responsible for the differences between variants.

Turnbull and Giblett (1961) have compared the amounts and rates of iron uptake of several variants. They found no evidence to suggest differences. Apart from its use as a marker in anthropological studies and work on its primary and higher structures, not much attention has been paid to transferrin in the last five years. A similar protein called lactoferrin exists in milk and other secretions. This probably has antimicrobial activity (MacGillivray and Brew, 1975).

3. Group-specific component (Gc) (vitamin-D-binding α-globulin, VDBG)

The Gc polymorphism was discovered by Hirschfeld (1959) using immunoelectrophoresis. Its principal interest until very recently has been as a marker for anthropological studies. It has been used in a large number of surveys of blood protein polymorphisms. Gc is an abbreviation for group-specific component, which is an α_2-globulin. Its molecular weight is about 50 000 (Bearn et al., 1964a) and it is produced in the liver (Prunier et al., 1964) beginning about the tenth to thirteenth week in gestation (Melartin et al., 1966). The variants of the protein differ in charge, so that their positions after electrophoresis can be shown by use of an appropriate antiserum. Hirschfeld (1959) found three patterns, a fast arc, a slow arc and a mixture of the two, i.e. a bimodal arc. Hirschfeld et al. (1960) called the three phenotypes Gc1-1, Gc2-2 and Gc2-1 respectively, and gave family data to support the hypothesis that two alleles, Gc^1 and Gc^2, were responsible for the differences. This hypothesis is supported by much further family data, the references to which are given by Reinskou (1968) and Giblett (1969), who have written comprehensive reviews of the system.

In addition to the immunoelectrophoretic techniques, Gc phenotypes may also be detected by starch gel electrophoresis (Schultze et al., 1962;

Arfors and Beckman, 1963; Parker *et al.*, 1963a; Bearn *et al.*, 1964b). The Gc proteins migrate as postalbumins, very close to the slow edge of the albumin. In haemolysed sera or in fluids where Gc concentration is low, AACE may be used (Sutcliffe and Brock, 1973). Each allele produces two bands, the faster of Gc^2 having the same mobility as the slower of Gc^1, so that Gc2-1 possesses three bands and the two homozygotes two each. The difference between the two bands produced by the same allele and between the Gc^1 and Gc^2 products has been investigated by Bearn *et al.* (1964a), Bowman and Bearn (1965) and Bowman (1967). It seems likely that both components consist of two very similar but not identical subunits, while the products of Gc^1 and Gc^2 differ by at least one amino acid substitution.

Besides the two common variants, there are a number of other types, which are mostly rare. Two alleles, Gc^{Chip} and Gc^{Ab}, which occur in American Indians and Australian Aborigines respectively, occur in low but apparently polymorphic frequencies (Giblett, 1969). Gc^{Ab} is associated with the bizarre neurological disease, kuru, which occurs in the Fore tribe of New Guinea (Kitchin *et al.*, 1973).

Weitkamp and his colleagues (Weitkamp *et al.*, 1970) have collected a great deal of data which show that the Gc locus and the structural locus for albumin (in which rare variants have been detected) are closely linked, the best estimate of the recombination fraction being 2·3%.

Until 1974 no biological function for the Gc protein was known. In that year Daiger *et al.* reported the interesting discovery that the Gc protein is in fact identical with vitamin-D-binding α-globulin (VDBG), which was first described in the same year as the Gc groups (Thomas *et al.*, 1959). A full account of their work has now appeared (Daiger *et al.*, 1975a, b). VDBG binds to the naturally occurring vitamins, D_2 and D_3, and to the metabolically more active forms, 25-hydroxyvitamin D and 1α,25-dihydroxyvitamin D. This discovery opens up the possibility of searching for possible selective differences between the various genotypes based upon this physiological role. An excess of heterozygotes occurs in the progeny from Gc2-2 × Gc1-2 matings, which suggests a possible balancing mechanism (Reinskou, 1968).

Daiger *et al.* (1975a, b) used labelling with radioactive D_3 followed by autoradiography to screen for genetic variants of VDBG. This was part of a general screening programme for genetic variants in various transport proteins in human plasma. The greatly increased sensitivity of this method of detecting the position of transport proteins after electrophoresis means that low-concentration components can be screened. As a result, two or possibly three other polymorphisms have been detected. (See below under thyroxine-binding globulin, transcobalamin, and testosterone-binding globulin.)

4. Caeruloplasmin

Caeruloplasmin is a copper binding α_2 globulin of the plasma first described by Holmberg and Laurell (1948). It is detectable after gel electrophoresis by its oxidase activity against o-dianisidine (Owen and Smith, 1961). Its biological function is not fully understood. It could be an oxidase since it reacts with dopa, epinephrine and serotonin (Laurell, 1960) or it could be concerned with copper metabolism (Laurell, 1960; Bearn, 1966). It is markedly reduced in Wilson's disease (Bearn, 1966) and some heterozygotes for this disease (Sternlieb et al., 1961). It is also reduced in some cases without any sign of accompanying abnormality of the subject (Cox, 1966). It is increased in several other conditions (for instance, in pregnancy) after the administration of steroid hormones, in fever and infections, in certain malignancies and after immunizations (Laurell, 1960). Using horizontal starch gel electrophoresis with a rather alkaline borate/sodium hydroxide buffer, Shreffler et al. (1967) detected polymorphism in American Blacks and American Caucasians. There are two forms, Cp^A and Cp^B. The two forms have very similar mobilities, which explains why they were not distinguished before. One might have expected them to have been, because caeruloplasmin also shows up on gels stained with o-dianisidine for the haptoglobin types (Owen and Smith, 1961). The Cp^A gene frequency is 0·053 in Negroes and 0·006 in Caucasians. A third polymorphic form CpTh has been found in Thais (Shokeir et al., 1968). The frequency of its allele is 0·157 and its mobility is very similar to but not identical with that of CpA. Several rare variants exist: CpC in Negroes (Shreffler et al., 1967), CpIF found in an Irish pedigree (McAlister et al., 1961) and CpNH (Shokeir et al., 1967). Other population studies have been performed (Kellerman and Walter, 1972; Vasiletz et al., 1974; and cf. p. 174).

Several suggestions have been made concerning its molecular structure. On the basis of more recently gathered data both Shokeir (1973) and Vasiletz et al. (1973) have proposed that it is a heterodimer, i.e. has one α and one β subunit of approximately equal molecular weights.

The Russian group has investigated the nature of the defect in Wilson's disease. They find that purified caeruloplasmin has lower specific oxidase activity and altered peptide patterns after trypsin hydrolysis (Neifakh et al., 1972). They also find that the amount of caeruloplasmin-forming liver polysomes is 10 to 20 times lower than in normal individuals, and a corresponding reduction in caeruloplasmin polypeptides is observed (Gaitshovski et al., 1975). Evidently the alteration to the structure of the gene for one (at least) of the polypeptides leads to a reduction in transcription or to a loss of stability of the mRNA.

5. Alpha-1-antitrypsin (protease inhibitor—Pi) system

Fågerhol and Braend (1965) used starch gel electrophoresis at acid pH to discover inherited variation in the prealbumin zone. Shortly afterwards AACE was used to show that this variation resided in the serum α_1-antitrypsin (AAT) component (Kueppers and Bearn, 1966; Fågerhol and Laurell, 1967). Since AAT is the principal protease inhibitor in human serum, the symbol Pi was chosen to designate the system. It has been reviewed by Fågerhol and Laurell (1970), Mittman (1972) and Mittman and Lieberman (1974). Useful short summaries are to be found in Kueppers (1973) and Janus and Carrell (1975). Over the last five years the Pi system has received more attention than any other plasma protein polymorphism. The literature on it has become dauntingly large. The principal reason for this is that variant AAT types predispose towards lung and liver disease. Z, the variant which is mainly involved, is also of interest because it appears that its lower level in plasma is due to a defect in its secretion from the liver cells, and not to any reduction in transcription. Several recent papers indicate that Z and perhaps other variants may also increase the likelihood that their bearer will produce chromosomally abnormal offspring, a finding that will undoubtedly be the subject of many more investigations.

Upon electrophoresis in starch or by AACE, a variety of different phenotypes may be found. The patterns observed can be explained in terms of a single locus for the molecule at which there are multiple alleles co-dominantly expressed. After starch gel electrophoresis at pH 4·9, each homozygote exhibits a pattern of three major and five minor bands (Fågerhol, 1969). The substitution of one allele for another leads to a step-wise change in the mobility of each band, which must therefore all be under the control of the one gene. Fågerhol and Laurell (1970) suggest that the eight bands may represent complexes with borate or different configurations of the same molecule. Fågerhol and Laurell also consider that it is unlikely that these eight forms are to be found in the native protein of the serum, a view which is supported by the fact that the molecule migrates as a single band in agar gel at more alkaline pHs (6 to 9). All eight bands can be demonstrated by AACE when the first electrophoresis is in starch at acid pH. The polymorphism can also be detected by isoelectric focusing on polyacrylamide gel slabs, but this method has not yet been used a great deal (Allen et al., 1974). AAT can be typed from post-mortem blood samples (Talamo and Thurlbeck, 1975).

A large body of family data supports the multiple allele hypothesis (early work reviewed in Fågerhol and Laurell, 1970; Cook, 1975). A total of 23 different alleles have been described, 22 electrophoretic variants and one null allele Pi^- (Blundell et al., 1974; Cox and Celhoffer, 1974; Cohen, 1974). One of these, Pi^M, is by far the most common in all populations which have been examined, generally having a frequency of at least 0·90. Only three alleles Pi^F, Pi^S and Pi^Z are polymorphic i.e. have frequencies greater than 0·01.

AAT inhibits the proteolytic action of trypsin, elastase, collagenase, leucocytic proteases, plasmin and thrombin. About 90% of the inhibitory capacity of serum for these enzyme lies in the a_1-antitrypsin component. Other bodily fluids also contain the protein where it presumably also serves the same function. The allele Pi^Z determines a mere 7% of the amount of protein determined by Pi^M while Pi^S causes a less marked reduction. The capacity of the phenotypes containing S and Z to inhibit trypsin is correspondingly reduced (Rynbrandt et al., 1975). Elastase is even less inhibited by the Z variant than is trypsin (Rowley et al., 1974) and may therefore be a better enzyme for detecting Pi^ZPi^Z homozygotes.

The molecule is a glycoprotein with a molecular weight of approximately 55 000 (Crawford, 1973; Chan et al., 1973). Part of the carbohydrate moiety includes sialic acid residues. A polypeptide which contains material immunologically related to AAT and which has a molecular weight of 18 000 can be isolated from the liver of Pi^MPi^M, Pi^MPi^2 and Pi^2Pi^2 individuals (Matsubara et al., 1974). The relationship of this molecule to the native protein found in the plasma has yet to be determined. If for example this polypeptide is represented twice in the native molecule, it is surprising that heterodimers are not found in heterozygotes. Rowley et al. (1974) have found evidence that the apoprotein accumulates in the hepatic parenchymal cells of Pi^ZPi^Z individuals. They argue that this suggests a defect in hepatic secretion as the basis for the lower amount of protein in such individuals. Cox (1975) has shown that the Z form lacks two terminal sialic acid residues in its carbohydrate chain, which may either cause difficulty in secretion, or else result in aggregation. A very detailed biochemical model of the difference between the carbohydrate moieties of M and Z and the effect of these differences upon secretion from the liver cells has been presented by Yunis et al. (1976). Owen and Carrell (1976) fingerprinted the S variant and find that a glutamic acid residue of the M type is replaced by a valine. They suggest that this substitution may lead to increased lability of the S protein, which would explain the reduction in its concentration in plasma.

Alpha-1-antitrypsin is increased in various stress situations: acute and chronic infection, cancer, pregnancy and in response to oestrogenic hormones. However, individuals homozygous or heterozygous for Pi^Z were unable to respond to the oestrogenic hormone diethylstilbestrol (Lieberman and Mittman, 1973). This also supports the notion that secretion of Z is deficient.

The literature on its association with lung and liver disease is too large to list in its entirety in this short account. The Pi^Z allele was recognized in the

6 INHERITED VARIATION IN PLASMA PROTEINS   293

homozygous state before the discovery of electrophoretic variation as a deficiency of α_1-antitrypsin associated with degenerative pulmonary disease (Laurell and Eriksson, 1963; Eriksson, 1964; Kueppers *et al.*, 1964). Despite earlier doubts a great deal of evidence now demonstrates that Pi^Z to a very marked degree and Pi^S and Pi^P to a lesser degree predispose towards emphysema (obstructive lung disease) in adulthood and cirrhosis of the liver in childhood (Mittman and Lieberman, 1974; Crawford *et al.*, 1974; Wilkinson *et al.*, 1974; Asarian *et al.*, 1975; Janus and Carrell, 1975; and earlier references mentioned therein). The homozygotes Pi^SPi^S and Pi^ZPi^Z were represented in a sample of 164 patients with obstructive lung disease 12 and 40 times greater than would be expected on the basis of their population frequencies (Mittman and Lieberman, 1974). There are also indications that Pi^S and/or Pi^Z may predispose towards other diseases, e.g. chronic pancreatitis (Novis *et al.*, 1975) gastric and duodenal ulcers (Andre *et al.*, 1974) and hepatic carcinoma (Schleissner and Cohen, 1975; Lieberman *et al.*, 1975). Because of this association with disease large-scale screening programmes have been initiated. Using rocket immunoelectrophoresis (see Techniques), Laurell and Sveger (1975) performed the feat of screening 108 000 newborn Swedish infants for α_1-antitrypsin deficiency! As discussed by Mittman and Lieberman (1974) the justification for the screening programmes must be that with knowledge of their Pi type individuals at risk can avoid harmful irritation to their lungs or liver. Such irritation may be caused by smoking, excessive alcohol consumption, frequent exposure to enzyme washing powders or occupations in dusty environments. Whether their prognosis will be sufficiently improved remains to be established. Chowdhury and Louria (1976) have shown that *in vitro* cadmium reduces α_1-antitrypsin in activity and concentration. They suggest that this may explain the emphysema commonly observed in industrial workers exposed to cadmium. The effect of cadmium upon the variant Pi types would be well worth a study.

Aarskog and Fågerhol (1970) and Fågerhol (1972) were the first to draw attention to the possibility of increased Pi heterozygosity in a sample of individuals with various chromosomal aberrations. A similar increase seemed to occur in their parents. Two further papers have supported this interesting notion (Kueppers *et al.*, 1975; Fineman *et al.*, 1976). The latter authors found the effect to be particularly marked when the mother of the chromosomally abnormal propositus was over 35 years of age. Since the variants concerned have lower inhibitory capacity, these findings raise the question whether proteolytic enzymes and their inhibitors are somehow concerned with the normal mechanisms of meiosis and mitosis.

Why has such manifestly deleterious genetic variation become established? Kueppers (1972) has put forward the hypothesis that high levels of antitrypsin may inhibit fertilization of the egg and that conversely lower levels might promote fertilization. If so, the Pi^Z frequency may represent a balance between increasing the frequency of fertilization on the one hand, and of decreased fitness of the resulting zygotes on the other.

Gedde-Dahl *et al.* (1972) have shown that the Gm and Pi groups are very likely linked, the best estimate for the two sexes combined being 25 cMs.

6. C3 component of complement

As is indicated in Fig. 8, C3 occupies a central position in the complement system. Its biological and physical properties have been summarized by Bokisch *et al.* (1975). A thermolabile β_1 plasma protein of molecular weight 190 000, C3 carries out many of the physiological functions necessary for the

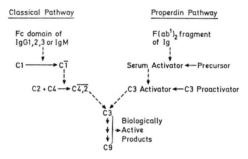

Fig. 8   The complement pathways. Complement is the name given to a group of plasma proteins which act sequentially to produce a variety of immunological and other biological reactions. The complement reactions are usually initiated by immunoglobulins, particularly after they have combined with antigen. One of two alternative sequences can begin the complement pathway. In the classical system the C1, C4, C2, C3, C5, C6, C7, C8 and C9 components act in that order. In the less well-characterized alternative or properdin system C1, C2 and C4 are by-passed and C3 is activated by properdin factor B (C3 activator). C3, which is by far the most concentrated component in the plasma, is thus the pivot of the whole system. This figure shows an outline of the probable relationships between the properdin and classical pathway

host to defend itself against pathogenic microorganisms and perhaps potential tumour cells. It occurs in plasma and other body fluids in inactive form. When activated by either the classical or properdin pathway it is split into two fragments, C3a the smaller and C3b the larger. C3a is one of the two anaphylotoxins which have been described. In very low concentrations this 9000 molecular weight polypeptide causes release of histamine from mast cells, chemotaxis of polymorphonuclear leucocytes, contraction of smooth muscle, and formation of cutaneous oedema and erythema. The

C3b fragment binds to the surface of cells and other large particles, particularly immune complexes, and enhances their ingestion by phago- cytes. C3b may also have a part in stimulating antibody-forming cells. Both C3a and C3b are inactivated by serum enzymes, the C3b component being split into fragments C3c and C3d. The molecule has two chains, a of 120 000 molecular weight, and β of 75 000 molecular weight which are united by three disulphide bands and non-covalent forces. The various fragments are formed by splitting the a-chain. Deficiency of C3 results in recurrent bacterial infection. It is probably made by the liver and its concentration is by far the highest of all complement components. The neonatal young makes its own C3 and there is no detectable placental transfer as judged from comparisons of mothers' and cord blood sera (Azen et al., 1969; Propp and Alper, 1968).

C3 is readily visualized after zone electrophoresis in agarose or starch using amido black. Two groups discovered polymorphism in it independ- ently. They were Alper and Propp (1968) using agarose and Azen and Smithies (1968) using high voltage starch gel electrophoresis. Rose and Geserick (1969) shortly afterwards discovered the same system in aged sera which accumulate C3c and described it as the Pt (post-transferrin) system. It was later realized that the variation discovered was the C3 system (Mauff et al., 1974; Alper, 1973).

Two variants now designated by international agreement as F and S are found in most populations. Their inheritance is autosomal codominant, the alleles being $C3^F$ and $C3^S$. Typing is usually done with high voltage agarose gel electrophoresis (Teisberg, 1970). Some 16 less common or rare variants exist in addition to the two common ones. The concentration of Ca^{2+} ions is critical in accurately typing some of these, since several variants have altered Ca^{2+} binding properties. Since all of the variants behave as alleles, it is probable that they are all in either the a- or β-chain although which one does not appear to have been discovered.

7. C6 complement component

Polymorphism in this component has recently been found by Hobart et al., 1975, cited in Lachmann, 1975. They used both electrophoresis and iso- electric focusing in polyacrylamide slab gels and detected the C6 bands by either immunofixation or an agarose overlay containing red cells coated with antibody and C6 deficient rabbit serum. Two common alleles and four rare ones have been found in Caucasians.

8. Properdin factor B (C3 proactivator or glycine rich glycoprotein)

Properdin is the name given to several factors which can activate comple- ment by an alternative pathway to the classical system (see Fig. 8). These are still in the process of characterization. Properdin factor B was discovered by Pillemer *et al.* (1954) as a heat-labile component of the properdin system. Since then it has been investigated in a number of different ways and given a variety of different names by various groups of workers who have only recently appreciated that they were dealing with different aspects of the same molecule. It has been isolated from serum by Haupt and Heide (1965) who described it as β_2-glycoprotein II and by Boenisch and Alper (1970) who called it glycine-rich β_2-glycoprotein (GBG). Neither of these pairs of workers realized its biological function at the time of isolation. Using the technique of high voltage agarose gel electrophoresis (Teisberg, 1970) followed by immunofixation (Alper and Johnson, 1969), Alper *et al.* (1971, 1972) described genetic polymorphism in GBG. Two common · alleles, now referred to as Bf^F and Bf^S were discovered together with two rare alleles Bf^{F1} and Bf^{S1}. Each allele specifies four equally spaced components separable by the electrophoretic technique. It was then realized that GBG was identical with factor B of the properdin system and the cobra factor-binding protein of human serum (Alper *et al.*, 1973) and with protein isolated by Götze and Müller-Eberhard (1971) and called by them as C3 proactivator. According to Ruley *et al.* (1973) it is also likely that it is identical with the Ce nephritis factor. Human factor B may exist in different forms in serum. It is rapidly converted to glycine-rich "γ globulin" and glycine-rich α-globulin upon incubation with antibody–antigen complexes or endotoxin (Alper *et al.*, 1971). The heparin in plasma apparently combines with it, rendering it unsuitable for genetic typing (Rittner *et al.*, 1975); this combination is reversible with protamine salts.

Further interest in this protein has been generated by the fact that it is linked to the major histocompatibility locus (HLA) which is on chromosome 6. This was first reported by Allen (1974). The loci for the C2 complement component (Fu *et al.*, 1974; Wolski *et al.*, 1975), the C4 complement component (Hauptmann and Rittner, cited by Meo *et al.*, 1975) and the properdin factor (Rittner *et al.*, 1975) are all linked to HLA. Rittner *et al.* (1975) suggest that this region of chromosome 6 may have other genes concerned with immunological functions, and drew a comparison with the H-2 region in the mouse. The mouse H-2 region has recently been shown to influence the levels of C3 (Ferreira and Nussenzweig, 1975). However, in man the *C3* locus itself is not linked to HLA (Gedde-Dahl *et al.*, 1974). The present probable map of the region is shown in Ch. 8, Fig. 1.

9. α₂-Acid glycoprotein (orosomucoid)

Johnson *et al.* (1969) have used immunofixation after agarose gel electrophoresis to define an autosomal codominant polymorphism determined by two alleles, Or^S and Or^F. They used neuraminidase-treated serum to obtain sialic acid-free orosomucoid. This presumably enables the phenotype to be distinguished more easily by eliminating electrophoretic heterogeneity due to varying numbers of sialic acid residues per molecule. The phenotypes are designated FF, SS and FS, corresponding to fast, slow and double electrophoretic patterns. Sera of homozygous S individuals have small amounts of material of identical electrophoretic mobility to F, and vice versa. Johnson et al. (1969) suggest the possibility that both F and S structural genes are present in all individuals and that the differences between patterns are the result of genetic variation in relative rates of synthesis. However, they regard it as more likely that conventional codominant inheritance is involved and that the different bands in homozygotes are due to secondary modification of the product of each allele. Family studies, twin pairs and population data support the two-allele hypothesis. Cord sera may differ from mothers' sera, hence transplacental passage of the protein probably does not occur. The gene frequency of Or^S ranges from 0·37 to 0·62 in various African populations; from 0·36 to 0·67 in Caucasian populations and is 0·47 in Chinese, 0·27 in Japanese, 0·44 in Indians from Calcutta and 0·54 in Mexican Indians.

10. Thyroxine binding α-globulin

Thyroxine (T_3 and T_4) in human plasma may be bound to three distinct proteins which are, in order of decreasing mobility on starch or acrylamide, thyroxine binding pre-albumin (TBPA), albumin itself, and thyroxine binding α-globulin (TBG). The major portion of protein bound thyroxine is found in the TBG fraction. TBPA appears to have the primary role of binding to a vitamin A-protein complex, with its transport of thyroxine being a secondary role. In rhesus monkeys TBPA variation is inherited in an autosomal co-dominant manner, and the molecule itself has a tetrameric structure (Alper et al., 1969; Bernstein et al., 1970). Binding to albumin occurs only after the other two are saturated.

Heritable low levels of TBG were first recorded by Nicoloff et al. (1964). Nikolai and Seal (1966, 1967) showed that this variation is X-linked. Low TBG does not lead to thyroid abnormality. A number of other such pedigrees also showing X-linkage have been recorded (Grant et al., 1974). No close linkages to other X-chromosome loci have been discovered. Shane et al. (1971) have described an X-linked elevated level of TBG. In this pedigree 13 of the 15 family members with high TBG had a goitre. Shane et al. concluded that the limited family data were best explained on the basis of two loci, one for the goitre and one for the high TBG.

During the course of the search for genetic variation in plasma transport proteins being carried out in Cavalli-Sforza's laboratory, Daiger and Rummel have discovered electrophoretic variation for TBG in American Blacks and Orientals (Daiger, 1976). The relatively high level, 15%, for the variant gene in American Blacks should make this locus a valuable one for further mapping of the human X. Whether the locus for electrophoretic variation and the locus for the deficiency are the same has yet to be determined.

11. Transcobalamin II (Tc, a vitamin B_{12}-binding protein)

Daiger et al. (1975) have described polymorphism in the transcobalamin II class of vitamin B_{12}-binding proteins. Using polyacrylamide gel electrophoresis of ^{57}Co vitamin B_{12}-labelled sera followed by autoradiography they are able to detect three co-dominant polymorphic alleles at an autosomal locus. Serum, EDTA plasma and citrate plasma all give good patterns after electrophoresis but those from heparin plasma are highly distorted, which suggests that heparin combines with vitamin B_{12}-binding proteins. Immunological evidence indicates that transcobalamin II is not haptoglobin, caeruloplasmin, transferrin, haemopexin, properdin factor B or a protein which binds free cobalt. Occasional pedigrees with hereditary transcobalamin II deficiency which results in megaloblastic anaemia early in life have been described (Hitzig et al., 1974).

12. Ag beta-lipoprotein (LDL) system

The Ag system is the name given to the low density serum lipoprotein polymorphism first recognized by Allison and Blumberg (1961). An isoimmune serum from a multiply transfused patient (serum de B.) reacted with a component later shown to be low density lipoprotein (Blumberg et al., 1962). Much further work has shown the system to be a complicated one, very like the Rh locus in the problems it presents in immunogenetic terms. The most recent reviews appear to be Hirschfeld (1971) and Bütler et al. (1971).

The antisera which define Ag types often come from individuals who suffer from thalassaemia major and have therefore been frequently transfused. Ag types may be determined either by an Ouchterlony test or else by a more sensitive method of passive haemagglutination of antigen-coated red cells devised by Bütler and Brunner (1966). The latter method is particularly useful for weak antibodies and when employed in an inhibition test, polyspecific anti-Ag sera (which occur frequently) can be used as monospecific reagents. Further details of the haemagglutination-inhibition technique may be found in Bütler et al. (1967). Hirschfeld (1968a) has discussed the reproducibility of the Ag determinations in the Ouchterlony test. This is influenced by the time of incubation, the strength of the antiserum employed, the number of specificities it contains, and the naturally occurring variation in the amount of antigen present. Stored sera may sometimes give negative reactions after being positive when first collected. Bradbrook (1971) has shown that the Ag antigens lie on the LDL lipoproteins in the density range 1·018 to 1·035 g ml^{-1}. Some purification procedures abolish the antigenicity of the Ag groups, so that they are probably somewhat unstable.

In conventional terms it may be hypothesized that there are eight antigens recognized by eight antibodies in the Ag system. The antigens are controlled by four pairs of alleles at very closely linked loci—indeed no recombinants have so far been recognized. Diagrammatically the situation is

$$\frac{\quad\quad\mid\quad\quad\quad\quad\mid\quad\quad\quad\quad\mid\quad\quad\quad\quad\mid\quad\quad}{}$$
$$Ag^x \quad Ag a^1 \quad Ag^z \quad Ag^c$$
$$Ag^y \quad Ag^d \quad Ag^t \quad Ag^q$$

where the order of the loci is arbitrary, and x/y, a_1/d, z/t, and c/q represent the four pairs of co-dominantly inherited antigens. This hypothesis allows 16 different chromosomes, 136 possible genotypes, and 81 different phenotypes; in 1971, 9 of the 16 chromosomes had been recognized (Hirschfeld, 1971). The population frequencies show markedly non-random associations (Hirschfeld, 1968b).

Anyone familiar with the long-drawn-out controversy over the nature of the *Rh* locus (Race and Sanger, 1975) will recognize that an alternative hypothesis would be that there were nine alleles at a single locus. In this interpretation each allele is conceived of as producing four antigens. In the absence of more genetic and biochemical data it is not really possible to distinguish between these two hypotheses—which one is preferred is largely a matter of taste.

The above two hypotheses assume that the antigen molecule may be complex, i.e. have more than one antigenic grouping, but that the antibodies are simple, i.e. have only one specificity. Hirschfeld (1965, 1971) has challenged this orthodoxy and explored the consequences of interpreting the data upon the basis of the opposite assumption, i.e. that the antigen molecules have only one specificity and each antibody molecule may have several. The reader is referred to Hirschfeld's papers for further details.

Fisher *et al.* (1975) have made the interesting discovery that the molecular weight of LDL varies considerably between individuals, from 2·4 to 3·9 × 10^6, but that within an individual it is confined to a narrow range. A strong heritable component is implicated in the determination of the individual differences. Preliminary family data are compatible with the hypothesis of a single locus with two co-dominant alleles. The differences in molecular weight are due to the amount of lipid associated with the apoprotein, which does not vary in size. These authors raise the question of the relationship of this variation to the Ag types, a question which deserves some further investigation.

13. Lp(a) lipoprotein

The Lp(a) antigen was first described by Berg (1963) using absorbed rabbit antisera against purified human β-lipoprotein. An Ouchterlony test was used to type individuals as reactors, Lp(a+) and non-reactors, Lp(a−). The early family data indicated that presence of Lp(a) is determined by a dominant autosomal gene which segregates independently from the low density lipoprotein Ag system, from which Lp(a) is immunologically distinct (Berg, 1968). More recent investigations have revealed more complex genetic control. These data show that the genetic variation is in levels of Lp(a) substance rather than its presence or absence (Ritter and Wichmann, 1967; Utermann and Weigandt, 1970; Ehnholm et al., 1971; Albers et al., 1974; Schultz et al., 1974; Sing et al., 1974). Albers and colleagues used radial immunodiffusion to measure Lp(a) concentration. In their data there is no detectable bimodality in the concentration. They conclude that the observed Lp(a) quantitative variation is compatible with a polygenic model of inheritance, although they do not rule out the possibility of major gene effects. The analysis carried out in Shreffler's laboratory (Schultz et al., 1974; Sing et al., 1974) offers a reconciliation between the hypotheses of Berg and Albers et al. The concentration was measured by a sensitive radioimmune inhibition technique. This shows that all individuals possess Lp(a) and that the distribution of concentrations has an obvious bimodality. In terms of "relative Lp(a) units", a complex measure of concentration derived from the original data, the range of values found was 0·021 to 1·0 i.e. the lowest is one-fiftieth of the highest. The lower mode is at 0·10, there is a trough at 0·20, with the larger values being spread in a rather flat distribution from 0·20 to 1·0. The threshold for distinguishing Lp(a+) from Lp(a−) in the Ouchterlony test is about 0·15 i.e. very near the trough. Sing et al. (1974) have carried out a searching statistical analysis of family and population data on concentrations measured by radioimmune inhibition. They conclude that the simplest hypothesis which can account for the data is of a single two allele locus whose effect is modified to a minor degree by both environmental and genetic factors. The dominant allele Lp^a determines one distribution of Lp(a) concentrations with the higher mean and the recessive allele another distribution with a lower mean. Under the influence of the modifying factors the spread of these two distributions is such that they overlap slightly. The Lp(a+) and Lp(a−) phenotypes are thus not unambiguously related to the genotypes at the locus. However, a truncation point can be selected such that a minimum amount of mis-classification occurs; at this point, 0·15 in relative Lp(a) units, 1·25% of individuals who are homozygous recessive react qualitatively as positives, i.e. have > 0·15, and 6·55% of positive genotypes behave as negative, i.e. have < 0·15. This interpretation, which is in essence consistent with Berg's original hypothesis, also explains other recent data. These include the occasional Lp(a+) offspring from Lp(a−) matings, the wide variation in frequencies of Lp(a+) individuals found by different investigators for the same population, and the class of "weak" Lp(a+) individuals found in some surveys.

The Lp(a+) molecule has been examined in several ways. Garoff et al. (1970) detected it using disc electrophoresis followed by staining with Sudan Black B. Dahlén et al. (1972) also detected a β-lipoprotein electro-phoretically which they called pre-β_1-lipoprotein. A later investigation showed that it was Lp(a) (Berg et al., 1974). Rider et al. (1970) isolated a "sinking pre-β-lipoprotein" in the ultracentrifuge, i.e. it did not float at a density of 1·006, as do low density lipoproteins (LDLs), and showed that it was Lp(a). However, other data indicate a close relationship to the LDL class of lipoprotein. Ehnholm et al. (1972) and Utermann et al. (1972) showed that upon storage or treatment with detergent, it dissociates into three fragments, LDL lipoprotein, Lp(a) apoprotein and albumin. It has some (non-polymorphic) antigens in common with LDL protein but none with high density lipoprotein (HDL) (Schultz et al., 1968; Weigandt et al., 1968). The native Lp(a) molecule has a different amino acid content, hydrated density, lipid/protein ratio, mean diameter, isoelectric point and carbohydrate content from LDL (Ehnholm et al., 1971). Harvie and Schultz (1973) have shown that some individuals have two Lp(a) molecules which are separable in the ultracentrifuge. The basis for this difference has yet to be determined. In view of its tendency to dissociate, typing is best done on fresh serum. Its instability may also explain why the antibody to Lp(a) can only be produced by intravenous injection, and not intramuscular injection, with or without adjuvant (Berg, 1965).

Lp(a) may be of medical importance in two respects. Berg et al. (1968) have shown that Lp(a) compatibility prolongs skin graft survival time. Berg (1971) has discussed the possibility that Lp(a) could be related to transplantation antigens. More recently several investigations have been undertaken on its relationship to heart disease. A fairly strong correlation between presence of Lp(a) and both coronary heart disease and sustained myocardial infarction has been found (Berg et al., 1974; Dahlén et al., 1975; Dahlén and Berg, 1975).

302   D. W. COOPER

14. a_2-Macroglobulin Xm system

During the course of his investigations on the Lp system, Berg discovered a comparatively weak antiserum made by a single rabbit injected
with Lp(a−) whole serum. After absorption, this antiserum reacted in the
Ouchterlony test with a lipid-free component of the serum proteins of some
but not all people. Tests with purified proteins showed the reaction to be
specific for a_2-macroglobulin. Positive individuals were called Xm(a+) and
negative Xm(a−). The "X" signifies the fact that the Xm (+) character
has proved to be inherited as an X-linked dominant characteristic (Berg and
Bearn, 1966a, b).

The main interest in the polymorphism derives from this fact. It has been
used chiefly to map the human X-chromosome. This work has been
summarized by Berg and Bearn (1968). The human X has proved to have a
long linkage map (p. 357). There are two groups of genes which are within
sufficient distance of each other to permit mapping. The orders within the
two groups are Xm—deutan—G6PD—protan—haemophilia B and
angiokeratoma—ocular albinism—ichthyosis—Xg. Edwards (1968) gives
the two most likely relationships between the two groups as Xg—ichthyosis
—Xm—colour-blindness—G6PD complex or Xg—ichthyosis—colour-
blindness—G6PD complex—Xm. A report on a pedigree showing transmission of an X with a secondary constriction near the end of its long arm
suggests that Xm is not located near the end of the long arm of the
X and so the first order is possibly more likely to be correct (Lubs,
1969). It is possible, although unlikely, that Xm and Xg are within measurable distance of each other but further data are required to establish this
point. Likewise it is also possible that Xm and X-linked Hunter's syndrome
may be within measurable distance of one another, but again further data
are required (Berg et al., 1968). Unfortunately the limited quantity of
antiserum has run out and no further example has been found.*

15. a_2-Macroglobulin (Al-M system)

In 1969 Kasukawa et al. found a weak antibody in a human plasma which
reacted with some but not all human sera, apparently with an a_1-globulin
component. The individual from whom the serum came had never been
transfused. Another example of the antiserum was then found and this
enabled Leikola et al. (1972) to show that the polymorphism resided in the
a_2-macroglobulin fraction, thus establishing a second a_2-macroglobulin
polymorphism. The presence of the antigen is under autosomal dominant
control. The authors consider it likely that it represents a new polymorphism. The allele for the antigen has a frequency of approximately 0·16 in
Japan.

* See addendum on p. 234.

16. β_2-Glycoprotein I

This protein was first isolated from plasma in 1961 by Schultze *et al.* Physical characterization of the molecule has been carried out by Heimburger *et al.* (1964) and Haupt *et al.* (1968). On polyacrylamide it migrates just ahead of slow (S)a_2-macroglobulin. So far no biological function has been found for it. Haupt *et al.* (1968) discovered a family in which two sibs totally lacked it and whose parents had half the normal quantity. This led Cleve (1968) to carry out a population and family study using the Mancini radial diffusion technique. He found that 94% of individuals had a concentration which ranged between 16 and 30 mg/100 ml with a mean of $21\pm3\cdot6$, while in the remaining 6% of sera the range was 6 to 14 mg/100 ml with a mean of $10\pm1\cdot3$. In these first data the distribution of all concentrations was sharply bimodal with no overlap between the high and low values. Cleve accordingly proposed that β_2-glycoprotein I concentrations are controlled by a pair of autosomal co-dominant alleles, Bg^N and Bg^D. $Bg^N Bg^N$ individuals have concentrations which fall in the higher range, $Bg^N Bg^D$ fall in the lower range. The two sibs with no detectable β_2-glycoprotein found by Haupt *et al.* represent the uncommon $Bg^D Bg^D$ homozygous genotype. Bg^N is thus a null allele. Cleve (1968) also showed that sex, pregnancy, age and various diseases affected its level. Its lower level in chronic liver disease was taken to indicate that the liver is the probable site of its synthesis. In a further investigation Cleve and Rittner (1969) gathered family data which showed that the hypothesis proposed was substantially correct but that some children with low concentrations occurred in families where both parents had normal levels. The same kind of results were obtained by Koppe *et al.* (1970). One assumes that the two distributions overlap slightly. Atkin and Rundle (1974) have presented data on the frequencies of concentration classes in an English sample; these are almost identical with the frequencies in the German samples of the above authors. The quantitative nature of the varia ion in β_2-glycoprotein I presents problems of interpretation which are similar to those encountered with the Lp(a) antigen. However, the codominant nature of the β_2-glyco-protein I variation, in contrast to the dominance of Lp(a), makes it clear beyond reasonable doubt that most of the variation is ascribable to a single locus.

17. Kappa light chains (now called Km, until 1975/1976 referred to in literature as Inv)

The Inv polymorphism was discovered by Ropartz *et al.* (1961a, b) using the inhibition agglutination test with incomplete anti-D. In order to conform to the nomenclature for the other two immunoglobulin polymorphisms, this symbol has recently been changed to Km (WHO meeting on

Human Immunoglobulin Allotypes, Rouen, 1974, cited by Piazza *et al.*, 1976). The antigen they discovered is now called Km(1) (previously Inv (1)). Later two other antigens now called Km(2) (Inv(a) or Inv(a)) and Km(3) (Inv(3) or Inv(b)) were found. Family and population studies showed that these are controlled by three alleles, $Km^{1,2}$, Km^2, and Km^3. Only the antiserum to the Km(1) antigen is readily available (Steinberg, 1969).

A series of investigations on the relation of the Km antigens to the immunoglobulins showed that they are on the kappa chains (Terry *et al.*, 1965). The Km antigens are accordingly found on all classes of immuno-globulins, unlike the other two globulin polymorphisms which are confined to a single class. Sequence analyses have shown that the Km antigens lie in the constant (C terminal) end of the kappa chain (see Chapter 12). The differences between the three chains as well as for the Bence Jones protein Cro (which does not react with any of the three antisera) have been given by Milstein *et al.* (1974) as:

Km allele	Residue	
	153	191
1, 2	Ala	Leu
3	Ala	Val
1	Val	Leu
Cro	Val	Val

For the three-dimensional structure of lambda light chains and therefore probably for the kappa chains as well, positions 153 and 191 are located on the surface of adjacent loops about one nanometer apart, as shown on p. 638. All three antisera are thought to recognize differences primarily at position 191. Both antisera Km(1) and Km(2) recognize leucine at 191 but anti-Km(2) is prevented from doing so when valine rather than the smaller molecule alanine is at 153. Anti-Km(3) recognizes valine at 191 if 153 is alanine. However, in the Bence Jones protein Cro 153 is valine, which indicates that anti-Km(3) is also sterically hindered by the larger molecule.

18. Immunoglobulin G heavy chains (Gm)

These were discovered by Grubb (1956), using the haemagglutination inhibition test. Grubb and Laurell (1956) showed that this inhibitory property was inherited in an autosomal dominant manner. Subsequent investigations by many workers using other rheumatoid sera have revealed a system of considerable complexity. Some 25 Gm antigens have been recognized (W.H.O., 1965; Giblett, 1969), some of which have proved to be identical. Reviews of the Gm groups may be found in Steinberg (1969), Grubb (1970), Natvig and Kunkel (1973) and Ropartz (1974). The antigenic variants (allotypes) lie in the constant position of the heavy chains of the IgG class. Four sub-classes of IgG designated IgG_1, IgG_2, IgG_3 and IgG_4 have been found with corresponding heavy chains γ_1, γ_2, γ_3 and γ_4. The inheritance of the Gm antigens indicates that these four chains are controlled by four closely linked loci. The antigens are inherited in tightly

Table 1

Immunoglobulin heavy chain allotypes and their immunoglobulin classes and sub-classes[a]

New (numerical W.H.O. recommendation)	Allotype nomenclature Original (letters)	Sub-class of heavy chain
Gm markers		
1	a	IgG1
2	x	IgG1
3	b^w or b^2	IgG1
4	f	IgG1
5	b and b^1	IgG3
6	c	IgG3
7	r	IgG1
8	e	non-marker?
9	p	non-marker?
10	b	IgG3
11	$b\beta$	IgG3
12	$b\gamma_3$	IgG3
13	b_4	IgG3
14	b^4	IgG3
15	s	IgG3
16	t	IgG3
17	z	IgG1
18	Rouen 2	IgG1
19	Rouen 3	?
20	San Francisco 2	IgG1
21	g	IgG3
22	y	non-marker?
23	n_0	IgG2
	b_5	IgG3
	b_3	IgG3
	c_5	IgG3
	c	IgG3
Am markers		
1	1 or +	IgA2

[a] Because the sub-class location was ignored in drawing up the numerical designation, many workers continue to use the lettering systems. (Adapted from Natvig and Kunkel, 1973.)

linked combinations, called haplotypes. In Europeans there are three common haplotypes ("alleles") $Gm^{1, 17; ..; 21}$ $Gm^{1, 2, 17; ..; 21}$ and $Gm^{3; 23;}$ [5,10,11,13,14]. If we take the first of these haplotypes, for example, 1 and 17 are found on the γ_1 chain, the two dots indicate that no antigenic differences on γ_2 are determined by this haplotype, and 21 occurs on γ_3. Twenty-eight allotypes have been described, eight on γ_1, one on γ_2, 14 on γ_3, none on γ_4, one unlocated and three occur on more than one chain. These last three are referred to as "non-markers" (Natvig and Kunkel, 1973) because they are markers which are shared by at least two sub-classes but show genetic variability within only one sub-class. Table 1 lists the allotypes and their sub-classes. It is hardly surprising that the complexities of this system have taken 20 years to unravel.

Different groups of polymorphic haplotypes characterize different racial groups. The degree of heterogeneity between races for the Gm groups is greater than for any other human polymorphism except perhaps HLA. Strong linkage disequilibrium between the allotypes which each haplotype controls is evident for all populations. Rare haplotypes are also found in well-studied populations and it is usually plausible to suppose that these occurred by crossing-over between two of the polymorphic haplotypes, although no undoubted instance of recombination has been found in any family data to date.

The IgG_4 sub-class has no regular genetic marker, but does share "non-markers" with other sub-classes. When IgG_4 is isolated from human serum, one pair of these "non-markers" can be shown to behave as allelic determinants for IgG_4. Its linkage to the other three chains is thus established (Kunkel et al., 1970).

Several examples of probable intergenic recombination between the γ_1-γ_4 loci have been described (Natvig, 1974; Natvig and Kunkel, 1973) one of which resulted in a Lepore-type $IgG_4 - IgG2 \gamma$ chain (cf. p. 646). The likely order of the chains appears to be $\gamma_4 \gamma_2 \gamma_3 \gamma_1$, with the γ_2 chain being duplicated in some American Blacks (cf. p. 655).

Walzer and Kunkel (1974) have shown that IgD levels are reduced in Gm(f) Gm(b) homozygotes. This may indicate linkage of the δ chains of IgD to the Gm locus.

By studying the Gm and Km types of premature infants Daveau et al. (1975) have confirmed that immunoglobulin is transferred to the foetus from the mother, although some foetuses do express their own genotypes.

19. Immunoglobulin A heavy chains (Am)

The first report of polymorphism in the IgA major class is that of Vyas and Fudenberg (1969a). They discovered the antibody defining it in a woman who had undergone an anaphylactoid reaction after blood transfusion.

Anti-IgA antibodies have been found in the sera of 86% of patients with anaphylactoid and urticarial transfusion reactions (Vyas *et al.*, 1969). The system of detection involves coupling red cells with an IgA paraprotein (a myeloma immunoglobulin of IgA specificity) using the $CrCl_3$ method (Gold and Fudenberg, 1967; Vyas *et al.*, 1968). Sera are typed by their capacity to inhibit the agglutination of those cells by the anti-IgA serum of Am_2 specificity. Two classes were defined, $Am_2(+)$, being positive for the antigen (inhibitory) and $Am_2(-)$ being negative. Subsequent work revealed that the antigen is in the IgA_2 sub-class (Vyas and Fudenberg, 1969b; Kunkel *et al.*, 1969). Accordingly the designation Am_2 seems the more appropriate, rather than the $Am(1)$ of Vyas and Fudenberg (1969a, b). The markers defined by the two laboratories are identical according to van Loghem (1974). This view is borne out by their frequencies in various racial groups. Vyas and Fudenberg (1969b) find frequencies for the gene for the antigen of 0·99 in Caucasians, 0·31 in American Negroes, 0·51 in Japanese and 0·34 in Chinese, while the figures for the same groups found by Kunkel *et al.* (1969) were 0·99, 0·17, 0·53 and 0·43. Kunkel *et al.* also found a gene frequency of 0·46 in Easter Islanders.

The data of Vyas and Fudenberg (1969a, b) did not reveal association with any other immunoglobulin type including Gm, but the studies of Kunkel *et al.* (1969b) and van Loghem *et al.* (1970) have shown that the Am_2 group is closely linked to the Gm groups (as shown in the Table on p. 437). Since the Am_2 polymorphism is confined to the IgA molecule, it is a heavy chain marker. We thus have the intriguing situation that the heavy chain cistrons of the IgG sub-classes and one of the IgA sub-classes are closely linked. Another feature of some interest is that $Am_2(+)$ types appear to lack disulphide bridges between the two heavy chains. The molecules will readily dissociate into LH monomers in acid or urea (Jerry *et al.*, 1970), unlike IgA_2 from $Am_2(-)$ individuals. The difference presumably involves the substitution of another amino acid in $Am_2(-)$ heavy chains for cysteine in $Am_2(+)$.

A second allotype at the *Am$_2$* locus has been described and designated $Am_2(2)$ (van Loghem *et al.*, 1973). The Am system has been reviewed by van Loghem (1974). The evolution of IgA has been discussed by Wang and Fudenberg (1974).

B. Some Possible Polymorphisms

1. Australia antigen (Au)
Whether presence or absence of Australia antigen reflects a genetic polymorphism or not is at present a matter of controversy. However, the

investigation of Blumberg's hypothesis that it does has been an instructive one, in that it demonstrates the difficulties involved in studying possible genetic variation in susceptibility to infectious agents in Man. The question posed is whether diseases with a familial distribution have a genetic basis or whether they are the result of infection by agents which are only spread as a result of close sustained contact of the kind which occurs between members of the same household. For that reason, the main facts are recounted here.

Australia antigen (Au) was first identified by Blumberg *et al.* (1965) in the sera of patients with leukaemia and was then shown to be present in patients with hepatitis B (Blumberg *et al.*, 1967). The appropriate antisera came from multiply transfused individuals, like the Ag antisera. Au has many of the properties of the agent which causes the disease and its presence identifies otherwise cryptic hepatitis B carriers. As a consequence blood bank donors throughout the world are routinely screened for Au in order to reduce the incidence of post-transfusion hepatitis B. Particles corresponding to the Au antigen can be identified with the electron microscope. It is also associated with diseases in which the immune system is impaired, such as the lymphocytic leukaemias, Hodgkin's disease, Down's syndrome, and hepatomas (Blumberg *et al.*, 1974). Whether the Au antigen itself is the infectious agent is uncertain. Unlike other viral agents it contains little if any nucleic acid and there are conflicting reports on its capacity to grow in cell culture systems (reviewed in Blumberg, 1974).

Contemporaneous with the discovery that Au was associated with hepatitis and other diseases was the collection of data indicating that it had a strongly familial pattern of distribution (Blumberg *et al.*, 1969; Ceppellini *et al.*, 1970). This led Blumberg *et al.* (1969) to propose that presence of Au antigen is an expression of a recessive gene (Au^1) for increased susceptibility to hepatitis, to other diseases, and in view of the number of asymptomatic carriers of Au in some populations, probably to non-pathogenic infectious agents as well. Thus on this hypothesis the presence or absence of Au antigen may be considered as a plasma protein polymorphism which becomes manifest after infection with certain agents, an interesting example of genotype–environment interaction producing a particular phenotype. The early evidence for and against the hypothesis has been reviewed in Blumberg (1974). Until then it appeared that with some reservations the genetic hypothesis was substantially in agreement with the data. Blumberg's reservations were that (a) there is a greater frequency of Au in males than in females, (b) some age groups are more susceptible than others, and the pattern of susceptibility with age varies from population to population, and (c) there appears to be some maternal–offspring transmission of the agent which leads to production of Au antigen.

Very recently Stevens and Beasley (1976) and Mazzur (1976) have published data which are incompatible with Blumberg's hypothesis. Stevens and Beasley find that mothers with Au antigen give roughly 60% offspring like themselves irrespective of whether the father has Au antigen or has anti-Au in his serum. Their data are from a Formosan population. Mazzur found a similar lack of agreement with the simple Mendelian hypothesis in a Melanesian population. Mazzur (1976) suggests that the distribution of Au antigen i.e. of hepatitis infection mimics the distribution of a genetically determined characteristic because it is only spread by very close contact, the possibility of which is greatest within members of the same family. In a brief review article, Zuckerman (1976) extends this argument. He suggests that other diseases with a familial distribution may also be caused by infectious agents which require sustained contact for their spread.

An association between HLA type and possession of Australia antigen has been demonstrated in Northern Europeans (Bertrams et al., 1974) and an Australian Aboriginal sample (Boettcher et al., 1975). In view of the recent discoveries that properdin factor B and the C2 and C4 complement fractions are linked to HLA (see above for references), it seems that it would be worthwhile investigating the possibility of a relationship between complement and Au. This speculation is supported by the fact that the Au particles contain serum proteins (Millman et al., 1971).

2. C4 complement component

The C4 component is a β_1 serum protein of molecular weight 230 000 and a carbohydrate content of 14%. Its concentration of 200 to 600 μg ml^{-1} is the second highest of the complement components but it is too low for ready detection with the usual stains used after zone electrophoresis. This problem was overcome by Rosenfeld et al. (1969) who used a specific antiserum in AACE patterns to demonstrate three separate components, A, A_1 and C, in order of decreasing anodic mobility. The level of each varies. Seven different AACE patterns can be defined and are reproducible in different runs and bleedings. The ratio of haemolytic activity to C4 protein concentration varies with pattern. It is highest for the C-alone pattern and lowest for the A_1-alone pattern. It is likely that A versus A_1 is inherited in an autosomal co-dominant manner but what determines the rest of the differences has still not been discovered.

Meo et al. (1975) have shown that the Ss-Slp protein in mouse, which is controlled by a locus within the mouse major histocompatibility region (H-2), is the murine homologue of C4. This raises the likelihood that the human C4 locus is linked to HLA. According to Meo et al., Hauptmann and Rittner have unpublished data which show that an association exists between C4 deficiency and HLA. However, the matter is not entirely resolved. Capra *et al.* (1975) have asserted that C2 rather than C4 may be homologous with Ss-S1, and from pedigree data on rare deficiency states C2 is also known to be linked to HLA (Fu *et al.*, 1974; Wolski *et al.*, 1975). It would seem therefore that a further investigation of the probable C4 polymorphism would be of some interest, particularly in order to resolve its relationship to the HLA system

3. Testosterone binding β-globulin

Daiger (personal communication, 1976) has examined small samples of Caucasians, American Blacks and Orientals using radioactive-labelled testosterone, polyacrylamide electrophoresis, and autoradiography. He finds variation which is suggestive of, but not conclusive proof for, an autosomal electrophoretic variant with a gene frequency of 1 to 5%. De Moor *et al.* (1975) have measured variation in the capacity of steroid binding β-globulin (SBβG) to bind dihydroxytestosterone in the plasma of normal twins and find indications of a high heritability.

4. Factors IX and X

Lester *et al.* (1972) have carried out a best-fitting normal distribution analysis on data for the activity of these two clotting factors. They conclude that there is evidence for bimodality (at least) and suggest that there may be a relatively high frequency polymorphism for factor X (Stuart factor) and a lower frequency for factor IX. Veltkamp *et al.* (1972) used twins to study sources of variation in factors I, II, V, VII, VIII, IX, X and XIII/(cf. Table 9 on p. 453). They find that the technical limitations in estimating the level of activity of each component are such that heritability estimates have wide standard errors. Their work illustrates the difficulty of recognizing heritable variation in the absence of reproducible assay methods for activity or variants which can be typed biochemically.

5. Heritability of immunoglobulin concentration

Several groups of workers have studied the heritability of IgA, IgG, and IgM concentrations. The most notable finding is that genes for the level of IgM appear to be carried on the X-chromosome (Rhodes *et al.*, 1969; Grundbacher, 1972; Price *et al.*, 1974). All three classes show high heritability for their levels, with those for IgA and IgM being interrelated, which suggests possible coordinate control (Grundbacher, 1974; see Table 9 on p. 453). Age and race influence all three but IgM and Km groupings do not affect any of them. The variation so far found is thus strictly quantitative, but the high heritability suggests the possibility of major gene effects.

6. Alpha-2 macroglobulin

Gallango and Castillo (1974, 1975) have presented preliminary evidence which indicates autosomally controlled variation in the electrophoretic mobility of this protein in the Venezuelan population. Using immuno-electrophoresis they define six types A, B, C, AB, AC and BC; they suggest three alleles and autosomal co-dominance to explain the differences. It is possible that the different forms have different capacities to bind trypsin and chymotrypsin (Gallango and Castillo, 1975). The relationship of this possible polymorphism to Xm and Al-M is not clear.

IV. RARE VARIATION

As Harris and Hopkinson (1972) and Harris (1974) have pointed out, electrophoretically detectable polymorphism occurs at only a fraction of enzyme or protein loci, but all or nearly all loci exhibit rare variation, or at any rate, rare electrophoretic variation. Much of this rare variation in Man is detected in the first instance because of its untoward effects upon the well being of its bearer. Table 2 lists some examples of the more thoroughly

Table 2

Some examples of rare variation in plasma proteins

Protein or group of proteins	Nature of variation	Mode of inheritance	Reference
Factor VIII (a) Haemophilia (b) von Willebrand's disease	deficiency in activity and CRM	X-linked recessive Autosomal recessive	Ekert and Firkin (1975), Aledort (1975)
Other clotting factors	deficiencies in activity	autosomal recessive	Ratnoff and Bennett (1973)
Albumin	electrophoretic	autosomal codominant	Weitkamp et al. (1973, 1974)
Complement factor concentrations	deficiencies	autosomal recessive	Polley and Bearn (1975), Lachmann (1975)
Liproprotein concentration	deficiency or excess	probably auto-somal recessive or not well defined	Lees et al. (1974)
IgA concentration	deficiency	multigenic	Koistinen (1976)

Table 3

A summary of plasma protein polymorphisms

Polymorphism (locus symbol)	Mode of detection[a]	Number of polymorphic[b] alleles	Approximate frequencies in Caucasian populations (mainly) Northern Europe)	Polypeptide formula	Molecular weight (daltons)
Haptoglobin ((Hp-a)	electrophoresis	3	$Hp^{1S} \simeq 0.26$, $Hp^{1F} \simeq 0.14$, $Hp^2 = 0.60$	$\alpha_2\beta_2$ in Hp 1-1 only	$\alpha_1 = 9000$ $\alpha_2 = 17\,000$ $\beta^2 = 42\,500$
Transferrin (Tf)	electrophoresis	5	$Tf^{B2} = 0.01$	one polypeptide	$77\,000$
Group-specific component (Gc) (vitamin D binding protein)	immunoelectrophoresis and electrophoresis	4	$Gc^1 = 0.75$ $Gc^2 = 0.25$	unknown	$\simeq 50\,000$
Caeruloplasmin (Cp)	electrophoresis	3	$Cp^A \simeq 0.01$	$\alpha\beta$	$148\,000$
Serum α_1-antitrypsin (Pi)	electrophoresis	4	$iP^M = 0.95$ $Pi^S = 0.02$ $Pi^2 = 0.02$ $Pi^F = 0.01$	not known	$50\,000$
C3 complement component	electrophoresis (IF)	3	$C3^S = 0.8$ and $C3^F = 0.2$	$\alpha\beta$	$\alpha = 120\,000$ $\beta = 75\,000$ TOTAL $= 195\,000$
C6 complement component	electrophoresis (IF)	2	two alleles, 0.63 and 0.36	not known	not known
Properdin factor B (Bf)	electrophoresis (IF)	2	$Bf^S = 0.71$, $Bf^F = 0.28$	not known	$80\,000$
Orosomucoid (Or)	AACE	2	see text	not known	$44\,100$
Thyroxine-binding globulin	electrophoresis (A)	2	monomorphic	not known	$55\,000$
Transcobalamin II (Tc)	electrophoresis (A)	2	$Tc^4 = 0.44$, $Tc^3 = 0.55$ $Tc^2 = 0.01$	not known	$59\,000$
Ag β-lipoprotein (Ag)	ouchterlony	6	$Ag^x = 0.21$, $Ag^{\alpha1} = 0.49$ $Ag^c = 0.30$, $Ag^r = 0.81$	not known	2.5–3.5×10^6
Lp β-lipoprotein (Lp)	ouchterlony	2	$Lp^a \simeq 0.2$	not known	4.8×10^6 (including lipid moiety)
α_2-macroglobulin (Xm)	ouchterlony	2	$Xm^a \simeq 0.26$	not known ⎫	⎫
α_2-macroglobulin ($A1$-M)	ouchterlony	2	not determined	not known ⎬	8×10^5 ⎬
β_2– glycoprotein I (Bg)	mancini	2	$Bg^N = 0.97$ $Bg^D = 0.03$	not known	$40\,000$
Kappa light chains (Km, previously Inv)	haemagglutination inhibition	3	$Km^1 < 0.01$ $Km^{1r2} \simeq 0.07$ $Km^3 \simeq 0.93$	found in all Ig classes	$K = 22\,000$
Ig G heavy chains (Gm)	haemagglutination inhibition	10	$Gm^{1,17;..;21} = 0.22$ $Gm^{1,2,17;..;21} = 0.16$ $Gm^{3;23;5,10,11,13,14} = 0.61$	$\gamma_2K_2, \gamma_2\lambda_2$	$\lambda = 53\,000$
Ig A heavy chains (Am)	haemagglutination inhibition	2	$Am^1_2 = 0.99$	$\alpha_2K_2, \alpha_2\lambda_2$	$\alpha = 64\,000$

[a]A   = autoradiography
AACE = antigen antibody crossed electrophoresis
IF   = immunofixation
[c]Not necessarily in the same population
[b]C = certain, M = modifying factors may cause exceptions to the simple genetic hypothesis, P = highly probable but only on or a few papers published on the subject

Primary structure analysis	Chromosomal location	Close linkages to:	Biological function	Plasma concentration (mg/100ml)	Status	Year of publication (or discovery)
chain sequenced chain import	16	lecithin: cholesterol acyltransferase (LCAT)	haemoglobin binding	50–220	C	1955
quencing progress	—	pseudo cholinesterase	iron transport	200–400	C	1957
eptide map	—	albumin	vitamin D transport	30–55	C	1959
eptide map	—	—	copper transport and/or oxidase activity	20–45	C	1967
eptide map	—	Gm (Am)	inhibits proteolytic enzymes	200–400	C	1965
eptide fragments olated	—	Lewis blood group?	complement reactions	30–140	C	1968
—	—	—	complement reactions		P	1975
—	6	HL-A	properdin reactions	10–20	C	1971
—	—	—	unknown	55–140	P	1969
—	X	—	thyroxine transport	1–2	P	1975
—	—	—	vitamin B_{12} transport	$3·5 \times 10^{-3}$	P	1975
—	—	—	lipid transport	290–950	C	1961
mino acid mposition	—	—	lipid transport		C(M)	1963
nuno acid mposition	X	—	binds trypsin, plasmin stimulates cell growth	150–350♂♂	P	1965
nino acid mposition	—	—	binds trypsin, plasmin stimulates cell growth	175–420♀♀	P	1969
nino acid mposition	—	—	unknown	6–32	C(M)	1968
quenced	—	—	antibody	—	C	1961
quenced	—	Am, (Ig, D?) (Pi)	antibody	IgG= 800–1800	C	1956
quenced	—	Gm, Pi	antibody	IgA=90–450	C	1969

investigated kinds of plasma protein variation. Haemophilia and von Willebrand's disease are distinct clinical and genetic entities, the former sex linked and the latter autosomal. However, they both seem to involve the same molecule, factor VIII or antihaemophiliac factor (AHF). The deficiencies of the various clotting factors have, as Chapter 13 will show, been useful in disentangling the complex sequence of reactions which constitutes the clotting mechanism. Likewise the various complement deficiencies have confirmed the validity of the complement-reaction sequence worked out in *in vitro* studies. Disturbances of lipoprotein concentrations lead to heart and other diseases and have therefore been the object of a great deal of study. The literature on rare variation in plasma proteins is as large as that on polymorphism, and cannot be surveyed here. The interested reader is referred to the reviews listed in Table 2 for introductions to the subject.

V. COMPARATIVE SUMMARY OF POLYMORPHISMS; PROSPECTS FOR FURTHER INVESTIGATION

Table 3 summarizes some of the properties of the human plasma protein polymorphisms. Much of what was said in the first edition of this book remains applicable. The discovery of genetic variation still outstrips the physico-chemical description of the molecules. While the physiological function for most molecules is known, there is little information on the significance of the differences between phenotypes as far as the organism as a whole is concerned. Only the Pi types are understood in this respect. The rate of discovery since 1955 is still about one polymorphism a year. The refinement of techniques which allow discovery of polymorphisms in low concentration components has continued. In addition to the more recently developed immunological techniques, the application of autoradiography to the detection of transport procedures has revealed two new polymorphisms, in thyroxine-binding globulin and transcobalamin, as well as revealing the biological function of the Gc protein. Thyroxine-binding globulin and transcobalamin are both in low concentration in plasma. If the trend to greater sensitivity of detection continues, it may become possible to search for polymorphism in the structure or level of hormone proteins. An understanding of the organization of the genes which control the protein hormones would be of great interest, given that their physiological function is so vital and so well worked out, and given that their molecular structure is now being elucidated.

REFERENCES

Aarskog, D. and Fågerhol, M. K. (1970). *J. Med. Genet.* **7**, 367.
Albers, J. J., Wahl, P. and Hazard, W. R. (1974). *Biochem. Genet.* **11**, 475.
Aledort, L. M. (Ed.) (1975). *Ann. N. Y. Acad. Sci.* **240**, 1.
Alepa, F. P. and Steinberg, A. G. (1964). *Vox. Sang.* **9**, 333.
Allen, C. R. (Ed.) (1974). "Electrophoresis and Isoelectric Focusing in Polyacrylamide Gel". Walter de Grayter, Berlin and New York.
Allen, F. H. (1974). *Vox. Sang.* **27**, 382.
Allen, R. C., Harley, R. A. and Talamo, R. C. (1974). *Am. J. Clin. Pathol.* **62**, 732.
Allison, A. C. (1959). *Nature, Lond.* **183**, 1312.
Allison, A. C. and Blumberg, B. S. (1961). *Lancet* **i**, 634.
Alper, C. A. (1973). *Vox. Sang.* **25**, 1.
Alper, C. A. and Johnson, A. M. (1969). *Vox. Sang.* **17**, 445.
Alper, C. A., Robin, N. I. and Refetoff, S. (1969). *Proc. Natn. Acad. Sci. USA* **63**, 775.
Alper, C. A., Boenisch, T. and Watson, L. (1971). *J. Immunol.* **107**, 323.
Alper, C. A., Boenisch, T. and Watson, L. (1972). *J. Exp. Med.* **135**, 68.
Alper, C. A., Grodofsky, I. and Lepow, I. H. (1973). *J. Exp. Med.* **137**, 424.
Andre, F., Andre, C., Lambert, R. M. and Descos, F. (1974). *Biomedicine* **21**, 222.
Arfors, K. E. and Beckman, L. (1963). *Acta Genet.* **13**, 231.
Arquembourg, P. C. (1975). "Immunoelectrophoresis: Theory, Methods, Identifications, Interpretations" (2nd revised ed. of "Primer of Electrophoresis"). S. Karger, Basel.
Asarian, J., Archibald, R. W. R. and Leberman, J. (1975). *J. Pediatr.* **86**, 844.
Atkin, J. and Rundle, A. T. (1974). *Humangenetik* **21**, 81.
Axelsen, N. H. (Ed.) (1975). *Scand. J. Immunol.* Suppl. No. 2.
Azen, E. A. and Smithies, O. (1968). *Science* **162**, 905.
Azen, E. A., Smithies, O. and Hiller, O. (1969). *Biochem. Genet.* **3**, 215.
Barnett, D. R. and Bowman, B. H. (1968). *Acta Genet.* **18**, 573.
Bearn, A. G. (1966). *In* "The Matabolic Basis of Inherited Disease". (J. B. Stanbury, J. B. Wyngaarden and D. S. Frederickson, Eds), 2nd ed, 761–779. McGraw-Hill, New York.
Bearn, A. G. and Franklin, E. C. (1958). *Science* **128**, 596.
Bearn, A. G., Bowman, B. H. and Kitchin, F. D. (1964a). *Cold Spring Harb. Symp. Quant. Biol.* **29**, 435.
Bearn, A. G., Kitchin, F. D. and Bowman, B. H. (1964b). *J. Exp. Med.* **120**, 83.
Beckman, L. (1962). *Nature, Lond.* **194**, 796.
Beckman, L. and Holmgren, G. (1963). *Acta Genet.* **13**, 361.
Berg, K. (1963). *Acta Path. Microbiol. Scand.* **59**, 369.
Berg, K. (1965). *Acta Path. Microbiol. Scand.* **63**, 142.
Berg, K. (1968). *Ser. Haemat.* **1**, 111.
Berg, K. (1971). *In* "Protides of the Biological Fluids". (H. Peeters, Ed.), Vol. 19, 169–177. Pergamon Press, London.
Berg, K. and Bearn, A. G. (1966a). *Trans. Ass. Am. Physns.* **79**, 165.
Berg, K. and Bearn, A. G. (1966b). *J. Exp. Med.* **123**, 379.
Berg, K. and Bearn, A. G. (1968). *Ann. Rev. Genet.* **2**, 341.
Berg, K., Danes, B. S. and Bearn, A. G. (1968a). *Am. J. Hum. Genet.* **20**, 398.

Berg, K., Ceppellini, R., Curtoni, E. S., Mattiuz, P. L. and Bearn, A. G. (1968b). *In* "Advances in Transplantation" (J. Dausset, J. Hamburger and J. Mathé Eds). Munksgaard, Copenhagen.

Berg, K., Dahlén, G. and Frick, M. H. (1974). *Clin Genet.* **6**, 230.

Bernstein, R. S., Robbins, J. and Rall, J. E. (1970). *Endocrinology* **86**, 383.

Bertrams, J., Reis, H. E., Kuwert, E. and Selmair, H. (1974). *Z. Immun-Forsch.* **146**, 300.

Black, J. Z. and Dixon, G. H. (1968). *Nature, Lond.* **218**, 736.

Black, J. Z., Chan, G. F., Hew, C. L. and Dixon, G. H. (1970). *Can. J. Biochem.* **48**, 123.

Blumberg, B. S. (1974). *In* "Genetic polymorphisms and diseases in man" (R. Ramot, A. Adam, B. Bonné, R. M. Goodman and A. Szeinberg, Eds) 311–317. Academic Press, London and New York.

Blumberg, B. S., Dray, S. and Robinson, J. C. (1962). *Nature, Lond.* **194**, 656.

Blumberg, B. S., Alter, H. J. and Visnich, S. (1965). *J. Am. Med. Ass.* **191**, 541.

Blumberg, B. S., Gerstley, B. J. S., Hungerford, D. A., London, W. T. and Sutnick, A. I. (1967). *Ann. Int. Med.* **66**, 924.

Blumberg, B. S., Friedlaender, J. S., Woodside, A., Sutnick, A. I. and London, W. T. (1969). *Proc. Natn. Acad. Sci. USA* **62**, 1108.

Blumberg, B. S., Hann, H.-W. L., London, W. T. and Yin, L.-K. (1974). *J. Immunogenet.* **1**, 83.

Blundell, G., Cole, R. B., Nevin, B. C. and Bradley, B. (1974). *Lancet* **ii**, 404.

Boenisch, T. and Alper, C. A. (1970). *Biochem. Biophys. Acta.* **221**, 529.

Boettcher, B., Hay, J., Watterson, C. A., Bashir, H., MacQueen, J. M. and Hardy, G. (1975). *J. Immunogenet.* **2**, 195.

Bokisch, V. A., Dierich, M. P. and Müller-Eberhard, H. J. (1975). *Proc. Natn. Acad. Sci. USA*, **72**, 1989.

Bowman, B. H. (1967). *Fedn. Proc.* **26**, 724.

Bowman, B. H. (1968). *Ser. Haemat.* **1**, 97.

Bowman, B. H. and Bearn, A. G. (1965). *Proc. Natn. Acad. Sci. USA* **53**, 722.

Brackenridge, C. J. (1973). *Hum. Hered.* **23**, 543.

Bradbrook, I. D. (1971). *In* "Protides of the Biological Fluids". (H. Peeters, Ed.) Vol. 19, 197–200. Pergamon Press, London.

Buettner-Janusch, J. (1970). *Ann. Rev. Genet.* **4**, 47.

Bütler, R. and Brunner, E. (1966). *Vox. Sang.* **11**, 738.

Bütler, R., Brunner, E., Politis, E. and Scaloumbacas, N. (1967). *Vox. Sang.* **13**, 508.

Bütler, R., Morganti, G. and Verucci, A. (1971). *In* "Protides of the Biological Fluids". (H. Peeters, Ed.) Vol. 19, 161–167. Pergamon Press, London.

Capra, J. D., Vitetta, E. S. and Klein, J. (1975). *J. Exp. Med.* **142**, 664.

Catsimpoolas, N. (Ed.) (1973). *Ann. N. Y. Acad. Sci.* **209**, 1.

Ceppellini, R., Bedarida, G., Carbonara, A. O., Trinchieri, G. and Filippi, G. (1970). *In* "Proceedings of the International Symposium on Australian Antigen and Viral Hepatitis, Milan" (cited by Blumberg, 1974).

Chan, S. K., Luby, J. and Wu, Y. C. (1973). *FEBS Lett.* **35**, 79.

Cheftel, R. I. and Moretti, J. M. (1966). *C.R. Hebd. Seanc. Acad. Sci., Paris* **262**, 1982.

Chowdhury, P. and Louira, D. B. (1976). *Science* **191**, 480.

Cleve, H. (1968). *Humangenetik* **5**, 294.

Cleve, H. and Deicher, H. (1965). *Humangenetik* **1**, 537.

Cleve, H. and Rittner, C. (1969). *Humangenetik* **7**, 93.
Cohen, B. H. (1974). *Am. J. Hum. Genet.* **26**, 775.
Connell, G. E., Dixon, G. H. and Smithies, O. (1962). *Nature, Lond.* **193**, 505.
Connell, G. E., Smithies, O. and Dixon, G. H. (1966). *J. Mol. Biol.* **21**, 225.
Cook, P. J. L. (1975). *Ann. Hum. Genet.* **38**, 275.
Cox, D. W. (1966). *J. Lab. Clin. Med.* **68**, 893.
Cox, D. W. (1975). *Am. J. Hum. Genet.* **27**, 165.
Cox, D. W. and Celhoffer, L. (1974). *Can. J. Genet. Cytol.* **16**, 297.
Crawford, I. P. (1973). *Arch. Biochem. Biophys.* **156**, 215.
Crawford, I. P., Dawson, A. and Stevenson, D. D. (1974). *Am. J. Med.* **57**, 210.
Crowle, A. J. (1961). "Immunodiffusion" Academic Press, New York and London.
Dahlén, G. and Berg, K. (1975). *Opusc. Med.* **20**, 127.
Dahlén, G., Ericson, C., Furberg, C., Lundkvist, L. and Svardsudd, K. (1972). *Acta Med. Scand.* Suppl. **531**, 17.
Dahlén, G., Berg, K., Gillnas, T. and Ericson, C. (1975). *Clin. Genet.* **7**, 334.
Daiger, S. P. (1976). Ph.D. Thesis. Stanford University.
Daiger, S. P., Schanfield, M. S. and Cavalli-Sforza, L. L. (1974). *Am. J. Hum. Genet.* **26**, 24a.
Daiger, S. P., Schanfield, M. S. and Cavalli-Sforza, L. L. (1975a). *Proc. Natn. Acad. Sci. USA* **72**, 2076.
Daiger, S. P., Labowe, M. L. and Cavalli-Sforza, L. L. (1975b). *Am. J. Hum. Genet.* **27**, 31a.
Daveau, M., Rivat, L., Ropartz, C. and Fessord, C. (1975). *Biomed. Express* **23**, 23.
Davis, B. J. (1964). *Ann. N. Y. Acad. Sci.* **121**, 404.
Edwards, J. H. (1968). Appendix to Berg and Bearn (1968).
Ehnholm, C., Simons, K. and Garoff, H. (1971). In "Protides of the Biological Fluids" (H. Peeters, Ed.) Vol. 19, 191–196. Pergamon Press, London.
Ehnholm, C., Garoff, H., Renkonen, O. and Simons, K. (1972). *Biochemistry* **11**, 3229.
Ekert, H. and Firkin, B. G. (1975). *Vox. Sang.* **28**, 409–421.
Eriksson, S. (1964). *Acta Med. Scand.* **175**, 197.
Fågerhol, M. K. (1969). *Scand. J. Clin. Lab. Invest.* **23**, 97.
Fågerhol, M. K. (1972). In "Pulmonary Emphysema and Proteolysis" (C. Miltman, Ed.) 123–137. Academic Press, New York and London.
Fågerhol, M. K. and Braend, M. (1965). *Science* **149**, 986.
Fågerhol, M. K. and Jacobsen, J. H. (1969). *Vox. Sang.* **17**, 143.
Fågerhol, M. K. and Laurell, C.-B. (1967). *Clin. Chim. Acta.* **16**, 199.
Fågerhol, M. K. and Laurell, C.-B. (1970). *Prog. Med. Genet.* **7**, 96.
Fantes, K. H. and Furminger, I. G. S. (1967). *Nature, Lond.* **215**, 750.
Ferguson, K. A. (1964). *Metabolism* **13**, 985.
Ferreira, A. and Nussenzweig, V. (1975). *J. Exp. Med.* **141**, 513.
Fineman, R. M., Kidd, K. K., Johnson, A. M. and Breg, W. R. (1976). *Nature, Lond.* **260**, 320.
Fisher, W. R., Hammond, M. C., Mengel, M. C. and Warmke, G. L. (1975). *Proc. Natn. Acad. Sci. USA* **72**, 2347.
Fu, S. M., Kunkel, H. G., Brusman, H. P., Allen, F. H. and Fotino, M. (1974). *J. Exp. Med.* **140**, 1108.
Gaitshovski, V. S., Kisselev, O. I., Moshkov, K. A., Puchkova, L. V., Shavlovski. M. M., Shulman, V. S., Vacharlovski, V. G. and Neifakh, S. A. (1975) *Biochem. Genet.* **13**, 533.

318   D. W. COOPER

Gallango, M. L. and Castillo, O. (1974). *J. Immunogenet.* **1**, 243.
Gallango, M. L. and Castillo, O. (1975). *Humangenetik* **26**, 71.
Garoff, H., Simons, K., Ehnholm, C. and Berg, K. (1970). *Acta Path. Microbiol. Scand.* **78**, 253.
Gedde-Dahl, T. Jr., Fågerhol, M. K., Cook, P. J. L. and Noades, J. (1972). *Ann. Hum. Genet.* **35**, 393.
Gedde-Dahl, T. Jr., Teisberg, P. and Thorsby, E. (1974). *Clin. Genet.* **6**, 66.
Giblett, E. R. (1969). "Genetic Markers in Human Blood". Blackwell Scientific Publications, Oxford and Edinburgh.
Giblett, E. R. and Steinberg, A. G. (1960). *Am. J. Hum. Genet.* **12**, 160.
Giblett, E. R., Hickman, C. G. and Smithies, O. (1959). *Nature, Lond.* **183**, 1589.
Gold, E. R. and Fudenberg, H. H. (1967). *J. Immunol.* **99**, 859.
Gordon, A. H. (1975). "Electrophoresis of Proteins in Polyacrylamide and Starch Gels". North Holland, Amsterdam.
Götze, O. and Müller-Eberhard, H. J. (1971). *J. Exp. Med.* **134**, 90s.
Grabar, P. and Burtin, P. (Eds) (1964). "Immunoelectrophoretic Analysis". Elsevier, Amsterdam.
Grabar, P. and Williams, C. A. Jr. (1953). *Biochem. Biophys. Acta* **10**, 193.
Grant, D. B., Michin-Clarke, H. G. and Putnam, D. (1974). *J. Med. Genet.* **11**, 271.
Greene, F. C. and Feeney, R. E. (1968). *Biochemistry* **7**, 1366.
Grubb, R. (1956). *Acta Path. Microbiol. Scand.* **39**, 195.
Grubb, R. (1970). "The Genetic Markers of Human Immunoglobulin". Chapman and Hall, London.
Grubb, R. and Laurell, A. B. (1956). *Acta Path. Microbiol. Scand.* **39**, 390.
Grundbacher, F. J. (1972). *Science* **176**, 311.
Grundbacher, F. J. (1974). *Am. J. Hum. Genet.* **26**, 1.
Hamaguchi, H. (1967). *Proc. Japan Acad.* **44**, 733.
Harris, D. C. and Aisen, P. (1975). *Nature, Lond.* **257**, 821.
Harris, H. (1974). *Sci. Prog.* **61**, 495.
Harris, H. (1975). "The Principles of Human Biochemical Genetics". (2nd ed.), North Holland Publishing Co., Amsterdam and London.
Harris, H. and Hopkinson, D. A. (1972). *Ann. Hum. Genet.* **36**, 9.
Harvie, N. R. and Schultz, J. S. (1973). *Biochem. Genet.* **9**, 235.
Haupt, H. and Heide, K. (1965). *Clin. Chim. Acta* **12**, 419.
Haupt, H., Schwick, H. G. and Storiko, K. (1968). *Humangenetik* **5**, 291.
Heimburger, N., Heide, K., Haupt, H. and Schultze, H. E. (1964). *Clin. Chim. Acta* **10**, 293.
Hess, M. and Bütler, R. (1962). *Vox. Sang.* **7**, 93.
Hirschfeld, J. (1959). *Acta Path. Microbiol. Scand.* **47**, 160.
Hirschfeld, J. (1960). *Sci. Tools* **7**, 18.
Hirschfeld, J. (1965). *Science* **148**, 968.
Hirschfeld, J. (1968a). *Ser. Haemat.* **1**, 38.
Hirschfeld, J. (1968b). *Vox. Sang.* **14**, 95.
Hirschfeld, J. (1971). *In* "Protides of the Biological Fluids" (H. Peeters, Ed.) Vol. 19, 157–160. Pergamon Press, London.
Hirschfeld, J., Jonsson, B. and Rasmuson, M. (1960). *Nature, Lond.* **185**, 931.
Hitzig, W. H., Dohmann, U., Pluss, H. J. and Vischer, D. (1974). *J. Pediatr.* **85**, 622.

Hobart, M. J., Lachmann, P. J. and Alper, C. A. (1975). *In* "Protides of the Biological Fluids" (H. Peeters, Ed.), Vol. 22, 575–580. Pergamon Press, London.

Holmberg, C. G. and Laurell, C.-B. (1948). *Acta Chem. Scand.* **2**, 550.

Janus, E. D. and Carrell, R. W. (1975). *New Zealand Med. J.* **81**, 461.

Javid, J. (1967). *Proc. Natn. Acad. Sci. USA* **57**, 920.

Jeppson, J. O. (1967). *Biochem. Biophys. Acta* **140**, 469.

Jeppson, J. O. and Sjöquist, J. (1966). Proc. 14th Colloq. Prot. Biol. Fluids, 87. Elsevier, Amsterdam.

Jerry, L. M., Kunkel, H. G. and Grey, H. M. (1970). *Proc. Natn. Acad. Sci. USA* **65**, 557.

Johnson, A. M., Schmid, K. and Alper, C. A. (1969). *J. Clin. Invest.* **48**, 2293.

Kasukawa, R., Yoshida, T. and Milgrom, F. (1969). *Int. Arch. Allerg.* **36**, 347.

Kellerman, G. and Walter, H. (1972). *Humangenetik* **15**, 84.

Kirk, R. L. (1968a). "The Haptoglobin Groups in Man". Monographs in Human Genetics, Vol. 4 (J. Beckman and M. Hauge, Eds), Karger, Basel.

Kirk, R. L. (1968b). *Acta Genet. Med. Gemell* **17**, 613.

Kirk, R. L. (1971). *Am. J. Hum. Genet.* **23**, 384.

Kirk, R. L., Kinno, H. and Morton, N. E. (1970). *Am. J. Hum. Genet.* **22**, 384.

Kitchin, F. D., Bearn, A. G., Alpers, M. and Gajdusek, D. C. (1973). *Amer. J. Hum. Genet.* **6**, s72.

Koistinen, J. (1976). *Vox. Sang.* **30**, 181.

Koppe, A. L., Walter, H., Chopra, V. P. and Bajatzadh, M. (1970). *Humangenetik* **9**, 164.

Kueppers, F. (1972). *In* "Pulmonary Emphysema and Proteolysis" (C. Mittman, Ed.), 133–137. Academic Press, New York and London.

Kueppers, F. (1973). *Am. J. Hum. Genet.* **25**, 677.

Kueppers, F. and Bearn, A. G. (1966). *Science* **154**, 407.

Kueppers, F., Briscoe, W. A. and Bearn, A. G. (1964). *Science* **146**, 1678.

Kueppers, F., O'Brien, P., Passarge, E. and Rüdiger, H. W. (1975). *J. Med. Genet.* **12**, 263.

Kunkel, H. G., Smith, W. K., Joslin, F. G., Natvig, J. B. and Litwin, S. D. (1969). *Nature, Lond.* **223**, 1247.

Kunkel, H. G., Yount, W. J. and Litwin, S. D. (1966). *Science* **154**, 1041.

Kunkel, H. G., Joslin, F. G., Penn, G. M. and Natvig, J. B. (1970). *J. Exp. Med.* **132**, 508.

Kurosky, A., Barnett, D. R., Rasco, M. A., Lee, T. H. and Bowman, B. H. (1974). *Biochem. Genet.* **11**, 279.

Lachmann, P. J. (1975). *J. Med. Genet.* **12**, 372.

Laurell, C.-B. (1960). *In* "The Plasma Proteins" (F. W. Putman, Ed.), 349–378, Academic Press, New York and London.

Laurell, C.-B. (1965). *Anal. Biochem.* **10**, 358.

Laurell, C.-B. (1966). *Anal. Biochem.* **15**, 45.

Laurell, C.-B. and Eriksson, S. (1963). *Scand. J. Lab. Clin. Invest.* **15**, 132.

Laurell, C.-B. and Sveger, T. (1975). *Am. J. Hum. Genet.* **27**, 213.

Lees, R. S., Wilson, D. E., Schonfeld, G. and Fleet, S. (1974). *Prog. Med. Genet.* **9**, 237.

Leikola, J., Fudenberg, H. H., Kasukawa, R. and Milgrom, F. (1972). *Am. J. Hum. Genet.* **24**, 134.

Lester, R. H., Elston, R. C. and Graham, J. B. (1972). *Am. J. Hum. Genet.* **24**, 168.

Lewis, L. A. (1970). *CRC Crit. Rev. Clin. Lab. Sci.* **1**, 233.

320   D. W. COOPER

Lieberman, J. and Mittman, C. (1973). *Am. J. Hum. Genet.* **25**, 610.
Lieberman, J., Silton, R. M., Agliozzo, C. M. and McMahon, J. (1975). *Am. J. Clin. Path.* **64**, 304.
Litwin, S. D. and Kunkel, H. G. (1966). *Transfusion* **6**, 140.
Lubs, H. A. (1969). *Am. J. Hum. Genet.* **21**, 231.
McAlister, R., Martin, G. M. and Benditt, E. P. (1961). *Nature, Lond.* **190**, 927.
Macdonald, J. L. and Papiha, S. S. (1974). *Hum. Hered.* **24**, 45.
MacGillivray, R. T. A. and Brew, K. (1975). *Science* **190**, 1306.
Malchy, B. and Dixon, G. H. (1973). *Can. J. Biochem.* **51**, 231.
Mancini, G., Vaerman, J. P., Carbonara, A. O. and Heremans, J. F. (1964). *In* "Protides of the Biological Fluids" (H. Peeters, Ed.), Proc. 11th Colloq. Bruges, 1963, p. 370. Elsevier, New York and Amsterdam.
Mann, K. G., Fish, W. W., Cox, A. C. and Tanford, C. (1970). *Biochemistry* **9**, 1348.
Manwell, C. and Baker, C. M. A. (1970). "Molecular Biology and the Origin of Species". Sidgwick and Jackson, London.
Margolis, J. (1969). *Anal. Biochem.* **27**, 319.
Margolis, J. and Kenrick, K. G. (1967a). *Nature, Lond.* **214**, 1334.
Margolis, J. and Kenrick, K. G. (1967b). *Biochem. Biophys. Res. Commn.* **27**, 68.
Margolis, J. and Kenrick, K. G. (1968). *Anal. Biochem.* **25**, 347.
Margolis, J. and Kenrick, K. G. (1969a). *Nature, Lond.* **221**, 1056.
Margolis, J. and Kenrick, K. G. (1969b). *Aust. J. Exp. Biol. Med. Sci.* **47**, 637.
Matsubara, S., Yoshida, A. and Lieberman, J. (1974). *Proc. Natn. Acad. Sci. USA* **71**, 3334.
Mauff, G., Potrafki, B. G., Freis, H. and Pulverer, G. (1974). *Humangenetik* **21**, 75.
Mazzur, S. (1976). *Nature, Lond.* **261**, 316.
Meera Khan, P. (1971). *Arch. Biochem. Biophys.* **145**, 470.
Melartin, L., Hirvonen, T., Kaarsalo, E. and Toivanen, P. (1966). *Scand. J. Haemat.* **3**, 117.
Meo, T., Krusteff, T. and Shreffler, D. C. (1975). *Proc. Natn. Acad. Sci. USA* **72**, 4536.
Millman, I., Hutanen, H., Merino, F., Bayer, M. E. and Blumberg, B. S. (1971). *Res. Comm. Chem. Path. Pharmacol.* **2**, 667.
Milstein, C. P., Steinberg, A. G., McLaughlin, C. L. and Solomon, A. (1974). *Nature, Lond.* **248**, 160.
Mittman, C. (Ed.) (1972). "Pulmonary Emphysema and Proteolysis" Academic Press, New York and London.
Mittman, C. and Lieberman, J. (1974). *In* "Genetic Polymorphisms and Diseases in Man". (A. R. Ramot, A. Adam, B. Bonné, R. M. Goodman and A. Szeinberg, Eds) 185–193. Academic Press, New York and London.
de Moor, P., Muelepas, E., Heyns, W., Derom, F. and Thierry, M. (1975). *Am. Endocrinol.* **36**, 87.
Moretti, J. M., Cheftel, R. I. and Cloarec, L. (1966). *Bull. Soc. Chim. Biol.* **48**, 843.
Natvig, J. B. (1974). *Ann. Immunol.* **125C**, 63.
Natvig, J. B. and Kunkel, H. G. (1967). *Nature, Lond.* **215**, 68.
Natvig, J. B. and Kunkel, H. G. (1968). *Ser. Haemat.* **1**, 66.
Natvig, J. B. and Kunkel, H. G. (1973). *Adv. Immunol.* **16**, 1.
Neifakh, S. A., Vasiletz, I. M. and Shavlovski, M. M. (1972). *Biochem. Genet.* **6**, 231.

Nerenberg, S. T. (1966). "Electrophoresis: A Practical Laboratory Manual". Davis and Co., Philadelphia.

Nicoloff, J. T., Dowling, J. T. and Patterson, D. D. (1964). *J. Clin. Endocrinol.* **24**, 294.

Nikolai, T. F. and Seal, U. S. (1966). *J. Clin. Endocrinol. Metab.* **26**, 835.

Nikolai, T. F. and Seal, U. S. (1967). *J. Clin. Endocrinol.* **27**, 1515.

Novis, B. H., Young, G. O., Bank, S. and Marks, I. N. (1975). *Lancet* **ii**, 748.

Nute, P. E. (1968). Ph.D. Thesis, Duke University, North Carolina, 203 pp.

Ogawa, A., Kageyama, S. and Kawamura, K. (1968). *Proc. Japan Acad.* **44**, 1054.

Ornstein, L. (1964). *Ann. N. Y. Acad. Sci.* **121**, 321.

Ouchterlony, O. (1948). *Ar. Kemi Miner. Geol.* **26B** (14), 1.

Ouchterlony, O. (1967). Immunodiffusion and Immunoelectrophoresis. *In* "Handbook of Experimental Immunology" (M. D. Weir, Ed.). Blackwell Scientific Publications, Oxford and Edinburgh.

Ouchterlony, O. and Nilsson, C. Å. (1973). *In* "Handbook of Experimental Immunology" (M. D. Weir, Ed.), 2nd ed., Ch. 9, pp. 19.1–19.39. Blackwell Scientific Publications, Oxford.

Owen, J. A. and Smith, H. (1961). *Clinica Chim. Acta.* **6**, 441.

Owen, J. A., Better, F. C. and Hoban, J. (1960). *J. Clin. Pathol.* **13**, 163.

Owen, M. C. and Carrell, R. W. (1976). *Br. Med. J.* **1**, 130.

Palmour, R. M. and Sutton, H. E. (1971). *Biochemistry* **10**, 4026.

Parker, W. C. and Bearn, A. G. (1962). *J. Exp. Med.* **115**, 83.

Parker, W. C., Cleve, H. and Bearn, A. G. (1963a). *Am. J. Hum. Genet.* **15**, 353.

Parker, W. C., Hagstrom, J. W. C. and Bearn, A. G. (1963b). *J. Exp. Med.* **118**, 975.

Peetom, F. (1963). "The Agar Precipitation Technique and Its Application as a Diagnostic and Analytical Method". Oliver and Boyd, London.

Piazza, A., van Loghem, E., de Lange, G., Curtoni, E. S., Ulizzi, L. and Terrenata, L. (1976). *Am. J. Hum. Genet.* **28**, 77.

Pillemer, L., Blum, L., Lepow, I. H., Ross, O. A., Todd, E. W. and Wardlaw, A. C. (1954). *Science* **120**, 279.

Polley, M. J. and Bearn, A. G. (1975). *Am. J. Med.* **58**, 105.

Polonovski, M. and Jayle, M. F. (1938). *C.R. Séanc. Soc. Biol.* **129**, 457.

Polonovski, M. and Jayle, M. F. (1940). *C.R. Hebd. Séanc. Acad. Sci., Paris* **211**, 517.

Price, W. H., Newlands, I. M. and Ferguson, T. (1974). *J. Immunogenet.* **1**, 221.

Prokop, O. and Uhlenbruck, G. (1969). "Human Blood and Serum Groups". Maclaren and Sons, London.

Propp, R. P. and Alper, C. A. (1968). *Science* **162**, 672.

Prunier, J. H., Bearn, A. G. and Cleve, H. (1964). *Proc. Soc. Exp. Biol. Med.* **115**, 1005.

Putnam, F. W. (Ed.) (1975). "Plasma Proteins; Structure, Function and Genetic Control" (Vol. 1, 2nd ed.). Blackwell, Oxford.

Race, R. R. and Sanger, R. (1975). "Blood Groups in Man" (6th ed.). Blackwell Scientific Publications, Oxford and Edinburgh.

Ramot, R., Adam, A., Bonné, B., Goodman, R. M. and Szeinberg, A. (Eds) (1974). "Genetic Polymorphisms and Diseases in Man". Academic Press, New York and London.

Ratnoff, O. D. and Bennett, B. (1973). *Science* **179**, 1291.

Raymond, S. and Weintraub, L. (1959). *Science* **130**, 711.

Reinskou, T. (1968). *Ser. Haemat.* **1**, 21–37.

Rhodes, K., Markham, R. L., Maxwell, P. M. and Monk-Jones, M. E. (1969). *Br. Med. J.* **3**, 439.

Rider, A. K., Levy, R. I. and Frederickson, D. S. (1970). Circulation XLI Suppl. III, 10.

Ritter, C. and Wichmann, D. (1967). *Humangenetik* **5**, 42–53.

Rittner, C., Grosse-Wilde, H., Rittner, B., Netzel, B., Scholz, S., Lorenz, H. and Albert, H. (1975). *Humangenetik* **27**, 173.

Robinson, J. C., Blumberg, B. S., Pierce, J. E., Cooper, A. J. and Hames, C. G. (1963). *J. Lab. Clin. Med.* **62**, 762.

Robson, E. B., Polani, P. E., Dart, S. J., Jacobs, P. A. and Renwick, J. H. (1969). *Nature, Lond.* **223**, 1163.

Ropartz, C. (1965). *Transfusion* **8**, 301–312.

Ropartz, C. (1974). *Ann. Immunol.* **125C**, 27.

Ropartz, C., Lenoir, J., Hemet, Y. and Rivat, L. (1960). *Nature, Lond.* **188**, 1120.

Ropartz, C., Rousseau, P.-Y., Rivat, L. and Lenoir, J. (1961a). *Rev. Etud. Clin. Biol.* **6**, 374.

Ropartz, C., Lenoir, J. and Rivat, L. (1961b). *Nature, Lond.* **189**, 586.

Rose, V. M. and Geserick, G. (1969). *Kriminal. Forens. Wiss.* **5**, 113.

Rosenfeld, S. I., Ruddy, S. and Austen, F.-K. (1969). *J. Clin. Invest.* **48**, 2283.

Rowley, P. T., Sevilla, M. L. and Schwartz, R. H. (1974). *Hum. Hered.* **24**, 472.

Ruley, E. J., Forristal, J., Davis, N. C., Andres, C. and West, C. D. (1973). *J. Clin. Invest.* **52**, 896.

Rundle, A. T., Atkin, J. and Sudell, B. (1975). *Humangenetik* **27**, 15.

Rynbrandt, D. J., Ihrig, J. and Kleinerman, J. (1975). *Am. J. Clin. Pathol.* **63**, 251.

Schleisner, L. A. and Cohen, A. H. (1975). *Am. Rev. Resp. Dis.* **111**, 863.

Schultz, J. S., Shreffler, D. C. and Harvie, N. R. (1968). *Proc. Natn. Acad. Sci. USA* **61**, 963.

Schultz, J. S., Shreffler, D. C., Sing, C. F. and Harvie, N. R. (1974). *Ann. Hum. Genet.* **38**, 39.

Schultze, H. E. and Heremans, J. (1966). "Molecular Biology of Human Proteins", Vol. 1, 424 pp. Elsevier, New York and Amsterdam.

Schultze, H. E., Heide, K. and Haupt, H. (1961). *Naturwissenschaften* **48**, 719.

Schultze, H. E., Biel, H., Haupt, H. and Heide, K. (1962). *Naturwissenschaften* **49**, 16.

Shane, S. R., Seal, U. S. and Jones, J. E. (1971). *J. Clin. Endocrinol.* **32**, 587.

Sheridan, J. W., Kenrick, K. G. and Margolis, J. (1969). *Biochem. Biophys. Res. Comm.* **35**, 474.

Shokeir, M. H. K. (1973). *Clin. Biochem.* **6**, 9.

Shokeir, M. H. K., Shreffler, D. C. and Gall, J. C. (1967). *Am. Soc. Hum. Genet. Meeting*, Toronto. Quoted in Giblett (1969), p. 262.

Shokeir, M. H. K., Rucknagel, P. L., Shreffler, D. C., Na-Nakorn, S. M. and Wasi, P. (1968). *Am. Soc. Hum. Genet. Meeting*, Abst. 79. Quoted in Giblett (1969). p. 579.

Shreffler, D. C., Brewer, G. J., Gall, J. C. and Honeyman, M. S. (1967). *Biochem. Genet.* **1**, 101.

Sing, C. F., Schultz, J. S. and Shreffler, D. C. (1974). *Ann. Hum. Genet.* **38**, 47.

Smithies, O. (1955a). *Nature, Lond.* **175**, 307.

Smithies, O. (1955b). *Biochem. J.* **61**, 629.

Smithies, O. (1957). *Nature, Lond.* **180**, 1482.

Smithies, O. (1959a). *Biochem. J.* **71**, 585.
Smithies, O. (1959b). *Adv. Prot. Chem.* **14**, 65.
Smithies, O. (1960). *In* "Genetics" (H. E. Sutton, Ed.), 130–134. Josiah Macy Jr. Foundation, New York.
Smithies, O. (1964). *Metabolism* **13**, 974.
Smithies, O. and Hiller, O. (1959). *Biochem. J.* **72**, 121.
Smithies, O. and Walker, N. F. (1955). *Nature, Lond.* **176**, 1265.
Smithies, O., Connell, G. E. and Dixon, G. H. (1962a). *Am. J. Hum. Genet.* **14**, 14.
Smithies, O., Connell, G. E. and Dixon, G. H. (1962b). *Nature, Lond.* **196**, 232.
Steinberg, A. G. (1962). *Prog. Med. Genet.* **2**, 1.
Steinberg, A. G. (1967). *In* "Advances in Immunogenetics". (T. J. Greenwalt, Ed.). Lipincott, Philadelphia and Toronto.
Steinberg, A. G. (1969). *Ann. Rev. Genet.* **3**, 25.
Sternlieb, I., Morell, A. G., Bauer, C. D., Comber, D., DeRobes-Sternberg, S. and Scheinberg, I. H. (1961). *J. Clin. Invest.* **40**, 707.
Stevens, C. E. and Beasley, R. P. (1976). *Nature, Lond.* **260**, 715.
Studd, J. W. W., Blainey, J. D. and Bailey, D. E. (1970). *J. Obstet. Br. C'wealth.* **77**, 42.
Sutcliffe, R. G. and Brock, D. J. H. (1973). *Biochem. Genet.* **9**, 63.
Sutton, H. E. (1970). *Prog. Med. Genet.* **7**, 163.
Sutton, H. E. and Karp, G. W. (1964). *Am. J. Hum. Genet.* **16**, 419.
Talamo, R. C. and Thurlbeck, W. M. (1975). *Am. Rev. Resp. Dis.* **112**, 201.
Teisberg, P. (1970). *Vox. Sang.* **19**, 47.
Teisberg, P. and Gjone, E. (1974). *Nature, Lond.* **249**, 550.
Terry, W. D., Fahey, J. L. and Steinberg, A. G. (1965). *J. Exp. Med.* **122**, 1087.
Thomas, W. C., Morgan, H. G., Conners, T. B., Haddock, L., Bills, C. E. and Hovord, J. E. (1959). *J. Clin. Invest.* **38**, 1078.
Turnbull, A. and Giblett, E. R. (1961). *J. Lab. Clin. Med.* **57**, 450.
Utermann, G. and Weigandt, H. (1970). *Humangenetik* **11**, 66.
Utermann, G., Lipp, K. and Weigandt, H. (1972). *Humangenetik* **14**, 142.
Van Loghem, E. (1974). *Ann. Immunol.* **125C**, 57.
Van Loghem, E., Natvig, J. B. and Matsumoto, H. (1970). *Ann. Hum. Genet.* **33**, 351.
Van Loghem, E., Wang, A. C. and Shuster, J. (1973). *Vox. Sang.* **24**, 481.
Van Oss, C. C. J. (1967). *In* "Advances in Immunology" (T. J. Greenwalt, Ed.), 11–13. Lippincott, Philadelphia and Toronto.
Vana, L. R. and Steinberg, A. G. (1975). *Am. J. Hum. Genet.* **27**, 224.
Vasiletz, I. M., Kushner, V. P., Moshkov, K. A. and Neifakh, S. A. (1973). *Dolk. Akad. Nauk. S. S. R.* **208**, 729.
Vasiletz, I. M., Mashkova, E. T., Teterina, Z. K., Rafalson, K. I., Shavlovski, M. M. and Neifakh, S. A. (1974). *Genetika* **10**, 144.
Veltkamp, J. J., Mayo, O., Motulsky, A. G. and Fraser, G. R. (1972). *Hum. Hered.* **22**, 102.
Vyas, G. N. and Fudenberg, H. H. (1969a). *Clin. Res.* **17**, 469.
Vyas, G. N. and Fudenberg, H. H. (1969b). *Proc. Natn. Acad. Sci. USA* **64**, 1211.
Vyas, G. N., Fudenberg, H. H., Pretty, H. M. and Gold, E. R. (1968). *J. Immunol.* **100**, 274.
Vyas, G. N., Holmdahl, L., Perkins, H. A. and Fudenberg, H. H. (1969). *Blood* **34**, 573.
Walzer, P. D. and Kunkel, H. G. (1974). *J. Immunol.* **113**, 274.

324 D. W. COOPER

Wang, A. C. and Fudenberg, H. (1974). *J. Immunogenet.* **1**, 3.
Wang, A. C. and Sutton, H. E. (1965). *Science* **149**, 435.
Wang, A. C., Sutton, H. E. and Riggs, A. (1966). *Am. J. Hum. Genet.* **18**, 454.
Wang, A. C., Sutton, H. E. and Howard, P. N. (1967). *Biochem. Genet.* **1**, 55.
Weigandt, H., Lipp, K. and Wendt, G. G. (1968). *Hoppe-Seylers Z. Physiol. Chem* **349**, 489.
Weitkamp, L. R., Renwick, J. H., Berger, J., Shreffler, D. C., Drachmann, O., Wuhrmann, F., Braend, M. and Franglen, G. (1970). *Hum. Hered.* **20**, 1.
Weitkamp, L. R., Salzano, F. M., Neel, J. V., Parta, F., Geerdink, R. A. and Tarnoky, A. L. (1973). *Ann. Hum. Genet.* **36**, 381.
Weitkamp, L. R., Yamonato, M. and Nishiyama, J. (1974). *Am. J. Hum. Genet.* **37**, 485.
Williams, J. (1974). *Biochem. J.* **141**, 745.
Williams, J. (1975). *Biochem. J.* **149**, 237.
Wilkinson, E. J., Raab, K., Browning, C. A. and Hosty, T. A. (1974). *J. Pediat.* **85**, 159.
Wolski, K. P., Schmid, F. R. and Mittal, K. K. (1975). *Science* **188**, 1020.
World Health Organization (1965). *Bull. Wld. Hlth. Org.* **33**, 721.
Yunis, E. J., Agostini, R. M. and Glew, R. H. (1976). *Am. J. Path.* **82**, 265.
Zuckerman, A. J. (1976). *Nature, Lond.* **261**, 275.

ADDENDUM

Berg has recently been quoted by Horne, Bohn and Towler as agreeing that the Xm protein is "related" to a_2- pregnancy associated globulin, or a_2-PAG for short (*in* "Plasma Hormone Assays In Evaluation of Foetal Well-being", 1976 (A. Klopper, Ed.) Churchill Livingstone, Edinburgh). a_2-PAG is the same as the Xh protein which was tentatively described as an X-linked protein (Bundschuh, G., 1966, *Acta Biol. Germ.* **17**, 785. Further data disproved this hypothesis (Dunston, G. M. and Gershowitz, H., 1973, *Vox Sang.* **24**, 343). The dependence of a_2-PAG level on age and sex is such that it could be mistaken for an X-linked polymorphism in a small body of data. Ritter (*Bibl. Haemat.* **38**, 393, 1971) has published data suggesting that the Xh and Xm antigens are closely related. It has therefore become necessary to establish whether the Xm protein is a_2-macroglobulin or a_2-PAG, and whether either may be affected by an X-linked locus.

7 Blood Group Antigens

B. BOETTCHER

Department of Biological Sciences, University of Newcastle, New South Wales, Australia

I. INTRODUCTION

The detection of blood group antigens needs the appropriate blood group antibodies, with which the antigens can interact to provide evidence of their presence. In obtaining antisera which recognize specific human blood group antigens, several procedures have been utilized. The ABO blood group system was discovered by Landsteiner (1900), who observed that serum from some individuals would agglutinate red cells from others. The antibodies, anti-A and anti-B, are "naturally occurring"; they are present in the serum when the antigen is not present on the red cells of the individual. Whether their presence is "natural" or whether their production is stimulated by an appropriate antigenic stimulus is not certain, though undoubtedly the bulk of opinion at present is that an antigenic stimulus is necessary for their expression, especially since Springer et al. (1959a, b) have shown that White Leghorn chicks, which normally develop an agglutinin for human red cells at a few weeks of age (commonly anti-B), do not develop antibody if they are hatched and raised in a germ-free environment. Further, many bacteria have ABO blood group antigenic activities (Springer, 1970a, b), and the administration of live or killed *Escherichia coli* 086, which has B antigen activity, leads to an increase in titre of anti-B in infants and adults (Springer and Horton, 1969).

ABO antibodies are the only human blood group antibodies that are routinely "naturally occurring", though sometimes antibodies in other systems have been detected without known antigenic stimulus and may be regarded as being "naturally occurring", e.g. in the MNSs system, anti-M (Speiser, 1956), anti-N (Hirsch et al., 1957; Stern et al., 1957; Masters and Vos, 1962), anti-M^g (Allen et al., 1958; Race and Sanger, 1962), anti-S (Constantoulis et al., 1955) and anti-Vw (Darnborough, 1957; Race and Sanger, 1962) have been reported in individuals who were not known to have received an appropriate antigenic stimulus.

The discovery of a new blood group or system usually involves the finding of a serum which gives a pattern of agglutination reactions different from the pattern given by reagents of known specificities. Sometimes the serum has arisen due to exchange transfusion or pregnancy, e.g. anti-D of the Rh system (Levine and Stetson, 1939). On other occasions the serum is the result of deliberate immunization. The MN and P blood group systems, discovered by Landsteiner and Levine, were the result of immunizing rabbits with human O red cells, absorbing the immune sera with samples of other red cells and then testing the specificities of the absorbed immune antisera. Some of the antisera produced gave agglutination patterns which were independent of the ABO system and from these the MN and P blood group systems were defined (Landsteiner and Levine, 1927a, b, 1928a, b).

Landsteiner and Wiener (1940) found that antisera prepared in rabbits immunized with rhesus monkey blood would agglutinate red cells from about 85 % of New York whites. The pattern of reactions of the anti-rhesus sera was later shown to be the same as that given by sera from some patients who had incompatible transfusion reactions after being transfused with blood of the correct ABO type (Wiener and Peters, 1940). Although the anti-rhesus sera and the incompatible transfusion sera appeared to have the same specificity, it is now recognized that they detect different antigenic determinants, LW and D (of the human Rh system), respectively (see Section VII).

An extremely valuable antiserum has been produced more recently by injecting a human volunteer with red cells from a specially selected donor. The blood groups of donor and recipient were carefully matched and the recipient was given intradermal injections of red and white cells, in the hope of stimulating the production of anti-leucocyte antibodies. However, his major response was of an antibody directed against the sex-linked blood group antigen, Xg^a (Shepherd et al., 1969).

With an antiserum which shows a pattern of reaction different from known antisera, it must be established that it is not a mixture of known antibodies. This is done by testing it against a panel of red cells classified for a wide range of antigens. (These are available commercially.) The antiserum may be detecting a "new" antigen of a known blood group system, e.g. Jk^b of the Kidd groups (Plaut et al., 1953); a new specificity of a known antigen, e.g. E^t of the E antigen in the Rh system (Vos and Kirk, 1962) or a new blood group system, e.g. Xg^a (Mann et al., 1962).

Table 1

ABO blood groups

Blood group	Genotypes	Antigens on cells	Antibodies in serum
A	*AA, AO*	A	anti-B
B	*BB, BO*	B	anti-A
AB	*AB*	A and B	neither
O	*OO*	neither	anti-A and anti-B

Blood group antigens and their respective antibodies are of practical importance in blood transfusions. If the transfusion recipient possesses antibodies against an antigen present on the transfused cells, a harmful (and possibly fatal) reaction can occur. Because of the relative volumes involved, the presence of an antibody in transfused blood, against an antigen on the patient's cells, is not as serious. In this context, group O blood, containing anti-A, can be given in certain circumstances to a group A person. Since ABO antibodies are "naturally occurring", and can give rise to severe reactions in incompatible transfusion (see Table 1), the ABO groups of transfusion recipients and donors are always determined. If possible, patients are given blood of the same ABO group as their own (see Table 2). Apart from ABO antibodies, the ones most likely to produce adverse reactions during transfusion are associated with the Rh and Kell blood group systems. This is because of the frequency and also the strengths of activities of such antibodies.

Table 2

Possible transfusions, based on ABO blood groups of donor and recipient

| | Blood group of recipient | | | |
	A (anti-B)	B (anti-B)	AB (—)	O (anti-A anti-B)
A	C	I	C	I
B	I	C	C	I
AB	I	I	C	I
O	C	C	C	C

C=groups are compatible with possible transfusion
I=groups are incompatible with possible transfusions

Blood group antibodies can also be of significance in pregnancy. If the mother possesses an antibody against an antigen on the red cells of the foetus, and if this antibody is IgG (the only class of immunoglobulin which is able to cross the placenta), then the maternally-derived antibodies can be harmful to the foetus. The source of the maternal antibodies may be "natural", as with ABO antibodies (which can sometimes be haemolytic at 37^0C and lead to foetal loss), or be formed in response to earlier transfusion, or pregnancy with a foetus which has a paternally-inherited antigen not possessed by the mother. The mother can then be sensitized by foetal cells transferred into her circulation during delivery.

It seems unlikely that many new major blood group systems will be found through antibodies resulting from pregnancy or transfusion, since many opportunities for the formation of such antibodies must have arisen during the last 30 years, the time when the specificities of such antibodies have been investigated intensively. Newly recognized blood group systems would most likely involve antigens of either very high or very low incidence, in which case the circumstances necessary for the stimulation of antibody production would be rare, or they would involve antigens which are only weakly immunogenic and it would be expected that rurther examples of the antisera would occur infrequently. The latter appears to be the position with antisera directed against the sex-linked blood group antigen, Xg^a. Since about 7% of all pregnancies in Britain and about 20% of all transfusions are incompatible with regard to Xg blood groups, many individuals capable of forming anti-Xg^a must have been exposed to the antigen, but fewer than ten cases of the occurrence of the antibody have been reported.

Among human genetic markers, the blood groups are the most useful. Virtually all populations are polymorphic for some antigens; the samples are easily transported and the tests are relatively simple to perform. The provision of good antisera in useful quantities appears to be the main obstacle. Why populations should be usefully polymorphic for about a dozen blood group systems (ABO, MNSs, P, Rh, Se, K, Lu, Jk, Le, Fy, Xg, Do, Sf, Au) is unknown. The role of the blood group compounds themselves is unknown. Since they are (usually) components of the red cell membrane, it is possible that they are structural compounds which are antigenic only secondarily. Within at least two systems, ABO and Rh, strong selection is operating, on the basis of the antigenic properties of the compounds, so that the antigenicity of the compounds themselves may be important. It has been proposed that resistance to disease based on ABO blood groups and possible protection against microorganisms by the associated anti-A and anti-B antibodies, may have influenced the world-wide distribution of ABO blood groups (Mourant, 1954). This concept has had its supporters and its challengers (Pettenkofer et al., 1962; Springer and Wiener, 1962; Harris et al., 1963; Azevêdo et al., 1964) and this still seems to be the situation. Some data from India indicate that individuals of blood group A are more likely to contract smallpox than individuals of other ABO groups and, further, among smallpox sufferers A individuals have a higher mortality rate (Vogel and Chakravartti, 1966; Chakravartti et al., 1966). But data from the Congo do not support these observations (Lambotte and Israel, 1967). However, rare individuals whose red cells lack ABO antigens, the phenotype being termed "Bombay", show no unusual physical properties of their red cells, nor any indication of haematological disorder (Bhatia et al., 1974). Since the ABO antigens appear not to be structurally important, it seems that their role is related to their well established antigenic properties.

However, the situation appears to be different with some antigens, where involvement in the structure of the red cell membrane is demonstrated by the unusual physical properties of red cells lacking them. Studies carried out on red cells lacking all Rh antigens, the phenotype being termed Rh_{null}, show that the cells are not uniform in size, they have an unusual appearance in blood films (stomatocytes), they show greater than usual osmotic fragility and have a decreased survival time in vivo (Bar-Shany et al., 1967; Hasekura and Boettcher, 1970; Sturgeon, 1970; Levine et al., 1973). Additionally, the rare individuals involved suffer a haemolytic anaemia

(Sturgeon, 1970; Schmidt and Holland, 1971). Cells lacking a very common antigen, En^a, have unusual expression of antigens in several blood group systems and a markedly reduced sialic acid content. It has been proposed that a modification of the red cell envelope is the basis of the unusual properties (Darnborough *et al.*, 1969; Furuhjelm *et al.*, 1969). Weak expression of red cell antigens in several blood group systems, presumably a heterozygous condition in a man and two sons, since antigen expression on cells from the wife and the daughter were normal, was accompanied by altered osmotic fragility of the cells (Simmons *et al.*, 1969).

Some of the most stimulating areas of research on blood group antigens at present are: the studies relating genetical and biochemical data in establishing the pathways for the genetically-controlled biosynthesis of the antigens; the inter-relationships between antigens in different blood group systems, which provide evidence of common biosynthetic pathways and/or related spatial arrangements on the red cell membrane; chromosomal locations and linkage relationships using blood groups as markers, and the studies of sex-linkage and sex-chromosome aneuploidy, using the Xg blood group locus as the reference marker. These topics will be discussed in this chapter. However, these are only part of the fascinating (and useful) study of blood group antigens. For further reading, refer to Race and Sanger's classic work "Blood Groups in Man" (1975) now in its sixth and, sadly, final edition and to Mollison's "Blood Transfusion in Clinical Medicine" (1972), fifth edition. An English translation of "Human Blood and Serum Groups", originally written in German by Prokop and Uhlenbruck (1969), has been published. This book provides discussion of a number of topics in a different manner from the other two.

II. METHODOLOGY

A. Qualitative Studies

The best treatment of this topic is given by Mollison (1972) but some general principles are discussed here.

The studies of blood group antigens on red cells invariably involve agglutination reactions. Agglutination is a second order immunological reaction, since combination of antibody with red cell antigens can occur without agglutination resulting; it involves the linking of cells which have antibody attached. Consequently, absence of agglutination cannot necessarily be evidence that an antigen–antibody interaction has not occurred. To detect antigen–antibody interactions which have not resulted in agglutination (i.e. reactions involving "incomplete" antibody), several techniques have been developed. Sometimes suspending red cells in albumin (Diamond and Denton, 1945), or other high molecular weight compounds such as gelatin (Fisk and McGee, 1947), dextran (Grubb, 1949) or polyvinylpyrrolidone (McNeil and Trentelman, 1951) is sufficient to permit agglutination to occur. Alternatively, the red cells can be incubated with proteolytic enzymes such as papain (Kuhns and Bailey, 1950), ficin, trypsin (Morton and Pickles, 1951) or bromelin before addition of the appropriate antiserum. These procedures alter the effects of surface charges on the cells which normally cause red cells to repel each other in solution. Albumin and the other high molecular weight compounds raise the dielectric constant of the medium, whereas enzymes remove molecules responsible for the surface charges of the erythrocytes (Pollack, 1965; Pollack et al., 1965). Additional antigenic sites are not exposed, but the rate constant for association is increased (Hughes-Jones et al., 1964).

An alternative type of test is the antiglobulin, or indirect Coombs test (Coombs et al., 1945). After incubation of red cells with the appropriate antiserum which attaches to the cells, but does not agglutinate them, an antiserum reacting with human gammaglobulin is added. If antibody (gammaglobulin) has attached to the red cells, the cells will agglutinate on the addition of anti-gammaglobulin.

Different procedures have been found to be satisfactory for different blood group systems and for different antibodies within the one blood group system, and different laboratories have preferences on which procedure they use to detect incomplete antibodies. In general, haemolytic tests do not play an important part in detecting human blood group antigens.

A variety of tests is used to detect antigens on cells other than red cells. The fluorescent antibody technique has been applied successfully to blood group antigens in body tissues (Szulman, 1960, 1962, 1964) and to the detection of minor cell populations (Cohen et al., 1960). The mixed cell agglutination (Coombs and Bedford, 1955) and the mixed anti-globulin (Coombs et al., 1956a) tests have proved very effective in certain situations (Coombs et al., 1956b; Jones and Silver, 1958; Chalmers et al., 1959; Edwards et al., 1964).

The mixed cell agglutination test involves incubating the cells under test with the appropriate antibody, washing them and then adding cells (or particles) possessing the antigen being tested for. If the cells being tested possess the antigen, mixed agglutination will occur, and agglutinates involving the test cells and the cells bearing the known antigen will be formed. In the mixed anti-globulin test, cells being examined are incubated with the appropriate antibody and then mixed with known cells that have been coated with the same antibody. If antibody has attached to the cells being examined (because they possess the relevant antigen), mixed agglutination of the two cell types occurs on the addition of the anti-globulin reagent.

Tests with human foetal kidney have shown that IgM-type anti-A will cause mixed agglutination and serological adhesion, whereas IgG-type antibody will not react in the mixed agglutination test (Högman and Killander, 1962). Because of this and the possibility of non-specific uptake of globulins by tissues, a certain amount of familiarity with the reagents and procedures is necessary before mixed agglutination tests can be performed reliably.

The presence of antigen on other than red cells, or in solution, can also be detected by an inhibition test, where some of the antibody in a solution is used up and then the amount of antibody remaining is back-titrated to the appropriate red cells.

Different types of antibodies with similar specificities can have different properties. Rh antibodies of the IgG class fail to agglutinate red cells suspended in saline, whereas IgM antibodies will agglutinate in saline (Campbell *et al.*, 1955; Fudenberg *et al.*, 1959; Adinolfi *et al.*, 1962). However, IgG anti-A will agglutinate A red cells suspended in saline, but such agglutination is enhanced in a serum medium (Polley *et al.*, 1963). Human serum antibodies to blood group antigens are mainly of the immunoglobulin classes IgG and IgM, whereas ABO antibodies in secretions, such as saliva and milk, are mainly of the IgA class (Tomasi *et al.*, 1965; Tomasi and Bienenstock, 1968). In the sera of Group B and A individuals, anti-A and anti-B are predominantly of the 19S (IgM) type, whereas in group O individuals there is usually an appreciable amount of 7S (IgG) antibodies as well (Fudenberg *et al.*, 1959; Kunkel and Rockey, 1963; Polley *et al.*, 1963; Kochwa *et al.*, 1960). In O sera, cross-reacting anti-AB, i.e. antibody which can be absorbed by and eluted from A cells and which can then agglutinate B cells (or vice versa), has been identified as a 7S (IgG) immunoglobulin (Rawson and Abelson, 1960). The anti-A and anti-B antibodies produced by B and A individuals in response to a known stimulus are predominantly 19S (IgM), but O individuals produce a considerable amount of 7S (IgG) antibody.

Some of the properties of IgG and IgM anti-A and anti-B are different. IgG-type antibodies differ from IgM-type in being less readily neutralized by soluble blood group substances, in being able to bind less complement and in having their activity enhanced when used in a serum medium (Polley *et al.*, 1963). However, since A and B individuals may fail to produce appreciable amounts of IgG antibodies after stimulation (although it is commonly held that antibody stimulation results mainly in an IgG response), tests for the above properties are not always useful in detecting immune antibodies.

The majority of Rh antibodies in human serum are "incomplete". They will not agglutinate cells in saline, and are of the IgG class (Adinolfi et al., 1962). But a high proportion of agglutinins of the IgM class may be produced in the early stages of the immune response (Diamond, 1947; Wiener and Sonn-Gordon, 1947). Among Rh antibodies of the IgG type, virtually all are of the IgG1 or IgG3 sub-classes, though sera containing anti-D of sub-class IgG4 have also been reported (Frame et al., 1970).

B. Quantitative Studies

As with many genetically determined characters, studies of blood group antigens have been mainly qualitative. However, quantitative studies can provide useful information unobtainable through other means. The simplest method for comparing antigen expression on different samples of red cells is to obtain a series of (usually doubling) dilutions of the appropriate antiserum, and determine the greatest dilution which will just agglutinate an appropriate dilution of the cells. Alternatively, a score can be given to the agglutination test with each antiserum dilution. It has been found convenient to give a score of from 10 to 0, according to the degree of agglutination seen (Race and Sanger, 1958). The sums of the scores recorded for each antiserum dilution can then be compared for different cell suspensions. This procedure has been used frequently and with important results in the elucidation of the genetic background of an Rh_{null} individual (Levine et al., 1965a).

More sophisticated and time-consuming quantitative studies have been performed. A series of studies of the interaction of anti-B with B red cells, made by counting in a haemocytometer the number of cells agglutinated under different conditions of cell concentration, temperature, etc., coupled with micro-Kjeldahl protein estimations and sedimentation velocity calculations, enabled Filitti-Wurmser et al. (1954) to calculate the heats of reaction and the free energy changes of the interactions between human B antigen and anti-B antibodies. In this work it was established that individuals of genotypes A_1O, A_1A_1 and OO produce anti-B's of different types. In addition, they were unable to show a difference in the number of B sites on red cells from individuals of genotypes A_1B or BO.

With the stimulus provided by the work of the Wurmser group, Wilkie and Becker (1955) developed a more practical quantitative haemagglutination technique. Duplicate mixtures of 0·5 ml of red cell suspension (12 500 to 14 500 cells/mm³) and 0·5 ml of the appropriate serum dilutions are incubated with shaking for two hours. Samples are removed from the mixtures and the numbers of free cells are determined in haemocytometers by direct viewing or by counting from photographs of the haemocytometer chambers. The percentage agglutination of the red cells is transformed into probits. The probits are then plotted against the logarithm of the antiserum concentration.

The plot of percentage agglutination against logarithm of the antiserum concentration is a sigmoidal shaped curve. However, when the percentage agglutination is transformed to probits, the curve becomes a straight line, since this is the principle involved in probit transformations (Finney, 1962). In comparisons between the regression lines given by the log probit assays of different cells samples with the one antiserum, differences in slope reflect some qualitative difference between the antigens on the different cells (Solomon, 1964; Solomon et al., 1965c).

The Wilkie and Becker technique was initially used by them for studies on the group B-anti-B system (as with the Wurmser group). It has been used also in studies of the A-anti-A and H-anti-H interactions within the ABO blood group system (Gibbs and Akeroyd, 1959a, b; Solomon, 1964). Salmon and his colleagues have used the procedure in several contexts, including cases involving unusual ABO alleles (Salmon et al., 1964, 1965; Salmon and Salmon, 1968; Reviron et al., 1967; Bouguerra-Jacquet et al., 1970). The technique has been modified slightly for studies in the Rh system (Silber et al., 1961a, b), where a higher cell concentration is used. The reactivity of different anti-D agglutinins, and the expression of the D antigen on cells from individuals of different genotypes have been studied (Silber et al., 1961a, b). The technique has been applied to studies on individuals suspected of being Rh_{null} heterozygotes (Gibbs, 1966; Hase-kura and Boettcher, 1970; Boettcher and Watts, 1978).

In our experience we have found that sometimes there can be an appreciable loss of red cells due to their haemolysis while in the haemocytometer. This can be overcome by adding bovine serum albumin (about 0.8%) to the red cell suspension. It is possible that the haemolysis is due to contact of the washed cells with a clean glass surface, or to an alteration in pH (Prankerd, 1961), which can be buffered by the albumin.

Electronic particle counters have been used to count unagglutinated cells in the Wilkie and Becker technique, with accuracy as great as that obtained with haemocytometers (Bowdler and Swisher, 1964; Gibbs et al., 1965). The particle counters make the work faster and less tedious.

Solomon et al. (1965a, b) have reviewed the development and application of methods in quantitative haemagglutination.

III. ABO ANTIGENS

The earliest recognized groups within the ABO blood group system were A, B, AB and O, which are determined, by the presence of absence of the A and B antigens on the red cells (Table 1). The sub-division of the A antigen into the two types A_1 and A_2 was made quite early after their initial discovery (von Dungern and Hirszfeld, 1911). Another sub-group, A_3, was described first in 1936 by Friedenreich. Further sub-division has occurred, and new variants of both the A and B antigens are still being reported. Some variants of the expression of A and B antigens are due to the effects of modifying genes at other loci, but some are inherited as alleles of the ABO system and a complete understanding of ABO blood group antigens requires a knowledge of the molecular basis for these sub-groups of the antigens A and B. Some workers consider that many of the variants are simply quantitative in origin (Rosenfield and Rubinstein, 1966), but this is not wholly satisfying since the concepts invoked to explain this would lead to expectation of a continuous range of antigenic expression, while the A and H antigen expressions of A_1 and A_2 cells show marked discontinuities (Solomon, 1964; Grundbacher, 1965). Biochemical information on the ABO blood group antigens has given no useful indication of the basis for the sub-groups (see page 344). Other sub-groups of A can be conveniently grouped into:

A_x type: where cells give negative or weak reaction with anti-A (B serum), but a good reaction with anti-A + B (O serum). Serum usually contains anti-A_1. If the individual is a secretor, H substance is present in the saliva, but not A, though Alter and Rosenfield (1964) have shown that anti-A, or anti-A + B, absorbed by and eluted from A_x red cells can be inhibited, at least partially, by saliva from an A_x individual.

A_m type: where cells give negative or weak reaction with anti-A and with anti-A + B. No anti-A is present in the serum. If the individual is a secretor, A and H substances are present in the saliva.

However, there are other rare variants of the A antigen which are due to alleles at the ABO locus, but which do not fit into these two general classes (Race and Sanger, 1975). Sub-groups of B, analogous to those of A have been reported, including the first case of B_x which is analogous to A_x (Yamaguchi et al., 1970).

In 1924 Bernstein proposed the three allele mode of inheritance of the ABO blood group which is now accepted. This was to take account of the antigens A and B and the lack of either, O. The theory has been extended to take into account the A_1 and A_2 antigens (Thomsen *et al.*, 1930) and, by implication, all other A and B variants. The concept is of a multi-allelic locus, with each allele being responsible for the presence of an A or a B antigen or neither, O. However, some pedigree studies have shown the existence of rare alleles A_2B and A_1B (Seyfried *et al.*, 1964; Yamaguchi *et al.*, 1965, 1966; Madsen and Heisto, 1968), termed *"cis-AB"* by Reviron *et al.* (1968), which determine both an A and B antigen. A consideration of these cases has suggested that A_2B alleles arise by intra-genic recombination (Boettcher, 1966). This is also considered to be the basis of the one A_1B allele reported (Reviron *et al.*, 1967). An explanation invoking inter-genic recombination with the *A* and *B* genes being at different (though adjacent) loci is not satisfying, since the frequency of the A_2 antigen is very low in Japan, while at least ten pedigrees involving an A_2B allele have been found there (Yamaguchi *et al.*, 1970; Kogure, 1971). Of 9280 blood samples tested in Osaka, only 35 showed the presence of the A_2 antigen. Since, as will be discussed later, the products of the *A* and *B* genes are considered to be enzymes responsible for sugar transferase activity, and the A and B antigenic specificities reside in sugar residues, it is possible to explain these unusual *"cis-AB"* alleles in terms of a single blood-group-synthesizing enzyme which can transfer both A and B specifying sugars on to the one blood group substance macromolecule. Such an enzyme would be a single mutational step away from either an A- or a B-specifying enzyme and there would seem to be little reason to expect that the frequencies of such mutations would be different in populations with different *A* and *B* allele frequencies.

However, if intra-genic recombination is involved, such recombination would be expected to occur more frequently in populations having high frequencies of both the *A* and *B* alleles, which are needed to participate in recombination. Since almost all of the *"cis-AB"* alleles that have been reported have come from populations with high frequencies of both *A* and *B* alleles, this observation would tend to be in favour of the recombination explanation.

Observations made on several cases of the "Bombay" phenotype, originally observed by Bhende *et al.* (1952), where the cells do not group as A, B, AB or O, and where the serum contains anti-A, anti-B and anti-H, have provided valuable insight into the genetical pathways leading to the synthesis of ABO blood group antigens. "Bombay" individuals who do not express A_1, A_2 or B genes (Aloysia *et al.*, 1961; Levine *et al.*, 1955b) but, nevertheless, transmit them to their offspring, have been identified. Although the cells do not show a normal ABO blood group, some form of antigen occurs on them, since Levine *et al.* (1955b) were able to demonstrate weak B antigen on the cells of a "Bombay" woman who transmitted a

B gene, and since Lanset *et al.* (1966) have demonstrated that anti-A and anti-B can be eluted from "Bombay" cells where these antigens are suppressed. It appears that the defect in the "Bombay" phenotype occurs before the synthesis of H antigen, and that this antigen is necessary as a precursor for the synthesis of A and B antigens. The *H* locus is independent of the *ABO* locus (Aloysia *et al.*, 1961). A way of visualizing the genetical control of the synthesis of H, A and B antigens is presented in Fig. 1. The relationships of this genetic information with the biochemical evidence will be discussed later.

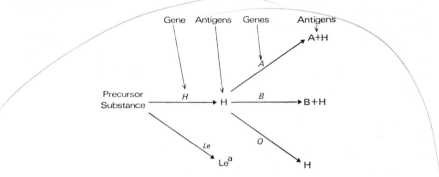

Fig. 1   The genetical control of the synthesis of the H, A, B and Le^a antigens. It is considered that H antigen is an essential precursor of A and B and that the "Bombay" phenotype involves defective H antigen

IV. SECRETORS AND NON-SECRETORS

The ABO antigens, as well as existing on red cells, can exist in water-soluble form in body fluids (Yamakami, 1926; Lehrs, 1930; Putkonen, 1930). The presence of the appropriate antigens in body fluids is controlled by the secretor gene, with secretion being a dominant character (Schiff and Sasaki, 1932), though a significant difference between the amounts of blood group substances secreted by homozygous and heterozygous secretors has been reported (Matsunaga, 1959; Kaklamanis *et al.*, 1964), but not supported (Chung *et al.*, 1965; Kelso, 1968). The initial studies and most of those since were carried out on ABO blood group substances in saliva, but other body tissues also are affected by the secretor gene (Hartmann, 1941).

   The secretor gene does not appear to control the presence of a known "new" immunological specificity. It determines the presence and activity of an enzyme, which is considered to be the product of the *H* gene in milk, submaxillary glands and stomach mucosa of human secretors (Shen *et al.*, 1968; Schenkel-Brunner *et al.*, 1972), and presumably in other fluids. The concept is that the secretor gene, *Se*, is a "switch" which is responsible for "switching on" the *H* gene. In this way the H antigen and, in appropriate individuals, the A and/or B antigens can be expressed.

Recent cases of dispermic chimaeras have shown that, for the appropriate blood group substances to appear in body fluids, the secretor gene needs to be in the same cell as the appropriate ABO gene which will express its antigen (Beattie *et al.*, 1964; Zuelzer *et al.*, 1964; Moores, 1966; Klinger *et al.*, 1968). This observation will obviously be important when the action of the secretor gene is fully understood.

Although no product of the secretor gene is recognized at present, the gene is known to influence interactions between the ABO and Lewis blood group antigens and the number of serum alkaline phosphatase isoenzymes seen after electrophoresis. In ABO non-secretors the Lewis blood group antigen, Le^a, can be present in the saliva and also detectable on the red cells. In ABO secretors the antigens Le^a and Le^b can be detected in the saliva but normally only the Le^b antigen is detectable on the red cells. However, some observations contrary to this have been reported, and individuals who have both Le^a and Le^b expressed on the red cells have been observed (Lewis *et al.*, 1957; Chandanayingyong *et al.*, 1967; Vos and Comley, 1967; Boettcher and Kenny, 1971).

With serum alkaline phosphatase, there are basically two enzyme patterns detected on starch gel electrophoresis. The most frequent pattern is one with a single band of activity, Pp1. Pp2, the less common two-band pattern, is observed more frequently in individuals of blood groups B and O than of blood group A and it is observed in a far lower proportion of non-secretors than secretors (Beckman, 1964, 1968). The electrophoretic technique of Beckman (1964) provided a clear cut differentiation, since all of the individuals whose sera gave a Pp2 pattern were secretors.

It has been reported that the concentration of proteins in salivas from secretors is on the average higher than in salivas from non-secretors (Hope *et al.*, 1968). Additionally, it has been shown that this increase in protein is due to proteins which are precipitated on boiling, rather than glycoproteins which will still remain in solution after boiling (Fig. 2). Thus, it appears that the secretor gene is responsible for the activity of an enzyme which can confer H specificity in secretions; the presence in some sera of a second alkaline phosphatase isoenzyme and an increase in the amount of protein present in the saliva.

Beckman (1968) has suggested, from a consideration of the interaction between secretor status and serum alkaline phosphatase type, that the secretor gene could be responsible for specific permeability of cell membranes. An explanation of the action of the secretor gene in terms of a permeability factor could explain the observations on alkaline phosphatase and protein concentrations and could also be consistent with the dispermy evidence. But, it seems difficult to fit into this scheme the observation of the "H enzyme" in the milk of secretors but not of non-secretors, since the present concept is that all individuals, even non-secretors, have precursor "blood group substance" molecules in their secretions (Watkins and Morgan, 1956/57) and that secretors have the H determinant added to these. An explanation involving intra-cellular permeability could satisfy all of the observations with, in non-secretors, the H enzyme being unable to contact appropriate substrate.

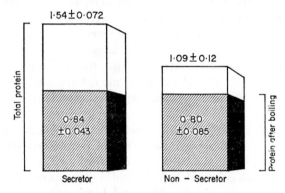

Fig. 2 Mean protein concentrations (mg ml^{-1}), and the standard errors of these means, in salivas from 144 secretors and 43 non-secretors. Protein concentrations were determined by the Lowry technique (Lowry et al., 1951) after the salivas had had mucinous compounds removed by freezing and centrifuging, and again after these had been boiled. The average protein concentrations before boiling (total protein) were significantly different ($t_{185} = 2.29$, $P < 0.05$) but the concentrations after boiling were not ($t_{185} = 0.23$, $P > 0.25$). Unpublished data of Boettcher and Kenny

However, it is clear that more work is needed before the role of the secretor gene can be fully understood, and whether or not it exerts its effect as a controlling element within the nucleus.

V. LEWIS ANTIGENS

There are interactions between the ABO and Lewis antigens and the secretor status of individuals. The Lewis antigens are regarded as being primarily antigens of body fluids and secondarily of the red cells. The Lewis gene, Le, is responsible for synthesis of Le^a specificity; its only known allele, le, is inactive. It is considered that, once formed, the Lewis antigens can be adsorbed by red cells from plasma. ABO antigens can also be adsorbed by red cells from plasma both *in vivo* (Renton and Hancock, 1962) and *in vitro* (Wherrett *et al.*, 1971), though sera from group O individuals is necessary to demonstrate the uptake of A or B antigens. However, this phenomenon is secondary to the intrinsic ABO antigens of the cells.

Since Lewis antigens are regarded as being primarily of body fluid origin this discussion will refer to body fluids rather than to red cells. The presence of Lewis antigen is a dominant condition. It appears that ABO and Lewis genes can act on the same precursor compound (Fig. 1) to convert it to ABO and/or Lewis substance, respectively. In individuals who are non-secretors the Lewis gene, *Le*, converts the precursor compound to Le^a substance. In individuals who are secretors the same conversion can occur but, in addition, the *H* gene converts some precursor substance to H substance. If both the *H* and Lewis genes act on the same chain of the precursor substance macromolecule, the Le^b antigen is formed. Thus, in secretors Le^a, H and Le^b antigenic determinants are formed (as well as A and B antigenic determinants) in individuals of the appropriate blood groups.

The ABO and Lewis antigenic determinants can occur on the same molecules, since ABO antigens in saliva can be precipitated by Lewis antisera (Brown *et al.*, 1959; Kaklamanis *et al.*, 1964), and since ABO antisera will precipitate molecules with ABO and Lewis specificities (Watkins, 1958). However, in the plasma, ABO and Lewis antigenic determinants have been found to reside on different molecular species (Andresen *et al.*, 1968). But studies on a twin chimaera pair have shown that a compound antigen, $A_1 Le^b$, is expressed on the red cells and in the plasma of the twin who possesses A_1, *Se* and *Le* genes, but not in the twin whose own cells are of group O, whose A_1 cells in the circulation were derived originally from her brother, and who formed anti-$A_1 Le^b$ (Crookston *et al.*, 1970).

VI. BIOCHEMICAL STUDIES ON ABO AND LEWIS ANTIGENS AND THE ROLE OF THE SECRETOR GENE

Although most genetical studies on secreted blood group substances have been performed on saliva, biochemical studies have utilized the large amounts of purified blood group substances which can be isolated from pseudomucinous ovarian cysts. In this work it has been assumed that the blood group compounds in the saliva and cyst fluids are under the same genetical control and there is no evidence to suggest that this is not a valid assumption. Nevertheless, markedly different quantitative relationships exist between ABH and Lewis antigens in different body fluids (Lawler, 1959; Denborough *et al.*, 1967; Boettcher, 1969) and it may be that there are unknown but important basic differences between blood group compounds in saliva and cyst fluids.

ABO and Lewis antigens in saliva are glycoproteins and ABO antigens on red cells are mainly glycolipids, but only the carbohydrate portions of these molecules will be considered here, since the antigenic specificities reside in the sugar residues of the carbohydrate chains and these appear to be the same in the respective glycoproteins and glycolipids.

The earliest observations indicating that the antigenic specificities of the ABH and Lewis antigens reside in carbohydrate were made by Watkins and Morgan (1952), who showed that L-fucose inhibited the agglutination of O

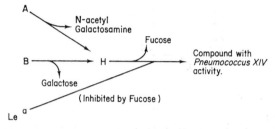

Fig. 3 The results of enzymic degradation of blood group substances. A diagrammatic summary of the work of several authors (see text). This figure should be compared with Fig. 1

cells by anti-H eel serum; by Côté and Morgan (1956) who showed that a disaccharide, α-N-acetylgalactosaminoyl-(1 → 3)-galactose isolated from human A substance would inhibit the agglutination of A cells by human anti-A; and by Kabat who showed that an α-D-galactosyl unit inhibited the B- anti-B interaction (Kabat and Leskowitz, 1955). Enzyme preparations which destroy blood group substances have been found. Their destructive activity could be inhibited by sugars which were N-acetylgalactosamine, D-galactose and L-fucose in the cases of A, B and H substances, respectively. Similarly, it was found that L-fucose inhibited the digestion of Le^a substance by an enzyme from *Trichomonas foetus*.

Progressive degradation of blood group substances, with the loss of one activity and the appearance of a new one, has provided biochemical evidence on the structure of blood group substances which links with the information obtained from ABO blood group studies (Fig. 3). Thus, if A or B substances are treated with A- or B-decomposing enzyme preparations, the A or B activity is lost and N-acetylgalactosamine or galactose, respectively, is released, with a concurrent increase in the H-activity of the degraded substance (Iseki and Masaki, 1953; Iseki and Ikeda, 1956; Watkins, 1962; Harrap and Watkins, 1964). Enzymic degradation of H substance releases fucose, and antigenic activity, identical to that of the capsule polysaccharide of *Pneumococcus* type XIV, is increased (Watkins and Morgan, 1955; Tyler and Watkins, 1960). Such degradation can be inhibited by fucose (Watkins and Morgan, 1955; Watkins, 1962). The degradation of Le^a substance leads to an increase in *Pneumococcus* type XIV activity. This degradation and the agglutination of Le(a+) cells by anti-Le^a sera can be inhibited by fucose (Watkins and Morgan, 1957; Watkins, 1962).

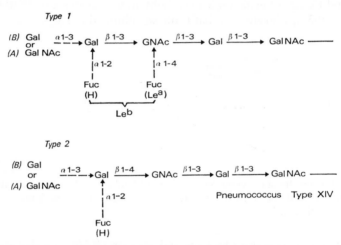

Fig. 4 Structures of Type 1 and Type 2 carbohydrate chains of the A, B, H and Le blood group substances. The possible sugar residues which can be added, and thereby confer H, A, B, Le^a or Le^b specificities are indicated by broken arrows

Gal = D-galactose   GNAc = N-acetylglucosamine
Gal NAc = N-acetylgalactosamine
Fuc = L-fucose

The results of these studies indicate that fucose is involved in both H and Le^a specificities. However inhibition studies, using oligosaccharides containing fucose, have provided evidence that the position of the H-determining fucose on blood group substance molecules is different from that of the Le^a-determining fucose (Watkins and Morgan, 1957; Watkins, 1962).

From the results obtained on the identification of the immunospecific sugars and of the oligosaccharides obtained from purified blood group substances isolated from pseudomucinous ovarian cysts, the structures proposed for the carbohydrate chains controlled by the *H, Le, A* and *B* genes have been detailed (Watkins, 1966; Lloyd and Kabat, 1968) (Fig. 4). Since the blood group specificities reside in sugar residues in carbohydrate chains, it is envisaged that the blood group genes "control the formation, or functioning, of specific glycosyl transferase enzymes that add sugar units from a donor substrate to the carbohydrate chains" (Watkins, 1966). There are two basic types of precursor chains, based on the linkage between the terminal galactose residue and the adjacent N-acetylgalactosamine residue, viz.:

Type 1 O-β-D-galactosyl-(1 → 3)-N-acetylglucosaminyl chain
Type 2 O-β-D-galactosyl-(1 → 4)-N-acetylglucosaminyl chain

From inhibition studies, it appears that Type 2 chains are more reactive with anti-*Pneumococcus* type XIV, than are Type 1 chains (Watkins and Morgan, 1962).

Le[a] specificity, due to an α-L-fucose residue bound by a 1 → 4 link to the subterminal N-acetylglucosamine residue can be added, under the influence of the *Le* gene, only to Type 1 chains since in Type 2 chains carbon atom 4 of the subterminal N-acetylglucosamine residue is already involved in a bond with the terminal galactosyl residue. H specificity, conferred by the addition of an α-L-fucose residue to the terminal galactosyl residue by a 1 → 2 link, can be added to both types of chains. Similarly, A or B specificity can be added to either type of chain in the form of an N-acetyl-D-galactosamine or a galactose unit, respectively, attached by an α-(1 → 3) link to the terminal galactosyl residue. Type 1 chains with both Le[a]- and H-specific fucose residues on adjacent sugars assume Le[b] specificity.

Enzymes with the anticipated specificities of glycosyl transferases specified by blood group genes, and from individuals of appropriate blood groups, have been detected by incubating radioactively-labelled sugar nucleotides with tissue extracts and identifying the labelled products. An L-fucosyltransferase with the properties of the postulated product controlled by the *H* gene has been found in serum (Schenkel-Brunner *et al.*, 1972) and in the milk and submaxillary gland extracts of secretors, but not of non-secretors (Shen *et al.*, 1968; Chester and Watkins, 1969). Another L-fucosyltransferase with the expected properties of the Le[a] enzyme has been detected in milk (Shen *et al.*, 1968) and in submaxillary gland extracts (Chester and Watkins, 1969).

A D-galactosyltransferase has been detected in individuals of groups B and AB, but not of groups A or O, in milk (Kobata *et al.*, 1968a), submaxillary gland (Race *et al.*, 1968), stomach mucosal linings (Poretz and Watkins, 1972), ovarian cyst fluids and linings (Hearn *et al.*, 1972) and plasma (Sawicka, 1971). Similarly, an N-acetyl-D-galactosyltransferase has been found in individuals of groups A and AB, but not of groups B or O, in similar locations (Kobata *et al.*, 1968b; Race *et al.*, 1968; Poretz and Watkins, 1972; Hearn *et al.*, 1972; Sawicka, 1971; Kim *et al.*, 1971).

From genetical studies on "Bombay" individuals, it is known that individuals with this phenotype do not express A or B genes (though they may possess them) due to the lack of function of a normal H gene. In terms of the biochemical concepts presented here, this would be interpreted as indicating that a terminal N-acetylgalactosamine or galactose residue could not be added to the terminal galactosyl residue of the preformed blood group carbohydrate chain, without the H-specific fucosyl residue having been added first. It is consistent and of interest, that in biochemical studies *in vitro* oligosaccharides having a terminal non-reducing galactosyl residue, but lacking the appropriate fucosyl residue, have not proved to be good acceptors (Kobata *et al.*, 1968a, b; Race *et al.*, 1968; Hearn *et al.*, 1968). However, such oligosaccharides show some ability to accept N-acetylgalactosamine (Hearn *et al.*, 1968) and this may be the basis of the very weak activity shown by cells from some "Bombay" individuals (Levine *et al.*, 1955b; Lanset *et al.*, 1966). Further, it has been demonstrated that "Bombay" individuals lack in their serum the L-fucosyltransferase considered to be controlled by the H gene, even though they have the relevant galactosyl- or N-acetyl- D-galactosyl-transferases controlled by their A or B genes (Schenkel-Brunner *et al.*, 1972; Race and Watkins, 1972), as had been expected from the earlier serological and genetical data.

Oligosaccharides with fucosyl residues attached to the terminal galactosyl and the sub-terminal N-acetylgalactosaminyl resides will not readily accept galactose or N-acetylgalactosamine residues (Kobata *et al.*, 1968b; Hearn *et al.*, 1968; Race *et al.*, 1968). The comparable observation *in vivo* would be that A or B specificity is not added to Le[b]-active structures (Kobata *et al.*, 1968b). Such a conclusion has been reached by Boettcher and Kenny (1971) from a consideration of the results of quantitative studies on A, H and Le[a] antigens in salivas from Australian aborigines, which showed a significantly greater expression of Le[a] in salivas from A_1 individuals than in those from O individuals. Additionally, oligosaccharides of structures analogous to Le[a] antigen cannot, *in vitro*, accept the H-determining fucosyl residue. But, oligosaccharides analogous to H antigen can accept the Le[a]-determining fucosyl residue (Hakamori and Kobata, 1974). Thus, these observations indicate that H, A and B genes express their effects in blood group substance carbohydrate chains before Le, as has been indicated in Fig. 5. If the transferase controlled by the A_1 gene were responsible for adding N-acetylgalactosamine residues to the terminal positions of Type 1 and Type 2 chains, whereas the transferase specified by the A_2 gene were responsible for adding residues to only one chain type (Lloyd and Kabat, 1968), then this would explain the difference between the A_1 and A_2 subgroups. However, *in vitro* studies indicate that transferases from A_2 individuals act on oligosaccharides of both Type 1 and Type 2

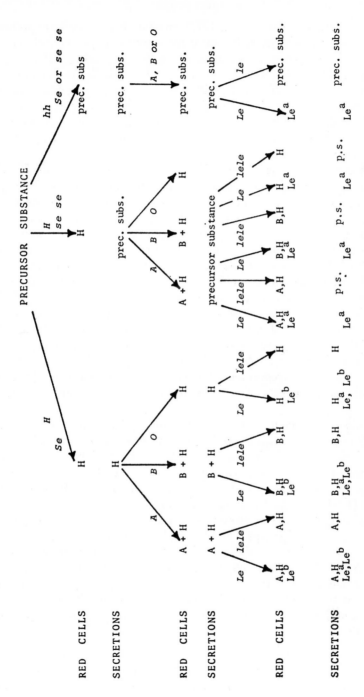

Fig. 5   Antigens detectable on red cells and in secretions determined by the secretor, H ABO and Lewis genes

(Watkins, 1970; Schachter *et al.*, 1973), making this attractive hypothesis unlikely, even though the activities of the enzymes might be different *in vivo*. But the enzymes are qualitatively different, having different pH optima and K_m values (Schachter *et al.*, 1973). A similar explanation for the action of the *"cis-AB"* allele, with A determinants being added to one chain type and B determinants to the other (Boettcher, 1972) also seems now to be unlikely. In biochemical terms it is now difficult to envisage the basis for the serological (and genetically-determined) differences of the sub-types of the A and B antigens.

The secretor gene, responsible for the presence of H, A and B substances in body secretions of the appropriate individuals, is considered to exert its influence in the synthesis of H substance from precursor, since Le[a] substance, using the same precursor, can be found in salivas of most non-secretors and since H substance is absent from the salivas of non-secretors. Because of the lack of immunological specificity associated with the presence of the *Se* gene, it has been suggested that the gene acts as a "switch" responsible for "switching on" the *H* gene in the cells producing water soluble ABO antigens. This concept is supported by lack of the appropriate L-fucosyltransferase in tissues producing ABH glycoproteins, but not in the serum, of non-secretors, as discussed earlier. Although the effect of the secretor gene is most noticeable in body fluids, glycoproteins isolated from erythrocyte stroma also show the same antigenic patterns, indicating that the glycoproteins but not the glycolipids of erythrocytes are influenced by effects of the secretor genes (Gardas and Kościelak, 1971).

The role of the secretor gene appears to be regulatory and, although the *A* and *B* genes are most commonly thought of as being structural (Hakamori and Kobata, 1974), it is also possible that they might be regulatory. This latter possibility receives support from the observations of the expression of red cell antigens in malignant tissues different from those on red cells (Levine, 1975). Specifically, A antigen has been demonstrated by immunofluorescence in cancerous tissues, but not normal tissues, of gastric carcinoma patients of groups B and O (Häkkinen, 1970). Such observations can be interpreted in terms of derepression of genes in the cancerous tissues, with *A*, *B* and *O* alleles normally being involved in repression of ubiquitous genes (Levine, 1975). If this were so, it would avoid the apparently unique situation at the ABO locus of the allelic genes *A* and *B* specifying the structures of different enzymes, which has been worrying some geneticists (Boettcher, 1966; Rostenberg, 1976).

VII. Rh ANTIGENS

The Rh blood group system derived its name from the reactions of rabbit anti-rhesus monkey antisera with human red cells (Landsteiner and Wiener, 1940). Those cells reacting with antisera exhaustively absorbed with selected bloods were termed Rh-positive and those which did not react, Rh-negative. The specificity of some immune antibodies formed by individuals who suffered transfusion reactions, or which caused erythroblastosis fetalis was shown to be identical (Wiener and Peters, 1940) or almost identical (Levine et al., 1941) with the rabbit anti-rhesus monkey antisera and both types of sera became known as anti-Rh. Shortly after the clinical importance of human anti-Rh was established, it was realized that the human Rh blood group system was not as simple as it appeared at first. The six editions of Race and Sanger's "Blood Groups in Man" detail the growth and the realization of the increasing complexity of the system.

At the commonly used level of consideration of the Rh groups, it is convenient to consider that the Rh antisera anti-D, anti-C, anti-c, anti-E and anti-e detect the D antigen or its absence, and the "allelic" antigens C/c and E/e. The Rh antigen, considered common to rhesus monkey and to human cells, became the D antigen in the CDE notation. The Rh antigens possessed by an individual are determined by two doses of some form of a sequence of D, C and E genes and these triplets are inherited as units. A well supported case of recombination or mutation within the Rh locus has been reported by Steinberg (1965).

The first suggestion that the human D antigen and the antigen on rhesus monkey cells are different was observed by Fisk and Foord (1942) who found that cells of newborn babies, whether Rh-positive or Rh-negative (as defined by human anti-D), reacted with guinea pig anti-rhesus monkey serum. The next step in showing the difference was made by Murray and Clark (1952), who found that heat extracts of human Rh-positive or Rh-negative red cells could stimulate the production of apparent anti-D in guinea pigs. It was realized that the specificity was not truly anti-D, especially since one sample of blood used to stimulate the production of the antibody came from an Rh-negative woman who had been immunized against the D-antigen during pregnancy. Levine et al. (1961) confirmed Murray and Clark's findings and established that the specificities of human anti-D and the anti-"D like" would still agglutinate human red cells that had been blocked by incomplete human anti-D. Further, Levine and Celano (1962a, b) demonstrated the presence of the "D-like" antigen on the cells of various monkeys including the rhesus monkey, but found that rhesus monkey red cells failed to absorb or elute specific antibodies when exposed to human anti-D.

With the existence of two antigens established, D and "D-like", the former being found only on human red cells and the latter being present on both human and monkey cells, Levine *et al.* (1963) proposed that the "D-like" antigen be termed LW, after Landsteiner and Wiener. Renaming the monkey antigen appeared to be easier than renaming the human antigen, since there was such a huge volume of literature on human Rh blood groups.

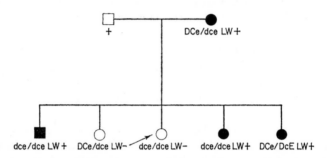

Fig. 6 Part of the pedigree of Mrs. V. W. (Swanson and Matson, 1964) which provides evidence that the *LW* locus is not closely linked with the *Rh* locus

Anti-LW has been found in sera from some individuals whose cells are Rh-positive but LW-negative. A family study in one of the cases has established the gene controlling the LW antigen is not part of the Rh complex locus (Fig. 6) (Swanson and Matson, 1964) and in another the evidence is most suggestive (Tippett, 1963).

The original confusion between Rh and LW antigens arose because Rh-positive human cells react strongly with anti-LW, whereas Rh-negative cells react weakly (if at all). These observations have been confirmed by Gibbs (1966) who observed a correlation between the D and the LW expressions of cells when using quantitative haemagglutination techniques.

The first human blood found to lack all known Rh antigens, the phenotype being described as Rh_{null} (Levine *et al.*, 1965a), was reported by Vos *et al.* in 1961. The cells were also found to be LW-negative (Levine *et al.*, 1962). A family with an Rh_{null}, LW-negative proposita was investigated by Levine *et al.* (1965a, b), who found that the proposita did not express *Rh* genes, though she transmitted them. It was proposed that the proposita was homozygous for alleles which prevented the synthesis of a compound which is the common precursor for both Rh and LW antigens (Levine *et al.*, 1965b), a situation similar to that of "Bombay" in the ABO system.

Further studies on relatives of the original Rh_{null} proposita (Hasekura and Boettcher, 1970) have shown that her genetical background is similar to that of the proposita studied by Levine et al. (1965a), and further cases have also been reported. Such cases are referred to as regulator-type Rh_{null}.

A case of the Rh_{null} phenotype, apparently due to homozygosity for amorphic Rh alleles, ---/---, was reported from Japan by Ishimori and Hasekura (1966, 1967). Based on the model for the genetically-controlled biosynthesis of the Rh and LW antigens proposed by Levine et al. (1965b), it would be expected that this Rh_{null} individual would be LW-positive. But he is LW-negative (Ishimori and Hasekura, 1966, 1967). Two other cases of amorphic-type Rh_{null} propositi have been reported (Seidl et al., 1972; Race and Sanger, 1975 p. 223). These also are LW-negative.

Schmidt et al. (1967) and Schmidt and Vos (1967) found additional unusual reactions of the cells from the two earliest Rh_{null} propositae. The U and s antigens were expressed weakly and the cells were agglutinated strongly by both anti-I and anti-i, which may have indicated "marrow stress", as had been described earlier (Giblett and Crookston, 1964; Hillman and Giblett, 1965). The series of unusual reactions suggested that the condition involved a complex heritable structural abnormality of the red cell (Schmidt et al., 1967). As further propositi have been studied, it is now recognized that the phenotypes of regulator-type Rh_{null} individuals are heterogeneous. Some are anaemic, others are not. Some show depression of Ss U antigens, and others do not. In some families the Rh antigens of obligate heterozygotes are depressed, in others they are not. The cells of one would fix and elute anti-D, whereas those of four others tested would not (Race and Sanger, 1975, p. 227). Because of this heterogeneity it is possible that other cases of severely depressed Rh antigens due to modifying genes unlinked with Rh (Chown et al., 1972; Rosenfield and McGuire, 1973) might involve alleles at the same locus as the regulator of Rh_{null}.

With the Japanese amorph-type Rh_{null}, no unusual MNSs reactions of the cells have been recorded (Ishimori and Hasekura, 1966, 1967), and the red cells of the parents of the propositus, who are obligate --- heterozygotes, do not show weakened expression of their Rh antigens (Hasekura, personal communication) though other --- heterozygotes have been found to have weak Rh antigen expression (Henningsen, 1959; Prokop and Schneider, 1960).

To accommodate the observations made on all Rh_{null} individuals, Giblett (1969) has proposed a model for Rh and LW antigen biosynthesis where the Rh antigen (presumably D) is an intermediate step in the biosynthesis of LW (Fig. 7). This would fit with the observations of Gibbs (1966) showing a correlation between the expression of D and LW on different cell types. However, as pointed out in Section XI, LW is phylogenetically older than Rh, yet this model has Rh antigens (or at least D) as an obligate precursor for LW, which does not appear to fit easily. Race and Sanger (1975, p. 213) are troubled by the exceptionally high consanguinity of parents of propositi who have only the D antigen present (D---/D--), and of the high rate of D-- sibs of such propositi, which suggests that we are not yet fully aware of the organization of the Rh gene complex. Perhaps further information will resolve these present doubts about the correctness of the proposed scheme for the Rh and LW antigen biosynthesis.

The chemical nature of the Rh antigenic determinants is not certain, though Green (1965, 1967a) working with lyophilized red cell ghosts,

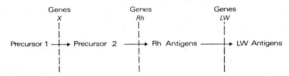

Fig. 7　Genetically determined biosynthetic pathways leading to the production of Rh and LW antigens. The Rh_{null} phenotype can arise from a defect in the biosynthesis of precursor for Rh antigens, or from amorphic *Rh* genes

obtained evidence that the D antigen is protein or peptide in nature and that at least one disulphide bond and at least one sulphydryl group are necessary for activity. Further studies showed that treatment of lyophilized ghosts with butanol destroyed all D activity, but that this could be restored by re-adding the butanol extract (Green, 1967b). The butanol extract from D cells would not confer D activity on D-negative ghosts, but a butanol extract of D-negative ghosts restored D activity on D ghosts extracted with butanol (Green, 1967b). Thus the D specificity resides in the red cell membrane, but it needs a butanol-soluble component, shown to be a phospholipid (Green, 1968), for its expression. Natural lecithin, phosphatidylethanolamine and phosphatidylglycerol were equally effective in restoring the D activity, but a number of lipids without phosphate groups were ineffective (Green, 1972). Weicker (1968) has isolated a peptide from Rh-positive red cell haemolysates which inhibits agglutination of Rh-positive cells with anti-D. The peptide has a molecular weight of 6000 to 12 000 and consists of 12 amino acids with a small amount of xylose but no other carbohydrate or lipid. From biochemical, serological and haematological observations, it seems that the Rh antigens are an integral part of the red cell membrane (Nicolson *et al.*, 1971; Levine *et al.*, 1973), and protein in nature. Under these circumstances, the great difficulties experienced in solubilizing them from red cell membranes (their only source) are understandable.

Whittemore *et al.* (1967) have observed that butanol extraction of red cell membranes destroys both ATPase and Rh activity. Giblett (1969) has speculated that Rh molecules may have ATPase activity, as has been suggested with the sheep blood group antigen M (Tosteson, 1963; Shreffler, 1967).

To say the least, in the Rh system, antibody-inhibition studies have not provided the same sort of valuable information that they have in the ABO system. A wide range of compounds has been reported to inhibit agglutination of Rh-positive cells by anti-D, but no clear pattern has emerged. Inhibiting compounds include sialic acids (Dodd *et al.*, 1960; Bigley *et al.*, 1963), a related compound, colominic acid (Boyd and Reeves, 1961), *N*-substituted glycosylamines (Chattoraj and Boyd, 1965), human brain gangliosides (Dodd *et al.*, 1964), amino acids (Chattoraj and Boyd, 1965; Hackel, 1964), ribonucleotides (Hackel, 1964) and para-amino-salicylic acid (Scott and Good, 1970). Bigley *et al.* (1963) found that a polysaccharide (glycopeptide) which had five amino acids attached, isolated from an unknown *Pseudomonas* species, inhibited anti-D and stimulated the production in a rabbit of antibodies which reacts with Rh-positive cells.

Perhaps these studies will be more meaningful when the structure of the D antigen is known.

VIII. MNSs ANTIGENS

The antigens M and N were originally found with and are still commonly detected by antisera prepared in rabbits (Landsteiner and Levine, 1927a, b). The correct two allele mode of inheritance of the groups, M, MN and N was proposed in 1928 (Landsteiner and Levine, 1928b). Both humans and rabbits seem to find it harder to form anti-N than anti-M but this may be related to the chemistry of the compounds.

An important extension of the MN system was the finding of the associated S (Walsh and Montgomery, 1947; Sanger and Race, 1947; Sanger *et al.*, 1948) and s antigens (Levine *et al.*, 1951), which extended the number of possible gene arrangements in the system to four, *MS, Ms, NS* and *Ns*. The frequencies of these gene complexes in Britain are about 0·25, 0·28, 0·08 and 0·39, respectively (Race and Sanger, 1975), indicating that there is not free recombination between the *M,N* and *S,s* genes. In this regard the system is similar to the Rh system. The MNSs blood group system is the most efficient at distinguishing between groups of unrelated individuals, because MN and Ss heterozygotes can be identified, and is therefore important in human genetics.

Further antigens have been found to be associated with the MNSs system, which is now recognized as being extremely complex. Anti-U serum (Wiener et al., 1953, 1954) reacts with almost all human red cells. U-negative people found are most commonly Negroes (Greenwalt et al., 1958; Fraser et al., 1966). Those cells which do not react with anti-U also do not react with anti-S nor with anti-s (Greenwalt et al., 1954). However, some S-negative, s-negative cells are U-positive (Race and Sanger, 1962, p. 92). The antigens Hu and He are recognized by antisera prepared in rabbits to red cells from Negroes of the names Hunter and Henshaw (Landsteiner et al., 1934; Ikin and Mourant, 1951). The antigens are more common in Negroes than Whites, but even in this group the frequencies are low (Chalmers et al., 1953). More recently, a further locus closely linked with MN and Ss has been reported, and termed Z. Only the Z antigen is recognized, with its alternative being the absence of Z (Booth, 1972).

Many other antigens have been found which can be regarded as being controlled by variants of the M,N or S,s genes or as antigens controlled by genes closely linked with M,N and S,s. M^g, an allele of M and N (Allen et al., 1958) does not react with anti-M or anti-N, though a little N activity was detectable on the one case of an M^gM^g homozygote studied (Metaxas-Bühler et al., 1966). M^k is an allele which produces no detectable M, N, S or s antigens (Metaxas and Metaxas-Bühler, 1964; Henningsen, 1966).

Anti-N reagents of rabbit, human or Vicia graminea origin can have almost all of their activity removed by absorption with M cells, though N cells will not exhaust anti-M (Landsteiner and Levine, 1928a; Hirsch et al., 1957; Levine et al., 1955a). The absorption of anti-N by M cells is increased by lowering the temperature or by trypsin treatment of cells, but is decreased by papain (Hirsch et al., 1957). The ability of M cells to react with anti-N is affected by the allele present at the S locus (Allen et al., 1960; Figur and Rosenfield, 1965). The order of reactivity is MS > Ms, with MS–s–U+ and MS–s–U− cells failing to react with anti-N.

The MN antigens appear to be carbohydrate in nature. Hohorst (1954) extracted MN antigens from erythrocyte ghosts using a phenol–water method and found the antigens in the polysaccharide fraction. Baranowski et al. (1956) arrived at the same conclusion and in 1959 reported that the MN substances had the composition of typical mucoids. Springer and Ansell (1958) and Mäkelä and Cantell (1958) found that neuraminidases destroyed MN antigens, and concluded that N-acetyl neuraminic acid (NANA) was involved in the M and N antigenic determinants. Uhlenbruck (1961) found that MN mucoids released from red cell stroma by proteolytic enzymes also had S activity. In 1963, Springer and Hotta showed that the NV_g determinant (detected by the anti-N lectin from Vicia graminea) did not have any NANA associated with it. It is destroyed by β-galactosidase or

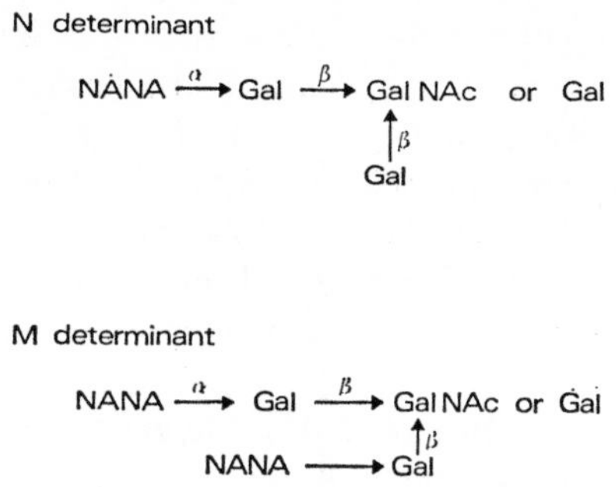

N determinant

$$\text{NANA} \xrightarrow{\alpha} \text{Gal} \xrightarrow{\beta} \text{Gal NAc} \quad \text{or} \quad \text{Gal}$$
$$\uparrow \beta$$
$$\text{Gal}$$

M determinant

$$\text{NANA} \xrightarrow{\alpha} \text{Gal} \xrightarrow{\beta} \text{Gal NAc} \quad \text{or} \quad \text{Gal}$$
$$\text{NANA} \longrightarrow \text{Gal} \uparrow \beta$$

Fig. 8   Structures proposed for N and M determinants (Springer and Tegtmeyer, 1972)

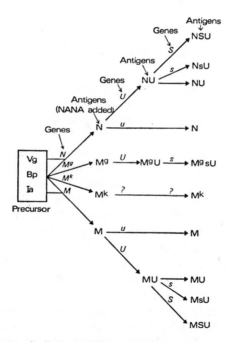

Fig. 9   Proposed genetically controlled pathways leading to the biosynthesis of MNSs antigens. Adapted from Uhlenbruck (1969). Vg, Bp, Ia refer to antigens detectable by *Vicia graminea, Bauhinia purpurea* and *Iberis amara* lectins

galactose oxidase (Uhlenbruck and Krupe, 1965; Springer *et al.*, 1966), implying that the D-galactopyranoside structure is important in the NV_g determinant. Romanowska (1964) found that a basic substance in the MN system was similar to N and consisted of a ground substance (which reacted with anti-N lectins) to which NANA had been added. This observation offered an explanation for the ability of M cells to remove anti-N activity from reagents. Springer and Tegtmeyer (1972) have proposed structures for M and N determinants which are consistent with this concept (Fig. 8).

Uhlenbruck (1969) has proposed possible genetically-controlled biosynthetic pathways leading to the synthesis of antigens associated with the MNSs Uu system (Fig. 9). The order of gene action should be *MN, U* and *Ss* alleles, which add specificity to a precursor which has determinants that react with anti-N lectins. A complication, considered with some reservation by Uhlenbruck, that Rh_{null} cells appear to be s-positive but U-negative (Schmidt and Vos, 1967), now appears not to be needed, since appropriate S, s and U antigens appear to be present on Rh_{null} cells though their expression may be somewhat unusual (Race and Sanger, 1968, p. 574).

IX. Xg ANTIGEN

In 1962, Mann *et al.*, recognized the first sex-linked human blood group system, using a serum from a multi-transfused man. With the frequency of the allele determining the dominantly inherited antigen, Xg^a, being approximately 0·65 and that of the silent allele 0·35, it was immediately recognized that this was the most useful sex-linked character yet demonstrated and that it could lead to satisfactory mapping of the X-chromosome, and could be used to determine the origins of X-chromosomes in cases of sex chromosome aneuploidy (Mann *et al.*, 1962; Noades *et al.*, 1966). Consequently, most of the work performed with the original anti-Xg^a serum has been devoted to studies on sex-linkage and sex-chromosome aneuploidy. Race and Sanger have devoted much of their time and energies to these questions for several years and have published excellent summaries of the work done in these fields (Frøland *et al.*, 1968; Race and Sanger, 1969, 1975; Race, 1970; see also p. 446).

The origin of sex chromosome aneuploidy can sometimes be determined by considering the Xg blood groups of the parents and the affected individual. For example, a girl with Turner's syndrome (XO sex chromosomes) who is Xg(a +) and who has an Xg(a +) father and an Xg(a –) mother has inherited her father's X-chromosome. With Klinefelter's syn-

drome (XXY sex chromosomes), an Xg(a +) Klinefelter male whose father
is Xg(a +) and whose mother is Xg(a –) has both an X and a Y-chromosome derived from his father and an X from his mother.

The frequencies of Xg^a groups among individuals with a single type of
sex-chromosome aneuploidy can sometimes be informative with regard to
the origin of these chromosomes. If the XXY chromosome complements of
Klinefelter males were composed of a maternal X and a paternal X and Y,
or the two maternal X's and the paternal Y or of a mixture of these two
possible origins, the Xg^a blood group frequencies of Klinefelter males
would be expected to be the normal female frequencies. If the XXY complement were formed from two copies of one of the maternal X's and the
male Y, the Xg frequencies of the Klinefelter males would be that of
normal males. However, the Xg distribution of Klinefelter males is significantly different from those of normal females and of normal males (Race
and Sanger, 1975) indicating that sometimes Klinefelter's two X chromosomes are copies of one of the maternal X's and sometimes two different
X's are involved.

It is estimated that in XXY males the extra X is of paternal origin in 39%
of cases; in 44% of cases it is due to both maternal chromosomes being
transmitted and in 17% of cases it is derived through duplication of one of
the maternal X's (Race and Sanger, 1975).

Pedigree data and Xg groups of individuals with Turner's (XO) syndrome indicate that the single X chromosome present can be of either
maternal or paternal origin. Statistical analysis of the Xg blood group
distribution among XO Turner individuals provides an estimate that in
77% of such individuals the X is of maternal origin and in 23% it is of
paternal origin (Race and Sanger, 1975). These results indicate that on
some occasions human spermatozoa lacking either an X or a Y chromosome
can function normally in fertilization, a situation comparable to that in
Drosophila where it has been shown quite clearly that spermatozoa with
abnormal chromosome complements behave normally (Muller and Settles,
1927).

Data from cases involving XXXY, XXXXY and XXYY individuals
indicate that on some occasions non-disjunction occurs at both the first and
the second meiotic divisions of spermatogenesis or oogenesis and that the
products are still functional.

It seems that the *Xg* locus is not involved in the sort of X-chromosome
inactivation proposed by Lyon (1962) when it is located on a structurally
normal X chromosome, but it is probably subject to inactivation when
carried on a structurally abnormal chromosome. This topic is discussed
elsewhere (p. 428). The *Xg* locus seems not to be on the distal third of the
long arm of the X chromosome. A woman with her father's normal X

carrying the silent allele, and her mother's X which lacks the distal third, expressed the Xg antigen fully (Clarke *et al.*, 1964).

A. Sex Linkage

Because the two recognizable phenotypes in the Xg system have been found to be in useful polymorphic frequencies in all populations so far tested, extensive sex-linkage studies have been performed, trying to link the *Xg* locus with those for rare sex-linked characters and the less rare traits, deutan and protan colour blindness, ichthyosis, glucose-6-phosphate dehydrogenase and the α_2-macroglobulin marker, Xm. Xg has been shown to be within measurable distance of the loci for a form of ichthyosis (Kerr *et al.*, 1964; Adam *et al.*, 1966; Wells *et al.*, 1966; Went *et al.*, 1969), ocular albinism (Fialkow *et al.*, 1967; Pearce *et al.*, 1968, 1971; Hoefnagel *et al.*, 1969), retinoschisis (Eriksson *et al.*, 1967; Ives *et al.*, 1970), and probably Fabry's disease which is also termed angiokeratoma (Opitz *et al.*, 1965; Spaeth and Frost, 1965; Johnston *et al.*, 1969; Malmqvist *et al.*, 1971). The loci for some other rare conditions may also be within measurable distance of Xg but since the conditions are rare the evidence, though suggestive, is not strong. The loci for glucose-6-phosphate dehydrogenase deficiency (Siniscalco *et al.*, 1960; Adam, 1961), deutan and protan colour blindness (Bell and Haldane, 1937; Porter *et al.*, 1962), haemophilia A (Whittaker *et al.*, 1962) and Becker's muscular dystrophy (Emery *et al.*, 1969; Skinner *et al.*, 1974; Zatz *et al.*, 1972) have been shown to be closely linked, but distant from the loci closely linked with *Xg* (Renwick and Schulze, 1964; Race and Sanger, 1975).

It seems possible that, in the near future, mouse–man cell hybridization will provide new information on linkage relationships on the X-chromosome (Siniscalco, 1974) since four enzymes controlled by X-linked loci can be detected in fibroblast cultures and, recently, it has been demonstrated that the Xg antigen can also be detected in cultured fibroblasts and short-term cultures of lymphocytes (Fellous *et al.*, 1974).

A genetic system involving an antigenic marker on α_2-macroglobulin molecules, controlled by a sex-linked locus designated *Xm* (Berg and Bearn, 1966; and see p. 302), still seems to offer the best chance of combining the above two clusters of X-linked genes into a single group. The estimates (which have wide confidence limits) of Xm from Xg and of Xm from G6PD are 0·35 and 0·16, respectively (Berg and Bearn, 1968), suggesting that Xm is a marker located between the two gene clusters. The Xm marker is in useful frequency—about 25 % of male and 50 % of female sera carry the marker—but there are some limitations since the original source of serum, a heteroimmune rabbit, has died and only fresh (not stored) sera can be grouped (Berg and Bearn, 1968). As with Xg, although the original source of serum has been lost, the value of anti-Xm reagents has been indicated and specific attempts to produce an anti-Xm reagent should be undertaken. It seems not unreasonable to hope that attempts to specifically induce anti-Xm will be successful, since Bundschuh (1966) appears to have

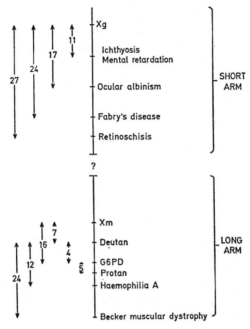

Fig. 10   An optimist's view of the X-chromosome; while the map distances (in cM) are the best estimates available (data taken largely from Race and Sanger (1975); Berg and Bearn (1968); and McKusick (1975)), the gene order is most uncertain

found a horse anti-human serum which contained anti-Xm as well as other antibodies and a related serum has been prepared in a goat (Rittner, 1967).

An idea of the gene clusters surrounding Xg and the colour blindness loci is given in Fig. 10.

B. Autosomal Linkage

Blood groups have provided data on autosomal linkage. Linkage between the loci controlling the Lutheran blood groups and secretor status was the first case of autosomal linkage reported in Man (Mohr, 1951; Sanger and

Race, 1958). Recently, the locus for myotonic dystrophy has been shown to be linked with Lutheran and secretor (Mohr, 1954; Renwick et al., 1971; Harper et al., 1972). Other linkages reported include ABO with nail–patella syndrome (Renwick and Lawler, 1955) and adenylate kinase (Rapley et al., 1968); Rh with elliptocytosis (Chalmers and Lawler, 1953), 6-phosphogluconate dehydrogenase (Weitkamp et al., 1971), phosphoglucomutase 1 (Renwick, 1971; Robson et al., 1973) and possibly enolase (Giblett et al., 1974); Chido and HLA (Middleton et al., 1974) and Lewis and the third component of complement (Weitkamp et al., 1974). There are indications that P is linked with the adenosine deaminase (Cook et al., 1970) and HLA (Edwards et al., 1972) loci. From workers interested in the major histocompatibility complex, of which the two HLA loci form part, there are claims that not only is HLA linked with Chido and P blood groups, but also with loci for the second and eighth complement factors, properdin B, pepsinogen, glyoxylase and the third form of phosphoglucomutase as well as the several other loci associated with the complex itself. If these linkages are substantiated, the investigational popularity of P and Chido will no doubt increase rapidly.

An interesting observation in autosomal linkage studies has been that of greater recombination frequencies in females than in males, as has been observed in some laboratory organisms, the extreme of which is Drosophila where recombination does not occur in the male. This phenomenon was observed initially in the Lutheran-secretor (Cook, 1965) and ABO–nail–patella (Renwick and Schulze, 1965) cases. It indicates that data involving doubly heterozygous fathers are more useful than those involving doubly heterozygous mothers, especially where there is loose linkage.

A blood group had the honour of being the first autosomal locus to be assigned to a chromosome. On the basis of studies performed in a family in which a number 1 chromosome with a clear secondary constriction is segregating, it seems that the Duffy (Fy) locus is on chromosome 1 (Donahue et al., 1968). Because the locus for a congenital cataract is linked with the Fy locus (Renwick and Lawler, 1963) it, too, was recognized as being on chromosome 1. Now, Rh and several other loci are also recognized as being on chromosome 1, though Rh and Fy are not within directly measurable distance of each other. As indicated earlier, Rh and PGM_1 are linked (Robson et al., 1973); PGM_1 is syntenic with peptidase C (van Cong et al., 1971) and peptidase C is syntenic with chromosome 1 (Ruddle et al., 1972), thereby allocating all of these loci to chromosome 1.

Tables 5 and 6 on pp. 436–437 show the recent state of the human autosomal map.

X. OTHER SYSTEMS

There are many blood group systems in Man, apart from those already mentioned in this chapter. A number of these have antigens which occur with polymorphic frequencies. Such systems are the P, Lutheran, Kell, Duffy, Kidd, Diego, Dombrock, Auberger, and Sid. Sections have not been devoted to these since much less is known about them than about those systems which have been discussed. Antibodies in these systems are seldom the cause of haemolytic disease of the newborn or of transfusion difficulties. Consequently, these systems have not drawn as much attention to themselves as have those to which more space has been devoted. However, they are certainly not without their interest and merit the attention of human geneticists.

The Kell system is now recognized as being complex. The antithetical antigens of three systems initially reported as Kell, Sutter and Penney (Stroup *et al.*, 1965; Allen and Lewis, 1957) are now all considered to be part of the Kell complex. Furthermore, it seems that the recently reported Weeks antigen (Strange *et al.*, 1974) belongs to the complex, and that the best way of considering the activities within the system is similar to that with Rh, but with four series of alternative antigens. Since Kell reagents are not as readily available as Rh, the intricacies of the system are not as well known. The Lutheran phenotype Lu (a–b–) is common in Negroes, but is rare in Whites. The Diego antigen Di^a is found practically only in Mongoloids, and is certainly rare in Whites. Conversely, the antithetical antigen, Di^b, is very common in Whites. The Sid antigen, Sd^a, is found on red cells and in the sorts of body secretions where ABO and Lewis antigens can be found, such as saliva and urine. However, the activity of the Sd^a antigen does not parallel those of A, B, or H (Morton *et al.*, 1970). The antigen appears to be carbohydrate, since it can be precipitated by ethanol without denaturation but is denatured by acid, alkali and periodate (Morton and Terry, 1970). Further, since agglutination can be inhibited by N-acetyl-D-galactosamine similarly to agglutination involving the A antigen (Sanger *et al.*, 1971), this sugar is important in the Sd^a antigen also. Because the antigen is present in aqueous fluids, it is in a convenient form for the types of studies which have been performed with such success on the ABH and Lewis antigens, and it is anticipated that more biochemical work will be carried out with Sd^a. The P blood group system involves at least three antigens, P_1, P^k and P, and combinations of these antigens can be detected in the groups, P_1, P_2, P_1^k and P_2^k with the above three antigens being absent in the p phenotype. The P_1 antigen, detected in ovarian cyst mucin, was shown to be a glycoprotein similar in nature to the ABH and

Lewis antigens (Morgan and Watkins, 1962). A glycosphingolipid with P_1 activity has been isolated from human erythrocyte stroma (Marcus, 1971) and the importance of an α-galactosyl linkage in the P_1 determinant has been confirmed by the destruction of this activity by coffee bean α-galactosidase (Anstee and Pardoe, 1973).

Cory et al. (1974), from the results of inhibition studies using anti-P_1, showed the P_1 determinant to have the structure:

D-galactosyl-α (1 $\rightarrow$ 4)-D-galactosyl-β(1 $\rightarrow$ 4)-N-acetyl-D-glucosamine. Recently, Watkins and Morgan (1976) have undertaken further studies, and have shown that the above trisaccharide and the disaccharide D-galactosyl-α(1 $\rightarrow$ 4)-D-galactose inhibit human anti-P^k sera equally effectively.

Naiki and Marcus (1974) showed that both the P and P^k antigens were glycosphingolipids, and proposed that P^k is the precursor of P. Watkins and Morgan (1976) found that, whereas anti-P_1 and anti-P^k sera were inhibited by structurally related oligosaccharides, with anti-P_1 reagents being sharply specific for the trisaccharide given above, anti-P sera were not inhibited by the tested compounds which had terminal D-galactosyl-α(1 $\rightarrow$ 4)-D-galactosyl residues. However, a glycolipid, N-acetyl-D-galactosaminyl-β (1 $\rightarrow$ 3)-D-galactosyl-α(1 $\rightarrow$ 4)-D-galactosyl-β(1 $\rightarrow$ 4)-glucosyl-ceramide, strongly inhibited anti-P sera.

The data are consistent with the antigen P^k being the precursor of P, and with P_1 and P^k being closely related. However, a satisfying scheme for the genetically-determined biosynthesis of these antigens and their inter-relationships in the five known phenotypes is difficult to devise at this stage. Nonetheless, it is envisaged that the genetical control of the biosynthesis of the P group antigens is effected through transferase enzymes, similarly to the ABH system.

In addition to these systems with antigens occurring in polymorphic frequencies, there are others where the antigen frequencies are either very high ("public" antigens) or very low ("private" antigens). In these systems the inherited nature of the rare condition is indicated by there being at least two sibs with the same phenotype. Examples of "public" antigens are I, Vel, Gerbich, Lan, Gregory, August, Jr[a], Knops, El, Dp and Gonsowski. An example of an antithetical relationship between a "public" and a "private" antigen, not recognized initially, is that of the antigens Sm and Bu[a] (Lewis et al., 1964, 1967) which are now joined in the Scianna blood groups. The antigen Co[a] was initially a "public" antigen, but Co[b] can now be recognized (Giles et al., 1970), giving rise to the Colton blood groups. A similar situation had earlier produced the Yt blood groups (Eaton et al., 1956; Giles and Metaxas, 1964).

The list of private antigens reported is quite long and probably there are many antigens which have been encountered but which have not been reported. Perhaps some of these bear antithetical relationships to "public" antigens, but these have yet to be demonstrated.

For further information on the above systems, the reader is referred to "Blood Groups in Man" (Race and Sanger, 1975) which provides an excellent introduction to the relevant literature.

XI. PRIMATES

Blood groups, of course, are not peculiar to Man nor are the blood group systems of Man peculiar to this one species. There is evidence of some similar (if not identical) antigens on the red cells of other primates and suggestions of the evolutionary development of some blood group systems through monkeys and apes to Man.

ABO antigens are present on red cells and the appropriate agglutinins are present in the sera from the apes. Individuals of groups O and A have been found in chimpanzees; A, B and AB in orang-utans; A, B and AB in gibbons; B in siamangs and A-like and B-like in gorillas. The A antigen of gibbons is similar to that of human A_1, that of chimpanzees is similar to human A_2 and orang-utan A in intermediate between these. The H expression of B gibbons appears to be stronger than that of humans of group B (Wiener and Moor-Jankowski, 1963, 1969; Wiener et al., 1968).

The red cells of baboons and most monkeys do not possess ABO group antigens, though A, B and H substances are found in their salivas and the complementary antibodies in their sera. They can thus be assigned to a particular ABO group. The red cells of New World monkeys possess a B-like antigen, which is independent of the salivary antigens (Moor-Jankowski et al., 1964), and marmosets have an A- and a B- like antigen on their red cells as well as an A salivary antigen (Gengozian, 1964). It would be interesting to determine whether the A- and B- like antigens of New World monkeys are chemically similar to the A and B antigens in Man, for in the former the antigens are determined by non-allelic genes whereas in the latter they are determined by allelic genes.

Almost all apes, baboons, monkeys and marmosets are salivary secretors of ABH antigens, though one non-secretor orang-utan has been reported (Moor-Jankowski et al., 1964). Lewis substance has been detected in saliva from apes, baboons and monkeys. Chimpanzees are either weak or non-secretors of Lewis, whereas orang-utans are Lewis secretors. Baboons show a range of activities, with some being strong, some weak and some non-secretors of Lewis substances. Based on small numbers of individuals tested (pigtail monkeys and Celebes black apes), two species of Old World monkeys are Lewis secretors, whereas two species of New World monkeys (*Cebus* and squirrel) are Lewis non-secretors (Moor-Jankowski *et al.*, 1964).

Thus, it seems that secretion of ABO antigens is a more primitive condition than their presence on red cells and the B-like antigen on cells of New World monkeys appears to be a development independent of the ABO system.

MN blood group antigens have been detected on red cells from apes. Landsteiner and Levine (1928a) identified the M antigen on chimpanzee cells and Levine *et al.* (1955a) were able to identify some form of N antigen on red cells from some chimpanzees using an anti-N reagent made from *Vicia graminea* seeds. This seed extract has been found to be valuable in determining group N in apes, because rabbit antisera to M and N can have different specificities. Rabbit anti-N's have given equivocal results and although selected rabbit anti-M's have given consistent results, some rabbit anti-M's which react with human M and MN cells react with all gibbons, some with no gibbons and others show a polymorphism (Wiener *et al.*, 1966). Groups M, MN and N have been identified in gibbons and M in siamangs (Wiener *et al.*, 1964a), M or not-M in orang-utans and MN and N in lowland gorillas (Wiener *et al.*, 1972).

As stated earlier, the LW antigen has been detected on red cells of apes, baboons and a variety of monkeys (Levine and Celano, 1962a, 1967; Levine *et al.*, 1962). No monkey with any of the Rh antigens has yet been found, though all species of apes have some form of the D and c antigens (Wiener *et al.*, 1964b; Boettcher *et al.*, 1965). Antigens of the E/e series have been found in Man only. Thus, since the Rh and LW groups are phenotypically related, it appears that the gene for the LW antigen is phylogenetically older than the *Rh* genes. The *Rh* genes may have arisen through a process of gene duplication and evolution (Boettcher, 1964; Race and Sanger, 1968, p. 512). If Giblett's model (Giblett, 1969) for the relationship between the biosynthetic pathways leading to the production of Rh and LW antigens is correct, viz. in humans LW is added to the Rh antigens, it seems difficult to visualize how evolution could occur to interpose Rh antigen biosynthesis before LW antigen biosynthesis in such a way that LW specificity is not lost and so that Rh antigen expression is obligatory before LW expression.

Among a number of species of apes and monkeys tested, including the chimpanzee, gorilla and orang-utan, the antigen Xg[a] was found only in the gibbon and in this species there is some suggestion that it might be sex-linked as in Man (Gavin *et al.*, 1964). It seems strange that the gibbon, considered not to be as closely related to Man as the other apes, should be the only ape to share this sex-linked blood group with him.

REFERENCES

Adam, A. (1961). *Nature, Lond.* **189**, 686.

Adam, A., Ziprkowski, L., Feinstein, A., Sanger, R. and Race, R. R. (1966). *Lancet* **i**, 877.

Adinolfi, M., Polley, H. J., Hunter, D. A. and Mollison, P. L. (1962). *Immunology* **5**, 566.

Allen, F. H. and Lewis, S. J. (1957). *Vox Sang.* **2**, 81.

Allen, F. H., Corcoran, P. A., Kenton, H. B. and Breare, N. (1958). *Vox Sang.* **3**, 81.

Allen, F. H., Corcoran, P. A. and Ellis, F. R. (1960). *Vox Sang.* **5**, 224.

Aloysia, M., Gelb, A. G., Fudenberg, H. H., Hamper, J., Tippett, P. and Race, R. R. (1961). *Transfusion* **1**, 212.

Alter, A. A. and Rosenfield, R. E. (1964). *Blood* **23**, 605.

Andresen, P. H., Goldman, F. and Henriksen, L. (1968). *Transfusion* **8**, 388.

Anstee, D. J. and Pardoe, G. I. (1974). *Eur. J. Biochem.* **19**, 149.

Azevêdo, E., Krieger, H. and Morton, N. E. (1963). *Am. J. Hum. Genet.* **16**, 451.

Baranowski, T., Lisowska, E., Morawiecki, A., Romanowska, E. and Strózecka, K. (1959). *Archism. Immun. Terap. Doswiad.* **7**, 15.

Baranowski, T., Lisowska, E. and Romanowska, E. (1956). Quoted in Prokop and Uhlenbruck (1969), 446.

Bar-Shany, S., Bastani, A., Cuttner, J. and Rosenfield, R. E. (1967). *Transfusion* **7**, 389.

Beattie, K. M., Zuelzer, W. W., McGuire, D. A. and Cohen, F. (1964). *Transfusion* **4**, 77.

Beckman, L. (1964). *Acta Genet.* **14**, 286.

Beckman, L. (1968). *Ser. Haemat.* **1**, 137.

Bell, J. and Haldane, J. B. S. (1937). *Proc. R. Soc. B* **123**, 119.

Berg, K. and Bearn, A. G. (1966). *J. Exp. Med.* **123**, 379.

Berg, K. and Bearn, A. G. (1968). *Ann. Rev. Genet.* **2**, 341.

Bernstein, F. (1924). *Klin. Wschr.* **3**, 1495.

Bhatia, H. M., Sathe, M., Gandhi, S., Mehta, B. E. and Levine, P. (1974). *Vox Sang.* **26**, 272.

Bhende, Y. M., Deshpande, C. K., Bhatia, H. M., Sanger, R., Race, R. R., Morgan, W. T. J. and Watkins, W. M. (1952). *Lancet* **i**, 903.

Bigley, N. J., Dodd, M. C., Randles, C. I., Goyer, V. B. and Lazer, A. G. (1963). *J. Immunol.* **90**, 526.

Boettcher, B. (1964). *Vox Sang.* **9**, 641.

Boettcher, B. (1966). *Vox Sang.* **11**, 129.

Boettcher, B. (1969). *In* "Immunology and Reproduction" (R. G. Edwards, Ed.), 148–154. International Planned Parenthood Federation, London.

Boettcher, B. (1972). *In* "The Biochemical Genetics of Man" 1st ed. (D. J. H. Brock and O. Mayo, Eds). 276. Academic Press, London and New York.

Boettcher, B. and Kenny, R. (1971). *Hum. Hered.* **21**, 334.

Boettcher, B. and Watts, S. (1978). *Vox Sang.* (In press).

Boettcher, B., Vos, G. H. and Hay, J. (1965). *Am. J. Hum. Genet.* **17**, 308.

Booth, P. B. (1972). *Vox Sang.* **22**, 524.

Bouguerra-Jacquet, A., Hrubisko, M., Salmon, D. and Salmon, C. (1970). *Nouv. Revue Fr. Hémat.* **10**, 173.

Bowdler, A. J. and Swisher, S. N. (1964). *Transfusion* 4, 153.

Boyd, W. C. and Reeves, E. (1961). *Nature, Lond.* 191, 511.

Brown, P. C., Glynn, L. E. and Holborow, E. J. (1959). *Vox Sang.* 4, 1.

Bundschuh, G. (1966). *Acta Biol. Med. Germ.* 17, 349.

Campbell, D. H., Sturgeon, P. and Vinograd, J. R. (1955). *Science* 122, 1091.

Chakravartti, M. R., Verna, B. K., Hanurav, T. V. and Vogel, F. (1966). *Humangenetik* 2, 78.

Chalmers, J. N. M. and Lawler, S. D. (1953). *Ann. Eugen.* 17, 267.

Chalmers, J. N. M., Ikin, E. W. and Mourant, A. E. (1953). *Brit. Med. J.* II, 175.

Chalmers, D. G., Coombs, R. R. A., Gurner, B. W. and Dausset, J. (1959). *Br. J. Haemat.* 5, 225.

Chandanayingyong, D., Sasaki, T. T. and Greenwalt, T. J. (1967). *Transfusion* 7, 269.

Chattoraj, A. and Boyd, W. C. (1965). *Vox Sang.* 10, 700.

Chester, M. A. and Watkins, W. M. (1969). *Biochem. Biophys. Res. Commun.* 34, 835.

Chown, B., Lewis, M., Kaito, H. and Lowen, B. (1972). *Am. J. Hum. Genet.* 24, 623.

Chung, C. S., Pitt, E. L. and Dublin, T. D. (1965). *Am. J. Hum. Genet.* 22, 194.

Clarke, G., Stevenson, A. C., Davies, P. and Williams, C. E. (1964). *J. Med. Genet.* 1, 27.

Cohen, F., Zuelzer, W. W. and Evans, M. M. (1960). *Blood* 15, 884.

Constantoulis, N. C., Paidoussis, M. and Dunsford, I. (1955). *Vox Sang.* 5, 143.

Cook, P. J. L. (1965). *Ann. Hum. Genet.* 28, 393.

Cook, P. J. L., Hopkinson, D. A. and Robson, E. B. (1970). *Ann. Hum. Genet.* 34, 187.

Coombs, R. R. A. and Bedford, D. (1955). *Vox Sang.* 5, 111.

Coombs, R. R. A., Mourant, A. E. and Race, R. R. (1945). *Brit. J. Exp. Path.* 26, 255.

Coombs, R. R. A., Marks, J. and Bedford, D. (1956a). *Br. J. Haemat.* 2, 84.

Coombs, R. R. A., Bedford, D. and Rouillard, L. M. (1956b). *Lancet* i, 461.

Cory, H. J., Yates, A. D., Donald, A. S. R., Watkins, W. M. and Morgan, W. T. J. (1974). *Biochem. Biophys. Res. Commun.* 61, 1289.

Côté, R. H. and Morgan, W. T. J. (1956). *Nature, Lond.* 178, 1171.

Crookston, M. C., Tilley, C. A. and Crookston, J. H. (1970). *Lancet* ii, 1110.

Darnborough, J. (1957). *Vox Sang.* 2, 362.

Darnborough, J., Dunsford, I. and Wallace, J. A. (1969). *Vox Sang.* 17, 241.

Denborough, M. A., Downing, H. J. and McRea, M. G. (1967). *Aust. Ann. Med.* 16, 320.

Diamond, L. K. (1947). *Proc. R. Soc. Med.* 40, 546.

Diamond, L. K. and Denton, R. L. (1945). *J. Lab. Clin. Med.* 30, 821.

Dodd, M. C., Bigley, N. J. and Geyer, V. B. (1960). *Science* 132, 1398.

Dodd, M. C., Bigley, N. J., Johnson, G. A. and McCluer, R. H. (1964). *Nature, Lond.* 204, 549.

Donahue, R. P., Bias, W. B., Renwick, J. H. and McKusick, V. A. (1968). *Proc. Natn. Acad. Sci. USA* 61, 949.

Eaton, B. R., Morton, J. A., Pickles, M. M. and White, K. E. (1956). *Br. J. Haemat.* 2, 333.

Edwards, J. H., Allen, F. H., Glenn, K. P., Lamm, L. U. and Robson, E. B. (1972). *In* "Histocompatibility Testing", 745. Munksgaard, Copenhagen.

Edwards, R. G., Ferguson, L. C. and Coombs, R. R. A. (1964). *J. Reprod. Fert.* **7**, 153.

Emery, A. E. H., Smith, C. A. B. and Sanger, R. (1969). *Ann. Hum. Genet.* **32**, 261.

Eriksson, A. W., Fellman, J. H., Vainio-Mattila, B., Sanger, R., Race, R. R., Krause, U. and Forsius, H. (1967). *Int. Congr. Neuro-Genet. and Neuro-Ophthal.* 360.

Fellous, M., Bengtsson, B. O., Finnegan, D. J. and Bodmer, W. F. (1974). *Ann. Hum. Genet.* **37**, 421.

Fialkow, P. J., Giblett, E. R. and Motulsky, A. G. (1967). *Am. J. Hum. Genet.* **19**, 63.

Figur, A. M. and Rosenfield, R. E. (1965). *Vox Sang.* **10**, 169.

Filitti-Wurmser, S., Jacquot-Armand, Y., Aubel-Lesure, G. and Wurmser, R. (1954). *Ann. Eugen.* **18**, 183.

Finney, D. J. (1962). "Probit Analysis", 2nd ed. Cambridge University Press, Cambridge.

Fisk, R. T. and Foord, A. G. (1942). *Am. J. Clin. Path.* **12**, 545.

Fisk, R. T. and McGee, C. A. (1947). *Am. J. Clin. Path.* **17**, 737.

Frame, M., Mollison, P. L. and Terry, W. D. (1970). *Nature, Lond.* **225**, 641.

Fraser, G. R., Giblett, E. R. and Motulsky, A. G. (1966). *Am. J. Hum. Genet.* **18**, 546.

Friedenreich, V. (1936). *Z. Immun. Forsch. Exp. Ther.* **89**, 409.

Frøland, A., Sanger, R. and Race, R. R. (1968). *J. Med. Genet.* **5**, 161.

Fudenberg, H. H., Kunkel, H. G. and Franklin, E. C. (1959). Proc. VII Cong. Int. Soc. Blood Transf., Rome, 1958, 124.

Furuhjelm, U., Myllylä, G., Nevanlinna, H. R., Nordling, S., Pirkola, A., Gavin, J., Gooch, A., Sanger, R. and Tippett, P. (1969). *Vox Sang.* **17**, 256.

Gardas, A. and Kościelak, J. (1971). *Vox Sang.* **20**, 137.

Gavin, J., Noades, J., Tippett, P., Sanger, R. and Race, R. R. (1964). *Nature, Lond.* **204**, 1322.

Gengozian, N. (1964). *Proc. Soc. Exp. Biol. Med.* **117**, 858.

Gibbs, M. B. (1966). *Nature, Lond.* **210**, 642.

Gibbs, M. B. and Akeroyd, J. H. (1959a). *J. Immunol.* **82**, 568.

Gibbs, M. B. and Akeroyd, J. H. (1959b). *J. Immunol.* **82**, 577.

Gibbs, M. B., Dreyfuss, J. N. and Aguilu, L. A. (1965). *J. Immunol.* **94**, 62.

Giblett, E. R. (1969). "Genetic Markers in Human Blood". Davis, Philadelphia.

Giblett, E. R. and Crookston, M. C. (1964). *Nature, Lond.* **201**, 1138.

Giblett, E. R., Chen, S.-H., Anderson, J. E. and Lewis, M. (1974). *Cytogenet. Cell Genet.* **13**, 91.

Giles, C. M. and Metaxas, M. N. (1964). *Nature, Lond.* **202**, 1122.

Giles, C. M., Darnborough, J., Aspinall, P. and Fletton, M. W. (1970). *Br. J. Haemat.* **19**, 267.

Green, F. A. (1965). *Vox Sang.* **10**, 32.

Green, F. A. (1967a). *Immunochemistry* **4**, 247.

Green, F. A. (1967b). *Blood* **30**, 870.

Green, F. A. (1968). *J. Biol. Chem.* **243**, 5519.

Green, F. A. (1972). *J. Biol. Chem.* **247**, 881.

Greenwalt, T. J., Sasaki, T., Sanger, R., Sneath, J. and Race, R. R. (1954). *Proc. Natn. Acad. Sci. USA* **40**, 1126.

Greenwalt, T. J., Sasaki, R., Sanger, R. and Race, R. R. (1958). Proc. 6th. Congr. Int. Soc. Blood Trans. 104.

Grubb, R. (1949). *J. Clin. Path.* **2**, 223.
Grundbacher, F. J. (1965). *Am. J. Hum. Genet.* **17**, 399.
Hackel, E. (1964). *Vox Sang.* **9**, 56.
Hakamori, S.-I. and Kobata, A. (1974). *In* "The Antigens" Vol. 2, 108.) M. Sela, Ed. Academic Press, New York.
Häkkinen, I. (1970). *J. Natn. Cancer Inst.* **44**, 1183.
Harper, P. S., Rivas, M. L., Bias, W. B., Hutchinson, J. R., Dyken, P. R. and McKusick, V. A. (1972). *Am. J. Hum. Genet.* **24**, 310.
Harrap, G. J. and Watkins, W. M. (1964). *Biochem. J.* **93**, 9.
Harris, R. C., Harrison, G. A. and Rondle, C. J. M. (1963). *Acta Genet.* **13**, 44.
Hartmann, G. (1941). "Group Antigens in Human Organs". Munksgaard, Copenhagen.
Hasekura, H. and Boettcher, B. (1970). *Proc. Jap. Acad.* **46**, 733.
Hearn, V. M., Smith, Z. G. and Watkins, W. M. (1968). *Biochem. J.* **109**, 315.
Hearn, V. M., Race, C. and Watkins, W. M. (1972). *Biochem. Biophys. Res. Commun.* **46**, 948.
Henningsen, K. (1959). Proc. VII Congr. Int. Soc. Blood Transf. 567.
Henningsen, K. (1966). *Acta Genet.* **16**, 239.
Hillman, R. S. and Giblett, E. R. (1965). *J. Clin. Invest.* **44**, 1730.
Hirsch, W., Moores, P. O., Sanger, R. and Race, R. R. (1957). *J. Haemat.* **3**, 134.
Hoefnagel, D., Allen, F. H. and Walker, M. (1969). *Lancet* **i**, 1072.
Högman, C. F. and Killander, J. (1962). *Acta Path. Microbiol. Scand.* **55**, 357.
Hohorst, H. J. (1954). *Z. Hyg.* **139**, 561.
Hope, R. M., Mayo, O. and Boettcher, B. (1968). *Vox Sang.* **15**, 70.
Hughes-Jones, N. C., Gardner, B. and Telford, R. (1964). *Vox Sang.* **9**, 175.
Ikin, E. W. and Mourant, A. E. (1951). *Br. Med. J.* **I**, 456.
Iseki, S. and Ikeda, T. (1956). *Proc. Jap. Acad.* **32**, 201.
Iseki, S. and Masaki, S. (1953). *Proc. Jap. Acad.* **29**, 460.
Ishimori, T. and Hasekura, H. (1966). *Proc. Jap. Acad.* **42**, 658.
Ishimori, T. and Hasekura, H. (1967). *Transfusion* **7**, 265.
Ives, E. J., Ewing, C. C. and Innes, R. (1970). *Am. J. Hum. Genet.* **22**, 17a.
Johnston, A. W., Frost, P., Spaeth, G. L. and Renwick, J. H. (1969). *Ann. Hum. Genet.* **32**, 369.
Jones, A. R. and Silver, S. (1958). *Blood* **13**, 763.
Kabat, E. A. and Leskowitz, S. (1955). *J. Am. Chem. Soc.* **77**, 5159.
Kaklamanis, E., Holborow, E. J. and Glynn, L. E. (1964). *Lancet* **i**, 788.
Kelso, J. (1968). *Vox Sang.* **14**, 282.
Kerr, C. B., Wells, R. S. and Sanger, R. (1964). *Lancet* **ii**, 1369.
Kim, Y. S., Perdomo, J., Bella, A. and Nordberg, J. (1971). *Proc. Natn. Acad. Sci. USA* **68**, 1753.
Klinger, R., Miggiano, V. C., Tippett, P. and Gavin, J. (1968). *Atti Accad. med. Lomb.* **23**, 1392.
Kobata, A., Grollman, E. F. and Ginsburg, V. (1968a). *Biochem. Biophys. Res. Commun.* **32**, 272.
Kobata, A., Grollman, E. F. and Ginsburg, V. (1968b). *Arch. Biochem. Biophys.* **124**, 609.
Kochwa, S., Rosenfield, R. E., Tallal, L. and Wasserman, L. R. (1960). *In* "Progress in Medical Genetics" (A. G. Steinberg, Ed.), Vol. 1, 120–157. Grune and Stratton, London.
Kogure, J. (1971). *Jap. J. Hum. Genet.* **16**, 69.

Kuhns, W. J. and Bailey, A. (1950). *Am. J. Clin. Path.* **20**, 1067.
Kunkel, H. G. and Rockey, J. H. (1963). *Proc. Soc. Exp. Biol.* **113**, 278.
Lambotte, C. and Israel, E. (1967). *Ann. Soc. Belge Med. Trop.* **47**, 405.
Landsteiner, K. (1900). *Zentbl. Bakt. Parasit Kde.* **27**, 357.
Landsteiner, K. and Levine, P. (1927a). *Proc. Soc. Exp. Biol.* **24**, 600.
Landsteiner, K. and Levine, P. (1927b). *Proc. Soc. Exp. Biol.* **24**, 941.
Landsteiner, K. and Levine, P. (1928a). *J. Exp. Med.* **47**, 757.
Landsteiner, K. and Levine, P. (1928b). *J. Exp. Med.* **48**, 731.
Landsteiner, K. and Wiener, A. S. (1940). *Proc. Soc. Exp. Biol.* **43**, 223.
Landsteiner, K., Strutton, W. R. and Chase, M. W. (1934). *J. Immunol.* **27**, 469.
Lanset, S., Ropartz, C., Rousseau, P.-Y., Guerbet, Y. and Salmon, C. (1966). *Transfusion* **9**, 255.
Lawler, S. D. (1959). Proc. 7th. Congr. Eur. Soc. Haemat., Part II, 1219.
Lehrs, H. (1930). *Z. Immun. Forsch. exp. Ther.* **66**, 175
Levine, P. (1975). *Wadley Med. Bull.* **5**. 174.
Levine, P. and Celano, M. J. (1962a). *Immunogenet. Let.* **2**, 66.
Levine, P. and Celano, M. J. (1962b). *Nature Lond.* **193**, 184.
Levine, P. and Celano, M. J. (1967). *Science* **156**, 1744.
Levine, P. and Stetson, R. E. (1939). *J. Am. Med. Ass.* **113**, 126.
Levine, P., Vogel, P., Katzin, E. H. and Burnham, L. (1941). *Science* **94**, 371.
Levine, P., Kuhnuckel, A. B., Wigod, M. and Koch, E. (1951). *Proc. Soc. Exp. Biol.* **78**, 218.
Levine, P., Ottensooser, F., Celano, M. J. and Pollitzer, W. (1955a). *Am. J. Phys. Anthrop.* **13**, 29.
Levine, P., Robinson, E., Celano, M. J., Briggs, P. and Falkinburg, L. (1955b). *Blood* **10**, 1100.
Levine, P., Celano, M. J., Fenichel, R., Pollack, W. and Singher, H. O. (1961). *J. Immunol.* **87**, 747.
Levine, P., Celano, M. J., Vos, G. H. and Morrison, J. (1962). *Nature, Lond.* **194**, 304.
Levine, P., Celano, M. J., Wallace, J. A. and Sanger, R. (1963). *Nature, Lond.* **198**, 596.
Levine, P., Chambers, J. W., Celano, M. J., Falkowski, F., Hunter, O. B. and English, C. T. (1965a). Proc. 10th. Congr. Int. Soc. Blood Transf., Stockholm, 350.
Levine, P., Celano, M. J., Falkowski, F., Chambers, J. W., Hunter, O. B. and English, C. T. (1965b). *Transfusion* **5**, 492.
Levine, P., Tripodi, O., Struck, J., jr., Zmijewski, C. M. and Pollack, W. (1973). *Vox Sang.* **24**, 417.
Lewis, M., Kaita, H. and Chown, B. (1957). *Am. J. Hum. Genet.* **9**, 274.
Lewis, M., Chown, B., Schmidt, R. P. and Griffiths, J. J. (1964). *Am. J. Hum. Genet.* **16**, 254.
Lewis, M., Chown, B. and Kaita, H. (1967). *Transfusion* **7**, 92.
Lloyd, K. O. and Kabat, E. A. (1968). *Proc. Natn. Acad. Sci. USA* **61**, 1470.
Lowry, O. H., Rosebrough, N. J., Farr, A. L. and Randall, R. J. (1951). *J. Biol. Chem.* **193**, 265.
Lyon, M. F. (1962). *Am. J. Hum. Genet.* **14**, 135.
McKusick, V. A. (1975). "Mendelian Inheritance in Man" 5th ed. Johns Hopkins Press, Baltimore.
McNeil, C. and Trentelman, E. F. (1951). *Proc. Soc. Exp. Biol.* **78**, 674.
Madsen, G. and Heisto, H. (1968). *Vox Sang.* **14**, 211.
Mäkelä, O. and Cantell, K. (1958). *Ann. Med. Exp. Biol. Fenn.* **36**, 366.

368   B. BOETTCHER

Malmquist, E., Ivemark, B. I., Lindsten, J., Maunsbach, A. B. and Martensson, E. (1971). *Lab. Invest.* **25**, 1.
Mann, J. D., Cahan, A., Gelb, A. G., Fisher, N., Hamper, J., Tippett, P., Sanger, R. and Race, R. R. (1962). *Lancet* **i**, 8.
Marcus, D. M. (1971). *Transfusion* **11**, 16.
Masters, P. L. and Vos, G. H. (1962). *Lancet* **ii**, 641.
Matsunaga, E. (1959). *Jap. J. Hum. Genet.* **4**, 173.
Metaxas, M. N. and Metaxas-Bühler, M. (1964). *Nature, Lond.* **202**, 1123.
Metaxas-Bühler, M., Cleghorn, T. E., Romanski, J. and Metaxas, M. N. (1966). *Vox Sang.* **11**, 170.
Middleton, J., Crookston, M. C., Falk, J. A., Robson, E. B., Cook, P. L. L., Batchelor, J. R., Bodmer, Julia, Ferrara, G. B., Festenstein, H., Harris, R. C., Kissmeyer-Nielsen, F., Lawler, S. D., Sachs, J. A. and Wolf, E. (1974). *Tissue Antigens* **4**, 366.
Mohr, J. (1951). *Acta Path. Microbiol. Scand.* **28**, 207.
Mohr, J. (1954). "A Study of Linkage in Man", Munksgaard, Copenhagen.
Mollison, P. L. (1972). "Blood Transfusion in Clinical Medicine", 5th ed. Blackwells, Oxford.
Moores, P. O. (1966). Quoted in "Blood Groups in Man" 5th ed. (R. R. Race and R. Sanger, Eds) 448. Blackwell, Oxford.
Moor-Jankowski, J., Wiener, A. S. and Rogers, C. M. (1964). *Nature, Lond.* **202**, 663.
Morgan, W. T. J. and Watkins, W. M. (1962). *Proc. 9th Congr. Int. Soc. Blood Transf.*, 1960, 225.
Morton, J. A. and Pickles, M. M. (1951). *J. Clin. Path.* **4**, 189.
Morton, J. A. and Terry, A. M. (1970). *Vox Sang.* **19**, 151.
Morton, J. A., Pickles, M. M. and Terry, A. M. (1970). *Vox Sang.* **19**, 472.
Mourant, A. E. (1954). *In* "The Distribution of the Human Blood Groups", 185. Blackwells, Oxford.
Muller, H. J. and Settles, F. (1927). *Z. Vererb Lehre* **43**, 285.
Murray, J. and Clark, E. C. (1952). *Nature, Lond.* **169**, 886.
Naiki, M. and Marcus, D. M. (1974). *Biochem. Biophys. Res. Commun.* **60**, 1105.
Nicolson, G. L., Masouredis, S. P. and Singer, S. J. (1971). *Proc. Natn. Acad. Sci. USA* **68**, 1416.
Noades, J., Gavin, J., Tippett, P., Sanger, R. and Race, R. R. (1966). *J. Med. Genet.* **3**, 162.
Opitz, J. M., Stiles, F. C., Wise, D., Race, R. R., Sanger, R., von Gemmingen, G. R., Kierland, R. R., Cross, E. G. and deGroot, W. P. (1965). *Am. J. Hum. Genet* **17**, 325.
Pearce, W. G., Sanger, R. and Race, R. R. (1968). *Lancet* **i**, 1282.
Pearce, W. G., Johnson, G. A. and Sanger, R. (1971). *Lancet* **i**, 1072.
Pettenkofer, H. J., Stoss, B., Helmbold, W. and Vogel, F. (1962). *Nature, Lond.* **193**, 445.
Plaut, G., Ikin, E. W., Mourant, A. E., Sanger, R. and Race, R. R. (1953). *Nature, Lond.* **171**, 431.
Pollack, W. (1965). *Ann. N. Y. Acad. Sci.* **127**, 892.
Pollack, W., Hager, H. J., Reckel, R., Toren, D. A. and Singher, H. O. (1965). *Transfusion* **5**, 158.
Polley, H. J., Adinolfi, M. and Mollison, P. L. (1963). *Vox Sang.* **8**, 385.
Poretz, R. D. and Watkins, W. M. (1972). *Eur. J. Biochem.* **25**, 455.

Porter, I. H., Schulze, J. and McKusick, V. A. (1962). *Nature, Lond.* **193**, 506.
Prankerd, T. A. J. (1961). *In* "The Red Cell", 92. Blackwells, Oxford.
Prokop, O. and Schneider, W. (1960). *Dt. Z. Ges. Gericht. Med.* **50**, 423.
Prokop, O. and Uhlenbruck, G. (1969). "Human Blood and Serum Groups", Maclaren and Sons, London.
Putkonen, T. (1930). *Acta Soc. Med. Fenn. "Duodecim"* **14**, 1.
Race, C., Ziderman, D. and Watkins, W. M. (1968). *Biochem. J.* **107**, 733.
Race, E. and Watkins, W. M. (1972). *FEBS Lett.* **27**, 125.
Race, R. R. (1970). *Phil. Trans. R. Soc. Lond. B.* **259**, 37.
Race, R. R. and Sanger, R. (1958). "Blood Groups in Man" 3rd ed. Blackwells, Oxford.
Race, R. R. and Sanger, R. (1962). "Blood Groups in Man" 4th ed. Blackwells, Oxford.
Race, R. R. and Sanger, R. (1968). "Blood Groups in Man" 5th ed. Blackwells, Oxford.
Race, R. R. and Sanger, R. (1969). *Br. Med. Bull.* **25**, 99.
Race, R. R. and Sanger, R. (1975). "Blood Groups in Man" 6th ed. Blackwells, Oxford.
Rapley, S. E., Robson, E. B., Harris, H. and Maynard Smith, S. (1968). *Ann. Hum. Genet.* **31**, 237.
Rawson, A. J. and Abelson, N. M. (1960). *J. Immunol.* **85**, 640.
Renton, P. H. and Hancock, J. A. (1962). *Vox Sang.* **7**, 33.
Renwick, J. H. (1971). *Nature, Lond.* **234**, 475.
Renwick, J. H. and Lawler, S. D. (1955). *Ann. Hum. Genet.* **19**, 312.
Renwick, J. H. and Lawler, S. D. (1963). *Ann. Hum. Genet.* **27**, 67.
Renwick, J. H. and Schulze, J. (1964). *Am. J. Hum. Genet.* **16**, 410.
Renwick, J. H. and Schulze, J. (1965). *Ann. Hum. Genet.* **28**, 379.
Renwick, J. H., Bundey, S. E., Ferguson-Smith, M. A. and Izatt, M. M. (1971). *J. Med. Genet.* **8**, 407.
Reviron, J., Jacquet, A., Delarue, F., Liberge, G., Salmon, D. and Salmon, C. (1967). *Nouv. Rev. Fr. Hémat.* **7**, 425.
Reviron, J., Jacquet, A. and Salmon, C. (1968). *Nouv. Rev. Fr. Hémat.* **8**, 323.
Rittner, C. (1967). *Vox Sang.* **13**, 29.
Robson, E. B., Cook, P. J. L., Corney, G., Hopkinson, D. A., Noades, J. and Cleghorn, T. E. (1973). *Ann. Hum. Genet.* **36**, 393.
Romanowska, E. (1964). *Vox Sang.* **9**, 578.
Rosenfield, R. E. and McGuire, D. (1973). Quoted in "Blood Groups in Man" 6th ed. (R. R. Race and R. Sanger, Eds) 227. Blackwells, Oxford.
Rosenfield, R. E. and Rubinstein, P. (1966). "Blood Groups in Immunogenetics". Proc. 11th Congr. Int. Soc. Haemat., Sydney, 244–250.
Rostenberg, I. (1976). Personal communication.
Ruddle, F. H., Ricciuti, F. C., McMorris, F. A., Tischfeld, J., Creagan, R. P., Darlington, G. J. and Chen, T. R. (1972). *Science,* **176**, 1429.
Salmon, C. and Salmon, D. (1968). *Bibthca. Haemat.* **29**, 189.
Salmon, C., Salmon, D. and Reviron, J. (1964). *Nouv. Rev. Fr. Hémat.* **4**, 739.
Salmon, C., deGrouchy, J. and Liberge, G. (1965). *Nouv. Rev. Fr. Hémat.* **5**, 631.
Sanger, R. and Race, R. R. (1947). *Nature, Lond.* **160**, 505.
Sanger, R. and Race, R. R. (1958). *Heredity* **12**, 513.
Sanger, R., Race, R. R., Walsh, R. J. and Montgomery, C. (1948). *Heredity* **2**, 131.

Sanger, R., Gavin, J., Tippett, P., Teesdale, P. and Eldon, K. (1971). *Lancet* **i**, 1130.

Sawicka, T. (1971). *FEBS Lett.* **16**, 346.

Schachter, H., Michaels, M. A., Tilley, C. A., Crookston, M. C. and Crookston, J. H. (1973). *Proc. Natn. Acad. USA* **70**, 220.

Schenkel-Brunner, H., Chester, M. A. and Watkins, W. M. (1972). *Eur. J. Biochem.* **30**, 269.

Schiff, F. and Sasaki, H. (1932). *Klin. Wschr.* **34**, 1426.

Schmidt, P. J. and Holland, P. V. (1971). *Bibl. Haemat.* **38**, 230.

Schmidt, P. J. and Vos, G. H. (1967). *Vox Sang.* **13**, 18.

Schmidt, P. J., Lostumbo, M. M., English, C. T. and Hunter, O. B. (1967). *Transfusion* **7**, 33.

Scott, J. S. and Good, W. (1970). *Obstet. Gynec.* **35**, 351.

Seidl, S., Spielmann, W. and Martin, H. (1972). *Vox Sang.* **23**, 182.

Seyfried, H., Walewska, I. and Werblinska, B. (1964). *Vox Sang.* **9**, 268.

Shen, L., Grollman, E. F. and Ginsburg, V. (1968). *Proc. Natn. Acad. Sci. USA* **59**, 224.

Shepherd, L. P., Feingold, E. and Shanbrom, E. (1969). *Vox Sang.* **16**, 157.

Shreffler, D. C. (1967). *Ann. Rev. Genet.* **1**, 163.

Silber, R., Gibbs, M. B., Jahn, E. F. and Akeroyd, J. H. (1961a). *Blood* **17**, 291.

Silber, R., Gibbs, M. B., Jahn, E. F. and Akeroyd, J. H. (1961b). *Blood* **17**, 282.

Simmons, R. T., Jakobowicz, R. and Young, N. A. F. (1969). *Med. J. Aust.* **I**, 339.

Siniscalco, M. (1974). *In* "Somatic Cell Hybridization" (R. L. Davidson and F. de la Cruz, Eds) 35. Raven Press, New York.

Siniscalco, M., Motulsky, A. G., Latte, B. and Bernini, L. (1960). *Atti. Acad. Naz. Lincei Rc.* **28**, 1.

Skinner, R., Smith, C. and Emery, A. E. H. (1974). *J. Med. Genet.* **11**, 317.

Solomon, J. M. (1964). *Transfusion* **4**, 3.

Solomon, J. M., Gibbs, M. B. and Bowdler, A. J. (1965a). *Vox Sang.* **10**, 54.

Solomon, J. M., Gibbs, M. B. and Bowdler, A. J. (1965b). *Vox Sang.* **10**, 133.

Solomon, J. M., Waggoner, R. and Leyshon, W. C. (1965c). *Blood* **25**, 470.

Spaeth, G. L. and Frost, P. (1965). *Arch. Ophthalmol.* **74**, 760.

Speiser, P. (1956). *Z. Immun. Forsch. Exp. Ther.* **113**, 165.

Springer, G. F. (1970a). *Naturwissenschaften* **57**, 162.

Springer, G. F. (1970b). *Ann. N. Y. Acad. Sci.* **169**, 134.

Springer, G. F. and Ansell, N. J. (1958). *Proc. Natn. Acad. Sci. USA* **44**, 182.

Springer, G. F. and Horton, R. E. (1969). *J. Clin. Invest.* **48**, 1280.

Springer, G. F. and Hotta, K. (1963). *Fedn. Proc. Fedn. Am. Socs. Exp. Biol.* **22**, 2261.

Springer, G. F. and Tegtmeyer, H. (1972). *Proc. Int. Congr. Blood Transfusion* 25.

Springer, G. F. and Wiener, A. S. (1962). *Nature, Lond.* **193**, 444.

Springer, G. F., Horton, R. E. and Forbes, M. (1959a). *J. Exp. Med.* **110**, 221.

Springer, G. F., Horton, R. E. and Forbes, M. (1959b). *Ann. N. Y. Acad. Sci.* **78**, 272.

Springer, G. F., Nagai, Y. and Tegtmeyer, H. (1966). *Biochemistry* **5**, 3254.

Steinberg, A. G. (1965). *Vox Sang.* **10**, 721.

Stern, K., Ellis, F. R. and Masaitis, L. (1957). *Am. J. Clin. Path.* **27**, 635.

Strange, J. J., Kenworthy, R. J., Webb, A. J. and Giles, C. M. (1974). *Vox Sang.* **27**, 81.

Stroup, M., MacIlroy, M., Walker, R. and Aydelotte, J. V. (1965). *Transfusion* 5, 309.

Sturgeon, P. (1970). *Blood* 36, 310.

Swanson, J. and Matson, G. A. (1964). *Transfusion* 4, 257.

Szulman, A. E. (1960). *J. Exp. Med.* 111, 785.

Szulman, A. E. (1962). *J. Exp. Med.* 115, 977.

Szulman, A. E. (1964). *J. Exp. Med.* 119, 503.

Thomsen, O., Friedenreich, V. and Worsaae, E. (1930). *Acta Path. Microbiol. Scand.* 7, 157.

Tippett, P. (1963). Quoted in "Blood Groups in Man" 6th ed. (R. R. Race and R. Sanger, Eds) 229. Blackwells, Oxford.

Tomasi, T. B. and Bienenstock, J. (1968). *Ad. Immun.* 9, 1.

Tomasi, T. B., Tan, E. M., Solomon, A. and Prendergast, R. A. (1965). *J. Exp. Med.* 121, 101.

Tosteson, D. C. (1963). *Fedn. Proc. Am. Socs. Exp. Biol.* 22, 19.

Tyler, H. M. and Watkins, W. M. (1960). *Biochem. J.* 74, 2P.

Uhlenbruck, G. (1961). *Zentbl. Bakt. Abt.* 181, 283.

Uhlenbruck, G. (1969). *Vox Sang.* 16, 200.

Uhlenbruck, G. and Krupe, M. (1965). *Vox Sang.* 10, 326.

van Cong, N., Billardon, C., Picard, J.-Y., Feingold, J. and Frezal, J. (1971). *C.R. Acad. Sci. Paris* 272, 485.

Vogel, F. and Chakravartti, M. R. (1966). *Humangenetik* 3, 166.

von Dungern, E. and Hirszfeld, L. (1911). *Z. Immun. Forsch. Exp. Ther.* 8, 526.

Vos, G. H. and Comley, P. (1967). *Acta Genet.* 17, 495.

Vos, G. H. and Kirk, R. L. (1962). *Vox Sang.* 7, 22.

Vos, G. H., Vos, D., Kirk, R. L. and Sanger, R. (1961). *Lancet* i, 14.

Walsh, R. J. and Montgomery, C. (1947). *Nature, Lond.* 160, 504.

Watkins, W. M. (1958). Proc. 7th. Congr. Int. Soc. Blood Transf., 692.

Watkins, W. M. (1962). *Immunology* 5, 245.

Watkins, W. M. (1966). *Science* 152, 172.

Watkins, W. M. (1970). *In* "Blood and Tissue Antigens" (D. Aminoff, Ed.) 200. Academic Press, New York and London.

Watkins, W. M. and Morgan, W. T. J. (1952). *Nature, Lond.* 169, 825.

Watkins, W. M. and Morgan, W. T. J. (1955). *Nature, Lond.* 175, 676.

Watkins, W. M. and Morgan, W. T. J. (1956/57). *Acta Genet.* 6, 521.

Watkins, W. M. and Morgan, W. T. J. (1957). *Nature, Lond.* 180, 1038.

Watkins, W. M. and Morgan, W. T. J. (1962). *Vox Sang.* 7, 129.

Watkins, W. M. and Morgan, W. T. J. (1976). *J. Immunogenet.* 3, 15.

Weicker, H. (1968). *Clin. Chim. Acta* 21, 513.

Weitkamp, L. R., Guttormsen, S. A. and Greendyke, R. M. (1971). *Am. J. Hum. Genet.* 23, 462.

Weitkamp, L. R., Johnston, E. and Guttormsen, S. A. (1974). *Cytogenet. Cell Genet.* 13, 183.

Wells, R. S., Jennings, M. C., Sanger, R. and Race, R. R. (1966). *Lancet* ii, 493.

Went, L. N., deGroot, W. P., Sanger, R., Tippett, P. and Gavin, J. (1969). *Ann. Hum. Genet.* 32, 333.

Wherrett, J. R., Brown, B. L., Tilley, C. A. and Crookston, M. E. (1971). *Clin, Res.* 29, 784.

Whittaker, D. L., Copeland, D. L. and Graham, J. B. (1962). *Am. J. Hum. Genet.* 14, 149.

Whittemore, N. B., Trabold, N., Weed, R. I., Rega, A., Reed, C. F. and Swisher, S. N. (1967). *Clin. Res.* **15**, 290.

Wiener, A. S. and Moor-Jankowski, J. (1963). *Science* **142**, 67.

Wiener, A. S. and Moor-Jankowski, J. (1969). *Ann. N. Y. Acad. Sci.* **162**, 37.

Wiener, A. S. and Peters, H. R. (1940). *Ann. Int. Med.* **13**, 2306.

Wiener, A. S. and Sonn-Gordon, E. B. (1947). *Am. J. Clin. Path.* **17**, 67.

Wiener, A. S., Unger, L. J. and Gordon, E. B. (1953). *J. Am. Med. Ass.* **153**, 1444.

Wiener, A. S., Unger, L. J. and Cohen, L. (1954). *Science* **119**, 734.

Wiener, A. S., Gordon, E. B. and Moor-Jankowski, J. (1964a). *J. Forens. Med.* **11**, 67.

Wiener, A. S., Moor-Jankowski, J. and Gordon, E. B. (1964b). *Am. J. Hum. Genet.* **16**, 246.

Wiener, A. S., Moor-Jankowski, J., Gordon, E. B., Riopelle, A. J. and Shell, W. F. (1966). *Transfusion* **6**, 311.

Wiener, A. S., Moor-Jankowski, J., Cadigan, F. C. and Gordon, E. B. (1968). *Transfusion* **8**, 235.

Wiener, A. S., Gordon, E. B., Moor-Jankowski, J. and Socha, W. W. (1972). *Haematologia* **6**, 419.

Wilkie, M. H. and Becker, E. L. (1955). *J. Immunol.* **74**, 199.

Yamaguchi, H., Okubo, Y. and Hazama, F. (1965). *Proc. Jap. Acad.* **41**, 316.

Yamaguchi, H., Okubo, Y. and Hazama, F. (1966). *Proc. Jap. Acad.* **42**, 512.

Yamaguchi, H., Okubo, Y. and Tanaka, M. (1970). *Proc. Jap. Acad.* **46**, 446.

Yamakami, K. (1926). *J. Immunol.* **12**, 185.

Zatz, M., Itskan, S. B., Sanger, R. and Fieve, R. R. (1972). *J. Am. Med. Ass.* **222**, 1624.

Zuelzer, W. W., Beattie, K. M. and Reisman, L. E. (1964). *Am. J. Hum. Genet.* **16**, 38.

8 Leucocyte Antigens

W. R. MAYR

Institute of Blood Group Serology, University of Vienna, Austria

I. INTRODUCTION

The interest of serologists was first directed to the antigens of leucocytes and thrombocytes when it became evident that some of them were important in the survival and function of grafted organs. If there are differences between donor and recipient in these antigens, called transplantation anti-gens or histocompatibility antigens, the graft induces an immune response resulting in the rejection of the transplanted organ.

The basic work in this field was done by Medawar (1944, 1945) who showed that the rejection of skin grafts is essentially due to immune re-actions and that allograft sensitivity (an allograft is a graft originating from the same species when donor and recipient are not genetically identical) can be induced by leucocyte injections (Friedman *et al.*, 1961; Medawar, 1946). This finding and the fact that skin grafts also produce circulating antibodies against leucocytes (Amos, 1953; Walford *et al.*, 1964) allowed the conclusion that transplantation antigens are present both in skin and leucocytes. This has been very important for research because leucocytes can easily be isolated for the purpose of serological investigation.

Dausset (1958) was the first to demonstrate a leucocyte antigen in Man, which he named Mac (now HLA-A2). He suggested that it might play an essential part in grafting. In spite of the importance of this finding, it was not until 1963 that van Rood and van Leeuwen described further leucocyte antigens (4a and 4b, now W4 and W6). Since then a great number of anti-gens have been found within a relatively short period of time. Modern genetical concepts of the antigenic systems of leucocytes and the elucidation of the genetical structure of the HLA system, the major histocompatibility complex of Man, are largely based on these serological findings.

II. METHODS

It would exceed the scope of this chapter to describe methods for the detection of leucocyte antigens comprehensively. Therefore, only the most widely used techniques are summarized. For details see Kissmeyer-Nielsen and Thorsby (1970); Manual of Tissue Typing Techniques (1974) and Mayr (1972a).

A. Lymphocytotoxicity Test

Lymphocytes are usually prepared from defibrinated or heparinized blood by a flotation technique, where the blood is layered upon a medium with specific gravity of 1·075. After centrifugation, the lymphocytes (specific gravity 1·060) are found at the interface between serum (or plasma) and medium. These cells are incubated in the presence of complement with specific test sera. The lymphocytes, having the antigen against which the antiserum is active, are killed. If they do not possess this antigen, they survive (Gorer and O'Gorman, 1956). Differentiation between living and dead lymphocytes is made by staining the dead cells. At present this technique is mainly used in a micromethod as described by Brand *et al.* (1970) (NIH standard technique of the microlymphocytotoxic test). Because of its excellent reproducibility this technique has become the most important method to detect the classical *s*erologically *d*efined (SD) antigens of the HLA system (i.e. tissue typing).

In addition to HLA SD specificities, the microlymphocytotoxic test also can be used for the determination of Ia type antigens, Le^a, and blood group A or B associated antigens on lymphocytes.

A serious disadvantage is the fact that the lymphocytes must be viable for this technique. Because of their limited lifespan, they should be used as promptly as possible. However, they can be kept living for approximately 10 days by adding some culture medium (Park and Terasaki, 1974) or for months by freezing them in preservative at −196°C (Fotino *et al.*, 1967).

B. Leucoagglutination

Leucocytes prepared by sedimenting erythocytes of acid-citrate-dextrose (ACD) or defibrinated blood are agglutinated by antibodies if they carry the corresponding antigen (Dausset and Nenna, 1952). It is very difficult and tricky to carry out the agglutination technique and its reproducibility is rather poor.

Although the first leucocyte antigens were defined by using the agglutination method, this technique has become less and less useful. It is still important for the definition of the granulocyte-specific antigens, the antigens of the Five system and some supertypic HLA SD specificities such as W4 or W6.

C. Complement Fixation on Platelets

Some antisera are able to fix complement when they are combined with antigens present on platelets which can easily be prepared from ACD or EDTA blood. The consumption of the added complement is tested by a haemolytic system (sensititized sheep red blood cells). If haemolysis is observed, no antigen–antibody reaction has occurred; this means that the platelets do not possess the corresponding antigen. If haemolysis is not seen, the complement has been fixed to the antigen–antibody complexes, demonstrating that the platelets carry this antigen. For this technique, a standardized micromethod has been introduced (Colombani *et al.*, 1971). Complement fixation on platelets can be used for the detection of platelet specific factors and for the determination of HLA SD antigens. The main advantage of this method is that the platelets can be stored for months at +4°C, but the fact that this technique is not very sensitive and that not all HLA SD specificities can be defined using complement fixation must be noted as a disadvantage.

III. LEUCOCYTE AND PLATELET ISOANTIGENS

A. Isoantigens Detectable on Leucocytes, Platelets and Tissues

1. ABO blood groups and associated antigens

The isoagglutinogens of the ABO blood groups, discovered by Landsteiner (1901), may be present on all cells of the organism, as discussed by Race and Sanger (1975).

On lymphocytes, however, the ABO isoagglutinogens are detectable by absorption studies, but not by lymphocytotoxic sera. Nevertheless, some determinants associated with the blood groups can be defined by lymphocytotoxicity. They show mainly an association with blood group A, but in rare cases also with blood group B. The sera found by Jeannet *et al.* (1973, 1974) may react with the specificities A_1 Le^b or B Le^b. The ones directed against A_1 Le^b (frequency of positive reactions in Caucasoids approximately 10%) could represent an equivalent of the red cell antibody Siedler (Seaman *et al.*, 1968) which reacts only with A_1 Le^b erythrocytes.

Another determinant, Atri (Dausset *et al.*, 1973; Marcelli-Barge *et al.*, 1976), is detectable on the lymphocytes of 2·3% of the French population; such individuals possess the genes *A* and *Se*. The antigen Atri is present in the serum of Atri positive people; the mode of inheritance is not yet clear, but it seems that in addition to the *H*, *ABO* and *Secretor* loci, at least one additional locus must be involved.

A further determinant associated with the ABO blood groups and ABH secretor status was described by Mayr and Mayr (1974a): RB. The frequency of lymphocytotoxic reactions of anti-RB was 31·4% in a German population. Family and population studies showed that the antibody activity was directed against cells of individuals possessing the genes *H*, A_1, *Se* or *H*, A_2, *Se*, *lele*. After incubation of RB positive lymphocytes in a medium containing 20% of RB negative serum for six days at +37°C, the cells lost their ability to react with anti-RB while RB negative cells incubated for the same period of time in 20% RB positive serum became RB positive. These results and other similar data (Mayr *et al.*, unpublished) indicate that RB is adsorbed from the serum on the lymphocyte surface as was suggested by Tilley *et al.* (1975). It is possible that RB represents a glycosphingolipid and that quantitative differences, which depend on the A subgroups and the *Le* and the *Se* loci (Tilley *et al.*, 1975), explain the reaction pattern of anti-RB.

2. The HLA system

The SD antigens of the HLA system (*H*uman *L*eucocyte Locus *A*) (Amos, 1968) are most probably, with quantitative differences, present on all nucleated cells of the organism. The largest amount can be detected in spleen, lung and liver, while brain and fatty tissue contain practically no antigens (Berah *et al.*, 1970). They are present on reticulocytes (Harris and Zervas, 1969) and can be found by very sensitive techniques (Doughty *et al.*, 1973; Nordhagen and Ørjasaeter, 1974; Perrault and Högman, 1971; Reekers *et al.*, 1975) even on erythrocytes. Furthermore, a close relation has been described between HLA-B7, BW17 and A28, and the erythrocyte antigens Bg^a, Bg^b and Bg^c, respectively (Morton *et al.*, 1969, 1971). For these reasons erythrocytes seem to carry small amounts of HLA SD antigens, although they possess no nucleus.

Besides the classical SD antigens, there exist other determinants coded by the HLA system which are the human equivalents of the Ia (*I* region *a*ssociated) antigens of the mouse. They are mostly present on B lymphocytes, macrophages, epidermal cells and spermatozoa (see Section IV.C).

The HLA system is the most important of all isoantigenic systems of leucocytes and will be described in detail in Section IV of this chapter.

3. The Five system

This system discovered by van Leeuwen *et al.* (1964) consists of two antigens, 5a and 5b, governed by autosomal codominant alleles having gene frequencies of 0·18 and 0·82, respectively in the Netherlands. Both antigens are only demonstrable by the agglutination technique. The Five system is inherited independently of the HLA system. Although its antigens have been demonstrated in several organs, it appears that they are of no importance in organ grafting (van Rood *et al.*, 1967).

4. Erythrocyte antigens other than ABO

The data on the presence of erythrocyte antigens other than ABO on leucocytes, platelets and tissues disagree (with one exception) to such an extent that no conclusions can be reached (Dausset and Tangün, 1965;

Thierfelder, 1968). Although it is claimed that some of them (e.g. P, D) are important in organ transplantation, there is as yet no definitive evidence for this (Ceppellini *et al.*, 1966; Dausset *et al.*, 1970b).

The only exception is the clearcut presence of the antigen Le[a] on lymphocytes, detected using the cytotoxic test with most human anti-Le[a] sera (Dorf *et al.*, 1972; Mayr, unpublished). As with the antigen RB, Le[a] is also probably adsorbed from the serum onto the lymphocyte surface (Mayr *et al.*, unpublished).

B. Isoantigens Detectable on Leucocytes Only

1. Granulocytes

The granulocyte-specific antigens and antibodies are of minor practical importance, but they possibly play some part in the pathogenesis of neonatal neutropenia and in febrile transfusion reactions (Kissmeyer-Nielsen and Thorsby, 1970; Lalezari and Bernard, 1965).

The Nine system. Only one antigen is known: 9a, which at present can be defined by a single agglutinating serum. It has a gene frequency of 0·41 in the Netherlands (van Rood *et al.*, 1965). The inheritance is autosomal dominant and independent of the HLA system (Workshop Data, Torino, 1967).

The NA system. This consists of two antigens, NA1 (=To1) and NA2 governed by autosomal co-dominant alleles with gene frequencies ot 0·34 and 0·66 respectively in Italy (Ceppellini *et al.*, 1967; Lalezari and Bernard, 1965; van der Weerdt and Lalezari, 1972). Its inheritance is independent of the HLA system.

The NB and Vaz system. Both systems consist of only one antigen, NB1 (Lalezari and Bernard, 1966) and Vaz (NC1) (Lalezari *et al.*, 1970) respectively. They are inherited as autosomal dominants and the corresponding gene frequencies are 0·83 and 0·72 in Caucasians in the U.S.A.

The G–A, G–B and G–C series. In this system 14 granulocyte antigens have been identified. They show autosomal dominant inheritance and may be coded for by three linked loci (Hasegawa *et al.*, 1975).

2. Lymphocytes

A short time ago there was only one lymphocyte-specific antigen known: Ly[D1], defined by complement fixation (Shulman *et al.*, 1964). Another lymphocyte-specific antigen, Ly-Co, probably present only on B lympho-cytes, has been found in the lymphocytotoxic test by Legrand and Dausset (1975a). Both antigens show autosomal dominant inheritance and their gene frequencies are both approximately 0·20. For this reason, they might be identical. *Ly–Co* segregates independently of *ABO, Rh, MNSs, Le, Fy, Ko, Hp, Gm, Inv, ISf, Lp, Ag, acP, PGM₁, PGM₂, AK₁ and HLA.*

C. Isoantigens Detectable on Platelets Only

Because HLA SD antigens can be detected easily on platelets, it should be mentioned that several systems of platelet-specific antigens are known (for details, see Majsky, 1970; Mayr, 1972a; Svejgaard, 1969a). Antibodies reacting with platelet-specific antigens can cause an extreme reduction of the survival time of transfused thrombocytes, isoimmune neonatal throm-bopenia or postransfusional purpura.

IV. THE HLA SYSTEM

The HLA system encompasses approximately one-thousandth of the human genome and consists of several loci, most of them being related functionally. For operational reasons the loci are sub-divided into different groups: loci coding for the classical SD antigens (these antigens are present on all nucleated cells), for the *l*ymphocyte *d*efined LD determinants, for the Ia type antigens, etc. The general notational scheme for the HLA system recommended by the WHO-IUIS terminology committee (1975) is as follows:

(a) HLA: designation for the system
(b) A, B, C, D, . . . : designation for the loci
(c) W: symbol to indicate a provisional specificity which is not yet defined in the best possible way. W may be dropped when the specificity is con-firmed
(d) 1, 2, 3, . . . : number identifying the specificities belonging to each locus.

The system prefix HLA must not be included, when it is clear from the context that loci of this system are being referred to. In the future, other loci situated in the *HLA* gene complex may obtain the designation *HLA-E*, *HLA-F* . . . when genetical, chemical and functional analyses demonstrate clearly that they belong to the HLA system.

A. Classical SD Antigens

1. Antigens

The SD antigens (J. Bodmer *et al.*, 1975) of the HLA system consist of three series corresponding to three loci: the series *A (SD-1, LA* or *First)*, the series *B (SD-2, FOUR* or *Second)* and the series *C (SD-3, AJ* or *Third)*. The specificities officially recognized are listed in Table 1 (for

Table 1

Officially recognized SD antigens of the HLA system

Locus *HLA-A*		Locus *HLA-B*		Locus *HLA-C*
A1		B5		CW1
A2		B7		CW2
A3		B8		CW3
A9	{ AW23	B12		CW4
	{ AW24	B13		CW5
A10	{ AW25	B14		
	{ AW26	B18		
A11		B27		
A28		BW15		
	A29	BW16	{ BW38	
	AW30		{ BW39	
AW19	AW31	BW17		
	AW32	BW21		
	AW33	BW22		
AW34		BW35		
AW36		BW37		
AW43		BW40		
		BW41		
		BW42		

comparison with the old nomenclatures, see Mayr, 1972a; Table of Equivalent Nomenclature, 1973; WHO-IUIS Terminology Committee, 1975). As shown in this table, some supertypic antigens (defined by sera with a broader reactivity) can be split into subtypic factors (defined by sera with a narrower reactivity). This phenomenon which is similar to the splitting of the blood group A into the subgroups A_1 and A_2 will be discussed later in connection with the cross-reactivity of antisera and the chemical structure of HLA SD antigens (Section IV.A.3).

AW34, AW36, AW43 and BW42 do not exist (or are extremely rare) in Caucasians; they are found in certain Mongoloid (AW34) or Negroid

(AW34, AW36, AW43 and BW42) populations (J. Bodmer *et al.*, 1975; see also Section IV.E.5).

Besides the officially recognized SD antigens, there are others less well-defined:

(a) HLA-A: subtypic factors of A28 (To54 and To55, Belvedere *et al.*, 1975c) and of AW33 (Degos *et al.*, 1975)

(b) HLA-B: subtypic factors of B5 (Richiardi *et al.*, 1975), of B12 (To50 and To51 = TT*, Kissmeyer-Nielsen *et al.*, 1975a; Richiardi *et al.*, 1974), of BW15 (To52 and To53, Richiardi *et al.*, 1974), of BW21 (BW21-SL-ET and BW21-ET*, Thorsby *et al.*, 1971) and of BW22 (Kissmeyer-Nielsen *et al.*, 1975b; Laundy and Entwistle, 1975), HR (Engelfriet *et al.*, 1973), HS (an antigen typical for Mongoloids, Payne *et al.*, 1975), IM2 (Gelsthorpe and Doughty, 1975), JA (Kissmeyer-Nielsen *et al.*, 1973) or 407* (Kissmeyer-Nielsen *et al.*, 1970)

(c) HLA-C: T6 (Mayr *et al.*, 1973) and T7 (Staub Nielsen *et al.*, 1975).

2. Antibodies

Antibodies against SD antigens of the HLA system are most likely to occur after immunization, as naturally occurring antibodies have been rarely described (Collins *et al.*, 1973; Lepage *et al.*, 1976). At present most antisera are of human origin and belong mainly to the IgG class of immunoglobulins (Ahrons and Glavind-Kristensen, 1971; Engelfriet, 1966; Walford *et al.*, 1965), but it is possible that immunization of animals will provide valuable typing reagents (Albert *et al.*, 1969; Balner *et al.*, 1971; Billing and Terasaki, 1974; Robb *et al.*, 1975; Sanderson and Welsh, 1973). Anti-HLA SD immunization is caused by blood transfusions (Dausset, 1954), pregnancies (Payne and Rolfs, 1958; van Rood *et al.*, 1958) or transplantations. Because of the close genetic relationship between mother and child, the antibodies produced by foeto–maternal immunization are in most cases mono- or oligospecific. Although they belong to the IgG class, they have no influence on the development of the foetus (Ahrons, 1971). Antibodies found in polytransfused or grafted patients are generally multispecific. The most promising method of antibody production is the immunization of volunteers with antigenic material from donors who are different from the recipient in one antigen only. In such cases, immunization can be produced by skin grafts, leucocyte injections or blood transfusions.

Unfortunately, it often happens that so-called "cross-reacting" antibodies are produced which react with the gene products of several alleles and cannot be separated by absorption (Dausset *et al.*, 1968; Svejgaard and Kissmeyer-Nielsen, 1968). The most frequently found cross-reacting antibodies are directed against the following groups of antigens ("cross-reacting groups"): A1 and A3 and A11, A1 and A10 and A11, A2 and A28, A2 and A9, AW23 and AW24, AW25 and AW26, A10 and AW19, AW25 and AW32, AW30 and AW31; B5 and BW35 and B18 and BW 15 and BW17 and BW21, B7 and B27 and BW22 and BW40, B8 and B14, B12 and B13, B12 and BW21, B13 and B27, B13 and BW40, BW40 and BW41 (Albert *et al.*, 1973; Colombani *et al.*, 1970a; Curtoni *et al.*, 1969; Kissmeyer-Nielsen and Thorsby, 1970; Mayr, 1971; Mittal and Terasaki, 1972, 1974; Müller-Eckhardt *et al.*, 1972).

W4 and W6 represent broad cross-reacting antisera including specificities of the B series (W4 = B5, B13, B27, BW17, BW21-SL-ET, BW37, BW38, To50, To52, HR, 407*; W6 = B7, B8, B14, B18, BW21-ET*, BW22, BW35, BW39, BW40, BW41, To51, To53 (J. Bodmer *et al.*, 1975). It is striking that the cross-reacting groups generally encompass antigens belonging to the same series. However, a few exceptions are known: A11 and W6 (Belvedere *et al.*, 1975b), A9 and W4 (Legrand and Dausset, 1975b; Scalamogna *et al.*, 1976); A11 and (BW22 and BW35 and B5 and BW15) (Tongio and Mayer, 1974).

3. Chemical definition

The HLA SD antigens are insolubly bound to the cell membrane. This fact explains the considerable difficulties encountered in the elucidation of their chemical structure. The methods of extraction and solubilization of antigens are manifold: crude membrane fragments are obtained by destroying mechanically the cell membrane of spleen cells, platelets or human lymphoid cells in long-term tissue culture. The solubilization of such fragments is carried out by autolysis or by using papain. Other methods are sonication, extraction with the aid of organic solvents, detergents or chaotropic agents (e.g. 3M KCl). The solubilization is followed by purification by gel filtration, ion-exchange chromatography, polyacrylamide disc electrophoresis or immune precipitation. The antigenic quality of the preparations is tested by inhibiting the effect of known test sera on positively reacting cells by the lymphocytotoxic test or by the complement fixation reaction.

The chemical investigations performed on these purified antigens show that they are glycoproteins (for references, see Transplantation Reviews, Vol. 21, 1974). Although small glycopeptides (10 000 daltons, Sanderson *et al.*, 1971) or carbohydrate-containing macromolecules (Springer *et al.*, 1973) are able to inhibit the activity of some anti-SD sera, it is highly probable that the HLA SD alloantigenicity resides only in the protein structure (Mann *et a* ., 1970; Reisfeld and Kahan, 1970; Transplantation Reviews, Vol. 21; 1974). The molecular weight of the antigens solubilized using papain is approximately 48 000. Exposure to mild acid treatment results in the dissociation of the molecule into two fragments: a glycoprotein with 33 000–35 000 daltons and a polypeptide with 11 000–12 000 daltons. The HLA SD alloactivity resides in the heavier chain, while the lighter chain, which is common to all the HLA molecules of the A, B and C series, is identical with β_2-microglobulin (Cresswell *et al.*, 1973; Grey *et al.*, 1973; Nakamuro *et al.*, 1973; Rask *et al.*, 1974, 1976; Tanigaki and Pressman, 1974; Tanigaki *et al.*, 1973). Solubilization with detergents yields not only dimers of the HLA molecule (Snary *et al.*, 1974) built up by two glycoprotein chains (molecular weight of each chain approximately 44 000 daltons, Strominger *et al.*, 1975) and two chains of β_2-microglobulin, but also monomers. The dimers are probably kept together by disulphide bridges (Cresswell and Dawson, 1975). In both preparations β_2-microglobulin is bound in a noncovalent way to the chains carrying the HLA SD allospecificity. Furthermore, it is probable that within the heavy chain there are three disulphide bridges, while there exists only one within the β_2-microglobulin (Strominger *et al.*, 1974). These facts and the extensive homology of β_2-microglobulin with the C_H3 domain of the IgG molecules (Cunningham and Berggård, 1974) point to an astonishing similarity of the structure of the HLA SD antigens and the constant part of IgG (Strominger *et al.*, 1974).

The close connection between the heavy chain with the HLA SD alloactivity and β_2-microglobulin can easily be demonstrated by cocapping experiments (Bernoco *et al.*, 1973): when lymphocytes are incubated under definite conditions with anti-β_2-microglobulin serum and, in a second step, with an antiserum against the IgG of the species providing the anti-β_2-microglobulin, the β_2-microglobulin molecules are aggregated on the pole of the cell containing the Golgi complex. This occurs, according to the fluid mosaic model of cell membranes (Singer and Nicolson, 1972), because the HLA antigens are free to move within the phospholipid bilayer constituting the lymphocyte membrane. All HLA SD alloantigens are also found within this cap (Bismuth *et al.*, 1973; Mayr, 1974a; Östberg *et al.*, 1974; Poulik *et al.*, 1973; Solheim and Thorsby, 1974). This can be demonstrated by two methods (Mayr *et al.*, 1973):

(a) Fluorostrip: the antiglobulin serum used to cap the β_2-microglobulin molecules must be conjugated with tetramethylrhodamine isothiocyanate (TRITC). After the capping, the cells are incubated with adequate anti-HLA SD sera and then with fluorescein isothiocyanate (FITC) conjugated antihumanglobulin serum. Using selective filter combinations in a fluorescence microscope, one can recognize that all the β_2-microglobulin molecules (stained red with TRITC) and all the HLA SD molecules (stained green with FITC) are cocapped, i.e. aggregated together in the same cap.

(b) Lysostrip: after capping the β_2-microglobulin, the cells are incubated with anti-HLA SD serum and complement. If the corresponding HLA SD antigens are aggregated in the cap, the complement is not able to lyse the cells which become resistant to the action of the specific antiserum and complement. By this technique, one can easily find that all the HLA SD antigens of the series A, B and C are cocapped by the treatment of the lymphocytes with anti-β_2-microglobulin.

In contrast to these findings, Mayr (1974a) could show that β_2-microglobulin is not coupled to RB or Le^a.

Other sources of HLA SD antigens are serum (Billing et al., 1973; Miyakawa et al., 1973; van Rood et al., 1970), seminal plasma (Singal et al., 1971) and mammary secretions (Dawson et al., 1974; Kachru and Mittal, 1975), where soluble antigens can be found. Highly purified material isolated from serum consists of only one chain of 33 000 daltons and contains no β_2-microglobulin (Oh et al., 1975).

Further biochemical studies showed clearly that the antigens of the A and B series can be separated by electrophoresis or by immune precipitation (Colombani et al., 1970b; Cresswell et al., 1974; Dautigny and Colombani, 1973; Mann et al., 1969; Snary et al., 1974; Thieme et al., 1974). Initial steps towards definition of the amino acid sequence of HLA SD specificities have already been taken, with the determination of the 16 N-terminal amino acids (Bridgen et al., 1976).

The chemical structure of the HLA SD antigens (polypeptide chains with a minor carbohydrate portion, in which the alloantigenicity is defined in principle by the primary sequence of the amino acids building up the molecule) and the fact that determinants on polypeptides may consist of only seven amino acids (Kabat, 1968) can easily explain the two postulated reasons for cross-reactivity (Ceppellini, 1971; Ceppellini and van Rood, 1974):

(a) the antisera may fit with different, but structurally similar determinants present on different gene products of the SD loci, and
(b) the antisera are directed against determinants shared by the gene products of different SD alleles. These determinants must be in extreme disequilibrium with the subtypic specificities carried by the same molecule.

These two explanations for the cross-reactivity have been verified by several experiments: Richiardi et al. (1973) showed that the F(ab')$_2$ fragments of antibodies against subtypic specificities were able to block only the reactivity of other antisera against the subtypic specificity, but not the effect of sera against the supertypic factors. If the $F(ab')_2$ fragments which lack Fc fragments are bound on their corresponding determinants, they cannot activate complement and therefore block the activity of intact antibody molecules directed against the same determinants. For instance, the $F(ab')_2$ fragments of an anti-A2 serum block the cytotoxicity of other anti-A2 sera, but not the lysis due to anti-(A2 and A28) sera. As the anti-A2 and the anti-(A2 and A28) sera provoke the capping of the same molecules, Richiardi et al. concluded that the anti-A2 and the anti-(A2 and A28) sera react with different determinants of the same gene product. Similar data were obtained by Legrand and Dausset (1973) using absorption–inhibition techniques. Further experiments showed, however, that the cross-reactivity due to a similar structure of the surface of determinants can also be found among HLA SD antigens (Legrand and Dausset, 1974).

The splitting of HLA SD specificities is mostly caused by the fact that one gene product carries both the supertypic and the subtypic determinant. This could be demonstrated for A10, AW25 and AW26 (Richiardi et al., 1973). The linkage of A10 with either AW25 or AW26 is extremely tight, because, so far, no individual has been found to be A10 positive, but AW25 and AW26 negative. However, some people have been reported who are A9 positive and AW23 and AW24 negative, so that the existence of a gene "*A9.3*" must be postulated which codes for the determinant A9, but not for AW23 or AW24 (J. Bodmer et al., 1975). The determinants defined by anti-W4 and anti-W6 also represent areas of the molecules which are different from the determinants corresponding to the subtypic specificities. The definition of W4 and W6, however, is very helpful in recognizing some subtypic specificities of the HLA-B series: for instance, BW38 is on the same molecule as W4 and BW39 on the same one as W6 (BW38 and BW39 are splits of BW16, Schreuder et al., 1975). A similar splitting of B12, BW15 or BW21 becomes easier by taking into consideration the reactivity of anti-W4 and anti-W6 (J. Bodmer et al., 1975; Bright, 1976; Richiardi et a!., 1974).

Furthermore, by immunizing rabbits with purified antigens of the A and B series, antisera were obtained with specificity for determinants present on all molecules of the A or B series, respectively (Cresswell and Ayres, 1976). It is possible that these determinants are related to the HLA common antigenic determinant described by Tanigaki et al. (1974). Thus, the HLA SD alloantigenic chain carries several determinants. According to the immunogenetic concepts discussed extensively by Hirschfeld (1972), the SD antigens of the HLA system and their antibodies are excellent examples of a "complex–complex" system.

4. Genetics

The "two locus" hypothesis. The genetics of the HLA SD antigens has long been explained by the postulate that one factor of the series A is inherited in close linkage with one factor of the series B on an autosomal chromosome (Amos and Ward, 1975; Bender *et al.*, 1975; Ceppellini, 1971; Ceppellini and van Rood, 1974; Dausset *et al.*, 1970a; Dausset, 1972; Kissmeyer-Nielsen and Thorsby, 1970; Mayr, 1974b; Piazza and Morton, 1973; Svejgaard *et al.*, 1970; Thorsby, 1974). This combination of genes situated on one chromosome and travelling together is called a haplotype. Because of the relatively high frequency of recombination between the two series and because of the fact that the chemically purified SD antigens of both series can be separated, it can be assumed that the antigens of the series A and B are governed by two different loci. This hypothesis has been confirmed by fluorostrip and lysostrip experiments showing that the molecules carrying determinants of the A and B series are independent on the lymphocyte surface (Bernoco *et al.*, 1973; Mayr, unpublished). The single antigens within both series are governed by allelic genes with co-dominant inheritance (the supertypic factors can be replaced in all cases by the subtypic ones, in analogy to the subgroups of the blood group A, where the gene A can be substituted by the genes A_1 and A_2). This conclusion is based on several findings: no individual has been found possessing more than two antigens of the loci A or B, the genes of each series segregate and no child has been observed to be positive for an A or a B antigen the parents lacked. Furthermore, the phenotype frequencies of both loci are in excellent agreement with the Hardy-Weinberg equilibrial expectations. Investigations of populations and families have shown that not all specificities of the loci A and B are yet serologically detectable. These unknown antigens are coded by genes called X_A and X_B. The increasing number of factors which have been identified suggest that silent alleles are infrequent at the loci A and B. The antigen frequencies and the gene frequencies determined with the aid of the gene-counting method of maximum likelihood (Yasuda and Kimura, 1968) which have been found in the Viennese population, are listed in Table 2 (Mayr, 1975a).

A comparison of the observed phenotypes with the expected values calculated from the gene frequencies shows a very good agreement both in the A locus ($\chi^2 = 30.86$, 31 d.f., $0.5 > p > 0.4$) and in the B locus ($\chi^2 = 42.35$, 45 d.f., $0.6 > p > 0.5$).

Assuming 14 and 18 SD antigens in the A and B series, respectively (see Table 2), and one X gene per locus, the polymorphism in Caucasians is already prodigious: 285 haplotypes, 40 755 genotypes and 18 232 phenotypes are possible.

Table 2

HLA-A and HLA-B antigen and gene frequencies in Vienna (n = 450)

	Antigen frequency (%)	Gene frequency
HLA-A: A1	28·89	0·1530
A2	50·00	0·2872
A3	25·78	0·1451
AW23	6·44	0·0331
AW24	16·00	0·0821
AW25	5·11	0·0256
AW26	8·22	0·0421
A11	9·56	0·0488
A28	8·89	0·0444
A29	3·56	0·0186
AW30	4·22	0·0220
AW31	5·56	0·0278
AW32	8·89	0·0444
AW33	4·67	0·0233
X_A		0·0025
		$\Sigma = 1·0000$
HLA-B: B5	16·89	0·0873
B7	26·00	0·1423
B8	18·00	0·0910
B12	20·89	0·1105
B13	5·11	0·0256
B14	6·89	0·0344
B18	13·11	0·0675
B27	8·44	0·0439
BW15	9·56	0·0478
BW16	8·89	0·0498
BW17	8·00	0·0400
BW21-SL-ET	3·11	0·0156
BW21-ET*	4·00	0·0207
BW22	5·56	0·0286
BW35	20·89	0·0195
BW37	2·00	0·1000
BW40	11·78	0·0607
BW41	1·78	0·0089
X_B		0·0059
		$\Sigma = 1·0000$

The haplotype frequencies calculated using the formula by Mattiuz *et al.* (1970) in the Viennese population are listed in Table 3. A comparison of these figures with the values obtained on the basis of the gene frequencies assuming a random association of *A* locus genes with *B* locus genes reveals a strong linkage disequilibrium: some haplotypes occur much more

Table 3

HLA-A/HLA-B haplotype frequencies in Vienna (n = 450; values $\times 10^4$)

HLA-A	B5	B7	B8	B12	B13	B14	B18	B27	BW15	BW16	HLA-B BW17	BW21-SL-ET	BW21-ET*	BW22	BW35	BW37	BW40	BW41	X_B	Σ
A1	148	49	684	24	13	15	34	19	33	1	132	22	1	33	139	32	121	11	19	1530
A2	260	448	58	480	114	1	163	189	283	196	158	25	50	1	211	12	192	11	20	2872
A3	57	473	74	20	26	47	110	69	60	39	37	18	18	41	211	26	98	20	7	1451
AW23	4	46	1	160	1	1	10	1	1	1	1	12	61	12	1	1	11	5	1	331
AW24	199	106	37	30	16	13	48	1	33	43	1	1	2	78	170	1	38	3	1	821
AW25	1	25	1	2	1	14	167	1	1	1	12	1	1	6	18	1	1	1	1	256
AW26	16	14	4	51	10	36	1	77	35	83	4	14	16	25	11	1	1	21	1	421
A11	85	60	7	22	8	13	15	1	1	44	10	9	2	10	189	1	4	6	1	488
A28	1	4	26	126	1	54	66	19	1	14	12	13	2	24	46	5	28	1	1	444
A29	7	22	1	64	5	13	1	1	3	15	3	8	6	7	26	1	1	1	1	186
AW30	17	29	12	21	56	4	12	11	12	1	1	7	13	5	1	6	10	1	1	220
AW31	44	54	1	20	2	1	1	20	3	31	8	5	1	14	10	1	60	1	1	278
AW32	31	79	1	80	1	1	40	22	1	27	11	13	32	23	35	5	40	1	1	444
AW33	1	12	1	4	1	130	5	7	10	1	9	7	1	5	26	6	1	5	1	233
X_A	2	2	2	1	1	1	2	1	1	1	1	1	1	2	1	1	1	1	2	25
Σ	873	1423	910	1105	256	344	675	439	478	498	400	156	207	286	1095	100	607	89	59	10000

frequently, other ones less frequently than expected. The most significant deviations are listed in Table 4 (positive association: $p < 0.001$; negative association: $p < 0.01$), together with the Δ values of gametic association (Bodmer and Payne, 1965) being a measure of the linkage disequilibrium between both genes involved (p. 162). Other genes with a less significant positive association are $A29$ and $B12$, as well as $AW26$ and $BW16$ ($p \simeq 0.01$). The haplotype frequencies and Δ values determined from family investigations give identical results. The linkage disequilibria listed in Table 4, especially the ones between $A1$ and $B8$, and $A3$ and $B7$ are typical for Caucasian populations (Joint Report, 1973).

Table 4

Positive and negative associations of *HLA-A* and *HLA-B* genes in Vienna (n = 450)

Positive association ($p < 0.001$)			Negative association ($p < 0.01$)		
	Δ	χ		Δ	χ^2
A1, B8	+0.0542	109.20	A2, B8	−0.0203	7.96
AW25, B18	+0.0182	84.38	A1, B7	−0.0182	7.83
AW33, B14	+0.0124	78.71	A3, B12	−0.0150	7.36
AW23, B12	+0.0153	31.81	A1, B12	−0.0152	6.75
AW30, B13	+0.0058	23.24			
A11, BW35	+0.0144	18.89			
A3, B7	+0.0248	17.12			

Table 5

HLA-A and HLA-B antigens of family Me045

		Phenotype		Genotype
		HLA-A	HLA-B	
Father		A1, A3	B7, B8	A1, B8/A3, B7
Mother		A2, A11	B5, B13	A2, B13/A11, B5
Children 1 and 4		A1, A2	B8, B13	A1, B8/A2, B13
Child 2		A1, A11	B5, B8	A1, B8/A11, B5
Child 3		A3, A11	B5, B7	A3, B7/A11, B5
Child 5		A2, A3	B7, B13	A3, B7/A2, B13

In order to illustrate the linked inheritance, an example is given in Table 5 (family Me045): the children 1 and 4 must have inherited the genes $A1$ and $B8$ from the father. Therefore, he must possess the haplotypes $A1, B8$

and *A3, B7*. Similarly, the haplotypes of the mother are *A2, B13* and *A11, B5*. These conclusions are confirmed by the phenotypes of the other children.

The data obtained in family 037a (Mayr and Mickerts, 1971) show an example of recombination between the loci *A* and *B* (Table 6): on account of the phenotypes of the children 1, 2 and 4, the genotype of the father can

Table 6

HLA-A and HLA-B antigens of family 037a

	Phenotype HLA-A	HLA-B	Genotype
Father	A3, A9	B12, B27	*A3, B27/A9, B12*
Mother	A1, AW33	B8, B14	*A1, B8/AW33, B14*
Child 1	A3, AW33	B14, B27	*A3, B27/AW33, B14*
Child 2	A9, AW33	B12, B14	*A9, B12/AW33, B14*
Child 3	A9, AW33	B14, B27	*A9, B27/AW33, B14*
Child 4	A1, A9	B8, B12	*A9, B12/A1, B8*

be determined as *A3, B27/A9, B12* and that of the mother as *A1, B8/ AW33, B14*. Child 3 shows the phenotype A9, AW33, B14, B27 which must correspond to the genotype *A9, B27/AW33, B14*. As no evidence of illegitimacy within this family has been found by investigating 18 systems which are used in cases of disputed paternity, it must be assumed that during meiosis a cross-over occurred in the father, resulting in the formation of the haplotype *A9, B27* which has been inherited by child 3.

The recombination frequency between the loci *A* and *B* in a large data set containing 4614 informative meioses is 0·87% (Belvedere *et al.*, 1975a). The maternal recombination fraction (1·14%) is higher by a factor of 1·6 than the paternal one (0·69%), but no effect of parental age on recombination could be demonstrated (Weitkamp *et al.*, 1973; Mayo, 1974).

The question of homozygosity or heterozygosity arising from the fact that there are antigens governed by each locus which are not yet serologically detectable, is already of minor importance due to the extremely low gene frequencies of X_A and X_B. If in one locus only one antigen is detectable, one can conclude with a high probability that it is present in homozygous form and not heterozygous with the gene product of an X gene. If a statistical statement on the probability of homozygosity is needed, the linkage disequilibrium must be considered. If e.g. an individual has the HLA type A3, B7, BW35, its genotypes could be *A3, B7/A3, BW35*; *A3, B7/X_A, BW35* or *X_A, B7/A3, BW35*. Based on the Viennese haplotype frequencies, the frequencies of these genotypes are $199\ 606 \times 10^{-8}$, 946×10^{-8} and 844×10^{-8}, respectively. The probability of homozygosity of $A3$ is therefore

$$\frac{199\ 606}{199\ 606 + 946 + 844} = 0 \cdot 9911$$

Furthermore, some hints for or against homozygosity can be given by investigations on dosage effects with lymphocytotoxic (Ahrons and Thorsby, 1970) or complement-fixing antibodies (Svejgaard, 1969b), by kinetic studies using a ^{51}Cr-release cytotoxic test (White et al., 1973) or by quantitative absorptions (Dumble and Morris, 1975; Schmid et al., 1974).

Broad antisera like anti-W4 or anti-W6 can also help to reach a decision concerning the zygosity. Anti-W4 contains, for example, antibodies reacting with the gene products of $B5$, while anti-W6 shows no activity against products of this gene (see Section IV.A.2). If a propositus is not only B5 positive, but also W4 and W6 positive, heterozygosity can be assumed, as anti-W6 does not contain antibodies against the gene product governed by $B5$.

Another way of determining the zygosity of a male person could be the typing of his spermatozoa. Fellous and Dausset (1970) stated that the HLA SD antigens probably have a haploid expression on these cells: a mixture of sera against antigens governed by the same haplotype killed approximately 40% of the spermatozoa, while a mixture of sera against antigens governed by different haplotypes killed 70%. The investigations of Piazza et al. (1969) gave no such clear-cut results, but might show an analogous trend. Another study by Halim et al. (1974) using double and single fluorescent labelling techniques also seemed to demonstrate the haploid expression of HLA SD antigens governed by the loci A and B on the head of the sperm.

In any case, the best way to demonstrate whether an antigen is present in homozygous or heterozygous form is to test the close relations of the propositus.

A third locus. Some sera have been found defining antigens which did not fit the "two locus" concept. The first of them was anti-AJ (now anti-CW1) (Sandberg et al., 1970), detecting a determinant highly correlated with B5, B27, BW15 and BW22. Later on, four other determinants (T2 = 170 = CW2; T3 = UPS = CW3; T4 = 315 = CW4 and T5 = CW5; Mayr et al., 1973; Svejgaard et al., 1973), all showing significantly positive associations with some B locus antigens, were discovered. In population studies, they behaved as though governed by alleles of an autosomal locus and in family investigations, they segregated together with the HLA system. In several families where crossing-over between the loci *A* and *B* had occurred these determinants were always inherited with the B gene product (Mayr, 1974c; Svejgaard *et al.*, 1973). Therefore, it was impossible to distinguish between two explanations for these results:

(a) the determinants defined by these antisera are coded by a third *SD* locus, very closely linked to *HLA-B*
(b) the antisera anti-AJ etc. . . . . determine subtypic specificities of several B locus antigens.

The problem was solved using the fluorostrip and lysostrip techniques: both methods demonstrated clearly the independence of the five determinants CW1, CW2, CW3, CW4 and CW5 from B locus gene products, thus indicating that the HLA system contains a third locus coding for SD antigens (Mayr *et al.*, 1973; Pierres *et al.*, 1975; Solheim *et al.*, 1973; Svejgaard *et al.*, 1973), called *HLA-C*. This molecular evidence for a third locus was confirmed by family studies which showed in two cases crossing-over between *HLA-A* and *HLA-C* on one side, and *HLA-B* on the other side (Hansen *et al.*, 1975; Löw *et al.*, 1974).

Table 7

HLA-C antigen and gene frequencies in Vienna (n = 450)

HLA-C:	Antigen frequency (%)	Gene frequency
CW1	8·22	0·0419
CW2	12·44	0·0641
CW3	19·11	0·1006
CW4	25·56	0·1377
CW5	9·11	0·0463
X_c		0·6094
		$\Sigma = 1·0000$

The sequence of the three loci on the chromosome is therefore *HLA-A:HLA-C:HLA-B*; the recombination fraction between *HLA-B* and *HLA-C* is approximately 0·2% (Mayr, 1975b; Staub Nielsen *et al.*, 1975). For the moment, five specificities of the *C* locus are well-defined. Their antigen and gene frequencies in Vienna are listed in Table 7 (Mayr, 1975a). As shown in this table, the frequency of X_C (the gene or genes coding for not yet detectable antigens) is rather high (approximately 60%). The fit with the Hardy-Weinberg equilibrium within the *HLA-C* locus is good ($\chi^2 = 7·87$, 7 d.f., $0·4 > p > 0·3$).

The computation of Δ values between *HLA-C* genes and genes of the other two *SD* loci give very high and statistically significant figures (Table 8), especially between *HLA-B* and *HLA-C* genes. This fact also indicates that the *C* locus is situated closer to *B* than to *A*.

Table 8

Positive and negative associations of *HLA-C* with *HLA-A* and *HLA-B* genes in Vienna (n = 450)

	Positive association (p < 0·001)	Δ	χ^2
HLA-C/HLA-A:	CW4, AW23	+0·0125	16·81
HLA-C/HLA-B:	CW4, BW35	+0·0851	265·30
	CW3, BW15	+0·0329	109·99
	CW2, B27	+0·0248	107·89
	CW5, B12	+0·0302	85·31
	CW2, BW40	+0·0230	66·18
	CW3, BW40	+0·0270	59·95
	CW1, BW22	+0·0118	55·41
	CW1, B27	+0·0131	44·98
	CW3, BW22	+0·0129	28·48
	CW1, B5	+0·0111	16·07

	Negative association (p < 0·01)	Δ	χ^2
HLA-C/HLA-A:	CW5, A3	−0·0107	7·83
HLA-C/HLA-B:	CW3, BW35	−0·0168	12·39
	CW3, B7	−0·0179	11·35
	CW4, BW40	−0·0145	11·26
	CW4, B27	−0·0104	7·86
	CW2, B7	−0·0120	6·85

The analysis of the genotypes of 225 Austrians confirms the close association of *HLA-C* with *HLA-B* genes (*CW1* with *B27, B5, BW22*; *CW2* with *B27, BW40*; *CW3* with *BW15, BW40, BW22*; *CW4* with *BW35*; *CW5* with *B12*; Mayr, 1975b). Because of the relatively small number of haplotypes carrying a detectable *C* gene (only 184 out of 450 haplotypes), it was not possible to perform statistical comparisons between the figures observed and the number of expected haplotypes based on a random distribution of the alleles of the three *SD* loci. However, it was striking that the haplotype *AW23, CW4, B12* occurred much more frequently than expected (Mayr, 1974c, 1975b). The high frequency of this haplotype was also remarked by Staub Nielsen *et al.* (1975).

Taking into account the five known specificities of the C locus, the polymorphism of the HLA SD antigens is even further increased: 1710 haplotypes, 1 462 905 genotypes and 291 712 phenotypes can potentially be differentiated.

B. LD Determinants

The LD (*l*ymphocyte *d*efined) determinants are detected by mixed lymphocyte culture, MLC (Bach and Hirschhorn, 1964; Bain and Lowenstein, 1964). The lymphocytes of two individuals are cultured together in a suitable medium (bidirectional MLC); if differences of LD determinants are present, the cells are changed in a characteristic way: the cytoplasm and the nucleus become larger and there are more mitoses. These changes correspond with an increased DNA production which can be measured using radioactive thymidine. The presence of stimulation gives evidence that both cell populations do not possess the same LD determinants. If stimulation does not occur, these determinants are identical. It is also possible to prepare the MLC in such a way that the DNA replication in one cell population is blocked (unidirectional MLC; Bach and Voynow, 1966). In this method, the stimulation of the lymphocytes from one individual (responder) can be measured in respect to the cells of another one (stimulator). As this technique gives much more information than the bidirectional MLC, it is used in most cases.

The stimulating LD determinants appear to be mainly expressed on B lymphocytes, while the responding cells mainly belong to the T cells; however, many questions regarding the function of the lymphocyte subpopulations in the MLC and the nature of the LD determinants remain to be answered. In addition to their presence on B lymphocytes, the LD determinants can be found on epidermal and endothelial cells, macrophages and spermatozoa (for review see Thorsby, 1974; Thorsby and Piazza, 1975).

The genetical basis of the MLC is better understood. As the MLC reactivity within the first families tested was identical with the results of the SD typing (no stimulation between SD identical siblings and strong stimulation between SD different siblings), it was believed that the cell interaction was controlled by the loci coding for SD antigens. However, in exceptional cases, there was a strong stimulation between SD identical siblings, but no stimulation between sibs differing in one haplotype. These investigations and tests performed in families with recombination between *HLA-A* and *HLA-B* led to the conclusion that the main reactivity detectable by the MLC is governed by a locus, now called *HLA-D* (*LD-1*), outside the region *HLA-A/HLA-B* and situated near *HLA-B* (recombination frequency *HLA-B/HLA-D* approximately 1%). Besides the *HLA-D* locus, there exists another one, *LD-2*, situated between *HLA-A* and *HLA-B*; disparities at this *LD-2* locus however, cause only relatively low stimulation in the MLC (for references see Thorsby, 1974).

Using different "typing cells" as stimulators, i.e. cells homozygous at the *HLA-D* locus, it was possible to define some HLA-D specificities: the LD determinant carried by the typing cells must be present on the responder if there is no, or only a weak response in the MLC. The typing cells originate from inbred families (first cousin marriages or incest (Jörgensen *et al.*, 1973; van den Tweel *et al.*, 1973)), but also from outbred families (Mempel *et al.*, 1973) in which the father and the mother share an *HLA-D* allele.

Up to now, eight specificities have been recognized: DW1, DW2, DW3, DW4, DW5, DW6, LD107 and LD108. The determinants DW1 to DW6 are better defined than LD107 and LD108 (Thorsby and Piazza, 1975).

Another way of defining the LD gene products is based on the finding that the HLA-D determinants might be expressed as the A or B gene products in haploid form on spermatozoa. By killing one population of the sperms using appropriate anti-SD sera and complement, it is probably possible to obtain spermatozoa carrying only one HLA-D determinant which can be used as stimulating cells (Halim and Festenstein, 1975). Results of genetical analysis of the individuals tested for the specificities DW1 to DW6, LD107 and LD108 are compatible with the hypothesis that they are governed by co-dominant alleles at the *HLA-D* locus. The sum of the gene frequencies of *DW1* to *DW6* in Caucasians equals 48%, so that the X_D allele (including the genes coding for LD107 and LD108) represents 52%. Furthermore, there is also a strong linkage disequilibrium between the *HLA-D* and the *HLA-B* genes: *DW1* with *BW35*, *DW2* with *B7*, *DW3* with *B8*, *DW4* with *BW15*, *DW5* with *BW16* and *LD107* with *B12*. The maximal $\varDelta$ value observed between *B* and *D* genes is $+0\cdot044$ for *B8/DW3* (Thorsby and Piazza, 1975). This $\varDelta$ is in the same order of magnitude as the linkage disequilibrium between *HLA-A* and *HLA-B*.

Although the genetical hypothesis assuming only two loci coding for the MLC reactivity is probably an oversimplification (Dupont *et al.*, 1975), it helps to understand a majority of the data and gives a good basis for a more sophisticated analysis of the genetical background of the MLC.

C. Ia Type Antigens

In the mouse, one central region of the H-2 system, the major histocompatibility complex of this species and therefore the equivalent of the HLA system, codes for the genetic control of virus susceptibility, of immune responsiveness and of cell interaction, while also governing serologically detectable antigens with a restricted tissue distribution. These are the Ia antigens (see Klein, 1975).

By means of lymphocytotoxic tests with B-cell enriched suspensions or with B-lymphocyte derived lines, as well as by more complicated methods (immunofluorescence, inhibition of the binding of aggregated IgG . . .), several groups have demonstrated that sera of individuals immunized against HLA SD gene products also contain antibodies against Ia type antigens. Up to now, several Ia type antigens have been described, but their mutual relationship is not yet known. Some of them show a very high correlation with LD determinants and it seems that these antigens are controlled by *HLA-D* or loci situated near *HLA-D* (Arbeit *et al.*, 1975; Ferrara *et al.*, 1975; Legrand and Dausset, 1975c; Mann *et al.*, 1975; van Rood *et al.*, 1975b; Solheim *et al.*, 1975; Terasaki *et al.*, 1975; Walford *et al.*, 1975; Wernet, 1976; Wernet *et al.*, 1975a, b; Winchester *et al.*, 1975a). However, it is probable that another locus (or loci) close to *HLA-A* is also coding for such determinants (Mann *et al.*, 1976a).

The Ia antigens are also polypeptides, do not contain β_2-microglobulin and are probably present on B lymphocytes, epidermal cells, macrophages and spermatozoa (W. F. Bodmer *et al.*, 1975; Fellous *et al.*, 1975; Wernet and Kunkel, 1975). This tissue distribution is nearly identical with the one observed for LD determinants and points also to a close relationship between the Ia type antigens and HLA-D gene products. An intensive investigation of the Ia type antigens will be performed during the 7th International Histocompatibility Workshop in 1976 and 1977, and it is hoped that the definition of some of these antigens will reach the goal of a serological detection of LD determinants or of gene products of loci governing immune responsiveness (or of very closely linked loci) (see Section IV.E.6).

D. Other Loci of the HLA System and Linked Loci

Investigations using somatic hybrids of human–Chinese hamster cells demonstrated that the loci *ME1* (NADP dependent cytoplasmic malate dehydrogenase), *SOD B* (tetrameric indophenol oxidase) and *PGM₃* (third locus of phosphoglucomutase) are situated on chromosome 6 (Jongsma *et al.*, 1973). The linkage of *PGM₃* with the HLA system (see below) indicated that the *HLA* loci are also localized on chromosome 6. This conclusion was confirmed by Lamm *et al.* (1974) who observed that in a family with a pericentric inversion of chromosome 6 (Inv(6) (p+q−)), the HLA SD antigens segregated in parallel with the marker chromosome. From the cytogenetic findings, the location of the *HLA* genes was deduced to be a little distal to the midpoint of one or other arm of chromosome 6. Another study in a family with a translocation t(6;21) (p22;q11) showed that the *HLA SD* genes might be situated on the long arm or proximal to 6p22 on the short arm of chromosome 6 (Borgaonkar and Bias, 1974). Autosomal deletion mapping of human chromosomes excluded the *HLA SD* loci from band 6p23 (Kreiger *et al.*, 1974).

At present, the following linkages between the HLA system and other loci are known: loci coding for enzymes (PGM$_3$, GLO, Pg), erythrocyte antigens (Chido), and plasma proteins (C'2, C'4, C'8, Bf = properdin

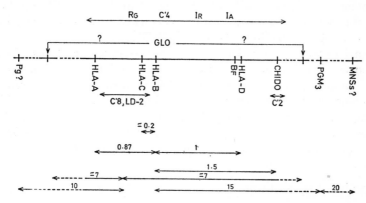

Fig. 1   The HLA system on chromosome 6. Distance in certimorgans (cM)

factor B). These linkage relationships are shown in Tables 5 and 6 on pp. 436–7; the currently most probable map of the region surrounding the HLA system is also shown in Fig. 1. The data used include those of Allen (1974), Day *et al.* (1975a), Fu *et al.* (1974), Friend *et al.* (1975), Hauptmann *et al.* (1976), Lamm *et al.* (1970, 1972, 1975, 1976), Mayr and Mayr (1974b), Mayr *et al.* (1975, 1976), Merritt *et al.* (1976), Middleton *et al.* (1974), Olaisen *et al.* (1976), Rittner *et al.* (1975), Weitkamp (1976), Weitkamp *et al.* (1975a, b) and Wolski *et al.* (1975, 1976). Another system which could belong to this linkage group is MNSs: the analysis of *PGM$_3$/MNSs* gave a lod score of +1·179 at $\theta = 0·20$ (Bissbort *et al.*, 1975). This result fits rather well with the lod score of +1·445 at $\theta = 0·30$ for *HLA/MNSs* (Mayr and Mayr, 1974b). However, as German and Chaganti (1973) postulated that *MNSs* is localized on chromosome 2 (the location listed in Chapter 9), these results have to be confirmed in other studies.

It is very interesting to note that there is a strong linkage disequilibrium existing between the gene controlling the synthesis of C'2 and the haplotype *A10, B18, DW2* (Friend *et al.*, 1975; Fu *et al.*, 1975), and also between the genes *B8* and *Bf^S* or *BW35* and *Bf^F* (Albert *et al.*, 1975; Olaisen *et al.*, 1975). Not all the loci determining components of the complement system are syntenic (Mayr and Mayr, 1974b; Day *et al.*, 1975b; Mittal *et al.*, 1976); see Chapter 6 for further discussion.

Recently, James *et al.* (1976) showed that 8 out of 9 Rg(a-) individuals (Rg = Rodgers is a blood group described by Longster and Giles, 1976) were probably *A1, B8* homozygous; the ninth person was A1, B8, BW35. From these data, one can conclude that the Rg system is also coded by a gene situated within the HLA complex and that there is a linkage disequilibrium between the haplotype *A1, B8* and the gene *Rg*.

Although β_2-microglobulin is firmly bound to the HLA SD gene products, the locus coding for this protein is situated on chromosome 15 and not on chromosome 6 (Goodfellow *et al.*, 1975).

By analogy with the situation found in the mouse, where a locus affecting morphogenesis (*T* locus) and loci controlling the immune responsiveness against various antigens (*Ir* loci) are linked with the major histocompatibility complex, the existence of similar loci has been investigated in Man. Although the presence of a human equivalent of the *T* locus has not been demonstrated unambiguously (Amos *et al.*, 1975; Bobrow *et al.*, 1975), *Ir* loci can be found with certainty in Man (see Section IV.E.6).

E. Practical Implications

1. Organ transplantation

Histocompatibility antigens. It can be stated without any doubt that the antigens A and B of the ABO blood groups are histocompatibility antigens (Dausset and Rapaport, 1966). It has been demonstrated that the survival time of skin grafts was less in cases of ABO incompatibility than in cases of ABO compatibility (Ceppellini *et al.*, 1965, 1969a). Furthermore, immunization against A or B in recipients of blood group O was followed by an accelerated rejection of skin grafts of blood group A or B, while skin grafts of blood group O had a normal survival time (Ceppellini *et al.*, 1969a; Dausset *et al.*, 1967, 1969). These observations show the pre-eminent importance of the isoagglutinogens of the ABO blood groups in organ grafting.

The presence of antibodies in the recipient against HLA SD antigens of the donor also causes an accelerated rejection of skin grafts (Dausset *et al.*, 1965; van Rood *et al.*, 1965). Therefore, the SD gene products are histocompatibility antigens. Furthermore, the effect of the HLA system on histocompatibility is non-immunized recipients could be demonstrated using skin grafts: in HLA SD identical siblings, the survival time was about 20 days, in non-identical sibs 13 to 14 days and in unrelated people only 12 days (Ceppellini *et al.*, 1969b). Transplantation in haploidentical situations, i.e. the recipient and the donor share one haplotype (mostly the skin of a child grafted to its father), showed a good correlation with the compatibility of SD antigens: the time of survival in cases of compatibility was 16·66 days, in cases of one SD incompatibility 14·35 days and in those of two SD incompatibilities 12·43 days (Dausset *et al.*, 1970b). In unrelated persons, no good correlation could be found between the numbers of SD incompatibilities and the survival time of skin grafts (Ceppellini *et al.*, 1969b; Walford *et al.*, 1969). Taking into account the gene products of the *HLA-D* locus, it seems that in unrelated people LD identity gives a higher mean survival time of skin transplants than SD identity; the best survival was found in SD and LD identical individuals (van Rood *et al.*, 1975b). The higher significance of the LD determinants could not be confirmed by Yunis *et al.* (1973) who showed that the gene (or genes) responsible for the rejection of skin grafts is inherited together with the *SD* loci in a family with a recombination between *HLA-B* and *HLA-D*.

From these data, one can conclude that the HLA gene complex is important for transplantation. The SD gene products are of relevance especially for immunized recipients; however, they are not the only determinants involved in the very complicated course of events occurring during rejection, as can be shown by the cell mediated lympholysis technique (CML), which represents an *in vitro* model of the rejection. In this test, lymphocytes are activated by a MLC and, after this first step, incubated with other target cells (Häyry and Defendi, 1970). The genetics of this reaction shows that lysis of the target cells occurs in most cases only after a positive MLC, i.e. if the stimulating cells carry LD determinants foreign to the effector cells. The stimulated effector cells react with determinants being coded by the *SD* loci or loci closely linked with them (M. L. Bach *et al.*, 1975). This rather simplified explanation, however, shows the complexity of the events, which are far from completely understood.

Transplantation. In order to avoid a hyperacute rejection of the grafted organ, the recipient must not have any antibodies against the histocompatibility antigens (ABO blood groups and HLA SD) of the donor (Kissmeyer-Nielsen *et al.*, 1966).

Besides this prerequisite, only the HLA complex is of importance for transplantation. This conclusion is based mainly on data from kidney grafts which showed that in HLA identical siblings the survival of grafts was nearly 100%, while HLA different organs were rejected frequently (Amos *et al.*, 1971; Descamps *et al.*, 1973; Hors *et al.*, 1974; Mickey *et al.*, 1971; Perkins *et al.*, 1973). In unrelated people, the situation is less clear: several transplantation centres are able to find a positive correlation between the survival of the kidney and histocompatibility (Eurotransplant, 1974; Festenstein *et al.*, 1976; Hors *et al.*, 1974; Scandiatransplant, 1975) while other centres cannot confirm these results (Cochrum *et al.*, 1975; Opelz *et al.*, 1974). In the first-mentioned centres, a more detailed analysis indicates that a compatibility for the HLA-D gene products might be more important for patients without antibodies against SD antigens (Eurotransplant, 1974; Festenstein *et al.*, 1976); recipients with HLA SD antibodies should be matched especially for HLA-B antigens in order to obtain a longer survival of the graft (Eurotransplant, 1974). Furthermore, it is possible that a large number of blood transfusions administered during haemodialysis has some good effect on the survival of the graft, probably due to enhancing antibodies (Festenstein *et al.*, 1976; Opelz *et al.*, 1973); however, this trend could not be observed in all centres (Hors *et al.*, 1974).

In spite of the discrepancies within these data, the joint analysis of the kidney grafts published by the Renal Transplantation Registry (1975) point to the importance of HLA SD compatibility. Therefore, a HLA SD identical kidney donor is at the moment the best one for grafting to an unrelated person. The discrepancies observed between the various centres are due to several factors: the patients are heterogeneous regarding their primary disease, the number of blood transfusions, immunosuppressive therapy, etc. Furthermore, it is almost impossible to include in the analyses the ability of the patients to produce an immune response against foreign histocompatibility antigens, which is known to vary widely amongst different individuals (Opelz *et al.*, 1972).

Due to the remarkable polymorphism of the HLA SD determinants, the chance of finding two unrelated SD identical persons is very small. This fact explains the establishment of national (Francetransplant, Londontransplant, Austrotransplant, Swisstransplant) or international (Eurotransplant, Scandiatransplant) programmes in which a large number of presumptive recipients is registered. The more recipients pooled, the higher is the chance of finding an SD identical patient should a donor be available.

For the transplantation of other organs, e.g. heart, lung or liver, the same criteria hold true, as these organs also possess ABO and HLA SD antigens. In transplantations of bone marrow, the dangers of rejection and of graft–versus–host reaction, which frequently causes the death of the patient, must be kept in mind. HLA SD and LD identity in recipient–donor pairs seems to give the best results in these cases (Floersheim and Storb, 1974; van Rood, 1974).

2. Platelet and granulocyte transfusion

The survival time of transfused platelets is significantly diminished if the recipient possesses antibodies against the donor's antigens (Bosch *et al.*,

1962). In these cases, the ABO, HLA SD and platelet-specific antigens must be considered. Similar considerations apply to the transfusion of granulocytes, in which antibodies against ABO, HLA SD 5a or 5b and granulocyte-specific determinants are of importance (Graw *et al.*, 1970).

3. Transfusion reactions

The transfusion reactions caused by antibodies against leucocytes are mostly febrile in nature (Brittingham and Chaplin, 1957). The antibodies can be directed against HLA SD antigens, against 5a or 5b, and against granulocyte-specific determinants.

4. Paternity testing

As is discussed further in Chapter 9, in order to use a genetical system in cases of disputed paternity, the following conditions must be fulfilled (Race and Sanger, 1975): the mode of inheritance must be known with certainty, the technique of determination must be reliable, the phenotype must reflect the genotype only, and the character must be developed at birth or soon thereafter. All these criteria apply to the products of the *HLA SD* loci (Mayr, 1974d).

If these antigens are utilized for paternity testing, only monospecific antisera of high titre must be used which have been evaluated in large family studies and which always give clear-cut results. Synergistic effects of minor antibody components or weak antibodies giving gene dosage effects (Ahrons and Thorsby, 1970) must not occur. The existence of weak variants (e.g. of HLA-A2; Mayr, 1973a) or of depressor genes (Bias *et al.*, 1974), and the possible influence of drugs which might alter the expression of the SD antigens (Ben-David *et al.*, 1973) should be borne in mind.

Three classes of exclusion can be distinguished:

(a) a man can be excluded if the child possesses an antigen lacked by the mother and the presumptive father. The chance of exclusion in false accusations of paternity within this class using the antigens coded by the loci *HLA-A* and *HLA-B* is 82%
(b) a man is excluded if the child possesses neither of both antigens of one locus demonstrable in the presumptive father. Such an exclusion is based on the allelism of the single genes governed by one locus. The chance of exclusion within this class amounts to 14%. The two classes jointly yield an exclusion chance of more than 91%
(c) a third constellation for possible exclusion is the one in which the child has inherited from the true father two genes which the presumptive father does not carry on one haplotype. However, because of the relatively high

402 W. R. MAYR

frequency of recombination, this class of exclusions should be used extremely cautiously.

Together with the other system used routinely in paternity testing (ABO, Rh, MNSs, P, K, Fy, Jk, Lu, Xg, Se, ACP, PGM_1, AK, ADA, GPT, Hp, Gm, Inv, Gc, C'3) which give an exclusion chance of 95·4%, the total chance of exclusion in false accusations of paternity is higher than 99·6%.

Formulae for the calculation of the chance of exclusion in single mother–child combinations (Mayr and Pausch, 1975) and for the determination of the plausibility of paternity according to Essen-Möller (Mayr, 1972b) have been published.

Practical experience with the HLA SD factors in paternity cases is excellent, especially in problematic ones, e.g. where several accused men could not be excluded by the classical systems or in cases where the accused man is not available, but his parents or other blood relations can be tested (Speiser *et al.*, 1974, 1975).

5. Anthropology

Because of its polymorphism and of the large variability of its gene frequencies amongst different populations, the HLA system is of great importance for anthropological investigations. In this paragraph, only the major differences between racial groups are mentioned; for details see the proceedings of the 5th International Histocompatibility Workshop Conference (Dausset and Colombani, 1973) which was mainly concerned with this problem.

The analysis of populations showed that A1, A3 and B8 are relatively frequent in Caucasians, while AW34, AW36, AW43 and BW42 do not exist in this race. Negroids frequently possess A28, AW30, AW33 and BW17. Mongoloids have a high occurrence of A9, A11, BW15 and BW40, but seem to lack B8, B14, CW2 and CW5. Furthermore, AW36, AW43 and BW42 are probably specificities typical for Negroids, and HS typical for Mongoloids. The linkage disequilibria between the loci also vary widely amongst the races (Joint Report, 1973). Even in Europe, there are considerable divergencies in the distribution of SD antigens: from North to South, the frequencies of A2, BW15 and BW40 decrease, while the frequencies of A10, B5 and BW35 increase significantly (Mayr, 1976).

The evolutionary trees based on the data of the 5th International Histocompatibility Workshop which were obtained using only the HLA SD polymorphism or using the SD antigens and other markers fit very well with the classical anthropological concepts of racial evolution (Piazza and Viganotti, 1973; Piazza *et al.*, 1975).

6. Association with diseases

In mice it has been observed that viruses which cause leukaemia bring about this disease if the animal possesses certain alleles of the H-2 system (see Klein, 1975). This finding suggested similar studies in Man, looking for associations between the HLA system and disease. Within the limited space of this chapter, it is not possible to discuss all these studies; for review, see: Dausset and Hors (1975), Dausset and Svejgaard (1976), Dausset et al. (1974), McDevitt and Bodmer (1974), van Rood et al. (1975a), Ryder et al. (1974), Svejgaard et al. (1975) and Vladutiu and Rose (1974).

Regarding the association between the HLA system and disease, there are many problems which must be considered: the genetical control of the disease is usually multifactorial, the effect of the gene (or genes) is not always expressed in all-or-none fashion, the influence of the environment cannot be measured, and the disease investigated may represent a hetero-geneous entity (cf. also p. 449). Furthermore, there are difficulties in the statistical evaluation of the results due to the large number of factors tested and due to problems concerning the definition of an appropriate control population. Some associations could also be caused by the correlation of certain gene products with a higher survival rate of the patients; errors of this kind can be avoided by prospective studies. Further heterogeneities may be due to the fact that the strength of the association is influenced by the age of the onset of the disease. This phenomenon is expressed in a very clear manner in juvenile diabetes mellitus (Cudworth and Woodrow, 1975; Schernthaner et al., 1976) and myasthenia gravis (Feltkamp et al., 1974).

The mechanisms of association between disease and the HLA system are not known; however, several hypotheses have been proposed (see McDevitt and Bodmer, 1974):

(a) the receptor hypothesis assumes that HLA gene products act as recep-tors for viruses or other pathogenic agents
(b) the molecular mimicry hypothesis postulates that the susceptibility to pathogenic agents represents a failure of the host to recognize some anti-gens as foreign and to produce an immune response against them
(c) the immune response hypothesis assumes that within the HLA com-plex immune response (Ir) genes control the ability of the host to produce an immune reaction against certain agents. This response, which can be quite variable, might bring on the symptoms of the disease.

As discussed by Dausset and Hors (1975), Klein (1975) or McDevitt and Bodmer (1974), the first two explanations are less likely than the third one, i.e. the existence of Ir genes. If these genes are not identical with the genes of the loci *HLA-A, HLA-B, HLA-C* or *HLA-D,* an association between them and the gene products of the four loci can only be found in unrelated people if a linkage disequilibrium exists. If there is no linkage disequilibrium, the presence of *Ir* genes responsible for a disease can only be based on family analyses which show parallel segregation of HLA haplotypes and the pathological condition.

This was the case in ragweed hay fever, where the genetic control of the production of antibodies against the antigen E segregated with HLA haplotypes (Blumenthal *et al.,* 1974; Levine *et al.,* 1972), in other allergic diseases (Geerts *et al.,* 1975), and in cutaneous hypersensitivity against several test antigens (Buckley *et al.,* 1973).

In contrast to these conditions, certain diseases show a clear-cut association with SD antigens. The first disease investigated was Hodgkin's disease (Amiel, 1967) which had an increased frequency of "4c" (a crossreacting group including B5, B18 and BW35). Further investigations showed that the highest risks of acquiring this disease are found for A1, B5 and B18 (Svejgaard *et al.,* 1975), thus confirming the results of Amiel. However, the relative risks (p. 450) are rather small. For other malignant diseases, the picture is not very clear and no definite associations have been observed (Svejgaard *et al.,* 1975).

In non-malignant diseases, the most significant and reliable associations are listed in Table 9 (based mainly on data in Caucasians summarized by Ryder *et al.,* 1974). Some of these diseases show stronger associations with other determinants controlled by the HLA system:

(a) coeliac disease and dermatitis herpetiformis with DW3 (Solheim *et al.,* 1976; Thomsen *et al.,* 1976) and a special Ia type antigen (Mann *et al.,* 1976b)
(b) juvenile diabetes mellitus with DW3 and DW4 (Thomsen *et al.,* 1975)
(c) Graves's disease with DW3 (Thorsby *et al.,* 1975)
(d) multiple sclerosis with DW2 (van den Berg-Loonen and Lucas, 1975; Bertrams *et al.,* 1975; Jersild *et al.,* 1975; Möller *et al.,* 1975) and an Ia type antigen, "Ag 7a" (Winchester *et al.,* 1975b).

In myasthenia gravis, there exists also an association with DW3, which is however less significant than the one with B8 (Möller *et al.,* 1976).

The disease with the highest relative risk, ankylosing spondylitis, is not highly correlated with a special HLA-D or HLA-C gene product (Möller *et al.,* 1975; Sachs *et al.,* 1975; Truog *et al.,* 1975). From these results, one may conclude that the *Ir* gene corresponding to anklylosing spondylitis shows a very strong linkage disequilibrium with *B27* and that its locus is situated near *HLA-B.* Similar considerations apply for the other associations.

Table 9

Association between diseases and SD antigens of the HLA system in Caucasians

Disease	HLA SD antigen	Approximate frequency (%) of the HLA SD antigen in patients	in controls	Relative risk
Ankylosing spondylitis	B27	90	7	120
Reiter's syndrome	B27	75	7	40
Acute anterior uveitis	B27	55	7	16
Coeliac disease	B8	80	24	13
Dermatitis herpetiformis	B8	60	24	5
Chronic aggressive hepatitis	B8	60	24	5
Myasthenia gravis	B8	60	24	5
Psoriasis vulgaris	B13	20	5	5
	BW17	25	8	4
Sjögren's syndrome	B8	55	22	4
Graves's disease	B8	45	24	3
Juvenile diabetes	B8	44	20	3
mellitus	BW15	30	14	3
	CW3	42	19	3
Multiple sclerosis	A3	36	27	2
	B7	36	27	2

Some of the correlations found in Caucasians can also be observed in other races, e.g. a very high frequency of B27 in Japanese with ankylosing spondylitis (Sonozaki et al., 1975). However, in most diseases the associations are different depending on the racial group tested, e.g. increase of B8 among Caucasians with Graves's disease, but association of this disorder with BW35 in Mongoloids (Grumet et al., 1975), or less significant increase of B27 in Blacks with ankylosing spondylitis or Reiter's disease (Good et al., 1976; Khan et al., 1976).

It is also remarkable that many of the organ specific autoimmune diseases show an increase of B8, which is particularly marked among the patients with detectable autoantibodies (Ludwig et al., unpublished; Thomsen et al., 1975). However, it is possible that an even stronger association will be obtained by studying LD determinants or Ia type antigens in these disorders.

Another disease in which a particular HLA constellation could be observed is choriocarcinoma. The tumour is made by the foetal tissue and

406   W. R. MAYR

might be considered as allograft. As demonstrated by Mogensen and Kiss-meyer-Nielsen (1968), the nature of the disease depends on the degree of HLA compatibility between mother and foetus. In generalized cases less incompatibilities have been observed than in localized ones (Mogensen et al., 1969). The frequency of choriocarcinoma is significantly higher in populations such as Eskimos of Greenland who have a restricted poly-morphism of SD antigens (Kissmeyer-Nielsen and Thorsby, 1970). How-ever, the results of the Scandinavian group could not be confirmed in other studies (Lawler et al., 1971; Tatra and Wolf, 1973) and should be further investigated.

Looking at the disease associations of the HLA system, it is striking that the disorders are either very rare or do not influence the reproduction so that they should not have any great significance regarding natural selection. In respect to this question, a possible influence of falciparum malaria, a disease exerting a strong selective pressure, was studied in two Sardinian populations of common ancestry: inhabitants of the highlands, who had never been infected with falciparum malaria, and inhabitants of the low-lands, where this disease was endemic for a long time (Piazza et al., 1973). The only differences which could not be explained by the effects of random drift were found in the frequencies of thalassaemia and G6PD deficiency (both known to be influenced by malaria), but also of HLA-B genes. From these analyses, the conclusion might be drawn that malaria selects for some special Ir genes closely linked with HLA-B genes. However, these results have to be verified in other populations with malaria infections.

Furthermore, data pointing to the importance of HLA linked Ir loci for the immune response against streptococci (Greenberg et al., 1975) and Au antigen (J. F. Bach et al., 1975; Boettcher et al., 1975) have been pub-lished. Such associations involving infectious agents could of course have an evolutionary significance.

V. CONCLUSIONS

In summary, it can be said that the leucocyte antigens are governed by various independent genetical systems. Up to now, it seems that only one of them is of major biological and clinical importance: the HLA system. In spite of progress in the definition of the classical HLA SD antigens, many questions remain to be solved, particularly in the field of the LD and Ia type determinants.

Furthermore, there is at the moment no clearcut answer to the question of how the tremendous polymorphism of the HLA system is maintained. Although Bodmer et al. (1973) were able to show that the HLA SD antigens are subjected to selection, several other authors could not find such effects (Bender *et al.*, 1976; Dausset *et al.*, 1970a; Joint Report, 1970; Mattiuz *et al.*, 1970; Mayr, 1973b), even when the segregation of the haplotypes with a strong positive linkage disequilibrium were analysed (Mayr, 1973b; Svejgaard *et al.*, 1971). It is, however, conceivable that the selective forces, acting in the past on the HLA complex (e.g. frequent epidemics) show no effect in the modern environment in which infectious diseases with a high lethality are extremely rare.

The hypotheses connecting the HLA polymorphism with antibody diversity (Jerne, 1971) or with a function to ensure the integrity of the body against the invasion by cells from other individuals of the same species (Burnet, 1973), might explain some aspects, but are rather difficult to verify. Other theories (Bodmer, 1972; Doherty and Zinkernagel, 1975) are also unable to solve all the problems.

In addition to this question, the significance of the linkage disequilibrium within the HLA system and the existence of "superhaplotypes" (e.g. *AW23, CW4, B12* or *A10, B18, DW2, C'2 deficiency*) are not yet understood.

It is striking that in other species ranging in phylogeny between *Xenopus laevis* (Du Pasquier *et al.*, 1975) and the chimpanzee (Balner and van Vreeswijk, 1975), e.g. the chicken, the mouse, the rat, the guinea pig, the pig, the dog or the rhesus monkey (for references see Klein, 1975), a chromosomal region can be found which is very similar to the HLA system. The extraordinary stability of this gene complex indicates that a very strong selective pressure must act on it which keeps the whole region intact over an evolutionary period covering more than 300 million years. It is probable that this selective pressure is connected with the function of the genes of the HLA system regarding the immune response (elimination of foreign histocompatibility antigens, cell to cell interaction, genetical control of complement components, etc.). The striking structural similarity of the HLA SD antigens with the constant part of the IgG molecule might also speak in favour of such a link. Nevertheless, present knowledge of the biological functions of the HLA system is rather modest and it is hoped that the new developments in this field (LD determinants, Ia type antigens, associations of HLA gene products with diseases) will lead to a better understanding of the significance of this gene complex.

ACKNOWLEDGEMENT

Parts of this work have been supported by the Hochschuljubiläumstiftung der Stadt Wien, Vienna.

408 W. R. MAYR

REFERENCES

Ahrons, S. (1971). *Tissue Antigens* 1, 129.
Ahrons, S. and Glavind-Kristensen, S. (1971). *Tissue Antigens* 1, 121.
Ahrons, S. and Thorsby, E. (1970). *Vox Sang.* 18, 323.
Albert, E. D., Kano, K., Abeyounis, C. J. and Milgrom, F. (1969). *Transplantation* 8, 466.
Albert, E. D., Mickey, M. R. and Terasaki, P. I. (1973). *Symp. Ser. Immunobiol. Standard.* 18, 156.
Albert, E. D., Rittner, C., Grosse-Wilde, H., Netzel, B. and Scholz, S. (1975). *In* "Histocompatibility Testing 1975" (F. Kissmeyer-Nielsen, Ed.) 941. Munksgaard, Copenhagen.
Allen, F. H. Jr. (1974). *Vox Sang.* 27, 382.
Amiel, J. L. (1967). *In* "Histocompatibility Testing 1967" (E. S. Curtoni, P. L. Mattiuz and R. M. Tosi, Eds) 79. Munksgaard, Copenhagen.
Amos, D. B. (1953). *Br. J. Exp. Path.* 34, 464.
Amos, D. B. (1968). *Science* 159, 659.
Amos, D. B. and Ward, F. E. (1975). *Physiol. Rev.* 55, 206.
Amos, D. B., Anderson, E. E., Glenn, J. F., Gunnells, J. C., Lancaster, S. L., McQueen, J. M., Robinson, R. R., Seigler, H. F., Stickel, D. L. and Ward, F. E. (1971) *Transplant. Proc.* 3, 993.
Amos, D. B., Ruderman, R., Mendell, N. R. and Johnson, A. H. (1975). *Transplant. Proc.* 7 Suppl. 1, 93.
Arbeit, R. D., Sachs, D. H., Amos, D. B. and Dickler, H. B. (1975). *J. Immunol.* 115, 1173.
Bach, F. H. and Hirschorn, K. (1964). *Science* 143, 815.
Bach, F. H. and Voynow, N. K. (1966). *Science* 153, 546.
Bach, J. F., Zingraff, J., Descamps, B., Naret, C. and Jungers, P. (1975). *Lancet* ii, 707.
Bach, M. L., Festenstein, H., Goulmy, E., Grunnet, N., Handwerger, B., Kristensen, T. and Röllinghoff, M. (1975). *In* "Histocompatibility Testing 1975" (F. Kissmeyer-Nielsen, Ed.) 898. Munksgaard, Copenhagen.
Bain, B. and Lowenstein, L. (1964). *Science* 145, 1315.
Balner, H. and Vreeswijk, W. van (1975). *Tranplant. Proc.* 7 Suppl. 1, 13.
Balner, H., Gabb, B. W., Dersjant, H., Vreeswijk, W. van, Leeuwen, A. van and Rood, J. J. van (1971). *Transplant. Proc.* 3, 1088.
Belvedere, M. C., Curtoni, E. S., Dausset, J., Lamm, L. U., Mayr, W. R., Rood, J. J. van, Svejgaard, A. and Piazza, A. (1975a). *Tissue Antigens* 5, 99.
Belvedere, M. C., Mattiuz, P. and Curtoni, E. S. (1975b). *Immunogenetics* 1, 538.
Belvedere, M. C., Richiardi, P., Luciani, G. and Curtoni, E. S. (1975c). *Tissue Antigens* 5, 108.
Ben-David, A., Orgad, S., Danon, Y. and Michali, D. (1973). *Tissue Antigens* 3, 378.
Bender, K., Mayerova, A. and Hiller, Ch. (1975). *In* "Humangenetik" Vol. I/3 (P. E. Becker, Ed.) p. 267. Thieme, Stuttgart.
Bender, K., Mayerova, A., Klotzbücher, B., Burckhardt, K. and Hiller, Ch. (1976). *Tissue Antigens* 7, 118.
Berah, M., Hors, J. H. and Dausset, J. (1970). *Transplantation* 9, 185.

Berg-Loonen, E. M. van den and Lucas, K. J. (1975). *In* "Histocompatibility Testing 1975" (F. Kissmeyer-Nielsen, Ed.) 773. Munksgaard, Copenhagen.

Bernoco, D., Cullen, S., Scudeller, G., Trinchieri, G. and Ceppellini, R. (1973). *In* "Histocompatibility Testing 1972" (J. Dausset and J. Colombani, Eds) 527. Munksgaard, Copenhagen.

Bertrams, J., Grosse-Wilde, H., Netzel, B., Mempel, W. and Kuwert, E. (1975). *In* "Histocompatibility Testing 1975" (F. Kissmeyer-Nielsen, Ed.) 782. Munksgaard, Copenhagen.

Bias, W. B., Hopkins, K. A., Hutchinson, J. R. and Hsu, S. H. (1974). *Tissue Antigens* 4, 36.

Billing, R. J. and Terasaki, P. I. (1974). *Transplantation* 17, 231.

Billing, R. J., Mittal, K. K. and Terasaki, P. I. (1973). *Tissue Antigens* 3, 251.

Bismuth, A., Neauport-Sautes, C., Kourilsky, F. M., Manuel, Y. and Greenland, T. (1973). *C.R. Acad. Sci. Paris* 277, 2845.

Bissbort, S., Kömpf, J. and Ritter, H. (1975). *Hum. Genet.* 28, 245.

Blumenthal, M. N., Amos, D. B., Noreen, H., Mendell, N. R. and Yunis, E. J. (1974). *Science* 184, 1301.

Bobrow, M., Bodmer, J. G., Bodmer, W. F., McDevitt, H. O., Lorber, J. and Swift, P. (1975). *Tissue Antigens* 5, 234.

Bodmer, J., Curtoni, E. S., Leeuwen, A. van, Mickey, M. R., Kjerbye, L., Degos, L., Botha, M. C., Wolf, E., Mayr, W. R., Staub Nielsen, L. and Piazza, A. (1975). *In* "Histocompatibility Testing 1975" (F. Kissmeyer-Nielsen, Ed.) 21. Munksgaard, Copenhagen.

Bodmer, W. F. (1972). *Nature, Lond.* 237, 139.

Bodmer, W. F. and Payne, R. (1965). *In* "Histocompatibility Testing 1965". Series Haematologica 11, 141. Munksgaard, Copenhagen.

Bodmer, W. F., Cann, H. and Piazza, A. (1973). *In* "Histocompatibility Testing 1972" (J. Dausset and J. Colombani, Eds) 753. Munksgaard, Copenhagen.

Bodmer, W. F., Jones, E. A., Young, D., Goodfellow, P. N., Bodmer, J. G., Dick, H. M. and Steel, C. M. (1975). *In* "Histocompatibility Testing 1975" (F. Kissmeyer-Nielsen, Ed.) 677. Munksgaard, Copenhagen.

Boettcher, B., Hay, J., Watterson, C. A., Bashir, H., MacQueen, J. M. and Hardy, G. (1975). *J. Immunogenetics* 2, 151.

Borgaonkar, D. S. and Bias, W. B. (1974). *Cytogenet. Cell Genet.* 13, 67.

Bosch, J., Jansz, A., Lammers, H. A., Leeuwen, A. van and Rood, J. J. van (1962). Proc. 8th Congr. Eur. Soc. Haemat., Vienna 1961. Paper No. 380. Karger, Basel and New York.

Brand, D. L., Ray, J. G. Jr., Hare, D. B., Kayhoe, D. E. and McClelland, J. D. (1970). *In* "Histocompatibility Testing 1970" (P. I. Terasaki, Ed.) 357. Munksgaard, Copenhagen.

Bridgen, J., Snary, D., Crumpton, M. J., Barnstable, C., Goodfellow, P. N and Bodmer, W. F. (1976). *Nature, Lond.* 261, 200.

Bright, S. (1976). *Tissue Antigens* 7, 23.

Brittingham, T. E. and Chaplin, H. Jr. (1957). *J. Am. Med. Assoc.* 165, 819.

Buckley, C. E. III., Dorsey, F. C., Corley, R. B., Ralph, W. B., Woodbury, M. A. and Amos, D. B. (1973). *Proc. Natn. Acad. Sci. USA* 70, 1257.

Burnet, F. M. (1973). *Nature, Lond.* 245, 359.

Ceppellini, R. (1971). *In* "Progress in Immunology" (B. Amos, Ed.) 973. Academic Press, New York and London.

Ceppellini, R. and Rood, J. J. van (1974). *Sem. Haematol.* 11, 233.

Ceppellini, R., Curtoni, E. S., Leigheb, G., Mattiuz, P. L., Miggiano, V. G. and Visetti, M. (1965). In "Histocompatibility Testing 1965". Series Haematologica 11, 13. Munksgaard, Copenhagen.

Ceppellini, R., Curtoni, E. S., Mattiuz, P. L., Leigheb, G., Visetti, M. and Colombi, A. (1966). Ann. N. Y. Acad. Sci. 129, 421.

Ceppellini, R., Curtoni, E. S., Mattiuz, P. L., Miggiano, V., Scudeller, G. and Serra, A. (1967). In "Histocompatibility Testing 1967" (E. S. Curtoni, P. L. Mattiuz and R. M. Tosi, Eds) 149. Munksgaard, Copenhagen.

Ceppellini, R., Bigliani, S., Curtoni, E. S. and Leigheb, G. (1969a). Transplant. Proc. 1, 390.

Ceppellini, R., Mattiuz, P. L., Scudeller, G. and Visetti, M. (1969b). Transplant. Proc. 1, 385.

Cochrum, K. C., Salvaterria, O. Jr., Perkins, H. A. and Belzer, F. O. (1975). Transplant Proc. 7 Suppl. 1, 659.

Collins, Z. V., Arnold, P. F., Peetom, F., Smith, G. S. and Walford, R. L. (1973). Tissue Antigens 3, 358.

Colombani, J., Colombani, M. and Dausset, J. (1970a). In "Histocompatibility Testing 1970" (P. I. Terasaki, Ed.) 79. Munksgaard, Copenhagen.

Colombani, J., Colombani, M., Viza, D. C., Degani-Bernard, O., Dausset, J. and Davies, D. A. L. (1970b). Transplantation 9, 228.

Colombani, J., d'Amaro, J., Gabb, B., Smith, G. S. and Svejgaard, A. (1971). Transplant. Proc. 3, 121.

Cresswell, P. and Ayres, J. L. (1976). Eur. J. Immunol. 6, 82.

Cresswell, P. and Dawson, J. R. (1975). J. Immunol. 114, 523.

Cresswell, P., Turner, M. J. and Strominger, J. L. (1973). Proc. Natn. Acad. Sci. USA 70, 1603.

Cresswell, P., Robb, R. J., Turner, M. J. and Strominger, J. L. (1974). J. Biol. Chem. 249, 2828.

Cudworth, A. G. and Woodrow, J. C. (1975). Br. Med. J. 3, 133.

Cunningham, B. A. and Berggård, I. (1974). Transplant. Rev. 21, 3.

Curtoni, E. S., Richiardi, P., Scudeller, G., Bernoco, D. and Ceppellini, R. (1969). Riv. Emoterap. Immunoemat. 16, 179.

Dausset, J. (1954). Vox Sang. 4, 190.

Dausset, J. (1958). Acta Haemat. 20, 156.

Dausset, J. (1972). Vox Sang. 23, 153.

Dausset, J. and Colombani, J., Eds. (1973). "Histocompatibility Testing 1972". Munksgaard, Copenhagen.

Dausset, J. and Hors, J. H. (1975). Transplant. Rev. 22, 44.

Dausset, J. and Nenna, A. (1952). C.R. Soc. Biol. (Paris) 146, 1539.

Dausset, J. and Rapaport, F. T. (1966). Ann. N. Y. Acad. Sci. 129, 408.

Dausset J. and Svejgaard, A., Eds. (1976). "HLA and Disease". Munksgaard, Copenhagen.

Dausset, J. and Tangün, Y. (1965). Vox Sang. 10, 641.

Dausset, J., Rapaport, F. T., Ivanyi, P. and Colombani, J. (1965). In "Histocompatibility Testing 1965", Series Haematologica 11, 63. Munksgaard, Copenhagen.

Dausset, J., Rapaport, F. T., Barge, A., Hors, J. H., Sasportes, M. and Santana, V. (1967). Presse Méd. 75, 1503.

Dausset, J., Colombani, J., Colombani, M., Legrand, L. and Feingold, N. (1968). Nouv. Rev. Franç. Hémat. 8, 398.

Dausset, J., Rapaport, F. T. and Barge, A. (1969). *Nouv. Rev. Franç. Hémat.* **9**, 339.

Dausset, J., Colombani, J., Legrand, L. and Fellous, M. (1970a). *In* "Histocompatibility Testing 1970" (P. I. Terasaki, Ed.) 53. Munksgaard, Copenhagen.

Dausset, J., Rapaport, F. T., Legrand, L., Colombani, J. and Marcelli-Barge, A. (1970b). *In* "Histocompatibility Testing 1970" (P. I. Terasaki, Ed.) 381. Munksgaard, Copenhagen.

Dausset, J., Colombani, J., Legrand, L., Lepage, V., Marcelli-Barge, A. and Dehay, C. (1973). *In* "Histocompatibility Testing 1972" (J. Dausset and J. Colombani, Eds) 107. Munksgaard, Copenhagen.

Dausset, J., Degos, L. and Hors, J. (1974). *Clin. Immunol. Immunopathol.* **3**, 127.

Dautigny, A. and Colombani, J. (1973). *C.R. Acad. Sci. Paris* **276**, 1641.

Dawson, J. R., Shasby, S. S. and Amos, D. B. (1974). *Tissue Antigens* **4**, 76.

Day, N. K., l'Espérance, P., Good, R. A., Michael, A. F., Hansen, J. A., Dupont, B. and Jersild, C. (1975a). *J. Exp. Med.* **141**, 1464.

Day, N. K., Rubinstein, P., de Bracco, M., Moncada, B., Hansen, J. A., Dupont, B., Thomsen, M., Svejgaard, A. and Jersild, C. (1975b). *In* "Histocompatibility Testing 1975" (F. Kissmeyer-Nielsen, Ed.) 960. Munksgaard, Copenhagen.

Degos, L., Joysey, V. C., Ferrara, G. B., Vives, J., Dastot, H., Dehay, C., Lethielleux, P., Reboul, M., Colombani, M. and Colombani, J. (1975). *In* "Histocompatibility Testing 1975" (F. Kissmeyer-Nielsen, Ed.) 247. Munksgaard, Copenhagen.

Descamps, B., Hinglais, N. and Crosnier, J. (1973). *Transplant. Proc.* **5**, 231.

Doherty, P. C. and Zinkernagel, R. M. (1975). *Lancet* ii, 1406.

Dorf, M. E., Eguro, S. Y., Cabrera, G., Yunis, E. J., Swanson, J. and Amos, D. B. (1972). *Vox Sang.* **22**, 447.

Doughty, R. W., Goodier, S. R. and Gelsthorpe, K. (1973). *Tissue Antigens* **3**, 189.

Dumble, L. and Morris, P. J. (1975). *Tissue Antigens* **5**, 103.

Du Pasquier, L., Chardonnens, X. and Miggiano, V. G. (1975). *Immunogenetics* **1**, 482.

Dupont, B., Yunis, E. J., Hansen, J. A., Reinsmoen, N., Suciu-Foca, N., Mickelson, E. and Amos, D. B. (1975). *In* "Histocompatibility Testing 1975" (F. Kissmeyer-Nielsen, Ed.) 547. Munksgaard, Copenhagen.

Engelfriet, C. P. (1966). "Cytotoxic Antibodies against Leucocytes", Thesis. Drukkerij Aemstelstad, Amsterdam.

Engelfriet, C. P., Veenhoven von Riesz, E., Kort-Bakker, M. and Berg-Loonen, P. M. van den (1973). *In* "Histocompatibility Testing 1972" (J. Dausset and J. Colombani, Eds) 475. Munksgaard, Copenhagen.

Eurotransplant Foundation (1974). Annual Report.

Fellous, M. and Dausset, J. (1970). *Nature, Lond.* **225**, 191.

Fellous, M., Mortchelewicz, F., Kamoun, M. and Dausset, J. (1975). *In* "Histocompatibility Testing 1975" (F. Kissmeyer-Nielsen, Ed.) 708. Munksgaard, Copenhagen.

Feltkamp, T. E. W., Berg-Loonen, P. M. van den, Oosterhuis, H. J. G. H., Nijenhuis, L. E., Engelfriet, C. P., Rossum, A. L. van and Loghem, J. J. van (1974). *In* "Recent Advances in Myology", Excerpta Medica Int. Congr. Ser. No. 360, 498. Excerpta Medica, Amsterdam.

Ferrara, G. B., Tosi, R. M., Antonelli, P, and Longo, A. (1975). *In* "Histocompatibility Testing 1975" (F. Kissmeyer-Nielsen, Ed.) 608. Munksgaard. Copenhagen.

412 W. R. MAYR

Festenstein, H., Sachs, J. A., Paris, A. M. I., Pegrum, G. D. and Moorhead, J. F. (1976). *Lancet* i, 157.

Floersheim, G. L. and Storb, R. (1974). *Schweiz. Med. Wschr.* 104, 3.

Fotino, M., Merson, E. J., Benoit, P., Rowe, A. W. and Allen, F. H. Jr. (1967). *In* "Histocompatibility Testing 1967" (E. S. Curtoni, P. L. Mattiuz and R. M. Tosi, Eds) 429. Munksgaard, Copenhagen.

Friedman, E. A., Retan, J. W., Marshall, D. C., Henry, J. L. and Merrill, J. P. (1961). *J. Clin. Invest.* 40, 2162.

Friend, P., Kim, Y., Handwerger, B., Reinsmoen, N., Michael, A. F. and Yunis, E. J. (1975). *In* "Histocompatibility Testing 1975" (F. Kissmeyer-Nielsen, Ed.) 928. Munksgaard, Copenhagen.

Fu, S. M., Kunkel, H. G., Brusman, H. P., Allen, F. H. Jr. and Fotino, M. (1974). *J. Exp. Med.* 140, 1108.

Fu, S. M., Stern, R. C., Kunkel, H. G., Dupont, B., Hansen, J. A., Day, N. K., Good, R. A., Jersild, C. and Fotino, M. (1975). *In* "Histocompatibility Testing 1975" (F. Kissmeyer-Nielsen, Ed.) 933. Munksgaard, Copenhagen.

Geerts, S. J., Pöttgens, H., Limburg, M. and Rood, J. J. van (1975). *Lancet* i, 461.

Gelsthorpe, K. and Doughty, R. W. (1975). *In* "Histocompatibility Testing 1975" (F. Kissmeyer-Nielsen, Ed.) 299. Munksgaard, Copenhagen.

German, J. L. and Chaganti, R. S. K. (1973). *Science* 182, 1261.

Good, A. E., Kawanishi, H. and Schultz, J. S. (1976). *New Eng. J. Med.* 294, 166.

Goodfellow, P. N., Jones, E. A., Heyningen, V. van, Solomon, E., Bobrow, M., Miggiano, V. and Bodmer, W. F. (1975). *Nature Lond.* 254, 267.

Gorer, P. A. and O'Gorman, P. (1956). *Transplant. Bull.* 3, 142.

Graw, R. G. Jr., Goldstein, I. M., Eyre, H. J. and Terasaki, P. I. (1970). *Lancet* ii, 77.

Greenberg, L. J., Gray, E. D. and Yunis, E. J. (1975). *J. Exp. Med.* 141, 935.

Grey, H. M., Kubo, R. T., Colon, S. M., Poulik, M. D., Cresswell, P., Springer, T., Turner, M. J. and Strominger, J. L. (1973). *J. Exp. Med.* 138, 1608.

Grumet, F. C., Payne, R. O., Konishi, J., Mori, T. and Kriss, J. P. (1975). *Tissue Antigens* 6, 347.

Halim, K. and Festenstein, H. (1975). *Lancet* ii, 1255.

Halim, K., Abbasi, K. and Festenstein, H. (1974). *Tissue Antigens* 4, 1.

Hansen, H. E., Ryder, L. P. and Staub Nielsen, L. (1975). *Tissue Antigens* 6, 275.

Harris, R. C. and Zervas, J. D. (1969). *Nature, Lond.* 221, 1062.

Hasegawa, T., Bergh, O. J., Mickey, M. R. and Terasaki, P. I. (1975). *Transplant. Proc.* 7 Suppl 1, 75.

Hauptmann, G., Sasportes, M., Tongio, M. M., Mayer, S. and Dausset, J. (1976). *Tissue Antigens* 7, 52.

Häyry, P. and Defendi, V. (1970). *Science* 168, 133.

Hirschfeld, J. (1972). *Nature, Lond.* 239, 385.

Hors, J. H., Busson, M. and Dausset, J. (1974). *Transplant. Proc.* 6, 421.

James, J., Stiles, P., Boyce, F. and Wright, J. (1976). *Vox Sang.* 30, 214.

Jeannet, M., Bodmer, J. G., Bodmer, W. F. and Schapira, M. (1973). *In* "Histocompatibility Testing 1972" (J. Dausset and J. Colombani, Eds) 493. Munksgaard, Copenhagen.

Jeannet, M., Schapira, M. and Magnin, C. (1974). *Schweiz. Med. Wschr.* 104, 152.

Jerne, N. K. (1971). *Eur. J. Immunol.* 1, 1.

Jersild, C., Dupont, B., Fog, T., Platz, P. J. and Svejgaard, A. (1975). *Transplant. Rev.* **22**, 148.

Joint Report of the Fourth International Histocompatibility Workshop (1970). *In* "Histocompatibility Testing 1970" (P. I. Terasaki, Ed.) 17. Munksgaard, Copenhagen.

Joint Report of the Fifth International Histocompatibility Workshop (1973). *In* "Histocompatibility Testing 1972" (J. Dausset and J. Colombani, Eds) 619. Munksgaard, Copenhagen.

Jongsma, A. P. M., Someren, H. van, Westerveld, A., Hagemeijer, A. and Pearson, P. L. (1973). *Hum. Genet.* **20**, 195.

Jörgensen, F., Lamm, L. U. and Kissmeyer-Nielsen, F. (1973). *Tissue Antigens* **3**, 323.

Kabat, E. A. (1968). "Structural Concepts in Immunology and Immunochemistry". Holt, Rinehart and Winston, New York.

Kachru, R. B. and Mittal, K. K. (1975). *In* "Histocompatibility Testing 1975" (F. Kissmeyer-Nielsen, Ed.) 404. Munksgaard, Copenhagen.

Khan, M. A., Kushner, I. and Braun, W. E. (1976). *Lancet* **i**, 483.

Kissmeyer-Nielsen, F. and Thorsby, E. (1970). *Transplant. Rev.* **4**, 3.

Kissmeyer-Nielsen, F., Olsen, S., Petersen, V. P. and Fjeldborg, O. (1966). *Lancet* **ii**, 662.

Kissmeyer-Nielsen, F., Staub Nielsen, L., Lindholm, A., Sandberg, L., Svejgaard, A. and Thorsby, E. (1970). *In* "Histocompatibility Testing 1970" (P. I. Terasaki, Ed.) 105. Munksgaard, Copenhagen.

Kissmeyer-Nielsen, F., Kjerbye, K. E., Lamm, L. U., Jörgensen, J., Bruun Petersen, G. and Gürtler, H. (1973). *In* "Histocompatibility Testing 1972" (J. Dausset and J. Colombani, Eds) 317. Munksgaard, Copenhagen.

Kissmeyer-Nielsen, F., Kjerbye, K. E. and Graugaard, B. (1975a). *In* "Histocompatibility Testing 1975" (F. Kissmeyer-Nielsen, Ed.) 277. Munksgaard, Copenhagen.

Kissmeyer-Nielsen, F., Kjerbye, K. E. and Graugaard, B. (1975b). *In* "Histocompatibility Testing 1975" (F. Kissmeyer-Nielsen, Ed.) 282. Munksgaard, Copenhagen.

Klein, J. (1975). "Biology of the Mouse Histocompatibility-2 Complex". Springer, Berlin, Heidelberg and New York.

Kreiger, D., Palmer, C. G. and Biegel, A. (1974). *Hum. Genet.* **23**, 159.

Lalezari, P. and Bernard, G. E. (1965). *In* "Histocompatibility Testing 1965", Series Haematologica **11**, 167. Munksgaard, Copenhagen.

Lalezari, P. and Bernard, G. E. (1966). *Fed. Proc.* **25**, 371.

Lalezari, P., Thalenfeld, B. and Weinstein, W. J. (1970). *In* "Histocompatibility Testing 1970" (P. I. Terasaki, Ed.) 319. Munksgaard, Copenhagen.

Lamm, L. U., Kissmeyer-Nielsen, F. and Henningsen, K. (1970). *Hum. Hered.* **20**, 305.

Lamm, L. U., Kissmeyer-Nielsen, F., Svejgaard, A., Bruun Petersen, G., Thorsby, E., Mayr, W. and Högman, C. F. (1972). *Tissue Antigens* **2**, 205.

Lamm, L. U., Friedrich, U., Bruun Petersen, G., Jörgensen, J., Nielsen, J., Therkelsen, A. J. and Kissmeyer-Nielsen, F. (1974). *Hum. Hered.* **24**, 273.

Lamm, L. U., Thorsen, I. L., Bruun Petersen, G., Jörgensen, J., Henningsen, K., Bech, B. and Kissmeyer-Nielsen, F. (1975). *Ann. Hum. Genet.* **38**, 383.

Lamm, L. U., Jörgensen, F. and Kissmeyer-Nielsen, F. (1976). *Tissue Antigens* **7**, 122.

Landsteiner, K. (1901). *Wien. Klin. Wschr.* **14**, 1132.

Laundy, G. J. and Entwistle, C. C. (1975). *In* "Histocompatibility Testing 1975" (F. Kissmeyer-Nielsen, Ed.) 278. Munksgaard, Copenhagen.

Lawler, S. D., Klouda, P. T. and Bagshawe, K. D. (1971). *Lancet* **ii**, 834.

Leeuwen, A. van, Eernisse, J. G. and Rood, J. J. van. (1964). *Vox Sang.* **9**, 431.

Legrand, L. and Dausset, J. (1973). *In* "Histocompatibility Testing 1972" (J. Dausset and J. Colombani, Eds) 441. Munksgaard, Copenhagen.

Legrand, L. and Dausset, J. (1974). *Tissue Antigens* **4**, 329.

Legrand, L. and Dausset, J. (1975a). *Transplant. Proc.* **7** Suppl. 1, 5.

Legrand, L. and Dausset, J. (1975b). *Transplantation* **19**, 177.

Legrand, L. and Dausset, J. (1975c). *In* "Histocompatibility Testing 1975" (F. Kissmeyer-Nielsen, Ed.) 665. Munksgaard, Copenhagen.

Lepage, V., Degos, L. and Dausset, J. (1976). *Tissue Antigens* **8**, 139.

Levine, B. B., Stember, R. H. and Fotino, M. (1972). *Science* **178**, 1201.

Longster, G. and Giles, C. M. (1976). *Vox Sang.* **30**, 175.

Löw, B., Messeter, L., Mansson, S. and Lindholm, T. (1974). *Tissue Antigens* **4**, 405.

Majsky, A. (1970). *In* "Fortschritte der Hämatologie" Vol. 1. (E. Perlick, W. Plenert and O. Prokop, Eds) 445. Barth, Leipzig.

Mann, D. L., Rogentine, G. N. Jr., Fahey, J. L. and Nathenson, S. G. (1969). *Science* **163**, 1460.

Mann, D. L., Fahey, J. L. and Nathenson, S. G. (1970). *In* "Histocompatibility Testing 1970" (P. I. Terasaki, Ed.) 461. Munksgaard, Copenhagen.

Mann, D. L., Abelson, L. D., Harris, S. and Amos, D. B. (1975). *J. Exp. Med.* **142**, 84.

Mann, D. L., Abelson, L. D., Harris, S. and Amos, D. B. (1976a). *Nature, Lond.* **259**, 145.

Mann, D. L., Katz, S. I., Nelson, D. L., Abelson, L. D. and Strober, W. (1976b). *Lancet* **i**, 110.

Manual of Tissue Typing Techniques. (1974). (J. G. Ray Jr., D. B. Hare, P. D. Pedersen and D. E. Kayhoe, Eds). DHEW Publication No. (NIH) 75–545.

Marcelli-Barge, A., Poirier, J. C., Benajam, A. and Dausset, J. (1976). *Vox Sang.* **30**, 81.

Mattiuz, P. L., Ihde, D., Piazza, A., Ceppellini, R. and Bodmer, W. F. (1970). *In* "Histocompatibility Testing 1970" (P. I. Terasaki, Ed.) 193. Munksgaard, Copenhagen.

Mayo, O. (1974). *Hum. Hered.* **24**, 144.

Mayr, W. R. (1971). *Schweiz. med. Wschr.* **101**, 361.

Mayr, W. R. (1972a). *In* "The Biochemical Genetics of Man" (D. J. H. Brock and O. Mayo, Eds) 1st edn, 301. Academic Press, London and New York.

Mayr, W. R. (1972b). *Z. Immun.-Forsch.* **144**, 18.

Mayr, W. R. (1973a). *Tissue Antigens* **3**, 212.

Mayr, W. R. (1973b). *Symp. Series Immunobiol. Standard.* **18**, 224.

Mayr, W. R. (1974a). *Z. Immun.-Forsch.* **148**, 92.

Mayr, W. R. (1974b). *In* "Fortschritte der Hämatologie" Vol. 3 (E. Perlick, W. Plenert, O. Prokop and H. Stobbe, Eds) 17. Barth, Leipzig.

Mayr, W. R. (1974c). *Fol. Haemat.* **101**, 401.

Mayr, W. R. (1974d). *Z. Rechtsmed.* **75**, 81.

Mayr, W. R. (1975a). *Wien. Klin. Wschr.* **87**, 488.

Mayr, W. R. (1975b). *In* "Histocompatibility Testing 1975" (F. Kissmeyer-Nielsen, Ed.) 330. Munksgaard, Copenhagen.

Mayr, W. R. (1976). *Homo* 27, 75.

Mayr, W. R. and Mayr, D. (1974a). *J. Immunogenet.* 1, 43.

Mayr, W. R. and Mayr, D. (1974b). *Hum. Genet.* 24, 129.

Mayr, W. R. and Mickerts, D. (1971). *Tissue Antigens* 1, 47.

Mayr, W. R. and Pausch, V. (1975). *Z. Immun.-Forsch.* 150, 447.

Mayr, W. R., Bernoco, D., de Marchi, M. and Ceppellini, R. (1973). *Transplant. Proc.* 5, 1581.

Mayr, W. R., Bissbort, S. and Kömpf, J. (1975). *Hum. Genet.* 28, 173.

Mayr, W. R., Mayr, D., Kömpf, J., Bissbort, S. and Ritter, H. (1976). *Hum. Genet.* 31, 241.

McDevitt, H. O. and Bodmer, W. F. (1974). *Lancet* i, 1269.

Medawar, P. B. (1944). *J. Anat.* 78, 176.

Medawar, P. B. (1945). *J. Anat.* 79, 157.

Medawar, P. B. (1946). *Br. J. Exp. Path.* 27, 15.

Mempel, W., Grosse-Wilde, H., Baumann, P., Netzel, B., Steinbauer-Rosenthal, I., Scholz, S., Bertrams, J. and Albert, E. D. (1973). *Transplant. Proc.* 5, 1529.

Merritt, A. D., Petersen, B. H., Biegel, A. A., Meyers, D. A., Brooks, G. F. and Hodes, M. E. (1976). *Cytogenet. Cell Genet.* 16, 331.

Mickey, M. R., Kreisler, M., Albert, E. D., Tanaka, M. and Terasaki, P. I. (1971). *Tissue Antigens* 1, 57.

Middleton, J., Crookston, M. C., Falk, J. A., Robson, E. B., Cook, P. J. L., Batchelor, J. R., Bodmer, J., Ferrara, G. B., Festenstein, H., Harris, R. C., Kissmeyer-Nielsen, F., Lawler, S. D., Sachs, J. A. and Wolf, E. (1974). *Tissue Antigens* 4, 366.

Mittal, K. K and Terasaki, P. I. (1972). *Tissue Antigens* 2, 94.

Mittal, K. K. and Terasaki, P. I. (1974). *Tissue Antigens* 4, 146.

Mittal, K. K., Wolski, K. P. Lim, D., Gewurz, A., Gewurz, H. and Schmid, F. R. (1976). *Tissue Antigens* 7, 97.

Miyakawa, Y., Tanigaki, N., Kreiter, V. P., Moore, G. E. and Pressman, D. (1973). *Transplantation* 15, 312.

Mogensen, B. and Kissmeyer-Nielsen, F. (1968). *Lancet* i, 721.

Mogensen, B., Kissmeyer-Nielsen, F. and Hauge, M. (1969). *Transplant. Proc.* 1, 76.

Möller, E., Link, H., Matell, G., Olhagen, B. and Stendhal, L. (1975). *In* "Histocompatibility Testing 1975" (F. Kissmeyer-Nielsen, Ed.) 778. Munksgaard, Copenhagen.

Möller, E., Hammarström, L., Smith, E. and Matell, G. (1976). *Tissue Antigens* 7, 39.

Morton, J A., Pickles, M. M. and Sutton, L. (1969). *Vox Sang.* 17, 536.

Morton, J. A., Pickles, M. M., Sutton, L. and Skov, F. (1971). *Vox Sang.* 21, 141.

Müller-Eckhardt, Ch., Heinrich, D. and Rothenberg, V. (1972). *Tissue Antigens* 2, 436.

Nakamuro, K., Tanigaki, N. and Pressman, D. (1973). *Proc. Natn. Acad. Sci. USA* 70, 2863.

Nordhagen R. and Ørjasaeter, H. (1974). *Vox Sang.* 26, 97.

Oh, S. K., Pellegrino, M. A., Ferrone, S., Sevier, E. D. and Reisfeld, R. A. (1975). *Eur. J. Immunol.* 5, 161.

Olaisen, B., Gedde-Dahl, T. Jr., Thorsby, E. (1976). *Hum. Genet.* **32**, 301.
Olaisen, B., Teisberg, P., Gedde-Dahl, T. Jr. and Thorsby, E. (1975). *Hum. Genet.* **30**, 291.
Opelz, G., Mickey, M. R. and Terasaki, P. I. (1972). *Lancet* **i**, 868.
Opelz, G., Sengar, D. P. S., Mickey, M. R. and Terasaki, P. I. (1973). *Transplant. Proc.* **5**, 253.
Opelz, G., Mickey, M. R. and Terasaki, P. I. (1974). *Transplantation* **17**, 371.
Östberg, L., Lindblom, J. B. and Peterson, P. A. (1974). *Nature, Lond.* **249**, 463.
Park, M. S. and Terasaki, P. I. (1974). *Transplantation* **18**, 520.
Payne, R. and Rolfs, M. R. (1958). *J. Clin. Invest.* **37**, 1756.
Payne, R., Radvany, R. and Grumet, F. C. (1975). *Tissue Antigens* **5**, 69.
Perkins, H. A., Kountz, S. L., Belzer, F. O., Kidd, K. K. and Payne, R. O. (1973). *Transplant. Proc.* **5**, 237.
Perrault, R. and Högman, C. F. (1971). *Vox Sang.* **20**, 356.
Piazza, A. and Morton, N. E. (1973). *Am. J. Hum. Genet.* **25**, 119.
Piazza, A. and Viganotti, C. (1973). *In* "Histocompatibility Testing 1972" (J. Dausset and J. Colombani, Eds) 731. Munksgaard, Copenhagen.
Piazza, A., Mattiuz, P. L. and Ceppellini, R. (1969). *Haematologica* **54**, 703.
Piazza, A., Belvedere, M. C., Bernoco, D., Conighi, C., Contu, L., Curtoni, E. S., Mattiuz, P. L., Mayr, W. R., Richiardi, P., Scudeller, G. and Ceppellini, R. (1973). *In* "Histocompatibility Testing 1972" (J. Dausset and J. Colombani, Eds) 73. Munksgaard, Copenhagen.
Piazza, A., Sgaramella-Zonta, L., Gluckman, P. and Cavalli-Sforza, L. L. (1975). *Tissue Antigens* **5**, 445.
Pierres, M., Fradelizi, D., Neauport-Sautes, C. and Dausset, J. (1975). *Tissue Antigens* **5**, 266.
Poulik, M. D., Bernoco, M., Bernoco, D. and Ceppellini, R. (1973). *Science* **182**, 1352.
Race, R. R. and Sanger, R. (1975). "Blood Groups in Man", 6th edn. Blackwell, Oxford, London, Edinburgh and Melbourne.
Rask, L., Östberg, L., Lindblom, B., Fernstedt, Y. and Peterson, P. A. (1974). *Transplant. Rev.* **21**, 85.
Rask, L., Lindblom, J. B. and Peterson, P. A. (1976). *Eur. J. Immunol.* **6**, 93.
Reekers, P., McShine, R. L., Boon, J. M. and Kunst, V. A. J. M. (1975). *In* "Histocompatibility Testing 1975" (F. Kissmeyer-Nielsen, Ed.) 398. Munksgaard, Copenhagen.
Reisfeld, R. A. and Kahan, B. D. (1970). *Fed. Proc.* **29**, 2034.
Renal Transplant Registry. (1975). *J. Am. Med. Assoc.* **233**, 787.
Richiardi, P., Carbonara, A. O., Mattiuz, P. L. and Ceppellini, R. (1973). *In* "Histocompatibility Testing 1972" (J Dausset and J. Colombani, Eds) 455. Munksgaard, Copenhagen.
Richiardi, P., Castagneto, M., d'Amaro, J., Schreuder, I., Vasalli, P. and Curtoni, E. S. (1974). *J. Immunogenet.* **1**, 323.
Richiardi, P., Belvedere, M. C., Castagneto, M., Fagiolo, U. and Curtoni, E. S. (1975). *In* "Histocompatibility Testing 1975" (F. Kissmeyer-Nielsen, Ed.) 270. Munksgaard, Copenhagen.
Rittner, C., Hauptmann, G., Grosse-Wilde, H., Grosshans, E., Tongio, M. M. and Mayer, S. (1975). *In* "Histocompatibility Testing 1975" (F. Kissmeyer-Nielsen, Ed.) 945. Munksgaard, Copenhagen.

Robb, R. J., Humphreys, R. E. and Strominger, J. L. (1975). *Transplantation* **19** 445.

Rood, J. J. van (1974). *Sem. Haematol.* **11**, 253.

Rood, J. J. van and Leeuwen, A. van (1963). *J. Clin. Invest.* **42**, 1382.

Rood, J. J. van, Eernisse, J. G. and Leeuwen, A. van (1958). *Nature, Lond.* **181**, 1735.

Rood, J. J. van, Leeuwen, A. van, Schippers, A. M. J., Vooys, W. H., Frederiks, E., Balner, H. and Eernisse, J. G. (1965). *In* "Histocompatibility Testing 1965", Series Haematologica **11**, 37. Munksgaard, Copenhagen.

Rood, J. J. van, Leeuwen, A. van, Schippers, A. M. J., Pearce, R., Blankenstein, M. van and Volkers, W. (1967). *In* "Histocompatibility Testing 1967" (E. S. Curtoni, P. L. Mattiuz and R. M. Tosi, Eds) 203. Munksgaard Copenhagen.

Rood, J. J. van, Leeuwen, A. van and Santen, M. C. T. van (1970). *Nature, Lond.* **226**, 366.

Rood, J. J. van, Hooff, J. P. van and Keuning, J. J. (1975a). *Transplant. Rev.* **22**, 75.

Rood, J. J. van, Leeuwen, A. van, Parlevliet, J., Termijtelen, A. and Keuning, J. J. (1975b). *In* "Histocompatibility Testing 1975" (F. Kissmeyer-Nielsen, Ed.) 629. Munksgaard, Copenhagen.

Ryder, L. P., Staub Nielsen, L. and Svejgaard, A. (1974). *Hum. Genet.* **25**, 251.

Sachs, J. A., Sterioff, S., Robinette, M., Wolf, E., Curry, H. L. F. and Festenstein, H. (1975). *Tissue Antigens* **5**, 120.

Sandberg, L., Thorsby, E., Kissmeyer-Nielsen, F. and Lindholm, A. (1970). *In* "Histocompatibility Testing 1970" (P. I. Terasaki, Ed.) 165. Munksgaard, Copenhagen.

Sanderson, A. R. and Welsh, K. I. (1973). *Transplant. Proc.* **5**, 471.

Sanderson, A. R., Cresswell, P. and Welsh, K. (1971). *Transplant. Proc.* **3**, 220.

Scalamogna, M., Mercuriali, F., Pizzi, C. and Sirchia, G. (1976). *Tissue Antigens* **7**, 125.

Scandiatransplant Report (1975). *Lancet* **i**, 240.

Schernthaner, G., Ludwig, H. and Mayr, W. R. (1976). *J. Immunogenet.* **3**, 117.

Schmid, M., Ercilla, G., Hors, J. H. and Dausset, J. (1974). *Transplantation* **17**, 427.

Schreuder, I., d'Amaro, J., Sandberg, L. and Rood, J. J. van (1975). *Tissue Antigens* **5**, 142.

Seaman, M. J., Chalmers, D. G. and Franks, D. (1968). *Vox Sang.* **15**, 25.

Shulman, N. R., Marder, V. J., Hiller, M. C. and Collier, E. M. (1964). *Progr. Haemat.* **4**, 222.

Singal, D. B., Berry, R. and Naipaul, N. (1971). *Nature New Biol.* **233**, 61.

Singer, S. J. and Nicolson, G. L. (1972). *Science* **175**, 720.

Snary, D., Goodfellow, P. N., Hayman, M. J., Bodmer, W. F. and Crumpton, M. J. (1974). *Nature, Lond.* **247**, 457.

Solheim, B. G. and Thorsby, E. (1974). *Tissue Antigens* **4**, 83.

Solheim, B. G., Bratlie, A., Sandberg, L., Staub Nielsen, L. and Thorsby, E. (1973). *Tissue Antigens* **3**, 439.

Solheim, B. G., Bratlie, A., Winther, N. and Thorsby, E. (1975). *In* "Histocompatibility Testing 1975" (F. Kissmeyer-Nielsen, Ed.) 713. Munksgaard, Copenhagen.

Solheim, B. G., Ek, J., Thune, P. O., Baklien, K., Bratlie, A., Rankin, B., Thoresen, A. B. and Thorsby, E. (1976). Tissue Antigens 7, 57.
Sonozaki, H., Seki, H., Chang, S., Okuyama, M. and Juji, T. (1975). Tissue Antigens 5, 131.
Speiser, P., Mayr, W. R., Pacher, M., Pausch, V., Bleier, I., Melzer, G., Weirather, M. and Groer, K. (1974). Vox Sang. 27, 379.
Speiser, P., Mayr, W. R., Pausch, V. and Pacher, M. (1975). Beitr. Gerichtl. Med. 33, 230.
Springer, G. F., Mittal, K. K., Terasaki, P. I., Desai, P. R., McIntire, F. C. and Hirata, A. A. (1973). Z. Immun.-Forsch. 145, 166.
Staub Nielsen, L., Ryder, L. P. and Svejgaard, A. (1975). In "Histocompatibility Testing 1975" (F. Kissmeyer-Nielsen, Ed.) 324. Munksgaard, Copenhagen.
Strominger, J. L., Cresswell, P., Grey, H. M., Humphreys, R. E., Mann, D. L., McCune, J., Parham, P., Robb, R. J., Sanderson, A. R., Springer, T. A., Terhorst, C. and Turner, M. J. (1974). Transplant. Rev. 21, 126.
Strominger, J. L., Chess, L., Herrman, H. C., Humphreys, R. E., Malenka, D., Mann, D. L., McCune, J. M., Parham, P., Robb, R. J., Springer, T. A. and Terhorst, C. (1975). In "Histocompatibility Testing 1975" (F. Kissmeyer-Nielsen, Ed.) 719. Munksgaard, Copenhagen.
Svejgaard, A. (1969a). "Iso-antigenic Systems of human Blood Platelets. A Survey". Series Haematologica II, 3. Munksgaard, Copenhagen.
Svejgaard, A. (1969b). Vox Sang. 17, 112.
Svejgaard, A. and Kissmeyer-Nielsen, F. (1968). Nature, Lond. 219, 868.
Svejgaard, A., Thorsby, E., Hauge, M. and Kissmeyer-Nielsen, F. (1970). Vox Sang. 18, 97.
Svejgaard, A., Hauge, M., Kissmeyer-Nielsen, F. and Thorsby, E. (1971). Tissue Antigens 1, 184.
Svejgaard, A., Staub Nielsen, L., Ryder, L. P., Kissmeyer-Nielsen, F., Sandberg, L., Lindholm, A. and Thorsby, E. (1973). In "Histocompatibility Testing 1972" (J. Dausset and J. Colombani, Eds) 465. Munksgaard, Copenhagen.
Svejgaard, A., Platz, P. J., Ryder, L. P., Staub Nielsen, L. and Thomsen, M. (1975). Transplant. Rev. 22, 3.
Table of Equivalent Nomenclature (1973). In "Histocompatibility Testing 1972" (J. Dausset and J. Colombani, Eds) 7. Munksgaard, Copenhagen.
Tanigaki, N. and Pressman, D. (1974). Transplant. Rev. 21, 15.
Tanigaki, N., Nakamuro, K., Appella, E., Poulik, M. D. and Pressman, D. (1973). Biochem. Biophys. Res. Comm. 55, 1234.
Tanigaki, N., Nakamuro, K., Natori, T., Kreiter, V. P. and Pressman, D. (1974). Transplantation 18, 74.
Tatra, G. and Wolf, E. (1973). Arch. Gynäk. 213, 341.
Terasaki, P. I., Opelz, G., Park, M. S. and Mickey, M. R. (1975). In "Histocompatibility Testing 1975" (F. Kissmeyer-Nielsen, Ed.) 657. Munksgaard, Copenhagen.
Thieme, T. R., Raley, R. A. and Fahey, J. L. (1974). J. Immunol. 113, 323.
Thierfelder, S. (1968). Klin. Wschr. 46, 1.
Thomsen, M., Platz, P. J., Ortved Andersen, O., Christy, M., Lyngsöe, J., Nerup, J., Rasmussen, K., Ryder, L. P., Staub Nielsen, L. and Svejgaard, A. (1975). Transplant. Rev. 22, 125.
Thomsen, M., Platz, P. J., Marks, J. F., Ryder, L. P., Shuster, S., Svejgaard, A. and Young, S. H. (1976). Tissue Antigens 7, 60.

Thorsby, E. (1974). *Transplant. Rev.* **18**, 51.

Thorsby, E. and Piazza, A. (1975). *In* "Histocompatibility Testing 1975" (F. Kissmeyer-Nielsen, Ed.) 414. Munksgaard, Copenhagen.

Thorsby, E., Bratlie, A., Mayr, W. R., Spärck, J., Svejgaard, A., Kjerbye, K. E. and Kissmeyer-Nielsen, F. (1971). *Tissue Antigens* **1**, 32.

Thorsby, E., Segaard, E., Solem, J. H. and Kornstad, L. (1975). *Tissue Antigens* **6**, 54.

Tilley, C. A., Crookston, M. C., Brown, B. L. and Wherett, J. R. (1975). *Vox Sang.* **28**, 25.

Tongio, M. M. and Mayer, S. (1974). *Transplantation* **18**, 163.

Transplantation Reviews Vol. 21. "β_2-Microglobulin and HL-A Antigens" (1974). Munksgaard, Copenhagen.

Truog, P., Steiger, U., Contu, L., Galfré, G., Trucco, M., Bernoco, D., Bernoco, M., Birgen, I., Dolivo, P. and Ceppellini, R. (1975). *In* "Histocompatibility Testing 1975" (F. Kissmeyer-Nielsen, Ed.) 788. Munksgaard, Copenhagen.

Tweel, J. G. van den, Blussé van Oud Alblas, A., Keuning, J. J., Goulmy, E., Termijtelen, A., Bach, M. L. and Rood, J. J. van (1973). *Transplant. Proc.* **5**, 1535.

Vladutiu, A. O. and Rose, N. R. (1974). *Immunogenetics* **1**, 305.

Walford, R. L., Gallagher, R. and Sjaarda, J. R. (1964). *Science* **144**, 868.

Walford, R. L., Gallagher, R. and Troup, G. M. (1965). *Transplantation* **3**, 387.

Walford, R. L., Colombani, J. and Dausset, J. (1969). *Transplantation* **7**, 188.

Walford, R. L., Gossett, T., Smith, G. S., Zeller, E. and Wilkinson, J. (1975). *Tissue Antigens* **5**, 196.

Weerdt, Ch. M. van der and Lalezari, P. (1972). *Vox Sang.* **22**, 438.

Weitkamp, L. R. (1976). *Tissue Antigens* **7**, 273.

Weitkamp, L. R., Rood, J. J. van, Thorsby, E., Bias, W. B., Fotino, M., Lawler, S. D., Dausset, J., Mayr, W. R., Bodmer, J., Ward, F. E., Seignalet, J., Payne, R. O., Kissmeyer-Nielsen, F., Gatti, R. A., Sachs, J. A. and Lamm, L. U. (1973). *Hum. Hered.* **23**, 197.

Weitkamp, L. R., May, A. G. and Johnston, E. (1975a). *Hum. Hered.* **25**, 337.

Weitkamp, L. R., Townes, P. L. and May, A. G. (1975b). *Am. J. Hum. Genet.* **27**, 486.

Wernet, P. (1976). *Transplant. Rev.* **30**, 271.

Wernet, P. and Kunkel, H. G. (1975). *In* "Histocompatibility Testing 1975" (F. Kissmeyer-Nielsen, Ed.) 731. Munksgaard, Copenhagen.

Wernet, P., Rieber, E. P., Winchester, R. J. and Kunkel, H. G. (1975a). *In* "Histocompatibility Testing 1975" (F. Kissmeyer-Nielsen, Ed.) 647. Munksgaard, Copenhagen.

Wernet, P., Winchester, R. J., Dupont, B. and Kunkel, H. G. (1975b). *In* "Histocompatibility Testing 1975" (F. Kissmeyer-Nielsen, Ed.) 637. Munksgaard, Copenhagen.

White, A. G., da Costa, A. J. and Darg, C. (1973). *Tissue Antigens* **3**, 123.

WHO-IUIS Terminology Committee (1975). *In* "Histocompatibility Testing 1975" (F. Kissmeyer-Nielsen, Ed.) 5. Munksgaard, Copenhagen.

Winchester, R. J., Dupont, B., Wernet, P., Fu, S. M., Hansen, J. A., Laursen, N. and Kunkel, H. G. (1975a). *In* "Histocompatibility Testing 1975" (F. Kissmeyer-Nielsen, Ed.) 651. Munksgaard, Copenhagen.

Winchester, R. J., Ebers, G., Fu, S. M., Espinosa, L., Zabriskie, J. and Kunkel, H. G. (1975b). *Lancet* **ii**, 814.

420 W. R. MAYR

Wolski, K. P., Schmid, F. R. and Mittal, K. K. (1975). *Science* **188**, 1020.
Wolski, K. P., Schmid, F. R. and Mittal, K. K. (1976). *Tissue Antigens* **7**, 35.
Workshop Data (1967). *In* "Histocompatibility Testing 1967" (E. S. Curtoni, P. L. Mattiuz and R. M. Tosi, Eds) 435. Munksgaard, Copenhagen.
Yasuda, N. and Kimura, M. (1968). *Ann. Hum. Genet.* **31**, 409.
Yunis, E. J., Seigler, H. F., Simmons, R. L. and Amos, D. B. (1973). *Transplantation* **15**, 435.

ADDENDUM

The analysis of a family segregating for a reciprocal translocation between the chromosomes 6 and 20, t(6; 20) (p21; p13) (Breuning *et al.* (1977), *Hum. Genet.* **37**, 131), and of cell hybrids between Chinese hamster cells and human fibroblasts with a balanced reciprocal translocation between the chromosomes 1 and 6, t(1; 6) (p32; p21) (Francke and Pellegrino (1977), *Proc. Natl. Acad. Sci. USA* **74**, 1147), showed that the HLA gene complex is localized on the short arm of chromosome 6 (near the transition of band 6p21 to 6p22). The sequence of the loci on 6p is most probably: centromere::: PGM_3::: GLO::: HLA–D:HLA–B:HLA–C: HLA–A.

During the last meeting of the HLA Nomenclature Committee ("Histocompatibility Testing 1977", W. F. Bodmer *et al.*, Eds, Munksgaard, Copenhagen, in press), the following changes of the HLA nomenclature were agreed upon:
(a) HLA–A, B, C:

New Representative equivalents		New Representative equivalents	
HLA–A25	HLA–AW25	HLA–BW46	HS, SIN2
HLA–A26	HLA–AW26	HLA–BW47	407*, MO66, CAS, Bw40C
HLA–B15	HLA–BW15	HLA–BW48	KSO, JA, Bw40.3
HLA–B17	HLA–BW17	HLA–BW49	Bw21.1, SL–ET
HLA–B37	HLA–BW37	HLA–BW50	Bw21.2, ET*
HLA–B40	HLA–BW40	HLA–BW51	B5.1
HLA–BW4	W4, 4a	HLA–BW52	B5.2
HLA–BW6	W6, 4b	HLA–BW53	HR
HLA–BW44	B12 (not TT*), To50	HLA–BW54	Bw22J, SAP1
HLA–BW45	TT*, To51	HLA–CW6	T7

(b) HLA–D: five new provisional specificities have been ascertained: HLA–DW7 (=LD107), HLA–DW8 (=LD108), HLA–DW9 (=TB9, OH), HLA–DW10 (=LD16) and HLA–DW11 (LD–17).
(c) Ia type antigens: the Ia type antigens closely associated with the already defined specificities of *HLA–D* are called DR determinants (DR = HLA–D related). The first seven HLA–DR specificities, which are yet provisional, are DRW1–DRW7. The numbers correspond to the relevant highly associated DW specificities.

9 Uses of Polymorphism

O. MAYO and D. J. H. BROCK

Biometry Section, Waite Agricultural Research Institute, Glen Osmond, South Australia and Department of Human Genetics, University of Edinburgh, Scotland

I. INTRODUCTION

In Chapter 4, the nature and extent of human polymorphism were briefly outlined, and the relationship of genetical variation to evolution was considered. In the four chapters that followed, the most important types of polymorphism were described in some detail. Attempts were made to show both the extraordinary diversity of human genetical and biochemical

422   O. MAYO AND D. J. H. BROCK

variation and also its importance in health and disease. However, for the most part, information about how polymorphism may be used was not specifically illustrated. In this chapter, we attempt to draw together studies in which polymorphic systems have been used to solve particular practical problems.

In many areas, such as seeking compatibility between donor and recipient for blood transfusion or tissue transplantation, the critical and immediate importance of knowledge of the various polymorphic systems should be by now apparent. However, there are further important areas where the relationship is not so direct, and still others where the polymorphisms themselves are of secondary importance in the process of interest. Thus, in paternity testing or establishing zygosity of twins it does not matter which loci are used, provided that the results are unequivocal. Overall, there are three main areas which we wish to emphasize: basic physiological investigations, such as determining the origin of normal cells or tumours, the relationship of genes and isozymes, or the time of X-chromosome inactivation; practical genetical problems, such as counselling, resolution of genetic heterogeneity and antenatal diagnosis; and population genetical problems, such as the association between polymorphism and disease. Many of the techniques and problems mentioned here have of course been referred to or used in earlier chapters, and will be used again later in the book, but here we collect together most of the applications of the knowledge of variation which we regard as important. We cannot pretend that a chapter on the uses of polymorphism has a logical structure, but we regard it as necessary nonetheless.

II. PHYSIOLOGICAL INVESTIGATIONS

A. Genes and Isozymes

Widespread use of starch gel electrophoresis followed by selective staining has shown that many enzymes exist in multiple molecular forms or isozymes, differing in molecular structure but not in functional activity. Multiple forms of a given activity-defined enzyme may be generated: (1) by control of the enzyme by more than one genetic locus, (2) through multiple allelism at a single genetic locus and (3) through secondary modifications of the enzyme which are not genetical in character. The existence of polymorphic markers is of great importance in resolving the reasons for the frequently found complex patterns of isozyme bands observed in electrophoretograms.

The classic case of a polymorphism revealing an enzyme controlled by

Table 1

Enzymes where multiple gene control has been revealed by polymorphism

Enzyme	Loci	Polymorphic loci	Reference
Phosphoglucomutase	PGM_1 PGM_2 PGM_3	PGM_1 PGM_3	Spencer et al. (1964) Hopkinson and Harris (1968)
Alcohol dehydrogenase	ADH_1 ADH_3 ADH_3	ADH_3	Smith et al. (1971, 1973)
Amylase	AMY_1 AMY_2	AMY_1 AMY_2	Kamaryt and Laxova (1965) Ward et al. (1971)
Serum cholinesterase	E_1 E_2	E_1 E_2	Kalow and Stavron (1957) Harris et al. (1962)
Carbonic anhydrase	CA_1 CA_2	CA_2	Moore et al. (1971)
Superoxide dismurate (indophenoloxidase)	SOD_A SOD_B	SOD_A	Beckman (1973)
Malate dehydrogenase	MDH_S MDH_M	MDH_M	Davidson and Cortner (1967)
Glutamate oxaloacetate transaminase	GOT_S GOT_M	GOT	Davidson et al. (1970)
Glutamate pyruvate transaminase	GPT_S GPT_M	GPT_S	Chen and Giblett (1971)
Malic enzyme	MOD_S MOD_M	MOD_M	Cohen and Omenn (1972)

multiple loci comes from studies on phosphoglucomutase (PGM). Three separate loci are responsible for the various isozyme bands seen on starch gel electrophoretograms (Chapter 5). At each locus multiple alleles have been observed, though only the PGM_1 and PGM_3 loci show polymorphic variation. The isozymes of each of the three loci are expressed in most tissues, with the exception of a limited expression of PGM_3 in red cells and skeletal muscle. Many other enzymes have now been shown to be controlled by two or more gene loci (Table 1), and presumably there are others where the absence of polymorphism has not allowed multiple gene control to be discerned. Of particular interest in Table 1 are those enzymes of which malate dehydrogenase (MDH) is the prototype (Davidson and Cortner, 1967), where separate loci control the expression of mitochondrial and cytoplasmic forms of the enzyme. Since mitochondria contain a moderate amount of DNA one may ask whether mitochondrial enzymes are under the control of nuclear or mitochondrial DNA. In Drosophila and some other insects a separate genetical system transmitted by cytoplasmic inheritance through the maternal parent controls mitochondrial enzymes (Sager, 1964). In Man, however, the mitochondrial forms of MDH, glutamate-oxalate transaminase and malic enzyme are under the control of nuclear genes (Davidson and Cortner, 1967; Davidson *et al.*, 1970; Cohen and Omenn, 1972).

Also of interest are the enzymes whose controlling genes are subject to developmental changes. The original and best-known example of this is LDH, where the relative expression of the *A* and *B* loci depends not only on the stage in development but the tissue in which the enzyme is being examined. A recent and even more complex example of the interaction between multi-gene enzymes, development stage and tissue expression is given by the alcohol dehydrogenase system. Here three different polypeptides controlled by separate loci appear at different stages in different tissues. Thus in liver there is a sequential switching on of loci, starting with the α-chain-synthesizing ADH_1 locus in early foetuses and working through the β-chain ADH_2 locus to the γ-chain controlling ADH_3 locus which is expressed during neonatal life. In lung, however, the ADH_2 locus predominates throughout development, while in gastrointestinal mucosa it is the ADH_3 locus. Added to this is a polymorphism resulting in two common alleles at the ADH_3 locus (Smith *et al.*, 1971, 1973).

In addition to the genetical control of isozyme expression there are occasional non-genetical factors contributing to the complexity of electrophoretic patterns. For example the enzyme purine nucleoside phosphorylase shows an extraordinarily complicated electrophoretic pattern in red cell lysates. Density centrifugation of red cells before haemolysis shows that this is related to the age of the cells, with older and heavier cells showing many more bands than younger and lighter cells (Edwards *et al.*, 1971). It is assumed that there is one polypeptide under direct control of the relevant gene, which is the basic unit from which other bands are generated by secondary changes. Another example where polymorphism has contributed to the clarification of complex isozyme patterns comes from the adenylate kinase (AK) polymorphism. Here a complex set of bands is observed in red cell lysates; an equally complex but slightly different set of bands is observed in homogenates from other tissues. At first sight it would appear that the enzyme in red cells is controlled by a different gene locus from that in other tissues. However, it has been shown that when the phenotype of AK changes in red cells it also changes in solid tissues even though the detailed isozyme pattern remains slightly different. In this case it has been established that the secondary modification is due to the presence of haemoglobin, which binds selectively with AK and disturbs its electrophoretic migration in blood-containing tissues (Brock, 1970). (The different patterns in solid tissues relate to the existence of two other non-polymorphic *AK* loci; see Chapter 5 and Wilson *et al.*, 1976.)

B. Origin of Cells

1. Normal cells

The mosaic composition of female tissues with respect to characters coded for by genes carried on the X-chromosome may be used in tracing the developmental origins of particular cellular types. Gandini *et al.* (1968) were the first to point out that if a woman were a carrier of G6PD deficiency the enzyme activities of her tissues would be similar if they were derived from a common precursor pool and dissimilar if derived from separate precursor pools. They examined G6PD activity in four women selected because their erythrocyte enzyme activity was in the deficient range and showed that in these patients granulocyte G6PD was also in the deficient range, arguing for a common primordial pool for these two cell types. A more sophisticated approach to correlation of mosaic expression (Gandini and Gartler, 1969) used measurement of the A:B isozyme ratios in G6PD A/B heterozygotes and was able to demonstrate with some certainty that not only erythrocytes and granulocytes but also lymphocytes share a common precursor pool. More recently Fialkow (1973) has extended these studies to lymph node, skin and skeletal muscle. When tissues from 42 A/B heterozygotes were examined, correlation coefficients approaching 1·0 were found between the isozyme ratios in the different nucleated cell types, and lower but still statistically significant correlations between red cells and nucleated cells. When tissues from 27 A⁻/B heterozygotes (carriers of the Negro deficient gene) were examined the correlations in A:B ratio were much less impressive. It seems likely that this difference could be attributed to selection against the deficient phenotype, known to occur in HGPRT deficiency, and that a conclusion of a common precursor pool for the six cellular types is valid.

2. Tumours

In theory neoplastic cells may arise from an initial precipitating event in a single precursor cell or from a series of such events in a population of susceptible cells. The former would mean that tumours had a unicellular or clonal origin, the latter that they were multicellular in derivation. The idea of using the polymorphism of a protein coded for by the X-chromosome in

MAN—R *

resolving this problem was first suggested by Linder and Gartler (1965) in their study of uterine leiomyomas. They pointed out that because of X-chromosome inactivation tumours from a woman heterozygous for the G6PD A/B polymorphism would express both enzyme forms if cells were multicellular in derivation and only one form if they were unicellular in derivation. Careful dissection of 184 tumours in 25 heterozygous females showed only a single enzyme phenotype, while the surrounding myometrium expressed both A and B forms (Linder, 1969; Fialkow, 1972). A single tumour had both A and B forms, but was reported to contain a large amount of non-tumour tissue (Fialkow, 1972). The ratio of A:B forms in the tumours was close to 1:1, suggesting a random single-cell-directed primary event and arguing against the alternative multicellular origin followed by selection for cells containing one particular type of active X-chromosome.

The uterine leiomyoma is a particularly suitable tumour for this type of study. In addition to being relatively common it is usually well-demarcated from the surrounding normal myometrium, and the two adjacent tissues can thus be studied at the same time (Fialkow, 1972). Many other solid invasive malignancies contain considerable quantities of non-tumour cells which makes interpretation of double enzyme phenotypes difficult. Thus Beutler et al. (1967) examined 28 metastatic nodules from a patient with a primary carcinoma of the colon and found seven with equal amounts of A and B enzyme and 21 with predominantly A or B enzyme. A conclusion of multi-cellular tumour origin was cautiously reached. Neoplasms of the haematopoietic system are more easily studied, since tissue is accessible and easily purified and can be repeatedly sampled. Thus both chronic myelo-cytic leukaemia and Burkitt's lymphoma appear to be definitely clonal in origin (Fialkow et al., 1967, 1970).

In a majority of the neoplasms studied to date the G6PD A/B poly-morphism has been used. Most seem to have a unicellular origin (Table 2). At first sight this suggests somatic mutation followed by proliferation of the transformed cell. However, both Burkitt's lymphoma and the common wart, verruca vulgaris, are believed to have a viral aetiology and yet they, too, appear clonally derived. Presumably the interpretation here is of integration of the viral genome into a target cell to produce a transformed cell capable of rapid replication, rather than infection of a whole patch of cells. Among the tumours of multicellular origin it is noteworthy that both trichoepithelioma and neurofibroma are inherited as autosomal dominants. It is suggested that in these cases all cells in the tissue of origin have "innate tumorigenic potential" (Fialkow, 1972; see also Chapter 14).

Another rather different type of tumour, the benign ovarian teratoma or dermoid cyst, has been investigated by Linder and Power (1970). They examined three autosomal loci, PGM_1, PGM_3 and $6PGD$, and one X-linked locus, $G6PD$, in tissues from ovarian cysts, and found that when the woman was a heterozygote, a majority of the tumour phenotypes showed homozygous expression. Origin of tumours from a single cell appears almost certain, but more importantly since gene suppression can be ruled out for reasons not relevant to the present discussion, it seems likely that these teratomas arise from a single germ cell which has undergone some stages of meiosis, with at least the reduction division (meiosis I) having occurred. Linder and Power's tentative conclusion was that the teratoma arises after meiosis I, followed either by suppression of meiosis II, suppression of the second polar body or re-entry of the second polar body.

Table 2

The use of G6PD polymorphism in distinguishing the cellular origin of tumours

Tumour	Cellular origin	Reference
Uterine leiomyoma	unicellular	Linder and Gartler (1965)
Chronic myelocytic leukaemia	unicellular	Fialkow et al. (1967)
Burkitt's lymphoma	unicellular	Fialkow et al. (1970)
Lymphosarcoma	unicellular	Beutler et al. (1967)
Multiple myeloma	unicellular	McCurdy (1968)
Carcinoma of cervix	unicellular	Park and Jones (1968)
Common wart	unicellular	Murray et al. (1971)
Carcinoma of the colon	multicellular	Beutler et al. (1967)
Carcinoma of the breast	multicellular	McCurdy (1967)
Carcinoma of the nasopharynx	multicellular	Fialkow (1972)
Multiple trichoepitheliomas	multicellular	Gartler et al. (1955)
Multiple neurofibromas	multicellular	Fialkow (1972)

Given this conclusion, which has been supported by later results (Ott et al., 1976), it is possible to map the position of gene loci relative to their centromeres. This is because the frequency of heterozygous tumours of heterozygous hosts is equivalent to the frequency of second (meiotic) division segregation, which has been used for many years in centromere mapping in fungi (where ordered products of meiosis can be recovered). See Barratt et al., 1954, for review. Ott et al. (1976) give in detail the method of estimation of distance from the centromere using teratoma phenotypes; new results has so far been very few, but the method has distinct promise.

Though the atherosclerotic plaque is not a tumour, its origins have also been investigated by using the G6PD polymorphism. Surprisingly, 30 plaques from four patients appeared in most cases to have a unicellular origin (Benditt and Benditt, 1973). This seems to rule out the formation of fibrous tissue by a process of wound-healing, since this is believed to occur by simultaneous proliferation of cells from a number of margins. Selection against one or other of the regenerating cell types appears unlikely in view of the occurrence of both A and B isozymes in different plaques. Benditt and Benditt (1973) argue that the primary event in the formation of an atheromatous lesion is the transformation of a cell either by somatic mutation or viral invasion, and that elevated concentrations of blood lipids merely accelerate the proliferation of a clone of mutant cells. Whether this interpretation is correct or not, the use of polymorphic markers in investigating atherosclerosis opens up new possibilities which have not been previously considered.

C. Chromosomal Inactivation

1. X-chromosomes

The essential premise of the Lyon hypothesis is that in somatic cells of mammalian females only one of the two X-chromosomes is genetically active (Lyon, 1961). Inactivation takes place early in embryonic life in an apparently random fashion so that either paternally-derived or maternally-derived X-chromosomes can be affected. Thereafter all descendant cells faithfully retain in the inactive configuration the same X-chromosome which has been marked out by this early and mysterious event. Consequently female tissues must be regarded as mosaics with regard to characters coded for by genes carried by the X-chromosome. Naturally this mosaicism is only observable when maternally- and paternally-derived X-chromosomes are non-identical, that is when suitable genetic markers are present.

Abundant physiological and biochemical evidence for the Lyon hypothesis has now been presented, though the mechanism of inactivation remains obscure (Lyon, 1971). In man the most convincing studies come from the use of X-linked polymorphic enzymes. Using the G6PD A/B polymorphism as marker, Davidson et al. (1963) cultured skin fibroblasts from heterozygous women, and by cloning techniques derived cell populations which expressed only one of the two enzyme forms. No clones showed both forms of the enzyme nor was there any indication of the A/B heteropolymer which may be seen in somatic cell hybrids of A-type and B-type cells (Migeon et al., 1974). Similar studies have been performed on cloned fibroblasts from heterozygous carriers of various X-linked recessive conditions (Table 3), using both enzyme assays and cellular indicators of abnormal metabolism. Indeed a demonstration of cellular mosaicism in a presumed carrier of an X-linked mutant is regarded as the definitive evidence of heterozygous status (Chapter 10). So widely is the Lyon hypothesis now accepted that it is used to explain females who manifest symptoms of known X-linked recessives, such as Duchenne muscular dystrophy (Emery, 1963) and Christmas disease (Revesz et al., 1972). It is assumed that in these women a large majority of normal X-chromosomes were inactivated during embryogenesis.

Table 3

Fibroblast cloning to demonstrate mosaicism in heterozygous carriers of X-linked recessives

Disorder	Mosaic character in clones	Reference
G6PD deficiency	G6PD	Davidson et al. (1963)
Lesch-Nyhan syndrome	HGPRT	Migeon et al. (1968)
Fabry's disease	α-galactosidase	Romeo and Migeon (1970)
Phosphoglycerate kinase deficiency	phosphoglycerate kinase	Deys et al. (1972)
Hunter's syndrome	metachromasia	Danes and Bearn (1967)
Chronic granulomatous disease	nitroblue tetrazolium staining	Windhorst et al. (1967)

Stability of inactivation. The stability of the inactivated X-chromosome through many rounds of mitotic divisions appears remarkable. Various attempts have been made to reactivate this chromosome *in vitro* using systems of cultured skin fibroblasts and searching for protein products by analysing for X-linked genetical variants. Comings (1966) grew fibroblasts from human uterine leiomyomas, which are known to be clonal in origin and which express only one form of G6PD in women heterozygous for the A/B polymorphism, but after exposing the cultures to physical and chemical agents capable of stimulating RNA synthesis, failed to find evidence of reactivation. Romeo and Migeon (1975) transformed clones of fibroblasts from a G6PD A/B heterozygote with SV40 virus to obtain heteroploid cultures, but these continued to express only the single phenotype of the original clone. Migeon (1972), using the observation that interspecific somatic cell hybridization may sometimes induce the formation of gene products not found in the parental cell, fused mouse cells with cloned fibroblasts from a G6PD A/B heterozygote expressing only the A phenotype. The resulting heterokaryons, however, showed only the G6PD electrophoretic patterns of the mouse, the human–murine heteropolymer and the human A-type isozyme.

The only evidence to date of a partial reactivation of an inactive human X-chromosome comes from the work of Kahan and DeMars (1975). They started with a human female fibroblast line in which one of the X-chromosomes carried a translocated segment of the long arm of chromosome No. 1 and was therefore cytologically distinctive. This line was heterozygous for HGPRT deficiency as a consequence of a mutation on the abnormal X-chromosome. Cultures of this line had the normal X-chromosome inactivated and were therefore HGPRT deficient. Fusion of these cells with a mouse cell line also deficient in HGPRT and selection of the resulting heterokaryons in a medium in which survival depended on the presence of cellular HGPRT activity produced a low yield of human–mouse hybrids. The hybrids expressed a HGPRT isozyme which was by immunological criteria human and had a morphologically normal human X-chromosome. This was taken as evidence of the reactivation of the X-chromosome demonstrated to be inactivated in the parental fibroblasts. However, it appeared to be a partial activation since no evidence could be found in the hybrids for two other known X-linked human enzymes, G6PD and phosphoglycerate kinase.

Special case of the Xg blood group. As discussed in Chapter 7, a red cell antigen, coded for by a gene carried on the X-chromosome, was first recognized by Mann *et al.* in 1962. It is weakly immunogenic, for although some 20% of blood transfusions should allow the recipient to make antibodies, only a handful of suitable antisera have been discovered (Race and Sanger, 1975). Of the two observed phenotypes, Xg (a$^+$) and Xg (a$^-$), the former is defined by anti-Xg^a antiserum and has dominant expression, while the latter is presumed to represent homozygosity for a silent allele, *Xg*. In European populations, some 66% of males and 89% of females are Xg (a$^+$), making this an extremely useful polymorphism. The antigen cannot be detected in serum or saliva but has been observed in cultured fibroblasts and short-term cultured lymphoid cell lines (Fellous *et al.*, 1974).

For some time controversy has raged over whether the *Xg* locus, like all other X-linked loci, is inactivated in human females. Gorman *et al.* (1963) devised a technique for demonstrating Xg (a$^+$) and Xg (a$^-$) cells in an artificial 1:1 mixture of red cells, but found that in heterozygous females all cells typed as Xg (a$^+$), suggesting either that both X-chromosomes were active in each erythrocyte precursor or that the antigen was synthesized elsewhere and taken up passively by the cell surface. The passive adsorption theory of Xg expression seems unlikely in view of the discovery of twin chimeras with separable mixtures of Xg (a$^+$) and Xg (a$^-$) cells in the same circulations (Ducos *et al.*, 1971). However, the alternative explanation of two active *Xg* loci received a set-back with the discovery in Salt Lake City of two sisters heterozygous for X-linked sideroachrestic anaemia, each of whom had two morphologically distinct populations of red cells (Lee et al., 1968). One population was microcytic and typed as Xg (a⁺), the other normal and typed as Xg (a⁻). This seemed to show a normal process of X-chromosome inactivation. However, a similar family examined by Weatherall et al. (1970) behaved quite differently, and recently Race and Sanger (1975) re-examined the Salt Lake City sisters and failed to confirm the original observations. The bulk of the evidence (Race and Sanger, 1975) now points firmly to the exclusion of the red cell Xg locus from the usual process of random early inactivation. In confirmation of this Fellous et al. (1974) demonstrated that in fibroblast cultures from a known Xg^aXg heterozygote, all the isolated clones typed as Xg (a⁺).

In contrast to this is the situation obtained when one of the X-chromosomes is structurally abnormal. Polani et al. (1970) pointed out that when the short arm of the X-chromosome is deleted some daughters of Xg (a⁺) fathers typed as Xg (a⁻), suggesting that the Xg locus is carried on the short arm of the X-chromosome. However, when the long arm of the X-chromosome was deleted a similar situation was found, pointing to localization of the Xg locus on the long arm. The only way of reconciling these differences is to assume that in the known preferential inactivation of abnormal X-chromosomes, the Xg locus is also involved.

Time of inactivation. In early human embryos studied by Park (1957) significant amounts of sex chromatin were observed at 32 days after the last menstrual period or about 18 days after fertilization. This suggested that X-inactivation might be occurring before the end of the third week of embryonic life, even though the correspondence between inactivation and the appearance of sex chromatin is not necessarily an exact one. Migeon and Kennedy (1975) examined the G6PD isozyme patterns in a variety of tissues from human embryos heterozygous for the A/B polymorphism and concluded that X-inactivation was complete and irreversible by at most the 5th week after conception. On the other hand, Nadler (1968) claimed to have observed twice the expected amount of G6PD activity in amniotic fluid cells cultured from a female foetus at 10 weeks gestation and suggested that this represented a pre-inactivation state. In the mouse creation of artificial chimaeras by injection of single cells into the blastocyst shows that inactivation of the X-chromosome is occurring at $4\frac{1}{2}$ days after ovulation (Gardner and Lyon, 1971), which suggests well before the 5th week in humans. Apart from Nadler's observations, which have not been confirmed, the evidence points to an early inactivation of the human X-chromosome. On the other hand, it is known that both X-chromosomes are active in the human oocyte (Gartler et al., 1972, 1973).

Since it is virtually impossible to obtain human embryonic material much before three weeks after conception, the study of the precise time of X-inactivation has been necessarily indirect. The general approach has been to examine the ratio of isozymes in the same tissue from a number of women known to be heterozygous for the G6PD A/B polymorphism. It is argued that if the tissue is derived from a single cell after X-inactivation, only one of the two possible isozymes will be found. If there were two progenitor cells, approximately one-quarter of the women would type as A, one-quarter would type as B and half as AB. If there were n cells at the time of inactivation, then $(\frac{1}{2})^n$ of the women would show hemizygous expression of the A isozyme, a similar number would be apparent hemizygotes for B and the remainder would type as heterozygotes. Obviously this approach has severe limitations if primordial pool size is larger than six, when only two out of 64 heterozygotes would have hemizygous manifestation. To circumvent this difficulty, Fialkow (1973) has measured the ratio of A to B isozymes in the same tissue (granulocytes) of a number of heterozygotes, making the assumption that if primordial pool size were large and there were no selection *in vivo* against either of the isozymes, a 1:1 ratio of isozymes would be found. Considerable deviations from the expected equivalence ratio were found and these were used to calculate a precursor pool of between 10 and 25 cells with a mean of 16. A similar approach using red cells made earlier by Nance (1964) had suggested X-inactivation occurred in a precursor pool of 18 cells. Small errors in the estimation of the A to B isozyme ratio can obviously affect the calculation considerably. On the other hand, the alternative approach used by Gandini *et al.* (1968) of calculating the number of heterozygotes for G6PD deficiency who show hemizygous expression of the normal allele is prone to error if the mutant allele is selected against *in vivo*. Such selection is known to occur in females heterozygous for HGPRT deficiency (Nyhan *et al.*, 1970).

2. Autosomes

The discovery that only one of the two X-chromosomes in human female cells produces protein products raises the question of whether members of any of the other 22 pairs of human chromosomes are similarly inactivated. Apart from obvious situations, such as partial chromosome deletions or where there are genetical indications of a silent allele, the only evidence of autosomal "allelic exclusion" comes from a study of the immunoglobulin loci. Here it seems fairly certain that in any given cell type only one of each pair of immunoglobulin alleles is active, while there is a further limitation to a single kind of heavy chain and a single kind of light chain.

The evidence for allelic exclusion within the immune system comes from three main sources. Immunofluorescence studies on the Gm phenotypes of IgG-secreting cells in human lymphoid tissue show only one genetic type per cell in heterozygous individuals (Curtain and Baumgarten, 1966;

Curtain and Golab, 1966). The myeloma protein of IgG sub-class in an individual heterozygous for Gm(1) and Gm(4) contains either Gm(1) or Gm(4) specificity, but not both (Fudenberg and Warner, 1970). And in Rh isoimmunization, the anti-Rh antibody has been shown to be predominantly of one Gm type (Martensson, 1964) as have antibodies to polysaccharide antigens (Fudenberg *et al.*, 1972). Although several ingenious hypotheses have been advanced to explain these findings the mechanism of this limited type of autosomal inactivation remains obscure. It is of some interest that in long-term cultured human lymphocytoid lines which synthesize IgG and IgM genic exclusion has not been observed (Bloom *et al.*, 1971).

D. Somatic Cell Hybridization

The fusion of mammalian cells *in vitro* to produce a hybrid with a single nucleus capable of further division was first described by Barski *et al.* in 1960. The importance of this discovery for somatic cell genetics was emphasized by the work of Ephrussi and Weiss (1965) who showed that fusion could occur between the cells of different species. At the same time Harris and Watkins (1965) introduced the use of inactivated Sendai virus to promote the wholesale formation of hybrids in mixed cell cultures. To date most of the applications of somatic cell hybridization in human genetics have utilized interspecific hybrids, particularly those between man and mouse and man and Chinese hamster, rather than intraspecific hybrids between different human lines. This has followed from the observation that in the interspecific heterokaryon human chromosomes are preferentially shed (Weiss and Green, 1967; Kao and Puck, 1970), so that the loss or retention of human gene products may be correlated with loss or retention of human chromosomes. Many of the linkage relationships and chromosome assignments in Table 5 have been derived by this method and somatic cell hybridization has been an invaluable complement to classical genetic procedures in linkage studies. Whereas classical segregation analysis depends on polymorphism and the occurrence of rare variants for detecting linkage (because it is by its nature intraspecific), the use of interspecific cell hybrids does not. The chromosomes of mouse and man are sufficiently different to be identifiable under the light microscope while the gene products are usually chosen because of characteristic electrophoretic or immunological differences. Thus in an ideal situation, where a man–mouse heterokaryon contained only a single residual human chromosome pair, an enzyme synthesized by the hybrid cell which had the expected electrophoretic mobility of the human prototype could be assigned unambiguously to the residual human chromosome. The enzyme thymidine kinase was located on chromosome 17 in this way (Miller *et al.*, 1971). This ideal situation is rarely observed in practice; usually linkage assignments are made by analysing a series of hybrid lines of different karyotype and correlating the concordant loss or retention of human enzymes and human chromosomes. Such analysis of course measures synteny (the presence of genes on the same chromosome) rather than true genetical linkage.

1. Intraspecific hybrids

Intraspecific heterokaryons, formed by the fusion of two different human cell lines, were first reported almost simultaneously by Siniscalco *et al.*, 1969 and Silagi *et al.*, 1969. Both groups of workers used half-selective systems to isolate their hybrids, i.e. a culture medium was chosen so that one of the two parental lines was killed off and the resulting culture was a mixture of hybrid cells and the other parental line. This illustrates one of the persistent problems in the technology of somatic cell hybridization; the difficulty of finding a medium which will select against parental cells and allow the proliferation of heterokaryons. The HAT medium (hypoxanthine, aminopterin and thymidine) of Littlefield (1966) which kills cells deficient in the enzyme hypoxanthine–guanine phosphoribosyltransferase (HGPRT) is one of the well-defined selective media for human cells, and HGPRT-deficient cells have thus frequently been used as one of the parental lines in human hybrids. In many other situations no attempt has been made to select the hybrids and analyses have been carried out on mixtures which may contain up to five different cell types, two diploid parental lines, two self-fused tetraploid lines and a heterokaryon line.

Where such mixtures have been of great value is in demonstration and resolution of genetical heterogeneity. In 1970 Nadler *et al.* performed pair-wise hybridizations of fibroblasts from seven different patients with the metabolic disorder galactosaemia. Each of the parent lines lacked detectable levels of the enzyme galactose-1-phosphate uridylyltransferase (GALT), the primary lesion in this disorder. Most of the fusion mixtures continued to be GALT-deficient, but of the 21 possible heterologous lines, three were found to contain appreciable levels of the missing enzyme. These three had in common one particular parental cell line, and it appeared that this line and the patient from whom it was derived differed genetically from the remaining six galactosaemic patients. Thus for the first time genetical heterogeneity had been demonstrated by cell fusion in a situation where the patient's clinical and biochemical characteristics were indistinguishable. It seemed that within the nucleus of the heterokaryon complementation had taken place to form a genetical unit capable of coding for the GALT enzyme. In view of independent evidence for dual locus control of the enzyme, intergenic rather than intragenic complementation seems the most likely explanation.

Table 4

Use of somatic cell hybridization in demonstrating genetical heterogeneity

Disorder	Enzyme	Reference
Galactosaemia	galactose-1-phosphate uridylyl transferase	Nadler *et al.* (1970)
Xeroderma pigmentosum	DNA repair enzyme	De Weerd-Kastelein *et al.* (1972) Bootsma (1974) Kraemer *et al.* (1975)
Maple syrup urine disease	α-keto acid decarboxylase	Lyons *et al.* (1973)
GM2-gangliosidosis	hexosamidase A and B	Galjaard *et al.* (1974b) Thomas *et al.* (1974)
GM1-gangliosidosis	β-galactosidase	Galjaard *et al.* (1975)
Methylmalonicacidaemia	methylmalonyl-CoA mutase	Gravel *et al.* (1975)

Somatic cell hybridization has now been used to demonstrate genetical heterogeneity in a number of other inborn errors of metabolism (Table 4). In most cases selective media have not been employed, since the test for complementation has been the appearance in the culture of the enzyme deficient in both parental lines. However, it should be clear that the failure of restoration of enzyme activity in fusion cultures is not necessarily a criterion for genetical identity of the parental lines, since technical problems in the form of a low rate of hybrid formation may prevent complementation between different genomes. Methods in which auto-radiographic examination of individual heterokaryons have been employed (as in the fusion of cells from different patients with Xeroderma pigmentosum) are probably a fairer test for genetical heterogeneity. Introduction of microtechniques for the assay of lysosomal enzymes in single cells (Galjaard *et al.*, 1974a) will also provide a more exacting test, and these have been used by Galjaard *et al.* (1975) in examining the heterogeneity of generalized gangliosidosis. (See Chapter 10 for a discussion of heterogeneity in genetical disease.)

III. FAMILY STUDIES

A. Linkage Analysis

1. Pedigree methods
Recombination experiments have permitted the calculation of linkage maps for the chromosomes of many diploid species. Among the species investigated in this manner, least perhaps is known about linkage in Man.

This is because matings cannot be specified as necessary and because of the large number of chromosomes present. A list of known human autosomal linkages is shown in Table 6, while chromosomal assignments are shown in Table 5.

Table 5

Assignments to chromosomes by pedigree studies, chromosomal anomalies and somatic cell hybridization[b]

Chromosome	Loci
1	*1qh Fy Cae AOD AMY$_1$ AMY$_2$ PGM$_1$ El$_1$ Rh PEPC Do An$_2$ ENO$_1$ PGD PPH RN5S FH$_1$ FH$_2$ St UGPP GUK$_1$ GUK$_2$ AK$_2$ AdV12-CMS-1 UMPK αFUC*
2	*ICD$_1$ MDH$_1$ ACP$_1$ If$_1$ MNSs^a Tys Gal$^+$-Act Hbα or Hbβa Ig^a AHH*
3	*GALT ACON$_M$*
4	*PGM$_2$ Hbα or Hbβ^a*
5	*If$_2$ HEXB DTS*
6	*ME$_S$ SOD$_B$ PGM$_3$ HLA-A HLA-B HLA-C HLA-D MLC Bf Chi Ir P Pg-5 GLO$_1$ Rg C2 C4 C8*
7	*XII HEX-C MDH$_M$ SV40T βGLU Coll I*
8	*GSR*
9	*ABO Np AK$_1$ Xp WS$_1$ ACON$_S$ AK$_3$ ASS*
10	*GOT$_1$ HK$_1$ GSS ADK PP*
11	*LDH-A ESA$_4$ ACP$_2$ AL Hbβ^a*
12	*TPI LDH-B PEPB GlyA$^+$ CS$_M$ GAPD γG ENO$_2$ SHM*
13	*RNr ESD RB$_1$*
14	*RNr NP TRPRS*
15	*RNr β$_2$M HEXA ICD$_M$ PK$_3$ MPI*
16	*APRT Hpα LCAT TK$_M$ Hbα^a*
17	*TK$_S$ GK AdV12-CMS-17*
18	*PEPA*
19	*GPI PolioRS E11S PEPD*
20	*ADA$_2$ DCE ITP*
21	*AVP RNr SOD$_1$ GAPS*
22	*RNr βGAL*

[a] inconsistent assignment

[b] Mayo, 1975; modified and extended by the Baltimore Conference, 1975; Brown *et al.*, 1976; Sturt, 1976; Lewis *et al.*, 1976; Bruns and Gerald, 1976; Teisberg *et al.*, 1975; Sykes and Solomon 1978 and Westerveld *et al.*, 1976, See Table 6 for linkage distances, where available, and Table 7 for description of loci

Table 6

Map distances, θ, between pairs of loci in centiMorgans[a]

| | Map distance | | | |
Male	Female	Neuter	Locus	Pair
26		25	Rh	PGM_1
13			Rh	PGD
		17	Rh	$UMPK$
		15	PGD	$PEPC$
		35	$PEPC$	Fy
		10	Fy, Cae	$1qh$
		10	Fy, Cae	AOD
20	34		Fy	AMY_1, AMY_2
<3		3	Rh	El_1
		27	Rh	Sc
0			Rh	PPH
		34	AMY	PGM_1
		25	$UMPK$	PGM_1
		0·8	$HLA\text{-}A$	$HLA\text{-}B$
		20	HLA	$Pg\text{-}5$
		4·5	$HLA\text{-}A$	Chi
		1·5	$HLA\text{-}B$	Chi
		1·9	$HLA\text{-}A$	Bf
		0	$HLA\text{-}B$	Bf
		0	$HLA\text{-}C$	Bf
		15	HLA	GLO_1
10		10	ABO	Np
		13	Np	AK_1
0	0	0	$Hp\alpha$	$LCAT$
10		13	Lu	Se
5		8	Se	Dm
		0	$Hb\beta$	Hb
		0	γG_2	γG_3
		0	γG_3	γG_1
		0	γG_1	γA_2
		13	γG	Pi
		15	Tf	E_1
		3	Alb	Gc
		4	MNS_s	Tys
		5	$EBS\ Ogna$	GPT
		15	Le	ACP_1
7			Le	$C'3$

[a] References as for Table 5

Table 7

Loci which have either been assigned to autosomes (Table 5) or to linkage groups (Table 6)

Locus	Description
ABO	ABO blood group system
ACON$_M$	aconitase, mitochondrial
ACON$_s$	aconitase, soluble
ACP$_1$	acid phosphatase$_1$
ACP$_2$	lysosomal acid phosphatase
ADA$_2$	adenosine deaminase$_2$
ADK	adenosine kinase
AdV12-CMS-1	adenovirus-12-chromosome modification site-1
AdV-12-CMS-17	adenovirus-12-chromosome modification site-17
AHH	aryl hydrocarbon hydroxylase
AK$_1$	adenylate kinase$_1$
AK$_2$	adenylate kinase$_2$
AK$_3$	adenylate kinase$_3$
AL	human species antigen (lethal; 3 loci)
Alb	albumin
αFUC	α-fucosidase
AMY$_1$	salivary amylase
AMY$_2$	pancreatic amylase
An$_2$	aniridia type II Baltimore
AOD	auriculo-osteodysplasia
APRT	adenosine phosphoribosyl transferase
ASS	argininosuccinate synthetase
AVP	antiviral protein
βGAL	β-galactosidase
βGLU	β-glucuronidase
β$_2$M	β$_2$ microglobulin
Bf	properdin factor B (glycine-rich-β-glycoprotein)
Cae	zonular pulverulent cataract
Chi	Chido antigen
CS	citrate synthase (mitochondrial)
C2	C2 component of complement
C'3	C3 component of complement
C4	C4 component of complement
C8	C8 component of complement
Coll I	Collagen type I
DCE	desmosterol to cholesterol enzyme
Dm	myotonic dystrophy
Do	Dombrock blood groups
DTS	diphtheria toxin sensitivity

Table 7—*contd.*

Locus	Description
E_1	pseudocholinesterase
EBS Ogna	epidermolysis bullosa simplex, Ogna type
El_1	elliptocytosis$_1$
El1S	echo 11 sensitivity
ENO_1	enolase$_1$
ENO_2	enolase$_2$
ESA_4	esterase A$_4$
ESD	esterase D
FH_1	fumarate hydratase$_1$
FH_2	fumarate hydratase$_2$
Fy	Duffy blood groups
GALT	galactose-1-phosphate uridylyl transferase
Gal$^+$-Act	galactose enzyme activator
γA_2	Am immunological region$_2$
γG	Gm system
γG_1	Gm immunological region$_1$
γG_2	Gm immunological region$_2$
γG_3	Gm immunological region$_3$
GAPD	glyceraldehyde-3-phosphate dehydrogenase
GAPS	phosphoribosylglycine amide synthetase
Gc	group specific component
GK	galactokinase
GLO_1	glyoxalase$_1$
GOT_1	soluble glutamic oxaloacetic transaminase$_1$
GPI	glucose-phosphate isomerase
GPT	erythrocyte glutamic-pyruvic transaminase
GSR	glutathione reductase
GSS	glutamate-γ-semialdehyde synthetase
GUK_1	guanylate kinase$_1$
GUK_2	guanylate kinase$_2$
Hb α	α haemoglobin chain structural locus
Hb β	β haemoglobin chain structural locus
Hb δ	δ haemoglobin chain structural locus
HEXA	hexosaminidase-A
HEXB	hexosaminidase-B
HEXC	hexosaminidase-C
HK_1	hexokinase$_1$
HLA-A	human leucocyte antigen system, locus A
HLA-B	human leucocyte antigen system, lucus B
HLA-C	human leucocyte antigen system, locus C
HLA-D	human leucocyte antigen system, locus D
Hpα	α-haptoglobin

Table 7—contd.

Locus	Description
ICD_1	isocitrate dehydrogenase$_1$
ICD_M	isocitrate dehydrogenase, mitochondrial
If_1	interferon$_1$
If_2	interferon$_2$
Ig	immunoglobulin heavy chains
Ir	immune response
ITP	inosine triphosphatase
LCAT	lecithin: cholesterol acyl transferase
LDH-A	lactate dehydrogenase A
LDH-B	lactate dehydrogenase B
Le	Lewis antigen system
Lu	Lutheran blood groups
MDH_1	malate dehydrogenase$_1$ (NAD-dependent)
MDH_M	malate dehydrogenase, mitochondrial
ME_1	malic enzyme$_1$ (soluble malate dehydrogenase (NAD dependent)
MLC	mixed lymphocyte culture antigen
MNSs	MNSs blood groups
MPI	mannose phosphate isomerase
Np	nail-patella syndrome
NP	nucleoside phosphorylase
1qh	uncoiler-1
P	P blood group system
PEPA	peptidase A
PEPB	peptidase B
PEPC	peptidase C
PEPD	peptidase D
PGD	6-phosphogluconate dehydrogenase
Pg 5	urinary pepsinogen$_5$
PGM_1	phosphoglucomutase$_1$
PGM_2	phosphoglucomutase$_2$
PGM_3	phosphoglucomutase$_3$
Pi	α1-antitrypsin
PK_3	white cell pyruvate kinase$_3$
Polio RS	polio receptor site
PP	inorganic pyrophosphatase
PPH	phosphopyruvate hydratase
RB_1	retinoblastoma$_1$
Rg	Rodgers blood group
Rh	Rhesus blood groups
RN5S	5S rRNA region
RNr	rRNA region

Table 7—*contd.*

Locus	Description
Sc	Scianna blood groups
Se	ABH secretion
SHM	serine hydroxymethylase
SOD_A	superoxide dismutase-1 (dimeric indophenol oxidase-A)
SOD_B	mitochondrial superoxide dismutase (tetrameric indophenol oxidase-B)
SV40-T	Simian virus 40-T antigen
Tf	transferrin
TK_M	thymidine kinase, mitochondrial
TK_S	thymidine kinase, soluble
TPI	triose phosphate isomerase
TRPRS	tryptophanyl-transfer RNA synthetase
Tys	sclerotylosis
UGPP	UDP-glucose pyrophosphorylase
UMPK	uridine monophosphate kinase
WS_1	Waardenburg syndrome, type 1
XII	Factor XII (Hageman)
XP	Xeroderma pigmentosum

Genetical mapping depends upon the characterization of a set of loci, so that recombination frequencies can be calculated for groups of genes having at least one locus in common, i.e. one set of genotypes at least must be known to be the result of allelic genes. Polymorphism at these loci is a helpful but not necessarily an essential feature of the method and a rare variant may permit the existence of a particular locus to be detected. Special methods of linkage analysis have had to be developed for man, and Bailey (1961) gives their history. Currently, most linkage investigations use likelihood ratio methods developed since about 1946. The general outline of the methods is as follows:

Consider a set of families $F_1, F_2, \ldots F_f$ from which information is being obtained and let θ and θ' be the recombination fractions in females and males respectively, with $\theta + \psi = 1$. The probability $Pr(F_r/\theta,\theta')$ that a family similar to the family F_r would arise if θ and θ' were the true frequencies of recombination is calculated and P_r called the likelihood of the family.

If all the f families are unrelated, the probability of obtaining the whole sample S is the product

$$P_r(S/\theta,\theta') = \prod_{II}^{f} P_r(F_r/\theta,\theta')$$

where Π indicates the multiple product of $P_r(F_1/\theta,\theta')$, $P_r(F_2/\theta,\theta')$ etc. This can be simplified by considering

$$L(\theta,\theta') = k_r P_r(F_r/\theta,\theta')$$

where

$$k_r = 1/P_r(F_r/\theta = \tfrac{1}{2}, \theta' = \tfrac{1}{2}),$$

the reciprocal of the likelihood when there is no linkage, that is the recombination fraction is $1/2$.

This $L(\theta,\theta')$ is a likelihood ratio. By setting

$$z(\theta,\theta') = \log_e L(\theta,\theta'),$$

the sum

$$Z(\theta,\theta') = \Sigma \log_e (k_r Pr(F_r/\theta,\theta'))$$
$$= \Sigma z(\theta,\theta', F_r)$$

is obtained. These z's are called log likelihoods or lods (log of odds) for short and can in many cases be tabulated, since human families are not large and many loci yield the same segregation patterns. To illustrate the method we examine a specific case.

Consider so-called double back-cross matings between one locus, G,g, and another locus T,t of the form $GgTt \times ggtt$. In such families, the doubly heterozygous individual (here, the mother) may be either GT/gt (coupling phase) or Gt/gT (repulsion phase) with probability of 0·5 of each initially (unless one knows of linkage disequilibrium, about which there is as yet little information in Man: see Chapter 4, Section II.A.2 for discussion). Now there will be four phenotypes for progeny, with observed numbers, a,b,c,d and expected frequencies as shown:

	GT	Gt	gT	gt	
observed numbers	a	b	c	d	$a+b+c+d = n$
expectation if mother is GT/gt	$\tfrac{1}{2}\psi$	$\tfrac{1}{2}\theta$	$\tfrac{1}{2}\theta$	$\tfrac{1}{2}\psi$	
expectation if mother is Gt/gT	$\tfrac{1}{2}\theta$	$\tfrac{1}{2}\psi$	$\tfrac{1}{2}\psi$	$\tfrac{1}{2}\theta$	

Since the heterozygote's genotype (including phase) is not known other than by her offspring (on account of dominance etc.), families scorable for crossing-over must have at least two children, who must contain one of these combinations of phenotypes: gt with any other; Gt with gT. As shown above, a sibship consists of a GT, b Gt, c gT, d gt, with likelihood

$$P_r(F_r/\theta) = \tfrac{1}{2}[(\tfrac{1}{2}\psi)^a(\tfrac{1}{2}\theta)^b(\tfrac{1}{2}\theta)^c(\tfrac{1}{2}\psi)^d + (\tfrac{1}{2}\theta)^a(\tfrac{1}{2}\psi)^b(\tfrac{1}{2}\psi)^c(\tfrac{1}{2}\theta)^d]$$
$$= (\tfrac{1}{2})^n (\psi^{a+d}\theta^{b+c} + \psi^{b+c}\theta^{a+d})$$

Hence

$$L_r(\theta) = 2^{n-1}(\psi^{a+d}\theta^{b+c} + \psi^{b+c}\theta^{a+d})$$

This $L_r(\theta)$ thus gives the probability of observing the family actually seen with any particular value of $\theta(\neq \frac{1}{2})$ relative to its chance of occurrence if the loci are independent $(\theta = \frac{1}{2})$; in other words, the odds, as noted earlier. The log of L can now be tabulated for all possible useful combinations of a, b, c and d and the size of z summed over all available informative families give a measure of confidence in the particular value of θ being the true value. Because not all families are informative, corrections must be made to the z's; tables for these corrections are to be found in Maynard-Smith *et al.* (1961).

What is being estimated is the frequency of recombination, which is a reflection of crossing-over in meiosis; Bailey (1961) discusses the relationship of the frequencies of recombination and crossing-over in Man. For small values of θ the two variables are very similar.

In the study of human chromosomes the most accurate and extensive mapping of linked genes has been accomplished for the X-chromosome. Recombinants can be directly observed in male offspring and X-recombination occurs only in females, the homogametic sex. Hence, loci for rare traits are more easily mapped than if they were autosomal. This is because one may specify, for example, that the father (and the maternal grandfather) of a propositus is normal with respect to two or more sex-linked anomalies and that the traits in question must be located upon the single maternally-derived X-chromosome. Thus, linkage phase can more easily be determined independently of the sibship of interest. A tentative map of the X-chromosome is shown in Fig. 7, Chapter 7.

Loci may be located by other methods, including the use of chromosomal aberrations (Section III.A.2) or somatic cell hybridization (p. 109). Centromeric mapping with ovarian teratomas (p. 426) is yet another curious and specialized method. Fluorescence staining of chromosomes has also provided a new method of linkage analysis. As discussed in Chapter 3, all human chromosomes display characteristic fluorescence patterns with quinacrine mustard, quinacrine hydrochloride and related compounds. Thus, chromosomal aberrations or even polymorphisms may be reliably identified and pedigree studies can yield extra information (Geraedts and Pearson, 1974).

Given a linkage map, one can use it in at least three ways: in the resolution of genetical heterogeneity; in assessing the meaning of an association; and in genetic counselling. In the first case, one of the earliest established linkages (Rh–El) was used to show that at least two kinds of elliptocytoses exist (Morton, 1956; and see Chapter 7). In the second case, known information about a linkage, e.g. ABO–AK, could be used to obtain information about an association, such as ABO–cholesterol (see p. 454). In the third case, which may be the most important in the intermediate future, many possibilities exist, but not all are equally promising. Let us first look at the existing map of chromosome 1. There are 18 loci assigned to it. Of these, three (elliptocytosis, zonular pulverulent cataract and auricular osteodysplasia) are abnormal traits, but in none of them would antenatal diagnosis be particularly helpful. Two (Duffy and Rhesus) are blood group systems and so can be determined antenatally, at least at the present time, only with difficulty because of the problems of placental aspiration and of typing mixed foetal and maternal blood (Kan et al., 1975). Of the others, the 5S-r RNA locus and the adenovirus-12-chromosome modification site-1 are not polymorphisms, and therefore would be unlikely to be useful in linkage investigations by present methods.

Chromosome 2, the next best mapped autosome after 1 and 6 (the latter being discussed in Chapter 8), has eight loci assigned to it, of which one (sclerotylosis) is a disease, while the Hb locus will be important in thalassaemia and other haemoglobin disease states (and foetal Hb phenotypes may be obtained directly in some cases, Kan et al., 1975). The MNSs system, like the other blood groups mentioned, would still present problems, while interferon-1 and the galactose enzyme activator would not be useful for linkage studies.

The relatively unsatisfactory balance of promise to performance for autosomal linkages has been a little redressed by progress achieved with X-linked traits. Here, for example, antenatal sex-determination can allow therapeutic abortion of males having a 50% risk of being affected with haemophilia, HGPRT deficiency or Duchenne muscular dystrophy. Problems exist in the application of this practice, since, for example, it will lead to an increase in the number of carrier women (if there is reproductive compensation for the aborted male foetuses; see Fraser, 1972, and Emery et al., 1971, for discussion). Nonetheless, it can perhaps be regarded as an advance in a world concerned with quality as well as quantity of life.

The resolution of genetical heterogeneity has also been more successful for X-linked traits. Linkage analyses have partly resolved the problem of the nature and number of the loci involved in colour vision defects (Fraser, 1969, 1970; Thuline et al., 1969), and the relatively mild Becker muscular dystrophy has shown to be determined by a different locus from that responsible for the severe Duchenne type (Emery et al., 1969).

One new area of interest in the use of linkage in human biochemical genetics is the role of chromosomal location in subcellular organization. Bruns and Gerald (1976) have pointed out that three of the enzymes of the Embden–Meyerhof pathway are located on chromosome 12. In attempting to investigate the significance of this finding, we have constructed Fig. 1. This shows all enzymes of this pathway for which structural loci have so far been assigned to chromosomes. In some cases, of course, other

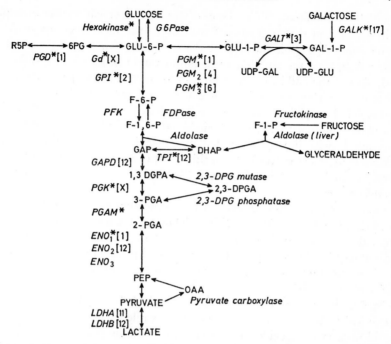

Fig. 1 The Embden-Meyerhof glycolytic pathway, showing chromosomal location and polymorphism of structural gene loci for the enzymes of the pathway.
Loci are shown in italics. Chromosomal assignments are indicated by symbols in
square brackets [], polymorphisms detected in one or more populations by an
asterisk*. Standard abbreviations have been used

isozymes coded for by as yet unassigned loci also catalyse particular steps
in the pathway, but even allowing this no clear pattern emerges.

In Fig. 1 we have also shown which loci are known to be polymorphic,
for comparison with Fig. 5 of Chapter 10. Once again, it is difficult to draw
any general conclusion from the pattern of normal as against deleterious
variation.

2. Aneuploidy

The nature and effects of aneuploidy have been extensively discussed in
Chapter 3. Here, we wish only to consider the use of polymorphic markers
in the detection and determination of the origin of aneuploidy, and the use
of aneuploidy in the mapping of polymorphic traits.

It was realized many years ago (Lang-Brown et al., 1953; Bateman,
1960) that for any marker aneuploidy could cause disturbed segregation. In
the case of a gross chromosomal defect, such as a trisomy, the aberration
would be recognized cytologically, and polymorphic markers could be

examined in the family for evidence of disturbed segregation. For example, if it were found that in a case of trisomy 21 the individual was of ABO blood group A_1B while the mother was of blood group O and the putative father A_1B, then the simplest explanation would be that non-disjunction had allowed the child to inherit both chromosomes 21 and that the ABO locus was on chromosome 21. Of course, as Lawler (1964) pointed out, matings of the type indicated are rare (less than 1 % of matings in England), but with many polymorphisms available, this should be less of a problem, particularly with linkage. As noted by Magenis et al. (1975a), the HLA system is the most powerful from this point of view, but as it has already been mapped on to Chromosome 6 (Table 5), it can be expected to be useful only rarely.

Goodman (1965) has considered the population genetics of trisomy in most detail, while Rundle (1973) has elaborated the techniques further and applied them to Down's syndrome. So far, no clear evidence of anomalous phenotypes has been presented; even the sophisticated analysis of centromeric segregation (Côté and Edwards, 1975) has not yet proved productive (Côté et al., 1975).

Smaller but still chromosomal defects such as large deletions could also allow loci to be mapped in the same way, while such loci, once mapped, could in turn yield evidence of deletions or duplications too small to be detected by the light microscope. In Table 5, many of the assignments to chromosomes were mapped through aneuploidy (cf. also Chapter 3). For example, assignment of the haptoglobin α ($Hp\alpha$) locus to chromosome 16 (Robson et al., 1969) was perhaps the first such achievement by means of linkage analysis of families in which chromosomal anomalies were segregating. In one family, a translocation between single arms of chromosomes 2 and 16 was segregating in coupling with the Hp^1 allele, with no recombinants in 11 possible cases. Thus, it seemed likely that Hp was on chromosome 2 or chromosome 16. In a second family Hp^2 was segregating in coupling with a break-point involving arms of chromosomes 1 and 16. Thus, it could be concluded logically that Hp lay on chromosome 16, as was soon confirmed (Magenis et al., 1970).

Fraser (1963, 1966) has considered the various sex chromosomal aneuploidies in detail, developing maximum likelihood methods to determine the parental sex chromosome contributions in each case, by the use of data on the Xg blood group system and the G6PD locus.

As an example of the general approach, consider the case of XO individuals. Suppose that there are two alleles, G and g, with frequencies of p and q, respectively. Then if a XO individuals arise by maternal non-disjunction and b by paternal ($a+b = 1$), in n XO individuals the offspring–parent combination will be as follows:

	Case I GG, Gg indistinguishable		Case II GG, Gg distinguishable	
class	frequency observed	frequency expected	frequency observed	frequency expected
$GG \times G{:}G$ ⎫ $Gg \times G{:}G$ ⎬	A	$np^2(1+aq)$	T $\mathcal{J}$	⎰np^3 ⎱$np^2q(1+a)$
$GG \times g{:}G$ ⎫ $Gg \times g{:}G$ ⎬	B	$npq(1-a)$	K	⎰$np^2q(1-a)$ ⎱$npq^2(1-a)$
$gg \times G{:}G$	C	npq^2a	M	npq^2a
$Gg \times G{:}g$	D	$np^2q(1-a)$	N	$np^2q(1-a)$
$GG \times g{:}g$ ⎫ $Gg \times g{:}g$ ⎬	E	$npq(a+a)$	P	⎰np^2qa ⎱$npq^2(1+a)$
$gg \times G{:}g$	F	$npq^2(1-a)$	R	$npq^2(1-a)$
$gg \times g{:}g$	H	nq^3	S	nq^3

In this table, the notation $GG \times G{:}G$ means a $GG \times G$ mating with XO offspring of genotype G, and so on. In case I, solution of

$$\frac{Aq}{1+aq} + \frac{C}{a} + \frac{E}{a+q} - \frac{B+D+F}{1-a} = 0$$

yields the maximum likelihood estimate of a, while in case II, a is obtained by solution of

$$\frac{\mathcal{J}+Q}{1+a} + \frac{M+P}{a} - \frac{K+L+N+R}{1-a} = 0$$

As Fraser (1963) noted, the expectations in Case I are such that loss of the maternal X-chromosome is extremely hard to demonstrate. However, this has now been achieved in a number of cases (Race and Sanger, 1975, Chapter 30, especially Table 107).

B. Paternity Testing

For many years, in some countries, human blood groups have been accepted legal tools for the identification of individuals, for deciding questions of paternity, and for testing whether babies have been unintentionally exchanged in hospitals. The most common application is disputed paternity, and in general these genetical data have only been used for the exclusion of paternity. That is, if, for example, a woman of a blood group O has a child of blood group A_1, then a putative father of blood group B cannot be the real biological father. A man of blood group A_1 could be the father, of course, but as blood group A_1 is by no means rare in most populations, this would hardly be compelling evidence that the man was the father to the child.

448   O. MAYO AND D. J. H. BROCK

The utility of the various blood group systems for paternity testing has recently been examined by Chakraborty *et al.* (1974), from the point of view that utility depends upon the probability of excluding a wrongly nominated male if the latter be randomly selected from the general population. In North American whites, these utilities range from about 0·31 for the MNSs system, and 0·27 for the Rhesus system, down through 0·18 for the Duffy system, 0·13 for ABO, to 0 for Diego, Henshaw and other systems which are not polymorphic in North American Whites. Overall, there is a cumulative probability of over 90% that a wrongly nominated man will be excluded from paternity by the 25 systems which Chakraborty *et al.* (1974) considered. Similar values hold for other well studied ethnic groups, such as North American blacks.

The law moves slowly, so that the large number of enzyme polymorphisms known (and discussed in Chapter 5) are in most parts of the world not yet admissible in this kind of court case, but Chakraborty *et al.* (1974) show that they are at least as informative as the blood group systems mentioned above. Thus, it seems important that these other systems should be accepted as being of equivalent merit to those already established, especially as the enzyme polymorphisms are regarded as being more readily assayed without error. (As Chakraborty *et al.* (1974) point out, they are of course no more immune to other kinds of error such as mislabelling than are blood group systems.)

The HLA system has peculiar merits for paternity testing, and these are discussed in Chapter 4, Section II.B.2.

The details of the application of polymorphic traits to these problems have been well established (Andresen, 1952; Fisher, 1951), and need not be repeated here. Chakraborty *et al.* (1974) give an excellent discussion of recent work in the field, as well as treating the points mentioned above, while Bucher and Elston (1975) and Langaney and Pison (1975) provide commentary on the work of Chakraborty *et al.* (1974).

IV. POPULATION STUDIES

A. Association of Polymorphism and Disease

In Chapter 4, the general problem of the association between polymorphism and disease was examined in detail, for the light that investigations of such associations might shed on the general problem of the maintenance of genetical variability. In Chapter 8, the relationship of the HLA system to disease was examined in detail, as this is the most important relationship yet discovered in this area. In this section, we merely outline a useful method for the detection and estimation of such associations, and discuss problems in such surveys.

In such surveys, the data will consist essentially of a set of phenotype frequencies for persons in whom a disease is present and those in whom the disease is absent (the control group). In assembling such data, the main problem will be to avoid bias. Controls and patients should obviously be matched for race and should belong to the same geographically located population. Racial admixture can create spurious associations, as noted on page 168. A less obvious source of error is the fact that the control population may be biased. For example, blood donors are frequently used as controls, and they are more healthy than the population in general, for obvious reasons. In one recent study, Sørensen and Dissing (1975) found that the C'3 alleles in blood donors are not present at the same frequencies as those in the general population. If this is the case for many polymorphisms, then data on blood donors should be used far more carefully than they have been in the past. (The South Australian control data in Table 8 are from blood donors.) In view of the rarity of age and sex differences in polymorphic gene frequencies, it would seem not worthwhile to match for age and sex, but nonetheless this too could introduce problems. Thus, there are many sound reasons for viewing reported associations with initial caution, as noted in Chapter 4, but when these become quite compelling, as in the cases discussed in Chapter 8, one can hope that further investigation will actually elucidate the associations. The analysis which follows, taken from Mayo (1975), illustrates two points about the investigation of associations: first, one should obtain a meaningful measure of the relative risk of those having different phenotypes; and secondly that similar findings from different populations will be helpful in deciding the reality of an association.

Woolf's method (1955) is as follows. Consider a set of data collected so that the numbers of people are classified in two ways: pair-wise by phenotype and pair-wise by presence or absence of disease. If there are several such sets of data, they can be set out thus:

sample	disease present phenotype 1 (h)	phenotype 2 (k)	disease absent (control) phenotype 1 (H)	phenotype 2 (K)				
1	h_1	k_1	H_1	K_1	x_1	y_1	w_1	$w_1 y_1^2$
2	h_2	k_2	H_2	K_2	x_2	y_2	w_2	$w_2 y_2^2$
.	.	.	.	.	.	.	.	.
.	.	.	.	.	.	.	.	.
n	h_n	k_n	H_n	K_n	x_n	y_n	w_n	$w_n y_n^2$

$$x = \frac{hK}{Hk} \qquad y = \ln x \qquad w = \left(\frac{1}{h} + \frac{1}{k} + \frac{1}{H} + \frac{1}{K} \right)^{-1} \qquad wy^2$$

Here, x is the relative risk of contracting the disease for phenotype 1 as against phenotype 2.

Setting

$$Y = \frac{\Sigma wy}{\Sigma w}$$
$$X = \text{antilog } Y$$

gives the overall relative risk, while the χ^2 analysis is as follows:

Source	degrees of freedom	χ^2
Y	1	$Y^2 \Sigma w$
heterogeneity	$n-1$	$\Sigma wy^2 - Y^2 \Sigma w$
TOTAL	n	Σwy^2

From this analysis of χ^2 one can see, first, whether X differs significantly from 1 and, secondly whether the different samples are homogeneous.

For example, consider the data shown in Table 8. The comparison of interest to Czechowicz and Pamnany (1972) was between individuals of blood group O and others.

Table 8

ABO blood group distribution in schizophrenics and controls

Source	Sample	Schizophrenic O	A	AB	B	TOTAL	Control O	A	AB	B	TOTAL
South Australia (Czechowicz and Pamnany, 1972)	1	31	46	5	15	97	534	409	33	95	1071
Lancashire (Masters, 1967)	2	31	31	0	5	87	334	243	17	58	652

Pooling groups A, B and AB, we obtain the following results:

Sample	Schizophrenic O(h)	not-O(k)	Control O(H)	not-O(K)	x	y	w	wy^2
1	31	66	534	537	0·47234	−0·75007	19·55416	11·00127
2	41	46	334	318	0·84861	−0·16415	19·13143	0·51550

Then

$$Y = \frac{\Sigma wy}{\Sigma w} = \frac{-17·80741}{38·68559} = -0·46031$$

and

$$X = \text{antilog } Y = 0·63111$$

i.e. the overall relative risk of being schizophrenic for persons of blood group O as against those of other blood groups is 0·631. The χ^2 analysis is as follows:

Source of variation	Degrees of freedom	χ^2
Y (relative risk ratio)	1	8·19691
heterogeneity	1	3·31986
TOTAL	2	11·51677

The interpretation of the two χ^2s is (1) that the risks of schizophrenia differ significantly as between those of blood group O and others, and (2) that the two sets of data are homogeneous.

B. Estimation of Genetical Distance

In Chapter 4, evolutionary divergence between separated human populations was considered, and it was pointed out that this could come about through adaptive change, i.e. natural selection, or random drift. The idea that a measure of this divergence could be obtained from data on gene frequencies at human polymorphic loci was introduced by Cavalli-Sforza and Conterio (1960), and the method most widely used is that of Cavalli-Sforza and Edwards (1964).

Let the allelic frequencies in two populations for a given m-allelic autosomal locus be $x_1, x_2, \ldots x_m$ and $y_1, y_2, \ldots y_m$ respectively

$$\left(\sum_{i=1}^{m} x_i = \sum_{i=1}^{m} y_i = 1. \right)$$

Then define an angle θ by

$$\cos \theta \sum_{i=1}^{m} \sqrt{x_i y_i} \qquad 0 \leq \theta \leq \frac{\pi}{2}$$

The genetical distance between the two populations is then given by θ. However, in most cases $\sqrt{2(1\text{-}\cos \theta)}$ is used instead, since data from k independent loci may then be combined by

$$G = [\sum_{K} 2(1\text{-}\cos \theta_K)]^{\frac{1}{2}}$$

The properties of $2(1\text{-}\cos \theta)$ are often regarded as critically important.

If the two populations diverged at a certain time, then by making the assumption that there is no selection, migration or mutation, if generation time is constant, expected values for θ and G may be derived. For example, with population size N, in the case of two alleles with frequencies p and q, after time t,

$$\Sigma[2(1\text{-}\cos\theta_K)]$$

$$= 4pq[1\text{-}e^{\frac{-t}{2N}} -pqe^{\frac{-2t}{2N}} +(p\text{-}q)^2e^{\frac{-3t}{2N}} +0(e^{\frac{-6t}{2N}})]$$

Even more approximately,

$$\Sigma[2(1\text{-}\cos\theta_K)] \simeq \frac{t}{4N}$$

Thus, on this simple model, genetical distance is approximately linearly dependent on the time since separation. Heuch (1975) discusses various possible elaborations and some of the limitations of this approach. Some of the limitations are overcome by Nei's model (Nei, 1975). For example, under fairly restrictive assumptions about the absence of selection, it shows a precisely linear relationship between genetical distance and time, and is apparently robust to variations in population size during divergence. Because of this, it is now being used quite widely. It is, however, closely related to the other measures of genetical distance, and all such measures will in general be sensitive to variation in gene frequency distribution (e.g. blood group and isozyme polymorphism data should not be combined (Yamazaki and Maruyama, 1974)), whether on account of selection or for other reasons (see also Ewens and Feldman, 1976).

C. Quantitative Variation

The classical techniques of quantitative genetics, introduced by Fisher (1918), allow the partitioning of the observed population (phenotypic) variance in a quantitative trait, V_P, into various meaningful components. Thus,

$$V_P = V_G+V_E$$

where V_G = genetical variance

and V_E = environmental variance,

while V_G may be further partitioned:

$$V_G = V_A+V_D+V_I$$

where V_A = variance due to the additive effects of gene affecting the character

V_D = variance due to dominance effects

and V_I = variance due to interaction between loci (epistasis).

Such partitioning is valid only in the absence of interaction between genotype and environment (Jinks and Fulker, 1970). In animal breeding, it is frequently of great interest to know what proportion of the observed phenotypic variance V_P is attributable to additive effects, i.e. the ratio V_A/V_P is important. It is defined as the heritability, h^2, of the trait in question.

Table 9

Heritability estimates, h^2, for several biochemical traits[d]

Trait	h^{2a}	Reference
Immunoglobulin A (serum level)	0.42 ± 0.05 ⎤	
Immunoglobulin G (serum level)	0.39 ± 0.05 ⎬	Grundbacher (1974)
Immunoglobulin M (serum level)	0.37 ± 0.06 ⎦	
Serum cholesterol	0.51 ± 0.17	Mayo et al. (1969)
Proportion of HbS in erythrocytes of HbA/HbS heterozygotes	0.88 ± 0.20	Nance and Grove (1972)
Phenylbutazone metabolism	0.65 ± 0.21	Whittaker and Price Evans (1970)
Serum trehalase activity	0.58 ± 0.22	Eze et al. (1970)
Erythrocyte acid phosphatase activity	0.82 ± 0.11	Eze et al. (1974)
β-amino-isobutyric acid excretion	1.00^b ⎤	
Lysine excretion	0.94^b ⎬	Gartler et al. (1955)
Taurine excretion	0.62^b ⎥	
Tyrosine excretion	0.60^b ⎦	
Threonine excretion	1.00^b	Berry et al. (1955)
Clotting factor V level	0.89^b ⎤	Veltkamp et al. (1972)
Clotting factor VII level	0.96^b ⎦	
Serum dopamine-β-hydroxylase activity	0.86 ± 0.25^c	Weinshilboum et al. (1975)

[a] Standard error, where available
[b] Twin study
   From full-sib correlation, so that the estimate is an upper limit (Falconer, 1960)
[d] Estimation by regression of offspring on midparent unless otherwise indicated

When quantitative human genetics was developed, it appeared to many that heritability would be a useful parameter to estimate, so that standard techniques (Falconer, 1960; Smith, 1975) were applied to variation in many biochemical traits; some of these results are shown in Table 9. In certain cases, it may indeed be relevant that there is genetical variation in a biochemical trait. For example, the rate of drug metabolism such as phenylbutazone hydroxylation may be relevant to a treatment regime, in that knowledge of the genetical contribution to variation in this rate may prevent iatrogenic disease. However, as has been forcefully and frequently emphasized by Edwards (1975) and originally by Fisher (1951), mere knowledge of

h^2 is in itself hardly helpful. The next step in analysis of the observed genetical variation is much harder but potentially of greater benefit.

It was mentioned in Chapter 4 that the phenotypes of the ABO blood group system are associated with variation in plasma cholesterol level. It is possible to partition this variability in such a way as to show that about 5% of the total variance (i.e. about 10% of the genetical component) is associated with the ABO system (Mayo et al., 1971). As this association is consistent across populations, it seems likely that it is not fortuitous, but this is in principle testable by the use of information about loci linked to ABO. Thus, since it appears (Fraser et al., 1974) that the adenylate kinase (AK) locus does not affect cholesterol level, an examination of the recombinant and non-recombinant offspring of a large number of matings of the form $\dfrac{AK^1A}{AK^2O} \times \dfrac{AK^1O}{AK^1O}$ would reveal whether a locus lying between ABO and AK affected cholesterol level. As the genome becomes better mapped, a process now gathering speed, such tests will become simple and precise, rather than far-fetched, complex and inaccurate as they may now appear.

A further important role for polymorphism in the elucidation of quantitative biochemical genetics comes from the actual variation in enzyme activity and its sequelae induced by known polymorphic loci. It is well known (Smith, 1975) that a very high heritability for a trait can be a hint that a single locus is the major influence on that trait. Thus, the value of 0·82 for erythrocyte acid phosphatase activity shown in Table 9, which was computed crudely from individuals unequally representing different age groups and different family sizes, might induce a suspicion that a major locus was influencing enzyme activity. This, of course, is the case; Eze et al. (1974) computed h^2 merely by way of illustration, having already determined the actual enzyme activities of five available phenotypes at the erythrocytic acid phosphatase locus, ACP (cf. Fig. 2, Chapter 5). Further analysis would allow the partitioning out of the effect of the AcP locus, so that the importance of age and other non-genetical factors could then be evaluated, allowing a far more meaningful understanding of quantitative biochemical variation. (Meanwhile, Magenis et al. (1975b) have used quantitative differences between ACP alleles to confirm the assignment of this locus to the short arm of chromosome 2.) Weinshilboum et al. (1975) have begun such an analysis for serum dopamine-β-hydroxylase, an enzyme tentatively suggested as a useful measure of the function of the sympathetic nervous system (Rush and Geffen, 1972).

Some detailed aspects of quantitative variation are considered in Chapters 5 and 6. For an extensive and original review of the topic, see Modiano (1976).

D. Pharmacogenetics

It is well known that different individuals taking a "normal" dose of a particular drug react rather differently. In most situations individual responses to the drug as measured in the clearance rates follow a continuous unimodal distribution. This may be interpreted as indicating that drug metabolism is a polygenically controlled trait and therefore only within the province of biochemical genetics to the limited extent discussed in the previous section. However, an alternative explanation of such distributions is that they represent unifactorial genetical control obscured by a strong environmental component. To distinguish between the two alternative explanations it is necessary to discover genetical variation which breaks the smooth distribution. Thus the observation that there are certain individuals who cannot metabolize hydrogen peroxide at all, and furthermore that their condition is inherited as an autosomal recessive and due to a deficiency of the enzyme catalase (Takahara, 1952), showed that hydrogen peroxide clearance distributions in "normal" people are probably environmentally controlled. Similarly suxamethonium sensitivity, primaquine sensitivity, diphenylhydantoin toxicity, warfarin resistance (Chapter 13 Section V) and malignant hyperthermia (Vesell, 1973) are other examples of Mendelian traits where occasional abnormal patients reveal the unifactorial nature of the drug metabolism.

Interesting as these unusual patients may be, from the point of view of clinical management those drug responses which show polymorphic variation, i.e. considerable numbers of people show variant responses, are of much greater importance. The classic example is the acetylator polymorphism, first revealed in response to the anti-tuberculostatic drug isoniazid (Hughes, 1953). Clearance rates of isoniazid, usually followed through urinary ratios of isoniazid to acetyl-isoniazid are bimodal, with different individuals being classed as either fast or slow acetylators. Acetylation is a Mendelian trait with fast being dominant over slow and fast acetylators thus comprising both heterozygote $(Ac^F Ac^S)$ and homozygote $(Ac^F Ac^F)$ genotypes (Price Evans et al., 1960). The enzymatic basis of this polymorphism is a microsomal liver acetyltransferase, whose difference between fast and slow individuals appears to be based on the amount of an enzyme protein rather than on structural variation (Jenne, 1965). Gene frequencies for the Ac^F and Ac^F alleles vary from one population to another, so that slow inactivators range from 50% of Caucasian populations to only 10% of Japanese people.

It might be anticipated that the acetylator polymorphism would be of considerable importance in deciding a strategy of isoniazid therapy. Despite the marked difference in the rate of clearance of the drug from the blood-stream, no systematic difference in response to the anti-tuberculostatic activity has been demonstrated between slow and fast inactivators (Price Evans, 1963). Slow inactivators, however, are more likely to develop side effects during prolonged therapy, of which peripheral neuropathy and a syndrome resembling systemic lupus erythematosis are the most important. This appears to be due to the ability of isoniazid to complex and sequester vitamin B_6, a property more apparent in those who have high levels of drug in their plasma for appreciable periods of time. It is standard practice to supplement isoniazid with vitamin B_6 during prolonged therapy, rather than determine the acetylator status of the patient.

Several drugs, whose chemical structures suggested that they might be metabolized by the same enzymatic system as isoniazid, have been sub-jected to clearance distribution tests. With the exception of sulphadimidine (sulphamethazine), sulphonamides are cleared unimodally, and since acetylation is the mechanism of their metabolism, a second and distinct acetylase system must exist in the liver (Mattila et al., 1969). Para-amino-benzoic acid and para-aminosalicylic acid are likewise removed at the same rate from the bloodstreams of both slow and fast inactivators as defined by the isoniazid system (Jenne et al., 1961; Price Evans, 1964).

Drugs whose acetylation like isoniazid is polymorphic include hydrala-zine, sulphadimidine, sulphapyridine, phenelzine, sulphasalazine and possibly serotonin (Price Evans and White, 1964; Schroder and Price Evans, 1972; Gelber et al., 1971; Johnstone and Marsh, 1973; Propping and Kopun, 1973). Phenelzine has attracted some attention because it is claimed that only patients who are slow inactivators respond to the use of this drug in the treatment of neurotic depression (Johnstone and Marsh, 1973), and that in fast inactivators placebo is equally effective. This finding requires confirmation in view of the fact that earlier Price Evans et al. (1965) were only able to observe an increased incidence of side effects in slow inactivators on phenelzine. Sulphasalazine (azulfidine, salicylazosylpha-pyridine) has also been carefully studied, because although highly effective in the treatment of ulcerative colitis and Crohn's disease it is often dis-continued because of unpleasant and persistent side effects. Das et al. (1973) have shown that this is largely restricted to slow inactivators. But the most important claim for polymorphic acetylation concerns the potent vasoconstrictor serotonin. On the basis of competitive inhibition studies in the acetylation of isoniazid it has been argued that serotonin is probably the natural substrate for the polymorphic N-acetyl-transferase in human liver microsomes (Schloot et al., 1969). Such a claim could have a considerable impact on the development of molecular behavioural genetics. However, White et al. (1969), using purified enzyme preparations from human post-mortem livers, could find no correlation between acetylation rates of isoniazid and serotonin or sulphadimidine and serotonin, while isoniazid and sulphadimidine were exactly correlated. They concluded that serotonin is not acted on by the polymorphic acetyltransferase.

The acetylator polymorphism is an example of pharmacogenetics at the level of practical importance. Another such example, though less clearly defined, follows from the original suggestion of Pare et al. (1962), that depressive states may be divided into two distinct types on the basis of drug response. Patients with "reactive" depressions respond to monoamine oxidase inhibitors like phenelzine, while those with "endogenous" depressions respond to the tricyclic antidepressants like imipramine and nortriptyline (Pare and Mack, 1971). Phenelzine, as we have seen, is involved in the polymorphic acetylase system, with slow acetylators showing major response (Johnstone and Marsh, 1973). Nortriptyline, on the other hand, is hydroxylated to 10-hydroxynortriptyline, and its clearance rate though unimodal shows an extraordinarily wide range (Sjoqvist et al., 1968). Intensive investigations of nortriptyline levels in plasma have tended to suggest that its rate of metabolism is polygenically controlled (Asberg et al., 1971). Similar conclusions have been reached for the other tricyclic antidepressants, imipramine, amitriptyline and desmethyl-imipramine (Propping and Kopun, 1973). The observations of Pare and Mack (1971) thus appear to be grounded in two different genetical systems, one Mendelian, the other polygenic. Further studies on the metabolic bases to these antidepressants are obviously of great importance in devising more rational approaches to the treatment of depression.

E. Sensory Perception

Since at least the time of the coining of the proverb, de gustibus non disputandum est, it has been widely recognized that individual sensory perceptions vary. Those defects related to vision, for example myopia and colour blindness, are perhaps the most obvious and have certainly been recognized for many centuries, but it is really only in the last half century that very wide variation in the ability to detect individual chemical substance by means of the senses has been observed. This applies particularly to the gustatory and olfactory senses.

The fact that the individual sensitivity to—that is ability to recognize the taste of—phenylthiocarbamide (PTC) has a bimodal frequency distribution was recognized in the 1920s and 30s (Fox, 1931), and indeed it was extended by 1939 to the larger anthropoid apes (Fisher et al., 1939). In later years, this ability was found to vary in the same way for certain other

Table 10

Specific inabilities to taste or smell certain substan

Sense	Substance	Population	Frequency Phenotype	Allele	Mode of inheritance	Comments	Reference
Taste	phenylthio-carbamide	London Fezzan, Libya Kikuyu, Kenya	0·37 0·18 0·02	0·61 0·42 0·14	autosomal recessive	well-established; see text and Amerine *et al.* (1965) for general discussion	Kalmus (1952) Sunderland and Rosa (1975)
Smell	"freesia scent"	Oxford (under-graduates)	0·05-0·08		autosomal recessive		McWhirter, cited by Amoore (1971)
	"verbena scent"	North America	0·33		unclear	one-third of sub-jects could smell pink verbena but not red, and *vice versa* the other two-thirds	Blakeslee (1918)
	HCN	Australia	0·20		X-linked recessive	not repeatable by Kirk and Stenhouse (1953) other workers (Brown and Robinette, 1967)

ketones	Hungary			recessive	not yet repeated; see text	Forrai et al. (1970)
isovaleric acid	North America —(Whites) —(Negroes)	0·01 0·09	0·10 0·30	autosomal recessive	similar results obtained for isobutyric acid (Amoore, 1970)	Whissel-Buechy and Amoore (1973)
exaltolide	North America —(Whites) —(Negroes) London	0·07 0 0·09	0·26 0·30	autosomal recessive		Kalmus and Seedburgh (1975)
n-butyl mercaptan		0·002		unclear	some of the families of anosmics showed autosomal recessivity for the trait, others autosomal dominance with late onset; not yet repeated	Patterson and Lauder (1948)

chemically related substances with bitter flavours, and the possibility that there are selective differences between the phenotypes has already been mentioned in Chapter 4. A possible antithetical sensitivity, to the bitterness of the berries of *Antidesma bunius*, has recently been reported (Henkin and Gillis, 1977).

While other taste and smell sensitivity variations have been detected, their biochemical elucidation still awaits a comprehensive and precise theory of sensory perception. (See Duncan and Sheppard, 1963, for a treatment of problems in sensory discrimination, and Amoore, 1970, for an extensive discussion of the sense of smell.) Amoore *et al.* (1968) were the first to carry out a systematic evaluation of specific anosmias, detecting a number of these, such as the inability to detect the sweaty odour of iso-valeric acid and isobutyric acid. Forrai *et al.* (1970) even reported that there was a bimodal distribution for the ability to perceive ketone compounds in a Hungarian population, though one of us (Mayo, unpublished) has been unable to show that the threshold is as precise as Forrai *et al.* (1970) reported.

In a more recent investigation, Kalmus and Seedburgh (1975) were able to confirm the existence of the bimodal odour threshold for an artificial musk, exaltolide, reported by Whissell-Buechy and Amoore (1973) and to show further that this related to at least one other synthetic musk, musk ambrette, though not to musk ketone.

Table 10 lists the cases described, together with a few others not so clearly defined. It can be seen that there is wide variation in individual abilities to perceive different compounds, and it should be recognized that these abilities may well contribute to the tendency, noted above, for individuals to vary in their preferences for different food and drink products. Thus, it is certain that the recognition and measurement of specific anosmias and taste variations will be of increasing importance in the food and drink industries, probably particularly in the wine industry. Comfort (1973) has pointed to their possible importance in sexual behaviour. Although this aspect has as yet received little scientific attention, Thomas (1975) has speculated that the major histocompatibility complex may have a role to play in mammalian mating preference.

REFERENCES

Amerine, M. A., Pangborn, R. M. and Roessler, E. B. (1965). "Principles of Sensory Evaluation of Food". Academic Press, New York and London.
Amoore, J. E. (1970). "The Molecular Basis of Odor". Charles C. Thomas, Springfield, Illinois.

Amoore, J. E. (1971). *In* "Handbook of Sensory Physiology Vol. 4, Olfaction". 246. Springer, Berlin.

Amoore, J. E., Venstrom, D. and Davis, A. R. (1968.) *Percept. Motor Skills* **26**, 143.

Andresen, P. H. (1952). "The Human Blood Groups: Utilized in Disputed Paternity Cases and Criminal Proceedings". Charles C. Thomas Springfield. Illinois.

Asberg, M., Price Evans, D. A. and Sjoqvist, F. (1971). *J. Med. Genet.* **8**, 129.

Bailey, N. T. J. (1961). "The Mathematical Theory of Genetical Linkage". Clarendon Press, Oxford.

Baltimore Conference (1975). Human Gene Mapping 3. 3rd Internat. Workshop on Human Gene Mapping. *Cytogenet. Cell Genet.* **16**, 1 (1976).

Barratt, R. W., Newmeyer, D., Perkins, D. D. and Garnjobst, L. (1954). *Adv. Genet.* **6**, 1.

Barski, G., Sorieul, S. and Cornefert, F. (1960). *C. R. Acad. Sci. (Paris)* **251**, 1825.

Bateman, A. J. (1960). *Lancet* **i**, 1293.

Beckman, G. (1973). *Hereditas* **73**, 305.

Bender, K. and Grzeschik, K-H. (1976). *Hum. Genet.* **31**, 341.

Benditt, E. P. and Benditt, J. M. (1973). *Prlc. Nat. Acad. Sci. USA* **70**, 1753.

Berry, H. K., Dobzhansky, Th., Gartler, S. M., Levene, H. and Osborne, R. H. (1955). *Am. J. Hum. Genet.* **7**, 93.

Beutler, E., Collins, Z. and Irwin, L. E. (1967). *New Eng. J. Med.* **276**, 389.

Blakeslee, A. F. (1918). *Science* **48**, 298.

Bloom, A. D., Choi, K. W. and Lamb, B. J. (1971). *Science* **172**, 382.

Bootsma, D. (1974). *In* "Somatic Cell Hybridisation" (R. L. Davidson and F. de la Cruz, Eds), 265, Raven Press, New York.

Brock, D. J. H. (1970). *Biochem. Genet.* **4**, 617.

Brown, K. S. and Robinette, R. R. (1967). *Nature, Lond.* **215**, 406.

Brown, S., Wiebel, F. J., Gelboin, H. V. and Minna, J. D. (1976). *Proc. Natl. Acad. Sci. USA* **73**, 4628.

Bruns, G. A. P. and Gerald, P. S. (1976). *Science* **192**, 54.

Bucher, K. D. and Elston, R. C. (1975). *Am. J. Hum. Genet.* **27**, 689.

Cavalli-Sforza, L. L. and Conterio, F. (1960). *Atti. Assoc. Genet. Ital.* **5**, 333.

Cavalli-Sforza, L. L. and Edwards, A. W. F. (1964). *Proc. XI Intern. Congr. Genet.* **3**, 923.

Chakraborty, R., Shaw, M. and Schull, W. J. (1974). *Am. J. Hum. Genet.* **26**, 477.

Chen, S. H. and Giblett, E. R. (1971). *Science* **173**, 148.

Cohen, P. and Omenn, G. S. (1972). *Biochem. Genet.* **7**, 303.

Comfort, A. (1973). *Nature, Lond.* **245**, 157.

Comings, D. E. (1966). *Lancet* **ii**, 1137.

Côté, G. B. and Edwards, J. H. (1975). *Ann. Hum. Genet.* **39**, 51.

Côté, G. B., Edwards, J. H., Chown, B., Giblett, E. R., Lewis, M., Moore, B. P. L., Steinberg, A. G. and Uchida, I. A. (1975b). *Ann. Hum. Genet.* **39**, 61.

Curtain, C. C. and Baumgarten, A. (1966). *Immunology* **10**, 499.

Curtain, C. C. and Golab, T. (1966). *Aust. J. Exp. Biol. Med. Sci.* **44**, 589.

Czechowicz, A. and Pamnany, L. (1972). *Med. J. Aust.* **1**, 1252.

Danes, B. S. and Bearn, A. G. (1967). *J. Exp. Med.* **126**, 509.

Das, K. M., Eastwood, M. A., McManus, J. P. A. and Sirans, W. (1973). *New Eng. J. Med.* **289**, 491.

Davidson, R. G. and Cortner, D. A. (1967). *Science* **157**, 1569.

Davidson, R. G., Nitowsky, H. M. and Childs, B. (1963). *Proc. Natn. Acad. Sci. USA* **50**, 481.

462   O. MAYO AND D. J. H. BROCK

Davidson, R. G., Cortner, D. A., Rattazzi, M. C., Ruddle, F. C. and Lubs, H. A. (1970). *Science* **169**, 391.
Denney, R. M. and Craig, I. W. (1976). *Biochem. Genet.* **14**, 99.
De Weerd-Kastelein, E. A., Keijzer, W. and Bootsma, D. (1972). *Nature, Lond.* **238**, 82.
Deys, B. P., Grzeschick, K. H., Grzeschick, A., Jaffe, E. R. and Siniscalco, M. (1972). *Science* **175**, 1002.
Ducos, J., Marty, Y., Sanger, R. and Race, R. R. (1971). *Lancet* **ii**, 219.
Duncan, C. J. and Sheppard, P. M. (1963). *Proc. R. Soc.* B. **153**, 343.
Edwards, J. H. (1975). *J. R. Stat. Assoc.* A, **138**, 131.
Edwards, J. H., Hopkinson, D. A. and Harris, H. (1971). *Ann. Hum. Genet.* **34**, 395.
Emery, A. E. H. (1963). *Lancet* **i**, 1126.
Emery, A. E. H., Smith, C. A. B. and Sanger, R. (1969). *Ann. Hum. Genet.* **32**, 261.
Emery, A. E. H., Nelson, M. M. and Mayo, O. (1971). *In* "Actualités de Pathologie Neuromusculaire" (G. Serratrice Ed.). Expansion Scientifique, Paris.
Ephrussi, B. and Weiss, M. C. (1965). *Proc. Natn. Acad. Sci. USA* **53**, 1040.
Ewens, W. J. and Feldman, M. (1976). *In* "Population Genetics and Ecology" (S. Karlin and E. Nero, Eds). Academic Press. New York and London.
Eze, L. C., Tweedie, M. C. K. and Price Evans, D. A. (1970). *J. Med. Genet.* **7**, 5.
Eze, L. C., Tweedie, M. C. K., Bullen, M. F., Wren, P. J. J. and Price Evans, D. A. (1974). *Ann. Hum. Genet.* **37**, 333.
Falconer, D. S. (1960). "Introduction to Quantitative Genetics". Oliver and Boyd Edinburgh.
Fellous, M., Bengtsson, B., Finnegan, D. and Bodmer, W. F. (1974). *Ann. Hum. Genet.* **37**, 421.
Fialkow, P. J. (1972). *In* "Advances in Cancer Research" (G. Klein and S. Weinhouse, Eds), Vol. 15, 191. Academic Press, New York and London.
Fialkow, P. J. (1973). *Ann. Hum. Genet.* **37**, 39.
Fialkow, P. J., Gartler, S. M. and Yoshida, A. (1967). *Proc. Natn. Acad. Sci. USA* **58**, 1468.
Fialkow, P. J., Klein, G., Gartler, S. M. and Clifford, P. (1970). *Lancet* **i**, 384.
Fisher, R. A. (1918). *Trans. R. Soc. Edinb.* **52**, 399. *In* "Collected Papers of R. A. Fisher" (J. H. Bennett, Ed.), Vol. 1. University of Adelaide, 1974.
Fisher, R. A. (1951a). *Br. Agric. Bull.* **4**, 217. *In* "Collected Papers of R. A. Fisher" (J. H. Bennett, Ed.), Vol. 5. University of Adelaide, 1974.
Fisher, R. A. (1951b). *Heredity* **5**, 95. *In* "Collected Papers of R. A. Fisher" (J. H. Bennett, Ed.), Vol. 5. University of Adelaide, 1974.
Fisher, R. A., Ford, E. B. and Huxley, J. (1939). *Nature, Lond.* **144**, 750.
Forrai, G., Szabados, T., Papp, E. S. and Bánkovi, G. (1970). *Humangenetik* **8**, 348.
Fox, A. L. (1931). *Science* **73** Suppl. 14.
Francke, U. (1976). *Am. J. Hum. Genet.* **28**, 357.
Fraser, G. R. (1963). *Ann. Hum. Genet.* **26**, 297.
Fraser, G. R. (1966). *Ann. Hum. Genet.* **29**, 323.
Fraser, G. R. (1969). *Am. J. Hum. Genet.* **21**, 593.
Fraser, G. R. (1970). *Am. J. Hum. Genet.* **22**, 692.
Fraser, G. R. (1972). *Am. J. Hum. Genet.* **24**, 359.
Fraser, G. R., Volkers, W. S., Bernini, L. F., de Greve, W. B., van Loghem, E., Meera Khan, P., Nijenhuis, L. E., Veltkamp, J. J., Vogel, G. P. and Went, L. N. (1974). *Hum. Hered.* **24**, 424.

Fudenberg, H. H. and Warner, N. L. (1970). "Advances in Human Genetics" (H. Harris and K. Hirschhorn Eds), Vol. 1 pp.131. Plenun, New York.
Fudenberg, H. H., Pink, J. R. L., Stites, D. P. and Wang, A. C. (1972). "Basic Immunogenetics". Oxford University Press, Oxford and New York.
Galjaard, H., Hoogeveen, A., Keijzer, W., de Wit-Verbeck, E. and Vlek-Noot, C. (1974a). *Histochem. J.* 6, 491.
Galjaard, H., Hoogeveen, A., de Wit-Verbeck, H. A., Reuser, A. J. J., Keijzer, W., Westerveld, A. and Bootsma, D. (1974b). *Exp. Cell Res.* 87, 444.
Galjaard, J., Hoogeveen, A., Keijzer, W., de Wit-Verbeck, H. A., Reuser, A. J. J., Ho, M. W. and Robinson, D. (1975). *Nature, Lond.* 257, 60.
Gandini, E. and Gartler, S. M. (1969). *Nature, Lond.* 224, 599.
Gandini, E., Gartler, S. M., Angioni, G., Argiolas, N. and Dell'Acqua, G. (1968). *Proc. Natn. Acad. Sci. USA* 61, 945.
Gardner, R. L. and Lyon, M. F. (1971). *Nature, Lond.* 231, 385.
Gartler, S. M., Dobzhansky, Th. and Berry, H. K. (1955). *Am. J. Hum. Genet.* 7, 108.
Gartler, S. M., Ziprkowski, L., Krakowski, A., Ezra, R., Szeinberg, A. and Adam, A. (1966). *Am. J. Hum. Genet.* 18, 282.
Gartler, S. M., Liskay, R. M., Campbell, B. K., Sparkes, R. and Cant, N. (1972). *Cell Differ.* 1, 215.
Gartler, S. M., Liskay, R. M. and Cant, N. (1973). *Exp. Cell Res.* 82, 464.
Gedde-Dahl, T., Cook, P. J. L., Fågerhol, M. K. and Pierce, J. A. (1975). *Ann. Hum. Genet.* 39, 43.
Gelber, R., Peters, J. H., Gordon, G. R., Glazko, A. J. and Levy, L. (1971). *Clin. Pharmacol. Therap.* 12, 225.
Geraedts, J. P. M. and Pearson, P. L. (1974). *Clin. Genet.* 6, 247.
Goodman, H. O. (1965). *Am. J. Hum. Genet.* 17, 111.
Gorman, J. G., Di Re, J., Treacy, A. M. and Cahan, A. (1963). *J. Lab. Clin. Med.* 61, 642.
Gravel, R. A., Mahoney, M. J., Ruddle, F. H. and Rosenberg, L. E. (1975). *Proc. Natn. Acad. Sci. USA* 72, 3181.
Grundbacher, F. J. (1974). *Am. J. Hum. Genet.* 26, 1.
Harris, H. and Smith, C. A. B. (1949). *Ann. Eugen. Lond.* 14, 309.
Harris, H. and Watkins, J. F. (1965). *Nature, Lond.* 205, 640.
Harris, H., Hopkinson, D. A. and Robson, E. B. (1962). *Nature, Lond.* 196, 1296.
Henkin, R. I. and Gillis, W. T. (1977). *Nature, Lond.* 265, 536.
Heuch, I. (1975). *Biometrics* 31, 685.
Hopkinson, D. A. and Harris, H. (1968). *Ann. Hum. Genet. Lond.* 31, 359.
Hughes, H. B. (1953). *J. Pharmacol. Exp. Theor.* 109, 196.
Jenne, J. W. (1965). *J. Clin. Invest.* 44, 1992.
Jenne, J. W., MacDonald, F. M. and Mendoza, E. (1961). *Am. Rev. Resp. Dis.* 84, 371.
Jinks, J. L. and Fulker, D. W. (1970). *Psychol. Bull.* 73, 311.
Johnstone, E. C. and Marsh, W. (1973). *Lancet* i, 567.
Kahan, B. and DeMars, R. (1975). *Proc. Natn. Acad. Sci. USA* 72, 1510.
Kalmus, H. (1952). *Ann. Hum. Genet.* 22, 222.
Kalmus, H. and Seedburgh, D. (1975). *Ann. Hum. Genet.* 38, 495.
Kalow, W. and Stavron, N. (1957). *Can. J. Biochem. Physiol.* 33, 1305.
Kamaryt, J. and Laxova, R. (1965). *Humangenetik* 1, 579.

Kan, Y. W., Golbus, M. S., Trecartin, R., Furbetta, M. and Cao, A. (1975). *Lancet* **ii**, 790.

Kao, F. T. and Puck, T. T. (1970). *Nature, Lond.* **228**, 329.

Kirk, R. L. and Stenhouse, N. S. (1953). *Nature, Lond.* **171**, 698.

Kraemer, K. H., Coon, H. G., Petinga, R. A., Barrett, S. F., Rahe, A. E. and Robbins, J. H. (1975). *Proc. Natn. Acad. Sci. USA* **72**, 59.

Langaney, A. and Pison, G. (1975). *Am. J. Hum. Genet.* **27**, 558.

Lang-Brown, H., Lawler, S. D. and Penrose, L. S. (1953). *Ann. Hum. Genet.* **17**, 307.

Lawler, S. D. (1964). *Hum. Biol.* **36**, 146.

Lee, G. R., McDiarmid, W. D., Cartwright, G. E. and Wintrobe, M. M. (1968). *Blood* **32**, 59.

Lewis, M., Kaita, H. and Chown, B. (1976). *Am. J. Hum. Genet.* **28**, 619.

Linder, D. (1969). *Proc. Natn. Acad. Sci. USA* **63**, 699.

Linder, D. and Gartler, S. M. (1965). *Science* **150**, 67.

Linder, D. and Power, J. (1970). *Ann. Hum. Genet.* **34**, 21.

Littlefield, J. W. (1966). *Exp. Cell Res.* **41**, 190.

Lyon, M. F. (1961). *Nature, Lond.* **190**, 372.

Lyon, M. F. (1971). *Nature New Biol.* **232**, 229.

Lyons, L. B., Cox, R. P. and Dancis, J. (1973). *Nature, Lond.* **243**, 533.

Magenis, R. E., Hecht, F. and Lovrien, E. W. (1970). *Science* **170**, 85.

Magenis, R. E., Overton, K., Wyandt, H., Bergstrom, T., Hecht, F. and Lovrien, E. W. (1975a). *Humangenetik* **27**, 91.

Magenis, R. E., Koler, R. D., Lovrien, E., Bigley, R. H., Du Val, M. C. and Overton, K. M. (1975b). *Proc. Natn. Acad. Sci. USA* **72**, 4526.

Mann, J. D., Cahan, A., Gelb, A. G., Fisher, N., Harper, J., Tippett, P., Sanger, R. and Race, R. R. (1962). *Lancet* **ii**, 8.

Martensson, L. (1964). *J. Exp. Med.* **120**, 1169.

Masters, A. B. (1967). *Br. J. Psychiat.* **113**, 1309.

Mattila, M. J., Tutinen, H. and Alhara, E. (1969). *Ann. Med. Exp. Fenn.* **47**, 308.

Maynard Smith, S., Smith, C. A. B. and Penrose, L. S. (1961). "Mathematical Tables for Research Workers in Human Genetics". Churchill, London.

Mayo, O. (1975). *In* "Textbook of Human Genetics" (G. R. Fraser and O. Mayo, Eds). Blackwell, Oxford.

Mayo, O., Fraser, G. R. and Stamatoyannopoulos, G. (1969). *Hum. Hered.* **19**, 86.

Mayo, O., Wiesenfeld, S. L., Stamatoyannopoulos, G. and Fraser, G. R. (1971). *Lancet* **ii**, 554.

McCurdy, P. R. (1967). *Clin. Res.* **15**, 65.

Migeon, B. R. (1972). *Nature, Lond.* **239**, 87.

Migeon, B. R. and Kennedy, J. F. (1975). *Am. J. Hum. Genet.* **27**, 233.

Migeon, B. R., Der Kaloustian, V. M., Nyhan, W. L., Young, W. J. and Childs, B. (1968). *Science* **160**, 425.

Migeon, B. R., Novum, R. A. and Corsaro, C. M. (1974). *Proc. Natn. Acad. Sci. USA* **71**, 937.

Miller, O. J., Allderdyce, P. W., Miller, D. A., Breg, W. R. and Migeon, B. R. (1971). *Science* **173**, 244.

Modiano, G. (1976). "Genetically Determined Quantitative Protein Variations in Man excluding Immunoglobulins" (Accadmia Nazionale dei. Lincei, Rome.

Moore, M. J., Funakoshi, S. and Deutsch, H. F. (1971). *Biochem. Genet.* **5**, 497.

Morton, N. E. (1956). *Am. J. Hum. Genet.* **8**, 80.

Murray, R., Hobbs, J. and Payne, B. (1971). *Nature, Lond.* **232**, 51.

Nadler, H. L. (1968). *Biochem. Genet.* **2**, 119.

Nadler, H. L., Chacko, C. M. and Rachmeler, M. (1970). *Proc. Natn. Acad. Sci. USA* **67**, 976.

Nance, W. E. (1964). *Cold Spring Harbor Symp. Quant. Biol.* **29**, 415.

Nance, W. E. and Grove, J. (1972). *Science* **177**, 716.

Nei, M. (1975). "Molecular Genetics and Evolution" North Holland Publishing Co., Amsterdam.

Nyhan, W. L., Bakay, B., Connor, J. D., Marks, J. F. and Kelle, D. K. (1970). *Proc. Natn. Acad. Sci. USA* **65**, 214.

Olaisen, B., Teisberg, P., Gedde-Dahl, Jr., T. and Thorsby, E. (1975). *Humangenetik* **30**, 291.

Ott, J., Linder, D., McCaw, B. K., Lovrien, E. W. and Hecht, F. (1976). *Ann. Hum. Genet., Lond.* **40**, 191.

Pare, C. M. B. and Mack, J. W. (1971). *J. Med. Genet.* **8**, 306.

Pare, C. M. B., Rees, L. and Sainsbury, M. J. (1962). *Lancet* **ii**, 1340.

Park, I. and Jones, K. W. (1968). *Am. J. Obstet. Gynecol.* **102**, 106.

Park, W. W. (1957). *J. Anat.* **91**, 369.

Patterson, B. M. and Lauder, B. A. (1948). *J. Hered.* **39**, 295.

Polani, P. E., Angell, R., Giannelli, F., de la Chapelle, A., Race, R. R. and Sanger, R. (1970). *Nature, Lond.* **227**, 613.

Price Evans, D. A. (1963). *Am. J. Med.* **34**, 639.

Price Evans, D. A. (1964). *Proc. R. Soc. Med.* **57**, 508.

Price Evans, D. A. and White, T. A. (1964). *J. Lab. Clin. Med.* **63**, 394.

Price Evans, D. A., Manley, K. E. and McKusick, V. A. (1960). *Br. Med. J.* **2**, 485.

Price Evans, D. A., Davidson, K. and Pratt, R. T. C. (1965). *Clin. Pharmacol. Therap.* **6**, 430.

Propping, P. and Kopun, M. (1973). *Humangenetik* **20**, 291.

Race, R. R. and Sanger, R. (1975). "Blood Groups in Man", 6th ed. Blackwell, Oxford.

Revesz, T., Schuler, D., Goldschmidt, B. and Elod, S. (1972). *J. Med. Genet.* **9**, 396.

Robson, E. B., Polani, P. E., Dart, S. J., Jacobs, P. A. and Renwick, J. A. (1969). *Nature, Lond.* **223**, 1163.

Romeo, G. and Migeon, B. R. (1970). *Science* **170**, 180.

Romeo, G. and Migeon, B. R. (1975). *Humangenetik* **29**, 165.

Rotterdam Conference (1974) "Human Gene Mapping 2". Second Internat. Workshop on Human Gene Mapping. *Cytogenet. Cell Genet.* **14**, 180 (1975).

Rundle, A. T. (1973). *Clin. Genet.* **4**, 520.

Rush, R. A. and Geffen, L. B. (1972). *Circ. Res.* **31**, 444.

Sager, R. (1964). *New Eng. J. Med.* **271**, 352.

Schloot, W., Tigges, F. J., Blaesner, H. and Goedde, H. W. (1969). *Hoppe-Seylers Z. Physiol. Chem.* **35**, 1353.

Schroder, H. and Price Evans, D. A. (1972). *J. Med. Genet.* **9**, 168.

Silagi, S., Darlington, G. and Bruce, S. A. (1969). *Proc. Natn. Acad. Sci. USA* **62**, 1085.

Siniscalco, M., Klinger, H. P., Eagle, H., Koprowski, H., Fujimoto, W. Y. and Seegmiller, J. E. (1969). *Proc. Natn. Acad. Sci. USA* **62**, 693.

Sjoqvist, F., Hammar, W., Idestrom, C. M., Lind, M., Tuck, D., Asberg, M. (1968). *Excerpta Med. Internat. Congr. Ser.* **145**, 146.

Smith, C. (1975). *In* "Textbook of Human Genetics" (G. R. Fraser and O. Mayo, Eds), Blackwell, Oxford.

Smith, M., Hopkinson, D. A. and Harris, H. (1971). *Ann. Hum. Genet.* **34**, 251.

Smith, M., Hopkinson, D. A. and Harris, H. (1973). *Ann. Hum. Genet.* **36**, 401.

Sørensen, H. and Dissing, J. (1975). *Hum. Hered.* **25**, 284.

Spencer, N., Hopkinson, D. A. and Harris, H. (1964). *Nature, Lond.* **201**, 299.

Sturt, E. (1976). *Ann. Hum. Genet.* **40**, 147.

Sunderland, E. and Rosa, P. J. (1975). *Hum. Biol.* **47**, 473.

Sykes, B. and Soloman, E. (1978) *Nature, Lond.* **272**, 548.

Takahara, S. (1952). *Lancet* **ii**, 1101.

Teisberg, P., Gjone, E. and Olaisen, B. (1975). *Ann. Hum. Genet.* **38**, 327.

Thomas, G. H., Taylor, H. A., Miller, C. S., Axelman, J. and Migeon, B. R. (1974). *Nature, Lond.* **250**, 580.

Thomas, L. (1975). Fourth Internat. Convoc. Immunol. (E. Neter and F. Milgrom, Eds) S. Karger, Basel, Buffalo, New York.

Thuline, H. C., Hodgkin, W. E., Fraser, G. R. and Motulsky, A. G. (1969). *Am. J. Hum. Genet.* **21**, 581.

Veltkamp, J. J., Mayo, O., Motulsky, A. G. and Fraser, G. R. (1972). *Hum. Hered.* **22**, 102.

Vesell, E. S. (1973). *In* "Progress in Medical Genetics" (A. G. Steinberg and A. G. Bearn, Eds), Vol. 9, 291. Grune and Stratton, New York.

Ward, J. C., Merritt, A. D. and Bixler, D. (1971). *Am. J. Hum. Genet.* **23**, 403.

Weatherall, D. J., Pembrey, M. E., Hall, E. G., Sanger, R., Tippet, P. and Gavin, J. (1970). *Lancet* **ii**, 744.

Weinshilboum, R. M., Schrott, H. G., Raymond, F. A., Weidman, W. H. and Elveback, L. R. (1975). *Am. J. Hum. Genet.* **27**, 573.

Weiss, M. C. and Green, H. (1967). *Proc. Natn. Acad. Sci. USA* **58**, 1104.

Weitkamp, L. R., Townes, P. L. and May, A. G. (1975). *Am. J. Hum. Genet.* **27**, 486.

Westerveld, A., Jongsma, A. P. M., Meera Khan, P., van Someren, H. and Bootsma, D. (1976). *Proc. Natn. Acad. Sci. USA* **73**, 895.

Whissell-Buechy, D. Y. E. and Amoore, J. E. (1973). *Nature, Lond.* **242**, 271.

White, T. A., Jenne, J. W. and Price Evans, D. A. (1969). *Biochem. J.* **113**, 721.

Whittaker, J. A. and Price Evans, D. A. (1970). *Br. Med. J.* **4**, 333.

Wilson, D. E., Povey, S. and Harris, H. (1976). *Ann. Hum. Genet.* **39**, 305.

Windhorst, D. B., Holmes, B. and Crod, R. A. (1967). *Lancet* **i**, 737.

Woolf, B. (1955). *Ann. Hum. Genet.* **19**, 251.

Yamazaki, T. and Maruyama, T. (1974). *Science* **183**, 1091.

Pathological Variation

10 Inborn Errors of Metabolism

D. J. H. BROCK

Department of Human Genetics, University of Edinburgh, Scotland.

I. INTRODUCTION

In the past twenty years inborn errors of metabolism have generated research activity out of all proportion to their importance as diseases. With the exception of a few haematological disorders such as the α and β-thalassaemias, sickle cell anaemia and glucose-6-phosphate dehydrogenase (G6PD) deficiency, where special reasons may be found for high frequencies (Chapter 4), the inborn errors rarely achieve incidences of more than one in ten thousand in any outbred population (Table 1). Most are far less common; indeed many are so uncommon that calculation of frequency or prevalence is not possible and it is a difficult task to establish genetic causation. Despite this comparative lack of importance in the mainstream of medicine, interest in inborn errors of metabolism has continued to accelerate. On the medical side, understanding of the molecular cause of an inborn error may provide rational grounds for attempts at therapy and assist the process of carrier detection and prenatal diagnosis. On the biological side, the insights which the study of single-protein defects has afforded into cellular metabolism have been enormous, since these "natural experiments" are often a valuable way for exploring human biochemical pathways. And perhaps most important of all, the fact that many inborn errors have the superficial appearance of discrete problems, soluble within the context of existing biological knowledge, has been a powerful stimulant to laboratory investigation.

Modern understanding of inborn errors of metabolism is based on a framework of ideas put forward by Sir Archibald Garrod at the turn of the century. In a classic paper (Garrod, 1902) he summarized his observations on the benign condition of alkaptonuria, and in a later series of lectures (Garrod, 1908) extended and refined his concept of an inborn error of metabolism to include cystinuria, albinism and pentosuria, but to exclude what he called diseases of metabolism, such as gout, diabetes and obesity. He observed that

> "as far as our present knowledge of them enable us to judge, they (inborn errors of metabolism) apparently result from failure of some step or other in the series of chemical changes which constitute metabolism" and "each successive step in the building up and breaking down, not merely of proteins, carbohydrates and fats in general, but even of individual sugars, is the work of special enzymes set apart for each particular purpose".

In other words Garrod saw quite clearly that inborn errors of metabolism were the consequences of metabolic blocks, themselves the consequences of enzyme deficiencies. Through Bateson, he became conversant with

Table 1

Inborn errors of metabolism at high frequencies in particular populations

Condition	Population	Frequency (per 10 000)	Reference
Phenylketonuria	Scotland	2·0–2·8	Lindsay (1970)
Tyrosinaemia	Chicoutimi region, Canada	1·5–1·8	Laberge (1969)
Tay-Sachs disease	Ashkenazi Jews, U.S.A.	1·7	Myrianthopoulos and Aronson (1966)
Gaucher's disease (adult type)	Ashkenazi Jews, Israel	4	Fried (1968)
Duchenne muscular dystrophy	N. E. England	3·3	Gardner Medwin (1970)
Myotonic dystrophy	New Zealand	9	Caughey and Barclay (1954)
Huntington's chorea	Tasmania	1·7	Brothers (1949)
Porphyria variegata	South African whites	30	Dean (1963)
α₁-antitrypsin deficiency	Scandinavia	8	Fågerhol and Laurell (1970)
Adrenogenital syndrome	Yupik Eskimos	20	Hirschfeld and Fleshman (1969)
Cystic fibrosis of the pancreas	Afrikaners, S.W. Africa	16	Super (1975)
Dubin-Johnson syndrome	Iranian Jews	7·7	Shani et al. (1970)

Mendelian laws, with dominant and recessive modes of inheritance and with the ratios of normal to affected persons to be expected from a recessively transmitted characteristic.

With hindsight it seems strange that the relationship between gene and enzyme, implicit in Garrod's work and writings, was not developed (Beadle and Tatum, 1941) or explicitly formulated for another 37 years (Beadle, 1945). Obviously it was an idea before its time and if Garrod had seen the consequences of his discoveries he might well have concluded that there was little point in speculating on a biological relationship which he lacked the tools to pursue. For it was not until Pauling et al. (1949) demonstrated an electrophoretic difference between normal haemoglobin and the haemoglobin of sickle cell anaemia, that it became apparent that inherited diseases in Man could be both investigated and understood in molecular terms. Once this had happened it became possible to apply the one gene–one enzyme concept to Garrod's earlier observations and to conclude that diseases obeying Mendelian genetics should be associated with specific enzyme deficiencies. In 1952 the first biochemical "solution" of an inborn error was reported, a deficiency of hepatic glucose-6-phosphatase in von Gierke's disease (Cori and Cori, 1952). In 1958 alkaptonuria itself yielded to the accelerating biochemical onslaught (La Du et al., 1958). At the time of writing well over 150 may be regarded as "solved", in that the primary enzyme or protein defect has been identified (Tables 11–14).

Over recent years the concept of inborn errors of metabolism has experienced a gradual enlargement. Garrod's four examples were inherited as autosomal recessives and he saw them as based on specific enzyme deficiencies. Alkaptonuria, where there is a defect in the enzyme homogentisic acid oxidase (La Du et al., 1958) and pentosuria, where there is a defect in xylitol dehydrogenase (Wang and Van Eys, 1970), have turned out to be just that and indeed where the molecular pathology is clear, a majority of inborn errors conform to the classic Garrodian ideas. However, the discovery of an abnormal haemoglobin molecule in sickle cell anaemia (Pauling et al., 1949) and in a range of other haemoglobinopathies (see Chapter 11) has shown that metabolic abnormalities which are inherited according to Mendelian principles may be based on proteins whose primary function is transport rather than catalysis. A further extension of ideas on the genetic control of metabolism derives from advances in the understanding of the molecular mechanism of renal tubular and intestinal transport phenomena. Although the protein species involved in the transport of individual molecules across cell membranes are as yet poorly characterized, the study of disorders such as cystinuria and iminoglycinuria (see Section III) has made it clear that specific protein carriers are involved and that defects in them lead to diseases which should certainly be classified as inborn errors of metabolism. Garrod's understanding of metabolism was remarkably clear for his time, but probably corresponds to what is now referred to a intermediary metabolism, the interconversion of small molecules by activity-defined enzymes. A modern view of metabolism should be broader than this and the definition of an inborn error extended to encompass any inherited deficiency of a protein nature, whether the protein's role be catalysis, transport or structure.

The one gene–one enzyme theory of Beadle and Tatum, 1941 (expressed in its modern form of one cistron–one polypeptide chain) has few serious challengers. This means that in principle any human phenotype showing Mendelian genetics can be based on a variant protein. Thus not only autosomal and X-linked recessives but also autosomal and X-linked dominant conditions may arise from mutant genes expressed through abnormal proteins. Because Garrod's original examples of inborn errors of metabolism were all inherited as recessives, there has been a tendency to regard dominantly inherited conditions as things apart, needing separate definition. To a large extent this tendency has been reinforced by the great difficulties encountered in finding molecular solutions to dominantly inherited disorders (see Section II). It has seemed that expression of a disease in a heterozygote has required a quite different molecular explanation from expression in a homozygote. It is quite possible that this is partly true and that some dominants do derive from new forms of gene–protein interaction or realms of protein, gene or chromosome chemistry not yet understood. But even if partly true, it is not universally true and advances in haemoglobin chemistry have already provided some examples of dominantly inherited conditions which can be readily explained in terms of protein pathology (Section II.B).

A broad definition of an inborn error of metabolism is therefore any condition of clinical significance which shows a Mendelian mode of inheritance. By far the most complete summary of such disorders is to be found in McKusick's catalogue (McKusick, 1975), which at the 4th edition listed 2336 genetic loci in man which had been defined by variation—most of it pathological variation. In only a small proportion of these has sufficient biochemical knowledge accumulated for the primary protein or enzyme lesions to be established, and it is on this group that this chapter will concentrate. Since the haemoglobinopathies, the immunoglobulinopathies and the coagulation disorders have become associated with a formidable technology of their own, they are dealt with in separate chapters (Chapters 11, 12 and 13).

II. MOLECULAR CONCEPTS

A. Structural and Control Genes

The one gene–one enzyme (or one cistron–one polypeptide chain) hypothesis is a statement of a relationship without any attempt to describe the
mechanism of the relationship. Modern concepts of molecular biology
suggest that genes exercise control over proteins either by specifying the
structure of the molecule or by specifying the rate at which it is produced.
In principle, therefore, inborn errors of metabolism could arise both
through deficiencies in enzyme structure and through deficiencies in
enzyme quantity. The earliest examples of disorders obeying Mendelian
principles in which the molecular defect was established were the haemoglobinopathies, in which the mutant proteins were identified electrophoretically and therefore in terms of structural change. This influenced thinking on other inborn errors of metabolism, which were seen as resulting from
structurally altered proteins and enzymes, unable to carry out normal
catalytic or transport functions in the cell.

The introduction of an elegant physical model for the genetic control of
quantitative aspects of protein synthesis (Jacob and Monod, 1961) caused a
revision in thinking. The Jacob–Monod model of regulator, operator and
structural genes and repressor substances was derived exclusively from the
study of bacterial mutations, but the resulting picture was so clear and so
reasonable that it was not long before it was being applied somewhat
injudiciously to human genetic disease. A major textbook of inborn errors
of metabolism (Stanbury et al., 1966) listed a majority of the biochemically
solved diseases as probable examples of regulator and operator gene mutations. But as these disorders were more intensively studied it became
apparent that in most cases evidence could be found for the presence of a
mutant protein rather than the absence of protein or reduction in quantity
of normal protein, and in a later edition (Stanbury et al., 1972) this distinction was dropped. Today the feeling is that there is considerable evidence
against the validity of using bacterial genetic concepts in explaining mammalian systems.

The importance of the Jacob–Monod theory for human genetics is that
it has forced recognition of the fact that in higher organisms there must be
more than one category of gene. The structural gene is comparatively un-
controversial; it contains the information which defines the amino acid
sequence of a particular polypeptide. A mutation leads to an alteration in
amino acid sequence and this may affect both physical and chemical

properties of the polypeptide. A change in charge or shape of the molecule can be reflected in altered solubility or electrophoretic mobility. If the polypeptide has enzymatic activity this may be totally destroyed, drastically reduced, moderately reduced or even increased. The affinity of an enzyme for its substrate, its thermostability, responsiveness to effector molecules (activators, inhibitors, co-factors) and other kinetic criteria are likewise dependent on amino-acid sequence and may be altered by this type of mutation. Most of the evidence for structural gene mutations has been based on a detailed examination of the physicochemical properties of residual enzyme activities in inborn errors of metabolism (Table 2). Demonstration that a protein retains its antigenicity while losing its biological activity, first shown in some cases of haemophilia A, is a newer and more laborious method of establishing a structural gene mutation.

Table 2

Types of evidence used in reaching a conclusion of a structural gene mutation responsible for an inborn error of metabolism

Experimental observation	Examples
Altered electrophoretic mobility	G6PD variants
Altered affinity for substrate	"K_m variants" in pyruvate kinase deficiency
Altered pH-activity profile	β-galactosidase in generalized gangliosidosis
Decreased stability	HGPRT variants
Altered response to inhibitors	pseudocholinesterase variants
Altered affinity for co-factors	vitamin-responsive aminoacidopathies
Decreased biological activity with normal antigen activity	factor VIII in haemophilia A

In contrast to the structural gene, other categories of gene are poorly understood in eukaryotic systems. Most commentators agree that some form of regulator gene is necessary to account for the characteristic phenomena of higher organisms, of which cellular differentiation is the most obvious. Models have proliferated but there is no agreement yet even on the number of regulator elements necessary to explain, for example, development. Britten and Davidson (1969) suggest no less than three separate categories of gene involved in the control of protein quantity; Waddington (1969) regarded this as a subminimal model and added a fourth. For the investigator of inborn errors of metabolism, it is sufficient at present to conclude that a category of control genes must exist and that situations will arise where metabolic diseases are a consequence of mutations within this category.

The problem is how to recognize such situations. This is more than pedantry, for there can be little doubt that a better understanding of the molecular origins of a disease will assist attempts to find a rational basis for therapy. At the moment it seems wise to regard inborn errors of metabolism as resulting from structural gene mutations, except where there are obvious misfits. Some such misfits tend to take the form of recessively inherited disorders where more than one discrete enzyme is impaired. The

Table 3

Inborn errors of metabolism with apparent defects in more than one enzyme

Disorder	Enzyme defects	Reference
Orotic aciduria (type I)	orotidylate decarboxylase orotidylate pyrophosphorylase	Huguley et al. (1959)
Sandhoff's disease	hexosaminidase A hexosaminidase B	Sandhoff et al. (1968)
Maple syrup urine disease	α-ketoisocaproate decarboxylase α-keto-β-methylvalerate decarboxylase α-ketoisovalerate decarboxylase	Dancis et al. (1960)
Congenital sucrose intolerance	sucrase isomaltase	Auricchio et al. (1965)
I-cell disease	multiple lysosomal enzyme deficiencies	LeRoy and DeMars (1967)

simultaneous deficiency of orotidylate pyrophosphorylase and orotidylate decarboxylase in one form of hereditary oroticaciduria (Huguley et al., 1959) is a well-known example, almost impossible to explain as a consequence of two discrete mutations. Other similar situations are shown in Table 3. Such inborn errors may, of course, be the consequence of a mutation affecting a polypeptide which is a common part of two distinct and different enzyme activities. In this context the recent findings on the hexosaminidase isozymes are instructive. A deficiency of hexosaminidase A is responsible for Tay-Sachs disease and of both hexosaminidase A and B for Sandhoff's disease. For some time it has seemed probable that the A and B isozymes shared a common polypeptide, i.e. A would be $\alpha\beta$ and B would be $\beta\beta$, so that a mutation in the α locus would damage the A isozyme and in the β locus both isozymes (Srivastava and Beutler, 1973). Evidence for this model has been produced by Beutler and Kuhl (1975) who showed that hexosaminidase A could be converted to the B isozyme by freezing and thawing and that all the antigenic determinants present in hexosaminidase

B could be found in the A form, but not vice versa. Interspecific somatic cell hybridization studies (Lalley *et al.*, 1974) supported this conclusion by showing that clonal expression of hexosaminidase B was independent of A, but that expression of A was dependent on B. More recently, however, Gilbert *et al.* (1975), also using somatic cell hybridization, have claimed that the A isozyme is coded for by a locus carried on chromosome 5 and the B isozyme by a locus on chromosome 15, and that the expression of the two forms is independent. The interpretation of enzyme types in hybrid clones is notoriously difficult and if the studies of Gilbert *et al.* (1975) are correct, one is forced to the conclusion that Sandhoff's disease arises either from a double mutation or a defect in an element controlling quantitative expression of the A and B isozymes. A double mutation is extremely improbable while a regulatory mutation is argued against by the observation that some Sandhoff's patients have antigenically detectable hexosaminidases (Srivastava and Beutler, 1974). Whatever the truth is, the conflicting reports on the molecular pathology of Tay-Sachs and Sandhoff's diseases show some of the difficulties in resolving apparent double enzyme deficiencies.

Despite information from DNA–DNA hybridization experiments which suggests that a large proportion of the eukaryotic genome is not involved in coding for proteins (i.e. is not structural gene material) and is therefore likely to be involved in regulation, evidence for control gene mutations remains sparse in man. Among the large number of G6PD variants, all that have been carefully examined have turned out to have altered physico-chemical properties and there are still no examples of an enzyme with completely deficient activity in all tissues. The α-haemoglobin locus has revealed 110 variants and the β-locus 140 variants (Chapter 11) and in each case there is evidence for a structural gene mutation. Even the α-thalassaemias, which at one time looked reasonable examples of regulator-type mutations, now appear to be the consequence of gene deletions (Taylor *et al.*, 1974). The absence of clear evidence for control gene mutations among the proliferating examples of human metabolic defects is an extremely puzzling phenomenon.

B. Dominance and Recessiveness

According to Mendel "characters which are transmitted entire, or almost unchanged in the hybridization, are termed the dominant, and those which become latent in the process recessive". Dominance and recessiveness are thus functions of characters or phenotypes and not genes, and the mode of inheritance may change when the method of observation is altered. In biochemical genetics the usual attempt is to study variant phenotypes at the level of altered protein, so that the mode of inheritance becomes by definition dominant. This may create confusion with medical observations which continue to concentrate on the full-blown syndrome. The classic example is sickle cell anaemia, a recessively inherited haemolytic disease, whose underlying molecular pathology based on the presence of haemoglobin S can be observed by any first year student with an electrophoresis set-up, to be expressed in the heterozygote. More complex examples are compound heterozygotes where an individual has inherited a different allelic gene from each parent. Haemoglobin SC disease is clinically a recessive, but does not fit the Mendelian definition of either dominant or recessive (McKusick, 1975). Similar examples of compound heterozygosity manifesting as recessively inherited syndromes have been observed in pyruvate kinase deficiency (Paglia et al., 1968) and the mucopolysaccharidoses (McKusick et al., 1972).

However, in clinical genetics diseases may usually be divided quite sharply into those inherited as dominants and those inherited as recessives. Heterozygotes for recessive diseases are often detectable by biochemical techniques (Section VI) and found to be completely free of the clinical symptoms seen in the affected homozygote. Heterozygotes for dominant diseases may be affected at birth (e.g. achondroplasia) or later in life (e.g. Huntington's chorea), but in both cases the disease is caused by a single dose of a mutant gene. It is obviously important to know whether the differences between diseases caused by double doses of mutant genes and those caused by single doses provide any clues to the primary molecular lesion. Such consideration must be necessarily speculative, for few dominantly inherited conditions are understood at the protein level. Guidelines to the types of molecular lesion likely to be found may be derived from two sources; consideration of abnormalities of protein function and properties and consideration of abnormalities in the gene control of protein quantity. The former is a moderately useful exercise, for knowledge of protein chemistry is quite extensive. The latter demands a model for gene regulation and at the moment must remain frustratingly speculative.

1. Protein function and properties

Major proteins. The most comprehensively understood group of dominantly-inherited disorders are the haemolytic anaemias arising from abnormal haemoglobins (Chapter 11). Examination of this group of diseases suggests that two separate conditions must be satisfied for heterozygote expression; the mutant protein must be quantitatively important in the tissue affected and the nature of the mutation must be such that the properties of the abnormal protein over-ride those of the normal protein present in the same cell. Haemoglobin is of course a special case in that it represents a large majority of the total protein of the red cell; nonetheless most haemoglobin abnormalities are observed clinically only in the homozygote. It is usually when the mutant protein is unstable and precipitates within the red cell causing inclusion bodies and disturbing red cell function and longevity that overt haemolysis and anaemia results. Of the 111 abnormal haemoglobins which have been described, only 17 have dominant expression as unstable haemoglobins (McKusick, 1975). In view of the evidence that the α-chain locus is duplicated in some populations (Chapter 11), it is not surprising that 13 of these are β-chain variants. Some are sufficiently unstable to produce a life-long haemolytic anaemia in the carrier (e.g. Hb Bibba, Hb Hammersmith), some produce moderate haemolytic disease improving after splenectomy (e.g. Hb Shepherd's Bush) while others are asymptomatic except after drug-induction (e.g. Hb Zurich). In addition the haemoglobins M and the decreased oxygen-affinity haemoglobins may produce cyanosis in the carrier while the increased oxygen-affinity haemoglobins may lead to erythrocytosis (Wintrobe, 1974).

In many ways akin to the haemoglobinopathies are the hereditary dysfibrinogenaemias. Fibrinogen is the major protein of the clotting process with a plasma concentration of 250 mg per 100 ml. Afibrinogenaemia, the congenital absence of fibrinogen, is inherited as an autosomal recessive and patients have a pronounced haemorrhagic tendency from birth (Wintrobe, 1974). A number of qualitative abnormalities of fibrinogen structure have been described in heterozygotes which lead to delayed release of fibrinopeptides following the enzymatic action of thrombin (e.g. fibrinogen Baltimore), delayed or disordered polymerization of fibrin monomers (fibrinogen Detroit) or deficient cross-linking of fibrin monomers in the presence of factor XIII (fibrinogen Oklahoma). In general symptoms are mild with some haemorrhagic diathesis, slow wound healing and occasional thromboembolic complications (Chapter 13). The relationship of the molecular structure of the variant fibrinogen to the degree of disability in the carrier has not yet been described in the same detail as in the haemoglobin variants.

Other major proteins of specialized tissues include myosin of muscle fibres, keratin in skin and collagen in connective tissue. The latter has long been suspected as underlying the molecular defects in the dominantly inherited disorders of connective tissue such as osteogenesis imperfecta, the Marfan syndrome and the Ehler–Danlos syndromes. Pentinnen et al. (1975) have shown that fibroblasts cultivated from skin of an infant that died at birth with a severe form of osteogenesis imperfecta had greatly decreased synthesis of type I collagen, the major collagen component of mature skin, bone and tendon. Priest et al. (1973) have reported that collagen isolated from fibroblasts of patients with Marfan syndrome is abnormally soluble and postulate a structural abnormality that interferes with normal cross-linking. Pope *et al.* (1975) have shown that type III collagen is absent from tissues of a patient with Ehler–Danlos syndrome type IV, and that fibroblasts from such patients synthesize only type I collagen. However, in the studies on osteogenesis imperfecta and Ehler–Danlos syndrome cited above, the evidence for dominant inheritance is not clear, while the exact nature of the soluble collagen in Marfan syndrome has yet to be established. The apparent absence of type I and III collagens (cited above) is easier to reconcile with a recessively inherited condition.

Membrane proteins. The idea that an abnormality in a membrane protein could give rise to a dominantly inherited disorder is an attractive one. Implicit in the hypothesis is the assumption that the variant protein should be sufficiently normal to be incorporated into the membrane matrix without difficulty, but should, once incorporated, function defectively. Such malfunction could take the form of a structural weakness or an inability to provide receptor sites for hormones or other molecular signals. In the dominantly inherited red cell defects of spherocytosis and elliptocytosis erythrocyte deformability suggests a structural weakness, though its nature has yet to be elucidated (Jandl and Cooper, 1972; Nozawa *et al.*, 1974). In the X-linked dominant pseudohypoparathyroidism there is a failure of receptor sites in bone and kidney to respond to parathyroid hormone, though again the nature of the receptor sites remains unknown (Potts, 1972).

Currently, the best evidence for a membrane defect in a dominantly inherited disorder comes from the study of familial hypercholesterolaemia (FH), also known as Fredrickson's type II hyperlipoproteinaemia. FH heterozygotes manifest hypercholesterolaemia, tendinous xanthomas and later in life signs of coronary heart disease (Fredrickson and Levy, 1972), while homozygotes manifest a much more severe clinical syndrome characterized by extreme elevations of serum cholesterol level, cutaneous planar xanthomas appearing within the first four years of life, and rapidly progressive coronary heart disease developing in childhood. Heterozygotes and homozygotes for FH may be distinguished clinically through the time of onset of xanthomas and other symptoms, and biochemically through the fasting plasma cholesterol level (Goldstein *et al.*, 1974). The disorder thus represents one of the most convenient prototypes for the study of the molecular pathology of dominant inheritance.

In a series of elegant studies, Brown, Goldstein and colleagues have shown that the rate of cholesterol formation is mediated through the control of the activity of the enzyme which catalyses the first step in its biosynthesis, 3-hydroxy-3-methylglutaryl coenzyme A reductase (HMG-CoA reductase). In cultured fibroblasts the level of HMG-CoA reductase is suppressed and the activity of the enzyme is inhibited by serum low density lipoprotein (LDL) which binds to receptor sites on the cell membrane. LDL also stimulates cholesterol esterification and after transfer into the cell is degraded to its constituent amino acids. Control of cholesterol and LDL concentrations in serum is thus mediated through a complex set of interactions involving enzyme suppression, inhibition and stimulation and binding to components on the cellular membrane (Brown et al., 1973; Brown and Goldstein, 1974a; Brown et al., 1975).

In cultured skin fibroblasts from patients with homozygous FH the activity of HMG-CoA reductase is increased 40 to 60 fold (Goldstein and Brown, 1973). This is due to the absence of functional binding sites on the cell surface for LDL and the consequent inability of the lipoprotein to pass into the cell and regulate HMG-CoA reductase activity (Brown and Goldstein, 1974a). The homozygous patient thus displays profound hypercholesterolaemia due to overproduction of cholesterol and a moderate hyperlipidaemia due to underdegradation of LDL. More recently Goldstein et al. (1975) have identified a second type of homozygous FH, where cell receptor sites are deficient rather than absent. These patients may be the same as the group of FH homozygotes identified by Breslow et al. (1975) as "therapy-responsive", whose plasma cholesterol may be lowered by dietary control and treatment with cholestyramine and nicotinic acid.

Since the mechanisms of the molecular defects in homozygous FH have been so well studied, it has been possible to investigate the reasons for clinical manifestations in the FH heterozygote. Brown and Goldstein (1974b) have shown that there is a large reduction in LDL receptor sites in the fibroblasts of these subjects. This in turn affects both the rate of LDL turnover and the control of the level of HMG-CoA reductase and cholesterol biosynthesis. The data presented by Brown and Goldstein (1974b) suggest strongly that the mechanism of dominant expression in the "receptor-absent" form of FH revolves around the stoichiometry of LDL-receptor binding. FH homozygotes have no binding which leads to no LDL turnover and uncontrolled HMG-CoA reductase activity, whereas FH heterozygotes have 40% of normal binding, deficient LDL turnover and partially controlled HMG-CoA reductase activity. This type of explanation will probably also hold for a different type of FH heterozygotes, where the molecular defect appears to lie in a mutant LDL incapable of normal binding to fully active cellular receptor sites (Higgins et al., 1975).

Rate-limiting enzymes. Most enzymes function at a tiny fraction of their real capacity and it is not uncommon to find that flux through a metabolic pathway under normal physiological conditions may be maintained by less than 1% of the available enzyme. This overcapacity is the main reason that few carriers of recessively inherited enzymopathies show any clinical symptoms at all; if the reaction in question can be maintained by 1% of the normal level of enzyme, 50% of enzyme represents a handsome excess. The exception to this general rule is a limited group of enzymes known as rate-limiting, whose tissue concentrations are such that they regulate the amount of metabolite passing through a reaction sequence. Deficiencies in these enzymes will obviously affect metabolic flux in a direct way and it is conceivable that a 50% reduction in a rate-limiting enzyme could produce a critical starving of tissue of the end-product of the sequence.

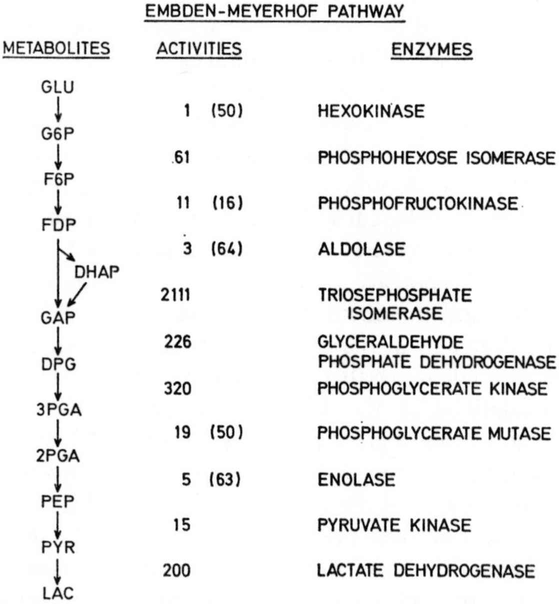

EMBDEN-MEYERHOF PATHWAY

METABOLITES	ACTIVITIES		ENZYMES
GLU			
↓	1	(50)	HEXOKINASE
G6P			
↓	.61		PHOSPHOHEXOSE ISOMERASE
F6P			
↓	11	(16)	PHOSPHOFRUCTOKINASE
FDP			
↘	3	(64)	ALDOLASE
DHAP			
↓↗	2111		TRIOSEPHOSPHATE ISOMERASE
GAP			
↓	226		GLYCERALDEHYDE PHOSPHATE DEHYDROGENASE
DPG			
↓	320		PHOSPHOGLYCERATE KINASE
3PGA			
↓	19	(50)	PHOSPHOGLYCERATE MUTASE
2PGA			
↓	5	(63)	ENOLASE
PEP			
↓	15		PYRUVATE KINASE
PYR			
↓	200		LACTATE DEHYDROGENASE
LAC			

Fig. 1   Embden-Meyerhof pathway of anaerobic glycolysis with enzyme activities in IU/g haemoglobin. Values in brackets are activities in the presence of low substrate concentration

Possibly the best-known of all metabolic sequences is the Embden-Meyerhof or anaerobic glycolytic pathway (Fig. 1). Consideration of the relative concentrations of the 11 enzymes necessary to produce lactate from glucose shows that the first enzyme, hexokinase, has the lowest capacity, and might confidently be expected to be rate-limiting. Haemolytic anaemia due to hexokinase deficiency has been described (Valentine *et al.*, 1967; Keitt, 1969; Necheles *et al.*, 1970a), but is apparently inherited as an autosomal recessive with heterozygous carriers being symptom-free. In this case the pronounced reticulocytosis which accompanies most enzyme-based haemolytic anaemias (Section III.B) is able to supply enough enzyme to maintain glycolysis at an adequate level even in the affected homozygote. However, a further problem is that it is not valid to reach conclusions about rate-limiting enzymes for inspection of relative activities, such as those shown in Fig. 1. These activities are obtained from *in vitro* systems where substrate, co-factor and ion concentrations are quite different from *in vivo*. Despite many studies it is still not clear whether hexokinase is rate-limiting in red cell glycolysis, or whether a number of different enzymes can become pace-makers as intracellular conditions change, or whether coenzyme supply (ADP, ATP, NAD, NADH) is the critical factor (Valentine *et al.*, 1968).

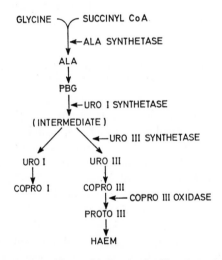

Fig. 2 Schematic pathway of haem biosynthesis. Uro I synthetase and copro III oxidase have been suggested as the primary enzyme lesions in acute intermittent porphyria and hereditary coproporphyria, respectively

In a quite different area of intermediary metabolism it has been claimed (McKusick, 1975) that deficiency of a rate-limiting enzyme in haem biosynthesis may be responsible for dominantly inherited acute intermittent porphyria (AIP). AIP is one of three types of hepatic prophyria, characterized chemically by urinary excretion of large amounts of porphobilinogen (PBG) and δ-aminolaevulinic acid (ALA). As in the other hepatic porphyrias the enzyme δ-aminolaevulinic acid synthetase is greatly increased (Fig. 2). However, low levels of the third enzyme in the sequence, uroporphyrinogen I synthetase, which uses PBG as substrate, have been reported in liver of patients with AIP (Strand *et al.*, 1970) and subsequently confirmed in erythrocytes (Meyer *et al.*, 1972). This would explain the increased excretion of PBG and ALA. Since conversion of PBG to uroporphyrinogen and subsequently to porphyrins and haem is depressed and since haem is known to be a feed-back inhibitor and repressor of δ-aminolaevulinic acid synthetase, the increased activity of this enzyme is also explained (Marver and Schmid, 1972). Sassa *et al.* (1974) have shown that erythrocyte levels of uroporphyrinogen I synthetase are approximately halved in AIP patients and non-symptomatic AIP carriers, and that the K_m for PBG is decreased. They also noted disturbed steroid hormone metabolism and concluded that though the uroporphyrinogen I synthetase defect is primary to AIP, other triggering factors are necessary. A 50% reduction in the activity of coproporphyrinogen oxidase (Fig. 2) in fibroblasts from patients with hereditary coproporphyria (Elder *et al.*, 1976) may also explain the dominant expression of this type of hepatic porphyria.

2. Control gene mutations

Interpreting dominantly inherited disorders in terms of control gene mutations is fraught with difficulty in view of the absence of an agreed model of gene control in the eukaryotic system. However, a dominant loss of function responding to known inducers of enzyme synthesis can be interpreted in terms of a control gene mutation, without too much attention being paid to the mechanism. One of the best-known examples of this system is the intermediate form of unconjugated hyperbilirubinaemia usually associated with the name of Arias (Arias, 1962). Patients have considerably reduced levels of hepatic glucuronyl transferase, and a chronic jaundice which responds dramatically to large doses of phenobarbital (Yaffe *et al.*, 1966). Phenobarbital is known to induce the synthesis of a number of microsomal enzymes including glucuronyl transferase, though in cases of Arias-type hyperbilirubinaemia increased enzyme synthesis has not been directly demonstrated. It is assumed that in this disorder a control gene malfunction leading to diminished output of the enzyme is overcome by the inductive effects of phenobarbital. In the recessively-inherited Crigler-Najjar syndrome, a form of hyperbilirubinaemia in which glucuronyl transferase is severely diminished or absent, phenobarbital is without effect (Arias *et al.*, 1969).

Another, and more thoroughly studied, dominant which has been interpreted in terms of a control gene mutation is hereditary angioneurotic oedema (HANE). The biochemical defect is in the inhibitor of the activated first component of complement, an α-2-neuraminoglycoprotein, referred to as the $C\bar{1}$-inhibitor (Donaldson and Evans, 1963). Both direct measurements and immunochemical assays show extremely low levels of

Cī-inhibitor (Shokeir, 1973). In a variant form of HANE, comprising per-haps 15% of patients, a functionally inactive but immunochemically nor-mal Cī-inhibitor has been described (Rosen *et al.*, 1965). But in both types the problem has been to explain the virtual absence of functional inhibitor activity in heterozygotes, who would be expected to carry one normal allele. A series of careful studies have shown that deficient inhibitor is not the product of increased degradation or clearance, but decreased hepatic syn-thesis (Johnson *et al.*, 1971).

A hypothesis to explain the molecular pathology of HANE, based on the operon model of Jacob and Monod (1961), has been proposed by Shokeir (1973). It envisages a mutation at the regulator site leading to an abnormal repressor, incapable of binding inducer. Such "super–repressor" mutations, leading to dominant loss of function, have been described in bacterial systems (Dreyfus, 1972). However, there is little evidence that the Jacob–Monod model can be applied to the variant form of HANE, for although there is a dominant loss of function, there are normal levels of impotent inhibitor in the circulation. A more satisfactory model should surely con-centrate on mechanisms whereby precursor inhibitor is rendered function-ally and immunochemically active.

III. EXPERIMENTAL APPROACH

A. General Considerations

The experimental approach to inborn errors of metabolism often proceeds in a series of well-defined stages. The first is the clinical description of the disease, with sometimes an indication in the naming that it belongs to a particular tissue or organ system. In the second stage, chemical and bio-chemical analysis may focus attention on an area of metabolism which seems to be deranged and should also lead to a refined clinical description of the disease and exclusion of cases which do not conform to the new criteria. The third stage has been a final one for many inborn errors—the proposal of a specific protein or enzyme lesion, which makes sense of the clinical and chemical observations and which is found in all patients with the disease. However, there are two further stages which are increasingly being seen in these disorders. One is the demonstration of heterogeneity within a disease defined quite clearly by a single genetic locus or by a speci-fied variant protein (Section V). The other stage, which will presumably remain the final one until the molecular analysis of DNA becomes feasible, is investigation of the nature of the protein defect. It is a process which has been extensively developed in the haemoglobinopathies, where the amino-acid changes in a large number of mutant haemoglobins are known (see Chapter 11) and where a confident statement may be made about a structural gene mutation. In other inborn errors, there have been as yet few attempts to sequence the mutant proteins, usually because residual enzyme activities are too low to allow sophisticated biochemistry. However, single amino acid differences have been demonstrated between the common B form of G6PD and both the A and Hektoen variants (Yoshida, 1967; Yoshida, 1970).

The primary biochemical objective in investigating an inborn error of metabolism is usually the location of a specific protein or enzyme defect. Different problems arise in different diseases, but the one which haunts many workers is the possibility that the error may lie in an area of metabolism as yet inadequately understood. It is possible that in heavily investigated diseases like Duchenne muscular dystrophy the defect involves a protein which has not yet been recognized in normal tissue. Thus the claim of Roses et al. (1975) to have identified a specific membrane protein kinase abnormality in this disorder is difficult to assess in the context of our current paucity of knowledge about individual membrane proteins and their functions. Even when the area of metabolism is better understood, certain problems are common to most or even all disease investigations. The difficulty of obtaining tissue confronts those who are not fortunate enough to choose a haematological disorder or an enzyme defect expressed in red cells (Section IV). Environmental effects may confuse findings in a number of ways, a principal source of difficulty being the existence of enzyme defects, themselves the product rather than the cause of tissue damage. Some enzymes are more susceptible than others to tissue necrosis: glucose-6-phosphatase deficiency, primary to von Gierke's disease, is also apparent in other liver diseases (Hers, 1959); muscle phosphorylase deficiency (McArdle's disease) has been reported in Duchenne muscular dystrophy (di Mauro et al., 1967) while lactase deficiency can be inherited or acquired. Glutathione reductase deficiency, for long a well-established red cell enzyme defect (Waller et al., 1965; Waller, 1968), has been shown to be at least partly dependent on the subject's riboflavin intake (Beutler, 1969a, b).

Even when environmental effects have been disentangled and sufficient tissue obtained, the investigator is faced by the knowledge that single enzyme defects can have multiple metabolic and physiological consequences. Pleiotropy, as this is called, means that working back from the clinical and chemical facts of the disease to the primary enzyme lesion demands sophisticated understanding of metabolic inter-relationships and a good deal of luck. Enzyme deficiencies can lead inter alia to product depletion, precursor accumulation, activation of normally minor pathways and interference with other biochemical sequences. If the enzyme is comparatively non-specific, as for example β-galactosidase in GM1 gangliosidosis, a variety of different substrates can accumulate in separate tissues (O'Brien, 1975). Many of these situations have been observed in inborn errors of metabolism and have provided clues leading to the enzyme defect. Nonetheless, the construction of diagrams of pleiotropic effects is only really possible when a primary defect has been proposed, and it then serves the purpose of allowing an assessment of the validity of the proposal. To arrive at a specific hypothesis, the investigator must proceed through various indirect stages, where an area of deranged metabolism is localized, to the direct stage where measurement of individual enzymes or proteins is undertaken.

B. Indirect Approach

1. Metabolite accumulation

Because high concentrations of metabolites may be self-indicating (homogentisic acid in alkaptonuric urine blackens on exposure to light) or apparent in routine clinical testing (uric acid in Lesch-Nyhan syndrome), accumulation of various chemical compounds has helped the elucidation of a number of inborn errors. The compound may be stored in solid tissues at an expected subcellular site (glycogen in liver cytoplasm in von Gierke's disease) or in an unexpected site (glycogen in liver and muscle lysosomes in Pompe's disease). It may be found in circulating erythrocytes (ceramide trihexoside in Fabry's disease), leucocytes (glycogen in Forbes's disease) or plasma (phenylalanine in phenylketonuria). It may appear in the CSF (citrulline in citrullinaemia) or be excreted in large quantities in the urine (orotic acid in oroticaciduria). Usually elevated levels of known compounds are found, but in a few cases previously unrecognized metabolites are observed (D-glyceric acid in hyperoxaluria type II), or known compounds with abnormal structure (glycogen in Andersen's disease). The compound may be proximal to the metabolic block (GM1 ganglioside in generalized gangliosidosis) or separated from it by several enzymatic steps (glycogen in von Gierke's disease).

Though many of these examples have been of great importance in focusing attention on a discrete area of biochemistry, the lessons of metabolite accumulation have been most valuable in the storage diseases and in the aminoacidopathies. In the typical storage diseases—the glycogenoses, the spingolipidoses and the mucopolysaccharidoses—chemical analysis of the stored compound has led to the elucidation of many of the primary enzyme defects (Tables 11, 15 and 16). In the aminoacidopathies, the presence of amino acids in circulating plasma and the easy recognition of abnormal concentrations through the improving technique of ion-exchange chromatography, has allowed rapid advances to be made in the delineation of a range of biochemically-defined inborn errors of metabolism (Table 14).

2. Metabolite depletion

Depletion of metabolites is often less obvious than their accumulation and may indeed be a secondary observation. The absence of urocanic acid in the sweat of patients with histidinaemia followed discovery of elevated serum and urinary histidine levels (La Du *et al.*, 1963); hypouricaemia in xanthine oxidase deficiency followed the discovery of urinary stones composed of xanthine (Dent and Philpot, 1954). In other diseases depletion is more obvious; well-known cases are the absence of melanin in oculocutaneous albinism and the inability of ischaemically exercising muscle to produce lactate in McArdle's disease (McArdle, 1951). Because of the interrelationships of metabolic pathways the lowered level of a common metabolite may not be entirely helpful; hypoglycaemia is found in galactosaemia, hereditary fructose intolerance, maple syrup urine disease and several of the glycogen storage diseases and it is possible to relate the finding to the primary enzyme deficience. It is also seen in a number of inborn errors where the enzyme lesion has yet to be located, such as leucine-sensitivity (Cochrane *et al.*, 1956) and ketotic hypoglycaemia of childhood (Pagliara *et al.*, 1971). Hypouricaemia is a primary finding in xanthinuria but also occurs in the Fanconi syndrome (Harrison, 1958), Wilson's disease (Bishop *et al.*, 1954) and haemochromatosis (Ayvazian, 1964).

3. Metabolic pathway analysis

In principle the integrity of a metabolic pathway may be tested by measuring the rate of product formation from a suitable precursor, often many enzymatic steps removed. When such studies are carried out *in vivo* they are complicated by alternative routes of metabolism of most compounds which can be safely administered to human subjects and of course by the ethics of experimentation on sick people. Nonetheless, such investigations are often both desirable and permissible in the search for specific enzyme deficiencies. Oral and intravenous administration of amino acids has been widely used in locating the metabolic blocks in a number of aminoacidopathies where accessible tissue (often liver) has been difficult to acquire. Thus in a patient with hydroxyprolinaemia, an oral load of hydroxyproline caused a three-fold increase in the plasma concentration of the amino acid (Efron *et al.*, 1965) without excretory response of the metabolites found in normal subjects (Scriver and Efron, 1972). This led to the suggestion of a specific deficiency of liver hydroxyproline oxidase, now generally accepted as the primary biochemical lesion, even though direct enzyme assay has not been carried out. Likewise, in a patient with isovaleric acidaemia, leucine loading was found to cause plasma elevation of isovalerate but not β-methylcrotonate, thus pointing to a block in the leucine degradative pathway between these two acids and implicating the enzyme isovaleryl-CoA dehydrogenase (Tanaka et al., 1966).

In vitro investigations of metabolic pathways, often easier to perform and interpret, may be compromised by the non-representative nature of the tissue selected. Glycolytic rates in the red cells of patients with hereditary fructose intolerance would not reveal the abnormality of the specific liver aldolase isoenzyme, even though aldolase is an essential glycolytic enzyme (Froesch, 1972). But in general this is probably not a serious objection, for the range of genetically distinct tissue isozymes seems small. They may be defined both by studies of polymorphisms (see Chapter 5) and by studies of inborn errors of metabolism (Table 7).

A more serious hazard could be the maintenance of metabolic integrity under artificial conditions. Even here, a growing body of experience with red cell, leucocyte and fibroblast preparations suggests that many pathways may be kept functional under carefully controlled conditions and the measurement of intermediate metabolite levels used to locate an abnormal enzymatic step. Thus in triosephosphate isomerase deficiency, incubation of isolated red cells in glucose-fortified plasma leads to an increase of dihydroxyacetone phosphate (Schneider et al., 1968) which is proximal to the enzyme deficiency (Fig. 1). In maple syrup urine disease, Dancis et al. (1960) found that incubation of leucocytes with leucine, isoleucine and valine failed to produce the expected yield of carbon dioxide, the degradative pathway being interrupted at the keto acid step. In Refsum's disease, Mize et al. (1969) showed that cultured fibroblasts failed to oxidize phytanic acid to carbon dioxide and then localized the defect in a specific phytanic acid oxidase. It seems probable that in future leucocyte and fibroblast metabolism will be increasingly investigated with the particular object of finding an interrupted biochemical pathway.

4. Measurements of active transport

In a number of recessively inherited disorders of metabolism (including one of Garrod's original group, namely cystinuria) the defect does not involve a conventionally-defined enzyme but rather a protein which mediates the transfer of small molecules across cell membranes. These proteins have enzymatic activity in the sense that they catalyse a transport process against a concentration gradient and have defined substrate specificity, but they are not amenable to normal studies with purified fractions in soluble systems. Since transport proteins or "permeases" are of great importance in the

Table 4

Transport disorders[a]

Disorder	Tissues affected	Metabolites involved
Cystinuria (three types)	kidney and intestine	cystine, lysine ornithine, arginine
Hypercystinuria	kidney	cystine
Iminoglycinuria (four types)	kidney and intestine	glycine, proline, hydroxyproline
Hartnup disease I	kidney and intestine	neutral mono
II	kidney	amino acids
Tryptophan malabsorption	intestine	tryptophan
Methionine malabsorption	intestine	methionine
Hyperdibasic aminoaciduria (two types)	kidney, intestine	lysine, ornithine, arginine
Glucose-galactose malabsorption	intestine, kidney	glucose, galactose
B_{12} malabsorption	intestine	vitamin B_{12}
Renal glycosuria (two types)	kidney	glucose
Iodide transport defect	thyroid	iodide
Congenital chloridorrhoea	colon	chloride
Hypophosphataemic rickets	kidney	phosphorus, calcium
Dicarboxylic aminoaciduria	kidney, intestine	glutamate, aspartate
Bartter syndrome	kidney, RBC	sodium ion
Pseudohyperparathyroidism	kidney, bone	parathyroid hormone (calcium)
Renal tubular acidosis I	kidney	hydrogen ion
II (X-linked)	kidney, intestine	bicarbonate
III	kidney	bicarbonate
Diabetes insipidus	kidney	antidiuretic hormone (water)

[a] Adapted from Rosenberg (1974), Rosenberg and Scriver (1974), and Scriver *et al.* (1976).

cellular control of metabolite concentrations, mutations may be expected to lead to aberrant phenotypes simulating inborn errors of metabolism, but perhaps more correctly called inborn errors of active transport. Many of the earlier examples involved transport of single amino acids or groups of amino acids in the kidney and were characterized by aminoaciduria in the absence of aminoacidaemia. Recently several other types of transport defect have been discovered (Table 4). Though the "permeases" responsible for the defects in Table 4 have not been isolated or purified, there seems little reason why these disorders should not be regarded as to some extent biochemically solved.

The conventional method of studying transport disorders is by measuring plasma, faecal and urinary metabolites after the patients have been given an appropriate oral load. A more direct method is to follow the uptake of radioactively-labelled compounds by biopsied intestinal mucosa or kidney cortex *in vitro*. These studies have shown that some transport sites are specific for more than one metabolite, such as cystine, lysine, ornithine and arginine at the "cystinuric" site and a range of neutral mono-amino mono-carboxylic amino acids at the "Hartnup" site. They have showed, conversely, that certain metabolites like glycine and glucose are absorbed in both kidney and intestine at more than one site. *In vitro* studies have also been used to confirm suggestions of genetic heterogeneity within a single defect of active transport, cystinuria (Rosenberg et al., 1966).

Attempts to extend the study of transport to more readily available tissues have not been very successful. Though white blood cells and fibroblasts share with intestinal mucosa and kidney cortex the ability to concentrate amino acids against concentration gradients, the appropriate mechanisms appear unimpaired in Hartnup disease (Tada et al., 1966a), cystinuria (Rosenberg and Downing, 1965) and iminoglycinuria (Tada et al., 1966b). In the glucose–galactose malabsorption syndrome the renal intestinal defect is not expressed in red cells (Meeuwisse, 1970), where a separate carbohydrate transport mechanism appears to function.

5. Cellular histochemistry

In the storage or deposition diseases excessive amounts of metabolites accumulate in various solid tissues of the body as well as spilling over into the plasma and urine. Glycogen is laid down in liver and muscle in the glycogen storage diseases, mucopolysaccharide and complex lipid is found in brain and visceral organs in the mucopolysaccharidoses and sphingolipid and mucopolysaccharide is deposited in a range of tissues in the sphingolipidoses. These metabolites may be detected by histochemical stains and the findings used both as an aid to diagnosis and as an indication of the type of metabolism which is deranged.

The discovery that the characteristic feature of some storage disorders is reflected in fibroblasts cultured from the skin of patients with the diseases has released investigators from the constraints of solid tissue biopsy. Activity has largely centred around the staining properites of the dye toluidine blue. When this is taken up by fibroblasts which contain abnormal amounts of polyanionic high molecular weight compounds, the resulting complex has a distinctive reddish-pink colour. Metachromasia, as this is called, is exhibited by fibroblasts cultured from patients with various forms of mucopolysaccharidosis and sphingolipidosis, as well as other less easily defined disorders (Table 5). Though it is perhaps unwise to be too definite about the molecular meaning of metachromasia, it is fair to say that it is probably an indication of excessive deposition of high molecular weight metabolites in the cell. Furthermore in those disorders exhibiting metachromasia where the primary enzyme defect is known, it has been found to be a deficiency of a degradative enzyme rather than an excessive activity of a biosynthetic one.

Table 5

Disorders detectable in fibroblast culture by metachromatic or alcianophilic staining

Metachromasia	
Hurler's[a]	Generalized gangliosidosis[a]
Hunter's[a]	Fabry's[a]
Sanfilippo[a]	Gaucher's[a]
Scheie[a]	Krabbe's[a]
Morquio[a]	Late infantile amaurotic idiocy[a]
Marfan's[a]	Juvenile amaurotic idiocy[a]
Pseudoxanthoma elasticum[a]	Larsen's[c]
Chediak-Higashi[b]	Lafora's[d]
Cystic fibrosis of the pancreas[f]	Pompe's[e]
Myotonic dystrophy[t]	Pelizaeus-Merzbacher[e]

Alcianophilia	
Hurler's[c]	Morquio[c]
Hunter's[c]	Maroteaux-Lamy[c]
Scheie[c]	Larsen's[c]
Generalized gangliosidosis[c]	Myotonic dystrophy[t]

[a] Matalon and Dorfman (1969
[b] Danes and Bearn (1967b)
[c] Danes et al. (1970a)
[d] Fluharty et al. (1970)
[e] Gertner et al. (1970)
[f] Swift and Finegold (1969)

Metachromasia has been criticized on the grounds that it lacks specificity and also because too many positive staining reactions are observed in cells from normal controls (Taysi et al., 1969; Milunsky and Littlefield, 1969). An amended technique which employs the affinity of another dye, alcian blue, for polyanionic molecules (alcianophilia) is more specific (Danes et al., 1970a) as a selective indicator of those disorders (mucopolysaccharidoses and sphingolipidoses) in which mucopolysaccharide accumulation occurs. However, both metachromasia and alcianophilia belong to an era when the primary enzyme defects in the mucopolysaccharidoses and sphingolipidoses were unknown and have fallen into disuse in recent years.

6. Pharmacogenetics

An early definition of pharmacogenetics states that it refers to hereditary disorders initially revealed by the action of drugs (Vogel, 1959). Though this has now been widened to encompass any situation where a drug response is modified by genetic factors (Kalow, 1962; and cf. Chapter 9), pharmacogenetics has been occasionally useful in uncovering unsuspected enzyme deficiency states which can quite properly be regarded as inborn errors of metabolism, since the person concerned can become severely ill under the influence of the wrong drug. The classic situation is glucose-6-phosphate dehydrogenase (G6PD) deficiency which came to be studied intensively when large-scale use of the antimalarial primaquine during World War II precipitated haemolytic crises in considerable numbers of Negro soldiers. The enzymatic basis of these haemolytic episodes was established by Carson et al. (1956) and it is now known that G6PD deficiency is the most common inborn error of metabolism in mankind. Its study has been seminal in the development of many aspects of medical genetics.

Despite the early definition of pharmacogenetics as a revelatory situation, unexpected drug responses have only been moderately useful as a tool in the investigation of inborn errors of metabolism. Acatalasia, the deficiency of the enzyme catalase (Takahara, 1952), and suxamethonium sensitivity, the deficiency of serum pseudocholinesterase (Evans et al., 1952), were brought to light by the action of hydrogen peroxide and succinyl choline, their respective substrates. The instability of haemoglobin Zurich was revealed by the agency of sulphonamides (Hitzig et al., 1960) and the deficiency of diphenylhydantoin hydroxylase by the use of the anticonvulsant diphenylhydantoin (Kutt et al., 1964). But many other pharmacogenetic episodes appeared after the molecular pathology of the inborn error was already understood, such as the oxidant drug-induced haemolyses in glutatione reductase deficiency (Waller, 1968), glutathione peroxidase deficiency (Necheles et al., 1970b), gluthatione synthetase deficiency (Prins et al., 1966) and methaemoglobin reductase deficiency (Szorady, 1973), the effect of aminopyrine on pentosuric patients (Hiatt, 1972), and the dangers of general anaesthesia to carriers of the sickle cell trait (Anon, 1970). Other abnormal drug responses, such as that of barbiturates in porphyric patients (Marver and Schmid, 1972), of alcohol in hypokalaemic and normokalaemic periodic paralyses (Pearson and Kalyanaraman, 1972) and of thiopentane anaesthesia in Huntington's chorea and myotonic dystrophy (Bush, 1968) may eventually provide the necessary clue to the molecular defect.

7. Response to vitamins

Since 1937, when Albright et al. reported a case of intractable rickets and hypophosphataemia which responded to 1000 times the normal dose of vitamin D, it has been known that some inborn errors of metabolism may be ameliorated by pharmacological (as opposed to physiological) levels of vitamins. Hunt *et al.* (1954), describing a child with seizures whose convulsions could be controlled with massive doses of pyridoxine but not normal anticonvulsants, termed this phenomenom vitamin-dependency, in contrast to the more usual avitaminoses which could be corrected with physiological levels of the appropriate vitamin or vitamin precursor. A number of vitamin-responsive inborn errors of metabolism have been described in the last ten years (Table 6) and vitamin-dependence has been useful in probing the origins of the molecular defect. The subject has been well reviewed (Mudd, 1971; Scriver, 1973; Rosenberg, 1976).

There are two main mechanisms underlying vitamin dependence. The first, which is characteristic of the pyridoxine, biotin and thiamine-responsive group, involves enzyme–coenzyme interactions. The mutation in the enzyme (styled apoenzyme to indicate that it is the protein moiety of the complex which is affected) impairs its ability to bind its specific coenzyme (the coenzymatically active form of the vitamin). The resulting complex has diminished or absent catalytic activity and is often indistinguishable from a mutant enzyme where the catalytic site has been directly affected. However, since apoenzyme–coenzyme binding is concentration-dependent, massive doses of the coenzyme or its vitamin precursor may push the equilibrium in the direction of the complex and restore catalytic activity. Frimpter (1965) showed that cystathioninase activity in liver homogenates from a patient with cystathioninuria could be restored by addition of pyridoxal phosphate, the coenzymatically active form of vitamin B_6, while Tada *et al.* (1967) produced the same effect with a defective kynurerinase from a patient with xanthurenic aciduria. However, in a majority of cases of pyridoxine-responsive homocystinuria the effect would appear to be slightly different, for little enhancement of enzyme activity is obtained in studies *in vitro* (Uhlendorf *et al.*, 1973). Instead it is suggested that the active form of the vitamin operates by stabilizing the low residual levels of the mutant enzyme (Mudd, 1971), providing enough catalytic activity to perform the appropriate metabolic function.

The second mechanism for vitamin-dependence which has been described results from defects in vitamin transport or coenzyme synthesis. The story of methylmalonic acidaemia is instructive here. First reported cases of this inborn error of metabolism were found to be due to a deficiency of the enzyme methylmalonyl-CoA mutase (Fig. 3) (Morrow *et al.*, 1969b). Since the enzyme was known to use cobalamin as co-factor attempts were made to use vitamin B_{12} in therapy. Some patients responded favourably suggesting that they had a mutation in the mutase apoenzyme with defective capacity for binding B_{12} (Rosenberg *et al.*, 1968). However, further studies on fibroblasts from responsive patients showed that the levels of the vitamin B_{12} coenzyme, 5′-deoxyadenosylcobalamin were

Table 6

Vitamin-responsive inborn errors of metabolism

Vitamin	Disorder	Enzyme affected	Reference
Thiamine (B$_1$)	maple-syrup urine disease	branched chain ketoacid decarboxylases	Scriver et al. (1971)
	pyruvic acidaemia Leigh's subacute necrotizing encephalopathy	pyruvate decarboxylase	Blass et al. (1970)
Pyridoxine (B$_6$)	cystathioninaemia	pyruvate carboxylase	Brunette et al. (1972)
		cystathioninase	Frimpter (1965)
	xanthurenic aciduria	kynureninase	Tada et al. (1967)
	homocystinuria	cystathionine synthetase	Barber and Spaeth (1967)
	hyperoxaluria type I	glyoxalate α-ketoglutarate carboligase	Smith and Williams (1967)
	infantile convulsions	glutamate decarboxylase	Yoshida et al. (1967)
Cobalamin (B$_{12}$)	methylmalonic acidaemia	methylmalonyl-CoA mutase	Rosenberg et al. (1968)
Folic acid	formiminotransferase deficiency	formimino transferase	Arakawa (1970)
	homocystinuria and hypomethioninaemia	N^5, N^{10}-methylene tetrahydrofolate reductase	Mudd et al. (1972)
	megaloblastic anaemia	dihydrofolate reductase	
Biotin	propionic acidaemia	propionyl-CoA carboxylase	Tauro et al. (1976)
	β-methylcrotonyl glycinuria	β-methylcrotonyl-CoA carboxylase	Barnes et al. (1970)
			Gompertz et al. (1971)

about 10% of normal and that supplementation of the coenzyme *in vitro* could restore full activity (Rosenberg *et al.*, 1969). The defect thus lies in the synthesis of the coenzymes and not in methylmalonyl CoA mutase itself (Fig. 3). It is now known that there are at least three different forms of B_{12}-responsive methylmalonic acidaemia involving apparently different steps in coenzyme synthesis (Mahoney *et al.*, 1975).

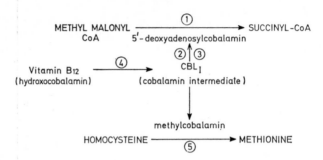

Fig. 3   Inborn errors of metabolism related to vitamin B_{12}: 1. B_{12}-independent methylmalonic acidaemia due to deficiency of methylmalonyl-CoA mutase; 2 and 3. B_{12}-dependent forms of methylmalonic acidaemia due to different defects in 5′-deoxyadenosyl cobalamin synthesis; 4. B_{12}-dependent methylmalonic acidaemia with disordered sulphur amino acid metabolism due to defect in cobalamin synthesis prior to CBL_1; 5. homocystinuria due to defective cystathionine synthetase (Rosenberg, 1976)

C. Direct Approach

When indirect approaches have yielded evidence of an area of deranged metabolism specific hypotheses must be tested by direct enzyme or protein assay. A number of problems arise in the interpretation of results and these are discussed below.

1. Tissue necrosis

In Duchenne muscular dystrophy extensive degeneration of muscle occurs before the onset of clinical symptoms and the invasion of fibres by fat, connective tissue and macrophages leads to a series of distorted enzyme values (Pennington, 1969). Most of these must be the consequence of the disease process, but it is just possible that one of the deficient enzymes is the causative agent. Tissue necrosis is a common finding in inborn errors of metabolism; it is a particularly difficult problem when the advancement of the disease is rapid and widespread and when the investigator is confronted by a range of obviously degenerate tissues and a series of abnormal enzyme values. If kinetic or electrophoretic analysis of a suspected enzyme is possible, this can provide valuable corroboratory evidence for the primacy of the lesion. Confidence in a hypothesis is gained if an enzyme deficiency is expressed in several tissues, some of which have minimal clinical involvement. In glycogen storage disease type III (Forbes's disease), the diminished levels of red cell amylo-1, 6-glucosidase (Van Hoof, 1967) confirmed that the previously observed hepatic enzyme deficiency (Hers, 1959) was not the consequence of liver degeneration. But unfortunately the reverse is not true, and a normal enzyme in one tissue may disguise its deficiency in another (Section 4).

2. More than one deficient enzyme

Even when the effects of tissue degeneration have been adequately controlled, more than one enzyme may appear to be genuinely depressed. Several such cases have been reported; most are difficult to assess as the opportunities for thorough biochemical analysis have not been available (Table 3).

3. Partial enzyme deficiencies

In a majority of biochemically solved inborn errors of metabolism some residual activity of the defective enzyme can be detected. It is a confident investigator who declares that he can find no trace of an enzyme activity and the statement is more credible when backed up by immunological demonstration of the absence of enzyme protein, as in muscle phosphorylase deficiency (Robbins, 1960) and phosphofructokinase deficiency (Layzer et al., 1967). But the wide acceptance of the fact of partial enzyme deficiencies poses the question as to when a diminished enzyme is severe enough to account for the clinical features of a disease. Common sense suggests that the finding should be confirmed in other individuals and perhaps other tissues. Electrophoretic analysis or Michaelis constants of the residual activity are evidence of a structural change in the suspected protein, but have been possible only in a limited number of cases. In other situations a claim to have identified the primary enzymatic lesion must be assessed in the light of the thoroughness of the accompanying biochemical findings and also whether the proposed solution makes sense of the clinical and chemical facts.

4. Elevated enzyme levels

A generalized increase in the specific activity of a group of enzymes may be a sign of haemapoietic compensation (see reticulocytosis, below) or tissue inflammation. The latter is characteristic of the sphingolipidoses and the mucopolysaccharidoses. Van Hoof and Hers (1968) have measured an enormous series of acid hydrolases in the biopsied livers of children with different types of the six classical muscopolysaccharidoses. Many of these enzymes are grossly elevated; it is presumed that this is related to the enlargement and proliferation of the lysosomes and may be regarded as another and special case of tissue necrosis. In I-cell disease there are strong increases in a whole series of lysosomal enzymes in the serum, but in this case the activity of the same enzymes is decreased in the lysosomes despite their enlargement and proliferation. The most plausible explanation is a defect in the permeability of the lysosomal membrane (Van Hoof, 1974).

In other disorders more limited and specific elevations of enzyme activity have been found. In the Hektoen variant of G6PD, the red cell enzyme value is four times that found in the normal B form, as a consequence of a single amino acid substitution (Yoshida, 1970). A pseudocholinesterase variant (E. Cynthiana) with two to three times normal activity is also thought to be the product of a structural gene mutation (Yoshida and Motulsky, 1969). In both cases immunological demonstration of increased protein quantity as well as enzyme activity suggests facilitated transcription or translation of the gene or gene product. On the other hand a mutant phosphoribosylpyrophosphate synthetase has been described with two- to three-fold higher activity per molecule, a structural gene mutation which has presumably led to a more efficient enzyme (Becker et al., 1973). In three dominantly inherited porphyrias, acute intermittent porphyria, hereditary coproporphyria and variegate porphyria, the first enzyme of the haem biosynthetic pathway, hepatic δ-aminolaevulinic acid synthetase (ALA synthetase), is greatly increased (Tschudy et al., 1965; Kaufman and Marver, 1970; Dowdle et al., 1967). It is difficult to envisage all three situations arising from structural gene mutations in this enzyme; a more probable explanation, though one still embroiled in controversy and uncertainty, invokes some form of enzyme derepression or de-inhibition. Recent studies suggest that underproduction of haem due to enzyme deficiencies in the pathway commencing at ALA (Fig. 2), may be responsible for the high levels of ALA synthetase.

5. Reticulocytosis

A special type of enhancement of enzyme activity is often observed in haematological disorders, where compensatory haematopoiesis leads to the production of increased amounts of younger red cells. In these reticulocytes, enzyme levels per cell are several fold higher than in older red cells, for protein chains are newly synthesized and have not yet begun the slow decay that characterizes their life-time in the enucleated erythrocyte. A partial enzyme defect will be marked by a high percentage of reticulocytes in the blood. In the three cases of hexokinase deficiency reported (Valentine *et al.*, 1967; Keitt, 1969; Necheles *et al.*, 1970a) the enzyme value measured by conventional techniques was within the normal range. The abnormality had to be revealed by assays in fractions enriched with older red cells, by comparisons with other red cell enzymes (which were elevated) and by analysis of enzyme kinetics. Unless the contribution of the reticulocyte fraction is recognized, it is easy to miss the enzyme deficiency in a congenital, non-spherocytic haemolytic anaemia.

IV. TISSUE DISTRIBUTION

Many of the sequences of intermediary metabolism are found in a wide range of body tissues. The enzymes of glycolysis, the citric acid cycle, fatty acid biosynthesis and catabolism, purine and pyrimidine metabolism and hydrolytic degradation are quite extensively distributed, while active glycolysis is even a feature of the enucleated red cell. This does not mean, however, that an enzyme deficiency in one of these sequences will necessarily be expressed in all the tissues using that metabolic pathway. A system of isozymes have evolved; these can be defined as alternate enzymes with the same broad catalytic function but with different molecular forms and kinetic characteristics. Knowledge of the isozyme structure of particular activity-defined enzymes is far from complete; it is not clear how many enzymes are composite groups of isozymes, how the isozymes are structurally related to one another and whether they are controlled by one or more genetic loci (Chapter 1). The investigation of inborn errors of metabolism has been particularly useful in regard to this last point. If a genetically determined enzyme deficiency is found in one group of tissues while the same enzyme is normal in another group of tissues, it can be concluded that the enzyme is coded for by more than one locus. Unfortunately it is not always possible to obtain a wide enough variety of tissues to be certain about the extent of the enzyme deficiency. The examples shown in Table 7 are therefore limited to a restricted number of cases where tissue availability has allowed enzymatic demonstration of genetically distinct isozyme differences. In other cases current evidence suggests a generalized deficiency; in glycogen storage disease type III (Forbes's disease) amylo-1, 6-glucosidase is almost completely inactive in muscle, liver, leucocytes and erythrocytes and it is reasonable to assume that this extends to all glycogenolytic tissues. The lysosomal disorders, inborn errors involving hydrolytic enzymes of the lysosomal fraction of the cell, are examples of metabolic defects where the enzyme deficiency may be measured in virtually all cells with exception of red cells.

Table 7

Genetically distinct isozymes revealed by inborn errors of metabolism

Disorder	Enzyme	Enzyme normal in	Enzyme deficient in
Hexokinase deficiency	hexokinase	wbc	rbc
Phosphofructokinase deficiency	PFK	rbc (1 isozyme)	muscle
Pyruvate kinase deficiency	PK	wbc	rbc
Fructose intolerance	aldolase	muscle, rbc	liver
McArdle's disease	phosphorylase	liver	muscle
Myeloperoxidase deficiency	myeloperoxidase	eosinophils	neutrophils, monocytes
Phosphorylase kinase deficiency	phosphorylase kinase	muscle	liver, wbc
Hers's disease	phosphorylase	muscle	liver, wbc
Oculocutaneous albinism	tyrosinase	adrenals	melanocytes
Hypophosphatasia	alkaline phosphatase	intestine	bone, liver kidney
Chronic granulomatous disease	glutathione peroxidase	rbc	wbc

A few metabolic sequences are known to be comparatively tissue-specific. One such is gluconeogenesis, the synthesis of glucose from pyruvate and lactate, which occurs only in liver, kidney and intestinal mucosa. A deficiency of a gluconeogenic enzyme, glucose-6-phosphatase, is responsible for von Gierke's disease and the defect is observable only in these tissues and not in others. Though it is not possible to generalize about the location of amino acid metabolizing enzymes, a few such as phenylalanine hydroxylase (absent in phenylketonuria), parahydroxyphenylpyruvate hydroxylase (deficient in tyrosinaemia) and proline hydroxylase (deficient in hyperprolinaemia) are demonstrable only in liver. Others, including many of the amino acid transaminases, though primarily important in the liver, are also expressed in more readily available leucocytes (e.g. valine transaminase in hypervalinaemia) or cultured skin fibroblasts (e.g. cystathionine synthetase in homocystinuria). As pointed out in Section III.B, transport defects also appear specific to kidney and intestine.

The subcellular distribution of an enzyme could conceivably be of importance in the investigation of an inborn error. Defects in glycolysis involve cytoplasmically localized enzymes; the defect in von Gierke's disease occurs in the microsomal fraction, that in propionicacidaemia in the mitochondrion and in Gaucher's disease in the lysosomes. If different isozymes were found in the various subcellular fractions, then tissue fractionation would be necessary before any conclusions could be reached about the possible role of an enzyme in the pathogenesis of a disorder. As far as it is possible to be certain, this does not appear to occur frequently and most enzyme deficiencies have been investigated in whole tissue homogenates. However, in hyperoxaluria type I, Koch *et al.* (1967) demonstrated a deficiency of α-ketoglutarate: glyoxalate carboligase in the cytoplasmic fraction, with a normal enzyme level in the mitochrondrial fraction of the cell. A similar finding has been made in a case of tyrosine: α-ketoglutarate amino-transferase deficiency (Fellman *et al.*, 1969).

A somewhat unpredictable distribution of enzymes in body tissues has presented many problems in the detailed study of inborn errors of metabolism. Visceral organ biopsy is hardly ethical unless precise diagnosis will assist management and treatment of the disease. A few accessible tissues have had limited usefulness; histidase (histidinaemia) can be assayed in stratum corneum, tyrosinase (oculocutaneous albinism) in hair root bulbs, β-galactosidase (generalized gangliosidosis) in urine and ceramide trihexosidase (Fabry's disease) in plasma. If the enzyme is expressed in red cells the investigator has few excuses for not producing a detailed study of residual activities. But apart from the haematological disorders only a few inborn errors may be monitored in this most available of tissues (Table 8).

Table 8

Non-haematological enzyme deficiency conditions which are expressed in red cells

Condition	Enzyme deficiency	Reference
Pentosuria	xylitol dehydrogenase (NADP-linked)	Wang and van Eys (1970)
Galactosaemia	galactose-1-phosphate uridylyl transferase	Kalckar *et al.* (1956)
Galactokinase deficiency	galactokinase	Gitzelmann (1967)
Forbes's disease	amylo-1, 6-glucosidase	Hers (1964)
PFK myopathy	phosphofructokinase	Tarui *et al.* (1965)
Acatalasia	catalase	Takahara (1952)
Argininosuccinicaciduria	argininosuccinase	Tomlinson and Westall (1964)
Hyperargininaemia	arginase	Terheggen *et al.* (1970a)
Oroticaciduria	orotidylate pyrophosphorylase orotidylate decarboxylase	Huguley *et al.* (1959)
Lesch-Nyhan syndrome	hypoxanthine-guanine phosphoribosyl transferase	Seegmiller *et al.* (1967)
Phosphorylase kinase deficiency	phosphorylase kinase	Huijing and Fernandes (1970)
Congenital erythropoietic porphyria	uroporphyrinogen III cosynthetase	Romeo *et al.* (1970a, b)

Developments in fibroblast culture have greatly assisted the detailed study of inborn errors of metabolism. Biopsy of a small patch of the patient's skin (usually so painless that no local anaesthesia is necessary) and subsequent growth of the tissue in well-described cell culture media allow the gradual generation of quite considerable quantities of cells over a period of months. The fibroblast-type cells which grow out from the original explant must be subcultured when they reach monolayer confluences and this does allow the possibility of contamination by other cells (particularly if the laboratory maintains stocks of HeLa cells) or by ever-present mycoplasma. Cells may be arrested in their growth at an early stage and stored with suitable preservative at liquid nitrogen temperatures until needed, when restoration of growth can be effected by returning them to cell culture conditions. Many laboratories around the world have gained extensive experience in skin fibroblast culture to the point that banks of mutant lines have been established, so that an investigator studying a particular disorder can gain access to additional samples without troubling the patient.

In general the phenotype of fibroblast lines appears to reflect that of the donor quite faithfully. The range of enzymes expressed encompasses most of the major pathways of intermediary metabolism. A useful rule-of-thumb is that if an enzyme is measurable in white blood cells it will be found in fibroblasts, and for that matter in cultured amniotic fluid cells, though specific activities may be quite different. A few tissue-specific enzymes such as fructose diphosphatase and glucose-6-phosphatase are apparently not detectable in fibroblasts and of course the haemoglobinopathies, immunoglobinopathies and coagulation disorders cannot be studied in this type of material. Various attempts have been made to induce enzymes and proteins by chemical manipulation of cultured cells but with little success. A claim by Uhlendorf and Mudd (1968) that cystathionine synthetase is absent from normal skin but present in cultured skin fibroblasts and has therefore been induced, has not been confirmed, and it seems more probable that the apparent absence of the enzyme in the starting material is due to the technical difficulties of the assay in small amounts of material. The best chance of enzyme induction comes through developments in somatic cell hybridization.

A tissue which has come increasingly into use in the study of inborn errors of metabolism is the long-term lymphoid cell line. Long-term as opposed to short-term cultivation of lymphoid cells was first reported in 1964 (Pulvertaft; Epstein and Barr) from solid tumours and tumour fluid aspirates from patients with African Burkitt's lymphoma. Cells grew in suspension culture, doubled their numbers every 24 to 28 hours and could be maintained for very long periods of time in active growth or suspended in liquid nitrogen.

Though it was claimed that long-term lymphoid lines could be established from large volumes of blood from normal individuals (Moore *et al.*, 1967), it was subsequently shown that Epstein-Barr virus was responsible for the proliferation of the cells (Miller *et al.*, 1971), and it is now accepted that viral transformation is necessary for long-term as opposed to short-term survival. Current methods for the establishment of long-term lines use a combination of Epstein-Barr virus and phytohaemagglutinin (Beratis and Hirschhorn, 1973) and have a high degree of setting-up success.

Though lymphoid lines have many properties which made them suitable for the study of inborn errors of metabolism, their basic biochemistry has yet to be adequately studied. From the incomplete information now available it appears that enzymes of glycolysis, of purine metabolism and of lysosomal scavenging are all present, though often at much lower specific activities than in cultured fibroblasts (Glade and Beratis, 1976). It would be surprising if the basic metabolism of the long-term lymphoid cell turned out to be very different to that of the untransformed leucocyte, and the list of inborn errors of metabolism diagnosable through lymphoid cells (Table 9) to date contains only examples of enzymes also expressed in isolated white blood cells.

Table 9

Inborn errors of metabolism diagnosable in long-term lymphoid cells lines[a]

Disorder	Enzyme
Phosphoglycerate kinase deficiency	phosphoglycerate kinase
Triosephosphate isomerase deficiency	triosephosphate isomerase
Glucosephosphate isomerase deficiency	glucosephosphate isomerase
Glucose-6-phosphate dehydrogenase deficiency	G6PD
Galactosaemia	galactose-1-phosphate uridylyl transferase
Galactokinase deficiency	galactokinase
Pompe's disease	α-1, 4-glucosidase
Tay-Sachs, Sandhoff's	hexosaminidase
Generalized gangliosidosis	β-galactosidase
β-glucuronidase deficiency	β-glucuronidase
Metachromatic leucodystrophy	arysulphatase A
Gaucher's disease	β-glucosidase
Lesch-Nyhan	HGPRT
Homocystinuria	cystathionine synthetase
Cystathioninaemia	cystathionase
Citrullinaemia	argininosuccinate synthetase
Hurler's syndrome	α-iduronidase

[a] Adapted from Glade and Beratis (1976)

V. HETEROGENEITY

After a genetic disease has been named, further analysis often reveals not a single disorder, but a collection of disorders with varying pathologies and perhaps even varying modes of inheritance. For example, four distinct types of intestinal polyposis, each with autosomal dominant inheritance, have been reported; on the other hand the disease known as retinitis pigmentosa shows autosomal dominant, autosomal recessive and X-linked recessive modes of inheritance (McKusick, 1975). Though little is known about the molecular origins of these diseases, the gradual increase in understanding of primary protein lesions in other inborn errors is beginning to reveal the different ways in which genetic heterogeneity can work. Basically there are two distinct situations, involving either a single locus (allelism) or separate loci (non-allelism). Information from genetic linkage may permit a decision on these alternatives, if the disorder is common enough to allow analysis of segregating markers, such as the Rhesus-linked and unlinked forms of hereditary elliptocytosis (Chapter 9). Biochemical investigations, however, can usually only determine whether a single protein or different proteins are involved. The main exception to this is in the haemoglobino-pathies where the polypeptide structure of the protein is well-known and amenable to detailed analysis.

It must also be remembered that the clinical expression of a disease, caused by a defined mutant protein, can be altered by the rest of the genotype in ways which at the moment cannot be resolved. This kind of heterogeneity is said to be the consequence of modifying genes, but it is not known how these influence the action of other genes at the molecular level. Another possible source of heterogeneity is environmental effects which in certain situations mimic genetic effects to produce what are called phenocopies. Thus lactase deficiency can be acquired as a result of intestinal disorders or may be inherited (Gray, 1972), while renal glycosuria, normally the product of a genetically-determined defect in the kidney transport mechanism for glucose, can be induced by phlorhizin poisoning (Bondy, 1969). The discovery that women with phenylketonuria may have severely mentally-retarded offspring because of the effects of high concentrations of plasma phenylalanine and its metabolites on the developing foetal brain (Huntly and Stevenson, 1969) is a special type of phenocopy in which the disorder is based on the genotype of the mother. Kacser et al. (1973), studying histidin-aemia in the mouse, have suggested that this phenomenon may be found in other aminoacidopathies. The role of drugs, other teratogens and viruses in the aetiology of other apparent genetic disorders is still very unclear.

A. Non-Allelic Genes

Mutant genes at separate and distinct loci may give rise to disorders which are clinically and even chemically very difficult to distinguish. If the abnormal genes control enzymes in a metabolic sequence, defects may manifest themselves primarily as failures to produce the ultimate product of the pathway. When the pathway is a multireaction one it can be difficult to locate the exact enzymatic site of breakdown. This point is well illustrated in the non-spherocytic haemolytic anaemias arising from deficiencies in the enzymes of red cell glycolysis. Hexokinase, phosphohexose isomerase, phosphofructokinase (red cell isozyme) and pyruvate kinase deficiencies all interfere with the smooth functioning of the major pathway of glucose utilization (Fig. 1), deplete the cell of an adequate source of adenosine triphosphate and lead to increased fragility and rapid breakdown of the red cell membrane. Other deficiencies in the related hexosemonophosphate sequence (e.g. certain forms of G6PD deficiency and glutathione reductase deficiency) can have similar results. Unless detailed metabolite analysis or direct enzyme assay is resorted to, the clinician is at a loss to place a precise diagnosis on these forms of haemolytic anaemia.

A somewhat different situation is found in the spectrum of congenital methaemoglobinaemias. In these rare disorders, characterized by persistent bluish-grey cyanosis, both dominant (Horlein and Weber, 1948) and recessive (Scott, 1960) modes of inheritance have been described. The dominantly transmitted condition results from the presence of a structurally abnormal haemoglobin molecule (Hb M), more stable in the oxidized (methaemoglobin) stage than in the reduced, and resistant to normal cellular mechanisms for methaemoglobin reduction (see Chapter 11). The recessively transmitted condition is caused by a defect in the principal cellular mechanism for reducing methaemoglobin, the enzyme methaemoglobin reductase (Gibson, 1948). In both situations there is a build-up in red cells of a molecule incapable of transporting oxygen. The two main types of methaemoglobinaemia are further complicated by the fact that several varieties of haemoglobin M are known (involving both the α- and the β-chains, and therefore the two major loci governing haemoglobin structure; see Chapter 11) and by the existence of different abnormal methaemoglobin reductases (Kaplan and Beutler, 1967; Bloom and Zarkowsky, 1969).

The use of pharmacological doses of vitamins in the management of certain inborn errors of metabolism (Table 6) has provided another method of probing for heterogeneity. Unresponsive patients may have a defect in the apoenzyme portion of the holoenzyme complex, so that the complex is incapable of catalytic activity even in the presence of adequate coenzyme (Section III.B). Alternatively the apoenzyme may be frankly deficient. Responsive patients, on the other hand, may have impared apoenzyme–coenzyme binding or a deficiency in the enzyme-mediated pathway from vitamin precursor to coenzymatically active vitamin. Response and lack of response to vitamin therapy is thus in itself a strong indicator of genetic heterogeneity within a defined inborn error of metabolism. This heterogeneity may be further subdivided in the case of vitamin-responsive patients, since several of the pathways from precursor to coenzymatically active vitamin are complex and involve several enzymatic steps. Three different vitamin B_{12}-dependent types of methylmalonic acidaemia have been described (Fig. 3), while similar defects have been described in the complex set of reactions involving the folic acid coenzymes (Erbe, 1975).

It will be apparent from the examples given that resolution of heterogeneity is greatly helped by knowledge of the protein defect. If the primary lesion is unknown the task is more difficult. Family studies which reveal different modes of inheritance for apparently identical disorders provide sound evidence for separate genes. Careful clinical examination of patients and chemical analysis of unusual metabolites is useful, but cannot ultimately discriminate between allelic and non-allelic heterogeneity. A tissue culture method, introduced by Neufeld and her colleagues (Fratantoni et al., 1969a, b; Neufeld and Fratantoni, 1970), has given promising results in distinguishing different types of mucopolysaccharidoses and may be generally applied where cellular metabolism is markedly deranged (Danes and Bearn, 1970). Neufeld and her colleagues showed that cultured fibroblasts from patients with clinically-defined Hurler's syndrome incorporated excessive amounts of radioactive sulphate into the mucopolysaccharide fraction; so did fibroblasts from patients with Hunter's syndrome. But when cells from the two types of patients were mixed, the abnormality disappeared. It seemed that each of the two cell types contained a factor which complemented the other's deficiency and restored normality by abolishing excessive radioactive sulphate uptake. The factors could be exchanged by cells growing together in culture, or by treating one line of cells with soluble extract from another line, and have now been identified as enzymes. Neufeld and Fratantoni (1970) proposed that this in vitro complementation could be used as a test for heterogeneity. If the fibroblasts from two patients with an apparently identical disease corrected each other's defects, then the same mutant gene was not being expressed and the disease was heterogeneous. On the basis of complementation tests the Sanfilippo syndrome, a clinically homogeneous entity, was divided into two distinct disorders, subsequently proved to be caused by different enzyme deficiencies (Kresse and Neufeld, 1972; O'Brien, 1972). Conversely, lack of *in vitro* complementation could be taken as evidence that a single genetic locus was involved. Fibroblast studies were used to show that type I and V mucopolysaccharidoses (Hurler and Scheie syndromes) did not complement and were either identical or allelic (Wiesmann and Neufeld, 1970). A deficiency of α-iduronidase has now been described in fibroblasts from both of these syndromes (Matalon and Dorfman, 1972; Sjoberg *et al.*, 1973). A new mucopolysaccharidosis, also based on α-iduronidase deficiency, is presumed to be a genetic compound between the allelic Hurler and Scheie genes (McKusick *et al.*, 1972; Danes, 1974).

The use of somatic cell hydridization in searching for heterogeneity has been discussed in Chapter 9. It is technically difficult since it requires skill in the growth of fibroblast cultures (with high efficiency in the formation of heterokaryons) and in microtechniques for enzyme assay on a limited numbers of cells. When complementation between parent lines fails to occur the conclusions are: (1) that the lines are identical or (2) that they contain allelic mutations or (3) that they contain non-allelic mutations but that no adequate heterokaryon formation has occurred. The last possibility can sometimes be excluded by further exhaustive hybridization experiments, but allelism and identity cannot be distinguished. When complementation does occur the usual conclusion is that the parent lines expressed non-allelic mutations. If the parent lines suffered the same enzyme deficiency this suggests that the enzyme is a polymer under the control of more than one genetic locus. However, the possibility of intragenic (interallelic) complementation in fibroblast hybrids cannot be completely excluded.

B. Allelic Genes

Unless the polypeptide structure and amino acid sequence of a protein is known, it is difficult to be certain whether observed examples of variation are the products of allelic genes or non-allelic genes. Among the haemoglobin variants, polypeptide separation and sequence analysis has placed individual mutations at the α or β loci. Among the multiple G6PD variants the weight of evidence is for single X-linked locus control and therefore for allelic variation. With the exception of the evidence from *in vitro* complementation and somatic cell hybridization, there is, for most other human enzymes and proteins, little definite knowledge about the number of genes controlling structure.

From the laboratory standpoint, it is probably more important to recognize heterogeneity deriving from different variants of the same protein as opposed to heterogeneity deriving from different proteins, than to be able to distinguish the number of loci involved. Knowledge of heterogeneity based on single proteins has accumulated slowly, to a large extent using the same criteria which have been applied to defining structural gene mutations (Section II.A). Provided that residual enzyme activity can be detected in the inborn error under consideration, the mutant protein may be examined and compared with other mutants in terms of electrophoretic mobility, kinetic properties and immunological status. The G6PD variants have been particularly instructive in such studies (Chapter 5), since it is unusual to find a complete enzyme deficiency, and red blood cells are an accessible tissue.

VI. HETEROZYGOTE DETECTION

Though most recessively inherited inborn errors of metabolism are quite rare, the frequencies of their heterozygotes are of course substantially higher. When the incidence of a disease such as cystic fibrosis is 1 in 2000 of the population, approximately 1 in 22 people will carry a single dose of the mutant gene and about 1 in 450 marriages will be at risk. The advantages of being able to detect clinically normal heterozygotes for potentially lethal diseases are obvious from the point of view of preventive genetic counselling. This is particularly so when a reliable technique for antenatal diagnosis exists (Section VII). One of the most impressive programmes in heterozygote detection in a large community is that organized by Kaback to find carriers of Tay-Sachs disease among the Jewish population of Baltimore (Kaback *et al.*, 1974). Tay-Sachs disease is common among Ashkenazi Jews, it is a particularly distressing lethal disorder, the primary enzyme lesion is known, carrier detection is possible from a small serum sample and the disorder can be diagnosed *in utero* early in pregnancy. By finding carriers before marriage it was possible to identify high risk pregnancies in advance and through antenatal monitoring to insure that they were normal or considered for termination. Few, if any, other inborn errors of metabolism satisfy all these criteria, though sickle cell anaemia and cystic fibrosis of the pancreas are not far away.

In X-linked recessive disorders a somewhat different situation is found, for the heterozygote frequency is now only twice the disease frequency. However, because a carrier female transmits the disease to the same proportion (a quarter) of her children irrespective of the man (or men) she marries, heterozygote detection is of even greater practical benefit. If the genic make-up of the daughters of a known carrier of a disease like haemophilia, Duchenne muscular dystrophy or Lesch-Nyhan syndrome can be established, then their risks of having affected children can be stated quite independently of their future mating patterns. The sons of carriers of X-linked recessives have a 50% chance of being affected, and since for many parents this is an unacceptably high risk, selective abortion of male foetuses is often carried out. However, recently a number of X-linked metabolic disorders have become diagnosable *in utero* from cultured amniotic fluid cells (Section VII).

Heterozygote detection has also been useful because it can contribute to basic knowledge on metabolic disorders. Estimation of carrier prevalence allows calculation of more accurate gene frequencies for recessively inherited disorders where there are problems of total ascertainment. For rare disorders which might be either sporadic (i.e. non-Mendelian) or Mendelian recessives, establishment of carrier status in the symptom-free parents is the best proof that one is dealing with a genetic condition and that parents should be counselled on risk of recurrence. In X-linked recessives carrier detection is useful in distinguishing between transmitted disorders and those which are the result of a new mutation.

A. Autosomal Recessive Transmission

In disorders transmitted as autosomal recessives, heterozygote detection is dependent on a dosage effect. The assumption is made that a single dose of mutant gene will produce half as much of the abnormal protein as a double dose. If the abnormal protein is an enzyme which has lost its catalytic function, then the expectation is of almost no activity in the affected homozygote, full activity in the unaffected homozygote and approximately half activity in the heterozygote. This expectation is closely followed in many disorders; indeed the success of the Tay-Sachs screening programme was dependent on the ability of laboratory personnel to distinguish heterozygotes from normal homozygotes on the basis of a serum enzyme assay. When the primary protein lesion in the disorder is unknown heterozygote detection is more difficult and sometimes only successful through the chance finding of a metabolite or protein which will serve as a marker.

Dosage effects have been found with some regularity in enzymological attempts to establish heterozygous status, from tissues as diverse as red cells, white cells, plasma, urine and cultured fibroblasts. Indeed heterozygotes for enzyme deficiency conditions have been discovered accidentally in the course of quantitative enzymological surveys aimed at other ends (Singer and Brock, 1971). The hazards of drawing facile conclusions about genetic constitution from enzyme assays should not, however, be underemphasized. The growing heterogeneity of many thoroughly investigated inborn errors of metabolism, and recognition of compound heterozygosity as underlying several recessively inherited conditions (McKusick, 1975), suggests that many parental pairs of heterozygotes will be dissimilar. This was first clearly shown in congenital haemolytic anaemia due to glucose-phosphate isomerase deficiency, where the parents of the patient were found to have starch gel electrophoretograms differing from normal and each other (Detter *et al.*, 1968). It is an obvious explanation for the complete absence of hexosaminidase A activity to synthetic substrate which has been described in a symptom—free obligate carrier of Tay-Sachs disease (Navon *et al.*, 1973; Vidgoff *et al.*, 1973). When compound heterozygosity and environmental effects on enzyme levels have been discounted, the various lists of enzyme tests for heterozygosity which have been compiled (Nitowsky, 1975; Kolodny, 1975) suggest that gene dosage is the most powerful single determinant of protein levels.

In some disorders difficulties of obtaining tissue make it impossible to screen for potential carriers by quantitative enzymology. Measurement of circulating or excreted metabolites is a useful alternative, though it is often impossible to distinguish between a heterozygote and a normal homozygote. In phenylketonuria, for example, plasma phenylalanine of carriers is elevated but overlaps with levels found in normal individuals. In situations like this where the enzyme responsible is expressed only in the liver, various loading tests have been devised to stress the metabolic pathway under consideration. The suspected carrier is given a standard dose of phenylalanine and the rate of disappearance of the amino acid from the plasma is followed over a period of hours and compared with rates in normal people and known phenylketonurics. Heterozygotes have tolerance curves which lie between those of normal and affected homozygotes. In cystinuria heterozygotes may be identified by their urinary excretion patterns of cystine and lysine without loading, while in cystathioninaemia a methionine load exaggerates the heterozygote's urinary cystathionine excretion (Rosenberg and Scriver, 1974). Other loading tests have been summarized by Hsia (1969).

Even when the primary lesion in a disorder is unknown, a suitable enzymatic or chemical marker may be found. The metachromatic staining of cultured fibroblasts by toluidine blue (Section III.B) has been widely used to detect storage disorders of lipid and mucopolysaccharide nature. It has been claimed that this technique will detect heterozygotes in a number of inborn errors of metabolism, both where the primary lesion is known and where it is unknown. In the latter group cystic fibrosis is the most important because of its high frequency in North European populations and the difficulties which have been experienced with bioassay systems such as the rabbit trachea, oyster cilia and the motile protozoan, *Proteus vulgaris*

(Danes, 1975). Detection of heterozygotes by cultured fibroblast meta-chromasia was first reported by Danes and Bearn in 1968. Though the technique has been criticized for its lack of specificity and reproducibility, Danes *et al.* (1975) have more recently repeated their claim for meta-chromasia. They assert that the most reliable method of heterozygote detection in cystic fibrosis is to show metachromasia without alcianophilia in cultured fibroblasts and to follow this with a demonstration of metabolic cooperation between the heterozygote and normal fibroblasts.

Though dosage effects usually ensure that the average level of enzyme activity in heterozygotes is significantly less than in normal homozygotes, there is always considerable variation in both groups, and the activity distributions often overlap. The mean values may be useful in confirming a postulated recessive mode of inheritance, but are not helpful if an individual's status as carrier has to be established. One possible way of improving the discriminatory power of enzyme assays has been suggested by Hirschhorn *et al.* (1969). They found that α-glucosidase activities in lymphocytes of people heterozygous for Pompe's disease overlapped the activities of controls (values were zero in two affected homozygotes). When lymphocyte transformation was induced with phytohaemagglutinin, α-glucosidase in controls increased sharply but in heterozygotes remained constant, so that there was no longer overlap between the two groups. Stimulation of lymphocytes by phytohaemagglutin leads to morphological changes, increased protein synthesis and mitosis, and is known to be associated with enhanced activities of several enzymes (Hirschhorn *et al.*, 1967). It is not clear, however, why the lymphocytes of a person heterozygous for α-glucosidase deficiency fail to show any increase in enzyme activity. A similar finding has been made with respect to lysosomal acid phosphatase deficiency (Nadler and Egan, 1970) and cystathionine synthetase in homo-cystinuria (Goldstein *et al.*, 1973).

B. X-Linked Recessive Transmission

The detection of female heterozygotes for disorders coded for by genes carried on the X-chromosome raises special problems. According to the Lyon hypothesis (Lyon, 1961; Chapter 9), early in embryonic life there is random inactivation of either the maternally-derived or paternally-derived X-chromosome. Once this has occurred in any cell the same X-chromosome continues to be inactivated in all daughter cells derived from it. Females may therefore be regarded as mosaics both for X-chromosomes and X-linked genes. This peculiar form of dosage compensation means that the activities of enzymes coded for by X-linked genes are no higher in females (with two X-chromosomes) than in males (with one X-chromosome). It also means that if one of the X-chromosomes of a female carries a mutant gene, the expression of the gene product can be highly variable. Tissues with cells derived largely from a precursor cell with a normal X-chromosome will be near normal both physiologically and enzymatically; tissue containing cells derived largely from an abnormal X-chromosomal precursor may deviate significantly from normality. It is difficult to know in advance to what extent a particular tissue of a female heterozygote will display the presence of a mutant X-linked gene. In a majority of cases, however, some indication of heterozygous status can be detected, probably because most tissues have multi-celled origins.

The first X-linked disorder to be solved at the protein level was G6PD deficiency (Carson et al., 1956). Soon after the Lyon hypothesis was put forward, Beutler et al. (1962) presented evidence for two distinct populations of red cells in female carriers of the disease and later achieved a partial separation of the fractions (Beutler and Baluda, 1964). Davidson et al. (1963) cultured fibroblasts from women known to be heterozygous for the enzyme deficiency (B/B⁻) and for the common Negro variant (A/B) and, by cloning and subculture, derived cell populations which expressed only one of the two enzyme forms. In B/B⁻ heterozygotes enzyme assays showed either very low activity or full activity in the separate clones; in A/B heterozygotes electrophoresis revealed either the A or B form in the clones. The relative expression of the products of normal and mutant genes in different tissues is highly variable. Stamatoyannopoulos et al. (1967) were able to detect only 65% of obligate heterozygotes for the Mediterranean form (B⁻) of G6PD deficiency by direct red cell enzyme assay, most of the remainder overlapping the normal range. Gandini et al. (1968) found four women in a large group of carriers who had no detectable red cell enzyme activity, but measurable activities in granulocytes and lymphocytes, and half normal activity in skin. The simplest and most reliable methods of carrier detection in G6PD deficiency are those that depend on histochemical demonstration of two populations of erythrocytes, using either methaemoglobin generation or tetrazolium reduction (Beutler, 1968; Pinsky, 1975).

An X-linked condition which has attracted a great deal of investigative effort is the Lesch-Nyhan syndrome, which results from a deficiency of hypoxanthine-guanine phosphoribosyl transferase (HGPRT). The enzyme is essentially absent in the red cells of affected males, but at or above normal levels in red cells (Henderson et al., 1969) and lymphocytes (Dancis et al., 1968) of heterozygotes. Mosaicism of the cultured fibroblasts of carriers was demonstrated by autoradiographic techniques (Rosenbloom et al., 1967) and later the cells cloned into enzyme-positive and enzyme-deficient populations (Salzmann et al., 1968; Migeon et al., 1968). Because normal cells have been found to predominate in fibroblast cultures from known carriers, a medium which selects against normal cells and allows growth of any mutant cells present is useful in ensuring that the correct decision is made about heterozygote status (Migeon, 1970).

Other disorders in which cloning techniques have been used in heterozygote detection are Fabry's disease, Hunter's syndrome, phosphoglycerate kinase deficiency and one form of chronic granulomatous disease (Chapter 9, Table 3). Though such in vitro demonstration of mosaicism are the most reliable methods of X-linked carrier detection, they are subject to two major problems—cell selection and metabolic cooperation. Cell selection is most easily visualized in the composition of the erythrocyte population of a carrier of G6PD deficiency. If the initial population consists of roughly equal proportions of both types of cells, those which are G6PD deficient will be at a disadvantage in situations of oxidative stress (e.g. in the presence of certain drugs) and will be rapidly removed from the circulation by haemolysis. Haemotopoiesis will replace deficient cells with the same proportion of both types of cells and the overall proportion of normal cells with gradually rise. It is therefore no surprise that the principal difficulty in detection of G6PD deficiency heterozygotes is overlap with the normal range of values (Beutler, 1968). The same problem, but in a more acute form, is found in the Lesch-Nyhan syndrome, where practically all heterozygous females have normal red cell enzyme activities. In this case intravascular haemolysis of enzyme-deficient cells is not the explanation, for survival times of red cells in affected males have been shown to be normal. Kelley et al. (1969) suggest that HGPRT-deficient red cell precursors are either destroyed preferentially or fail to mature properly in the marrow of heterozygotes. However, this explanation does not account for the normal population of red blood cells in affected hemizygotes. That only one type of red cell is present in HGRPT-deficiency carrier's circulation has been demonstrated rather neatly by Nyhan et al. (1970). They screened the cells of two women heterozygous both for HGPRT-deficiency and the G6PD A/B polymorphism and found hemizygous expression of the G6PD phenotype in red cells and heterozygous expression in fibroblasts.

In cultured fibroblasts where heterozygous detection of Lesch-Nyhan carriers must be performed, the problem of metabolic cooperation arises. This was originally defined by Subak-Sharpe et al. (1969) as modified cellular metabolism caused by the exchange of factors between deficient and normal cells in contact, in contrast to in vitro complementation where no contact is required. Dancis et al. (1969a) invoked metabolic cooperation to explain the observation that as cell density increases the number of HGRPT-positive cells in a culture derived from a heterozygote (or from an artificial mixture) also increases. Fujimoto and Seegmiller (1970) have confirmed this observation and Migeon (1970) has explained the predominance of normal cells in heterozygote cultures as being due to "cross-feeding". It is not yet clear which of several possible factors is being transferred from positive to negative cells; in the ascertainment of women who are carriers of the Lesch-Nyhan gene, metabolic cooperation further complicates a problem already make difficult by random X-inactivation and cell selection. Though in principle this type of concealment of cellular mosaicism could be disastrous for tissue culture methods of heterozygote detection, the only other disorder in which it has been reported is Hunter's syndrome. Here uncloned fibroblasts have normal incorporation of radioactive sulphate into mucopolysaccharides, as would be predicted from studies on in vitro complementation (Pinsky, 1975). What is not yet known is whether levels of the deficient enzyme, sulphoiduronate sulphatase, are also affected.

Culturing and cloning of fibroblasts is a time-consuming and arduous process unsuitable for routine studies of heterozygote status. The observation that hair follicles exhibit a degree of clonal growth (Gartler et al. 1971) allows the possibility of plucking hairs from the scalp of a presumed carrier and examining them for cellular or biochemical mosaicism. Heterozygotes for the Lesch-Nyhan syndrome have been diagnosed in this way by direct quantitative assay of HGPRT in single follicles (Gartler et al., 1971; Silvers et al., 1972) and by electrophoretic separation and histochemical staining of the enzyme zone on the gel (Francke et al., 1973). As expected, three different types of follicle were observed; HGPRT positive, HGPRT negative and mixed (presumably due to non-clonal growth). The use of a reference enzyme in such studies is important, since many follicles will be enzyme-negative due to atrophied sheath cells. Follicle analysis can in theory be applied to any X-linked disorder expressed in fibroblasts.

Where an inborn error is not expressed in fibroblasts the difficulties of reliable carrier detection are considerable. A majority of females heterozygous for haemophilia A have depressed Factor VIII coagulant activity, but a substantial minority have activities within the normal range, so that a definitive statement on the risks of an individual woman bearing affected sons cannot always be made. Discovery that a majority of haemophiliacs, though lacking coagulant activity, have levels of Factor VIII which may be measured immunologically with an appropriate antiserum (Bennett and Huehns, 1970), has suggested a new approach to carrier detection in X-linked recessives. The method is appropriate only in CRM (cross-reactive material) positive deficiencies, that is situations where a mutant has abolished biological activity but not antigenic activity in the hemizygote. In these cases the woman who is heterozygous for the mutation will possess two X-chromosomes both capable of directing synthesis of protein with normal antigenic potential but only one X-chromosome capable of directing synthesis of protein with normal biological potential. Carrier status will be revealed by a distorted ratio of biological to immunological activity. Though such ratio measurements have so far been applied only to carrier detection in haemophilia A (Bennett and Huehns, 1970; Zimmerman *et al.*, 1971; Bennett and Ratnoff, 1973) the principle could be applied to other X-linked recessives, where the mutation is CRM-positive and an appropriate antiserum has been prepared.

VII. PRENATAL DETECTION

Though biochemical techniques for the detection of carriers of a single dose of a mutant gene have made impressive progress in recent years, heterozygote detection remains a procedure with strictly limited use. Few inborn errors of metabolism are sufficiently common (see Table 1 for exceptions) to be accompanied by high heterozygote frequencies. Even fewer quantitative biochemical assays are easy or cheap enough to be applied at the population level. The first indication of heterozygous status in a person is nearly always the appearance of a disorder, known to be inherited recessively, in one of his children. At this point the precise establishment of heterozygosity for the disorder in the parents is largely redundant, for the fact of a diseased child will usually outweigh theories about parental heterozygosity.

Normal medical practice is to inform parents of the high risk of the recurrence of the disorder in subsequent children and to advise against further pregnancies through effective contraception or sterilization. This may seem harsh to those desperately wanting a normal child and to whom a 25% risk may be an ineffective deterrent. The frequency with which genetic advice is deliberately or accidentally ignored (Carter *et al.*, 1971) accounts to some extent for the rapid rise of interest in procedures for the prenatal detection of genetic disorders. High-risk pregnancies which can be monitored in the early stages and treated or terminated if the foetus is shown to be affected, provide a ready solution to the problem of heterozygous parents wanting their own normal child. Though it is difficult to see prenatal detection becoming an aspect of routine medical care in large populations, it has been pioneered in countries with facilities for management in depth of individual patients and has already achieved considerable momentum. The first demonstration of a foetus shown to be affected with an inborn error of metabolism was carried out in 1965 (Jeffcoate *et al.*, 1965); the topic has developed rapidly in the past ten years and been the subject of comprehensive reviews (Brock, 1973; Milunsky, 1973, 1975).

A. Techniques

The standard method for diagnosing an inborn error of metabolism pre-natally is by removing an aliquot of amniotic fluid by amniocentesis and culturing the cells to a point where biochemical measurements are possible. Amniotic puncture is usually performed at around the 15th or 16th gestational week when the volume of fluid is sufficient to provide a reasonable target for the amniocentesis needle (possibly guided by ultrasonar scan). Between 10 and 30 ml of fluid is removed, the cells gently centrifuged and plated out on glass or plastic for conventional tissue culture procedures. Amniotic fluid cells, derived largely from the foetal skin and amnion, are already in a dispersed state and will grow more readily than the fibroblasts of an adult skin explant. Morphologically, both epithelioid-like and fibroblast-like cells appear in early cultures, though the latter tend to overgrow the former as the culture persists. Accumulating sufficient cells for conventional macroscopic enzyme assay takes longer than acquiring enough cells for karyotype analysis, and usually several subcultures are necessary. This introduces the danger of contamination by ever-present mycoplasma or by rapidly-proliferating heteroploid human cells such as the HeLa line. It also subjects the patient to the strain of waiting 4 to 8 weeks between amniocentesis and the reporting of a biochemical diagnosis (Littlefield, 1971; Sutherland and Bain, 1973).

Diagnosis of inborn errors of metabolism by enzymatic or metabolite assay on either uncultured amniotic fluid cells or the fluid supernatant itself could greatly reduce the waiting time. However, uncultured cells are not reliable since they consist of a mixture of viable and non-viable cells as well as meconium and other debris (Sutcliffe and Brock, 1971). Quantitative and qualitative analysis of mucopolysaccharides in the supernatant fluid has met with mixed success (Matalon et al., 1970, 1972; Brock et al., 1971), and can no longer be justified now that the primary enzyme lesions in the muco-polysaccharidoses are known. Though several amniotic fluid lysosomal enzymes, such as hexosaminidase A (O'Brien et al., 1971), β-galactosidase (Lowden et al., 1973) and arylsulphatase A (Borressen and van der Hagen, 1973), appear to reflect the foetal phenotype, there is always a danger of contamination by maternal enzymes. Most of the protein of the amniotic fluid is maternal in origin (Sutcliffe and Brock, 1973) and at least one lyso-somal enzyme, α-1, 4-glucosidase, can be dangerously misleading if used to diagnose Pompe's disease (Cox et al., 1970). Somewhat unexpectedly, measurement of amniotic fluid α-fetoprotein, introduced as a tool for the diagnosis of spina bifida and anencephaly (Brock and Sutcliffe, 1972), has turned out to be useful in prenatal detection of the recessively inherited condition of congenital nephrosis (Kjessler *et al.*, 1975). Since α-fetoprotein enters the fluid via foetal urination (Brock, 1976) it may well prove valuable in other nephrotic syndromes.

The introduction by Galjaard and his co-workers (Galjaard *et al.*, 1973, 1974b; Niermeijer *et al.*, 1975, 1976) of elegant microtechniques for enzyme assay in cultured cells shows promise of releasing prenatal diagnosis from the constraints of massive cell culture. Amniotic fluid cells are grown in small plastic-bottomed dishes to a point where a number of clones become visible. The cells are freeze-dried and groups of clones dissected out under the microscope for fluorimetric assay in small volumes. For lysosomal enzymes 10 to 14 days is sufficient to accumulate a series of clones each containing 100 to 300 cells. Since the cells are being inspected during the dissection it is possible to choose clones showing the most vigorous growth and to count individual clones so that they contain equal numbers of cells. At this early stage of cellular growth most of the clones contain mainly epithelioid-like cells rather than the fibroblast-like cells which predominate in later cultures. Controls are therefore chosen amongst normal cells grown in a similar way and the use of reference enzymes is recommended. Microtechniques have been employed mainly in lysosomal storage disorders where commercial fluorogenic substrates are available, but could in theory be much more widely applied.

B. Results

In theory any inborn error of metabolism in which the enzyme defect is expressed in cultured fibroblasts should be diagnosable by appropriate assay on cultured amniotic fluid cells. The actual list of disorders is now quite extensive (Table 10). Quantitative experience is also increasing and in 1976 Lindsten *et al.* collected a total of 280 diagnoses of 27 disorders in 10 European countries. Heading the list was Tay-Sachs disease (56 diagnoses), the only condition in which (because of several programmes of prospective heterozygote detection) diagnoses have been made before the birth of an index affected child. Conspicuously absent are the two common disorders cystic fibrosis of the pancreas and phenylketonuria.

In the latter case recent observations that phenylalanine hydroxylase is expressed at low levels in cultured fibroblasts (Hoffbauer and Schrempf, 1976; Schlesinger *et al.*, 1976) suggest that this disorder may soon be diagnosable *in utero*. In cystic fibrosis there is also now a prospect of prenatal diagnosis. Hösli *et al.* (1976) have found that when fibroblasts from patients are cultured in the presence of Tamm-Horsfall protein the activity of alkaline phosphatase increases dramatically. Stimulation of alkaline phosphatase by feeding cultured fibroblasts with endogenous macromolecules has

Table 10

Inborn errors of metabolism which have been diagnosed prenatally[a]

Disorder	Enzyme or metabolite assayed
Lipidoses	
Generalized gangliosidosis	β-galactosidase
Tay-Sachs disease	hexosaminidase A
Sandhoff's disease	hexosaminidase A and B
Gaucher's disease	β-glucosidase
Fabry's disease	α-galactosidase
Niemann-Pick disease	sphingomyelinase
Metachromatic leucodystrophy	arylsulphatase A
Krabbe's disease	galactosylceramidase
Wolman's disease	acid lipase
Mucopolysaccharidoses	
Hurler and Scheie syndromes (MPS I and V)	α-iduronidase
Hunter syndrome (MPS II)	sulphoiduronate sulphatase
Sanfilippo A syndrome (MPS IIIA)	heparan sulphate sulphatase
Sanfilippo B syndrome (MPS IIIB)	α-N-acetylglucosaminidase
Maroteaux-Lamy (MPS VI)	arylsulphatase B
I-cell disease	lysosomal enzymes
Fucosidosis	α-fucosidase
Carbohydrate disorders	
Galactosaemia	galactose-1-phosphate uridylyl transferase
Pompe's disease (GSD II)	α-glucosidase
Andersen's disease (GSD IV)	amylo-(1, 4→1, 6) transglucosidase
Pyruvate decarboxylase deficiency	pyruvate decarboxylase
Aminoacidopathies	
Argininosuccinic aciduria	arginino succinase
Citrullinaemia	arginino succinate synthetase
Cystinosis	cystine uptake
Homocystinuria	cystathionine synthetase
Maple syrup urine disease	α-ketoacid decarboxylase
Methylmalonic acidaemia	methylmalonyl-CoA mutase
Propionic acidaemia	propionyl-CoA carboxylase
Miscellaneous	
Lesch-Nyhan syndrome	HGPRT
Adenosine deaminase deficiency	adenosine deaminase
Lysosomal acid phosphatase deficiency	acid phosphatase
Xeroderma pigmentosum	DNA repair enzyme
Menke's disease	copper uptake
Congenital nephrosis	α-fetoprotein concentration
Hypophosphatasia	alkaline phosphatase
Acute intermittent porphyria	uroporphyrinogen I synthetase

[a] Adapted from Milunsky (1973, 1975) and Brock (1973)

been noted in other inborn errors and is therefore not specific (Hösli, 1973). However, since the prenatal diagnostic decision in cystic fibrosis is a limited one (the foetus is either normal, heterozygous or homozygous for the particular disorder), Hösli's procedure may well prove practical.

In a majority of the conditions in Table 10, diagnoses have been based on direct enzyme assays of cultured amniotic fluid cells. In others assay has been indirect, using for example radioactive sulphates in the mucopolysaccharidoses or radioactive citrulline incorporation in citrullinaemia. In I-cell disease where the primary enzyme lesion remains obscure, assay of a variety of lysosomal hydrolases forms the basis for apparently reliable diagnoses (Aula *et al.*, 1975; Matsuda *et al.*, 1975). In cystinosis and Menke's disease, incorporation of cystine (Schneider *et al.*, 1974) and copper (Horn, 1976), respectively, into cultured cells allows early prenatal diagnosis of these as yet unsolved disorders.

An entirely new approach to prenatal diagnosis has been introduced recently by Kan *et al.* (1976b). Earlier they had showed that the molecular defect in homozygous α-thalassaemia was due to deletion of the gene that determines the structure of the α-globin chain (Taylor *et al.*, 1974). This had been achieved by molecular hybridization studies using synthetic cDNA whose structure was complementary to the normal α-globin gene (Chapter 2). Since all nucleated cells contain a full complement of genes, whether these are expressed or not, Kan *et al.* (1976b) showed that it is possible to probe for the number of α-globin genes in cultured amniotic fluid cells using the specially prepared cDNA. On the basis of the percentage of DNA annealing to cultured fibroblasts and cultured amniotic fluid cells they were able to distinguish between homozygous α-thalassaemia (hydrops fetalis), haemglobin H disease, heterozygous α-thalassaemia and normals. The method as currently used is only suitable for disorders in which gene deletions have been demonstrated. Apart from α-thalassaemia this at present comprises only hereditary persistence of foetal haemoglobin and $\delta^0\beta^0$ thalassaemia.

C. Future Prospects

The major limitation to the advancing prospects of early prenatal diagnosis stems from the apparent non-expression of certain gene products in cultured amniotic fluid cells. Three possible mechanisms exist for circumventing this difficulty. The first, the use of linked markers, has been discussed in Chapter 9. The second, foetal tissue biopsy, is already a limited reality. The third, enzyme induction through hybridization of amniotic fluid cells with other differentiated cells, remains at present a very distant prospect.

The technique of foetal tissue biopsy is dependent on the development of a safe method for visualizing the foetus *in utero* early in pregnancy. The introduction of fibre optic amnioscopes (foetoscopy) into the womb before the 20th gestational week is a delicate procedure, all too easily precipitating miscarriages (Scrimgeour, 1973). However, the possibilities that it offers in terms of foetal blood sampling for the diagnosis of haemoglobinopathies, coagulation disorders and immunoglobinopathies or foetal liver biopsy for the diagnosis of Von Gierke's disease, account for the persistence of its proponents. Further impetus for the development of foetoscopy comes from the observation that foetal reticulocytes synthesize appreciable quantities of haemoglobin A (in addition to the abundant haemoglobin F) from about the 9th week of gestation (Hollenberg *et al.*, 1971). This means that haemoglobinopathies based on mutations in the β locus, such as sickle cell anaemia and β-thalassaemia, which are among the most common of human Mendelian disorders, should be diagnosable early enough in pregnancy to allow a safe termination. Since foetal red cells are a great deal more active than maternal red cells in globin chain synthesis, a mixed placental blood aspirate can be used to detect a foetal abnormality (Kan *et al.*, 1972). If the proportion of foetal red cells in the mixture is low they can be concentrated by agglutination with anti-i antiserum (Kan *et al.*, 1974).

Prenatal diagnoses of homozygous β-thalassaemia were reported by two groups of workers in 1975 (Kan *et al.*, 1975a, b; Alter *et al.*, 1975). Blood sampling was conducted both with foetoscopic visualization and "blind". The latter procedure is claimed to be satisfactory if the placenta lies in the anterior position (Kan *et al.*, 1974) or even if it is posterior, provided ultrasound localization is used (Kan *et al.*, 1975a). Prenatal diagnosis of homozygous sickle cell anaemia (Kan *et al.*, 1976) and of a foetus homozygous for β^S and heterozygous for $\alpha^{G-Philadelphia}$ (Alter *et al.*, 1976) have been reported. Since the foetus also has detectable coagulation factors including factor VIII from twelve weeks of life, prenatal diagnosis of haemophilia A should now also be possible (Holmberg *et al.*, 1974).

The use of somatic cell hybridization in unlocking the hidden potential of amniotic fluid cells is an intriguing possibility. It is known that fused cells can express gene products not found in either of the parental lines (Harris, 1970). Peterson and Weiss (1972) hybridized rat hepatoma and mouse fibroblast cells and detected mouse albumin in the heterokaryon, a product which the parental mouse fibroblasts could not synthesize. Darlington *et al.* (1973) hybridized mouse hepatoma cells and human leucocytes and found human albumin in the products of the hybrid—again a protein which human leucocytes do not make. Of some relevance to prenatal diagnosis of haemoglobinopathies is the work of Deisseroth *et al.* (1975a, b, 1976) who fused mouse erythroleukaemia cells and human marrow cells to form hybrids which on exposure to dimethylsulphoxide could be induced to synthesize both mouse and human globins. As yet there have been no reports of somatic cell hybrids between amniotic fluid cells and other differentiated cells, but there seems little doubt that this will be feasible in the near future. The question remains as to whether the complicated technology involved will prove more amenable to diagnostic use than the equally complicated (and probably more dangerous) technology necessary for intrauterine foetal biopsy.

VIII. CLASSIFICATION OF INBORN ERRORS IN METABOLISM

The Tables (11–24) which follow this section have been classified biochemically. This is not always convenient to the clinician who might prefer a classification based on symptoms, but as the numbers of solved inborn errors increase it seems inevitable that attempts to maintain order in the subject will have to start from the basic protein lesion. There are two other disadvantages in this approach. Some disorders have their origins in areas of biochemistry where the Tables overlap. Thus it is an open question as to whether Wolman's disease should be placed among the lysosomal disorders or the defects in lipid metabolism, for it appears to be both. Other disorders have powerful alternative claimed solutions; Leigh's subacute necrotizing encephalomyelopathy has been suggested as originating in a defect in gluconeogenesis and a disorder of thiamine triphosphate synthesis—two rather different areas of metabolism. It is just possible that heterogeneity will justify both claims eventually.

A large majority of the inborn errors listed are inherited (or believed to be inherited) as autosomal recessives. Where there is evidence for X-linkage or dominant transmission this has been indicated. An attempt has been made to give a reference to the first study localizing the protein or enzyme defect, and to a recent study containing review material. Most of the enzyme and protein deficiencies have been directly demonstrated, though in some cases powerful indirect methods have established the lesion.

522   D. J. H. BROCK

Table 11

Glycogen storage diseases[a]

Disorder	Deficient enzyme and tissues involved	Reference
GSD type I (von Gierke's disease)	glucose-6-phosphatase (liver, kidney)	Cori and Cori (1952) Howell (1972)
GSD type II (Pompe's disease and mild variant)	α-1, 4-glucosidase (generalized)	Hers (1963) Howell (1972) Hudgson et al. (1968) Askanas et al. (1976)
GSD type III (Forbes's disease, limit dextrinosis)	amylo-1, 6-glucosidase; "debrancher enzyme" (generalized and rbc)	Illingworth et al. (1956) Van Hoof and Hers (1967) Howell (1972)
GSD type IV (Andersen's disease)	amylo- (1, 4→1, 6)-transglucosidase; "brancher enzyme" (generalized)	Brown and Brown (1966) Howell (1972)
GSD type V (McArdle's disease)	muscle phosphorylase (muscle)	Mommaerts et al. (1959) Schmid et al. (1959) Howell (1972)
GSD type VI (Hers's disease)	liver phosphorylase (liver)	Hers (1959) Hers and Van Hoof (1968)
GSD type VII	muscle phosphofructokinase (muscle, partial deficiency in rbc)	Tarui et al. (1965) Layzer et al. (1967)
GSD type VIII (X-linked)	phosphorylase kinase (generalized)	Hug et al. (1969) Huijing and Fernandes (1969)
Phosphorylase kinase phosphokinase deficiency	cAMP-dependent phosphorylase kinase phosphokinase (muscle, liver)	Hug et al. (1970b)
Glycogen synthetase deficiency	glycogen synthetase (liver)	Lewis et al. (1963) Dykes and Spencer-Peet (1972)

[a] Fig. 4

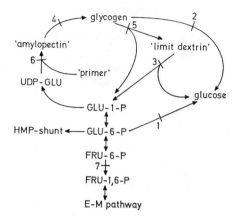

Fig. 4   Glycogen storage diseases: 1. glucose-6-phosphatase deficiency (von Gierke's); 2. lysosomal α-1, 4-glucosidase deficiency (Pompe's); 3. amylo-1, 6-glucosidase deficiency (Forbes'); 4. amylo-(1,4→1,6)-transglucosidase deficiency (Andersen's); 5. muscle and liver phosphorylase deficiency (McArdle's and Hers), respectively; 7. phosphofructokinase deficiency (GSD VII). Standard abbreviations

Table 12[a]

Disorders related to glycolysis and gluconeogenesis

Disorder	Deficient enzyme and tissues involved	Reference
Hexokinase deficiency	hexokinase (rbc)	Valentine et al. (1967), Piomelli and Corash (1976)
Glucosephosphate isomerase deficiency	glucosephosphate isomerase (phosphohexose isomerase) (generalized)	Baughan et al. (1968), Piomelli and Corash (1976)
Phosphofructokinase deficiency	phosphofructokinase (rbc)	Miwa et al. (1972), Waterbury and Frenkel (1972)
Triosephosphate isomerase deficiency	triosephosphate isomerase (generalized)	Schneider et al. (1965, 1968), Piomelli and Corash (1976)
Phosphoglycerate kinase deficiency (X-linked)	phosphoglycerate kinase (generalized)	Kraus et al. (1968), Piomelli and Corash (1976)
Pyruvate kinase deficiency	pyruvate kinase (rbc, liver)	Valentine et al. (1961), Paglia et al. (1976)
Pyruvate kinase deficiency	pyruvate kinase (neutrophils)	Burge et al. (1976)
Diphosphoglycerate mutase deficiency	diphosphoglycerate mutase (rbc)	Schroter (1965), Cartier et al. (1972)
Glucose-6-phosphate dehydrogenase deficiency (X-linked)	glucose-6-phosphate dehydrogenase (generalized)	Carson et al. (1956), Beutler (1972)
Fructose-1, 6-di-phosphatase deficiency	fructose-1, 6-diphosphatase (liver, wbc)	Baker and Winegrad (1970), Newcombe (1975)
Galactosaemia	galactose-1-phosphate uridylyltransferase (generalized)	Isselbacher et al. (1956), Segal (1972), Tedesco et al. (1975)
Galactokinase deficiency	galactokinase (generalized)	Gitzelmann (1967), Segal (1972)
Galactose epimerase deficiency	UDP-galactose-4-epimerase (rbc)	Gitzelmann (1972)
Hereditary fructose intolerance	fructose-1-phosphate aldolase (aldolase B) (liver, intestinal mucosa)	Hers and Joassim (1961), Newcombe (1975)
Essential fructosuria	fructokinase (liver)	Schapira et al. (1961), Froesch (1972)
Pyruvate carboxylase deficiency	pyruvate carboxylase (liver)	Hommes et al. (1968b), Tada et al. (1969)

[a] Fig. 5

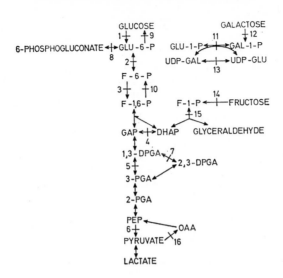

Fig. 5   Disorders related to anaerobic glycolysis and gluconeogenesis: 1. hexo-kinase deficiency; 2. glucosephosphate isomerase deficiency; 3. phosphofructo-kinase deficiency; 4. triosephosphate isomerase deficiency; 5. phosphoglycerate kinase deficiency; 6. pyruvate kinase deficiency; 7. diphosphoglycerate mutase deficiency; 8. glucose-6-phosphate dehydrogenase deficiency; 9. glucose-6-phosphatase deficiency (von Gierke's disease); 10. frutose-1, 6-diphosphatase deficiency; 11. galactose-1-phosphate uridylyltransferase deficiency (galactosaemia); 12. galactokinase deficiency; 13. galactose epimerase deficiency; 14. fructokinase deficiency (essential fructosuira); 15. aldolase B deficiency (hereditary fructose intolerance; 16. pyruvate carboxylase deficiency. Standard abbreviations

Table 13

Other disorders of carbohydrate metabolism

Disorder	Deficient enzyme and tissues involved	Reference
Congenital sucrose intolerance	sucrase and isomaltase (maltases IA and IB) (intestinal mucosa)	Dahlquist (1962), Gray et al. (1976)
Congenital lactose intolerance	lactase (intestinal mucosa)	Auricchio et al. (1963), Freiburghaus et al. (1976)
Pyruvate decarboxylase deficiency	pyruvate decarboxylase (generalized)	Blass et al. (1970, 1971), Farrell et al. (1975)
Pentosuria	NADP-linked xylitol dehydrogenase (rbc)	Wang and van Eys (1970), Hiatt (1972)

Table 14[a]

Disorders of amino acid metabolism

Disorder	Deficient enzyme and tissues involved	Reference
Phenylketonuria and hyperphenylalaninaemia	phenylalanine hydroxylase (liver, kidney, fibroblasts), dihydropteridine reductase (generalized)	Jervis (1953), Cotton and Danks (1976), Kaufman et al. (1976)
Tyrosinaemia	p-hydroxyphenylpyruvate oxidase (liver)	Gentz et al. (1965), Rosenberg and Scriver (1974)
Tyrosinosis	tyrosine: α-ketoglutarate aminotransferase (liver)	Campbell et al. (1967), Rosenberg and Scriver (1974)
Alkaptonuria	homogentisic acid oxidase (liver, kidney)	La Du et al. (1958), La Du (1972)
Oculocutaneous albinism (tyrosinase negative)	tyrosine hydroxylase (skin, hair-root bulbs)	Witkop et al. (1970), King and Witkop (1976)
Maple syrup urine disease	branched chain α-keto acid decarboxylase (generalized)	Dancis et al. (1960), Dancis and Levitz (1972)
Leucine/isoleucinaemia	leucine/isoleucine transaminase (wbc)	Jeune et al. (1970)
Hypervalinaemia	valine: α-ketocaproate aminotransferase (valine transaminase) (wbc)	Wada et al. (1963), Dancis and Levitz (1972)
Isovaleric-acidaemia	isovaleryl-CoA dehydrogenase (wbc)	Tanaka et al. (1966), Dancis and Levitz (1972), Tanaka (1973)
Methylmalonic-acidaemia	methylmalonyl-CoA mutase (isomerase) (generalized), methylmalonyl-CoA racemase (fibroblasts), conversion of vitamin B_{12s} to 5-deoxyadenosyl-cobalamin (fibroblasts)	Morrow et al. (1969a, b), Rosenberg (1976). Kang et al. (1972), Rosenberg and Mahoney (1973) Rosenberg et al. (1969), Rosenberg (1976)
Methylmalonic-acidaemia with homocystinuria and cystathioninaemia	defect in the formation of coenzymatically active Vitamin B_{12} (fibroblasts)	Mudd et al. (1969), Rosenberg (1976)

Disorder	Deficient enzyme and tissues involved	References
α-methylaceto-aceticaciduria	β-ketothiolase (fibroblasts)	Daum *et al.* (1971)
Propionicacidaemia (ketotic hyperglycinaemia)	propionyl-CoA carboxylase (generalized)	Hommes *et al.* (1968a), Gompertz (1973)
β-hydroxyisovaleric aciduria	β-methylcrotonyl-CoA decarboxylase (liver, kidney)	Eldjarn *et al.* (1970), Gompertz *et al.* (1971), Stokke *et al.* (1973)
Homocystinuria I	cystathionine synthetase (generalized)	Mudd *et al.* (1964), Rosenberg (1976)
Homocystinuria II (folate responsive)	N-(5, 10)-methylene tetra-hydrofolate reductase (fibroblasts)	Mudd *et al.* (1972)
Cystathioninaemia	cystathionase (generalized)	Frimpter (1965, 1972)
Sulphite oxidase deficiency	sulphite oxidase (generalized)	Laster *et al.* (1967), Mudd *et al.* (1967), Frimpter (1972)
Hypermethioninaemia	methionine adenosyl transferase (liver)	Gaull and Tallan (1974)
Argininosuccinic aciduria	argininosuccinase (generalized)	Tomlinson and Westall (1964)
Citrullinaemia	argininosuccinate synthetase (generalized)	Mohyuddin *et al.* (1967)
Ornithine trans-carbamylase deficiency (X-linked)	ornithine transcarbamylase (liver)	Russell *et al.* (1962), Ricciuti *et al.* (1976)
Carbamylphosphate synthetase deficiency	carbamylphosphate synthetase (liver, brain)	Freeman *et al.* (1964)
Hyperargininaemia	arginase (rbc, liver)	Terheggen *et al.* (1969), Colombo *et al.* (1973)
Hyperprolinaemia I	proline oxidase (liver)	Efron (1965), Scriver and Efron (1972)
Hyperprolinaemia II	Δ'-pyrroline-5-carboxy-late dehydrogenase (liver)	Scriver and Efron (1972), Valle and Phang (1974)
Hydroxyprolinaemia	hydroxyproline oxidase (liver)	Efron *et al.* (1965), Scriver and Efron (1972)
Non-ketotic hyperglycinaemia	glycine-cleavage enzyme (brain)	Yoshida *et al.* (1967), Perry *et al.* (1975)
Hyperoxaluria I	α-ketoglutarate: glyoxalate carboligase (liver, spleen, kidney)	Koch *et al.* (1967), Williams and Smith (1972)
Hyperoxaluria II	D-glycerate dehydrogenase (wbc)	Williams and Smith (1968, 1972)

Disorder	Deficient enzyme and tissues involved	References
Sarcosinaemia	sarcosine dehydrogenase (liver)	Gerritsen and Waisman (1966), Scott *et al.* (1970)
Infantile ketoacidosis	succinyl-CoA: 3-ketoacid-CoA transferase (brain, muscle, kidney, fibroblasts)	Tildon and Cornblath (1972)
Urocanic aciduria	urocanase (liver)	Yoshida *et al.* (1971b)
Histidinaemia	histidase (liver, skin)	Zannoni *et al.* (1962), Levy *et al.* (1974)
Carnosinaemia	carnosinase (serum)	Perry *et al.* (1967), Scriver and Perry (1972)
Formimino transferase deficiency	formimino glutamate tetrahydrofolate formimino transferase (liver)	Arakawa (1970), Niederwieser *et al.* (1974)
Hyperlysinaemia	lysine: α-ketoglutarate NADP oxidoreductase (saccharopine synthetase) (fibroblasts)	Dancis *et al.* (1969b)
Hyperlysinaemia with hyperammonaemia	lysine: NAD oxidoreductase (liver)	Colombo *et al.* (1967)
Xanthurenic aciduria	kynureninase (liver)	Tada *et al.* (1967)
Glutamic decarboxylase deficiency	glutamic decarboxylase (kidney)	Yoshida *et al.* (1971a)
Enterokinase deficiency	enterokinase (duodenal mucosa and juice)	Hadorn *et al.* (1969), Tarlow *et al.* (1970)
Trypsinogen deficiency	trypsinogen (duodenal juice)	Townes *et al.* (1967)
Glutathione synthetase deficiency I haemolytic anaemia II pyroglutamic acidaemia	glutathione synthetase (rbc, generalized)	Oort *et al.* (1961), Mohler *et al.* (1970), Wellner *et al.* (1976), Marstein *et al.* (1976)
Haemolytic anaemia due to glutamylcysteine synthetase deficiency	γ-glutamylcysteine synthetase (rbc)	Konrad *et al.* (1972)
Glutathione peroxidase deficiency	glutathione peroxidase (wbc)	Necheles *et al.* (1968, 1970b)
Glutathione reductase deficiency	glutathione reductase (rbc, wbc)	Waller (1968), Loos *et al.* (1976)

a Figs 6–12

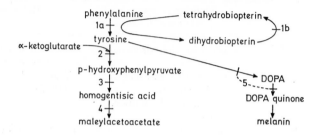

Fig. 6 Disorders of phenylalanine metabolism: 1a. PKU due to phenylalanine hydroxylase deficiency; 1b. PKU due to dihydropteridine reductase deficiency; 2. tyrosinosis; 3. tyrosinaemia; 4. alkaptonuria; 5. oculocutaneous albinism

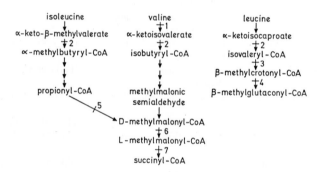

Fig. 7 Disorders of branched-chain amino acid metabolism: 1. valine transaminase deficiency (hypervalinaemia); 2. α-keto acid decarboxylase deficiency (maple syrup urine disease); 3. isovaleryl-CoA dehydrogenase deficiency (isovalericacidaemia); 4. β-methylcrotonyl-CoA decarboxylase deficiency (β-hydroxyisovaleric aciduria); 5. propionyl-CoA carboxylase deficiency (propionicacidaemia); 6. methylmalonyl-CoA mutase deficiency (methylmalonicacidaemia); 7. methylmalonyl-CoA racemase deficiency (methylmalonicacidaemia)

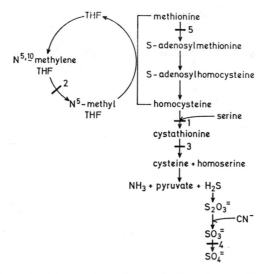

Fig. 8   Disorders of sulphur amino acid metabolism: 1. cystathionine synthetase deficiency (homocystinuria I); 2. N-(5,10)-methylene tetrahydrofolate reductase deficiency (homocystinuria II); 3. cystathionase deficiency (cystathioninaemia); 4. sulphite oxidase deficiency

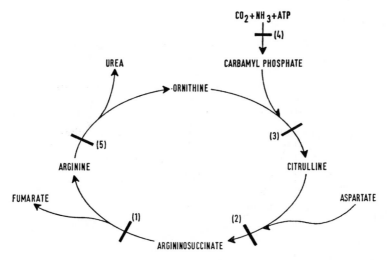

Fig. 9   Disorders of the urea cycle: 1. argininosuccinase deficiency (argininosuccinic aciduria); 2. argininosuccinate synthetase deficiency (citrullinaemia); 3. ornithine transcarbamylase deficiency (hyperammonaemia I); 4. carbamylphosphate synthetase deficiency (hyperammonaemia II); 5. arginase deficiency (hyperargininaemia)

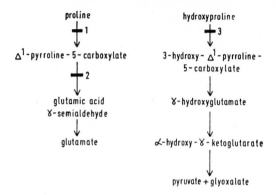

Fig. 10   Disorders of proline and hydroxyproline metabolism: 1. proline oxidase deficiency (hyperprolinaemia I); 2. Δ^1-pyrroline-5-carboxylate dehydrogenase deficiency (hyperprolinaemia II); 3. hydroxyproline oxidase deficiency (hydroxyprolinaemia)

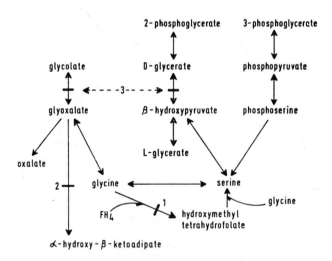

Fig. 11   Disorders related to glycine metabolism. Deficiencies of: 1. glycine cleavage enzyme (non-ketotic hyperglycinaemia); 2. glyoxalate: α-ketoglutarate carboligase (hyperoxaluria I); 3. D-glycerate dehydrogenase (hyperoxaluria II)

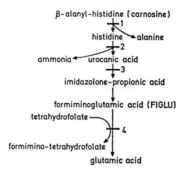

Fig. 12   Disorders of histidine metabolism: 1. carnosinase deficiency (carnosin-aemia); 2. histidase deficiency (histidinaemia); 3. urocanase deficiency (urocani-caciduria); 4. FIGLU: tetrahydrofolate formimino transferase deficiency

Table 15

Sphingolipidoses[a]

Disorder	Deficient enzyme[b]	Reference
Generalized gangliosidosis (GMI gangliosidosis)	β-galactosidase (non-specific)	Van Hoof and Hers (1968), Okada and O'Brien (1968), Dacremont and Kint (1968), O'Brien (1975)
Tay-Sachs disease (GM2 type I)	N-acetyl hexosaminidase A	Okada and O'Brien (1969), Sandhoff (1969, 1974)
Globoside storage disease (GM2 type II; Sandhoff's disease)	N-acetyl hexosaminidase A and B	Sandhoff et al. (1968), Sandhoff (1969, 1974)
Fabry's disease (angiokeratoma corporis diffusum) (X-linked)	ceramide trihexosidase (α-galactosidase)	Brady et al. (1967), Mapes et al. (1970a), Kint (1970), Brewster et al. (1974)
Gaucher's disease	glucocerebrosidase (β-glucosidase)	Patrick (1965), Brady et al. (1965), Brady (1974)
Niemann-Pick disease	sphingomyelinase	Brady et al. (1966), Schneider and Kennedy (1967), Brady (1974)
Metachromatic leucodystrophy	arylsulphatase A	Austin et al. (1963), Mehl and Jatzkewitz (1965), Brady (1974)
Multiple sulphatase deficiency	arylsulphatase A, B and C	Murphy et al. (1971)
Krabbe's disease (globoid leucodystrophy)	galactocerebroside β-galactosidase	Suzuki and Suzuki (1970a, 1971), Wenger and Riccardi (1976)
Farber's disease	ceramidase	Sugita et al. (1972)
GM3 gangliosidosis	UDP-N-acetylgalactosamine: GM3-N-acetyl-galactosaminyl transferase	Fishman et al. (1975).

[a] Fig. 13
[b] All lysosome-containing tissues involved

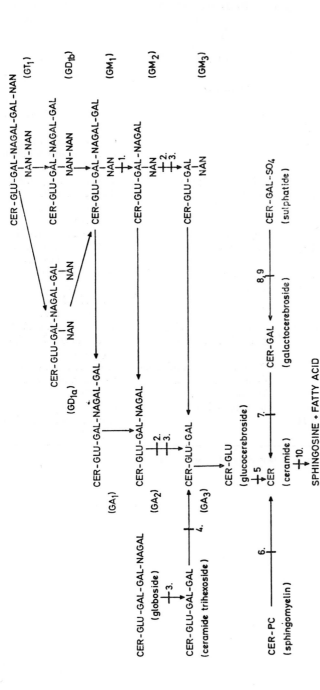

Fig. 13 Sphingolipidoses. Deficiencies of: 1. lysosomal non-specific β-galactosidase (generalized gangliosidosis); 2. N-acetyl-hexosaminidase A (Tay-Sachs); 3. N-acetylhexosaminidase A and B (globoside storage disease); 4. ceramide trihexosidase (Fabry's); 5. glucocerebrosidase (Gaucher's); 6. sphingomyelinase (Niemann-Pick); 7. galactocerebroside β-galactosidase (Krabbe's); 8. arylsulphatase A (metachromatic leucodystrophy); 9. arylsulphatase A, B and C (multiple sulphatase deficiency). Abbreviations: CER ceramide; GLU glucose; GAL galactose; NAGAL N-acetylgalactosamine; NAN N-acetylneuraminic acid

Table 16

Mucopolysaccharidoses

Disorder	Deficient enzyme[a]	Reference
Hurler syndrome (MPS I) and Scheie syndrome (MPS V)	α-L-iduronidase	Matalon and Dorfman (1972), Bach et al. (1972), Dorfman and Matalon (1976)
Hunter syndrome (MPS II) (X-linked)	L-iduronosulphate sulphatase	Sjoberg et al. (1973), Bach et al. (1973), Dorfman and Matalon (1976)
Sanfilippo syndrome type A (MPS IIIA)	heparansulphate sulphatase	Kresse and Neufeld (1972), Dorfman and Matalon (1976)
Sanfilippo syndrome type B (MPS IIIB)	N-acetyl-α-D-glucosaminidase	O'Brien (1972), Von Figura and Kresse (1972), Dorfman and Matalon (1976)
Morquio syndrome (MPS IV)	N-acetylhexosaminidase-6-sulphate sulphatase	Matalon et al. (1974), Dorfman and Matalon (1976)
Maroteaux-Lamy syndrome (MPS VI)	N-acetylhexosaminidase-4-sulphate sulphatase (arylsulphatase B)	Fluharty et al. (1974), Dorfman and Matalon (1976)
Mucopolysaccharidosis type VII (MPS VII)	β-glucuronidase	Hall et al. (1973), Dorfman and Matalon (1976)

[a] Lysosomal enzyme defects involve all tissues except rbc

Table 17

Other disorders involving lysosomal enzymes

Disorder	Deficient enzyme[a]	Reference
Fucosidosis	α-fucosidase	Van Hoof and Hers (1968), Beratis et al. (1975)
Mannosidosis	α-mannosidase	Ockerman (1967, 1969), Autio et al. (1973)
Xylosidosis	β-xylosidase	Payling-Wright and Evans (1970)
Acid phosphatase deficiency	acid phosphatase	Nadler and Egan (1970)
Wolman's disease	acid lipase	Patrick and Lake (1969), Hirschhorn and Weissman (1976)
I-cell disease	multiple lysosomal enzyme deficiencies	LeRoy and DeMars (1967), Holmes et al. (1975)
Aspartylglucosaminuria	N-aspartyl-β-glucosaminidase	Pollitt et al. (1968), Aula et al. (1973)
Chronic granulomatous disease	glutathione peroxidase	Holmes et al. (1970)
Chronic granulomatous disease (X-linked)	NADH oxidase (CN⁻ insensitive)	Baehner and Karnovsky (1968), Hirschhorn and Weissman (1976)
Myeloperoxidase deficiency	myeloperoxidase	Lehrer and Cline (1969)

[a] All lysosome-containing tissues involved

Table 18

Disorders of lipid metabolism

Disorder	Deficient protein or enzyme and tissues involved	Reference
Refsum's disease	phytanic acid α-hydroxylase (fibroblasts)	Steinberg et al. (1967a, b), Herndon et al. (1969), Mize et al. (1969), Steinberg (1972)
Familial lecithin: cholesterol acyltransferase deficiency	lecithin: cholesterol acyltransferase (plasma)	Norum and Gjone (1967), Torsvik (1970), Glomset et al. (1970),
Hyperlipoproteinaemia (type I, fat induced)	heparin-induced lipo-protein lipase (plasma, adipose tissue)	Havel and Gordon (1960), Harlan et al. (1967), Fredrickson and Levy (1972)
Pancreatic lipase deficiency	lipase (pancreas)	Sheldon (1964), Frezal and Rey (1970)
Familial hypercholesterolaemia (dominant)	membrane receptor sites for low density lipoprotein (fibroblasts)	Brown and Goldstein (1974, 1976)

Table 19[a]

Disorders of porphyrin and haem metabolism

Disorder	Deficient enzyme and tissues involved	Reference
Acute intermittant porphyria (dominant)	uroporphyrinogen I synthetase (rbc, liver, fibroblasts)	Strand et al. (1970), Meyer et al. (1972)
Hereditary coproporphyria (dominant)	coproporphyrinogen oxidase (wbc, fibroblasts)	Elder et al. (1976)
Congenital erythrocytic porphyria (Gunther's disease)	uroporphyrinogen III cosynthetase (generalized)	Romeo et al. (1970a, b), Tschudy (1974)
Crigler-Najjar syndrome	glucuronyl transferase (liver)	Arias et al. (1969), Schmid (1972)
Gilbert syndrome	glucuronyl transferase (liver)	Black and Billing (1969), Schmid (1972)

[a] See Fig. 2

Table 20

Disorders of steroid hormone metabolism

Disorder	Deficient enzyme[a]	Reference
Congenital adrenal hyperplasias (adrenogenital syndromes)	21-hydroxylase	Bongiovanni (1958, 1972)
	11 β-hydroxylase	Eberlein and Bongiovanni (1955)
	17-hydroxylase	Biglieri et al. (1966), Bongiovanni (1972)
	3β-hydroxysteroid dehydrogenase	Bongiovanni (1961, 1972)
	20, 21-desmolase	Prader and Gurtner (1955), Bongiovanni (1972)
Male pseudo-hermaphroditism	5α-reductase	Imperato-McGinley et al. (1974)
	17-ketosteroid dehydrogenase	Saez et al. (1971)
	17, 20-desmolase (X-linked)	Zachmann et al. (1972)

[a] Tissues involved are presumed to be adrenals and gonads

Table 21[a]

Disorders of purine and pyrimidine metabolism

Disorder	Abnormal enzyme and tissues involved	Reference
Lesch-Nyhan syndrome and some forms of gouty arthritis (X-linked)	hypoxanthine-guanine phosphoribosyl transferase (HGPRT) (generalized + rbc)	Seegmiller et al. (1967), Rosenbloom et al. (1967, 1968a) Seegmiller (1976)
Xanthinuria	xanthine oxidase (liver, intestine)	Watts et al. (1963), Watts (1974)
Adenine phosphoribosyl transferase deficiency	adenine phosphribosyl transferase (rbc)	Kelley et al. (1969), Watts (1974)
Haemolytic anaemia due to adenylate kinase deficiency	adenylate kinase (rbc)	Szeinberg et al. (1969)
Immune deficiency with deficiency of adenosine deaminase	adenosine deaminase (generalized + rbc)	Giblett et al. (1972), Trotta et al. (1976)
Immune deficiency with defect in purine nucleoside phosphorylase	purine nucleoside phosphorylase (rbc, fibroblasts)	Giblett et al. (1975), Cohen et al. (1976)
Haemolytic anaemia due to adenosine triphosphatase deficiency	adenosine triphosphatase (rbc)	Harvald et al. (1964)
Purine over-production syndrome	phosphoribosylpyrophosphate sythetase (increased) (rbc)	Becker et al. (1973)
Oroticaciduria type I	orotidylic pyrophosphorylase orotidylic decarboxylase (generalized + rbc)	Huguely et al. (1959), Smith et al. (1972)
Oroticaciduria type II	orotidylic pyrophosphorylase (generalized + rbc)	Fox et al. (1969), Smith et al. (1972)
Haemolytic anaemia due to pyrimidine-5'-nucleotidase deficiency	pyrimidine-5'-nucleotidase (rbc)	Valentine et al. (1974), Vives-Corrons et al. (1976)
Uridine monophosphate kinase deficiency	uridine monophosphate kinase (rbc)	Giblett et al. (1974)

[a] Fig. 14

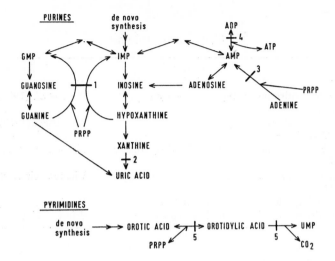

Fig. 14   Disorders of purine and pyrimidine metabolism. Deficiencies of: 1. hypoxanthine–guanine phosphoribosyl transferase (Lesch–Nyhan); 2. xanthine oxidase (xanthinuria); 3. adenine phosphoribosyl transferase; 4. adenylate kinase; 5. orotidylic pyrophosphorylase and decarboxylase (oroticaciduria). Standard abbreviations but with PRPP phosphoribosyl pyrophosphate

Table 22

Collagen disorders

Disorder	Defect	Reference
Ehler-Danlos syndrome type IV	type III collagen	Pope *et al.* (1975)
Ehler-Danlos syndrome type V (X-linked)	lysyl oxidase	McKusick (1975)
Ehler-Danlos syndrome type VI	collagen lysyl hydroxylase	Krane *et al.* (1972)
Ehler-Danlos syndrome type VII	procollagen peptides	Lichtenstein *et al.* (1973)
Osteogenesis imperfecta	type I collagen	Pentinnen *et al.* (1975)
Marfan syndrome	soluble collagen	Priest *et al.* (1973)

Table 23

Disorders of serum proteins

Disorder	Deficient protein	Reference
Drug-induced apnoea	pseudocholinesterase	Bourne *et al.* (1952)
Pulmonary emphysema and/or liver cirrhosis	α-1-antitrypsin	Laurell and Eriksson (1963)
Vitamin B_{12} deficiency	transcobalamin II	Hakami *et al.* (1971), Hitzig *et al.* (1974)
Thyroxine-binding globulin deficiency (X-linked)	thyroxine-binding globulin	Nikolai and Seal (1966)
Wilson's disease	caeruloplasmin	Shokeir and Shreffler (1969)
Analbuminaemia	albumin	Bennhold *et al.* (1954)
Atransferrinaemia	transferrin	Heilmeyer *et al.* (1961)
Analphalipoproteinaemia (Tangier disease)	high-density α-lipoprotein	Fredrickson *et al.* (1961)
Abetalipoproteinaemia (acanthocytosis)	β-lipoprotein	Bassen and Kornzweig (1950)
Familial hypercholesterolaemia	defective low density lipoprotein	Higgins *et al.* (1975)
Complement deficiencies	complement components C1R, C1S, C2, C3, C4, C5, C6, C7, C8, C1 inhibitor, C3b inactivator	cf. Lachmann (1975)

Table 24

Miscellaneous disorders

Disorder	Deficient enzyme and tissues involved	Reference
Acatalasia	catalase (generalized + rbc)	Takahara (1952), Aebi and Suter (1972)
Methaemoglobinaemia	NADH-diaphorase (cytochrome b₅reductase) (rbc)	Scott and Griffith (1959), LeRoux et al. (1975)
Methaemoglobinaemia with mental retardation	NADH-diaphorase (cytochrome b₅ reductase) (generalized)	LeRoux et al. (1975)
NADPH-linked methaemoglobin reductase deficiency	NADPH-diaphorase (rbc)	Sass et al. (1967), Bloom and Zarkowsky (1969)
Hypophosphatasia	alkaline phosphatase (serum, bone)	Rathburn (1948), Bartter (1972)
Carbonic anhydrase deficiency	carbonic anhydrase B (rbc)	Shapira et al. (1974)
Xeroderma pigmentosum	DNA excision repair mechanisms (fibroblasts)	Cleaver (1968), Lehmann et al. (1975), Kraemer et al. (1975), Sutherland et al. (1975)
Thyroid hormone deficiency type II	iodide peroxidase (thyroid)	Haddad and Sidbury (1959), Stanbury (1972)
Thyroid hormone deficiency type IV	iodotyrosine iodinase (thyroid, wbc)	Kusakabe and Myake (1964) Stanbury (1972)

544 D. J. H. BROCK

REFERENCES

Aebi, H. and Suter, H. (1972). In "The Metabolic Basis of Inherited Disease" (J. B. Stanbury, J. B. Wyngaarden and D. S. Fredrickson, Eds), 3rd ed. 1710–1729. McGraw-Hill, New York.

Albright, F., Bugler, A. M. and Bloomberg, E. (1937). Am. J. Dis. Child. 54, 529.

Alter, B. P., Modell, C. B. and Chang, H. (1975). Pediat. Res. 9, 320.

Alter, B. P., Friedman, S., Hobbins, J. C., Mahoney, M. J., Sherman, A. S., McSweeney, J. F., Schwartz, E. and Nathan, D. G. (1976). New Eng. J. Med. 294, 1040.

Anon (1970). Br. Med. J. ii, 98.

Arakawa, T. (1970). Am. J. Med. 48, 594.

Arias, I. M. (1962). J. Clin. Invest. 41, 2233.

Arias, I. M., Gartner, L. M., Cohen, M., Ezzer, J. B. and Levi, A. J. (1969). Am. J. Med. 47, 395.

Askanas, V., Engel, W. K., DiMauro, S., Brooks, B. R. and Mehler, M. (1976). New Eng. J. Med. 294, 573.

Aula, P., Nanto, V., Laipio, M. L. and Autio, S. (1973). Clin. Genet. 4, 297.

Aula, P., Rapola, J., Autio, S., Raivio, K. O. and Karjalainen, O. (1975). J. Pediat. 87, 221.

Auricchio, S., Rubino, A., Landolt, M., Semenza, G. and Prader, A. (1963). Lancet ii, 324.

Auricchio, S., Rubino, A., Prader, A., Rey, J., Jos, J., Frézal, J. and Davidson, M. (1965). J. Pediat. 66, 555.

Austin, J. H., Balasubramanian, A. S., Pattabiraman, T. N., Saraswathi, S., Basu, D. K. and Bachhawat, B. K. (1963). J. Neurochem. 10, 805.

Autio, S., Norden, N. E., Ockerman, P. A., Reikkinen, P., Rapola, J. and Louhimo, T. (1973). Acta. Pediat. Scand. 62, 555.

Ayvazian, J. H. (1964). New Eng. J. Med. 270, 18.

Bach, G., Friedman, R., Weissman, B. and Neufeld, E. F. (1972). Proc. Natn. Acad. Sci. USA 69, 2048.

Bach, G., Eisenberg, F., Cantz, M. and Neufeld, E. F. (1973). Proc. Natn. Acad. Sci. USA 70, 2134.

Baehner, R. L. and Karnovsky, M. L. (1968). Science 162, 1277.

Baker, L. and Winegrad, A. (1970). Lancet ii, 13.

Barber, G. W. and Spaeth, G. L. (1967). Lancet i, 337.

Barnes, N. D., Hull, D., Balgobin, L. and Gompertz, D. (1970). Lancet ii, 244.

Bartter, F. C. (1972). In "The Metabolic Basis of Inherited Disease" (J. B. Stanbury, J. B. Wyngaarden and D. S. Fredrickson, Eds), 3rd ed., 1295–1304. McGraw-Hill, New York.

Bassen, F. A. and Kornzweig, A. L. (1950). Blood 5, 381.

Baughan, M. A., Valentine, W. N., Paglia, D. E., Ways, P. O., Simons, E. R. and DeMarsh, Q. B. (1968). Blood 32, 236.

Beadle, G. W. (1945). Chem. Rev. 37, 15.

Beadle, G. W. and Tatum, E. L. (1941). Proc. Natn. Acad. Sci. USA 27, 499.

Becker, M. A., Kostel, P. J., Meyer, L. J. and Seegmiller, J. E. (1973). Proc. Natn. Acad. Sci. USA 70, 2749.

Bennett, E. and Huehns, E. R. (1970). Lancet, ii, 956.

Bennett, E. and Ratnoff, O. D. (1973). New Eng. J. Med. 288, 342.

Bennhold, H., Peters, H. and Roth, E. (1954). *Verh. Dt. Ges. Inn. Med.* **60**, 630.
Beratis, N. G. and Hirschhorn, K. (1973). *Mamm. Chrom. Newslett.* **14**, 114.
Beratis, N. G., Turner, B. M. and Hirschhorn, K. (1975). *J. Pediat.* **87**, 1193.
Beutler, E. (1968). *In* "Hereditary Disorders of Erythrocyte Metabolism" (E. Beutler, Ed.), 114–125. Grune and Stratton, New York.
Beutler, E. (1969a). *J. Clin. Invest.* **48**, 1957.
Beutler, E. (1969b). *Science* **165**, 613.
Beutler, E. (1972). *In* "The Metabolic Basis of Inherited Disease" (J. B. Stanbury, J. B. Wyngaarden, D. S. Fredrickson, Eds), 3rd ed., 1358–1388, McGraw-Hill, New York.
Beutler, E. and Baluda, M. C. (1964). *Lancet* **i**, 189.
Beutler, E. and Kuhl, W. (1975). *Nature, Lond.* **258**, 262.
Beutler, E., Yeh, M. and Fairbanks, V. F. (1962). *Proc. Natn. Acad. Sci. USA* **48**, 9.
Biglieri, E. G., Herron, M. A. and Brust, N. (1966). *J. Clin. Invest.* **45**, 1946.
Bishop, C., Zimdahl, W. T. and Talbott, J. H. (1954). *Proc. Soc. Exp. Biol. Med.* **86**, 440.
Black, M. and Billing, B. H. (1969). *New Eng. J. Med.* **280**, 1266.
Blass, J. P., Avigan, J. and Uhlendorf, B. W. (1970). *J. Clin. Invest.* **49**, 423.
Blass, J. P., Lonsdale, D., Uhlendorf, B. W. and Horn, E. (1971). *Lancet* **i**, 1302.
Bloom, G. E. and Zarkowsky, H. S. (1969). *New Eng. J. Med.* **281**, 919.
Bondy, P. K. (1969). *In* "Duncan's Diseases of Metabolism" (P. K. Bondy, Ed.) 6th ed., 199–294. Saunders, Philadelphia.
Bongiovanni, A. M. (1958). *J. Clin. Invest.* **37**, 1342.
Bongiovanni, A. M. (1961). *J. Clin. Endocrinol.* **21**, 860.
Bongiovanni, A. M. (1972). *In* "The Metabolic Basis of Inherited Disease" (J. B. Stanbury, J. B. Wyngaarden, D. S. Fredrickson, Eds), 3rd ed., 857–885, McGraw-Hill, New York.
Borressen, A. L. and van der Hagen, C. B. (1973). *Clin. Genet.* **4**, 442.
Bourne, J. G., Collier, H. O. J. and Somers, G. F. (1952). *Lancet* **i**, 1225.
Brady, R. O. (1974). *In* "Duncan's Diseases of Metabolism: Genetics and Metabolism" (P. K. Bondy and L. E. Rosenberg, Eds), 7th ed., 445–464, Saunders, Philadelphia.
Brady, R. O., Kanfer, J. N. and Schapiro, D. (1965). *Biochem. Biophys. Res. Commun.* **18**, 221.
Brady, R. O., Kanfer, J. N., Mock, M. B. and Fredrickson, D. S. (1966). *Proc. Natn. Acad. Sci. USA* **55**, 366.
Brady, R. O., Gal, A. E., Bradley, R. M., Martensson, E., Warshaw, W. L. and Laster, L. (1967). *New Eng. J. Med.* **276**, 1163.
Breslow, J. L., Spaulding, D. R., Lux, S. E., Levy, R. I. and Lees, R. S. (1975). *New Eng. J. Med.* **293**, 900.
Brewster, M. A., Whaley, S. A. and Kane, A. C. (1974). *Clin. Chem.* **20**, 383.
Britten, R. J. and Davidson, E. H. (1969). *Science* **165**, 349.
Brock, D. J. H. (1973). *In* "Antenatal Diagnosis of Genetic Disease" (A. E. H. Emery, Ed.), 82–112, Churchill Livingstone, Edinburgh.
Brock, D. J. H. (1976). *Lancet* **ii**, 345.
Brock, D. J. H. and Sutcliffe, R. G. (1972). *Lancet* **ii**, 197.
Brock, D. J. H., Gordon, H., Seligman, S. and Lobo, E. (1971). *Lancet* **ii**, 1324.
Brothers, C. R. D. (1949). *Proc. R. Australas. Coll. Phys.* **4**, 48.
Brown, B. I. and Brown, D. H. (1966). *Proc. Natn. Acad. Sci. USA* **56**, 725.
Brown, M. S. and Goldstein, J. L. (1974a). *Proc. Natn. Acad. Sci. USA* **71**, 788.

Brown, M. S. and Goldstein, J. L. (1974b). *Science* **185**, 61.

Brown, M. S. and Goldstein, J. L. (1976). *New Eng. J. Med.* **294**, 1386.

Brown, M. S., Dana, S. E. and Goldstein, J. L. (1973). *Proc. Natn. Acad. Sci. USA* **70**, 2162.

Brown, M. S., Dana, S. E. and Goldstein, J. L. (1975). *Proc. Natn. Acad. Sci. USA* **72**, 2925.

Brunette, M. G., Delvin, E., Hazel, B. and Scriver, C. R. (1972). *Pediatrics* **50**, 702.

Burge, P. S., Johnson, W. S. and Hayward, A. R. (1976). *Br. Med. J.* **1**, 742.

Bush, G. H. (1968). *Proc. R. Soc. Med.* **61**, 171.

Campbell, R. A., Buist, N. R. M., Jacinto, E. Y., Koler, R. D., Hecht, F. and Jones, R. T. (1967). *Proc. Soc. Pediat. Res. Atlantic City, N.J.*, p. 80.

Carson, P. E., Flanagan, C. L., Ickes, C. E. and Alving, A. S. (1956). *Science* **124**, 484.

Carter, C. O., Roberts, J. A. F., Evans, K. A. and Buck, A. R. (1971). *Lancet* **i**, 281.

Cartier, P., Labie, P., Leroux, J. P., Najman, A. and de Maugre, F. (1972). *Nouv. Rev. Fr. Haematol.* **12**, 269.

Caughey, J. E. and Barclay, J. (1954). *Aust. Ann. Med.* **3**, 165.

Cleaver, J. R. (1968). *Nature, Lond.* **218**, 652.

Cochrane, W. A., Payne, W. W., Simkiss, M. J. and Woolf, L. I. (1956). *J. Clin. Invest.* **35**, 411.

Cohen, A., Doyle, D., Martin, D. W. and Ammann, A. J. (1976). *New Eng. J. Med.* **295**, 1449.

Colombo, J. P., Burgi, W., Richterich, R. and Rossi, E. (1967). *Metabolism* **16**, 910.

Colombo, J. P., Terheggen, H. G., Lowenthal, A., van Sande, M. and Rogers, S. (1973). *In* "Inborn Errors of Metabolism" (F. A. Hommes and C. J. van den Berg, Eds), 239–254, Academic Press, New York and London.

Cori, G. T. and Cori, C. F. (1952). *J. Biol. Chem.* **199**, 661.

Cotton, R. G. H. and Danks, D. M. (1976). *Nature, Lond.* **260**, 63.

Cox, R. P., Douglas, G., Hutzler, J., Lynfield, J. and Dancis, J. (1970). *Lancet*, **i**, 893.

Dacremont, G. and Kint, J. A. (1968). *Clin. Chim. Acta* **21**, 421.

Dahlquist, A. (1962). *J. Clin. Invest.* **41**, 463.

Dancis, J. and Levitz, M. (1972). *In* "The Metabolic Basis of Inherited Disease" (J. B. Stanbury, J. B. Wyngaarden and D. S. Fredrickson, Eds), 3rd ed., 426–439, McGraw-Hill, New York.

Dancis, J., Hutzler, J. and Levitz, M. (1960). *Biochem. Biophys. Acta* **43**, 342.

Dancis, J., Berman, P. H., Jansen, V. and Balis, M. E. (1968). *Life Sci.* **7**, 587.

Dancis, J., Cox, R. P., Berman, P. H., Jansen, V. and Balis, M. E. (1969a). *Biochem. Genet.* **3**, 609.

Dancis, J., Hutzler, J., Cox, R. P. and Woody, N. C. (1969b). *J. Clin. Invest.* **48**, 1447.

Danes, B. S. (1974). *Lancet*, **i**, 680.

Danes, B. S. (1975). *J. Med. Genet.* **12**, 405.

Danes, B. S. and Bearn, A. G. (1967). *Lancet* **ii**, 65.

Danes, B. S. and Bearn, A. G. (1970). *Proc. Natn. Acad. Sci. USA* **67**, 357.

Danes, B. S., Scott, J. E. and Bearn, A. G. (1970). *J. Exp. Med.* **132**, 764.

Danes, B. S., Beck, B. and Flensborg, E. W. (1975). *Clin. Genet.* **8**, 85.

Darlington, G. J., Bernhard, H. P. and Ruddle, F. H. (1973). *In Vitro* **8**, 444.

Daum, R. S., Lamm, P. H., Maner, O. A. and Scriver, C. R. (1971). *Lancet* **ii**, 1289.

Davidson, R. G., Nitowsky, H. M. and Childs, B. (1963). *Proc. Natn. Acad. Sci. USA* **50**, 481.

Dean, G. (1963). "The Porphyrias: A Story of Inheritance and Environment" Pitman Medical Publishing Co., London.

Deisseroth, A., Barker, J., Anderson, W. F., Nienhuis, A. (1975a). *Proc. Natn. Acad. Sci. USA* **72**, 2682.

Deisseroth, A., Burk, R. R., Picciano, D., Minna, J., Anderson, W. F., Nienhuis, A. (1975b). *Proc. Natn. Acad. Sci. USA* **72**, 1102.

Deisseroth, A., Velez, R. and Nienhuis, A. (1976). *Science,* **191** 1262.

Dent, C. E. and Philpot, G. R. (1954). *Lancet* **i**, 182.

Detter, J. C., Ways, P. O., Giblett, E. R., Baughan, M. A., Hopkinson, D. A., Povey, S., Harris, H. (1968). *Ann. Hum. Genet.* **31**, 329.

Di Mauro, S., Angelini, C. and Catani, C. (1967). *J. Neurol. Neurosurg. Psychiat.* **30**, 411.

Donaldson, V. H. and Evans, R. R. (1963). *Am. J. Med.* **35**, 37.

Dorfman, A. and Matalon, R. (1976). *Proc. Natn. Acad. Sci. USA* **73**, 630.

Dowdle, E. B., Mustard, P. and Eales, L. (1967). *S. Afr. Med. J.* **41**, 1093.

Dreyfus, J. C. (1972). *Biochimie* **54**, 559.

Dykes, J. R. W. and Spencer-Peet, J. (1972). *Arch. Dis. Child.* **47**, 558.

Eberlein, W. R. and Bongiovanni, A. M. (1955). *J. Clin. Invest.* **15**, 1531.

Efron, M. L., Bixby, E. M. and Pryles, C. V. (1965). *New Eng. J. Med.* **272**, 1299.

Elder, G. H., Evans, J. O., Thomas, N., Cox, R. P., Brodie, M. J., Moore, M. R., Goldberg, A. and Nicholson, D. C. (1976). *Lancet* **ii**, 1217.

Eldjarn, L., Jellum, E., Stokke, O., Pande, H. and Waaler, P. E. (1970). *Lancet* **ii**, 521.

Epstein, M. A. and Barr, Y. M. (1964). *Lancet* **i**, 252.

Erbe, R. W. (1975). *New Eng. J. Med.* **293**, 807.

Evans, F. J., Gray, P. W. S., Lehmann, H. and Silk, E. (1952). *Lancet* **i**, 1229.

Fågerhol, M. K. and Laurell, C. B. (1970). *In* "Progress in Medical Genetics" (A. G. Steinberg and A. G. Bearn, Eds), Vol. 7, 96–111. Heinemann, London.

Farrell, D. F., Clark, A. F., Scott, C. R. and Wennberg, R. P. (1975). *Science* **187**, 1082.

Fellman, J. H., van Bellinghen, P. J., Jones, R. T. and Koler, R. D. (1969). *Biochemistry* **8**, 615.

Fishman, P. H., Max, S. R., Tallman, J. F., Brady, R. O., MacLaren, N. K. and Cornblath, M. (1975). *Science* **187**, 68.

Fluharty, A. L., Porter, M. T., Hirsh, G. A., Perida, E. and Kihara, H. (1970). *Lancet* **ii**, 109.

Fluharty, A. L., Stevens, R. L., Saunders, D. L. and Kihara, H. (1974). *Biochem. Biophys. Res. Commun.* **59**, 455.

Forget, B. G., Benz, E. J., Skoultchi, A., Baglioni, C. and Housman, D. (1975). *Nature, Lond.* **247**, 379.

Fox, R. M., O'Sullivan, W. J. and Firkin, B. G. (1969). *Am. J. Med.* **47**, 332.

Francke, U., Bakay, B. and Nyhan, W. L. (1973). *J. Pediat.* **82**, 472.

Fratantoni, J. C., Hall, C. W. and Neufeld, E. F. (1969a). *Science* **162**, 570.

Fratantoni, J. C., Hall, C. W. and Neufeld, E. F. (1969b). *Proc. Natn. Acad. Sci. USA* **64**, 360.

Fredrickson, D. S. and Levy, R. I. (1972). *In* "Metabolic Basis of Inherited Disease" (J. B. Stanbury, J. B. Wyngaarden and D. S. Fredrickson, Eds), 3rd ed., 545–614. McGraw-Hill, New York.

Fredrickson, D. S., Altrocchi, P. H., Avioli, L. V., Goodman, D. S. and Goodman, H. C. (1961). *Ann. Intern. Med.* **55**, 1016.

Freeman, J. M., Nicholson, J. F., Masland, W. S., Roland, L. P. and Carter, S. (1964). *J. Pediat.* **65**, 1039.

Freiburghaus, A. U., Schmitz, J., Schnidler, M., Rotthauwe, H. W., Kuitunen, P., Launiala, K. and Hadorn, B. (1976). *New Eng. J. Med.* **294**, 1030.

Frezal, J. and Rey, J. (1970). *In* "Advances in Human Genetics" (H. Harris and K. Hirschhorn, Eds), Vol. 1, 275–376. Plenum Press, New York.

Fried, K. (1968). Private communication.

Frimpter, G. W. (1965). *Science* **149**, 1095.

Frimpter, G. W. (1972). *In* "The Metabolic Basis of Inherited Disease" (J. B. Stanbury, J. B. Wyngaarden and D. S. Fredrickson, Eds), 3rd ed., 413–425, McGraw-Hill, New York.

Froesch, E. R. (1972). *In* "The Metabolic Basis of Inherited Disease" (J. B. Stanbury, J. B. Wyngaarden and D. S. Fredrickson, Eds), 3rd ed., 131–148. McGraw-Hill, New York.

Fujimoto, W. Y. and Seegmiller, J. E. (1970). *Proc. Natn. Acad. Sci. USA* **65**, 577.

Galjaard, H., Makes, M., de Josselin de Jong, J. and Niermeijer, M. F. (1973). *Clin. Chim. Acta* **49**, 361.

Galjaard, H., Van Hoogstraten, J. J., de Josselin de Jong, J. and Niermeijer, M. F. (1974a). *Histochem. J.* **6**, 409.

Galjaard, H., Niermeijer, M. F., Hahnemann, N., Mohr, J. and Sørenson, S. A. (1974b). *Clin. Genet.* **5**, 368.

Gandini, E., Angioni, G., Gartler, S. M., Argiolas, N. and Dell'Acqua, G. (1968). *Proc. Natn. Acad. Sci. USA* **61**, 945.

Gardner Medwin, D. (1970). *J. Genet.* **7**, 334.

Garrod, A. E. (1902). *Lancet* **ii**, 1606.

Garrod, A. E. (1908). *Lancet* **ii**, 1, 73, 142, 214.

Gartler, S. M., Scott, R. C., Goldstein, J. L., Campbell, B. and Sparkes, R. S. (1971). *Science* **172**, 572.

Gaull, G. E. and Tallan, H. H. (1974). *Science* **186**, 56.

Gentz, J., Jagenburg, R. and Zetterstrom, R. (1965). *J. Pediat.* **66**, 670.

Gerritsen, T. and Waisman, H. A. (1966). *New Eng. J. Med.* **275**, 66.

Gertner, M., Zalay, E. and Hirschhorn, K. (1970). *Clin. Genet.* **1**, 28.

Giblett, E. R., Anderson, J. E., Cohen, F., Pollara, B. and Meuwissen, H. J. (1972). *Lancet* **i**, 1067.

Giblett, E. R., Anderson, J. E., Chen, S. H., Teng, Y. S. and Cohen, F. (1974). *Am. J. Hum. Genet.* **26**, 627.

Giblett, E. R., Ammann, A. J. and Wara, D. W. (1975). *Lancet* **i**, 1010.

Gibson, Q. H. (1948). *Biochem. J.* **42**, 13.

Gilbert, F., Kucherlapati, R., Creagan, R. P., Murnane, M. J., Darlington, G. J. and Ruddle, F. H. (1975). *Proc. Natn. Acad. Sci. USA* **72**, 263.

Gitzelmann, R. (1967). *Pediatr Res.* **1**, 14.

Gitzelmann, R. (1972). *Helvet. Pediat. Acta* **27**, 125.

Glade, P. R. and Beratis, N. G. (1976). *In* "Progress in Medical Genetics: New Series" (A. G. Steinberg, A. G. Bearn, A. G. Motulsky and B. Childs, Eds), Vol. 1, 1–48, Saunders, Philadelphia.

Glomset, J. A., Norum, K. R. and King, W. (1970). *J. Clin. Invest.* **49**, 1827.

Goldstein, J. L. and Brown, M. S. (1973). *Proc. Natn. Acad. Sci. USA* **70**, 2804.

Goldstein, J. L., Campbell, B. K. and Gartler, S. M. (1973). *J. Clin. Invest.* **52**, 218.
Goldstein, J. L., Harrod, M. J. E. and Brown, M. S. (1974). *Am. J. Hum. Genet.* **26**, 199.
Goldstein, J. L., Dana, S. E., Brunschede, G. Y. and Brown, M. S. (1975). *Proc. Natn. Acad. Sci. USA* **72**, 1092.
Gompertz, D. (1973). *In* "Inborn Errors of Metabolism" (F. A. Hommes and C. J. van den Berg, Eds), 291–302. Academic Press, New York and London.
Gompertz, D., Draffan, G. H., Watts, J. L. and Hull, D. (1971). *Lancet* **ii**, 22.
Gray, G. M. (1972). *In* "The Metabolic Basis of Inherited Disease" (J. B. Stanbury, J. B. Wyngaarden and D. S. Fredrickson, Eds), 3rd ed., 1453–1464. McGraw-Hill, New York.
Gray, G. M., Conklin, K. A. and Townley, R. R. W. (1976). *New Eng. J. Med.* **294**, 750.
Haddad, H. M. and Sidbury, J. B. (1959). *J. Clin. Endocrinol.* **19**, 446.
Hadorn, B., Tarlow, M. J., Lloyd, J. K. and Wolff, O. H. (1969). *Lancet* **i**, 812.
Hakami, N., Neiman, P. E., Canellos, G. P. and Lazerson, J. (1971). *New Eng. J. Med.* **285**, 1163.
Hall, C. W., Cantz, M. and Neufeld, E. F. (1973). *Arch. Biochem. Biophys.* **155**, 32.
Harlan, W. R., Wineseh, P. S. and Wasserman, A. J. (1967). *J. Clin. Invest.* **46**, 239.
Harris, H. (1970). "Cell Fusion", Harvard University Press, Cambridge, Massachusetts.
Harrison, H. E. (1958). *J. Chronic Dis.* **7**, 346.
Harvald, B., Hanel, K. H., Squires, R. and Trap-Jensen, J. (1964). *Lancet* **i**, 18.
Havel, R. J. and Gordon, R. S. (1960). *J. Clin. Invest.* **39**, 1777.
Heilmeyer, L., Keller, W., Vivell, O., Keiderling, W., Betke, K., Wohler, F. and Schultze, H. E. (1961). *Dt. Med. Wschr.* **86**, 1745.
Henderson, J. F., Kelley, W. N., Rosenbloom, F. M. and Seegmiller, J. E. (1969). *Am. J. Hum. Genet.* **21**, 61.
Herndon, J. H., Steinberg, D., Uhlendorf, B. W. and Fales, H. M. (1969). *J. Clin. Invest.* **48**, 1017.
Hers, H. G. (1959). *Rev. Int. Hepat.* **9**, 35.
Hers, H. G. (1963). *Biochem. J.* **86**, 11.
Hers, H. G. (1964). *Adv. Metabol. Dis.* **1**, 1.
Hers, H. G. and Joassim, G. (1961). *Enzymol. Biol. Clin.* **1**, 4.
Hers, H. G. and van Hoof, F. (1968). *In* "Carbohydrate Metabolism and its Disorders" (F. Dickens, P. J. Randle and W. J. Whelan, Eds), Vol. 2, 151–168, Academic Press, New York and London.
Hiatt, H. H. (1972). *In* "The Metabolic Basis of Inherited Disease" (J. B. Stanbury, J. B. Wyngaarden and D. S. Fredrickson, Eds), 3rd ed., 119–130, McGraw-Hill, New York.
Higgins, M. J. P., Lecamwasam, D. S. and Galton, D. J. (1975). *Lancet* **ii**, 737.
Hirschfeld, A. J. and Fleshman, J. K. (1969). *J. Pediat.* **75**, 492.
Hirschhorn, K., Nadler, H. L., Waithe, W. I., Brown, B. I. and Hirschhorn, R. (1969). *Science* **166**, 1632.
Hirschhorn, R. and Weissman, P. (1976). *In* "Progress in Medical Genetics" (A. G. Steinberg, A. G. Bearn, A. G. Motulsky and B. Childs, Eds), Vol. 1 49–102. Saunders, Philadelphia.
Hirschhorn, R., Hirschhorn, K. and Weismann, G. (1967). *Blood* **30**, 84.
Hitzig, W. H., Frick, P. G., Betke, K. and Huisman, T. H. J. (1960). *Helv. Pediat. Acta* **15**, 499.

Hitzig, W. H., Dohmann, U., Pluss, H. J. and Vischer, D. (1974). *J. Pediat.* **85**, 622.
Hoffbauer, R. W. and Schrempf, G. (1976). *Lancet* **ii**, 194.
Hollenberg, M. D., Kaback, M. M. and Kazazian, H. H. (1971). *Science* **174**, 698.
Holmberg, L., Henriksson, P., Ekelund, H. and Astedt, B. (1974). *J. Pediat.* **85**, 860.
Holmes, B., Park, B. H., Malawista, S. E., Quie, P. G., Nelson, D. L. and Good, R. A. (1970). *New Eng. J. Med.* **283**, 217.
Holmes, E. W., Miller, A. L. and Frost, R. G. (1975). *Am. J. Hum. Genet.* **27**, 719.
Hommes, F. A., Kuipers, J. R. G., Elema, J. D., Jansen, J. F. and Jonxis, J. H. P. (1968a). *Pediat. Res.* **2**, 519.
Hommes, F. A., Polman, H. A. and Reerink, J. D. (1968b). *Arch. Dis. Child.* **43**, 423.
Horlein, H. and Weber, G. (1948). *Dt. Med. Wschr.* **73**, 476.
Horn, N. (1976). *Lancet* **i**, 1156.
Hösli, P. (1973). *In* "Birth Defects" (A. G. Molulsky and W. Lenz, Eds), 226–233. Excerpta Medica, Amsterdam.
Hösli, P., Erickson, R. P. and Vogt, E. (1976). *Biochem. Biophys. Res. Commun.* **73**, 209.
Howell, R. R. (1972). *In* "The Metabolic Basis of Inherited Disease" (J. B. Stanbury, J. B. Wyngaarden and D. S. Fredrickson, Eds), 3rd ed., 149–173, McGraw-Hill, New York.
Hsia, D. Y. Y. (1969). *Med. Clin. N. Am.* **53**, 857.
Hudgson, P., Gardner Medwin, D., Worsford, M., Pennington, R. J. T. and Walton, J. N. (1968). *Brain* **91**, 435.
Hug, G., Schubert, W. K. and Chuck, G. (1969). *J. Clin. Invest.* **48**, 704.
Hug, G., Schubert, W. K. and Chuck, G. (1970). *Biochem. Biophys. Res. Commun.* **40**, 982.
Huguley, C. M., Bain, J. A., Rivers, S. L. and Scoggins, R. B. (1959). *Blood* **14**, 615.
Huijing, F. and Fernandes, J. (1969). *Am. J. Hum. Genet.* **21**, 275.
Huijing, F. and Fernandes, J. (1970). *Am. J. Hum. Genet.* **22**, 484.
Hunt, A. D., Stokes, J., McCrory, W. W. and Stroud, H. H. (1954). *Pediatrics* **13**, 140.
Huntly, C. C. and Stevenson, R. E. (1969). *Obstet. Gynaecol.* **34**, 694.
Illingworth, B., Cori, G. T. and Cori, C. F. (1956). *J. Biol. Chem.* **218**, 123.
Imperato-McGinley, J., Guerrero, L., Gautier, T. and Peterson, R. E. (1974). *Science* **186**, 1213.
Isselbacher, K. J., Anderson, E. P., Kurahashi, K. and Kalckar, H. M. (1956). *Science* **123**, 635.
Jacob, F. and Monod, J. (1961). *J. Mol. Biol.* **3**, 318.
Jandl, J. H. and Cooper, R. A. (1972). *In* "The Metabolic Basis of Inherited Disease" (J. B. Stanbury, J. B. Wyngaarden and D. S. Fredrickson, Eds), 3rd ed., 1323–1337. McGraw-Hill, New York.
Jeffcoate, T. N. A., Fliegner, J. R. H., Russell, S. H., Davis, J. C. and Wade, A. P. (1965). *Lancet* **i**, 732.
Jervis, G. A. (1953). *Proc. Soc. Exp. Biol.* **82**, 514.
Jeune, M., Collombel, C., Michel, M., David, M., Guibault, P., Guerrier, G. and Albert, J. (1970). *Sem. Hop. (Ann. Pediat.)* **17**, 85.
Johnson, A. M., Alper, C. A., Rosen, F. S. and Craig, J. M. (1971). *Science* **173**, 553.

Kaback, M. M., Zeiger, R. S., Reynolds, L. W. and Sonneborn, M. (1974). *In* "Progress in Medical Genetics" (A. G. Steinberg and A. G. Bearn, Eds), Vol. 10, 103–104. Grune and Stratton, New York.

Kacser, H. K., Bulfield, G. and Wallace, M. E. (1973). *Nature, Lond.* **244**, 77.

Kalckar, H. M., Anderson, E. P. and Isselbacher, K. J. (1956). *Biochem. Biophys. Acta* **20**, 262.

Kalow, W. (1962). "Pharmacogenetics. Heredity and Response to Drugs". Saunders, Philadelphia.

Kan, Y. W., Dozy, A. M., Alter, B. P., Frigoletto, F. D. and Nathan, D. G. (1972). *New Eng. J. Med.* **287**, 1.

Kan, Y. W., Valenti, C., Carnazza, V., Guidotti, R. and Rieder, R. F. (1974). *Lancet* **i**, 79.

Kan, Y. W., Golbus, M. S., Klein, P. and Dozy, A. M. (1975a). *New Eng. J. Med.* **292**, 1096.

Kan, Y. W., Golbus, M. S. and Trecartin, R. (1975b). *Lancet* **ii**, 790.

Kan, Y. W., Golbus, M. S. and Trecartin, R. (1976a). *New Eng. J. Med.* **294**, 1039.

Kan, Y. W., Golbus, M. S. and Dozy, A. M. (1976b). *New Eng. J. Med.* **295**, 1165.

Kang, E. S., Snodgrass, P. J. and Gerald, P. S. (1972). *Pediat. Res.* **6**, 875.

Kaplan, J. C. and Beutler, E. (1967). *Biochem. Biophys. Res. Commun.* **29**, 605.

Kaufman, L. and Marver, H. S. (1970). *New Eng. J. Med.* **283**, 954.

Kaufman, S., Holtzman, N. A., Milstein, S., Butler, I. J. and Krumholz, A. (1976). *New Eng. J. Med.* **293**, 785.

Keitt, A. S. (1969). *J. Clin. Invest.* **48**, 1997.

Kelley, W. N., Levy, R. I., Rosenbloom, F. M., Henderson, J. F. and Seegmiller, J. E. (1968). *J. Clin. Invest.* **47**, 2281.

Kelley, W. N., Greene, M. L., Rosenbloom, F. M., Henderson, J. F. and Seegmiller, J. E. (1969). *Ann. Intern. Med.* **70**, 155.

King, R. A. and Witkop, C. J. (1976). *Nature, Lond.* **263**, 69.

Kint, J. A. (1970). *Science* **167**, 1268.

Kjessler, B., Johansson, S. G. O., Sherman, M., Gustavsson, K. H. and Hultqvist, G. (1975). *Lancet* **i**, 432.

Koch, J., Skokstad, E. L. R., Williams, H. E. and Smith, L. H. (1967). *Proc. Natn. Acad. Sci. USA* **57**, 1123.

Kolodny, E. H. (1975). *In* "The Prevention of Genetic Disease and Mental Retardation" (A. Milunsky, Ed.), 182–203. Saunders, Philadelphia.

Konrad, P. N., Richards, F., Valentine, W. N. and Paglia, D. E. (1972). *New Eng. J. Med.* **286**, 557.

Kraemer, K. H., Coon, H. G., Petinga, R. A., Barrett, S. F., Rahe, A. E. and Robbins, J. H. (1975). *Proc. Natn. Acad. Sci. USA* **72**, 59.

Krane, S. M., Pinnell, S. R. and Erbe, R. W. (1972). *Proc. Natn. Acad. Sci. USA* **69**, 2899.

Kraus, A. P., Langston, M. F. and Lynch, B. L. (1968). *Biochem. Biophys. Res. Commun.* **30**, 173.

Kresse, H. and Neufeld, E. F. (1972). *J. Biol. Chem.* **247**, 2164.

Kusakabe, T. and Myake, T. (1964). *J. Clin. Endocrinol.* **24**, 456.

Kutt, H., Wolk, M., Scherman, R. and McDowell, F. (1964). *Neurology* **14**, 542.

Laberge, C. (1969). *Am. J. Hum. Genet.* **21**, 36.

Lachmann, P. (1975). *J. Med. Genet.* **12**, 372.

La Du, B. (1972). *In* "The Metabolic Basis of Inherited Disease" (J. B. Stanbury,

552   D. J. H. BROCK

J. B. Wyngaarden and D. S. Fredrickson, Eds), 3rd ed., 308–325. McGraw-Hill, New York.

La Du, B. N., Zannoni, V. G., Laster, L. and Seegmiller, J. E. (1958). *J. Biol. Chem.* **230**, 251.

La Du, B. N., Howell, R. R., Jacoby, G. A., Seegmiller, J. E., Sober, E. K., Zannoni, V. G., Canby, J. P. and Ziegler, L. K. (1963). *Pediatrics* **32**, 216.

Lalley, P. A., Rattazzi, M. C. and Shows, T. B. (1974). *Proc. Natn. Acad. Sci. USA* **71**, 1569.

Laster, L., Irreverre, F., Mudd, S. H. and Heizer, W. D. (1967). *J. Clin. Invest.* **46**, 1082.

Laurell, C. B. and Eriksson, S. (1963). *Scand. J. Clin. Lab. Invest.* **15**, 132.

Layzer, R. B., Rowland, L. P. and Ranney, H. M. (1967). *Arch. Neurol.* **17**, 512.

Lehmann, A. R., Kirk-Bell, S., Arlett, C. F., Paterson, M. C., Lohman, P. H. M., de Weerd-Kastelein, E. A. and Bootsma, D. (1975). *Proc. Natn. Acad. Sci. USA* **72**, 219.

Lehrer, R. I. and Cline, M. J. (1969). *J. Clin. Invest.* **48**, 1478.

LeRoux, A., Junier, C., Kaplan, J. C. and Bamberger, J. (1975). *Nature, Lond.* **258**, 619.

LeRoy, J. G. and DeMars, R. (1967). *Science* **157**, 804.

Levy, H. L., Shih, V. E. and Madigan, P. M. (1974). *New Eng. J. Med.* **289**, 1214.

Lewis, G. M., Spencer-Peet, J. and Stewart, K. M. (1963). *Arch. Dis. Child.* **38**, 40.

Lichtenstein, J. R., Martin, G. R., Kohn, L. D., Byers, P. H. and McKusick, V. A. (1973). *Science* **182**, 298.

Lindsay, G. (1970). *In* "Disorders of Phenylalanine, Thyroxine and Testosterone Metabolism" (W. Hamilton and F. P. Hudson, Eds) 8–9, Livingstone, Edinburgh.

Lindsten, J., Zetterstrom, R., and Ferguson-Smith, M. A. (1976). "Prenatal Diagnosis of Genetic Disorders of the Foetus". Inserm, Paris.

Littlefield, J. W. (1971). Birth Defects. Original Article Series **7** (5), 15.

Loos, H., Loos, D., Weeing, R. and Houwerzijl, J. (1976). *Blood* **48**, 53.

Lowden, J. A., Cutz, E., Conen, P. E., Rudd, N. and Dorran, T. A. (1973). *New England. J. Med.* **288**, 225.

Lyon, M. F. (1961). *Nature, Lond.* **190**, 372.

McArdle, B. (1951). *Clin. Sci.* **10**, 13.

McKusick, V. A. (1975). "Mendelian Inheritance in Man", 4th ed. Johns Hopkins Press, Baltimore.

McKusick, V. A., Howell, R. R., Hussells, I. E., Neufeld, E. F. and Stevenson, R. E. (1972). *Lancet* **i**, 993.

Mahoney, M. J., Hart, A. C., Steen, V. D. and Rosenberg, L. E. (1975). *Proc. Natn. Acad. Sci. USA* **72**, 2799.

Mapes, C. A., Anderson, R. L. and Sweeley, C. C. (1970). *FEBS Lett.* **7**, 180.

Marstein, S., Jellum, E., Halpern, B., Eldjarn, L. and Perry, T. L. (1976). *New Eng. J. Med.* **295**, 406.

Marver, H. S. and Schmid, R. (1972). *In* "The Metabolic Basis of Inherited Disease" (J. B. Stanbury, J. B. Wyngaarden and D. S. Fredrickson, Eds) 3rd ed., 1087–1140. McGraw-Hill, New York.

Matalon, R. and Dorfman, A. (1969). *Lancet* **ii**, 838.

Matalon, R. and Dorfman, A. (1972a). *Biochem. Biophys. Res. Commun.* **42**, 340.

Matalon, R. and Dorfman, A. (1972b). *Biochem. Biophys. Res. Commun.* **47**, 959.

Matalon, R., Dorfman, A., Nadler, H. L. and Jacobson, C. B. (1970). *Lancet* **i**, 83.

Matalon, R., Dorfman, A. and Nadler, H. L. (1972). *Lancet* i, 798.
Matalon, R., Arbogast, B., Justice, P., Brandt, I. K. and Dorfman, A. (1974). *Biochem. Biophys. Res. Commun.* **61**, 759.
Matsuda, I., Arashima, S., Mitsuyama, S., Oka, Y., Ikenchi, T., Kaneko, Y. and Ishikawa, M. (1975). *Humangenetik* **30**, 69.
Meeuwisse, G. W. (1970). *Scand. J. Clin. Lab. Invest.* **25**, 145.
Mehl, E. and Jatzkewitz, H. (1965). *Biochem. Biophys. Res. Commun.* **19**, 407.
Meyer, U. A., Strand, L. J., Doss, M., Rees, A. C. and Marver, H. S. (1972). *New Eng. J. Med.* **286**, 1277.
Migeon, B. R. (1970). *Biochem. Genet.* **4**, 377.
Migeon, B. R., Der Kaloustian, V. M., Nyhan, W. L., Young, W. J. and Childs, B. (1968). *Science* **160**, 425.
Miller, G., Lisco, H., Kohn, H. I. and Still, D. (1971). *Proc. Soc. Exp. Biol. Med.* **137**, 1459.
Milunsky, A. (1973). *In* "The Prenatal Diagnosis of Hereditary Disorders" 62–114. Thomas, Springfield, Illinois.
Milunsky, A. (1975). *In* "The Prevention of Genetic Disease and Mental Retardation" (A. Milunsky, Ed.), 221–263. Saunders, Philadelphia.
Milunsky, A. and Littlefield, J. W. (1969). *New Eng. J. Med.* **281**, 1128.
Miwa, S., Sato, T. and Murao, H. (1972). *Acta Haematol. Japan* **35**, 113.
Mize, C. E., Herndon, J. H., Blass, J. P., Milne, G. W. A., Follansbee, C., Laudat, P. and Steinberg, D. (1969). *J. Clin. Invest.* **48**, 1033.
Mohler, D. N., Majerus, P. W., Minnich, V., Hess, C. E. and Garrick, M. D. (1970). *New Eng. J. Med.* **283**, 1253.
Mohyuddin, F., Rathbun, J. C. and MacMurray, W. C. (1967). *Am. J. Dis. Child.* **113**, 142.
Mommaerts, W. F. H., Illingworth, B., Pearson, C. M., Guillory, R. J. and Seraydarian, K. (1959). *Proc. Natn. Acad. Sci. USA* **45**, 791.
Moore, G. E., Gerner, R. E. and Franklin, H. A. (1967). *J. Am. Med. Assoc.* **199**, 519.
Morrow, G., Barness, L. A., Auerbach, V. H., Di George, A. M., Ando, T. and Nyhan, W. L. (1969a). *J. Pediat.* **74**, 680.
Morrow, G., Barness, L. A., Cardinale, G. J., Abeles, R. H. and Flaks, J. G. (1969b) *Proc. Natn. Acad. Sci. USA* **63**, 191.
Mudd, S. H. (1971). *Fed. Proc.* **30**, 970.
Mudd, S. H., Finkelstein, J. D., Irreverre, F. and Laster, L. (1964). *Science* **143**, 1443.
Mudd, S. H., Irreverre, F. and Laster, L. (1967). *Science* **156**, 1599.
Mudd, S. H., Levy, H. L. and Abeles, R. H. (1969). *Biochem. Biophys. Res. Commun.* **35**, 121.
Mudd, S. H., Uhlendorf, B. W., Freeman, J. M., Finkelstein, J. D. and Shih, V. E. (1972). *Biochem. Biophys. Res. Commun.* **46**, 905.
Murphy, J. V., Wolfe, H. J., Balazs, E. A. and Moser, H. W. (1971). *In* "Lipid Storage Diseases: Enzymatic Defects and Clinical Implications" (J. Bernsohn and H. J. Grossman, Eds), 67–110, Academic Press, New York and London.
Myrianthopoulos, N. C. and Aronson, S. M. (1966). *Am. J. Hum. Genet.* **18**, 313.
Nadler, H. L. and Egan, T. J. (1970). *New Eng. J. Med.* **282**, 302.
Navon, R., Padeh, B. and Adam, A. (1973). *Am. J. Hum. Genet.* **25**, 287.
Necheles, T. F., Boles, T. A. and Allen, D. M. (1968). *J. Pediat.* **72**, 319.

Necheles, T. F., Rai, U. S. and Cameron, D. (1970a). *J. Lab. Clin. Med.* **76**, 593.
Necheles, T. F., Steinberg, M. H. and Cameron, D. (1970b). *Br. J. Haematol.* **19**, 605.
Neufeld, E. F. and Fratantoni, J. C. (1970). *Science* **169**, 141.
Newcombe, D. S. (1975). "Inherited Biochemical Disorders and Uric Acid Metabolism", H. M. and M., Aylesbury.
Niederwieser, A., Giliberti, P., Matasovic, A., Pluznik, S., Steinmann, B. and Baerlocher, K. (1974). *Clin. Chim. Acta* **54**, 293.
Niermeijer, M. F., Koster, J. F., Jahodova, M., Fernandes, J., Heukels-Dully, M. J. and Galjaard, H. (1975). *Pediat. Res.* **9**, 498.
Niermeijer, M. F., Sachs, E. S., Jahodova, M., Tichelaar-Klepper, C., Kleijer, W. J. and Galjaard, H. (1976). *J. Med. Genet.* **13**, 182.
Nikolai, T. F. and Seal, U. S. (1966). *J. Clin. Endocrinol.* **26**, 835.
Nitowsky, H. M. (1975). *In* "The Prevention of Genetic Diseases and Mental Retardation" (A. Milunsky, Ed.), 114–133. Saunders, Philadelphia.
Norum, K. R. and Gjone, E. (1967). *Scand. J. Lab. Invest.* **20**, 231.
Nozawa, Y., Noguchi, T., Iida, H., Fukushima, H., Skiya, T. and Ito, Y. (1974). *Clin. Chim. Acta* **55**, 81.
Nyhan, W. L., Bakay, B., Connor, J. D., Marks, J. F. and Keele, D. K. (1970). *Proc. Natn. Acad. Sci. USA* **65**, 214.
O'Brien, J. S. (1972). *Proc. Natn. Acad. Sci. USA* **69**, 1720.
O'Brien, J. S. (1975). *Clin. Genet.* **8**, 303.
O'Brien, J. S., Okada, S., Fillerup, D. L., Veath, M. L., Adornato, B., Brenner, P. H. and LeRoy, J. G. (1971). *Science* **172**, 61.
Ockerman, P. A. (1967). *Lancet* **ii**, 239.
Ockerman, P. A. (1969). *J. Pediat.* **75**, 360.
Okada, S., and O'Brien, J. S. (1968). *Science* **160**, 1002.
Okada, S. and O'Brien, J. S. (1969). *Science* **165**, 698.
Oort, M., Loos, J. A. and Prins, H. K. (1961). *Vox Sang.* **6**, 370.
Paglia, D. E., Valentine, W. N., Baughan, M. A., Miller, D. R., Reed, C. F. and McIntyre, O. R. (1968). *J. Clin. Invest.* **47**, 1429.
Paglia, D. E., Gray, G. R., Growe, G. H. and Valentine, W. N. (1976). *Br. J. Haematol.* **34**, 61.
Pagliara, A., Karl, I., Devivo, D., Feigin, R. and Kipnis, D. (1971). *J. Clin. Invest.* **50**, 73 A.
Patrick, A. D. (1965). *Biochem. J.* **97**, 17C.
Patrick, A. D. and Lake, B. D. (1969). *Nature, Lond.* **222**, 1067.
Pauling, L., Itano, H. A., Singer, S. J. and Wells, I. C. (1949). *Science* **110**, 543.
Payling-Wright, C. R. and Evans, P. R. (1970). *Lancet* **ii**, 43.
Pearson, C. M. and Kalyanaraman, K. (1972). *In* "The Metabolic Basis of Inherited Disease" (J. B. Stanbury, J. B. Wyngaarden and D. S. Fredrickson, Eds), 3rd ed., 1180–1203, McGraw-Hill, New York.
Pennington, R. J. (1969). *In* "Disorders of Voluntary Muscle" (J. N. Walton, Ed.), 385–410. Churchill, London.
Pentinnen, R. P., Lichtenstein, J. R., Martin, G. R. and McKusick, V. A. (1975). *Proc. Natn. Acad. Sci. USA* **72**, 586.
Perry, T. L., Hansen, S., Tischler, B., Bunting, R. and Berry, K. (1967). *New Eng. J. Med.* **277**, 1219.
Perry, T. L., Urquhart, N. and MacLean, J. (1975). *New Eng. J. Med.* **290**, 1269.
Peterson, J. A. and Weiss, M. C. (1972). *Proc. Natn. Acad. Sci. USA* **69**, 571.

Pinsky, L. (1975). *In* "The Prevention of Genetic Disease and Mental Retardation" (A. Milunsky, Ed.), 134–181, Saunders, Philadelphia.

Piomelli, S. and Corash, L. (1976). *In* "Advances in Human Genetics" (H. Harris and K. Hirschhorn, Eds), Vol. 6, 165–240. Plenum, New York.

Platter, H. and Martin, G. M. (1966). *Proc. Soc. Exp. Biol. Med.* **123**, 140.

Pollitt, R. J., Jenner, F. A. and Merskey, H. (1968). *Lancet* **ii**, 253.

Pope, F. M., Martin, G. R., Lichtenstein, J. R., Pentinnen, R. P., Gerson, B., Rowe, D. W. and McKusick, V. A. (1975). *Proc. Natn. Acad. Sci. USA* **72**, 1314.

Potts, J. T. (1972). *In* "The Metabolic Basis of Inherited Disease" (J. B. Stanbury, J. B. Wyngaarden and D. S. Fredrickson, Eds), 3rd ed., 1305–1322. McGraw-Hill, New York.

Prader, A. and Gurtner, H. P. (1955). *Helv. Pediat. Acta.* **10**, 397.

Priest, R. E., Moinuddin, J. F. and Priest, J. H. (1973). *Nature, Lond.* **245**, 265.

Prins, H. K., Oort, M., Loos, J. A., Zürcher, C. and Beckers, T. (1966). *Blood* **27**, 145.

Pulvertaft, R. J. V. (1964). *Lancet* **ii**, 552.

Rathburn, J. C. (1948). *Am. J. Dis. Child.* **75**, 822.

Ricciuti, F. C., Gelehrter, T. D. and Rosenberg, L. E. (1976). *Am. J. Hum. Genet.* **28**, 332.

Robbins, P. W. (1960). *Fed. Proc. Fed. Am. Soc. Exp. Biol.* **19**, 193.

Romeo, G., Glenn, B. L. and Levin, E. Y. (1970a). *Biochem. Genet.* **4**, 719.

Romeo, G., Kaback, M. M. and Levin, E. Y. (1970b). *Biochem. Genet.* **4**, 659.

Rosen, F. S., Charache, P., Pensky, J. and Donaldson, V. H. (1965). *Science* **148**, 649.

Rosenberg, L. E. (1974). *In* "Duncan's Diseases of Metabolism. Genetics and Metabolism" (P. K. Bondy and L. E. Rosenberg, Eds), 7th ed., 31–106, Saunders, Philadelphia.

Rosenberg, L. E. (1976). *In* "Advances in Human Genetics" (H. Harris and K. Hirschhorn, Eds), Vol. 6, 1–74. Plenum, New York.

Rosenberg, L. E. and Downing, S. J. (1965). *J. Clin. Invest.* **44**, 1382.

Rosenberg, L. E. and Mahoney, M. J. (1973). *In* "Inborn Errors of Metabolism" (F. A. Hommes and C. J. van den Berg, Eds), 303–320. Academic Press, New York and London.

Rosenberg, L. E. and Scriver, C. R. (1974). *In* "Duncan's Diseases of Metabolism: Genetics and Metabolism" (P. K. Bondy and L. E. Rosenberg, Eds), 7th ed., 465–653. Saunders, Philadelphia.

Rosenberg, E. L., Downing, S. J., Durant, J. L. and Segal, S. (1966). *J. Clin. Invest.* **45**, 365.

Rosenberg, L. E., Lilljeqvist, A. C. and Hsia, Y. E. (1968a). *New Eng. J. Med.* **278**, 1319.

Rosenberg, L. E., Lilljeqvist, A. and Hsia, Y. E. (1968b). *Science* **162**, 805.

Rosenberg, L. E., Lilljeqvist, A. C., Hsia, Y. E. and Rosenbloom, F. M. (1969). *Biochem. Biophys. Res. Commun.* **37**, 607.

Rosenbloom, F. M., Kelley, W. N., Henderson, J. F. and Seegmiller, J. E. (1967). *Lancet* **ii**, 305.

Rosenbloom, F. M., Henderson, J. F., Caldwell, I. C., Kelley, W. N. and Seegmiller, J. E. (1968). *J. Biol. Chem.* **243**, 1166.

Roses, A. D., Herbstreith, M. H. and Appel, S. H. (1975). *Nature, Lond.* **254**, 350.

Russell, A., Levin, B., Oberholzer, V. G. and Sinclair, L. (1962). *Lancet* **ii**, 699.

556 D. J. H. BROCK

Saez, J. M., De Peretti, E., Morera, A. M., David, M. and Bertrand, J. (1971). *J. Clin. Endrocrinol. Metabol.* **32**, 604.
Salzmann, J., DeMars, R. and Beneke, P. (1968). *Proc. Natn. Acad. Sci. USA* **60**, 545.
Sandhoff, K. A. (1969). *FEBS Lett.* **4**, 351.
Sandhoff, K. A. (1974). *J. Clin. Pathol.* **27** (Suppl. 8), 94.
Sandhoff, K. A., Andreae, U. and Jatzkewitz, H. (1968). *Life Sci.* **7**, 283.
Sass, M. D., Caruso, C. J. and Farhargi, M. (1967). *J. Lab. Clin. Med.* **70**, 760.
Sassa, S., Cranick, S., Bicker, D. R., Bradlow, H. L. and Kappas, A. (1974). *Proc. Natn. Acad. Sci. USA* **71**, 732.
Schapira, F., Schapira, G. and Dreyfus, J. C. (1961). *Enzymol. Biol. Clin.* **1**, 170.
Schlesinger, P., Cotton, R. G. H. and Danks, D. M. (1976). *Lancet* ii, 1245.
Schmid, R. (1972). In "The Metabolic Basis of Inherited Disease" (J. B. Stanbury, J. B. Wyngaarden and D. S. Fredrickson, Eds), 3rd ed., 1141–1179. McGraw-Hill, New York.
Schmid, R., Robbins, P. W. and Traut, R. B. (1959). *Proc. Natn. Acad. Sci. USA* **45**, 1236.
Schneider, A. S., Valentine, W. N., Hattori, M. and Heins, H. L. (1965). *New Eng. J. Med.* **272**, 229.
Schneider, A. S., Valentine, W. N., Baughan, M. A., Paglia, D. E., Shore, N. A. and Heins, H. L. (1968). In "Hereditary Disorders of Erythrocyte Metabolism" (E. Beutler, Ed.) 265–272. Grune and Stratton, New York.
Schneider, J. A., Verroust, F. M., Kroll, W. A., Garvin, A. J., Horger, E. O., Wong, V. G., Spear, G. S., Jacobson, C., Pellett, O. L. and Becker, F. L. A. (1974). *New Eng. J. Med.* **290**, 878.
Schneider, P. B. and Kennedy, E. P. (1967). *J. Lipid Res.* **8**, 202.
Schroter, W. (1965). *Klin. Wschr.* **43**, 1147.
Scott, C. R., Clark, S. H., Teng, C. C. and Svedberg, K. R. (1970). *J. Pediat.* **77**, 805.
Scott, E. M. (1960). *J. Clin. Invest.* **39**, 1176.
Scott, E. M. and Griffith, I. V. (1959). *Biochim. Biophys. Acta* **34**, 584.
Scrimgeour, J. B. (1973). In "Antenatal Diagnosis of Genetic Disorders" (A. E. H. Emery, Ed.), 40–57. Churchill-Livingstone, Edinburgh.
Scriver, C. R. (1973). *Metabolism* **22**, 1319.
Scriver, C. R. and Efron, M. L. (1972). In "The Metabolic Basis of Inherited Disease" (J. B. Stanbury, J. B. Wyngaarden and D. S. Fredrickson, Eds), 3rd ed., 351–369. McGraw-Hill, New York.
Scriver, C. R. and Perry, T. L. (1972). In "The Metabolic Basis of Inherited Disease" (J. B. Stanbury, J. B. Wyngaarden and D. S. Fredrickson, Eds), 3rd ed., 476–492. McGraw-Hill, New York.
Scriver, C. R., McKenzie, S., Clow, C. L. and Delvin, E. (1971). *Lancet* i, 310.
Scriver, C. R., Chesney, R. W. and McInnes, R. R. (1976). *Kidney Internat.* **9**, 149.
Seegmiller, J. E. (1976). In "Advances in Human Genetics", (H. Harris and H. Hirschhorn, Eds), Vol. 6, 75–163. Plenum, New York.
Seegmiller, J. E., Rosenbloom, F. M. and Kelley, W. N. (1967). *Science* **155**, 1682.
Segal, S. (1972). In "The Metabolic Basis of Inherited Disease" (J. B. Stanbury, J. B. Wyngaarden and D. S. Fredrickson, Eds), 3rd ed., 174–195. McGraw-Hill, New York.
Shapira, E., Ben-Yoseph, Y., Eyal, G. and Russell, A. (1974). *J. Clin. Invest.* **53**, 59.

Shani, M., Seligsohn, U., Gilon, E., Sheba, C. and Adam, A. (1970). *Quart. J. Med.* **39**, 549.

Sheldon, W. (1964). *Arch. Dis. Child.* **39**, 268.

Shokeir, M. H. K. (1973). *Clin. Genet.* **4**, 494.

Shokeir, M. H. K. and Shreffler, D. C. (1969). *Proc. Natn. Acad. Sci. USA* **62**, 867.

Silvers, D. N., Cox, R. P., Balis, M. E. and Dancis, J. (1972). *New Eng. J. Med.* **286**, 390.

Singer, J. D. and Brock, D. J. H. (1971). *Ann. Hum. Genet.* **35**, 109.

Sjoberg, I., Fransson, L. A., Matalon, R. and Dorfman, A. (1973). *Biochem. Biophys. Res. Commun.* **54**, 1125.

Smith, L. H. and Williams, H. E. (1967). *Modern Treatment* **4**, 522.

Smith, L. H., Huguley, C. M. and Bain, J. A. (1972). *In* "The Metabolic Basis of Inherited Disease" (J. B. Stanbury, J. B. Wyngaarden and D. S. Fredrickson, Eds), 3rd ed., 1003–1029 McGraw-Hill, New York.

Srivastava, S. K. and Beutler, E. (1973). *Nature, Lond.* **241**, 463.

Srivastava, S. K. and Beutler, E. (1974). *J. Biol. Chem.* **249**, 2054.

Stamatoyannopoulos, G., Papayannopoulou, T., Bakopoulos, G. and Motulsky, A. G. (1967). *Blood* **29**, 87.

Stanbury, J. B. (1972). *In* "The Metabolic Basis of Inherited Disease" (J. B. Stanbury, J. B. Wyngaarden and D. S. Fredrickson, Eds), 3rd ed., 223–265. McGraw-Hill, New York.

Stanbury, J. B., Wyngaarden, J. B. and Fredrickson, D. S. (1966). *In* "The Metabolic Basis of Inherited Disease" (J. B. Stanbury, J. B. Wyngaarden and D. S. Fredrickson, Eds), 2nd ed., 3–20. McGraw-Hill, New York.

Stanbury, J. B., Wyngaarden, J. B. and Fredrickson, D. S. (1972). *In* "The Metabolic Basis of Inherited Disease" (J. B. Stanbury, J. B. Wyngaarden and D. S. Fredrickson, Eds), 3rd ed., 3–28. McGraw-Hill, New York.

Steinberg, D. (1972). *In* "The Metabolic Basis of Inherited Disease" (J. B. Stanbury, J. B. Wyngaarden and D. S. Fredrickson, Eds), 3rd ed., 833–856. McGraw-Hill, New York.

Steinberg, D., Herndon, J. H., Uhlendorf, B. W., Mize, C. E., Avigan, J. and Milne, G. W. A. (1967a). *Science* **156**, 1740.

Steinberg, D., Mize, C. E., Avigan, J., Fales, H. M., Eldjarn, L., Try, K., Stokke, O. and Refsum, S. (1967b). *J. Clin. Invest.* **46**, 313.

Stokke, O., Jellum, E. and Eldjarn, L. (1973). *In* "Inborn Errors of Metabolism" (F. A. Hommes and C. J. van den Berg, Eds), 321–336. Academic Press, New York and London.

Strand, L. J., Felsher, B. W., Redeker, A. G. and Marver, H. S. (1970). *Proc. Natn. Acad. Sci. USA* **67**, 1315.

Subak-Sharpe, H., Burk, R. R. and Pitts, J. D. (1969). *J. Cell Sci.* **4**, 353.

Sugita, M., Dulaney, J. T. and Moser, H. W. (1972). *Science* **178**, 1100.

Super, M. (1975). *S. Afr. Med. J.* **49**, 818.

Sutcliffe, R. G. and Brock, D. J. H. (1971). *Clin. Chim. Acta* **31**, 363.

Sutcliffe, R. G. and Brock, D. J. H. (1973). *J. Obstet. Gynaecol. Br. Commonw.* **79**, 902.

Sutherland, B. M., Rice, M. and Wagner, E. K. (1975). *Proc. Natn. Acad. Sci. USA* **72**, 103.

Sutherland, G. R. and Bain, A. D. (1973). *Humangenetik* **20**, 251.

Suzuki, K. and Suzuki, Y. (1970). *Proc. Natn. Acad. Sci. USA* **66**, 302.

Suzuki, K., Schneider, E. L. and Epstein, C. J. (1971). *Biochem. Biophys. Res. Commun.* **45**, 1363.

Swift, M. R. and Finegold, M. J. (1969). *Science* **165**, 294.

Szeinberg, A., Kahana, D., Gavendo, S., Zaidman, J. and Benezzer, J. (1969). *Acta Haematol.* **42**, 111.

Szorady, I. (1973). *In* "Pharmacogenetics. Principles and Pediatric Aspects". Akademiai Kiado, Budapest.

Tada, K., Morikawa, T. and Arakawa, T. (1966a). *Tohoku J. Exp. Med.* **90**, 337.

Tada, K., Morikawa, T. and Arakawa, T. (1966b). *Tohoku J. Exp. Med.* **90**, 189.

Tada, K., Yokoyama, Y., Nakasawa, H., Yoshida, T. and Arakawa, T. (1967). *Tohoku J. Exp. Med.* **93**, 115.

Tada, K., Yoshida, T., Konno, T., Wada, Y., Yokayama, Y. and Arakawa, T. (1969). *Tohoku J. Exp. Med.* **97**, 99.

Takahara, S. (1952). *Lancet* **ii**, 1101.

Tanaka, K. (1973). *In* "Inborn Errors of Metabolism" (F. A. Hommes and C. J. Van den Berg, Eds), 269–290. Academic Press, New York and London.

Tanaka, K., Budd, M. A., Efron, M. L. and Isselbacher, K. J. (1966). *Proc. Natn. Acad. Sci. USA* **56**, 236.

Tarlow, M. J., Hadorn, B., Artherton, M. W. and Lloyd, J. K. (1970). *Arch. Dis. Child.* **45**, 651.

Tarui, S., Okono, G., Ikura, Y., Tanaka, T., Suda, M. and Nishikawa, M. (1965). *Biochem. Biophys. Res. Commun.* **13**, 517.

Tashian, R. E., Riggs, S. K. and Yu, Y. S. (1966). *Arch. Biochem. Biophys.* **117**, 320.

Tauro, G. P., Danks, D. M., Rowe, P. B., van der Weyden, M. B., Schwarz, M., Collins, V. L. and Neal, B. W. (1976). *New Eng. J. Med.* **294**, 466.

Taylor, J. M., Dozy, A., Kan, Y. W., Varmus, H. E., Lie-Injo, L. E., Ganesan, J. and Todd, D. (1974). *Nature, Lond.* **251**, 392.

Taysi, K., Kistenmacher, M. L., Punnett, H. H. and Mellman, W. J. (1969). *New Eng. J. Med.* **281**, 1108.

Tedesco, T. A., Wu, J. W., Boches, F. S. and Mellman, W. J. (1975). *New Eng. J. Med.* **292**, 737.

Terheggen, H. G., Schwenk, A., Lowenthal, A., van Sande, M. and Colombo, J. P. (1969). *Lancet* **ii**, 748.

Terheggen, H. G., Schwenk, A., Lowenthal, A., van Sande, M. and Colombo, J. P. (1970). *Z. Kinderheilk.* **107**, 298.

Tildon, J. T. and Cornblath, M. (1972). *J. Clin. Invest.* **51**, 493.

Tomlinson, S. and Westall, R. G. (1964). *Clin. Sci.* **26**, 261.

Torsvik, H. (1970). *Clin. Genet.* **1**, 310.

Townes, P. L., Bryson, M. F. and Miller, G. (1967). *J. Pediat.* **71**, 220.

Trotta, P. P., Smithwick, E. M. and Balis, M. E. (1976). *Proc. Natn. Acad. Sci. USA* **73**, 104.

Tschudy, D. P. (1974). *In* "Duncan's Diseases of Metabolism: Genetics and Metabolism" (P. K. Bondy and L. E. Rosenberg, Eds), 7th ed., 775–824. Saunders, Philadelphia.

Tschudy, D. P., Perlroth, M. G., Marver, H. S., Collins, A., Hunter, G. and Rechcigl, M. (1965). *Proc. Natn. Acad. Sci. USA* **53**, 841.

Uhlendorf, B. W. and Mudd, S. H. (1968). *Science* **160**, 1007.

Uhlendorf, B. W., Conerley, E. B. and Mudd, S. H. (1973). *Pediat. Res.* **7**, 645.

Valentine, W. N., Tanaka, K. R. and Miwa, S. (1961). *Trans. Assoc. Am. Phys.* **74**, 100.

Valentine, W. N., Oski, F. A., Paglia, D. E., Baughan, M. A., Schneider, A. S. and Naiman, J. L. (1967). *New Eng. J. Med.* **276**, 1.

Valentine, W. N., Oski, F. A., Paglia, D. E., Baughan, M. A., Schneider, A. S. and Naiman, J. L. (1968). *In* "Hereditary Disorders of Erythrocyte Metabolism" (E. Beutler, Ed.), 288–302. Grune and Stratton, New York.

Valentine, W. N., Fink, K., Paglia, D. E., Harris, S. R. and Adams, W. S. (1974). *J. Clin. Invest.* **54**, 866.

Valle, D. L. and Phang, J. M. (1974). *Science* **185**, 1053.

Van Hoof, F. (1967). *Eur. J. Biochem.* **2**, 271.

Van Hoof, F. (1974). *J. Clin. Pathol.* **27** (Suppl. 8), 64.

Van Hoof, F. and Hers, H. G. (1967). *Eur. J. Biochem.* **2**, 265.

Van Hoof, F. and Hers, H. G. (1968). *Eur. J. Biochem.* **7**, 34.

Vidgoff, J., Buist, N. R. M. and O'Brien, J. S. (1973). *Am. J. Hum. Genet.* **25**, 372.

Vives-Corrons, J. L., Montserrat-Costa, E. and Rozman, C. (1976). *Hum. Genet.* **34**, 285.

Vogel, F. (1959). *Ergebn. Inn. Med. Kinderheilk.* **12**, 52.

Von Figura, K. and Kresse, H. (1972). *Biochem. Biophys. Res. Commun.* **48**, 262.

Wada, Y., Tada, K., Minagawa, A., Yoshida, T., Morikawa, T. and Okamura, T. (1963). *Tohoku J. Exp. Med.* **81**, 46.

Waddington, C. H. (1969). *Science* **166**, 639.

Waller, H. D. (1968). *In* "Hereditary Disorders of Erythrocyte Metabolism" (E. Beutler, Ed.), 185–208. Grune and Stratton, New York.

Waller, H. D., Löhr, G. W., Zysno, E., Gerok, W., Vos, D. and Strauss, G. (1965). *Klin. Wschr.* **43**, 8.

Wang, Y. M. and Van Eys, J. (1970). *New Eng. J. Med.* **282**, 892.

Waterbury, L. and Frenkel, E. P. (1972). *Blood* **39**, 415.

Watts, R. W. E. (1974). *J. Clin. Pathol.* **27** (Suppl. 8), 48.

Watts, R. W. E., Engelman, K., Klinenberg, J. R., Seegmiller, J. E. and Sjoerdsma, A. (1963). *Nature, Lond.* **201**, 395.

Wellner, V. P., Sekura, R., Meister, A. and Larsson, A. (1976). *Proc. Natn. Acad. Sci. USA* **71**, 2505.

Wenger, D. A. and Riccardi, V. M. (1976). *J. Pediat.* **88**, 76.

Wiesmann, U. N. and Neufeld, E. F. (1970). *Science* **169**, 72.

Williams, H. E. and Smith, L. H. (1968). *New Eng. J. Med.* **278**, 233.

Williams, H. E. and Smith, L. H. (1972). *In* "The Metabolic Basis of Inherited Disease" (J. B. Stanbury, J. B. Wyngaarden and D. S. Fredrickson, Eds), 3rd ed., 196–222. McGraw-Hill, New York.

Wintrobe, M. M. (1974). "Clinical Haematology", 7th ed. Lea and Febriger, Philadelphia.

Witkop, C. J., Nance, W. E., Rawls, R. F. and White, J. G. (1970). *Am. J. Hum. Genet.* **22**, 55.

Yaffe, S. J., Levy, G., Matsuzawa, T. and Baliah, T. (1966). *New Eng. J. Med.* **275** 1461.

Yoshida, A. (1967). *Proc. Natn. Acad. Sci. USA* **57**, 835.

Yoshida, A. (1970). *J. Mol. Biol.* **52**, 483.

Yoshida, A. and Motulsky, A. G. (1969). *Am. J. Hum. Genet.* **21**, 486.

Yoshida, T., Tada, K. and Arakawa, T. (1967). *Tohoku J. Exp. Med.* **104**, 195.

Yoshida, T., Kikuchi, G., Tada, K., Narisawa, K. and Arakawa, T. (1969). *Biochem. Biophys. Res. Commun.* **35**, 577.
Yoshida, T., Tada, K. and Arakawa, T. (1971a). *Tohoku J. Exp. Med.* **104**, 195.
Yoshida, T., Tada, K., Honda, Y. and Arakawa, T. (1971b). *Tohoku J. Exp. Med.* **104**, 305.
Zachmann, M., Vollmin, J. A., Hamilton, W. and Prader, A. (1972). *Clin. Endocrinol.* **1**, 369.
Zannoni, V. G., Seegmiller, J. E. and La Du, B. N. (1962). *Biochem. J.* **88**, 160.
Zimmerman, T. S., Ratnoff, O. D. and Powell, A. E. (1971). *J. Clin. Invest.* **50**, 244.

11 Haemoglobin Variation

J. M. WHITE

Department of Haematology, King's College Hospital Medical School, London, England

I. INTRODUCTION

The haemoglobin molecule and its genetic variation form not only one of the most studied subjects but also one of the most written about (Baglioni, 1963; Jonxis, 1963; Huisman, 1963; Braunitzer et al., 1964; Rossi-Fannelli et al., 1964; Schroeder and Jones, 1965; Antonini, 1965; Riggs, 1965; Huehns and Shooter, 1965; Lehmann and Huntsman, 1966; Weatherall and Clegg, 1968; Lehmann and Carrell, 1969; Huehns, 1970; Huehns, 1973; Lehmann and Huntsman, 1975). This chapter reviews the clinical implications of the genetic variations of the structure and synthesis of haemoglobin in the light of recent data. Since the first edition of this book six years ago there have been major advances in the understanding of the genetic control of haemoglobin. Also, several new types of haemoglobin mutation have been found. In this chapter the emphasis will be placed upon the more recent advances and how such mutations apply to medicine

A. Haemoglobin as a Protein

The mammalian haemoglobin molecule, like many proteins, is a multi-chain unit; a tetramer of four monomeric polypeptide chains (subunits), two α-chains and two non α-chains. The function of the molecule—the reversible carriage of oxygen—depends upon its four haem groups, one attached to each of the four globin chains. The structure of the globin chain has evolved to meet this function, namely, to create and maintain a constant hydrophobic environment for its haem and also to maintain the molecule in a soluble state.

The tetrameric molecule also exhibits the property of allosterism, that is, it undergoes conformational changes between oxygenation and deoxygena-tion. However, the changes which take place do not affect the overall con-formation of each chain (thus maintaining a constant haem environment) but rather the relationship of one chain to another. In this respect haemoglobin differs from many other proteins which probably also undergo large conformational changes during function. The tetramer also has the ability to bind hydrogen ions and intracellular organic phosphates. Both these properties have an important physiological role in oxygen delivery by the molecule. They will be considered later.

In Man, the non α-chains can be either ε, γ, β or δ, depending on the stage of embryonic, foetal or adult development (Huehns and Shooter, 1965). The tertiary conformation of each of these chains is identical; the only difference between them is in the sequence of the amino acids and the fact that the α-chains have 141 amino acids and all the others 146. The sequence of amino acids in the α-chain shows 84 differences from that of the β-chain (Braunitzer et al., 1961; Konigsberg and Hill, 1962; Schroeder et al., 1962). The number of differences in sequence between the β and γ and the β- and δ-chains is much less: 37 (Schroeder, 1963) and 10 respectively (Ingram and Stretton, 1962a, b; Hill and Kraus, 1963; Jones, 1964). The tertiary conformation of the five human globin chains is not only common to each of them but also common to the globin chains of all species, even to myoglobin, and it is in relation to the latter that this conformation is best examined (Perutz, 1965). Another non-α-chain, the ζ-chain (Capp et al., 1970) has also been described. This appears to be an embryonic chain and will be considered in more detail later (p. 575).

II. STRUCTURE OF NORMAL HAEMOGLOBIN

A. The Globin Monomer

Detailed information concerning the structure of the globin monomer has resulted from studies of the amino acid sequence of the human α- and β-chains (Braunitzer et al., 1961; Konigsberg and Hill, 1962; Schroeder et al., 1962) and from studies of the three-dimensional conformation by X-ray crystallography of myoglobin (Kendrew et al., 1960, 1961) and horse haemoglobin (Perutz et al., 1960, 1968; Cullis et al., 1962; Perutz, 1965; Perutz, 1976). As stated above, it is accepted that this tertiary conformation of the globin monomer chains is identical for all species (Fig. 1). Each chain consists of a long sequence of polar and non-polar amino acids, 75% of which are in the α-helical conformation. The chain is then folded on itself to give rise to eight helical segments which are connected by short non-helical segments. In order to make the position of each amino acid in the chain meaningful, the helical segments of myoglobin were designated A to H (from the amino end to carboxy end), the intra-helical segments

AB, BC, etc., and each amino acid numbered according to the position in its helix (Perutz, 1965). Using this notation, for example, the haem-linked histidines of all globin chains are always the eighth amino acid in the F helix (F8).

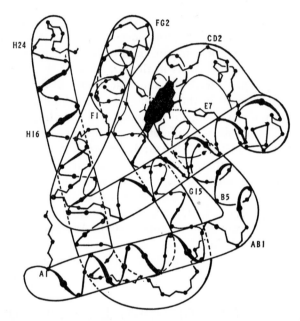

Fig. 1   A diagrammatic representation of the tertiary configuration of sperm whale myoglobin, showing helical segments, non-helical segments and haem group in haem pocket. Adapted from Dickerson (1964), see also p. 566

Perutz (1965) pointed out that one of the most consistent findings in all globin chains was that every third or fourth amino acid is non-polar. Therefore, because of the periodicity of the helix (one turn every 3·6 amino acids), the amino acids pointing internally are always non-polar and those pointing externally, polar. The aliphatic and aromatic side chains of the internally sited amino acids form van der Waal's bonds either with other non-polar amino acids on adjacent helical segments or with the haem group. These side chains create the necessary hydrophobic environment for the haem group, whilst the binding forces between them act as the major factor stabilizing the tertiary configuration of the chain. The side chains of the polar amino acids are, almost without exception, externally sited where they hydrogen-bond with the water of the cell. They are therefore only responsible for the solubility of the chain. (As will be appreciated when the genetic variants of haemoglobin are examined, those that result from a replacement of an internal (non-polar) amino acid generally produce marked alterations in either function or stability, whereas those that result from replacement of an externally sited amino acid (polar) generally do not.)

B. The Haem Group

The haem group lies in a pocket formed by folds in each chain and is held in position by bonds between non-polar amino acids, mainly from the F, E and CD helical segments and its non-polar methyl and vinyl side chains which are placed internally. The polar propionic acid side chains point externally and make at least one contact in the α-chain and two in the β-chain. However, their contribution to the binding of the haem group is not clear. In mammalian haemoglobins the haem contact amino acids, 19 in the α-chain and 21 in the β-chain, are invariant. Therefore, their contribution to the function and stability of the molecule is important. This is further supported by the fact that four of them are invariant in the globin chains of all species. The haem group in both chains is covalently bonded through the iron atom to the F8 histidines and lies opposite the E7 histidines. The small space between the iron atom and the imidazole group of histidine E7—the ligand site—is occupied by the oxygen in oxyhaemoglobin. Watson (1966) pointed out that the tertiary and secondary conformation of the chain was dependent upon the firm binding of the haem group. This is supported experimentally: for example, if the chain loses its haem, the helical content falls from 75% to 50% due to unfolding of the F and E helical segments (Hrkal and Vodrazka, 1967). In such a state the protein is very unstable.

C. The Haemoglobin Tetramer

The quaternary structure of the haemoglobin tetramer and the atomic contacts which bind the four chains together are now known in detail (Perutz et al., 1968). Haemoglobin is a globular protein (64Å $\times$ 55Å $\times$ 50Å) and consists of two symmetrically arranged β-chains. There are two pairs of major areas of contact, holding unlike α and β-chains together. One pair is designated the $\alpha_1\beta_1$ and $\alpha_2\beta_2$ and the other pair the $\alpha_1\beta_2$ and $\alpha_2\beta_1$. (A schematic diagram of the tetramer and contact regions is shown in Fig. 2.) The $\alpha_1\beta_1$ (or $\alpha_2\beta_2$) contact is large, hydrophobic and stable. It is made up of 34 amino acids, from the α- and β-chains, mainly those between G10 and H9, with some contribution from the B and D segments. During the conformational changes between oxygenation and deoxygenation, the movement across this contact is very small and the contact stabilizes the $\alpha\beta$ dimer unit.

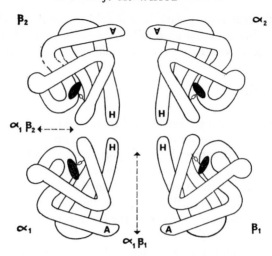

Fig. 2   Diagrammatic representation of the relationships between unlike subunits in haemoglobin tetramer (β_1 is equivalent to the configuration in Fig. 1). The approximate regions are shown for the contacts $\alpha_1\beta_1$ and $\alpha_1\beta_2$. For more complete detail of the quaternary structure of the tetramer see Perutz (1965) and Perutz *et al.* (1968)

Under many conditions the tetramer will readily dissociate into two symmetrical $\alpha\beta$ dimers and this is followed by a slow dissociation into monomers (Vinograd and Hutchinson, 1960; Huehns and Shooter, 1965).

$$\alpha_2\beta_2 \rightleftharpoons 2\alpha\beta \longrightarrow 2\alpha + 2\beta$$

This dissociation takes place across the $\alpha_1\beta_2$ contact, whilst the $\alpha_1\beta_1$ contact remains stable and prevents further dissociation of the dimer into unstable monomers (Rosemeyer and Huehns, 1967). It is not clear whether or not dissociation takes place within the red cell *in vivo* and has any functional importance. However, it has been suggested that dissociation of the tetramer into $\alpha\beta$ dimers is essential for oxygenation and deoxygenation of the molecule (Benesch *et al.*, 1965). This has been questioned by Gibson and Parkhurst (1968) on kinetic data and by Perutz (1969) on X-ray crystallographic data, both of whom consider the tetramer as the functional unit.

The $\alpha_1\beta_1$ contact is a smaller and smoother contact than the $\alpha_1\beta_1$. It comprises 19 amino acids, mainly from the FG and CD segments of the α- and β-chains. Perutz *et al.* (1968) have pointed out that the bonding between amino acids on this contact is such that one chain can move in relation to another without breaking it. During the change from the oxy- to deoxy-state the displacement across this contact is large and this movement is probably responsible for the normal allosteric changes which take place in the tetramer. (This is supported by observations on haemoglobin variants in which the site of substitution is on this contact. They all show marked alterations in the allosteric properties (p. 589).) In the tetramer there is an internal cavity lined by polar residues, mainly serines and threonines, and present evidence indicates that this cavity is the site of the binding of 2, 3-diphosphoglycerate (2, 3-DPG) and hydrogen ions to the molecule.

III. THE FUNCTION OF HAEMOGLOBIN

The structure–function relationships of the haemoglobin molecule are better understood than those of any other multi-chain protein. The information acquired over many years has given insight into how haemoglobin functions, and more recently it has been realized that the various functional properties of the molecule, especially oxygen affinity, have important implications in relation to the clinical morbidity of carriers of some haemoglobin variants (Bellingham and Huehns, 1968). It is no longer sufficient to consider the concentration of haemoglobin in the blood as a criterion of health or disease; one must now consider how haemoglobin functions in terms of its capacity to deliver oxygen to the tissues.

The function of the molecule is the transport of oxygen, a property which can be considered under three headings: the reversible oxygen: haem reaction, the cooperative effect of haem:haem interaction and the oxygen affinity of the molecule. Since 1968 (Perutz and Lehmann, 1968), a vast amount of knowledge has been obtained concerning these various properties. This has recently been summarized by Perutz (1976).

A. The Oxygen: Haem Reaction

The essential requirement that the haem environment must be fixed and hydrophobic for the preservation of the reversible oxygen:haem reaction is explained when this reaction is examined. Nobbs et al. (1966) showed that in oxymyoglobin the oxygen molecule occupied the 6th coordinate position in the Fe atom and that on reduction this site was vacated. This supported the earlier work of Weiss (1964a, b) and Viale et al. (1964) who postulated that the oxyhaemoglobin state approaches the ferric state, in that the electron of the Fe atom is partially transferred to oxygen, i.e. it is delocalized. This has since been confirmed (Lang and Marshall, 1966; Bearden et al., 1965). It was also shown by Nobbs et al. (1966) that in the absence of a ligand (water) the electron returns to the Fe atom which resumes the ferrous state when oxygen is released. However, the greater ability of the Fe atom to form a hydrogen bond with water rather than a weak ionic link with oxygen means that if water were to gain access to the haem group the spare electron would be rapidly donated to it and the Fe atom would become stabilized in the ferric state.

The interaction of oxygen with the iron atom has a profound effect on the tertiary structure of haemoglobin. In deoxyhaemoglobin the atomic radius of the iron atom increases and it is displaced out of the plane of the porphyrin ring. In oxyhaemoglobin the atomic radius shrinks and the atom is pulled into the plane of the ring. Since it is covalently bonded to histidine F8 the F helix is displaced; this slight movement has a profound effect on the quaternary conformation of the tetramer and is responsible for the very large movements which take place during deoxygenation and oxygenation of the molecule, the so-called R–T transition.

B. Respiration of Haemoglobin

In the oxy-state the haemoglobin tetramer is smaller than in the deoxy-state. Apart from the inter-subunit contacts there are no other bonds between the chains. Using the notation of Monod *et al.* (1965) it is in the R (relaxed) state. Upon deoxygenation, due to the displacement of the iron atom, the tertiary conformation of each chain alters and new contacts are formed between the four subunits. The net effect is to "prize open" the molecule, especially the β-chains, and to hold it in the T (tense) state. For exact details refer to Perutz (1972). The oxygen affinity of haemoglobin and the cooperative effect (p. 569) is due to this R–T transition. In the R state, the molecule has a very high affinity for oxygen; in the T state the affinity is very low.

The constant finding of a histidine opposite the Fe atom in position E7 in the globin chain of all species* has not been explained. It appears not to be essential for the oxygen:haem reaction, for Hb Zurich, in which histidine is replaced by arginine, can still reversibly combine with oxygen. Perutz and Lehmann (1968) suggested that its presence may in some way "assist" in maintaining the Fe atom in the ferrous state.

It was believed that it prevented the iron atom from undergoing spontaneous auto-oxidation. However, the recent experiments of Wallace *et al.* (1976) indicate that it only prevents the oxidation caused by drugs. As Perutz has pointed out, drugs are a modern invention and the explanation for its presence remains unclear, though of course in the environment there are many oxidants, e.g. nitrates in water.

* The CTT-III component of the haemoglobin of *Chironomus* is an exception in that the E7 histidine is replaced by glutamic acid (Braunitzer, 1970)

C. Cooperative (Haem:Haem) Interaction

The specific advantage of the globin chains of bony vertebrates (including the coelacanth) is that they undergo ordered association into polymers (a_2x_2) which possess the properties of linked oxygenation (cooperative effect or haem:haem interaction) and proton binding (Bohr effect). These allosteric phenomena are responsible for the sigmoid oxygen dissociation curve of the tetramer in contrast to the hyperbolic curve of the monomer (e.g. myoglobin) (Fig. 3).

The precise atomic events which take place during these allosteric

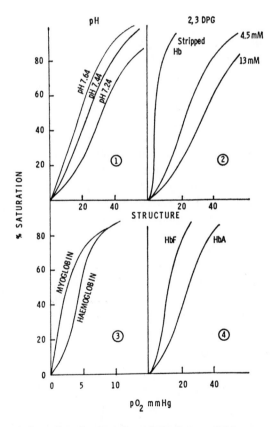

Fig. 3   Factors influencing the function of the haemoglobin tetramer: (1) proton binding (Bohr effect), (2) 2, 3-DPG binding, (3) comparison between oxygen dissociation curves of myoglobin (monomer, $n=1$) and haemoglobin (tetramer, $n=3$), and (4) comparison between oxygen dissociation curves of Hb F and Hb A

reactions are now understood (Perutz, 1970). The degree of conformational change in the molecule between the oxy- and deoxy- state has been measured (Perutz et al., 1968). It is confined to the displacement of one chain relative to another, mainly an outward displacement of the β-chains relative to the α-chains across the $\alpha_1\beta_2$ contact. The position of the α-chains relative to each other remains very close. The tetramer undergoes a reversed respiratory movement, i.e. upon deoxygenation the molecule opens up and on oxygenation it closes again. How this change in conformation of the tetramer increases the ability of the haem groups to combine with, or give up, oxygen is not understood. However, the sigmoid shape of the dissociation curve indicates that each haem does not function independently, but that the rate of reaction of one haem with oxygen is dependent on the state of oxygenation of the others.

The kinetics of the reaction are still uncertain (for review see Manwell and Baker, 1970). To explain them, Hill (1910) formulated his equation on the assumption that n molecules of oxygen combined with one molecule of haemoglobin. However, if this equation is correct, the n value for haemoglobin should be 4. In fact, it is 2·5 to 3·0. Adair (1927) explained this discrepancy by postulating that there were four different equilibrium constants for the reaction of oxygen with each haem ($K_1 < K_2 < K_3 < K_4$). Roughton (1949, 1965) showed that the major interaction takes place after three haems are oxygenated. The Hill equation, which Manwell and Baker (1970) regard as an allosteric approximation, only holds true between partial oxygen saturations of 15 to 80% (Roughton, 1949, 1965). Antonini (1965) suggested that at half saturation of the molecule the number of haems that interact must be equal to or greater than three.

The normal haem:haem interaction ($n = 3$) is dependent on a tetramer of two pairs of unlike chains (α_2x_2), as tetramers of like chains, e.g. Hb H (β_4), Hb Barts (γ_4) and Hb β_4^s have n values $= 1$. Their haem groups therefore function as four independent units (Benesch, 1966). It is also dependent on the amino acids on the $\alpha_1\beta_2$ contact, for those haemoglobin variants where substitution is on this site usually have abnormal n values.

D. Oxygen Affinity

There are three major factors which influence the degree of saturation of haemoglobin for any given oxygen concentration (pO_250)* (Fig. 3). First, the ability of the tetramer to bind hydrogen ions (Bohr effect); secondly, the binding of 2, 3-DPG; thirdly, the binding of CO_2; and finally, the

* The oxygen affinity of any haemoglobin is expressed as the pO_2 50, i.e. the partial pressure of O_2 which is required to half saturate the molecule

amino acid structure of certain regions of the molecule (for a detailed review see Kilmartin, 1976).

1. Bohr effect

The ability of the molecule to bind protons (H^+), which results in a decrease in the molecule's oxygen affinity, has an important physiological consequence, namely, that in the tissues low pH increases the quantity of oxygen delivered from the molecule. At the low pH of the tissues the inter-subunit bonds are stabilized; thus, there is a shift in the R–T equilibrium to the right. Also, once these bonds are formed the effect is to increase the pK of weak bases which will then combine more readily with protons. To date, 60% of the Bohr effect can be accounted for with 20% coming from the α amino group of $\alpha 1$ valine and 40% from the $\beta 146$ histidine (Kilmartin, 1973). The residues responsible for the remaining 40% have not been identified.

2. 2, 3-Diphosphoglycerate

The importance of this intracellular phosphate which binds to haemoglobin and decreases its oxygen affinity was first realized by Benesch et al. (1968) (for review, see Benesch and Benesch, 1969; Benesch and Benesch, 1974). These workers found that one mole of 2, 3-DPG would bind to one molecule of the haemoglobin tetramer with an association constant of $10^5 M^{-1}$. 2, 3-DPG binds mainly to the β-chains of haemoglobin when the molecule is in the deoxy form. In this state it binds mainly to the α amino groups and the C-terminal groups of the β-chain. Lysine EF6 also appears to be important. However, the internal cavity of the β-chain has an array of positively charged groups which would assist in the binding of 2, 3-DPG.

This simple organic compound has a marked effect on the oxygen affinity of haemoglobin in that it acts in simple terms as a spacer and shifts the R–T transition to the right. What has not been explained is at what point in this conformational transition 2, 3-DPG works. Since the binding constant for deoxyhaemoglobin is 100 times greater than oxyhaemoglobin, one might assume that 2, 3-DPG only binds to deoxyhaemoglobin and lowers the oxygen affinity for that molecule. Thus, there may well be a third state of haemoglobin: namely, the R state, the T state and the "S" state. The last is when 2, 3-DPG combines with haemoglobin.

3. The binding of CO₂

One of the functions of haemoglobin is the chemical reaction with CO_2 and, thus, its removal from the tissues. The sites of interaction are still not clearly defined but the α-amino groups of the α- and β-chains appear to be the major sites. An important effect of the binding of CO_2 is to lower the oxygen affinity of the molecule and, thus, enhance the delivery of oxygen to the tissues where the pCO_2 is high.

4. Amino acid structure

As previously stressed, the amino acids of the $\alpha_1\beta_2$ contact are extremely important both for the function of haem:haem interaction and the oxygen affinity of the molecule. However, there are several variants in which the site of substitution is distant from this contact and which show marked alterations in the oxygen affinity of the molecule. For example, Hb Seattle: E20 Ala→Glu (Huehns *et al.*, 1970) has a very low oxygen affinity and Hb Shepherd's Bush:E8 Gly→Asp (White *et al.*, 1970) has a high oxygen affinity with a reduced haem:haem interaction. There is no clear explanation for these altered functional properties but the observations underline the complexity of the atomic change which takes place.

IV. VARIATION IN THE STRUCTURE OF HAEMOGLOBIN

A. Phylogenetic Variation (Invariant and Variant Amino Acids)

The structure of the human haemoglobin molecule represents the current end-product of the evolution of haemoglobin. It is likely that haemoglobin developed to meet the oxygen requirements of primitive animals and that during evolution the molecule's efficiency increased progressively by mutation parallel with the increasing oxygen demand of higher vertebrates. This concept, however, is difficult to prove (by studying the structure and function of the haemoglobins of the accepted evolutionary steps of the human species). Analysis of the haemoglobins of different invertebrate and vertebrate species has shown that there is very little change in the chemical structure of the haem group (only one variant is known, Chlorocruorin in certain primitive worms). Major alterations, however, have taken place in the amino acid sequence of the globin chains. Thus, only five amino acids identical in type and position (invariant) have been found in the haemoglobin chains of all species so far examined (Perutz *et al.*, 1968). These appear to be of critical importance for the function of the globin monomer as a transporter of oxygen. However, when comparison is made between mammalian haemoglobin α- and β-chains many residues are invariant. These appear to be critically important for the structure and function of the molecule as a tetramer (i.e. for the allosteric phenomena). They form contacts with the haem group or unlike subunits.

The haemoglobin tetramer (α_2x_2) is common to all species from cartilagenous fishes upwards. Below them, the molecule shows major differences, e.g. lamprey haemoglobin is dimeric, and that of the hagfish monomeric. Thus, as Ingram (1963) pointed out, the haemoglobin molecule emerged probably at the time of development of the bony fishes. The α-chain of man still has 30 amino acids in common with the dimeric chain of the lamprey. The non-homologous areas in chains of different species must represent acceptable mutations which have taken place during evolution (Zuckerkandl and Pauling, 1962).

B. Developmental Variation of Human Haemoglobins

The six globin chains of man, α, β, γ, δ, ε, ζ are thought to arise from gene duplication and mutation of a common precursor (? α-chain). They form haemoglobins suitable for the environmental variations of the embryo (ε), foetus (γ), and adult (α, γ, δ). This is supported by the large degree of similarity of the amino acid sequence between all the chains (Itano, 1957; Gratzer and Allison, 1960; Braunitzer et al., 1961; Zuckerkandl and Pauling, 1962) and some estimate as to when gene duplication took place is given by the degree of concordance. (The greater the difference the farther back the gene duplication.) There are 84 amino acid differences between the α- and β-chains, 30 between the β-and γ-chains, but ony 10 between the β- and δ-chains. Thus the β-chain probably underwent duplication recently in evolutionary time (44 million years ago, according to Zuckerkandl and Pauling, 1962). It is unfortunate that the amino acid sequence of the human embryonic ε-chain is still unknown, for this might provide information as to the common precursor of all the haemoglobin chains. Ingram (1963) postulated that the human haemoglobin chains developed in the following sequence: α, γ, β and δ (Fig. 4). His assumption that the γ-chain preceded the β-chain is, however, very speculative. Both chains differ from the α-chain to a similar extent: the γ-chain by 86 amino acids and the β-chain by 89. An alternative hypothesis is that both the β- and γ-chains developed "simultaneously" as the foetus came to be clearly differentiated from the adult in evolution (Fig. 4).

Little insight into the problem can be obtained from the functions of the various haemoglobins, Gower I (ϵ_4), Gower II ($\alpha_2\epsilon_2$), Hb F ($\alpha_2\gamma_2$) and Hb A_2 ($\alpha_2\delta_2$), for these are still unclear. The ε-chains have usually disappeared after 10 weeks of intrauterine life, shortly after which the γ-chain appears, reaching maximal amounts at 30 weeks. The β-chain can be found in trace amounts of 16 weeks; thereafter its concentration steadily rises. At term the γ-chain comprises about 80% of the total secondary chain and the β-chain the remaining 20%. Shortly after birth the synthesis of the

γ-chain is switched off rapidly and this is accompanied by a rapid increase in the amount of β-chain synthesized. The δ-chain appears just before 40 weeks and thereafter is maintained at 2% of the total haemoglobin (Huehns and Shooter, 1965).

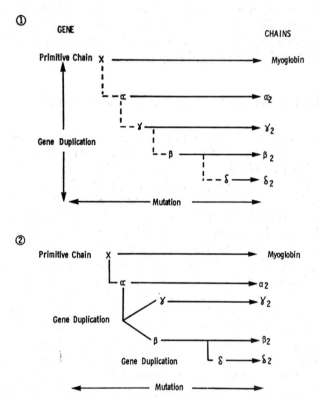

Fig. 4   Proposed mechanisms of development of human haemoglobin chains
(1) Postulate of Ingram (1963) that the developmental sequence was α, γ, β, δ
(2) Alternate mechanism of simultaneous development of β and γ chains when the foetal-maternal stage in evolution became defined

The major functional difference between Hb F and Hb A is that Hb F is unable to bind organic phosphates (2, 3-DPG) and is therefore a high oxygen-affinity haemoglobin, although the properties of haem:haem interaction and Bohr effect are preserved (Fig. 3). (The need for such a high affinity haemoglobin is still unexplained.)

Many vertebrate species form a foetal haemoglobin. However, the generally accepted view that the foetus will benefit because the high affinity of foetal haemoglobin facilitates the transport of the maternal oxygen from the mother to the foetus seems naïve since the same high affinity will deprive the foetal tissues of oxygen. In the author's opinion, the inability of the γ-chain to bind 2, 3-DPG (Benesch and Benesch, 1969) means that it does not interfere with foetal red cell metabolism. However, whether or not the foetal red cell benefits from this is speculative.

The function of Hb A_2 is also not clear. Kinderlerer et al. (1970) have, however, presented evidence that it increases the thermostability of Hb A. Hb A_2 is found only in higher primates (Kunkel et al., 1957) and Ingram (1963) has speculated that it may represent a more efficient haemoglobin chain destined (? in the remote future) to replace the β-chain of Hb A.

The relationship of the ζ-chain (Capp et al., 1970) to the development of the adult haemoglobin chains is as yet unclear. This chain has been shown to differ in its amino acid composition from the α-, β-, γ-, δ- and ϵ-chains. It has been established that it is the non α-chain component of the haemoglobin tetramer, Portland I, which can now be written as $\zeta_2\gamma_2$. This haemoglobin is detected in substantial amounts in infants with chromosomal abnormalities (Capp, 1966; Capp et al., 1967; Hecht et al., 1967), and in stillborn infants with the Hb Barts hydrops fetalis syndrome (Weatherall et al., 1970; Todd et al., 1970) and in small amounts in normal infants (Hecht et al., 1967). As yet it is unknown how similar the amino acid sequence of this chain is to the other chains. However, the recent work of Kamuzora and Lehmann (1975) indicates that it is very similar to the α-chain.

C. Structural Variation of Human Haemoglobins

There appears to be no general polymorphism of Hb A ($\alpha_2\beta_2$). Thus Ingram (1963) studied the haemoglobins of 120 individuals of different ethnic groups and did not find any differences in peptide sequence on partial amino acid analysis. His results have received support from similar studies over a number of years in many laboratories. Lehmann and Carrell (1969) have, however, pointed out that the methods used for analysis may miss 20% of the possible amino acid variation and that minor variation might therefore remain undetected. Von Ehrenstein (1966), Kilmartin and Clegg (1967) and Popp (1967) have shown that such minor variation does exist in animal haemoglobins, as in some species a particular amino acid site can be occupied by one of two amino acids, the relative proportions of which indicate that they are not alleles (i.e. not 1:1). They postulated that such amino acid variation could result from ambiguous reading of the genetic code, i.e. one gene could determine a family of polypeptide chains rather than a single chain. However, subsequent studies of the amino acid structure of the human γ-chain (Schroeder and Huisman, 1970) have suggested an alternative cause for the variation, namely, the presence of non-allelic genes, i.e. two genes could determine one polypeptide chain. This will be discussed in detail later (p. 601).

Table 1

Types of variation of human haemoglobin

	Remarks
Developmental	
Hb Gower I (ϵ_4)	embryonic haemoglobin
Hb Gower II ($a_2\epsilon_2$) or ($\gamma_2\epsilon_2$)	
Hb Portland ($\gamma_2\delta_2$)	
Hb F　　　　($a_2\gamma_2$)	
Hb A　　　　($a_2\beta_2$)	
Hb A$_2$　　　($a_2\delta_2$)	
Non-mutational (chemical)	
Hb A$_3$	fast migrating Hb, possibly due to oxidized β F9 SH groups
Hb A 1a, 1b, 1c, 1d, 1e	normal constituents of Hb A detected by IRC-50 chromatography
Hb F$_1$	normal constituent of foetal Hb
Hb Koelicher	fast migrating Hb found in plasma after intra-vascular haemolysis
Hb Pb	fast migrating Hb found in some children suffering from lead poisoning
Mutational	
Single point mutations	majority of Hb variants e.g. Hb's Freiburg, Leiden and Gun Hill
Intergenic non-homologous crossing-over	e.g. Hb's Lepore and anti Lepore
Stop codon mutations	e.g. Hb Constant Spring, Hb Tak, Hb Icaria
Frameshift mutations	e.g. Hb Wayne
Insertion mutations	e.g. Hb Grady

1. Types of variation (Table 1)

Non-mutational. Not all haemoglobin variants arise from substitution of amino acids. Haemoglobin A$_3$, a minor component, comprising up to 10% of normal haemoglobin (Kunkel and Wallenis, 1955), is believed to be normal Hb A in which the two β F9 cysteines are oxidized (Huisman and Dozy, 1962). (This has been questioned by Winterhalter and Birchmeier, 1970.) Haemoglobin F$_1$ (Allen *et al.*, 1958), which amounts to 10% of the total Hb F, differs from normal Hb F in that the terminal amino group of one of the γ-chains is acetylated, the chain having the formula $a_2\gamma^F\gamma^{Acet}$ (Schroeder *et al.*, 1962). Several minor compounds of Hb A have been detected by chromatography on IRC50; namely, Hb A 1a, 1b, 1c, 1d, 1e (Schnek and Schroeder, 1961). Haemoglobin A 1c, the major component (5% of the total) results from an aldehyde or a ketone forming a Schiff base with the amino terminal end of one β-chain (Holmquist and Schroeder, 1964; Schroeder and Holmquist, 1968). Another example of a chemically altered haemoglobin is the fast migrating Hb Koelicher found in plasma following intravascular haemolysis, which results from the removal of the terminal arginine of the β-chain by carboxypeptidase (Marti et al., 1967). A fast haemoglobin is found too, in lead poisoning, "Hb Pb" (Charache and Weatherall, 1966) but the nature of the chemical alteration has not yet been determined.

Point mutations. The vast majority of abnormal haemoglobins result from a single point mutation of the base codon of a single amino acid. The two types of base changes which can occur, transitions (purine to purine: e.g. A$\longleftrightarrow$G) and transversions (pyrimidine to purine: e.g. C$\longleftrightarrow$A) have been shown to have taken place equally frequently in the production of the known haemoglobin variants (Lehmann and Carrell, 1969; Lehmann and Kynoch, 1976); Lehmann and Carrell, however, also showed that the G$\longrightarrow$A mutation accounted for 30% of the variants, which is greater than expected, and postulated that a specific mechanism was responsible for the A$\longrightarrow$G mutation in DNA and the G$\longrightarrow$A in RNA transcription. The finding of equal numbers of transversion and transition base changes is interesting because random mutation of any one base should result in one transition to two transversions. When single amino acid substitutions have taken place there are no exceptions to a single point mutation of the codon triplet within the recognized genetic code. Analysis of the structure of haemoglobin variants has in fact confirmed the universality of the code.

It is noteworthy that twice as many variants of the β-chain as of the α-chain have been described while very few have been detected in the γ- and δ-chains. These differences remain unexplained. However, the apparent low mutational rate of the γ- and δ-chains may be due to technical difficulties in their recognition and isolation. The base changes in the variants have affected the first and second bases equally frequently but the third base in only a few instances. However, it is accepted that mutation of the third base in the codon will in most instances not alter the code for that particular amino acid.

Not all genetic variants arise as a single base point mutation in one chain. For example, Hb C Harlem: β A3 Glu$\longrightarrow$Val, E17 Asp$\longrightarrow$Asn (Bookchin et al., 1966) rises from a second point mutation in the β gene while Hb Memphis: α B4 Glu$\longrightarrow$Gln, β A3 Glu$\longrightarrow$Val (Kraus et al., 1966) is a variant caused by two single point mutations (one in each chain).

Another interesting observation by Steadman *et al.* (1970) has come from the interpretation of the amino acid substitution in Hb Bristol (β E11 Val—→Asp). Two other haemoglobins are known which result from the substitution of the Val β E11: namely, Hb Sydney:β E11 Val——→Ala and Hb M Milwaukee: β E11 Val——→Glu. Valine has four triplet codons: GUA, GUG, GUU and GUC. A single point mutation of any of these codons would result in the codon for alanine (GCA, GCG, GCU and GCC). However, the codons for glutamic acid are GAA and GAG and aspartic acid GAU and GAC; therefore, these two new amino acids could not have arisen from single point mutations of the same codon. Glutamic acid could only arise from a point mutation of the codons GUA or GUG and aspartic acid from a point mutation of GUU or GUC. As pointed out by the authors this must indicate a polymorphism for the valine codon in the population.

Deletion of the base codon. Eleven haemoglobins are known which results from deletions of the base codons (Lehmann and Kynoch, 1976). Hb Freiburg: β A5 Val deleted (Jones *et al.*, 1966) and Hb Leiden: A3 or A4 Glu deleted (de Jong *et al.*, 1968) arise from deletion of one base codon, whereas Hb Gun Hill: F7-FG2 (or F8-FG5 or F4-FG4) deleted, arises from a deletion of five of the base codons on the F and FG segments (Bradley *et al.*, 1967).

The above variants are examples of a "clean" intracistronic deletion, for the amino acid sequence of the carboxy terminal part of the peptide chain remains unchanged. Hb Leiden, however, may not be an example of deletion but be caused by non-homologous crossing-over of the δ- and β-chain loci. The first eight residues of these chains are identical (Lehmann and Carrell, 1969). Clean deletion of the base codon may explain the differences in length of the α- and β-chains (Ingram, 1963). For example, following gene duplication, there may have been clean deletions of 12 bases from the α-chain locus. Single nucleotide deletions ("dirty" deletions) have not yet been recorded. However, as they would result in a shift of the entire nucleotide frame of the peptide, the resultant molecule would not be expected to function.

Non-homologous crossing-over (fusion mutations). The genetic segregation of the δ- and β-chain variants (Boyer *et al.*, 1963) which demonstrate that the loci for the corresponding genes are closely linked, has been substantiated by the detection of the Hb's Lepore. These variants are formed from the N terminal portion of a δ-chain and the C terminal portion of a β-chain. Haemoglobins Lepore (Boston, Washington and The Bronx) are δ-like for the first 85 to 115 amino acids (Labie *et al.*, 1966) but Hb Lepore Hollandia is δ-like only up to and between the first 25 to 50 amino acids

(Barnabas and Muller, 1962; Curtain, 1964). Hb Pylos (Fessas *et al.*, 1962) may be another example of this type of variant. Baglioni (1962) suggested on this structural evidence that these hybrid haemoglobins arose from a misplaced synapse leading to non-homologous crossing-over of β and δ genetic loci. The recombinant hybrid haemoglobin—βδ—appears to be extremely rare but two such examples, Hb P Congo (Dherte *et al.*, 1959) and Hb Miyada (Yanase *et al.*, 1969) have been reported. In heterozygotes for Hb Miyada the minor haemoglobin (17%) has been shown to have non-α-chains which are β-like from the 12th to 22nd amino acid, the remaining carboxy end being δ-like. Of great interest is the recent finding of two variants, the non-α-chains of which consist of fusions between the λ-chains and the β-chains. They are Hb Kenya (Huisman *et al.*, 1972) and Hb Steinham (Huisman *et al.*, 1975). These variants prove that the non-α-chains are linked on the same chromosome and aligned λ-δ-β (p. 600).

Chain termination variants. Four new haemoglobins have recently been described which result from additions of amino acids to either the α- or the β-chains. The first was Hb Constant Spring (Milner *et al.*, 1971). This is an α-chain variant which has 31 extra residues. In this variant the α-chain stop codon UAA has a single point mutation which now codes for glutamine. Thus the 3′ end of the RNA is now read resulting in 31 extra residues (p. 29). Others which have arisen in a similar way are Hb Icaria, Hb Koya Dora and Hb Seal Rock (Lehmann and Kynoch, 1976).

Frameshift mutations. Hb Wayne is the first example of a "dirty" deletion. In this haemoglobin there has been a deletion of one of the bases in the triplet which codes for serine at position 138 of the α-chain. The sequence of subsequent codons is moved one place to the left, including the stop codon which now codes for an amino acid (p. 29). It is obvious that such a mutation could only take place at the C terminal end of a globin chain since if it occurred at the N terminus the whole protein would be non-viable as a haemoglobin chain. Two other β-chain variants, Hb Tak and Hb Cranston, are also thought to be the result of frameshift mutations (Lang and Lorkin, 1976).

Insertion mutations. In Hb Grady three extra amino acids are inserted in the α-chain at positions 120 to 122. The tripeptide, Phe–Thr–Pro, is a reiteration of the three immediately preceding amino acids at positions 117 to 119. The mechanism of this insertion is unclear (Huisman *et al.*, 1974).

2. Frequency of variation

The frequency of haemoglobin variation is difficult to establish with certainty; it depends to some extent on the population studied For example, in certain geographical regions Hb's S, C, D and E are endemic, while in one Pacific island an α-chain variant, Hb Tongariki, is present in 10% of the population (Gajdusek *et al.*, 1967), illustrating how in an isolated area a single but harmless mutation may reach a high frequency (cf. p. 172). The only widespread survey is that of Sick *et al.* (1967) whose study by paper electrophoresis (pH 8·6) of the haemoglobins of 8000 Europeans detected 10 new variants. However, as pointed out, of the 2217 possible amino acid substitutions only 700 would result in a change in charge and be detectable by the techniques used. Also, since a change in polarity is impossible for 25% of the amino acids (those internally sited), it was likely that only 25% of the possible number of haemoglobin variants were detected by such a survey. Lehmann and Carrell (1969) concluded that probably about 1 in 200 Europeans carried an abnormal haemoglobin and the probability of the same variant arising *de novo* in two individuals is about one in a million. This estimate is supported by the frequency of identical unstable haemoglobins and haemoglobins M (Gerald and Scott, 1966) in populations in which sibship could be excluded.

The frequency with which single point mutations occur in haemoglobin chain loci must exceed that which is demonstrated by currently described haemoglobin variants, for certain base codon mutations can occur without resulting in the production of an alternative amino acid. Thus the codon for leucine could undergo nine point mutations, four of which would still code for leucine. Also, with the exception of methionine and tyrosine the codon for all the other amino acids can undergo at least one single point mutation and still code for the same amino acid. This means that even a complete amino acid analysis of the chain would fail to demonstrate 20% of possible codon mutations. Lehmann and Carrell (1969) have pointed out, however, that a much more accurate estimate of the mutational rate of the haemoglobin loci may be obtained by studying the frequency in a population of the same (highly) unstable haemoglobin. These haemoglobins, which can result even from neutral substitutions of invariant amino acids, produce haemolysis in the heterozygote. In the case of Hb Hammersmith, β CD1 Phe——→Ser, they calculated that the mutational rate was 10^{-7} per generation. Taking into account that the mutation for Hb Hammersmith is only one of 2200 possible mutations, the true frequency of mutation would be about 10^{-4} per generation. These estimates are in general agreement with those for other human genetic loci, about 10^{-5} per generation (Livingstone, 1967).

3. Implications of genetic variation

On present evidence mutations in haemoglobin chains are constant in frequency but occur randomly (no one site in the chain(s) appears to be more frequently affected than another). The persistence of a particular variant will be determined primarily by the effect, if any, on the function of the haemoglobin molecule and the environment of the carrier. For example, in Hb J Baltimore the substitution of glycine A13 by aspartic acid does not result in any defect in the haemoglobin (therefore no abnormality in the subject affected) and it could therefore persist as an acceptable mutation. Harmful variants (e.g. Hb S) may, nevertheless, persist because of the advantage to the carrier in terms of his environment. For instance, there is epidemiological evidence that the heterozygous state for Hb's S, C, D and E protects the patients from falciparum malaria (an example of balanced polymorphism; see p. 164). The mechanism of protection is, however, uncertain. In contrast, a mutation which results in an abnormality of either function or stability of the protein (without any compensatory benefit) would tend not to persist in the population.

V. CLINICAL AND MOLECULAR IMPLICATIONS OF STRUCTURAL VARIATION

Since the isolation and analysis of Hb S (Pauling *et al.*, 1949) many other abnormal haemoglobins have been sought for and found in all parts of the world and the total now surpasses 250. Fortunately, the majority appear to be harmless in the heterozygous state, but studies on their structure and functional properties have been invaluable to the understanding of the structure–functional relationship of haemoglobin and the genetic control of protein synthesis.

It was fortunate that concurrent with the detection of the abnormal haemoglobins the three-dimensional structure of haemoglobin was being determined in such detail that an atomic model of the tetramer could be built (Perutz *et al.*, 1968). It was thus possible to insert the new amino acid into the model and to determine what degree of stereochemical alteration would result and to correlate this with the laboratory and clinical effects of the particular haemoglobin variant. Thus Perutz and Lehmann (1968) showed clearly that certain rules of the amino acid structure of the molecule had to be obeyed to preserve its structural–functional properties. Substitutions of amino acids on the exterior of the molecule were found to be harmless (with certain exceptions); however, substitutions of internal amino acids or those acting as haem contacts, helical contacts or subunit contacts lead to marked stereochemical alterations, which in some instances explained the abnormality associated with the haemoglobin variant. Evolution illustrates these rules in that the important structural–functional areas of the chains contain many amino acids invariant in most species; their size and dimensions must be important. In contrast, many of the amino acids on the surface of the chain are variant in most species. The surface is of lesser importance and substitutions in this region can therefore persist.

In general terms, the severity of abnormality associated with substitution can be correlated with the variance or invariance of the amino acid which is replaced, for example:

(a) Five sites in the globin chains of all species are occupied by the same amino acid. These must be of critical importance and substitution of any one of them by a different amino acid would seriously affect the function or stability of the globin monomer.

(b) In mammalian α- and β-chains the amino acids which make haem contacts, helical contacts and contacts between unlike subunits are invariant. Replacement of any one of these is likely to alter the stability or function of the tetramer.

(c) Polar amino acids are always excluded from the hydrophobic centre of the molecule in all species. Therefore, the insertion of a polar amino acid into this region would be expected to lead to a serious defect.

(d) The polar amino acids on the surface vary markedly between species. Their substitution would probably therefore not affect the stability or function of the molecule.

Recent studies of known variants support these general rules.

A. Replacement of External Residues

Over 60 variants due to replacement of externally sited amino acids have been described (α-chain 22, β-chain 29, γ-chain 4 and δ-chain 2). Of the 57 different substitutions described by Lehmann and Carrell (1969), 45 affected amino acids variant in most species and 12 were essentially invariant. Only five of these variants are associated with disease; the others cause no abnormality and are therefore acceptable mutations. The five which are clinically harmful only result in serious disease in the homozygous state. They are Hb S, Hb C, Hb I, Hb K Ibadan and Hb Hofu. Their effects are discussed below.

1. Hb S: β(A3 Glu→Val)

Haemoglobin S was the first genetically determined structural variation of a protein to be described (Pauling *et al.*, 1949), although sickle-cell disease, now known to result from the homozygous inheritance of the abnormal gene, had been recognized 39 years previously (Herrick, 1910). The geographical distribution of the gene is similar to that of falciparum malaria; its prevalence is believed to reflect the environmental selection of a harmful mutation because of the advantages it affords the carrier against malaria (p. 164). The gene is endemic in Central Africa but is found sporadically in the Mediterranean region, South Arabia and India. In some parts of Africa (Amba in East Africa, for instance) the gene frequency reaches 40%. In the West Indies and the United States of America, the frequency in the major cities is about 10% in the Negro population, making the sickling trait one of the most common congenital abnormalities.

Homozygosity for the gene Hb β^S results in a severe chronic haemolytic anaemia, on top of which are superimposed sickle-cell (microvascular thrombotic) or aplastic crises. The haemolysis is due to sickling of the red cells in the capillaries where the oxygen tension is low; if the sickling is prolonged, the cell membranes sustain irreversible damage and the rigid sickled cells are then removed from the circulation by reticuloendothelial cells. Sickle-cell crises are caused by a gross degree of sickling cutting off the blood supply to an organ with the resultant localized necrosis or infarction. The organs affected vary with age: in a young child the bones are the most common site, whereas in the adult pulmonary infarction is most frequent.

The molecular basis of the sickling phenomenon remains unclear. Oxy-Hb A and oxy-Hb S have the same solubility. However, in the deoxy-state Hb S becomes 50 times less soluble than Hb A and concentrated solutions of Hb S undergo gelling (Singer and Singer, 1953) and tactoid formation (Harris, 1950). Murayama (1966) and Stetson (1966) have reported the presence of tubular structures in deoxygenated solutions of Hb S, and Murayama (1966) suggested that they were due to the consequence of Val–Val bridges between β^S-chains of adjacent molecules. Perutz and Lehmann (1968) offered an alternative explanation. The externally sited, non-polar, valine is unbonded in the oxy-Hb state; however, the conformational changes which take place on deoxygenation would create a "sticky" or complementary site in the next molecule (? α-chain) on to which valine could bond. This hypothesis not only explains the formation of tubules in deoxy Hb S solutions but also the modifying effect of the presence of another (normal or abnormal) haemoglobin, for on deoxygenation the intermolecular bonding would be interrupted by the second haemoglobin and only short tubules formed. This explanation is supported by the work of Charache and Conley (1964) who compared the concentration of Hb S required for gelling in mixtures of Hb S and Hb D and of Hb C and Hb F with the clinical severity of the associated disease. Hb's D, C and F all modified gelling, in that higher concentrations of Hb S were required in the mixture for gelling to take place than when Hb S was present alone. This was most marked with mixtures of Hb S and Hb F. These *in vitro* findings agree with and partly explain the milder degrees of haemolysis in patients with Hb S/C, Hb S/D and Hb S/F phenotypes.

Two variants are known in which the mutation for sickling is associated with a second mutation in the molecule: haemoglobin C Harlem; βA3 Glu——$\rightarrow$Val, E17 Asp——$\rightarrow$Asn (Bookchin *et al.*, 1966) and Hb Memphis; αB4 Glu——$\rightarrow$Gln, βA3 Glu——$\rightarrow$Val (Kraus *et al.*, 1966). Both of these variants show less gelling and are associated with mild degrees of haemo-lysis. The modifying effect of the second mutation is illustrated best with Hb Memphis. The two affected patients were homozygous for the β-chain variant but heterozygous for the α-chain variant (genotype $Hb\alpha\ Hb\alpha^{\text{Memphis}}$ $Hb\beta^{s}\ Hb\beta^{s}$). The cells carrying this haemoglobin were shown to be much less rigid in the deoxy state than cells containing Hb S alone. It thus seems that small changes in other parts of the molecule can markedly affect the "complementary" site required for tubular aggregation. This view is supported by the studies of Bookchin *et al.* (1970) who examined the modi-fied gelling properties of Hb C Harlem by mixing solutions of Hb S with solutions of Hb Korle-Bu: β E17 Asp——$\rightarrow$Asn, which has only the second mutation. Solutions of Hb Korle-Bu inhibited the gelling. These authors pointed out that the E17 region is near the complementary bond site of the A3 valine, and also that the gelling of Hb S results from a three-dimen-sional network with cross reactions between the linear polymers which are formed. It was suspected that the substitution of E17 Asp——$\rightarrow$Asn affected with secondary cross-reaction rather than the primary linear aggregation. (See May and Huehns, 1976.)

2. Haemoglobin C (βA3 Glu$\rightarrow$Lys)

This gene is found in high frequency in West Africa. It appears to be virtually confined to the populations living west of the river Niger where it can reach a frequency as high as 30%. Rucknagel and Neel (1961) have reported that the frequency of the *Hb C* gene is inversely related to the frequency of the *Hb S* gene and it has been postulated that heterozygous carriers are afforded the same protection against falciparum malaria as are Hb S carriers (Motulsky, 1964a, b). The homozygous state ($\alpha_2\beta_2^{c}$) results in a chronic haemolytic anaemia associated with splenomegaly which is much more benign than sickle-cell disease. The molecular pathology (like that of Hb S) remains unclear, but Charache *et al.* (1967) have shown that Hb C is less soluble than Hb A in red cells, haemolysates and dilute phos-phate buffers and that red cells containing Hb C/C are more rigid than normal red cells. This is attributed to a precrystalline state of solutions of Hb C at low oxygen tensions; such crystals can be seen by light microscopy within the red cells exposed to very low oxygen tension (Kraus and Diggs, 1956; Wheby et al., 1956).

3. Other pathological variants due to external substitutions

Hb I:α A14 Lys——→Glu (Beale and Lehmann, 1965) is associated with a mild haemolytic anaemia. This is thought to result from the instability of the abnormal α-chain since Lys A14 stabilizes the A and GH segments (Perutz and Lehmann, 1968). Hb Hofu:β H4 Val——→Glu (Miyaji et al., 1968) is also associated with a mild haemolytic anaemia. This is probably due to the fact that valine H4 (invariant in most β-chains) makes a non-polar contact with β A11 leucine and phenylalanine β GH5; its replacement by the polar glutamic acid would make the β-chain less stable (Perutz and Lehmann, 1968). A mild haemolytic anaemia is associated with Hb K Ibadan: β CD5 Gly——→Glu (Allan et al., 1965), and this is attributed to distortion of the CD segment by the insertion of glutamic acid (Perutz and Lehmann, 1968).

B. Replacement of Internal Amino Acids

As stressed earlier, the internally sited amino acids which act as haem contacts, helical stabilizing contacts or contacts between unlike subunits, are invariant in the α- and β-chains of most mammalian haemoglobins. Their invariance underlines their importance to the normal structure and/or function of the haemoglobin chain and tetramer. This view is supported by the abnormalities of structure or function of the abnormal haemoglobins which result from replacement of these amino acids. They have so far only been found in heterozygous combination with Hb A. This is probably because the molecules are so abnormal that the homozygous state would not support life. The isolation and study of these variants has yielded much information on the normal structure and function of the molecule. The existence of these variants in Man provides a fine illustration of molecular disease; they show how amino acid precision in certain regions of a protein is critical for the preservation of the structure and function of the molecule.

Studies on the three-dimensional model of haemoglobin (Perutz and Lehmann, 1968) have shown the stereochemical effects of the substitutions in some of these haemoglobin variants and have partly explained the functional abnormalities which are associated with them. More recently, X-ray crystallographic studies of some of them have illustrated in even greater detail how a single substitution in a critical region of the molecule can result in gross distortion of its tertiary structure even in regions distant from the site of the substitution (Perutz and Greer, 1970).

MAN—Y*

These variants resulting from replacement of internal amino acids can be considered under three categories: the haemoglobins M, the altered affinity haemoglobins and the unstable haemoglobins. In several of them more than one abnormality in the protein is found.

C. The Haemoglobins M

The haemoglobins M are genetic variants in which the Fe atoms of either the α- or β-chains are stabilized in the ferric state. As a result only half the affected haemoglobin tetramer can combine with oxygen. This results in tissue hypoxia which leads clinically to cyanosis (methaemoglobinaemia) and secondary polycythaemia. The haemoglobins M (of which so far five variants have been characterized) have been found in many parts of the world and in all ethnic groups (Jaffé and Heller, 1964; Gerald and Scott, 1966) but only in heterozygous combination with Hb A.

1. The molecular pathology

With one exception, the molecular pathology is the same in all five: a replacement of either the distal (E7) or proximal (F8) haem-linked histidine of the α- or β-chain by tyrosine. In the exception, Hb M Milwaukee, valine β E11 is replaced by glutamic acid (Table 2). Perutz and Lehmann (1968) have shown that when the distal histidines are replaced, the phenolic oxygen of the tyrosine comes to within 2 to 2·5Å of the Fe atom of the haem and is therefore within the predicted distance (2·04Å) within which an ionic link is possible. When the proximal α or β histidines are replaced, the iron atom is covalently bonded to histidine E7 and the phenolic oxygen of the tyrosine F8 forms the ionic link with the Fe atom. The occupation of the ligand site in these haemoglobins by ionic links (Fig. 5) would theoretically inhibit ligand formation with KCN or CO. However, this is not observed when the distal (E7) histidines are replaced, probably because of displacement of the tyrosine; on the other hand it has been observed when the tyrosine occupies the proximal (F8) position, where its displacement would be difficult (Perutz and Lehmann, 1968). In Hb M Milwaukee the internal projection of the polar glutamic acid into the haem pocket would seem to be potentially disastrous. However, its side chain is long enough to reach the Fe atom and to form an ionic link, thereby neutralizing its charge.

There are marked differences in the functional properties of the various haemoglobins M. All of them show a marked reduction in haem:haem interaction ($n = 1·1$ to $1·3*$). In the α-chain variants, Hb M Boston: α E7

* Ranney (1970) has questioned the significance of this since in the normal haemoglobin tetramer, in which all four haems are reactive, the n value is only three

Table 2

The haemoglobins M

Name	Substitution	Oxygen affinity	"n" value	Bohr effect	Reference
M Boston	α E7 His⟶Tyr	decreased	1·2	decreased	Suzuki et al. (1965)
M Iwate	α F8 His⟶Tyr	decreased	1·1	decreased	Hayashi et al. (1966)
					Ranney et al. (1968b)
M Saskatoon	β E7 His⟶Tyr	normal	1·2	present	Suzuki et al. (1966)
M Hyde Park	β F8 His⟶Tyr	normal	1·3	present	Ranney et al. (1968b)
					Hayashi et al. (1968)
M Milwaukee	β E11 Val⟶Glu	decreased	1·2	present	Udem et al. (cited in Ranney, 1970)

His——→Tyr (Gerald and Efron, 1961) and Hb M Iwate: α F8 His——→Tyr
(Miyaji *et al.*, 1963), the oxygen affinity and Bohr effect are both decreased
(Suzuki *et al.*, 1965; Hayashi *et al.*, 1966; Ranney *et al.*, 1968a). However in
the β-chain variants, Hb M Saskatoon: β E7 His——→Tyr (Gerald and
Efron, 1961) and Hb M Hyde Park: β F8 His——→Tyr (Heller *et al.*, 1966),
the oxygen affinity and the Bohr effect are normal (Suzuki *et al.*, 1966;
Ranney *et al.*, 1968b; Hayashi *et al.*, 1968).

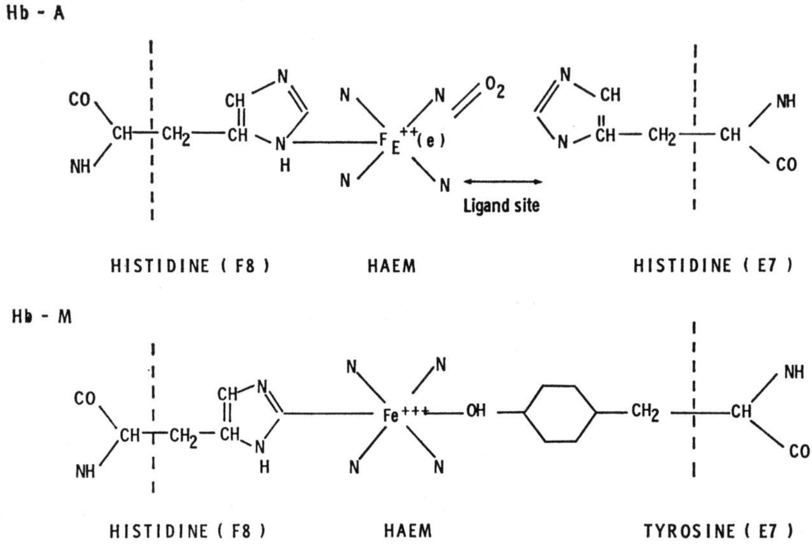

Fig. 5 Diagrammatic representation of stereochemical effect of replacement of E7
histidine by tyrosine in the haemoglobins M. The phenolic oxygen of tyrosine is
within bonding distance of the Fe^{2+} atom. When histidine F8 is replaced the
reverse situation is seen

The significance of the observed differences in the allosteric properties
of the haemoglobin tetramer according to whether the α- or β-chain is
abnormal is not clear. It appears that the preservation of the α-chain Fe
atom in the ferrous state is more important to haem:haem and proton bonding than is preservation of the β-chain Fe atom. However, the functional
changes associated with Hb M Milwaukee, which has a normal oxygen
affinity but an abnormal Bohr effect (Udem *et al.*, cited in Ranney, 1970),
do not support this concept. It should be added, however, that the stereochemical changes which result from the substitution in Hb M Milwaukee
may result in a more marked change in the quarternary structure of the
tetramer than the other β-chain Hb M's. The effect of the oxygenation of

the α-chain haems on the ionization of the terminal histidines of the β-chain (thought to be responsible for half the normal Bohr effect) has yet to be assessed.

The methaemoglobinaemia due to one of the Hb M's can be readily distinguished by changes in the spectral properties of acid methaemoglobin from that due to other causes, e.g. methaemoglobin reductase deficiency or the chronic ingestion of oxidative drugs. Those associated with the Hb M's not only differ from that of met-Hb A but are specific for each variant (Gerald and George, 1959). The abnormal haemoglobin is best demonstrated by starch gel electrophoresis at pH 7·0 of a haemolysate after conversion to acid methaematin (Gerald, 1958).

The clinical consequence resulting from the heterozygous inheritance of one of the haemoglobins M is slight. Cyanosis is usually evident at birth in the α-chain variants (met-Hb F) but does not appear until after the first three months in the β-chain variants. A blue colour (methaemoglobinaemia) is often the only presenting complaint and the patient seeks advice for cosmetic reasons. A mild haemolytic anaemia has been reported with Hb M Saskatoon (Josephson et al., 1962) and Hb M Milwaukee (Pisciotta et al., 1959), and although all the Hb M's are more heat-labile than Hb A in vitro, this does not appear to result in a severe degree of red cell destruction in vivo.

D. The Altered Affinity Haemoglobins

These are haemoglobin variants which, because of a single amino acid substitution in either the α- or β-chain, have marked alterations in the functional properties of the molecule, i.e. haem:haem interaction, Bohr effect or oxygen affinity. This class does not include the haemoglobin tetramers comprised of four like chains, Hb H (β_4) and Hb Barts (γ_4), which also show no allosteric properties; nor does it include the unstable haemoglobins, many of which are functionally abnormal as well as being unstable. Although the separation is artificial, a study of how single point mutations can alter the allosteric properties of the molecule is of great interest.

The first altered affinity haemoglobin variant described was Hb Chesapeake: α FG4 Arg——→Leu (Charache et al., 1966), which was detected in a family suffering from polycythaemia. At the present time 27 abnormal variants are known; 24 are high affinity haemoglobins (Table 3) and three low affinity haemoglobins (Bellingham, 1976).

1. Molecular pathology

It was stressed previously that the change in the configuration of the tetramer between the oxy- and deoxy-state centred around the $\alpha_1\beta_2$ ($\alpha_2\beta_1$)

Table 3

Haemoglobins with increased oxygen affinity (Bellingham, 1976)

Haemoglobin	Substitution	Site molecule affected	Whole-blood P_{33}	Bohr effect	n	Concentration[e] g dlitre^{-1} Female	Male	DPG inter-action	rbc DPG
Hb Chesapeake	FG4 α92(Arg→Leu)	$\alpha_1\beta_2$ contact	19·0	N	1·8	16·3	17·8	—	N
Hb J Capetown	FG4 α92(Arg→Gln)	$\alpha_1\beta_2$ contact	(↑)	(N)	(2·2)	13·7	15·6	—	—
Hb Malmö	FG4 β97(His→Gln)	$\alpha_1\beta_2$ contact	(↑)	(N)	(1·58)	—	19·5	—	—
Hb Yakima	G1 β99(Asp→His)	$\alpha_1\beta_2$ contact	12·0	N	1·1	—	17·2	N	N
Hb Kempsey	G1 β99(Asp→Asn)	$\alpha_1\beta_2$ contact	(↑)	(N)	1·1	19·5	20·9	—	—
Hb Ypsilanti	G1 β99(Asp→Tyr)	$\alpha_1\beta_2$ contact	(↑)	—	—	16·0	18·5	—	—
Hb Radcliffe	G1 β99(Asp→Ala)	$\alpha_1\beta_2$ contact	12·0	N	1·1	—	18·1	—	N
Hb Brigham	G2 β100(Pro→Leu)	$\alpha_1\beta_2$ contact	19·6	(N)	(↓)	15·8	18·9	N	N
Hb Georgia	G2 α95(Pro→Leu)	$\alpha_1\beta_2$ contact	(↑)	(? ↓)	(1·3)	10·8[a]	—	—	—
Hb Rampa	G2 α95(Pro→Ser)	$\alpha_1\beta_2$ contact	(↑)	(? ↓)	(→)	—	—	—	—
Hb Denmark Hill	G2 α95(Pro→Ala)	$\alpha_1\beta_2$ contact	(↑)	(? ↓)	1·8–2·4	12·8[d]	—	—	—
Hb Rainier	β145(Tyr→Cys)	C-terminal	12·9	−0·42	1·1	16·4	21·0	—	—
Hb Bethesda	β145(Tyr→His)	C-terminal	12·8	→	1·1	—	20·5	→	N

Hb			↑	→	(2·2)	16·0			
Hb Hiroshima	β146(His→Asp)	C-terminal	14·0	→	N	—	—	—	—
Hb Andrew Minneapolis	β144(Lys→Asn)	C-terminal	(↑)	→	—	—	19·8	N	—
Hb Abruzzo[b]	β143(His→Arg)	DPG-β contact	(↑)	—	(N)	—	—	—	—
Hb Little Rock	β143(His→Gln)	DPG-β contact	11·0	→	(1·1)	19·2	22·5	→	N
Hb Syracuse	β143 (His→Pro)	DPG-β contact	18·0	→	(N)	—	23·8	→	N
Hb Rahere	β82 (Lys→Thr)	DPG-β contact	{↑}	{N}	{1·2}	16·3	19·0	→	N
Hb Heathrow	β103(Phe→Leu)	haem pocket	16·4	(N)	(2·1)	17·1	21·0	—	N
Hb San Diego	β109(Val→Met)	α₁β₁ contact	(↑)	(↓)	(1·5)	12·0	—	N	—
Hb Hirose	β37(Tyr→Ser)	uncertain	18·6	N	2·3	—	13·3	—	—
Hb Olympia	β20(Val→Met)	uncertain	(↑)	N	(1·2)	—	19·4	N	N
Hb Creteil	β89 (Ser→Asn)	uncertain	(↓ ↓)	(↓)	(1·2)	—	19·5	→	N

Normal whole-blood P_{50} at pH 7·4 and 37°c = 27±2mm Hg; normal Bohr effect = −0·48±0·03; normal n value = 2·7±0·2

Abbreviations: N: normal; rbc: red blood cell; (): studies on haemolysates; {}: studies on whole cells; ↑ : increased; ↓ : decreased; ↓ ↓ : markedly decreased.

[a] Recorded as 10·8, but iron deficient blood film
[b] Associated with β-thalassaemia and raised oxygen affinity of doubtful clinical significance
[c] Hb concentrations are mean of recorded values for adults
[d] Pregnant when studied

contact. The displacement across this contact is large and is therefore probably essential for the allosteric properties of the tetramer. This view is supported by a loss or reduction of these properties in haemoglobin variants in which the site of substitution is on this contact (Table 3).

In Hb Chesapeake and HbJ Capetown (Botha *et al.*, 1966), arginine (a_1) FG4 is replaced by leucine and glutamine, respectively. Perutz *et al.* (1968) have shown that arginine (a_1) FG4 is in van der Waal's contact with (β_2) C6 and also that its guanidinium group makes a hydrogen bond with a recipient group in the β_2-chain during the change from the oxy- to deoxy-state. In these two variants in which this arginine is replaced the oxygen affinities of the molecules are high and the normal haem:haem interactions are lost. In Hb Chesapeake, however, the Bohr effect is preserved.

In Hb Yakima (Jones *et al.*, 1967; Novy *et al.*, 1967) and Hb Kempsey (Reed *et al.*, 1968) aspartic acid (β_1) FG6 is replaced by histidine and asparagine, respectively. The (β_2) FG6 aspartic acid forms a hydrogen bond with glutamic acid (β_2) G3 and could also make a new contact with (a_1) G3 valine (Perutz *et al.*, 1968); its replacement by histidine or asparagine would break these contacts. Both haemoglobins have high oxygen affinities with loss of haem:haem interaction.

In Hb Kansas (Reissman *et al.*, 1961; Bonaventura and Riggs, 1968) the site of substitution is also on the contact; (β_2) G4 asparagine is replaced by threonine. Normally asparagine (β_2) G4 forms a hydrogen bond with aspartic acid (a_1) G1 and the insertion of threonine would break this contact (Perutz *et al.*, 1968). This haemoglobin has a very low oxygen affinity and a low haem:haem interaction, but a large Bohr effect. The insertion of threonine into the contact also results in the molecule dissociating more readily into dimers, and decreases its thermostability. The two other low oxygen affinity variants are Hb Yoshizuka and Hb Agenogi (Bellingham, 1976).

Two altered affinity haemoglobins deserve further consideration. Hb Rainier: α HC2 Tyr——→His (Stamatoyannopoulos *et al.*, 1968b) and Hb Hiroshima: β H21 His——→Asp (Imai, 1968; Hamilton *et al.*, 1969) have their substitution site at the terminal portion of the H helix, not at the contact. The former haemoglobin has a high oxygen affinity and there is no haem:haem interaction. Though the site of substitution does not lie on the $a_1\beta_2$ contact, tyrosine HC2 forms a hydrogen bond with valine FG5 (a contact amino acid) and also acts as a spacer between the H and FG segment. Histidine, however, would be too short to make this contact (Perutz and Lehmann, 1968). Hb Hiroshima has a normal haem:haem interaction but has no Bohr effect and does not bind 2, 3-DPG; an observation which provides the first direct evidence as to the latter's binding site to haemoglobin.

The heterozygous carriers of a high affinity haemoglobin have poly-cythaemia and high haemoglobin levels resulting from tissue hypoxia. In contrast, carriers of Hb Kansas have cyanosis but low haemoglobin levels. Neither group, however, suffers from major disabilities and the other haematological indices are normal.

E. The Unstable Haemoglobins

The unstable haemoglobins are the largest group of haemoglobin variants which cause disease in the heterozygous state. Over 30 different variants are now known (Carrell and Lehmann, 1968; White and Dacie, 1971). Heterozygosity for one of these haemoglobins results in a congenital Heinz body haemolytic anaemia (CHBHA), the degree of severity depending on the degree of instability of the particular haemoglobin.

The isolation and analysis of these haemoglobins provided the first demonstration of the way in which a single change in either the size or charge of an amino acid in a chain could interfere with the stability of the tertiary structure of the molecule. In the CHBHA's this leads to instability of the affected protein within the red cell and its precipitation as insoluble inclusions or Heinz bodies.

CHBHA has been recognized since 1950 (Cathie), but the first indication that it might be due to an unstable haemoglobin came with the isolation of Hb Zurich: E7 His——>Arg (Muller and Kingman, 1961). Heterozygotes for this haemoglobin were found to suffer from acute haemolysis following sulphonamide ingestion. The first indication that a similar condition could result from a haemoglobin variant in which the abnormality was of a more subtle nature came from Grimes and Meisler (1962), who showed that 30% of the haemoglobin of a patient with severe CHBHA was heat labile, although no new haemoglobin could at the time be demonstrated by elec-trophoresis. A similar heat-labile haemoglobin was later shown to be present in several other affected patients (Dacie et al., 1964). The problem was resolved by the isolation and analysis of one of these haemoglobins, Hb Köln (Carrell et al., 1966). In this haemoglobin, valine β FG5 was replaced by methionine. It has been pointed out that although there was no change in charge, valine FG5 was a haem-contact amino acid invariant in all mammalian α- and β-chains; it was suggested that the instability of the protein was due to the fact that methionine was 2Å larger than valine and that its insertion into the FG segment would result in gross distortion which would weaken the binding of the haem group to the chain.

After the isolation of Hb Köln many other unstable haemoglobins were isolated from patients with CHBHA. All showed a common molecular pathology, namely, that the amino acid replaced was invariant in the chains of mammalian haemoglobins and had an important function within the molecule and that the amino acid which replaced it, either because of size or charge, weakened the tertiary folding of the chain or binding of the haem group.

1. The stability of haemoglobin

This depends basically on the combined van der Waal's forces of the internally sited non-polar amino acids which stabilize the helical segments or bind to the haem group. These forces are dependent on four factors:

(a) that the α-helical conformation of the helical segments is preserved;
(b) that all internally sited amino acids are non-polar;
(c) that the haem contact amino acids remain unchanged;
(d) that the amino acids on the $\alpha_1\beta_1$ ($\alpha_2\beta_2$) contact remain unchanged.

In all the known unstable haemoglobins one (or more) of these factors have been interfered with. To describe the molecular pathology of each of the unstable haemoglobins is beyond the scope of this chapter and only examples of each of the above categories will be given.

Substitution of a haem-contact amino acid. These variants generally result in the most severe instability and consequently the most severe degrees of haemolysis.

(a) Hb Hammersmith: β CD1 Phe—→Ser (Dacie *et al.*, 1967)

This is one of the most unstable haemoglobins. Phenylalanine CD1 is invariant in the globin chains of all species and makes an important haem contact. Serine is too short to make this contact and because its side chain is polar its presence allows the introduction of water into the haem pocket. The α-chain counterpart to Hb Hammersmith, Hb Torino: α CD1 Phe—→Val (Beretta *et al.*, 1968) is not as unstable, for the non-polar side chain of valine, although not in contact with the haem, preserves the hydrophobic nature of the haem pocket.

(b) Hb Sydney: β E11 Val—→Ala (Carrell *et al.*, 1967)

Valine E11 is an invariant haem contact: the side-chained alanine, though non-polar, is too short to reach the haem group. The loss of one haem contact results in a moderately unstable haemoglobin and a moderately severe haemolytic anaemia. In contrast, when the same valine is replaced by aspartic acid, as in Hb Bristol: E11 Val—→Asp, severe instability results (Steadman *et al.*, 1970). Aspartic acid is approximately of the same dimensions as valine and its polar side chain is too short to be swung out of the haem pocket and has to be neutralized internally, probably by histidine β E7. This results in gross distortion of the E helix and accounts for the severity of haemolysis.

Non-polar to polar substitution. The insertion of a polar residue for an internally sited non-polar residue would be energetically unfavourable to the molecule unless one of two things could take place; namely, its side chain could swing out on to the surface of the molecule or could be neutralized internally. Both would result in distortion of the helical segment, especially the latter.

(a) Hb Shepherd's Bush: β E18 Glu——→Asp (White *et al.*, 1970)

This is a moderately severe abnormality. Glycine E11 is an invariant non-polar residue. Aspartic acid is probably swung on to the surface or neutralized by histidine E17; it would not, however, distort the lower end of the E and F helices.

(b) Hb Riverdale Bronx: β B6 Gly——→Asp (Ranney *et al.*, 1968a)

This haemoglobin results in a mild degree of haemolysis. Glycine E6 is an invariant non-polar residue and the insertion of aspartic acid results in marked distortion of the segment.

Proline insertions. Proline is an imino acid and though it could be accommodated in the first three positions in the α-helical segment its insertion thereafter would disrupt the α-helical configuration (Perutz *et al.*, 1968). There are four unstable haemoglobins in which proline substitutes for leucine. In three of them, Hb Santa Ana: F4 Leu——→Pro (Opfell *et al.*, 1968) Hb Sabine: β F7 Leu——→Pro (Schneider *et al.*, 1969) and Hb Bibba: β H19 Leu——→Pro (Kleihauer *et al.*, 1968), the leucine is also an invariant haem contact. In Hb Genoa: β10 Leu——→Pro (Sansone *et al.*, 1967), however, leucine is not a haem contact and the instability of this haemoglobin illustrates how disruption of the helix by proline can result in instability of the molecule.

Substitution of stabilizing residues. These variants are usually only slightly unstable and result in a mild degree of haemolysis.

Hb Dakar: α G19 His——→Glu (Rosa *et al.*, 1968)

Histidine G19 stabilizes the G and B helices.

Hb Philly: β C1 Tyr——→Phe (Rieder *et al.*, 1969)

This substitution lies on the $\alpha_1\beta_1$ contact. Tyrosine β C1 forms a hydrogen bond with aspartic α H9. This bond is broken by the insertion of phenylalanine and the resultant haemoglobin dissociates readily into unstable α- and β-chain monomers.

Deletions. The three haemoglobins mentioned earlier, Hb Freiburg: β A5 Val deleted (Jones *et al.*, 1966); Hb Leiden: β A3 or A4 Glu deleted (de Jong *et al.*, 1968) and Hb Gun Hill (see p. 578; Rieder and Bradley, 1968), all illustrate how the breaking of the α-helical sequence by a deletion disrupts the ordered internal bonding of the affected segment and results in unstable molecule.

Clinical features. The heterozygous carriers of an unstable haemoglobin suffer from a congenital haemolytic anaemia, the severity of which correlates with the severity of the molecular lesion. There are no characteristic physical signs except the excretion of dark urine dipyrroluria) and occasionally cyanosis. The haematological findings are those of a congenital non-spherocytic haemolytic anaemia, associated in splenectomized patients with Heinz bodies in many of the circulating red cells. Heinz bodies can, however, be generated in the red cells of unsplenectomized patients by a variety of redox dyes or by sterile incubation of the blood for 24 hours at 37^0C. The specific diagnostic test for these haemoglobins is the demonstration of a heat-labilte haemoglobin and the temperature and rapidity at which precipitation takes place correlates with the instability resulting from the molecular lesion and with the clinical severity of the disease.

One advance in this field has been the recognition that superoxide ion plays an important part in the normal autoxidation of haemoglobin, and that it may damage the red cell membrane in the presence of unstable haemoglobins. Weiss (1964a, b) first suggested that the iron in oxyhaemoglobin may be in the ferric state, because an electron is transferred from it to the oxygen, which is thus changed into a superoxide ion. On transition to methaemoglobin this ion would be released into solutions and its position at the iron atom taken up by a water molecule. As many as 10^7 superoxide ions per red cell are formed daily. Within the normal cell these are rendered innocuous by the enzymes shown below, and the methaemoglobin formed on release of superoxide ion is reduced by the methaemoglobin reductase system; but in the presence of unstable haemoglobins superoxide ion, which is a powerful oxidizing agent, may be released in excess, causing irreversible oxidation of haemoglobin and of the unsaturated phospholipids of the red cell membrane.

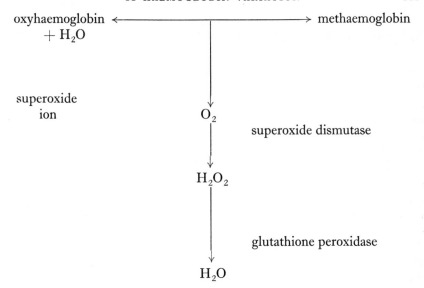

However, if the haemoglobin molecule is unstable the superoxide formed will increase to such amounts that it will oxidize haemoglobin and the unsaturated fatty acids of the membrane.

VI. GENETIC VARIATION OF GLOBIN SYNTHESIS

Over recent years studies of the genetic segregation of the haemoglobin variants and of the synthesis *in vitro* of mammalian haemoglobin chains have led to a clearer understanding of the mechanisms controlling haemoglobin synthesis (Hunt, 1976; Paul, 1976). However, there are still some aspects which await clarification. Several disorders which result from a genetically determined defect of the synthesis of globin are now recognized but, in contrast to the structural variants, the precise molecular pathology of these disorders is far from clear.

The genetic abnormalities of globin synthesis are classified according to alterations of the red cell indices, alterations of the proportions of the minor haemoglobin components in the cell (Hb A_2 and Hb F) and alterations of the rates of synthesis of the α- and β-chains. This section will consider these abnormalities but unfortunately will add little to the information provided by previous reviewers (Weatherall, 1965, 1969; Huehns and Shooter, 1965; Weatherall and Clegg, 1968; Fessas, 1968; Clegg and Weatherall, 1976).

A. Synthesis of Normal Haemoglobins

The synthesis of haemoglobin within the red cell precursor is a finely controlled process which involves the synthesis of the globin chains on cytoplasmic ribosomes, the synthesis of haem in the mitochondria and the
incorporation of iron into the cell. The synthesis of haem and the mechanisms responsible for iron incorporation are beyond the scope of this work
and only the synthesis of the globin chain will be considered.

Haemoglobin begins to be synthesized in the late basophilic or early poly-
chromatic normoblast, approximately 60 hours after differentiation from
the primitive haematocytoblast. Synthesis is maximal at about 90 hours but
then declines rapidly. The enucleated reticulocyte, however, is still capable
(at 180 hours) of synthesizing 10% of the cell's haemoglobin (Granick and
Levere, 1968). In this last respect the erythroid cell seems to be unique, in
that protein (globin chains) continues to be synthesized long after the synthesis of DNA has been terminated (80 hours); thereafter the synthesis of
the chains is dependent on the stability of the mRNA. The mechanism(s)
responsible for the initiation of haemoglobin synthesis in the maturing cell
is not known. However, it has been shown that within the chick blastoderm
globin chains are not synthesized until adequate amounts of amino laevul-
inic acid (ALA) are present (Granick and Levere, 1968). These authors
suggested that the initiating step may be the synthesis of mRNA for ALA
synthetase (the rate-limiting enzyme for haem synthesis). The synthesis of
ALA in the chick blastoderm has been shown to be induced by certain
steroids, in particular 5 β-H-androstane. These are thought to act as dere-
pressors for the operator gene controlling the synthesis of mRNA for ALA
synthetase. Such steroids, however, have not been shown to be active in
mammalian red cell precursors (Levere et al., 1967).

B. Molecular Biology of Haemoglobin Synthesis

With the isolation and purification of globin mRNA and the preparation
from this of a complementary DNA (cDNA) it has become possible to
study in some detail the expression of the globin gene which could, in
theory, be regulated at the levels of replication, transcription, processing,
translation or degradation of either nucleic acid or protein.

In a few instances, it is known that specific components of the genome
can be amplified beyond the other components (Perkowska et al., 1968).
One possibility is, therefore, that in the course of erythroid differentiation
the globin gene is replicated to a greater extent than the rest of the genome,

with the consequence that erythroid cells contain more globin genes than non-erythroid cells. This problem can be directly studied by measuring the number of globin genes using globin cDNA as a probe. When this was done with mouse tissues, it was found that the concentration of globin genes in sperm DNA, in total foetal DNA and in DNA from mouse foetal liver was exactly the same (Harrison *et al.*, 1976). This observation excludes hypotheses based on segregation of the globin genes or independent amplification of them.

To study the role of transcription, two kinds of experiments have been performed. On the one hand, the concentration of globin mRNA molecules in erythroid and non-erythroid tissues has been examined. This has shown that, whereas erythroid tissues have a very high concentration of globin mRNA, ranging from a few thousand molecules in a late erythroblast up to 150 000 molecules in the reticulocyte, globin mRNA is either totally absent, or nearly so, in other tissues. Generally, this kind of observation would argue in favour of regulation at the level of transcription of the globin gene. There is some evidence that even non-erythroid cells may have a very small number, perhaps two or three, mRNA molecules in each cell and, therefore, this control may not be absolute. It has also to be borne in mind that we have no satisfactory evidence about the rate of turnover of newly transcribed RNA, and until this has been obtained one cannot exclude the alternative possibility that there is a selective breakdown of certain species of RNA.

The alternative approach to this problem has been to study chromatin isolated from erythroid and non-erythroid cells. In several studies of this kind (Gilmour and Paul, 1973; Paul *et al.*, 1973), evidence has been obtained that the globin gene is accessible to transcription by bacterial RNA polymerase (RNA nucleotidyltransferase) in chromatin from erythroid cells, but not in chromatin from non-erythroid cells. These studies are confused by the facts that chromatin from erythroid cells often contains some globin mRNA as contamination and the amounts of globin mRNA produced are very small. Nevertheless, where these complications have been taken into account, there seems to be substantial agreement that this observation is correct. It provides very good evidence for a control at the transcriptional level determined by the structure of chromatin itself.

There is also much evidence (though it is not conclusive) that, after most RNA has been synthesized in the nucleus, it is subsequently cleaved and only a small component of the primary transcript gives rise to mRNA. In relation to globin RNA synthesis, there is very little precise experimental evidence about processing of a precursor molecule. The mRNA itself is subsequently modified by the attachment of a polyadenylic sequence to the 3' end and a modification called "capping" at the 5' end. It is not, at present, possible to state whether controls at these levels occur. At the next level, that of translation, there is substantial evidence for rate-regulating factors. There is no clear information as to whether there is normally regulation of haemoglobin synthesis by way of the translational machinery. However, if one lumps together processing and translation and considers them as post-transcriptional stages, then there is recent evidence that controls can, indeed, operate there.

C. Genetic Control

The chromosomes carrying the structural Hb loci for the α-, β-, γ- and δ-chains have not been identified with certainty, although there is some evidence to implicate chromosomes 2 and 4 (Price et al., 1972; and cf p. 436). There is also good evidence that the α- and non-α-chain loci are on different chromosomes (Deisseroth et al., 1976). This is reinforced by the segregation of the different variants, especially on certain critical families in which both types of variants were segregating. From such studies it appears that the α- and β-chains are each controlled by separate genes on different loci and that the chains are synthesized independently and associate in the cytoplasm to form haemoglobin. This concept is compatible with the three possible patterns of inheritance of α- and β-chain structural variants which have been found. The recent evidence that each mammalian γ-chain and probably α-chain are under the control of more than one locus (vide infra) does not alter this concept radically.

D. Linkage of Loci

1. α and β-chain loci

As stated above, there is very good evidence that the α- and β-chain loci reside on different chromosomes.

2. β-, γ and δ-chain loci

As stated earlier the discovery of hybrid or fusion haemoglobins (Lepore, anti-Lepore, Kenya and anti-Kenya) have provided conclusive evidence of the linkage of β-, γ- and δ-chain genes. In the haemoglobins Lepore, the non-α-chains have N-terminal δ-chain sequences and C-terminal β-chain sequences. In the two structurally analysed anti-Lepore haemoglobins

(Hb P Nilotic and Hb Miyada) the non-α-chains have *N*-terminal β-chain sequences and *C*-terminal δ-chain sequences (Lehmann and Huntsman, 1974). This strongly suggests contiguity of the δ- and β-chain genes. In Hb Kenya the non-α-chain has the *N*-terminal sequence of a γ-chain for the first 80 residues and a β-chain sequence for residues 87 to 146. Heterozygotes for Hb Kenya have haematological characteristics of hereditary persistence of foetal haemoglobin and the resulting Hb F is composed entirely of ^{G}γ-chains (γ-chains in which position 136 is occupied by glycine rather than alanine; *vide infra*). This indicates deletion of both the δ-chain gene and the γ-chain gene containing alanine at position 136 (^{A}γ). Thus the probable linkage arrangement is ^{G}γ ^{A}γδβ (Lang and Lorkin, 1976).

3. Non-allelic gene control of human γ and α-chains

It has been assumed for some time that each mammalian globin chain was under the control of a pair of allelic genes. That this assumption was incorrect has become evident from the analytical studies initially made on some animal haemoglobins and more recently on the γ- and the α-chains of human haemoglobins.

It had been shown earlier that the α-chain of mouse haemoglobin (Rifkin *et al.*, 1966; Popp, 1967), rabbit haemoglobin (von Ehrenstein, 1966) and horse haemoglobin (Kilmartin and Clegg, 1967) were composed of a mixture of α-chains, in that at least one position in the chain has a non-unique sequence of amino acids. There were three alternative explanations for this finding, namely: non-allelic genetic control, allelic genetic control and ambiguity in the reading of the base codons for these α-chains. That the first explanation (non-allelic control) was probably correct was suggested by studies of the structure of the α-chain of the goat (Huisman *et al.*, 1967). Goat Hb A contains two α-chains: ^{1}α and ^{11}α, which show 4 differences in their amino acid sequence (Huisman *et al.* 1968). Goat Hb B has an α-chain which is a single point mutation of ^{1}α (^{1}α′). It was shown that in a goat carrying both Hb's A and B, all three types of α-chain were present: ^{1}α, ^{11}α, and ^{1}α′.

Lehmann and Carrell (1968) were the first to consider that one of the human haemoglobin chains might also be under the control of a pair of non-allelic genes, i.e. the human α-chain. However, the evidence was indirect (*vide infra*). The first direct evidence for non-allelic genetic control of human chains has come from the studies on the amino acid composition of the human γ-chains. This evidence has already been extensively reviewed (Schroeder and Huisman, 1970) and only the relevant points will be discussed.

Following cyanogen bromide cleavage of foetal haemoglobin ($α_2γ_2$) the terminal peptide of the γ-chain (γCB3) can be isolated. The integral values of the amino acids of this peptide were originally thought to be: Ala_2, Arg_1, Gly_1, His_1, Leu_1, Ser_3, Thr_1, Tyr_1, Val_2 (Schroeder *et al.*, 1963). However, after careful analysis the integral values for Gly and Ala were found to be $0.7:2.3$ instead of $1:2$. The most likely explanation for this finding was that each γ-chain was under the control of multiple non-allelic genes (? four), and the observed integral values were due to mixtures of the two types of γ-chain. The alternative explanation—heterozygosity for a γ-chain variant—was excluded by finding identical ratios in over 40 different cord blood samples. Schroeder and Huisman (1970) pointed out that since alanine was present in two other positions, the integral value of glycine at position 3 was the important indicator as to the type of γ-chain present; e.g. if glycine $= 1$, glycine occupies position 3 in the γCB3 peptide ($^{G}\gamma$); if glycine $= O$, then alanine occupies this position ($^{A}\gamma$). Since the integral value for glycine was found to be 0.75, then a mixture of the two γ-chains must have been present with the ratio of $^{G}\gamma:^{A}\gamma$ being $3:1$. The final confirmation that the γ-chain was controlled by non-allelic genes was the finding that the γ-chain structural variants were single point mutations of either the $^{G}\gamma$ or $^{A}\gamma Hb$ loci (Schroeder *et al.*, 1968a). This has since been confirmed (Cauchi *et al.*, 1969; Ahern *et al.*, 1970).

Examination of the γCB3 peptide of the γ-chains of the foetal haemoglobin associated with neonates, adults and patients suffering from the thalassaemia syndromes and HPFH have shown marked differences in the Gly:Ala ratios. The significance of this will be discussed in the appropriate section.

The evidence for non-allelic control of the human α-chain is still indirect. Lehmann and Carrell (1968) and Lehmann (1970) drew attention to the fact that in heterozygotes for an α-chain variant, the abnormal haemoglobin usually only comprised 25% of the total, whereas the analogous β-chain variant comprised nearly 50% of the total. The authors interpreted this observation as good evidence that the human α-chain was under the control of four genes, two of which were probably non-allelic, and that each variant arose from a single point mutation of one of these genes. Kattamis and Lehmann (1970a, b) also postulated that non-allelic control for the α-chain would explain the clinical and biochemical heterogeneity of the α-thalassaemia syndromes. This, however, is debatable and will be discussed in more detail when the α-thalassaemia syndromes are considered. Stronger evidence for non-allelic control of the α-chain has come from the finding of a family carrying two α-chain variants. In this family two members were found who carried the two abnormal α-chains as well as the normal α-chain of Hb A (Brimhall *et al.*, 1970). Abramson *et al.* (1970) reported the finding of a similar family without detailed evidence. There is evidence that the situation may not be universal, for in a Melanesian population an α-chain variant, Hb Tongariki:α 115 Ala—→Asp, comprises 50% fo the total haemoglobin in heterozygotes and in the family studied by Abramson et al. (1970) two homozygotes were found. It should be added, however, that one of the parents of each homozygote was not available for study and although the α-thalassaemia/Hb Tongariki genotype of these homozygotes was extremely unlikely, it could not be definitely excluded (Abramson et al., 1970). There are two other α-chain variants; Hb G Chinese and Hb G Bangkok which also comprise 50% of the total haemoglobin. Lehmann has suggested that the structural gene for these mutants might be linked with the gene responsible for α-thalassaemia.

E. Developmental Control

As stated previously, five different globin chains are synthesized during embryonic, foetal and adult development. Though the anatomical site of erythropoiesis varies with the stage of development (12 weeks in liver and spleen: 40 weeks in the bone marrow), there is no evidence to indicate that any one type of globin chain is synthesized by a specific red cell precursor or from any specific site. Evidence would indicate that each red cell precursor is capable of synthesizing α-, γ- or β-chains (Dan and Hagiwara, 1967). However, this may not be true for the switch over from embryonic to foetal haemoglobin (Wood, 1976). Evidence indicates that the synthesis of the embryonic haemoglobin is confined to the yolk-sac (Kazazian, 1974).

The factors responsible for switching on and off the γ- and β-chains are unknown. The switching appears to be related only to the stage of development and is independent of such factors as maternal erythropoietin (Stohlman, 1967), prematurity and hypoxia (Thomas et al., 1960). Allen and Jandl (1960) and Necheles et al. (1965) have shown that in reticulocytes of cord blood in vitro the syntheses of Hb A and F respond independently to low oxygen tension and glucose deprivation. The significance of these findings is not clear. It also seems meaningless at present to relate the factors responsible for the switching from one haemoglobin type to another in response to anaemia in some animals, for instance, sheep (Hb C) and goats (Hb II), to the control mechanisms responsible for the γ–β switching in Man. No such change is known to exist (except Hb F in juvenile chronic myeloid leukaemia), unless the new haemoglobin is a structural variant indistinguishable from Hb A by routine techniques.

The ability of a new cell line developing in patients suffering from haematological abnormalities to synthesize Hb F is interesting but as yet unexplained. It is certainly not an invariable finding but has been obtained in several patients suffering from disorders such as aplastic anaemia, myelo-sclerosis or paroxysmal noctural haemoglobinuria (PNH). In PNH it may be related to the age of the patient, i.e. it is most frequent in young patients (White *et al.*, unpublished data).

Also of interest is the observation that in 20 to 25% of pregnant women, during the second trimester there is an increase in Hb F (Pembrey and Weatherall, 1971). The Hb F appeared to be of maternal origin and its peak level corresponded to the peak excretion of choriogonadotrophin.

The mechanism responsible for the switching from γ- to β-chain at birth not only involves these two chains but also the type of β-chains synthesized. Schroeder and Huisman (1970) have shown that at birth the Hb F γ-chain has a Gly:Ala ratio of 3:1. As the percentage of Hb F falls so the ratio alters and at five months when the level of Hb F is 2% the Gly:Ala ratio is 2:3 where it remains throughout adult life. (This finding is interesting because a Gly:Ala ratio of 2:3 could not arise if a mixture of only four γ-chains was present and suggests that the γ-chain may be under the control of more than four genes unless each gene is expressed differently.) There is no evidence to indicate that the γ-chains of the Hb F of the foetus are under the same genetic control as the γ-chains of the Hb F of the adult. The alterations in the Gly:Ala ratio would indicate that they are not. Indeed, in those haema-tological disorders associated with increased levels of the Hb F the Gly:Ala ratio of the γ-chain reverts back to the foetal type (3:1), indicating that the genetic loci for the foetal γ-chains can be reactivated under certain circum-stances. A more recent review of the problems involved is given by Wood (1976), whose final paragraph is summarized:

> The problem of the control of the switch needs to be understood at both the molecular and cellular levels. At the molecular level we know very little about the regulation of the $^{G}\gamma-^{A}\gamma-\delta-\beta$ gene complex. Several hypotheses have been proposed to explain what occurs during the switch and in those genetic dis-orders associated with the production of Hb F in adult life. These hypotheses include the presence of a single promoter locus at the beginning of the (ϵ?)–$^{G}\gamma-^{A}\gamma-\delta-\beta$ complex, with a progression in the actual gene transcribed being controlled by a process of intra-chromosomal crossing-over (Kabat, 1972). Alternative ideas invoke a series of controller genes between the globin struc-tural genes, with a series of overlapping deletions to explain the observed phenotypes (Huisman, 1974; Nigon and Godet, 1976). The rapidly develop-ing techniques of nucleic acid hybridization, gene amplification and nucleic acid sequencing, coupled with the development of an *in vitro* transcriptional system, may soon lead to new insights into the molecular aspects of the switch.
>
> At the cellular level there are several important questions requiring answers. We need to know at which level of haemopoietic stem-call differentiation the type of haemoglobin to be synthesized is determined and whether this can be altered during the further maturation of the cell. Secondly, we need to know

what factors are responsible for initiating the switch—humoral factors localized tissue agents (cell-cell interaction) or possibly some form of an internal clock? Evidence that can be interpreted in favour of a humoral factor, and against localized cell effects, comes from the increase in maternal Hb F synthesis during the first trimester of pregnancy (Pembrey *et al.*, 1973) and the apparent synchrony in the switch from Hb F to Hb A in different erythropoietic sites (Wood and Weatherall, 1973). The question of whether it is possible to reverse the switch in an individual stem cell, or whether the programming of a cell for β and δ-chain production is a permanent commitment, is extremely important in considering the exciting therapeutic potential of manipulating the switch to increase the production of Hb F in those severe haemoglobinopathies with defective β-chain production. If the switch is reversible in an individual stem cell, then the problem of increasing Hb F production lies at the molecular level, in increasing the transcription of the γ-chain genes relative to the β-chain gene. If, as seems more likely, the switch is irreversible, augmenting Hb F production in adults would involve increasing the contribution of those clones of cells in the adult which retain the capacity to produce Hb F (Weatherall and Clegg, 1972). Clearly these are two quite different propositions. The difficulties which would appear to surround augmenting the contribution of F-cell clones in adults make it likely that the best approach to this problem would be to attempt to arrest the switch from Hb F to Hb A production in the newborn, while the proportion of Hb F produced is still high.

F. Cytoplasmic Control

1. Gene level

Little is known about the nuclear (gene) control of chain synthesis. As discussed already in Chapters 2 and 10, there is no evidence that the concept of the operon (Jacob and Monod, 1961), thought to be the basic unit of regulation of protein synthesis in bacteria, is relevant to mammalian protein synthesis. Indeed, as Weatherall and Clegg (1968) have indicated, the structure of the Lepore haemoglobins must throw doubt on the validity of such a mechanism in Man. They argue that if the αβ locus is the structural gene of the operon, closely linked to an operator gene, it would then seem reasonable to assume that a single polycistronic mRNA is produced for these chains. The δ-chain would be read first and then the γ-chain. This is difficult to accept because of the differences in assembly of these two chains, namely the amount of δ-chain is assembled at a slower rate and the synthesis of the δ-chain is switched off much earlier than that of the β-chain (Winslow and Ingram, 1966). The only evidence for an operon-type mechanism in the erythroid precursor is in the synthesis of the mRNA for ALA synthetase (Granick and Levere, 1968), where in the chick blastoderm haem has been shown to act as a co-repressor and certain steroids as a derepressor for the synthesis of ALA.

2. Kinetics of chain assembly

The assembly of the globin chains in human and rabbit erythrocyte precursors is understood in greater detail than some other aspects of chain synthesis. However, major advances have been made over the last few years. (For comprehensive reviews see Proudfoot and Brownlee, 1976; Hunt, 1976.) For instance, the structure of the mRNAs and tRNAs required for globin chains in Man has now been sequenced. The α-chain messenger contains 640 nucleotides and the β-chain 740. Thus, there are two regions, coding and non-coding. The difference in the size of the two messengers lies in the non-coding regions (α-chain 220 and β 300). The 3' end of the mRNA of both chains is rich in adenine (poly A). This is probably put into the mRNA after it has been transcribed from the nuclear DNA (Birnboim et al., 1973).

More information is available concerning the nature of the mammalian ribosomes. They exist free within the cytoplasm of the erythroid precursor as aggregated units or polysomes. Polysome profiles have shown that the commonest aggregates are comprised of 4 to 6 ribosomes, but aggregates consisting of 10 ribosomes down to single ribosomes are also found. Hunt et al. (1968) have shown that β-chain synthesis is associated with the 6 to 8 unit polysomes, whilst α-chain synthesis is associated with the 4 to 6 unit polysomes. It appears likely that whilst the rate of assembly of α- and β-chains are the same, the ribosomes initiate synthesis of β-chains more readily than α-chains. However, the final product, i.e. α-chains and β-chains, are produced in equal amounts. This implies that there is more α-chain mRNA than β mRNA. It seems likely that disaggregation of polysomes represents one of the control mechanisms for terminating haemoglobin synthesis though the factors responsible are not known. Changes in polysome profile are seen during the in vitro incubation of reticulocytes (Knopf and Lamfrom, 1965), in that during the first 20 minutes there is an increasing aggregation of polysomes, but after 60 minutes disaggregation begins to take place. Disaggregation of polysomes also takes place in vivo during the maturation of the cell, for reticulocytes contain five times fewer polysomes than the nucleated precursors (Rifkind et al., 1964).

The role of formyl-methionine (important in E. coli) in the initiation of chain synthesis in mammalian cells (Clark and Marcker, 1966) has not yet been clearly defined but it is important. The globin chains are assembled from the N-terminal end by a stepwise addition of amino acids (Dintzis,

1961; Naughton and Dintzis, 1962), and the rate of assembly is linear. In the rabbit reticulocyte the rate of assembly of the chains is five to seven amino acids per second (Knopf and Lamfrom, 1965). Although the mechanism responsible for chain termination is not known, the codons UAG, UGA or UAA, thought to be responsible for *E. coli* chain termination, are probably also responsible in mammalian cells (Garen, 1968).

Short-term incubations with radioactive amino acids of human and rabbit reticulocytes have shown that a small α-chain pool exists in the cytoplasm, associated with the ribosomes (Baglioni and Colombo, 1964; Weatherall *et al.*, 1965). From these data it has been suggested that α-chains are released first, and then returned to the β-chain ribosomes to "pick up" nascent β-chains (Baglioni and Campana, 1967). Though this hypothesis is attractive, α-chains are not essential for β-chain synthesis and/or release, for in the absence of α-chains (α-thalassaemia) the β- and γ-chains continue to be synthesized and accumulate within the cell as haemoglobin tetramers.

The final stage in haemoglobin synthesis is the insertion of haem into the globin chains. Where this insertion takes place in relation to the assembly of the chains is unknown, but it most likely takes place after the chains are released into the cytoplasm rather than when they are being assembled in the ribosome (Felicetti *et al.*, 1966). Winterhalter *et al.* (1969) have postulated that the insertion of haem into the αβ dimer is an ordered process involving four steps as shown (H = haem):

$$\alpha\alpha\beta\beta \longrightarrow \alpha^H\alpha\beta\beta \longrightarrow \alpha^H\alpha^H\beta\beta \longrightarrow \alpha^H\alpha^H\beta^H\beta \longrightarrow \alpha^H\alpha^H\beta^H\beta^H$$

It should be added, however, that there is no reason on thermodynamic grounds why haem should not be inserted randomly into each chain of the tetramer after formation of the haem pocket.

The currently accepted theory that haem regulates the synthesis of the globin chain is of great interest but a full discussion is beyond the scope of this chapter. However, there have been many major advances which are summarized by Hunt (1976). Undoubtedly, in the absence of haem, globin chain synthesis is quantitatively retarded and the only direct explanation for this is a dissociation of the polysome units. *In vitro* haem will stimulate globin synthesis though the mechanism has not been elucidated. Such factors as stabilization of mRNA on ribosomes, increase in chain assembly time and facilitation of the initiation of chain synthesis have been thought to be responsible, but precise evidence is lacking. The recent observation that haem stops the function of an inhibitor of globin synthesis may be a more likely explanation. There is now good evidence that metabolites within the red cell, for instance oxidized glutathione, will inhibit protein synthesis. Of more interest is that even in the absence of haem cyclic AMP will stimulate the production of both chains, the reason for which is unclear (Hunt, 1976).

VII. GENETIC DISORDERS OF GLOBIN SYNTHESIS

There are three main groups of disorders which result from a defective synthesis of one or more of the globin chains, namely the α- and β-thalassaemia syndromes, the Hb Lepore syndromes and the hereditary persistence of foetal haemoglobin (HPFH). Each represents a heterogeneous group of genetically determined abnormalities which can be subdivided into distinct syndromes on clinical, haematological and biochemical grounds. Within the last few years major advances have been made in the understanding of these disorders and they can now be defined at a molecular level. For a recent appraisal of the subject readers are referred to Clegg and Weatherall (1976) and Weatherall (1977).

A. The Thalassaemia Syndromes

It was first suggested by Itano (1957) that the defect responsible for the condition called thalassaemia was a reduction in the rate of globin synthesis. This hypothesis was extended by Ingram and Stretton (1959) who postulated that there were probably two main types, one affecting the rate of synthesis of the β-chain (β-thalassaemia) and the other the rate of synthesis of the α-chain (α-thalassaemia). Several workers have now confirmed this by measuring the rates of synthesis of the α- and β-chains *in vitro* in patients suffering from thalassaemia (Heywood *et al.*, 1964; Bank and Marks, 1966; Weatherall *et al.*, 1965; Bargellesi *et al.*, 1968).

The defect of chain synthesis is genetically determined and is associated with alterations in the haematological indices, the red cell morphology and the proportions of the minor haemoglobin components of the cell (Hb A_2 and Hb F). The recognized genetic variants of the thalassaemia syndromes have already been extensively reviewed (Motulsky, 1964a; Nathan and Gunn, 1966; Weatherall, 1965, 1967, 1969; Fessas, 1968; Weatherall and Clegg, 1972; Clegg and Weatherall, 1976; Weatherall, 1977) and therefore they will only be summarized in this section.

1. The β-thalassaemia syndromes

These syndromes result from a genetically determined defect of the synthesis of the β-chains (or β- and δ-chains). Since this chain is under the control of a pair of allelic genes, affected patients can be either heterozygous or homozygous for the abnormal gene. The current classification of the β-thalassaemia syndromes is based on changes in the levels of Hb A_2 and Hb F in the heterozygote, and on the amount of β-chain synthesized in the homozygote (Table 4).

Table 4

β-Thalassaemia syndromes[a]

Genetic variant	Heterozygous state	Homozygous state
β°-Thalassaemia	increased Hb A_2 only	severe disease; Hb F and Hb A_2 only
β°-Thalassaemia (mild)	increased Hb A_2; increased Hb F (5 to 15%)	moderate disease; Hb F and Hb A_2 only
β±-Thalassaemia	increased Hb A_2 only	severe disease: high Hb F; some Hb A
βδ°-Thalassaemia (F-thalassaemia)	normal or low Hb A_2; increased Hb F (5 to 15%)	severe to moderate disease; Hb F only
βδ+-(?Silent thalassaemia gene)	normal Hb A_2 and Hb F	not described
Hb Lepore syndromes	low Hb A_2; Hb F increased; Hb Lepore 15%	severe disease: Hb F and Hb Lepore only (30%)
δ-Thalassaemia		normal levels of Hb A_2 and Hb F; thalassaemia-like picture

[a] Compiled from data cited in Weatherall and Clegg (1968), Weatherall (1969) and Fessas (1968)

β-Thalassaemia. This genetic variant results in a defective synthesis of the β-chain only. In the heterozygote (β-thalassaemia trait) it is associated with an increase in the amount of Hb A_2, usually to levels of 3·5 to 7%, with only a moderate increase (if any) in the level of Hb F. In the homozygote it results in a severe anaemia associated with a marked increase in the amount of Hb F but only minor increases in the level of Hb A_2. This genetic variant can be further sub-divided into two types; the non-β-chain producers and the β-chain producers:

(a) Complete suppression of β-chain synthesis ($β^0$)
There is good evidence that the gene involved completely suppresses β-chain synthesis. This has been obtained from the finding that double heterozygotes for β-thalassaemia and Hb S or Hb C do not synthesize any

Hb A and also that some patients homozygous for β-thalassaemia only synthesize Hb F and Hb A_2 (Fessas and Karaklis, 1962; Conconi et al., 1968;
Schokker et al., 1966). Here is some evidence that there is another genetic
variant of this type which only results in a mild degree of anaemia.

(b) Partial suppression of β-chain synthesis $(\beta\pm)$

The finding of Hb A in patients homozygous for β-thalassaemia, and double
heterozygotes for β-thalassaemia and Hb S or Hb C, indicates that there is
at least one variant of the β-thalassaemia gene which does not cause complete suppression of β-chain synthesis. The evidence that this gene is
distinct from the β^0-thalassaemia gene is convincing. Fessas (1968) has
reported on the amount of Hb A in double heterozygotes for β-thalassaemia
and a haemoglobin structural variant. These patients could be separated
into two groups; those that produce 20% or more of Hb A and those that
produce only 0 to 5%. The amount of Hb A produced by any one patient
was found to be constant within their family, which strongly indicates that
separate genes are involved. It is not known at present whether or not
separate genes are also to be found amongst the group who synthesize 0 to
5% of Hb A.

$\beta\delta$-Thalassaemia (F-thalassaemia). This genetic variant is recognized by a
thalassaemic-like picture, associated in the heterozygote with low or normal
levels of Hb A_2 and raised levels of Hb F (5 to 20%) and in the homozygote
with an absence of Hb A and Hb A_2 (Stamatoyannopoulos et al., 1969a).

(a) Suppression of β-chain synthesis

Evidence for complete suppression of $\beta\delta$-chain synthesis has resulted from
the findings of a complete absence of Hb A in double heterozygotes for this
gene and a β structural variant (Silvestroni and Bianco, 1964; Stamatoyannopoulos et al., 1968a); a complete absence of Hb A_2 in double heterozygotes for this gene and a δ-chain structural variant (Comings and Motulsky,
1966) and an absence of Hb A and A_2 in homozygotes (Stamatoyannopoulos et al., 1969a). The molecular defects responsible for this type of
thalassaemia are not known but three alternative mechanisms have been
postulated. (First, there is the suggestion that unequal crossing-over
between δ and β loci (i.e. Hb's Lepore) results in a highly unstable $\delta\beta$
mRNA or a highly unstable $\delta\beta$ hybrid haemoglobin. The second possibility
is a coupling of the β-thalassaemia gene with the δ-thalassaemia gene, which
because of the close linkage of the structural loci would always segregate
together. Thirdly, the $\delta\beta$ loci may have been deleted. The last of these
alternatives (indeed $\beta\delta$-thalassaemia), would be analogous to the postulated genetic defect for hereditary persistence of foetal haemoglobin and

will be discussed in more detail in that section (Stamatoyannopoulos *et al.*, 1969b).

(b) Partial suppression of $\beta\delta$-chain synthesis
The existence of this genetic variant has been suspected on purely haematological evidence for a number of years, for it was known that some children with typical thalassaemia major had one parent in whom the levels of Hb A_2 and Hb F were normal. The existence of this gene has been proven by Schwartz (1969). We have recently seen two further cases.

δ-*Thalassaemia*. This appears to be a rare genetic variant and it is uncertain if δ-chain deficiency *per se* would result in alteration of the haematological indices and red cell morphology (Fessas and Stamatoyannopoulos, 1962). Thompson *et al.* (1965) have reported a patient heterozygous for HPFH in whom no Hb A_2 was found and assumed that he was also homozygous for δ-thalassaemia. Ohta *et al.* (1970) have reported one homozygote for this gene in which Hb A_2 was absent, but it was not associated with any clinical abnormality.

Haemoglobin–Lepore syndromes. These syndromes which are thought to be due to a misplaced synapse and non-homologous crossing-over of the $\delta\beta$ loci during meiosis, result in a thalassaemia-like state and a hybrid haemoglobin. The structure of these hybrid haemoglobins has already been discussed and only how the abnormal fused $\delta\beta$ locus results in defective synthesis of β- and δ-chains will be considered.
Heterozygotes for this gene have a thalassaemic disease associated with a low or normal level of Hb A_2, a raised level of Hb F and about 10 to 15% of Hb Lepore which migrates on electrophoresis like Hb S. The homozygous state results in a severe thalassaemic disorder associated with only Hb F and Hb Lepore (30%) and an absence of Hb A and Hb A_2. Further evidence that the abnormal haemoglobin locus cannot synthesize β-chains or δ-chains is the absence of Hb A in double heterozygotes for this gene and Hb S or Hb C (Stamatoyannopoulos and Fessas, 1963; Silvestroni *et al.*, 1965a; Ranney and Jacobs, 1964). The reason for the defective synthesis of this hybrid haemoglobin appears to be due to the fact that the hybrid haemoglobins have the 5' part of the mRNA of the δ which is unstable. However, the anti-Lepore haemoglobins are also synthesized in small amounts which indicates that the 3' end also is implicated in the stability of the total mRNA. It is likely that there is a stereochemical interval between the 5' and the 3' end of the mRNA which protects it from degradation by enzymes within the cell.

2. The α-thalassaemia syndromes

The molecular pathology of α-thalassaemia has now been defined at least in patients from South East Asia. It has been shown that it is due to a deletion of one or more of the α-chain genes (p. 619). However, the genetics of the α-thalassaemia syndromes are far from clear at the present time. This results from a lack of knowledge concerning the number of loci controlling α-chain synthesis in different populations, the paucity of α- chain structural variants and the heterogeneity of the α-thalassaemia syndromes in different ethnic groups.

One concept of the genetics of the α-thalassaemia syndromes is that formulated by Wasi et al. (1964) which is based on the assumption that the α-chain was under the control of two genes, either of which could be altered, giving two distinct types of thalassaemia, designated α1-thal and α2-thal. These two types of α-thalassaemia genes were distinguished by the proportion of Hb Barts (γ_4) in cord bloods of neonates and the clinical and haematological findings of adults. The various α-thalassaemia syndromes were determined by which of these genes were inherited (Table 5). Although this genetic pattern appears to explain the heterogeneity of the α-thalas-saemia syndromes found in South East Asia, it may not be true for all racial groups. For example, the α-thalassaemia syndromes in Greece would seem to be equally well explained by assuming that the α-chain was under the control of four genes (Kattamis and Lehmann, 1970a, b) and that the various α-thalassaemia syndromes were determined by the number of genes affected. The problem becomes even more complex when the α-thalas-saemia syndromes found in Negro populations are considered, for as Weatherall (1963) showed, though α-thalassaemia trait is common, Hb H disease is extremely rare.

This section will consider the classical concept of Wasi et al. (1964) that there are two α-thalassaemia genes the segregation of which (e.g. α1 thal/α1 thal; α1 thal/α2 thal; α1 thal/α; α2 thal α) determines the clinical picture. This will then be evaluated in terms of the four genes per α-chain hypothesis of Kattamis and Lehmann (1970a, b).

Genotype α1 thal/α1 thal. This genotype results in the so-called Hb Barts hydrops fetalis syndrome, a common cause of foetal mortality in South East Asia. Affected foetuses are usually stillborn, and severely hydropic between 28 to 36 weeks of intrauterine development, although some survive until term but die shortly after. Examination of the haemoglobins of the cord blood of these infants shows a complete absence of Hb A; the only haemo-globins found are Hb Barts (γ_4), Hb H (β_4) and a new haemoglobin, Hb Portland (the structure of which has not been identified). The parents of

Table 5

The α-thalassaemia syndromes

Genetic pattern according to Wasi et al. (1964) (two genes per α-chain) Genotype	Clinical syndromes		Genetic pattern according to Kattamis and Lehmann (1970a,b) (four genes per α-chain) Genotype
	Neonatal findings	Adult findings	
α1 thal/α1 thal	Hb Barts hydrops fetalis syndrome, infant stillborn 28 to 40 weeks with only Hb Barts (γ_4) and Hb H (β_4)		four genes affected
α1 thal/α2 thal	moderate anaemia; 10 to 25% Hb Barts	moderate anaemia associated with 25% Hb H, majority of red cells contain Hb H	three genes affected
α1 thal/α	no anaemia: 5 to 10% Hb Barts	no anaemia, Hb H found in 1 to 3/1000 red cells	two genes affected
α2 thal/α	no anaemia; 1 to 2% Hb Barts	not detectable	one gene affected
α2 thal/α2 thal	not described	not described	no genes affected

the affected infants usually have the characteristic findings associated with the α1 thal trait (α1 thal/α), and the child is considered to be homozygous for the α1 thal gene. Hb Barts hydrops fetalis has been reported amongst Greeks but not amongst Negroes, although in the latter race the α1 thal/α genotype is common.

Genotype α1 thal/α2 thal (Hb H disease). This is defined by a moderate anaemia in the newborn associated with 25% Hb Barts in the cord blood. In adult life it results in a moderate thalassaemic disorder associated with 10 to 15% of Hb H. Though the genetics remain unclear it may be interpreted as representing the double heterozygous state for the α1 thal and α2 thal genes (Wasi et al., 1964, 1969; Huehns, 1965). Usually one of the parents will be found to have the typical haematological findings associated with α1 thal trait; however, the other parent is usually found to have normal haematological indices with no Hb H and is thought to be heterozygous for the α2 thal gene; Schwartz et al. (1969) have presented biochemical evidence to support this. They studied the rates of synthesis of the α- and β-chains in a family with two siblings of which one had Hb H disease. The ratio of α:β chain synthesized in the siblings with Hb H disease was approximately 0·4, that of the parent with α1 thal trait 0·8, and that of the α2 thal trait 0·85 to 0·9. That Hb H disease can be transmitted directly from the parent to siblings indicates that these two genes are not allelic or, alternatively, that the frequency of one of them must be high in populations where Hb H is common.

Genotype α1 thal/α. This is the so-called classical α1 thal trait. It is recognized in the neonate by the detection of 5 to 10% of Hb Barts in the cord blood. In adult life it results in a very mild thalassaemic condition associated with only minor changes in red cell indices and osmotic fragility, though Hb H inclusions can be detected in 1 to 3 in 3000 red cells. This type of α-thalassaemia is usually found in one of the parents of a child affected with Hb H disease and in both parents of children with Hb Barts hydrops fetalis.

Genotype α2 thal/α. This condition is only recognized in the neonate by the finding of 1 to 2% of Hb Barts. In adult life it is undetectable. One of the major objections in regard to this concept of α-thalassaemia is that the homozygous state for this gene (α2 thal/α2 thal), which according to Wasi et al. (1969) should be found in 1 in 3000 cases in Thailand, has not been described. These authors postulated that the abnormality associated with this genotype is probably too mild to detect. However, this is difficult to accept if in the heterozygous state it can be detected by an increase in the amount of Hb Barts in the cord blood of neonates.

Though the above genetic pattern of α-thalassaemia may explain the clinical syndromes found in South East Asia, geographical differences may exist and the alternative postulate of Kattamis and Lehmann (1970a, b) of four α-chain loci to explain the clinical heterogeneity of α-thalassaemia must be considered. The four recognized types of thalassaemia would be explained by how many of the four genes were affected in any one patient, i.e. Hb Barts hydrops fetalis—all four genes affected, Hb H disease—three genes affected, α1 thalassaemia trait—two genes affected and α2 thalassaemia trait—one gene affected. This situation is certainly compatible with the genetic pattern of thalassaemia in the Mediterranean and with DNA:DNA hybridization studies (p. 618).

3. Linkage of α- and β-thalassaemia loci with structural loci
The linkage between the genes responsible for structural mutation of the globin chains is a common occurrence where these two genes exist in high frequency.

β-Thalassaemia. The evidence of Rucknagel and Neel (1961) and Motulsky (1964b) indicates that the gene responsible for β-thalassaemia is closely linked to the β-chain structural gene, at least for Hb's S and C, but as would be expected the α-thalassaemia locus shows no close linkage. Where β-thalassaemia and β-chain variants are segregating, the effect is a reduction in the synthesis of the normal β-chain with an increase in the synthesis of the mutant haemoglobin and foetal haemoglobin. The clinical effect appears to be dependent on the amount of Hb A synthesized. Africans, doubly heterozygous for Hb S and β-thalassaemia, can be divided broadly into two groups; those who synthesize 25% of Hb A (mild disease) and those who synthesize 0 to 5% (severe disease). In the Far East (Hb E) β-thalassaemia is a fairly common condition, which results in an anaemia as severe as thalassaemia major.

α-Thalassaemia. There are several instances reported of linkage of the α-thalassaemia gene with α-chain variants (Hb I and Hb Q), β-chain variants (Hb S and Hb C), hereditary persistence of foetal haemoglobin and βδ-thalassaemia. The gene for α-thalassaemia is closely linked to that of the α-chain structural gene in that, where α-thalassaemia has been inherited with Hb I and Hb Q, very little Hb A is produced, the major haemoglobin being Hb Q or Hb I usually associated with some Hb H. Where the α-thalassaemia gene is found in association with a β-chain variant the result is a decrease in the amount of the mutant haemoglobin synthesized. This has been explained by the greater affinity of the normal β-chains than mutant β-chain for the relatively few α-chains. The combination of α- and β-thalassaemia is interesting. Fessas (1961) reported three such cases in which the clinical abnormality did not appear to be any more severe than that seen when only one of these genes was inherited. Nathan and Gunn (1966) measured the rates of synthesis of the α- and β-chains in such affected patients and showed that the α- and β-chains were synthesized in equal amounts *in vitro*. The authors postulated that the benign clinical abnormality could be explained by the fact that there was no imbalance of chain synthesis. This, however, must now be questioned in the light of the more recent finding that in two such affected patients, though the rates of synthesis of the chains were equal, both had a severe thalassaemic disorder (Wetherall, personal communication).

4. Hereditary persistence of foetal haemoglobin

This condition is considered with the thalassaemia syndromes because there is a defective synthesis of the β- and δ-chains; however, in contrast to thalassaemia, γ-chain synthesis is increased so as to compensate fully for the deficiency of the β- and δ-chains, i.e. there is no imbalance of chain synthesis.

The clinical condition resulting from the heterozygous or homozygous inheritance of this gene is extremely benign. Even in the homozygous state the only abnormality is a mild polycythaemia due to the high oxygen affinity of the red cells. The homozygous state is rare but one such patient has been well documented (Charache and Conley, 1969). The patient had a mild polycythaemia which was shown to result from the high oxygen content of his cells. These authors did not determine the origin of this but it is likely to be due to the inability of foetal haemoglobin to bind 2, 3-DPG. The cases which have been reported have haematocrits between 40 and 45%. One distinguishing feature of this condition is that the Hb F is distributed evenly throughout the red cells thus making this condition readily distinguishable from other casues of high haemoglobin F, namely thalassaemia.

Genetic defect. The gene is found in both Negro and Greek populations, the incidence in the former being 2%, but has also been detected sporadically in other ethnic groups (Charache and Conley, 1969). The evidence that this gene can be associated with a complete suppression of the β- and δ-chains stems from the findings of a complete absence of Hb A and Hb A_2 in the homozygous state and an absence of Hb A in double heterozygotes for this gene and Hb S and Hb C. There is now substantial evidence for heterogeneity of this genetic disorder. At a simple level there is a distinct subdivision between the Greek and Negro variety. The majority of Greeks have only 10 to 20% of Hb F whereas the Negro patients usually have higher levels, 20 to 35% (although some Negroes with only 10 to 20% have been reported). Also an American Caucasian family who appear to be another sub-class with low levels of Hb F has also been reported (Schroeder and Huisman, 1970). Further evidence for genetic heterogeneity has come from studies on the Gly:Ala ratios of the γ-chains of affected patients. Using this information Schroeder and Huisman (1970) were able to divide this condition into six sub-classes. In those patients with low levels of Hb F (10 to 20%) the γ-chains were shown to be either $^G\gamma$, $^A\gamma$, or $^A\gamma$ with small amounts of $^G\gamma$. In the high Hb F group (20 to 30%) the $^G\gamma$-and $^A\gamma$-chains are found in approximately equal amounts.

It is of interest that homozygous and double heterozygotes have only been described in the high Hb F group 20 to 35%), and it is not known whether some β- and δ-chains are synthesized in the low Hb F group.

The genetic defects responsible for this condition are not clear but three have been postulated: first, abnormalities in the operator or regulator genes; secondly, some form of allelism and thirdly, deletion of the δβ Hb loci. Nance (1963) has postulated that the δβ genes were aligned on the chromosome as -γδβ-; in the light of recent evidence of two γ-chain genes the arrangement would have to be $^G\gamma^A\gamma\delta\beta$ or $^A\gamma^G\gamma\delta\beta$. If the δβ loci were deleted only the $^G\gamma$ and $^A\gamma$ loci only would remain active. This would explain the equal amounts of $^G\gamma$ and $^A\gamma$ foetal haemoglobins in the group which synthesize high amounts of Hb F (20 to 35%) but complete deletion of these loci in the lower Hb F groups is difficult to explain on this concept. Since it seems likely there is some synthesis of β- and δ-chains complete deletion of these loci could not have taken place.

As mentioned previously, the condition is benign in that even in the homozygous state there is a full compensation of γ-chain production. Though the reasons for this are completely unknown, the obvious correlation between this genetic abnormality and βδ-thalassaemia is interesting and many important answers as to the aetiology of the thalassaemias must lie in the differences of the respective molecular pathologies of these two genetic disorders.

The gene is inherited as a single autosomal factor and is closely linked to the Hb S and Hb C genes. It has been described in the homozygous state (Hb F/F), heterozygous state (Hb A/F) and with Hb S, Hb C, β-thalassaemia and α-thalassaemia. Its combination with Hb S and C requires further consideration. This genotype results in a benign condition and though the haemoglobin patterns are similar to those associated with the genotypes Hb S/S, or Hb S/β thalassaemia, there is no evidence of anaemia or increased haemolysis. The benign clinical state appears to be adequately explained by the in vitro findings of Charache and Conley (1964) in that a solution of Hb F will reduce the gelling properties of Hb S.

MAN—Z*

5. The biochemical defect in thalassaemia

The studies of the relative rates of synthesis *in vitro* of the α- and β-chains in reticulocytes and bone marrows of patients with thalassaemia have shown conclusively that the basic abnormality in this condition is a defective synthesis of either one of these chains.

In homozygous β-thalassaemia the relative excess of α-chain over β- and γ-chains has been found to be greater by the order of three to five fold (Bank and Marks, 1966; Bargellesi *et al.*, 1968; Heywood *et al.*, 1964; Weatherall *et al.*, 1965). In certain instances a complete absence of β-chain synthesis has been demonstrated (Bargellesi *et al.*, 1967; Bank *et al.*, 1969). The rate of synthesis of the α-chain has also been shown to exceed the β-chain in the heterozygous state, with α:β ratios ranging between 1·6 to 1·9.

It has been demonstrated *in vitro* that the excess α- chains become associated rapidly with the red cell stroma (Bargellesi *et al.*, 1968; Weatherall *et al.*, 1965, 1969). *In vivo* these excess α-chains can be seen as red cell inclusions in bone marrow smears (Fessas, 1961) and in the circulating red cells of splenectomized patients when the cells are stained with cresyl violet. The kinetics of the α-chain pool have been closely studied *in vitro* (Modell *et al.*, 1969) and the α-chains have been shown to exist as a pool of dimers indicating that they are probably newly synthesized α-chains intermediate in the synthesis of Hb A and Hb F and not due to dissociation of the haemoglobin tetramer. The half-life of the α dimer pool is probably short and rapid dissociation into unstable monomers takes place.

In α-thalassaemia the reverse situation is found, i.e. a deficiency in the synthesis of the α-chain relative to the β-chain (Weatherall and Clegg, 1969). In reticulocytes of neonates with homozygous α1 thal Weatherall *et al.* (1970) were unable to demonstrate any synthesis of α-chains and small α-chain peptides were looked for but not found. In Hb H disease the synthesis of β-chain exceeds that of the α by a factor of two to three (Clegg and Weatherall, 1967). Schwartz *et al.* (1969) have shown that even in α1 thalassaemia trait and α2 thalassaemia trait genotypes, imbalance of chain synthesis can be demonstrated with α/β ratios of 0·8 and 0·8 to 0·9 respectively. The deficiency of α-chains with a relative excess of β- and γ-chains results in the accumulation of these chains in the form of Hb H (β_4) and Hb Barts (γ_4).

Over the past few years the understanding of the molecular causes of the thalassaemia syndromes has been partly clarified. Major advances have been made using the techniques of DNA:DNA and DNA:RNA hybridization. DNA which is complementary to globin mRNAs (and therefore also complementary to one of the strands of nuclear DNA from which the mRNAs are transcribed) may be prepared in cell-free systems using a viral reverse transcriptase. These complementary DNAs (cDNA) are then used as molecular probes in somewhat complicated hybridization experiments both to count the actual numbers of globin genes and to search for the presence of globin mRNA molecules.

Using a cDNA transcript of α globin mRNA, Ottolenghi et al. (1974) and Taylor et al. (1974) showed the absence of α globin genes in liver isolated from an infant dying of Hb Barts hydrops foetalis syndrome. DNA complementary to β globin mRNA hybridized as efficiency as normal in these experiments, thus suggesting that the homozygous α1 thal syndrome is due to a gene deletion. In an extension of the same approach Kan et al. (1975a) demonstrated that globin cDNA hybridized to nuclear DNA from a patient with HbH disease (presumptive genotype α1 thal/α2 thal; Table 5) at one-quarter of the rate that it hybridized to normal human DNA and at the same rate that it hybridized to an artificial mixture consisting of three parts of hydrops foetalis DNA and one part of normal DNA. This not only suggested gene deletion in HbH disease but also for the first time provided direct evidence that the α-chain locus is duplicated in some populations. The experiments of Kan et al. (1975a) suggest that HbH disease is caused by a deletion of three of the four globin genes.

DNA complementary to β-chain mRNA has so far not revealed gene deletions in the classical β-thalassaemias. However, Ramirez et al. (1976) have shown the absence of the β-chain gene in patients homozygous for δβ-thalassaemia and Kan et al. (1975b) have shown β gene deletion in the Negro form of HPFH, where the patients have Hb F of typ $^{G}\gamma$ $^{A}\gamma$ and no Hb A$_2$. These experiments as yet provide no evidence for the inadequate synthesis of γ-chains in homozygous δβ-thalassaemia when compared to that seen in HPFH.

When β-chain cDNA is used to probe for the presence of mRNA in the β-thalassaemias a complex picture emerges. Tolstoshev et al. (1976) showed less than 1% of β mRNA in two patients of Italian and Pakistani origin, even though β globin genes were present. On the other hand Kan et al. (1975b) found substantial amounts of β mRNA in bone-marrow and reticulocytes from two Chinese patients with β⁰-thalassaemia, though this material could not be induced to synthesize β globin chains in various cell-free systems. This distinguishes this form of β⁰-thalassaemia from the so-called Ferrara thalassaemia, where β globin chain synthesis can be induced in vivo by blood transfusion or in vitro by post-ribosomal supernatant from normal reticulocyte supernatants (Conconi et al., 1972). In Ferrara thalassaemia the β mRNA appears to decay faster than normal (Ottolenghi et al., 1977).

As summarized by Clegg and Weatherall (1976), the molecular basis of the classical thalassaemias and the thalassaemic-like syndromes is complex and only partly resolved. The conditions shown in Table 6 represent those were biochemical understanding is most advanced.

Table 6

Possible molecular defects in some thalassaemias

Type of thalassaemia	Possible molecular defect
Hb Barts hydrops fetalis	deletion of four α-chain genes
HbH disease	deletion of three α-chain genes
Homozygous δβ-thalassaemia	deletion of β-chain gene, absence of δ-mRNA
HPFH (Negro type)	deletion of β-chain gene
β^o-Thalassaemia (1)	absent β-chain mRNA, β-chain genes present
β^o-Thalassaemia (2)	β-chain mRNA present, non-functional
β^o-Thalassaemia (Ferrara)	β-chain mRNA present, can be induced to synthesize β globin chains, mRNA probably unstable

6. Anaemia of thalassaemia

The homozygous state for thalassaemia usually results in an extremely severe disease although a spectrum of severity is seen. It seems likely that the severity of the homozygous state will be determined by the degree of chain suppression. This is not easy to prove and such factors as environment, nutrition and infection must dictate the morbidity associated with these syndromes.

β-Thalassaemia. The heterozygous state for any of the β-thalassaemia genes results in a mild to moderate anaemia which often worsens during pregnancy or infection but does not usually require any specific treatment. The homozygous state results in an extremely severe anaemia associated with hepatosplenomegaly. The mortality is virtually 100%. Death can occur in the first year of life and is maximal in the second decade though survivals into the third and fourth decades have been recorded. The commonest cause of death is cardiac failure due to iron overload.

The anaemia has three major components: (a) ineffective erythropoiesis, (b) red cell haemolysis and (c) hypersplenism. The major component is the ineffective erythropoiesis which results in marked expansion of the bones. This is made worse by the high oxygen affinity of the foetal haemoglobin which is the major haemoglobin synthesized. It is likely that the chain imbalance, which leads to an accumulation of unstable α-chains which precipitate to form insoluble inclusions, will result in death of the cell within the marrow and will also be responsible for the haemoytic component of the red cells which get out into the circulation.

α-*Thalassaemia*. The homozygous state is incompatible with life. Hb H disease, however, results in a mild to moderate degree of anaemia associated with ineffective erythropoiesis and red cell haemolysis. It does not require any special therapy. The α-thalassaemia traits are clinically benign.

7. Hb F in thalassaemia

The amount of Hb F in the homozygous state of β-thalassaemia (and in α-thalassaemia) is increased by 10 to 100%. The factors responsible for the increased synthesis of the γ-chains are, however, not completely understood. The distribution of Hb F is in a separate population of cells and it is likely that within the marrow the cells that can synthesize Hb F will be the only ones to survive to reach the circulation. It is uncertain whether the γ-chain synthesis indicates that the cells are capable of synthesizing γ-chains in the deficiency of β- and δ-chain synthesis on the same loci or whether it comes from a separate clone of cells in an extremely anaemic child beyond the neonatal period. The finding that the Hb F is present in a separate population supports the latter explanation. Unfortunately, whether or not the level of β-chain activation remains constant or the switching off is just delayed is difficult to prove, for sooner or later these children have to be transfused because of falling haemoglobin. The type of γ-chain synthesized (Gly:Ala ratios) has been found to vary considerably amongst the thalassaemias and is another index for genetic heterogeneity of these syndromes.

VIII. CONCLUDING REMARKS

The implications of the molecular variation of a protein will differ according to the scientific interest of the student. Nevertheless, a study of the variation of haemoglobin satisfies all interests. Such a study not only defines clearly the meaning of molecular disease but also gives rise to greater understanding of the genetic control, the structure–function relationship and the synthesis of proteins.

The knowledge now acquired concerning the haemoglobin molecule is immense, with few gaps remaining and the clinician is now in a position at least to understand the cause of the patient's disease, albeit unable to cure it.

REFERENCES

Abramson, R. K., Rucknagel, D. L., Shreffler, D. C. and Soave, J. J. (1970). *Science* **169**, 194.

Adair, G. S. (1927). *Proc. Soc. Lond.* A **120**, 573.

Ahern, E. J., Jones, R. T., Brimhall, B. and Gray, R. M. (1970). *Br. J. Haemat.* **18**, 369.

Allan, N., Beale, D., Irvine, D. and Lehmann, H. (1965). *Nature, Lond.* **208**, 658.

Allen, D. W. and Jandl, J. H. (1960). *J. Clin. Invest.* **39**, 1107.

Allen, D. W., Schroeder, W. A. and Balog, J. (1958). *J. Am. Chem. Soc.* **80**, 1628.

Antoni, E. (1965). *Physiol. Rev.* **45**, 123.

Baglioni, C. (1962). *Proc. Natn. Acad. Sci USA* **48**, 1880.

Baglioni, C. (1963). *In* "Molecular Genetics" (J. H. Taylor, Ed.), 405. Academic Press, New York and London.

Baglioni, C. and Campana, T. (1967). *Eur. J. Biochem.* **2**, 480.

Baglioni, C. and Colombo, B. (1964). *Cold Spring. Harb. Symp. Quant. Biol.* **29**, 347.

Bank, A. and Marks, P. A. (1966). *Nature, Lond.* **212**, 1198.

Bank, A., Braverman, A. S. and Marks, P. A. (1969). *Ann. N. Y. Acad. Sci.* **165**, 231.

Bargellesi, A., Pontremoli, S. and Conconi, F. (1967). *Eur. J. Biochem.* **1**, 73.

Bargellesi, A., Pontremoli, S., Menoni, C. and Conconi, F. (1968). *Eur. J. Biochem.* **3**, 364.

Barnabas, J. and Muller, C. J. (1962). *Nature, Lond.* **194**, 431.

Beale, D. and Lehmann, H. (1965). *Nature, Lond.* **207**, 259.

Bearden, A. J., Moss, T. H., Caughey, W. S. and Beaudrew, C. A. (1965). *Proc. Natn. Acad. Sci. USA* **53**, 1246.

Bellingham, A. J. (1976). *Br. Med. Bull.* **32**, 234.

Bellingham, A. J. and Huehns, E. R. (1968). *Nature, Lond.* **218**, 924.

Benesch, R. (1966). *Proc. 11th Congr. Int. Soc. Haemat.* 406.

Benesch, R. and Benesch, R. E. (1969). *Nature, Lond.* **221**, 618.

Benesch, R. E. and Benesch, R. (1974). *Adv. Protein Chem.* **28**, 211.

Benesch, R. E., Benesch, R. and Macduff, G. (1965). *Proc. Natn. Acad. Sci. USA* **54**, 535.

Benesch, R., Benesch, R. E. and Yu, C. I. (1968). *Proc. Natn. Acad. Sci. USA* **54**, 526.

Beretta, A., Prato, V., Gallo, E. and Lehmann, H. (1968). *Nature, Lond.* **217**, 1014.

Birnboim, H. C., Mitchel, R. E. J. and Straus, N. A. (1973). *Proc. Natn. Acad. Sci. USA* **70**, 2189.

Bonaventura, J. and Riggs, A. (1968). *J. Biol. Chem.* **243**, 980.

Bookchin, R. M., Nagel, R. L. and Ranney, H. M. (1967). *J. Biol. Chem.* **242**, 248.

Bookchin, R. M., Nagel, R. L. and Ranney, H. M. (1970). *Biochem. Biophys. Acta* **221**, 373.

Botha, M. C., Beale, D., Isaacs, W. C. and Lehmann, H. (1966). *Nature, Lond.* **217**, 792.

Boyer, S. H., Rucknagel, D. L., Weatherall, D. J. and Watson-Williams, E. J. (1963). *Am. J. Hum. Genet.* **15**, 438.

Bradley, T. B. Jr., Brawner, J. N. III and Conley, C. L. (1961). *Bull. Johns Hopkins. Hosp.* **108**, 242.

Bradley, T. B. Jr., Wohl, R. C. and Rieder, R. F. (1967). *Science* **157**, 1581.

Braunitzer, G. (1970). *Biochem. J.* **119**, 31P.

Braunitzer, G., Gehring Mueller, G., Hilschmann, N., Hilse, K., Hobom, G., Rudloff, V. and Liebold Wittman, B. (1961). *Hoppe-Seylerps Z. Physiol. Chem.* **325**, 283.

Braunitzer, G., Hiese, K., Rudloff, V. and Hilschmann, N. (1964). *Adv. Protein Chem.* **19**, 1.

Brimhall, R., Hollan, S., Jones, R. T., Koler, R. D. and Szelenyi, J. G. (1970). *Clin. Res.* **18**, 184.

Capp, G. L. (1966). Thesis. University of Oregon.

Capp, G. L., Rigas, D. A. and Jones, R. T. (1967). *Science* **157**, 65.

Capp, G. L., Rigas, D. A. and Jones, R. T. (1970). *Nature, Lond.* **228**, 278.

Carrell, R. W. (1969). M. D. Thesis, Cambridge University.

Carrell, R. W. and Lehmann, H. (1968). *Semin. Haemat.* **6**, 116.

Carrell, R. W., Lehmann, H. and Hutchinson, H. E. (1966). *Nature, Lond.* **210**, 915.

Carrell, R. W., Lehmann, H., Lorkin, P. A., Raik, E. and Hunter, E. (1967). *Nature, Lond.* **215**, 626.

Cathie, I. A. B. (1950). *Gt. Ormond St. J.* **3**, 43.

Cauchi, M. N., Clegg, J. B. and Weatherall, D. J. (1969). *Nature, Lond.* **223**, 311.

Charache, S. and Conley, C. L. (1964). *Blood* **24**, 25.

Charache, S. and Conley, C. L. (1969). *Ann. N. Y. Acad. Sci.* **165**, 37.

Charache, S. and Weatherall, D. J. (1966). *Blood* **28**, 377.

Charache, S., Weatherall, D. J. and Clegg, J. B. (1966). *J. Clin. Invest.* **45**, 813.

Charache, S., Conley, C. L., Waugh, D. F., Ugoretz, R. J. and Spurrell, J. R. (1967). *J. Clin. Invest.* **46**, 1795.

Clark, B. F. C. and Marcker, K. A. (1966). *J. Mol. Biol.* **17**, 394.

Clegg, J. B. and Weatherall, D. J. (1967). *Nature, Lond.* **215**, 1291.

Clegg, J. B. and Weatherall, D. J. (1976). *Br. Med. Bull.* **32**, 262.

Comings, D. E. and Motulsky, A. G. (1966). *Blood* **28**, 54.

Conconi, F., Bargellesi, A., Pontremole, S., Vigi, V., Volpato, S. and Gaburro, D. (1968). *Nature, Lond.* **217**, 359.

Conconi, F., Rowley, P. T., Del Senno, L., Pontremoli, S. and Volpato, S. (1972). *Nature New Biol.* **238**, 83.

Cullis, A. F., Muirhead, H., Perutz, M. F., Rossmann, M. G. and North, A. C. T. (1962). *Proc. R. Soc.* A. **265**, 161.

Curtain, C. C. (1964). *Aust. J. Exp. Biol. Med. Sci.* **42**, 89.

Dacie, J. V., Grimes, A. J., Meisler, A., Steingold, L., Hemsted, E. H., Beaven, G. H. and White, J. M. (1964). *Br. J. Haemat.* **10**, 388.

Dacie, J. V., Shintan, N. K., Goffney, P. J. and Carrell, R. W. and Lehmann, H. (1967). *Nature, Lond.* **216**, 663.

Dan, M. and Hagiwara, A. (1967). *Exp. Cell. Res.* **46**, 596.

Deisseroth, A., Velez, R., Anderson, W. F., Nienhuis, A., Kucherlapati, R. and Ruddle, F. H. (1976). *Blood* **46**, 1049.

Dickerson, R. E. (1964). *In* "The Proteins" (N. Neurath, Ed.) 603. Academic Press, New York and London.

Dintzis, H. M. (1961). *Proc. Natn. Acad. Sci. USA* **47**, 247.

Edington, G. M. and Watson-Williams, E. J. (1964). *In* "C.I.O.M.S. Symposium on Abnormal Haemoglobins and Enzyme Deficiency", 33. Blackwells, Oxford.

Ehrenstein, G. von (1966). *Cold Spring Harb. Symp. Quant. Biol.* **31**, 705.

Felicetti, L., Colombo, B. and Baglioni, C. (1966). *Biochem. Biophys. Acta* **129**, 380.

Fessas, P. (1961). *In* "Haemoglobin Colloquium" (H. Lehmann and K. Betke, Eds) 74. Thieme, Verlag, Stuttgart.

Fessas, P. (1968). *Proc. 12th Congr. Int. Soc. Haemat.* 52.

Fessas, P. and Karaklis, A. (1962). *Clin. Chim. Acta* **7**, 133.

Fessas, P. and Stamatoyannopoulos, G. (1962). *Nature, Lond.* **195**, 1215.

Fessas, P., Stamatoyannopoulos, G. and Karaklis, A. (1962). *Nature, Lond.* **19**, 1.

Gajdusek, D. C., Guiant, J., Kirk, R. L., Carrell, R. W., Irvine, D., Kynoch, P. A. M. and Lehmann, H. (1967). *J. Med. Genet.* **4**, 1.

Garen, A. (1968). *Science* **160**, 149.

Gerald, P. S. (1958). *Blood* **13**, 936.

Gerald, P. S. (1966). *In* "Metabolic Basis of Inherited Disease" (J. B. Stanbury, J. B. Wyngaarden and D. S. Fredrickson, Eds) 1069. McGraw-Hill, New York.

Gerald, P. S. and Efron, M. L. (1961). *Proc. Natn. Acad. Sci. USA* **47**, 1758.

Gerald, P. S. and George, P. (1959). *Science* **129**, 393.

Gerald, P. S. and Scott, E. M. (1966). *In* "Metabolic Basis of Inherited Disease" (J. B. Stanbury, J. B. Wyngaarden and D. S. Fredrickson, Eds) 1090. McGraw-Hill, New York.

Gibson, Q. H. and Parkhurst, L. I. (1968). *J. Biol. Chem.* **243**, 5521.

Gilmour, R. S. and Paul, J. (1973). *Proc. Natn. Acad. Sci. USA* **70**, 3440.

Granick, S. and Levere, R. D. C. (1968). *Proc. 12th Congr. Int. Soc. Haemat.* 276.

Gratzer, W. B. and Allison, A. C. (1960). *Biol. Rev.* **35**, 359.

Grimes, A. J. and Meisler, A. (1962). *Nature, Lond.* **194**, 190.

Hamilton, H. B., Iuchi, I., Miyaji, T. and Shibata, S. (1969). *J. Clin. Invest.* **48**, 525.

Harris, J. W. (1950). *Proc. Soc. Exp. Biol. Med.* **75**, 197.

Harrison, P. R., Affara, N., Conkie, D., Rutherford, T., Sommerville, J. and Paul, J. (1976). *J. Mol. Biol.* **84**, 539.

Hayashi, N., Motokawa, Y. and Kikuchi, G. (1966). *J. Biol. Chem.* **241**, 79.

Hayashi, A., Suzuki, T., Shimizu, A., Imai, K., Morimoto, H., Miyaji, T. and Shibata, S. (1968). *Arch. Biochem. Biophys.* **125**, 895.

Hecht, F., Jones, R. T. and Koler, R. D. (1967). *Ann. Hum. Genet.* **31**, 215.

Heller, P., Coleman, R. D. and Yakulis, V. (1966). *J. Clin. Invest.* **45**, 102.

Herrick, J. B. (1910). *Arch. Int. Med.* **6**, 517.

Heywood, J. D., Karon, M. and Weissman, S. (1964). *Science* **146**, 530.

Hill, A. V. (1910). *J. Physiol.* **40**, 4.

Hill, R. J. and Kraus, A. P. (1963). *Fedn. Proc. Fedn. Am. Socs Exp. Biol.* **22**, 597.

Holmquist, W. R. and Schroeder, W. A. (1964). *Biochim. Biophys. Acta* **82**, 639.

Hrkal, Z. and Vodrazka, Z. (1967). *Biochim. Biophys. Acta* **133**, 527.

Huehns, E. R. (1965). *Post-grad. Med. J.* **41**, 718.

Huehns, E. R. (1970). *Ann. Rev. Med.* **21**, 157.

Huehns, E. R. (1973). *In* "Blood and its Disorders" (Hardisty and Weatherall, Eds) 526. Blackwell Scientific Publications, Oxford.

Huehns, E. R. and Shooter, F. M. (1965). *J. Med. Genet.* **2**, 48.

Huehns, E. R., Hecht, F., Yoshida, A., Stamatoyannopoulos, G., Hartman, J. and Motulsky, A. G. (1970). *Blood* **36**, 209.

Huisman, T. H. J. (1963). *Adv. Clin. Chem.* **6**, 231.

Huisman, T. H. J. (1974). *Ann. N. Y. Acad. Sci.* **232**, 107.

Huisman, T. H. J. and Dozy, A. M. (1962). *J. Chromat.* **7**, 180.

Huisman, T. H. J., Wilson, J. B. and Adams, H. R. (1967). *Arch. Biochem. Biophys.* **121**, 528.

Huisman, T. H. J., Brandt, G. and Wilson, J. B. (1968). *J. Biol. Chem.* **243**, 3675.

Hunt, R. T., Hunter, A. R. and Monroe, A. J. (1968). *Nature, Lond.* **220**, 481.

Hunt, T. (1976). *Br. Med. Bull.* **32**, 257.

Imai, K. (1968). *Arch. Biochem. Biophys.* **127**, 543.

Ingram, V. M. (1963). "The Haemoglobins in Genetics and Evolution", 97. Columbia University Press, New York.

Ingram, V. M. and Stretton, A. O. W. (1959). *Nature, Lond.* **184**, 1903.

Ingram, V. M. and Stretton, A. O. W. (1962a). *Biochim. Biophys. Acta* **62**, 456.

Ingram, V. M. and Stretton, A. O. W. (1962b). *Biochim. Biophys. Acta* **63**, 20.

Itano, H. A. (1957). *Adv. Protein. Chem.* **12**, 215.

Jacob, F. and Monod, J. (1961). *J. Mol. Biol.* **3**, 318.

Jaffé, E. R. and Heller, P. (1964). *Prog. Haemat.* **4**, 48.

Jones, R. T. (1964). *Cold Spring Harb. Symp. Quant. Biol.* **29**, 297.

Jones, R. T., Brimhall, B., Huisman, T. H. J., Kleihauer, E. and Betke, K. (1966). *Science* **154**, 1024.

Jones, R. T., Osgood, E. E., Brimhall, B. and Koler, R. D. (1967). *J. Clin. Invest.* **46**, 1840.

Jong, W. W. W. de, Went, L. N. and Bernini, L. F. (1968). *Nature, Lond.* **220**, 788.

Jonxis, J. H. P. (1963). *Ann. Rev. Med.* **14**, 297.

Josephson, A. M., Wainstein, H. G., Yakulis, V. J., Singer, L. and Heller, P. (1962). *J. Lab. Clin. Med.* **59**, 918.

Kabat, D. (1972). *Science* **175**, 134.

Kamuzora, H. and Lehmann, H. (1975). *Nature, Lond.* **256**, 511.

Kan, Y. W., Dozy, A. M., Varmus, H. E., Taylor, J. M., Holland, J. P., Lie-Injo, L. E., Ganesan, J. and Todd, D. (1975a). *Nature, Lond.* **255**, 255.

Kan, Y. W., Holland, J. P., Dozy, A. M., Charache, S. and Kazazian, H. H. (1975b). *Nature, Lond.* **258**, 162.

Kattamis, C. and Lehmann, H. (1970a). *Hum. Hered.* **20**, 156.

Kattamis, C. and Lehmann, H. (1970b). *Lancet* **ii**, 635.

Kazazian, H. H., Jr. (1974). *Semin. Haematol.* **11**, 525.

Kendrew, J. C., Dickerson, R. E., Strandberg, B. E., Hart, R. G., Davies, D. R., Phillips, D. C. and Shore, V. C. (1960). *Nature, Lond.* **185**, 422.

Kendrew, J. C., Watson, H. C., Strandberg, B. E., Dickerson, R. E., Phillips, D. C. and Shore, V. C. (1961). *Nature, Lond.* **190**, 666.

Kilmartin, J. V. (1973). *Biochem. J.* **133**, 725.

Kilmartin, J. V. (1976). *Br. Med. Bull.* **32**, 209.

Kilmartin, J. V. and Clegg, J. B. (1967). *Nature, Lond.* **213**, 269.

Kilmartin, J. V., Fogg, J., Luzzana, M. and Rossi Bernardi, L. (1973). *J. Biol. Chem.* **248**, 7039.

Kinderlerer, J., Lehmann, H. and Tipton, K. F. (1970). *Biochem. J.* **119**, 66.

Kleihauer, E. F., Reynolds, C. E., Dozy, A. M., Wilson, J. B., Moores, R. R., Berenson, M. P., Wright, C. S. and Huisman, T. H. J. (1968). *Biochim. Biophys. Acta* **154**, 220.

Knopf, P. M. and Lamfrom, H. (1965). *Biochim. Biophys. Acta* **95**, 398.

Konigsberg, W. H. and Hill, R. J. (1962). *J. Biol. Chem.* **237**, 2547.

Kraus, A. P. and Diggs, L. W. (1956). *J. Lab. Clin. Med.* **47**, 700.

Kraus, L. M., Miyaji, T., Iuchi, I. and Kraus, A. P. (1966). *Biochemistry* **5**, 3701.

Kunkel, H. G. and Wallenis, G. (1955). *Science* **122**, 288.

Kunkel, H. G., Ceppellini, R., Müller-Eberhard, V. and Wolf, J. (1957). *J. Clin. Invest.* **36**, 1615.

Labie, D., Schroeder, W. A. and Huisman, T. H. J. (1966). *Biochim. Biophys. Acta* **127**, 428.

Lang, A. and Lorkin, P. A. (1976). *Br. Med. Bull.* **32**, 239.

Lang, G. and Marshall, W. (1966). "Heme and Hemoproteins" (B. Chance, R. Estabrook and T. Yonetani, Eds) 15. Academic Press, New York and London.

Lehmann, H. (1970). *Lancet* **ii**, 78.

Lehmann, H. and Carrell, R. W. (1968). *Br. Med. J.* **4**, 748.

Lehmann, H. and Carrell, R. W. (1969). *Br. Med. Bull.* **25**, 14.

Lehmann, H. and Huntsman, R. G. (1974). "Man's Haemoglobins including the Haemoglobinopathies and their Investigation" (2nd ed.) North Holland Publishing Company, Amsterdam.

Lehmann, H., Huntsman, R. G. and Agar, J. A. M. (1966). *In* "The Metabolic Basis of Inherited Disease" (Stanbury, J. B., Wyngaarden, J. B. and Fredrickson, D. S., Eds) 1100. McGraw-Hill, New York.

Lehmann, H. and Kynoch, P. A. M. (1976). "Human Haemoglobin Variants and their Characteristics". North Holland Publishing Company, Amsterdam.

Levere, R. D. C., Kappas, A. and Granick, S. (1967). *Proc. Natn. Acad. Sci. USA* **58**, 985.

Livingstone, F. B. (1967). "Abnormal Haemoglobins in Human Populations". Aldine, Chicago.

Manwell, C. and Baker, C. M. A. (1970). "Molecular Biology and the Origin of Species", 226. Sedgwick and Jackson, London.

Marti, H. R., Beale, D. and Lehmann, H. (1967). *Acta Haemat.* **37**, 174.

May, A. and Huehns, E. R. (1976). *Br. Med. Bull.* **32**, 223.

Milner, P. F., Clegg, J. B. and Weatherall, D. J. (1971). *Lancet* **i**, 729.

Miyaji, T., Iuchi, I., Shibata, S., Takeda, I. and Tamura, A. (1963). *Acta Haemat. Jap.* **26**, 538.

Miyaji, T., Ohta, V., Yammomoto, K., Shibata, S., Iuchi, I. and Takareka, M. (1968). *Nature, Lond.* **217**, 89.

Modell, C. B., Benson, A. M. and Huehns, E. R. (1969). *Ann. N. Y. Acad. Sci.* **165**, 238.

Monod, J., Wyman, J. and Changeux, J. P. (1965). *J. Mol. Biol.* **12**, 88.

Motulsky, A. G. (1964a). *Am. J. Trop. Med.* **13**, 147.

Motulsky, A. G. (1964b). *Cold Spring Harb. Symp. Quant. Biol.* **29**, 399.

Muller, C. J. and Kingman, S. (1961). *Biochim. Biophys. Acta* **50**, 595.

Murayama, M. (1966). *J. Cell Physiol.* **67**, 21.

Nance, W. E. (1963). *Science* **141**, 123.

Nathan, D. G. and Gunn, R. B. (1966). *Am. J. Med.* **41**, 815.

Naughton, M. A. and Dintzis, H. M. (1962). *Proc. Natn. Acad. Sci. USA* **48**, 1822.

Necheles, T. F., Sheehan, B. G. and Meyer, H. J. (1965). *Proc. Soc. Exp. Biol. Med.* **119**, 1207.

Nigon, V. and Godet, J. (1976). *Int. Rev. Cytol.* **46**, 79.

Nobbs, C. L., Watson, H. C. and Kendrew, J. C. (1966). *Nature, Lond.* **209**, 331.

Novy, M. J., Edwards, M. J. and Metcalf, J. (1967). *J. Clin. Invest.* **46**, 1848.

Ohta, Y., Yamaoka, K., Sumida, I., Fujita, S., Fujimura, T., Hanada, M. and Yanase, T. (1970). *Proc. 13th Congr. Int. Soc. Haemat.* 233.

Opfell, R. W., Lorkin, P. A. and Lehmann, H. (1968). *J. Med. Genet.* **5**, 292.

Ottolenghi, S., Lanyon, W. G., Paul, J., Williamson, R., Weatherall, D. J., Clegg, J. B., Pritchard, J., Pootrakul, S. and Wong, J. B. (1974). *Nature, Lond.* **251**, 389.

Ottolenghi, S., Comi, P., Giglioni, B., Williamson, R., Vullo, G. and Conconi, F. (1977). *Nature, Lond.* **266**, 231.

Paul, J. (1976). *Br. Med. Bull.* **32**, 277.

Paul, J., Gilmour, R. S., Affara, N., Birnie, G., Harrison, P., Hell, A., Humphries, S., Windass, J. and Young, B. (1973). *Cold Spring Harbor Symp. Quant. Biol.* **38**, 885.

Pauling, L., Itano, H. A., Singer, S. J. and Wells, I. C. (1949). *Science* **110**, 543.

Pembrey, M. E. and Weatherall, D. J. (1971). *Abst. Br. Soc. Haemat.* 13.

Pembrey, M. E., Weatherall, D. J. and Clegg, J. B. (1973). *Lancet* **i**, 1350.

Perkowska, E., Macgregor, H. C. and Birnstiel, M. L. (1968). *Nature, Lond.* **217**, 649.

Perutz, M. F. (1965). *J. Mol. Biol.* **13**, 646.

Perutz, M. F. (1969). *Proc. R. Soc.* B. **173**, 113.

628

J. M. WHITE

Perutz, M. F. (1970). *Nature, Lond.* **228**, 726.

Perutz, M. F. (1972). *Nature, Lond.* **237**, 495.

Perutz, M. F. (1976). *Br. Med. Bull.* **32**, 195.

Perutz, M. F. and Greer, J. (1970). *Biochem. J.* **119**, 31.

Perutz, M. F. and Lehmann, H. (1968). *Nature, Lond.* **219**, 902.

Perutz, M. F., Rossman, M. G., Cullis, A. F., Muirhead, H., Will, G. and North, A. C. T. (1960). *Nature, Lond.* **185**, 416.

Perutz, M. F., Muirhead, H., Cox, J. M. and Goaman, L. C. G. (1968). *Nature, Lond.* **219**, 131.

Perutz, M. F., Muirhead, H., Mazzarella, L., Growther, R. Z., Greer, J. and Kilmartin, J. V. (1969). *Nature, Lond.* **222**, 1243.

Pisciotta, A. V., Ebbe, S. N. and Hinz, J. E. (1959). *J. Lab. Clin. Med.* **54**, 73.

Popp, R. A. (1967). *J. Mol. Biol.* **27**, 9.

Price, P. M., Conover, J. N. and Hirschhorn, K. (1972). *Nature, Lond.* **237**, 340.

Proudfoot, N. J. and Brownlee, G. G. (1976). *Br. Med. Bull.* **32**, 251.

Ramirez, F., O'Donnell, J. V., Marks, P. A., Bank, A., Musumeci, S., Schilino, G., Pizzarelli, G., Russo, G., Luppis, B. and Gambino, R. (1976). *Nature Lond.* **263**, 461.

Ranney, H. M. (1970). *New Engl. J. Med.* **282**, 144.

Ranney, H. M. and Jacobs, A. S. (1964). *Nature, Lond.* **204**, 163.

Ranney, H. M., Jacobs, A. S., Udem, L. and Zalusky, R. (1968a). *Biochem. Biophys. Res. Commun.* **33**, 1004.

Ranney, H. M., Nagel, R. L., Heller, P. and Udem, L. (1968b). *Biochim. Biophys, Acta* **160**, 112.

Raper, A. B. (1956). *Br. Med. J.* **1**, 965.

Reed, C. S., Hampson, R., Gordon, S., Jones, R. T., Novy, M. J., Brimhall, B., Edwards, M. J. and Koler, R. D. (1968). *Blood* **31**, 623.

Reissmann, K. R., Ruth, W. E. and Nomura, T. (1961). *J. Clin. Invest.* **40**, 186.

Rieder, R. F. and Bradley, T. B. (1968). *Blood* **32**, 355.

Rieder, R. F., Oski, F. A. and Clegg, J. B. (1969). *J. Clin. Invest.* **48**, 1627.

Rifkin, D. B., Rifkin, M. R. and Konigsberg, W. (1966). *Proc. Natn. Acad. Sci. USA* **55**, 586.

Rifkind, R. A., Luzzatto, L. and Marks, P. A. (1964). *Proc. Natn. Acad. Sci. USA* **52**, 1227.

Riggs, A. (1965). *Physiol. Rev.* **45**, 619.

Rosa, J., Oudart, J. C., Pagnier, J., Relhkodja, O., Boigne, J. M. and Labie, D. (1968). *Proc. 12th Congr. Int. Soc. Haemat.* 72.

Rosemeyer, M. A. and Huehns, E. R. (1967). *J. Mol. Biol.* **25**, 253.

Rossi-Fanelli, A., Antonini, E. and Canuto, A. (1964). *Adv. Protein Chem.* **19**, 74.

Roughton, F. J. W. (1949). *In* "Haemoglobins", 83. Interscience, New York.

Roughton, F. J. W. (1964). *J. Gen. Physiol.* **49**, 105.

Rucknagel, D. L. and Neel, J. V. (1961). *Prog. Med. Genet.* **1**, 158.

Sansone, G., Carrell, R. W. and Lehmann, H. (1967). *Nature, Lond.* **214**, 877.

Schneider, R. G., Ueda, S., Alperin, J. B., Brimhall, B. and Jones, R. T. (1969). *New Engl. J. Med.* **280**, 739.

Schnek, A. G. and Schroeder, W. A. (1961). *J. Am. Chem. Soc.* **83**, 1472.

Schokker, R. C., Went, L. N. and Bok, J. (1966). *Nature, Lond.* **209**, 44.

Schroeder, W. A. (1963). *Ann. Rev. Biochem.* **32**, 301.

Schroeder, W. A. and Holmquist, W. R. (1968). *In* "Structural Chemistry and Molecular Biology" (A. Rich and C. Davidson, Eds) 238. Freeman and Co., San Francisco.

Schroeder, W. A. and Huisman, T. H. S. (1970). *Proc. 12th Congr. Int. Soc. Haemat.* 26.

Schroeder, W. A. and Jones, R. T. (1965). "Fortschritte de Chemie Organischer Naturstaffe", 115. Zedimeister, Vienna.

Schroeder, W. A., Cua, J. T., Matsuda, G. and Fenninger, W. D. (1962). *Biochim. Biophys. Acta* **63**, 532.

Schroeder, W. A., Shelton, J. R., Shelton, J. B., Cormick, J. and Jones, R. T. (1963). *Biochemistry* **2**, 992.

Schroeder, W. A., Huisman, T. H. J., Shelton, J. R., Shelton, J. B., Kleihauer, E. F., Dozy, A. M. and Robberson, B. (1968a). *Proc. Natn. Acad. Sci. USA* **60**, 537.

Schwartz, E. (1969). *New Engl. J. Med.* **281**, 1327.

Schwartz, E., Kan, Y. W. and Nathan, D. G. (1969). *Ann. N. Y. Acad. Sci.* **165**, 288.

Sick, K., Beale, D., Irvine, D., Lehmann, H., Goodhall, P. T. and MacDougall, S. (1967). *Biochim. Biophys. Acta* **140**, 241.

Silvestroni, E. and Bianco, I. (1964). *Progr. Med.* **20**, 509.

Silvestroni, E. Bianco, I. and Baglioni, G. (1965a). *Blood* **25**, 457.

Singer, K. and Singer, L. (1953). *Blood* **8**, 1008.

Stamatoyannopoulos, G. and Fessas, P. (1963). *J. Lab. Clin. Med.* **62**, 193.

Stamatoyannopoulos, G., Sofroniadou, K. and Akrivakis, A. (1968a). *Blood* **30**, 772.

Stamatoyannopoulos, G., Yoshida, A., Adamson, J. and Heinenberg, S. (1968b). *Science* **159**, 741.

Stamatoyannopoulos, G., Fessas, P. and Papayannopoulou, T. (1969a). *Am. J. Med.* **47**, 194.

Stamatoyannopoulos, G., Papayannopoulou, T., Fessas, P. and Motulsky, A. G. (1969b). *Ann. Acad. Sci.* **165**, 25.

Steadman, J. H., Yates, A. and Huehns, E. R. (1970). *Br. J. Haemat.* **18**, 435.

Stetson, C. A. Jr. (1966). *J. Exp. Med.* **123**, 341.

Stohlman, F. Jr. (1967). *Semin. Hemat.* **4**, 304.

Suzuki, T., Akira, H. and Yamamura, Y. (1965). *Biochem. Biophys. Res. Commun* **19**, 691.

Suzuki, T., Hayashi, A., Shimizu, A. and Yamamura, Y. (1966). *Biochim. Biophys. Acta* **127**, 280.

Taylor, J. M., Dozy, A. M., Kan, Y. W., Varmus, H. E., Lie-Injo, L. E., Ganesan, J. and Todd, D. (1974). *Nature, Lond.* **251**, 392.

Thomas, E. D., Lochte, H. L. Jr., Greenough, W. B. III, and Wales, M. (1960). *Nature, Lond.* **185**, 396.

Thompson, R. B., Warrington, R., Odom, J. and Bell, W. N. (1965). *Acta Genet.* **15**, 190.

Todd, D., Lai, M. C. S., Beaven, G. H. and Huehns, E. R. (1970). *Br. J. Haemat.* **19**, 27.

Tolstoshev, P., Mitchell, J., Lanyon, G., Williamson, R., Ottolenghi, S., Comi, P., Giglioni, B., Masera, G., Modell, B., Weatherall, D. J. and Clegg, J. B. (1976). *Nature, Lond.* **259**, 95.

Udem, L., Ranney, H. M. and Bunn, H. F., cited in Ranney (1970).

Viale, R. O., Maggiora, G. M. and Ingraham, L. L. (1964). *Nature, Lond.* **293**, 183.

Vinograd, J. R. and Hutchinson, W. D. (1960). *Nature, Lond.* **187**, 216.

Wallace, W. J., Volne, J. A., Maxwell, J. C., Caughey, W. S., Charache, S. (1976). *Biochem. Biophys. Res. Commun.* **68**, 1379.

Wasi, P., Na-Nakorn, S. M. and Suingdumrong, A. (1964). *Nature, Lond.* **204**, 907.

Wasi, P., Na-Nakorn, S., Pootrakul, S., Sookanek, M., Disthasongchan, P., Pornpatkul, M. and Panich, V. (1969). *Ann. N. Y. Acad. Sci.* **165**, 288.

Watson, H. C. (1966). "Hemes and Hemoproteins" (B. Chance, R. Estabrook and T. Yonetani, Eds). Academic Press, New York and London.

Weatherall, D. J. (1963). *Br. J. Haemat.* **9**, 265.

Weatherall, D. J. (1965). "The Thalassaemia Syndromes". Blackwell Scientific Publications, Oxford.

Weatherall, D. J. (1967). *In* "Progress in Medical Genetics" (A. G. Steinberg and A. G. Bearn, Eds) 8. Grune and Stratton, New York.

Weatherall, D. J. (1969). *Br. Med. Bull.* **25**, 24.

Weatherall, D. J. (1977). *Clin. Sci. Mol. Med.* **52**, 223.

Weatherall, D. J. and Clegg, J. B. (1968). *Prog. Haemat.* **6**, 261.

Weatherall, D. J. and Clegg, J. B. (1969). *Ann. N. Y. Acad. Sci.* **165**, 242.

Weatherall, D. J. and Clegg, J. B. (1972). "The Thalassaemia Syndrome" (2nd ed.). Blackwell, Oxford.

Weatherall, D. J., Clegg, J. B. and Naughton, M. A. (1965). *Nature, Lond.* **208**, 1061.

Weatherall, D. J., Clegg, J. B., Na-Nakorn, S. and Wasi, P. (1969). *Br. J. Haemat.* **16**, 251.

Weatherall, D. J., Clegg, J. B. and Boon, W. H. (1970). *Br. J. Haemat.* **18**, 357.

Weiss, J. J. 1964a). *Nature, Lond.* **202**, 83.

Weiss, J. J. (1964). *Nature, Lond.* **202**, 182.

Wheby, M. S., Thorup, O. A. and Leavell, B. S. (1956). *Blood* **11**, 266.

White, J. M. and Dacie, J. V. (1971). *Prog. Haematol.* **7**, 69.

White, J. M., Brain, M. C., Larkin, P. A., Lehmann, H. and Smith, M. (1970). *Nature, Lond.* **225**, 939.

Winslow, R. M. and Ingram, V. M. (1966). *J. Biol. Chem.* **241**, 1144.

Winterhalter, K. H. and Birchmeier, W. (1970). *Proc. 12th Congr. Int. Soc. Haemat.* 9.

Winterhalter, K. H., Heywood, J. D., Huehns, E. R. and Finch, C. A. (1969). *Br. J. Haemat.* **16**, 523.

Wood, W. G. (1976). *Br. Med. Bull.* **32**, 282.

Wood, W. G. and Weatherall, D. J. (1973). *Nature, Lond.* **244**, 162.

Yanase, T., Hanada, M., Seita, M., Ohya, I., Ohta, T., Imamura, T., Fujimura, T., Kawasaki, K. and Yamaoka, K. (1969). *Jap. J. Hum. Genet.* **13**, 40.

Zuckerkandl, E. and Pauling, L. (1962). *In* "Horizons in Biochemistry" (M. Kasha and B. Pullman, Eds) 189. Academic Press, New York and London.

12 The Immunoglobulinopathies

M. W. TURNER

Department of Immunology, Institute of Child Health, London, England

I. INTRODUCTION

Healthy animals are protected from the harmful effects of microorganisms by two major defence mechanisms. These are usually referred to as innate (non-specific) and acquired (specific) immunity. Innate mechanisms are effective against a wide range of organisms and are mainly genetically controlled. Thus wide species and strain variations exist in terms of susceptibility to various organisms. Individual genetic factors, age and sex are also important. There are, in addition, other factors determining innate immunity such as physical barriers (skin, mucous membranes etc.) and a range of anti-bacterial and anti-fungal secretions.

Acquired specific immunity is based on a defence system which had already evolved some 400 million years ago among primitive vertebrates (Good *et al.*, 1967). Essentially, this type of immunity is based on the presentation of antigenic fragments from invading microorganisms (or any foreign material) to the cells of the immune system. These cells (macrophages and lymphocytes) are then responsible for the initiation of a specific immune response. The recognition of antigen and its processing constitute the so-called afferent pathway. The subsequent proliferation of immunocompetent cells results in the release of antibodies (humoral immunity) and/or sensitized lymphocytes (cell-mediated immunity)—the whole constituting the efferent pathway.

(a) Humoral immunity

The characteristic feature of humoral immunity is the large-scale production of antibody molecules or immunoglobulins. These proteins are secreted by plasma cells which have developed from B-lymphocytes. In addition to reacting specifically with the antigen which stimulated their production they give rise to a number of secondary effects such as bacterial lysis and phagocytosis.

(b) Cell-mediated immunity

In cell-mediated immunity there is a proliferation of lymphoid cells which are specifically sensitized to the inducing antigen. The cells involved are mainly T-lymphocytes and the inflammatory changes are due to pharmacologically active substances (lymphokines) released by these cells.

It is now known that T-lymphocytes are involved in many antibody responses (in a cooperative role with B-lymphocytes and macrophages) and it is possible that immunoglobulins are centrally involved in cell-mediated responses (as receptors on the surface membranes of lymphocytes and macrophages). The interaction of the cellular components of the immune system is outside the scope of this chapter but the interested reader may pursue this topic further in Roitt (1974) and Hobart and McConnell (1975).

The first part of this chapter will be concerned with a description of the immunoglobulins and their characteristics in healthy individuals. This lays the groundwork for the subsequent consideration of the immunoglobulinopathies which are discussed under three headings—namely the hyper-, the para- and the hypoimmunoglobulinopathies. Unlike the haemoglobinopathies, aberrant immunoglobulin molecules account for only a small number of the immunoglobulinopathies and the majority of these syndromes arise from faulty quantitative control of immunoglobulin synthesis. This chapter will attempt to describe and classify these diseases using the criteria available for the immunoglobulin populations of healthy individuals.

II. THE IMMUNOGLOBULINS

A. General Introduction

The immunoglobulins are a family of structurally related proteins with the major function of mediating circulating antibody responses. The name "immunoglobulins" was given in 1959 by Heremans to IgG, IgA and IgM. (Ig = Immunoglobulin). These three proteins differed in size and electrophoretic charge yet each clearly carried antibody activity. Two more classes have now been added to the list of human immunoglobulins (IgD and IgE) and it is probable that all five classes occur in most higher mammals. Table I gives the present nomenclature recommended by W.H.O.* and some of the synonyms which have been used in the past.

Table 1

The human immunoglobulins

Present nomenclature	Abbreviation	Previous nomenclature
Immunoglobulin G	IgG	γ-G globulin, 7Sγ-globulin
Immunoglobulin A	IgA	γ-A globulin, β_2A-globulin
Immunoglobulin M	IgM	γ-M globulin, 19Sγ-globulin
Immunoglobulin D	IgD	—
Immunoglobulin E	IgE	Reagin, IgND

Electrophoretically the immunoglobulins show a unique range of heterogeneity which extends from the γ- to the α-fractions of normal serum. In general it is the IgG class which exhibits most charge heterogeneity, the other classes having a more restricted mobility in the slow β- and fast γ-region (see Fig. 1).

Predictably, the heterogeneity of immunoglobulins has made characterization of the normal (polyclonal) antibodies an extremely difficult task.

* This and other aspects of immunoglobulin nomenclature are contained in publications by the Committees on Nomenclature for Human Immunoglobulins, *Bull. Wld. Hlth. Org.* (1964) **30**, 447; (1965) **33**, 721; (1966) **35**, 953; (1968) **38**, 151; (1969) **41**, 975

Indeed, much of our understanding of these proteins has come from studies of the homogeneous immunoglobulins found in the blood and urine of patients with myelomatosis and Waldenström's macroglobulinaemia. These proteins are the products of a proliferating neoplastic plasma cell clone and are pathological counterparts of normal immunoglobulins. Their availability in large amounts has permitted their isolation, classification and characterization not only in Man but also in the mouse and rat. In addition, the

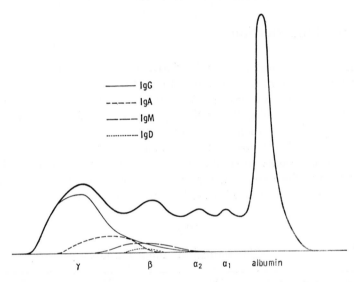

Fig. 1 Electrophoresis of normal human serum showing the distribution of the four major immunoglobulins. IgE is quantitatively insignificant on this scale but has a β-mobility similar to IgD

existence of the IgD and IgE classes was definitively established after the description of "untypeable" myeloma proteins of each class (Rowe and Fahey, 1965; Johansson and Bennich, 1967). A further contribution of myeloma proteins to our general knowledge of the immunoglobulins is in the field of sub-classes. In Man sub-classes of both IgG and IgA are known to occur and are likely for other classes also. Most mammals studied appear to have at least two sub-classes of IgG although these are probably relatively recent evolutionary events which have arisen independently in each species. The four sub-classes of human IgG (IgG1, IgG2, IgG3 and IgG4) are antigenically and chemically distinguishable, differ in their serum concentration and have different biological properties.

In the last 15 years impressive advances have been made in our understanding of immunoglobulin structure. In 1959 Porter showed that rabbit IgG antibodies could be split by the plant protease papain into three large fragments which could be separated by ion-exchange chromatography. Two of these fragments were identical and retained univalent antigen binding capacity (now called Fab: *f*ragment *a*ntigen *b*inding) whereas the third fragment was distinct and could be crystallized (now called Fc: *f*ragment *c*rystalline). The latter fragment has subsequently been shown to be associated with various so-called "effector" or "adjunctive" functions such as Clq binding, macrophage binding and placental transmission.

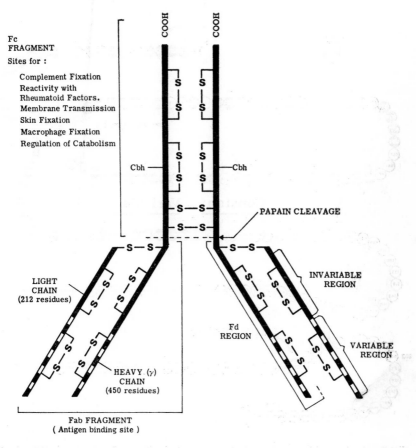

Fig. 2   Schematic diagram of human immunoglobulin G (IgG) molecule. The four chain structure was originally suggested by Porter (1962), the location of the heavy–light chain bridges was deduced by Pink and Milstein (1967) and the distribution of the intra-chain disulphide bonds was determined by Milstein (1964, 1966) and Frangione *et al.* (1969)

Chemical methods of separating the constituent peptide chains of immunoglobulins were described by Edelman and Poulik (1961) and Fleischman *et al.* (1961); shortly afterwards Porter (1962) proposed a four-chain model for immunoglobulin molecules based on two distinct types of polypeptide chain. The smaller (light) polypeptide chain has a molecular weight of 25 000 and is common to all classes of immunoglobulin whereas the larger (heavy) chain has a molecular weight of 50 000 to 77 000 and is structurally distinct for each class, e.g. the γ-chain is the heavy chain found in the IgG molecule, whereas the μ-chain is characteristic of the IgM molecule.

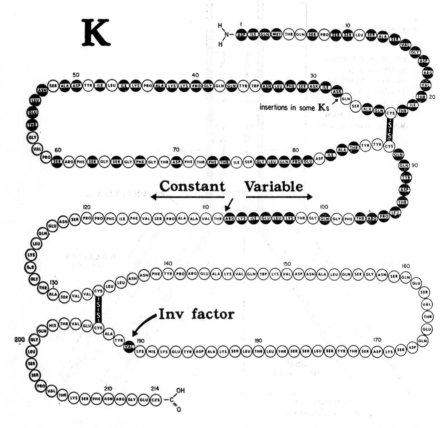

Fig. 3   Amino acid sequence of a human κ chain Bence–Jones protein. The black circles indicate positions where different amino acids were found in other human κ Bence–Jones proteins. Some regions (residues 25 to 34, 50 to 56 and 89 to 97) are known to be hypervariable and to contribute to the antigen binding region. This protein belongs to the Vκ₁ sub-group (see Fig. 4). Courtesy of Dr. F. W. Putnam

The polypeptide chains of immunoglobulins are linked together by both covalent and non-covalent forces to give a four-chain structure based on pairs of identical heavy and light chains (see Fig. 2). IgG, IgD and IgE occur only as monomers of the four-chain unit but IgA occurs in both monomeric and polymeric forms whereas IgM occurs as a pentamer with five four-chain subunits linked together.

The light chains of most vertebrates have been shown to exist in two antigenically distinct forms called kappa (κ-type) and lambda (λ-type). In any one molecule both light chains are of the same type and hybrid molecules have never been observed. In a single individual both types are usually present overall in the ratio 60:40 (κ:λ).

The elucidation of the relationship between myeloma proteins and Bence–Jones proteins on the one hand and normal intact immunoglobulins and free light chains on the other (Edelman and Gally, 1962) led to the application of protein sequencing techniques to these proteins. The work of Hilschmann and Craig (1965) and others established the principle that when light chains of the same type are sequenced they are found to consist of two distinct regions. The carboxyterminal half of the chain (approximately 107 amino acid residues) is constant except for certain allotypic and isotypic variations and is called the C_L-region whereas the amino-terminal half of the chain shows much sequence variability and is known as the V_L-region (see Fig. 3). The sequence variability is not evenly distributed throughout the variable region. Some residues (particularly in the regions 25 to 34, 50 to 56 and 89 to 98) are hypervariable and are thought to be directly involved in the formation of the antigen binding site.

In addition to the hypervariable residues the amino-terminal half of light chains contains residues showing a more restricted variation from protein to protein. When light chain sequences are examined for evidence of homology it is found that the variable regions of both κ- and λ-chains are divisible into sub-groups. Human κ-chains are divisible into three sub-groups and human λ-chains into five sub-groups. Prototype sequences of the first 20 residues of these sub-groups are shown in Fig. 4.

Sequence studies of heavy chains of monoclonal human immunoglobulins have revealed that, in common with light chains, there is a V-region at the N-terminus of the peptide chain. Generally V_H-regions are slightly longer than V_L-regions and contain 118 to 124 amino acid residues. Four regions of hypervariability have been delineated between residues 31 to 37, 51 to 68, 84 to 91 and 101 to 110 respectively. The heavy chain V-regions are not class-specific but are shared by all classes. Three V_H-sub-groups are recognized V_HI, V_HII and V_HIII and, within a sub-group, sequence homology is of the order of 80 to 90% (excluding hypervariable regions). The genetic implications of the variable region sub-groups are discussed later.

As shown in Fig. 2 there are two intrachain disulphide bridges in a light chain but there are four such bridges in each γ- and α-chain and five in each μ- and ϵ chain. Each bridge encloses a peptide loop of 60 to 70 amino acid residues and if the sequences of these loops are compared within a given heavy chain a striking degree of homology is revealed. Each loop represents

SUB GROUPS	SEQUENCE
	κ-CHAINS
VκI	Asp - Ile - Gln - Met - Thr - Gln - Ser - Pro - Ser - Ser - Leu - Ser - Ala - Ser - Val - Gly - Asp - Arg - Val - Thr
VκII	Glu - Ile - Val - Leu - Thr - Gln - Ser - Pro - Gly - Thr - Leu - Ser - Leu - Ser - Pro - Gly - Glu - Arg - Ala - Thr
VκIII	Asp - Ile - Val - Met - Thr - Gln - Ser - Pro - Leu - Ser - Leu - Pro - Val - Thr - Pro - Gly - Glu - Pro - Ala - Ser
	λ-CHAINS
VλI	*Glp - Ser - Val - Leu - Thr - Gln - Pro - Pro - ( ) - Ser - Val - Ser - Gly - Ala - Pro - Gly - Gln - Arg - Val - Thr
VλII	Glp - Ser - Ala - Leu - Thr - Gln - Pro - Ala - ( ) - Ser - Val - Ser - Gly - Ser - Pro - Gly - Gln - Ser - Ile - Thr
VλIII	( ) - Tyr - Val - Leu - Thr - Gln - Pro - Pro - ( ) - Ser - Val - Ser - Val - Ser - Pro - Gly - Gln - Thr - Ala - Ser
VλIV	Glp - Ser - Ala - Leu - Thr - Gln - Pro - Pro - ( ) - Ser - Ala - Ser - Gly - Ser - Pro - Gly - Gln - Ser - Val - Thr
VλV	( ) - Ser - Glu - Leu - Thr - Gln - Pro - Pro - ( ) - Ala - Val - Ser - Val - Ala - Leu - Gly - Gln - Thr - Val - Arg

* The symbol Glp indicates the residue derived from pyrrolid - 2 - one - 5 - carboxylic acid

Fig. 4   Prototype amino acid sequences at the aminotermini of human κ- and λ-chains belonging to different variable region sub-groups

the central portion of what is known as a homology region or "domain" comprising some 110 amino acid residues. In the γ heavy chain the amino terminal variable region (called V_H) is followed by three homology regions comprising the constant part of the chain and these are called the $C_\gamma 1$, $C_\gamma 2$ and $C_\gamma 3$ regions. The four constant region domains of IgM are known as $C_\mu 1$, $C_\mu 2$, $C_\mu 3$, $C_\mu 4$ and the corresponding regions of IgE are $C_\epsilon 1$, $C_\epsilon 2$, $C_\epsilon 3$ and $C_\epsilon 4$. Edelman and Gall (1969) proposed that each homology region is folded into a compact globular structure and linked to neighbouring domains by more loosely folded stretches of peptide chain. Furthermore, Edelman (1970) has suggested that each homology region has evolved to fulfil a specific function e.g. antigen binding is the major role of the V_L-and V_H-domains. There is considerable support for this view.

In both IgG and IgA molecules there exists a section of peptide chain between the first and second constant region domains which is not homologous with any other part of the molecule and is rich in cysteine and proline

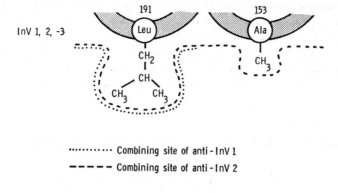

InV 1, 2, -3

191 — Leu
153 — Ala

·············· Combining site of anti - InV 1
– – – – – Combining site of anti - InV 2

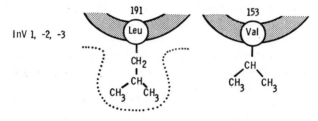

InV 1, -2, -3

191 — Leu
153 — Val

·············· Combining site of anti - InV 1

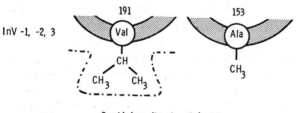

InV -1, -2, 3

191 — Val
153 — Ala

– ·· – ·· – ·· – Combining site of anti - InV 3

Fig. 5   Inv (or Km) antigens and amino acid sequence. Antisera that detect the Inv 1 (Km (1)) antigen probably interact with a leucine residue at position 191 but do not encompass residue 153. In contrast, antisera to Inv 2 (Km (2)) recognize the leucine residue at position 191 but also encompass an alanine residue at position 153 (Top). If the alanine residue is replaced by valine the anti-Inv 2 (Km (2)) reagent is no longer able to recognize the complete antigenic determinant, possibly because of steric hindrance by the larger valine residue (Middle). The third Inv allotype (Inv 3 or Km (3)) is expressed when a valine residue occurs at position 191 and appears to be independent of residue 153 although this is uncertain (Bottom)

amino acid residues. The possible flexibility contributed by the high density of proline residues led to the use of the expression "hinge region" to describe this part of the molecule.

The major physicochemical properties of the classes and sub-classes of human immunoglobulin are shown in Table 2 and the effector functions associated with these proteins are summarized in Table 3.

B. Immunoglobulin Genetic Markers (Allotypes) in Man

All healthy individuals have the sub-class variants of IgG and IgA in their serum and the antigenic markers of these proteins are termed isotypic markers. Certain genetic markers are also identifiable. These are inherited as autosomal co-dominant factors called allotypic markers and are found on γ-chains (Gm antigens), α-chains (Am antigens) and κ light chains (Inv antigens, now Km; see p. 303). An allotypic marker on human IgM, designated Mml, has been described by Wells et al. (1973) but little is known about its structural location.

The Km 1 and Km 3 antigens of κ light chains were the first such allotypic markers to be correlated with specific amino acid substitutions at position 191 (Hilschmann and Craig, 1965). More recently Milstein et al. (1974) have shown that the expression of Km 2 is dependent on the amino acid at position 153. The relationship of Km antigenicity to amino acid sequence is illustrated schematically in Fig. 5.

A large number of allotypic antigens (called Gm markers) have now been described for human γ-chains. These are listed in Table 4 together with the sub-class to which the antigen is restricted. A feature of the Gm system, which has not yet been described in any other animal allotype system, is the association of so-called "iso-allotypes" with many Gm antigens. Such antigens are shared by two or more sub-classes and are structurally antithetic to a Gm marker in one sub-class only. The most widely studied example is nG1m(a), an antigen present on all IgG2 and IgG3 proteins but only on IgG1 molecules lacking the G1m(a) antigen. It is thought that Gm antigens have arisen following mutation in a portion of the C-region gene controlling a sub-class specific segment of the γ-polypeptide chain; G1m (2) and G1m (f) at residue number 214 of the γ_1-chain are examples of such classic allelic antigens. In contrast, mutations in a part of the gene controlling regions which are common to other sub-classes may give rise to a genetic marker in one sub-class (e.g. G1m (a) in IgG1), but the antithetic marker on other IgG1 molecules (nG1m(a)) is shared with two other sub-classes (IgG2 and IgG3).

Table 2

Physicochemical properties of human immunoglobulins

Immunoglobulin	Heavy chain	Sedimentation constant	Molecular weight	Molecular weight of heavy chain	Number of heavy chain domains	Carbohydrate (percentage)
IgG1	γ_1	7S	146 000	51 000	4	2–3
IgG2	γ_2	7S	146 000	51 000	4	2–3
IgG3	γ_3	7S	170 000	60 000	4[a]	2–3
IgG4	γ_4	7S	146 000	51 000	4	2–3
IgM	μ	19S	970 000	65 000	5	12
IgA1	α_1	7S	160 000	56 000	4	7–11
IgA2	α_2	7S	160 000	52 000	4	7–11
sIgA[b]	α_1 or α_2	11S	385 000	52 000–56 000	4	7–11
IgD		7S	184 000	69 700	5	9–14
IgE		8S	188 000	72 500	5	12

[a] The hinge region of IgG3 may incorporate an intra-chain S–S bridge but this would not constitute a domain of 110 residues
[b] Secretory IgA

Table 3

Major effector functions of human immunoglobulins

Immunoglobulin	Mean serum concentration (mg/ml)	Classical complement fixation	Alternate pathway complement activation	Placental transfer	Binding to mononuclear cells	Binding to mast cells and basophils	Reactivity with staphylococcal protein A
IgG1	9	++	−	++	+	−	++
IgG2	3	+	−	+++	−	−	++
IgG3	1	+++	−	++	+	−	−
IgG4	0·5	−	−ᵃ	+	+	(?)	+
IgM	1·5	+++	−	−	−	−	−
IgA1	3·0	−	++	−	−	−	−
IgA2	0·5	−	+	−	−	−	−
sIgA	0·05	−	−	−	−	−	−
IgD	0·03	−	−	−	−	−	−
IgE	0·00005	−	−ᵇ	−	−ᵇ	+++	−

ᵃ Aggregated molecules may activate complement by alternate pathway
ᵇ Human IgE has been reported to bind to macrophages

A combination of peptide mapping, mild proteolytic fragmentation and amino acid analysis has permitted the partial structural localization of several antigens. Table 5 summarizes some of these data.

Table 4

Established Gm allotypes of human immunoglobulin G

Original nomenclature	New nomenclature	IgG sub-class
a	G1m(a)	IgG1
x	G1m(x)	IgG1
f	G1m(f)	IgG1
z	G1m(z)	IgG1
n	G2m(n)	IgG2
g	G3m(g)	IgG3
b^0	G3m(b^0)	IgG3
b^1	G3m(b^1)	IgG3
b^3	G3m(b^3)	IgG3
b^4	G3m(b^4)	IgG3
b^5	G3m(b^5)	IgG3
c^3	G3m(c^3)	IgG3
c^5	G3m(c^5)	IgG3
s	G3m(s)	IgG3
t	G3m(t)	IgG3

Table 5

Probable structural location of some human γ-chain allotypes and iso-allotypes

Allotype or iso-allotype	Chain	Homology region	Sequence	Amino-acid
G1m(a)	γ_1	Cγ3	355–358	Arg-*Asp*-Glu-*Leu*
nG1m(a)	$\gamma_1,\gamma_2,\gamma_3$	Cγ3	355–358	Arg-Glu-Glu-Met
G1m(f)	γ_1	Cγ1	214	Arg
G1m(z)	γ_1	Cγ1	214	Lys
nG4m(a)	$\gamma_1,\gamma_3,\gamma_4$	Cγ2	309	Val-Leu-His
nG4m(b)	γ_2,γ_4	Cγ2	309	Val-His
G3m(g)	γ_3	Cγ2	296^d	Tyr
nG3m(g)	γ_2,γ_3	Cγ2	296^a	Phe
G3m(b^0)	γ_3	Cγ3	436^b	Phe
nG3m(b^0)	$\gamma_1,\gamma_2,\gamma_3$	Cγ3	436^b	Tyr

[a] and [b] Correlative sequence differences observed in incompletely sequenced chains

IgA2 molecules exist in two allotypic forms called $A_2m(1)$ and $A_2m(2)$. The $A_2m(1)$ molecules have an unusual molecular structure in which the light chains are disulphide bonded to each other but not to the α-chains (Grey et al., 1968). The molecules are stabilized by strong non-covalent forces between the light and heavy chains. A second Am allotype which appears to be located in the F_C region of IgA2 molecules has been described by Wang et al. (1973).

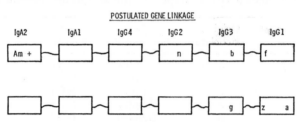

Fig. 6   Postulated genetic linkage between genes coding for the constant regions of IgG and IgA sub-classes. The two main Caucasian complexes of genetic markers are illustrated. Each horizontal row is inherited together as part of a closely linked H chain linkage group. The two complexes are antithetic to each other and are inherited as if controlled by two allelic gene clusters

The properties of the Gm and Am markers suggest that the gene coding for the constant region of a particular heavy chain sub-class (the so-called C gene) is probably not duplicated in a haploid set of chromosomes. The genetic markers at the $C\gamma$ loci rarely recombine with one another and are presumed to be closely linked. There also seems to be a close genetic linkage with the A_2m marker of the IgA2 heavy chain (Kunkel et al., 1969) and it is possible that all the C genes for heavy chains are linked (see Fig. 6).

There is good evidence that the serum levels of certain human immunoglobulins are related to the allotypic antigens expressed by the individual. Yount et al. (1967) showed that healthy subjects with the G3m (b) antigen marker on their IgG3 proteins had twice the serum level of this sub-class as individuals with the G3m(g) antigen. Similarly Steinberg et al. (1973) showed the concentration of IgG4 in the serum is significantly higher in individuals homozygous for the IgG2 allotype G2m (n) than in individuals heterozygous for this allotype and similarly the level of IgG4 in the heterozygotes is significantly higher than in individuals lacking G2m (n). Since the G2m (n) antigen appears to be closely linked to the IgG4 marker, nG4m (b), the authors suggested that the observed relation to IgG4 is probably with the nG4m (b) marker.

More recently Walzer and Kunkel (1974) measured serum IgD concentrations in normal Caucasian blood donors homozygous for either the G1m(a)–G3m(g) or the G1m(f)–G3m(b) linkage groups. By three methods of quantitation the G1m(f)–G3m(b) group had a significantly lower serum IgD level than did the G1m(a)–G3m(g) group indicating a close linkage between IgD and IgG heavy chain C-region genes.

It is also of interest that persons of G1m(a)–G3m(g) phenotype are reported to be better responders to injected flagellin from *Salmonella adelaide* than are individuals with the G1m(f)–G3m(b) phenotype (Wells *et al.* 1971).

C. The Immunogenetic Basis for Antibody Diversity

The origin of antibody diversity is poorly understood at present. Structural studies on immunoglobulin light and heavy chains suggest that the hypervariable regions of the V_L-and V_H-regions come together at the N-terminal surface of the molecule and constitute the "contact residues" of the antigen binding site. One of the major problems of present-day immunogenetics is explaining the origins of this hypervariability. Is it basically achieved by somatic or germ-line processes? The heat generated by controversies for and against both views serve as reminders of the extraordinary developments in this field within the last decade. For example, it seems astonishing to record that the suggestion by Dreyer and Bennett in 1965, that two genes control the synthesis of each immunoglobulin chain, although genetically unorthodox at the time, is now generally accepted. According to this view one gene codes for the constant part of the chain and another for the variable portion. The genes are brought together by an unknown mechanism and thereafter function as a single unit. Studies in several species suggest that there are three distinct linkage groups of immunoglobulin structural genes. These determine the primary structure of κ, λ and heavy chains (see Fig. 7). Within any linkage group any V-region sub-group gene can associate with any C-region gene (Kohler *et al.*, 1970) although some associations may be preferred (e.g. IgA-V_HIII since a high proportion of IgA proteins studied (75%) seem to belong to the V_HIII sub-group).

Recently it has been suggested by Capra and Kindt (1975) that three structural genes may interact to produce each immunoglobulin peptide chain. According to this theory a limited number of "group one" genes code for the relatively invariant portions of the variable region (perhaps one gene for each sub-group); a larger number of "group two" genes code for hypervariable regions; and a small number of "group three" genes would code for the constant regions.

The attraction of the theory proposed by Capra and Kindt (1975) is that it explains the association of identical idiotypes with V_H-regions of different allotype in the rabbit as observed by Kindt *et al.* (1974). However, an alternative explanation of these findings is that parallel evolution of identical

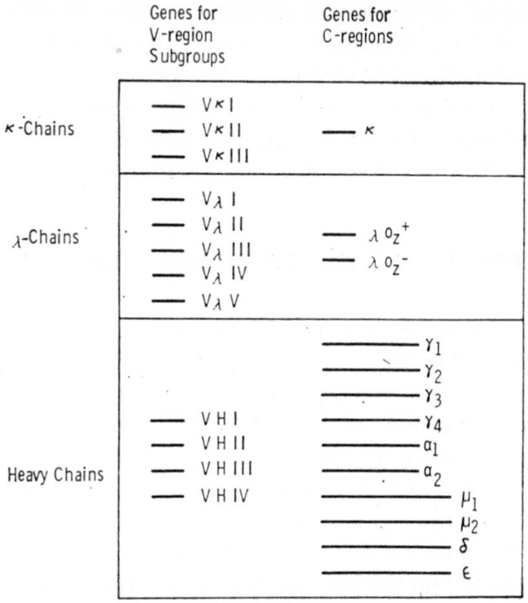

Genes for
V-region
Subgroups

Genes for
C-regions

κ-Chains
— $V\kappa$ I
— $V\kappa$ II
— $V\kappa$ III

— κ

λ-Chains
— V_λ I
— V_λ II
— V_λ III
— V_λ IV
— V_λ V

— $\lambda\, o_z{}^+$
— $\lambda\, o_z{}^-$

Heavy Chains
— V H I
— V H II
— V H III
— V H IV

———— γ_1
———— γ_2
———— γ_3
———— γ_4
———— α_1
———— α_2
———— μ_1
———— μ_2
———— δ
———— ϵ

Fig. 7 Minimum number of genetic systems for human immunoglobulin synthesis. Each box encloses a set of genes (represented by bars) which are probably linked. One V-region gene and one C-region gene contribute to the cistron for each immunoglobulin chain. Other genes (not shown) may contribute to the hypervariable regions. The heavy chain, κ-chain and λ-chain systems appear to be three unlinked systems. In the figure genes are labelled by the name of their protein product

or similar hypervariable regions has occurred and this view receives support from Cohn (1974) and Williamson and McMichael (1975).

Even to attempt to summarize the theories which have been proposed to account for the generation of antibody diversity is far beyond the scope of this chapter but the following reviews are recommended for the interested reader: Cohn (1974); Williamson and Fitzmaurice (1976); Williamson (1976).

D. Genetics of the Immune Response

There is abundant evidence that the immune response is subject to genetic control at several levels. One of the earliest indications of this was obtained by Scheibel (1943). He immunized randomly bred guinea pigs with diphtheria toxoid and divided the animals into good and poor responders to this antigen. Good responders were then bred with good responders and poor with poor until, after several generations, there were populations which were uniformly good or uniformly poor responders to diphtheria toxoid. This experimental approach has recently been exploited successfully in a series of investigations in mice bred for high and low responses to complex immunogens (reviewed by Biozzi et al., 1975). This group of workers has shown that selective breeding for quantitative antibody response to a complex multi-determinant immunogen produces lines of mice endowed with high or low responsiveness to a range of other, unrelated, immunogens. These authors conclude that general immune responsiveness is polygenic in character and probably determined by a group of about 10 independent loci. Biozzi and his colleagues have demonstrated that among this group of genes there is one linked with the H-2 (major histocompatibility) locus and another one with immunoglobulin allotype.

In addition to the Biozzi concept of a polygenic control of general immune responsiveness there exists a vast amount of data in animal systems suggesting specific antigen recognition by the products of individual dominant immune response (Ir) genes located in the genome in close relationship with genes of the major histocompatibility complex (McDevitt and Benacerraf, 1969; Benacerraf and McDevitt, 1972; and cf. p. 395).

T-cell regulation of immune responses is now thought to be mediated through factors acting on both T-cells and B-cells. Such factors may either enhance or suppress responses and may be either specific or non-specific. Katz and Benacerraf (1975) have proposed the existence of a distinct class of molecules present on the surface membranes of T-cells, B-cells and macrophages. It was suggested that these molecules are concerned with effective cell interactions and are coded for by genes in the so-called I region of the H-2 complex. Molecules of molecular weight 40 000 to 50 000 daltons have been obtained from activated T-cells, shown to carry determinants of gene products of the I region, and to have either helper or suppressor activity in both cell mediated and humoral immune responses. In a recent review of this field Benacerraf and Katz (1975) summarized their data as follows: (a) there is a class of molecules distinct from immunoglobulins that is capable of interacting specifically with antigen and which is composed, at least partially, of gene products of the major histocompatibility complex; (b) these molecules are the products of activated T-cells; their functions are to interact with appropriate cell interaction molecules on other T-cells, B-cells and macrophages to control the differentiation of immunocompetent cells in immune responses; (c) the antigen specificities of such molecules may be relevant to the process of selective concentration on to other cells which have bound antigen.

A full discussion of this field is inappropriate in this chapter since, at the present time, there is no evidence of association between Ir gene products and the immunoglobulins, and there is no direct evidence of the existence of *Ir* genes in Man. Nevertheless this is a rapidly developing area of immunogenetics and eventually these concepts may be expected to contribute significantly to our understanding of the aetiology of many diseases (see Chapter 8).

E. Biosynthesis and Metabolism of Immunoglobulins

Lymphoid cells actively secreting immunoglobulins have a network of rough endoplasmic reticulum and synthesize light and heavy chains on polyribosomes of different sizes. L-chain synthesis occurs on polyribosomes sedimenting at about 200S whereas H-chains are synthesized on larger polyribosomes sedimenting at about 300S. In the assembly of four chain immunoglobulin molecules various intermediates are produced and play a role in the control of synthesis. Freshly synthesized light chains are released from the polyribosomes and contribute to an intracisternal pool of free L-chains which are then used to assemble H_2L_2 molecules via various covalent intermediates such as H_2, H_2L or HL. These intermediates vary from class to class depending on the order of disulphide bond formation between H-chains or between H- and L-chains. After the H_2L_2 molecules are released into the cisternae of the endoplasmic reticulum they pass to the exterior via the Golgi apparatus. As this occurs oligosaccharides are added in a definite sequence by membrane bound glycosyl transferases and, in the case of polymeric IgA and IgM, J-chain is also incorporated before polymerization takes place (Parkhouse, 1974).

Much of the data on immunoglobulin biosynthesis has been obtained by studying neoplastic plasma cell tumours maintained in long-term culture or serially transplanted in mice. Many defective products have arisen spontaneously which are identical or analogous to the naturally occurring paraproteins considered in Section III.

Some data on the metabolic characteristics of the immunoglobulins, based mainly on turnover studies using purified radiolabelled proteins, are summarized in Table 6. Each class appears to be metabolically distinct with different synthetic and different fractional catabolic rates. The serum concentration of any one protein is determined by the balance between these two parameters. Thus although serum IgG and IgA have similar synthetic rates the serum concentration of IgA is much less because of the protein's higher fractional catabolic rate. For some of the immunoglobulins the catabolic rate is known to be closely linked to the serum concentration. Thus, in the case of IgG the fractional catabolic rate is directly proportional to the serum concentration and in patients with G-myeloma proteins the protein is degraded much more rapidly than in patients with sex-linked hypogammaglobulinaemia. In the case of IgD and IgE there is apparently an inverse relationship between serum concentration and catabolic rate so that individuals with high serum levels of these proteins also exhibit low fractional catabolic rates for the proteins. For IgM and IgA the fractional catabolic rate is independent of the serum concentration.

Table 6

Metabolic characteristics of human immunoglobulins

Immuno-globulins	Half-life (days)	Distribution (percentage intravascular)	Fractional catabolic rate (percentage intravascular pool catabolized per day)	Synthetic rate (mg/kg/day
IgG1	21		7	
IgG2	20	45	7	33
IgG3	7		17	
IgG4	21		7	
IgM	10	80	8·8	3·3
IgA1	6	42	25	24
IgA2	6			
sIgA	—	—	—	—
IgD	3	75	37	0·4
IgE	2	50	71	0·002

In contrast to the increasing body of knowledge about biosynthetic processes we know very little about the sites of immunoglobulin catabolism. Even their location is unknown. The liver, the gastrointestinal tract and the kidney have been implicated but Waldmann and Strober (1969) have also suggested that diffuse catabolism occurs in the endothelial system all over the body.

F. Development of Immunoglobulins Before and After Parturition

Foetal spleen and liver cell suspensions have been shown to synthesize both IgM and IgG as early as 10 to 12 weeks gestational age (Gitlin and Biasucci, 1969). At 13 weeks cells with intracellular immunoglobulin (IgM) may be identified by immunofluorescence and IgG containing cells may be observed after 20 weeks. IgM may be detected in foetal serum after 17 weeks and rises steadily to about 10 to 20% of adult levels at parturition. Using a sensitive radioimmunoassay technique Faulkner and Borella (1970) found measurable levels (0·15 mg% to 2·15 mg%; mean 0·8) of IgA in 60 specimens of cord blood.

IgG differs from the other immunoglobulin classes in that there exists a highly selective mechanism for the transmission of this protein across the placenta (Dancis *et al.*, 1961; Gitlin *et al.*, 1964). IgG is transported in preference to other immunoglobulins and Gitlin *et al.* (1964) were able to show that the Fcγ fragment of IgG was the most rapidly transported molecule out of the following series; α_1-acid glycoprotein, albumin, transferrin, fibrinogen, IgM, IgG, Fcγ fragment and Fabγ fragment.

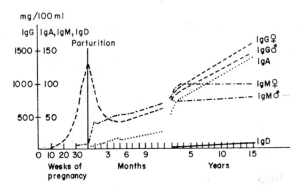

Fig. 8   Development of human immunoglobulin levels from birth to adulthood. By courtesy of Dr. S. G. O. Johansson

Using the Gm genetic markers to study the concentrations of IgG1, IgG2 and IgG3 molecules in cord serum Mellbye *et al.* (1970) found no difference between these sub-classes of IgG as regards placental transmission. The concentrations observed were similar to or somewhat higher than those of maternal sera. Virella *et al.* (1972) have also shown that all four sub-classes cross the placenta with only small differences in the relative proportions of the proteins in matched mother–cord pairs.

After birth the concentration of IgG appears to decrease during the first three months of life and this has been widely interpreted as being due to the catabolism of transplacentally acquired maternal IgG. However, Trevorrow (1959) has shown that the amounts of immunoglobulin are rather constant if the expanding plasma volume (and consequent serum protein dilution) is taken into account.

The level of IgM rises most rapidly after birth to about 75 % of the adult level at one year of age. Thereafter the increase is less pronounced and adult values are reached at about 15 years of age. IgG levels increase less rapidly to reach adult levels in early adolescence. After one year there is a significant sex difference in the levels of both IgG and IgM with females having higher levels of each protein (Berg and Johansson, 1969) (see Fig. 8).

IgA and IgD levels increase relatively slowly with age and no sex difference has been observed. IgE levels in non-allergic individuals rise steadily to reach a maximum at about four years but the range of normal values is very wide (< 1 to > 100 I.U. per ml; mean approximately 20 I.U. per ml).

III. THE IMMUNOGLOBULINOPATHIES

A. Classification and Definition of Terms

The heterogeneity of clinical syndromes associated with abnormalities of the immunoglobulin system has resulted in a complex and confusing terminology. We shall adopt a system proposed by Engle and Wallis (1969). This defines the immunoglobulinopathies (a sub-division of proteinopathies) as those diseases involving either a qualitative or quantitative abnormality in the immunoglobulins. There are three categories of immunoglobulinopathy (see Table 7). Syndromes characterized by a general elevation of one or more of the immunoglobulin classes are termed hyperimmunoglobulinopathies. In some diseases the elevation of immunoglobulin levels is due to the production of an electrophoretically homogeneous monoclonal protein in large amounts. Multiple myeloma and Waldenström's macroglobulinaemia are examples of such disturbances and the term paraimmunoglobulinopathies is suggested for this group. Finally there is a category of diseases characterized by a reduction of one or more classes of immunoglobulins and these are termed the hypoimmunoglobulinopathies.

B. Hyperimmunoglobulinaemia

Diffuse or isolated hyperimmunoglobulinaemia may be of either a primary or a secondary nature. The primary group are usually familial or have an ethnic basis whereas the secondary group are either associated with autoimmune disease and malignancies or are secondary to infections.

A direct genetic aetiology probably underlies all the primary syndromes but the situation with the secondary group is less clear. There now seems little doubt (as discussed earlier) that the immune response is under genetic control at several levels and therefore that hyperimmunoglobulinaemia secondary to infections or autoimmune disease derives in some way from this control. Unfortunately, direct evidence for this assertion is difficult to obtain in Man and we have to rely on indirect data. A familial

Table 7

Classification of Immunoglobulinopathies

		Proteinopathies	
	Immunoglobulinopathies		Disturbances in other plasma proteins
A	B	C	
Hyperimmunoglobulinopathies	Paraimmunoglobulinopathies	Hypoimmunoglobulinopathies	
Diffuse increase in some or all immunoglobulin classes	Characterized by presence of large amount of a monoclonal immunoglobulin in the serum	Deficiency of one or more immunoglobulin classes	
(1) primary	(1) myelomatosis	(1) primary defective synthesis	
(2) secondary	(2) macroglobulinaemia	(2) secondary defects	
	(3) multiple paraimmuno-globulinopathy		
	(4) heavy chain diseases		
	(5) other paraimmunoglobulin fragments		
	(6) associated with other diseases		

predisposition to autoimmune disease has long been known and has been cited by Fudenberg (1967) as an indication of genetic control.

1. Primary hyperimmunoglobulinaemia

Ethnic variations. It has been known for some years that the serum gamma globulin concentration of Negroid Africans is higher than that of Caucasians. Rowe *et al.* (1968) studied more than 1100 individuals from Gambia (West Africa) and showed that compared with a British group IgG and IgM levels were high at all ages. It has been suggested that this difference may be related to the wide range of antigenic stimuli (bacteria, viruses, protozoa and helminths) to which tropical people are repeatedly exposed. In support of this notion Schofield (1957) found that in West Africans resident for several years in the United Kingdom the IgG concentration falls progressively. Nevertheless Cohen and McGregor (1963) showed that such West Africans continue to synthesize IgG at almost twice the rate observed in healthy Europeans. This finding suggests that genetic factors may be significant.

Familial hyperimmunoglobulinaemia. This ill-defined syndrome has been observed in the relatives of patients with rheumatoid arthritis, systemic lupus erythematosus, scleroderma and dermatomyositis. Wilson *et al.* (1967) have described three families with a polyclonal increase in serum IgG, IgM and IgA; in each case a member of the family exhibited anaemia, hepatosplenomegaly, renal tubular acidosis and high latex fixation titres.

Table 8

Hyperimmunoglobulinaemia associated with tropical diseases[a]

Immunoglobulin class	Malaria	Trypanosomiasis	Leishmaniasis
IgG	increased	increased	greatly increased
IgM	increased	greatly increased	increased
IgA	slightly increased	possibly increased	increased
Rheumatoid-factor like globulins	sometimes increased	sometimes increased	often increased

[a] Modified from Houba and Allison (1966)
Data from Cohen and McGregor (1963); Curtain *et al.* (1964); Abele *et al.* (1965); Tobie (1965); Mattern *et al.* (1961) and Lumsden (1965)

2. Secondary hyperimmunoglobulinaemia

Secondary to infections. Infections are often followed by a rise in the level of one or more classes of immunoglobulin. In association with this rise, or possibly even without it, there is a rise in antibody to the infecting organism. Diffuse hyperimmunoglobulinaemia is frequently seen in the tropics as a result of chronic long-standing infections or repeated antigenic stimulation (see earlier). Extensive studies have been made on three such infections; malaria, trypanosomiasis and visceral leishmaniasis (kalaazar). Table 8 presents some of the data on the immunoglobulins and antibodies detected in these diseases. High levels of antibody to an invading microorganism are of obvious advantage to the host and the frequently observed rheumatoid factor-like antiglobulins in trypanosomiasis and leishmaniasis may represent an attempt by the host to enlarge the size of immune complexes and facilitate their removal by phagocytosis.

C. The Paraimmunoglobulinopathies

The paraimmunoglobulinopathies are neoplastic diseases involving the immunocytes, i.e. cells secreting immunoglobulins. In general, these conditions are characterized by the presence of excessive quantities of homogeneous immunoglobulin (the M-component) in the serum and urine. Table 9 classifies the major forms of paraimmunoglobulinopathy to be discussed.

Table 9

Classification of paraimmunoglobulinopathies

Multiple myeloma	Waldenström's macroglobulin-aemia	Multiple paraimmuno-globulins	Heavy chain diseases	Other para-immuno-globulin fragments
		(a) γ–HCD		
		(b) α–HCD		
		(c) μ–HCD		

The aetiological basis of the paraimmunoglobulinopathies is obscure. Irradiation (Potter and Fahey, 1960; Lewis, 1963), viruses (Parsons *et al.*, 1961; Sorenson, 1961) and mineral oils (Potter and Boyce, 1962) have all been cited as possible factors. In addition to these environmental influences the genetic susceptibility of the individual must also be considered. Thus Williams *et al.* (1967) have reported several monoclonal M-components in healthy members of some families in which the index case had myelomatosis. Similarly, Seligmann *et al.* (1967) have reported M-components in healthy relatives of patients with Waldenström's macroglobulinaemia.

Long-term follow-up of some cases of benign M-components have indicated the late development of myeloma (Kyle and Bayrd, 1966), but there seems to be no correlation between the classes of immunoglobulin represented by the different M-components. It appears that genetic factors may be involved with multifactorial expression.

Brown *et al.* (1967) have reported an extra chromosome in 5 to 20% of metaphases from peripheral blood lymphocytes and marrow smears taken from patients with macroglobulinaemia. The presence of this abnormally large chromosome may be a useful diagnostic aid, especially in studies of those individuals with "benign" M-components.

1. Incidence

In order to evaluate the incidence of the paraimmunoglobulinopathies, Axelsson *et al.* (1966) investigated 6995 persons living in Varmland, a county of Sweden. This represented 60% of the population over the age of 25. Of this number, a total of 64 (seven of whom were over 80) showed an M-component on serum electrophoresis. Further investigations indicated that the M-components belonged to the following immunoglobulin classes: IgG (39), IgA (17) and IgM (5). In two individuals there was evidence for biclonal M-components. This incidence is very similar to that in a larger series investigated by Hobbs (1969) (see Section 2).

2. Multiple myeloma

Multiple myeloma is a malignant neoplastic disease of plasma cells usually manifested by invasive destruction of the skeleton and associated with the presence of an abnormal protein in the blood. Paper or cellulose acetate electrophoresis (see Fig. 9) will readily reveal such an M-component and immunochemical investigations with specific antisera will identify which of the immunoglobulin classes (IgG, IgA, IgM, IgD or IgE) is involved. In many patients light chain synthesis is excessive and free light chains (Bence–Jones proteins) are secreted by the proliferating plasma cells in addition to the complete immunoglobulin. In such cases the myeloma protein and the Bence–Jones protein have the same light chain specificity, i.e. either type K or type L since both the myeloma protein and the Bence–Jones protein have the same light chain (Edelman and Gally, 1962).

Hobbs (1969) has analysed the immunochemical features of 212 consecutive cases of clinical myelomatosis and the incidence of the immunoglobulin classes in this series are given in Table 10. Such a small series

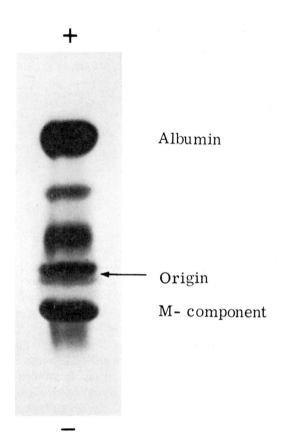

Fig. 9 Cellulose acetate electrophoresis of serum from a patient with multiple myelomatosis

Table 10

Incidence of immunoglobulin classes in 212 cases of myeloma[a]

	Percentage of total cases	Percentage of each with Bence–Jones protein
IgG	53	62
IgA	25	70
IgD	1	100
Biclonal	2	—
Bence–Jones only (light chain disease)	19	100

[a] Data from Hobbs (1969)

would not be expected to include an IgE myeloma (world total to date seven cases) and could also fail to include an IgD myeloma (frequency 2%). In general the frequencies of the various classes of paraprotein is in proportion to the serum concentrations of non-pathological immunoglobulin. Hobbs found that the age of clinical presentation of myelomata was similar to that reported by Innes and Newall (1961) with a main peak at 60 to 70 years (mean age 62). There appears to be little sex difference.

3. Waldenström's macroglobulinaemia

Waldenström (1944) was the first to describe this neoplastic condition arising from the proliferation of lymphoid cells. The clinical picture is usually one of lymphocytosis, enlargement of lymph nodes and spleen, bone marrow infiltration and the presence of abnormal levels of IgM in the blood. Bone pain is infrequent and X-rays of bones rarely reveal the punched-out lesions characteristic of myelomatosis.

Macroglobulinaemic patients are in an older age group than patients with myelomatosis. Thus the peak of age of incidence of myelomatosis is 50 to 65 years whereas that of macroglobulinaemia is 60 to 80 years (Martin, 1970). Another difference is that whereas in myelomatosis both males and females are equally affected, in macroglobulinaemia there is a ratio of about two males for every female. Finally, the prognosis of Waldenström's macroglobulinaemia is strikingly better than that of multiple myeloma. Cohen et al. (1966) have reported a survival range of 23 to 152 months from the estimated date of clinical onset of the disease.

4. Multiple paraimmunoglobulins

Hobbs (1969) found more than one paraprotein in 2% of cases of clinical myelomatosis. This is somewhat similar to the figures reported by Sanders et al. (1969), Imhof et al. (1966) and Bachmann (1965). (See Table 11.)

Sanders et al. (1969) have presented clinical, structural and cellular studies in three such patients with lymphoproliferative disorders. One patient had three paraproteins (G-myeloma protein, A-myeloma protein and Waldenström macroglobulin) in the serum plus Bence–Jones protein in the urine. Two patients had both G-myeloma protein and Waldenström macroglobulin in the serum. Two of the patients presented with clinical features similar to macroglobulinaemic lymphoma whilst the third patient had an oral plasmacytoma. Histological studies indicated lymphocytic–plasmocytic proliferative disorders indistinguishable from Waldenström's macroglobulinaemia. Immunofluorescent studies in the patient with three paraproteins indicated that single cells were capable of producing at least

Table 11

Frequencies of multiple immunoglobulin[a] abnormality

Series	Total number of paraproteins	Percentage of multiple paraproteins	Percentage frequency						
			M+G	M+A	G+A	D+G	G+G	M+G+A
National Cancer Institute, USA	802	0·9	0·4	—	0·2	0·1	0·1	0·1
Imhof et al. (1966)	390	1·3	0·3	0·3	0·5	—	0·3	—
Bachmann (1965)	585	1·2	—	—	1·2	—	—	—

[a] Modified from Sanders et al. (1969)

two of the proteins. This can be interpreted as evidence that separate proteins may be produced in a single neoplastic clone rather than by separate tumours.

More recently a patient (Til) has been described (Wang *et al.*, 1969, 1970) with a double myeloma (IgG2 and IgM) present simultaneously and with both proteins sharing idiotypic determinants, identical κ-chains and identical V_H-region amino acid sequences in the γ_2- and μ-chains. The shared V_H-region suggests that two sets of plasma cells arose from a single precursor with the potential to express both C_γ and C_μ genes. A second case of double IgG and IgM myeloma has also been reported by Penn *et al.* (1970) to exhibit shared idiotypic determinants.

5. Heavy chain diseases (HCD)

γ-Heavy chain disease

(a) Clinical and laboratory features. In 1964 Franklin *et al.* and, independently, Osserman and Takatsuki (1964) described a total of five patients with clinical evidence of a malignant lymphoma. The main symptoms were a generalized painful lymphadenopathy associated with hepatosplenomegaly and pyrexia. Laboratory investigations revealed the presence of an abnormal γ-heavy chain in the serum of each patient and led to the recognition of a distinct clinical syndrome associated with the production of heavy chain fragments (γ-HCD). At the time of reviewing these diseases in 1972, Frangione and Franklin (1973) reported a world total of 30 patients with γ-HCD. Clinical details were available for 24 patients and were rather varied although usually resembling malignant lymphoma rather than multiple myeloma. The disease occurs more commonly in males than in females and is most often seen in the elderly. In addition to the features associated with the first case, anaemia and a normal bone X-ray have also been frequently noted. Unlike myelomatosis Bence–Jones protein is not usually present in the urine although a heavy proteinuria with a protein excretion of 4 to 20 g/24 hours may be observed. Examination of the patient's plasma proteins usually reveals a reduced albumin concentration, hypogammaglobulinaemia and an M-component in the β-γ region. Once the disease is suspected on clinical grounds the final diagnosis can only be made by careful immunochemical characterization of the serum and urine proteins.

(b) Immunochemical investigations. The γ-HCD proteins usually have sedimentation coefficients of 3·5 to 4·0S and a molecular weight of 45 000 to 55 000 although some larger proteins (molecular weight 80 000) are also recorded. The heavy chains of the 26 proteins reviewed by Frangione and Franklin (1973) were related to all four sub-classes; 20γ_1, one γ_2, four γ_3 and one γ_4.

On the basis of structural studies from several laboratories Frangione and Franklin (1973) have proposed that the known γ-HCD proteins be classified into four types. Three of these are illustrated in Fig. 10.

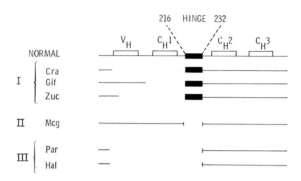

Fig. 10   Three types of γ-heavy chain disease protein compared to normal γ-chain. In Type I proteins the C_H1 region is deleted and part of the V_H-region may also be absent. A normal amino acid sequence resumes at position 216. In Type II proteins only the hinge region is deleted and in Type III proteins both the hinge and the C_H1 regions are absent together with much of the V_H-region

Type I. These proteins consist of γ-chains with a partial deletion of the Fd-region. They have both a normal hinge and a normal Fc-region. Protein Cra (Franklin (1964)) is a $\gamma_1\ V_H$III protein with only 10 to 11 V_HIII residues from the N-terminus linked directly to residue 216 (glutamic acid). Unlike normal γ_1-heavy chains it has three inter-heavy chain S–S bridges (see Fig. 11) since the cysteines normally linked to light chains have formed an extra bridge. Protein Gif (Cooper *et al.*, 1972) is a γ_2 protein lacking approximately 100 residues corresponding to the $C_\gamma1$ region and protein Zuc is a γ_3 protein with a gap of about 200 residues starting 17 to 18 residues after the N-terminus (Prahl, 1967; Frangione and Milstein, 1969).

Type II. A single protein, McG, represents this type of γ-HCD. It was described by Deutsch and Suzuki (1971) and shown by Fett *et al.* (1973) to lack 15 amino acid residues from the hinge region (residues 216 to 232).

Type III. Two proteins of this type (Par (γ_1) and Hal $(_2\gamma)$) have been described (Calvanico *et al.*, 1972; Frangione *et al.*, 1972). Both of these proteins have a molecular weight of about 55 000 in neutral solvents but in acid conditions this falls to about 25 000. Chemical investigations suggest that most of the Fd-region and all of the hinge region (with the inter-heavy chain S–S bridges) are absent. The two portions of γ-chain exist as a dimer stabilized by non-covalent bonds.

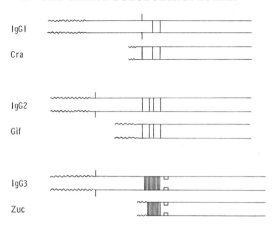

Fig. 11   Details of the structure of three Type I γ-HCD proteins showing the peptide chains and inter-chain S–S bonds. Normal immunoglobulins of the same sub-class are included for comparison. Intra-chain S–S bridges are omitted for the sake of clarity. Note the absence of the H–L bond on all γ-HCD proteins. In the case of protein Cra the cysteines normally forming such bonds have joined together to give an additional inter-heavy chain bond

Type IV. Proteins which may have arisen by enzymatic cleavage of a larger chain are grouped together as Type IV proteins. For example some of these proteins have N-terminal residues in the hinge region e.g. two proteins reported by Terry and Ein (1971) and others in the inter-domain region between the V_H and C_H1-regions. Usually there is insufficient data to establish whether such proteins represent undegraded gene products or are the result of postsynthetic enzymic degradation. Protein Cra (Type I) is also known to be rather heterogeneous at the N-terminus and may also have been subject to postsynthetic degradation.

(c) Genetic implications of structural studies. Frangione and Franklin (1973) suggest three possible explanations for the observed deletions of γ-HCD proteins: (i) deletion of a section of a gene, perhaps arising during cell division by mispairing of DNA strands; (ii) deletion of a section of a gene, reinitiation at the beginning of another gene and recombination or fusion of the transcribed genes; and (iii) DNA breakage at a hypothetical branch point and some place distal to it, with rejoining of the end of the distal portion of DNA back to the branch point (a mechanism first proposed by Smithies *et al.*, 1971).

The frequent observation of a deletion ending at position 216 led Franklin and Frangione (1971) to suggest that the codon(s) specifying glutamic acid, at residue 216, could represent the beginning of another gene implying that the Fc-region ($C_\gamma2$ and $C_\gamma3$ homology regions) and the

$C_\gamma 1$ region are under separate genetic control. If this is so the H-chain would be under the control of at least three structural genes since the V_H-region is known to be separately controlled (see also p. 647). The deletion of the hinge region in protein Mcg has been interpreted by some authors as evidence for yet another gene controlling the hinge region (Franklin and Frangione, 1971; Fett et al., 1973).

Structural studies on the Type III protein Hal have shown that the IgG4 sequence commences at residue 252 (methionine) and Frangione et al. (1973) have suggested that in the case of the Hal protein the AUG codon is a site for reinitiation of the chain and that in normal individuals a separate cistron begins at position 252.

(d) Is there a normal counterpart of γ-ECD? Lam and Stevenson (1973) have reasoned that the proteins secreted by many tumours of the lymphoid system have subsequently been identified at low levels in normal individuals. Bence–Jones protein, IgD and IgE are well-known examples of this. Accordingly these authors undertook a search for γ-HCD in normal plasma using three criteria for identification (i) a molecular weight appreciably lower than that of IgG, (ii) antigenic identity with Fc_γ but not Fab_γ and (iii) the presence in the N-terminal peptide of PCA (pyrrolidone carboxylic acid) which is characteristic of many human γ-chains. The authors isolated 2 mg of Fc-γ like protein from 12 litres of human plasma by a combination of gel filtration, ion exchange chromatography and immunoadsorption. The isolated protein sedimented at 3·9S reacted with anti-Fc_γ but not with anti-Fab_γ or anti-light chain and was shown to contain N-terminal PCA. Thus the isolated protein appeared to satisfy the original criteria for the identification of partially deleted γ-chain. The authors concluded that deletions in the γ-HCD proteins were not attributable to some characteristic of the neoplastic process but were probably the result of a genetic accident which could occur in either normal or neoplastic lymphocytes. The frequency of such events is clearly very low and the products are usually only recognized when neoplastic cell clones are involved.

α–Heavy chain disease

(a) Clinical and laboratory features. To date over 50 cases of α-chain disease have been recorded and two major clinical forms of the disease are now recognized. The first case to be described was a patient who presented with abdominal lymphoma and diffuse lymphoplasmocytic infiltration of the small intestine (Seligmann et al., 1968; Rambaud et al., 1968). This form of the disease is the most common and has been known for several years as "Mediterranean lymphoma" although it is not confined to this geographical area. Patients usually have severe malabsorption and diarrhoea and the disease generally pursues a progressive and fatal course. The second type of α-HCD was described by Stoop *et al.* (1971) in a Dutch child. Instead of intestinal involvement the patient presented with a diffuse lymphocytic plasmacytic reticular cellular infiltrate limited to the respiratory tract. A second child with a similar syndrome has also been described by Rosen (cited by Frangione and Franklin, 1973).

The diagnosis of α-HCD is particularly difficult and depends on careful laboratory investigations to show the presence of the α-chain fragment in the serum without associated light chain. This is made difficult by the well-known tendency of many IgA myeloma proteins to fail to react with anti-light chain antisera. In addition the α-HCD proteins are rarely detectable in the urine, a feature of γ-HCD which makes diagnosis considerably easier.

(b) Immunochemical investigations. The molecular weights of α-HCD proteins have been reported to range from 36 900 to 56 000 (Dorrington *et al.*, 1970; Frangione and Franklin, 1973) (see Table 12) and the proteins

Table 12

Molecular weights of α-HCD proteins[a]

	Monomer	Polypeptide without carbohydrate
$α_1$-chain	56 600	52 200
Al	38 600	34 200
AZ	37 000	31 600
DE	36 900	29 300
TL	34 500	29 500
KH	42 000	29 000

[a] Data reproduced from Frangione and Franklin (1973) with permission

generally have a large amount of carbohydrate, notably N-acetyl-galactosamine. All of the 50 proteins studied to date belong to the α1 sub-class and, in common with γ-HCD, light chains are invariably absent. Antigenic analyses have indicated the presence of an intact $Fc_α$-region in several proteins (Dorrington *et al.*, 1970) but the N-terminus of most proteins appears to be heterogeneous (Frangione, 1975) which poses two alternatives; the α-HCD may be either degraded normal α-chains or defective (deleted) α-chains which have undergone postsynthetic degradation. Wolfenstein-Todel *et al.* (1974) have investigated the α-HCD protein DEF and found a short heterogeneous N-terminal stretch of peptide probably corresponding to part of a V-region, absence of the rest of the V-region and the $C_α1$ domain and initiation of the α-chain sequence at a valine residue just before the hinge region (see Fig. 12). The authors postulated that the protein was

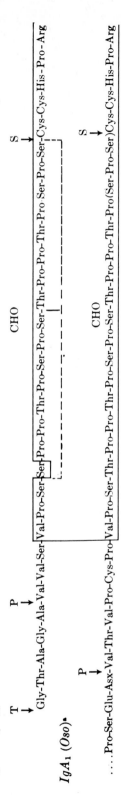

Fig. 12 Comparison of amino acid sequences in the hinge regions of α-HCD protein DEF and the normal α_1-chain from myeloma OSO. The sequences are essentially identical after the valine residue as indicated by the boxed in section. The amino acid sequence of α_1HCD (Def) which is shown on the N-terminal side of this valine residue is probably a portion of the V-region. Arrows indicate the sites of pepsin (P) trypsin (T) and subtilisin cleavage. The dashed line indicates a duplicated region and CHO indicates a carbohydrate moiety. (Reproduced with permission from Wolfenstein-Todel *et al.*, 1974)

synthesized as an internally deleted αl-heavy chain and subsequently underwent N-terminal proteolysis. Frangione (1975) has suggested that the valine residue before the α_1 hinge region may be the counterpart of the glutamic acid residue at position 216 of the γ-chain i.e. the translation product marking a switch point from one gene to another.

μ-Heavy chain disease
   (a) Clinical and laboratory features. Up to mid-1975 only six additional cases of μ-HCD had been reported since the first case was described by Forte *et al.* (1970). The general clinical picture is one of long standing chronic lymphocytic laukaemia involving spleen, liver and abdominal lymph nodes but with little peripheral lymphadenopathy.

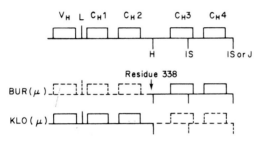

Fig. 13   Two variants of μ-HCD (BUR and KLO) compared to the homology regions and disulphide bridges of a normal μ-chain. H is the inter-heavy chain S–S bridge; IS is the inter-subunit S–S bridge and J the μ-chain–Jchain S–S bridge. The dotted lines indicate deletions. Reproduced with permission from Frangione (1975)

   The diagnosis of μ-HCD is suggested by immunoelectrophoretic identification of μ-chain antigenicity in the α_1 position. The abnormal protein is usually present in low concentrations in the serum and is generally not found in the urine. In contrast to γ-HCD and α-HCD, light chains have been found to be synthesized and secreted in five of the seven cases.
   (b) Immunochemical investigations. Most of the μ-HCD proteins appear to exist in the serum as pentamers resembling the $(Fc_\mu)_5$ fragment. However, the proteins have various molecular weights from 180 000 to 300 000 daltons and J-chain is present in some but not in others. Despite this heterogeneity alanine has been reported to be the N-terminal residue of each protein (Frangione, 1975). A partial amino acid sequence has been published for one of the proteins (Labreton *et al.*, cited by Frangione 1975) and has identified the N-terminus of the molecule as residue 338 (Ou sequence of Putnam *et al.*, 1972, 1973). Thus the deletion comprises the whole of the V-region and both the $C_\mu 1$ and $C_\mu 2$ domains (see Fig. 13).

In contrast the μ-chains of another protein described by De Coteau *et al.* (1973) appeared to lack only the $C_\mu 3$ and $C_\mu 4$ domains to give a molecular weight of 42 000 instead of 68 000 daltons. This protein was synthesized and secreted as $F(ab')_2\mu$ fragments with a molecular weight of 130 000 daltons.

6. Other paraimmunoglobulin fragments

Solomon *et al.* (1966) showed that in one-third of patients with typical Bence–Jones proteinuria there were also present fragments equal to half a light chain (i.e. molecular weight 11 000). Hobbs and Jacobs (1969) cited two cases in which such half chains were the only detectable paraprotein but there was no information on the origin of such fragments which may have represented post biosynthetic degradation products. Hobbs and Jacobs (1969) have also described a half-molecule IgG-plasmacytoma (molecular weight 75 000) which may have arisen as a result of a deletion in the heavy chain region responsible for heavy–heavy S–S bridge formation. The clinical picture resembled HCD with extensive plasmacytoma throughout the lymph nodes, liver and spleen but no certain bone involvement.

Another aberrant protein (SAC) was described by Lewis *et al.* (1968) and further characterized by Smithies *et al.* (1971). This protein sedimented at 5·4S and had a molecular weight of 125 000 daltons. Structural studies showed an internal deletion of the light chain (residues 19 to 99) and the absence of a V_H-region. It was suggested by Parr *et al.* (1972) that the latter arose from proteolysis facilitated by the structural abnormalities in the light chain. Isobe and Osserman (1974) have described an IgG myeloma protein (SM) with a similar defect.

Despont *et al.* (1974) and Despont and Abel (1974) have described a monomeric IgA1 myeloma protein (VO) with a normal light chain but with an α-chain of low molecular weight (42 000 instead of 58 000). Structural studies indicated the absence of the C-terminal $C_\alpha 3$ domain.

D. Hypoimmunoglobulinopathies

The hypoimmunoglobulinopathies are a fruitful area for the geneticist since practically all the well defined diseases in this category have a genetic basis. Nevertheless there are few examples of structural gene defects and in most cases the disease appears to result from the failure of a regulatory process.

Progress in recognizing and understanding immunodeficiency diseases, with or without a genetic aetiology, has been slow and the classification of immunity deficiency states has been a difficult task with few suggestions meeting universal approval.

Some of the immunodeficiencies for which there are clear genetic aetiologies are discussed in Section 1 below. In addition, patients with immunodeficiency secondary to hypercatabolism have been described and a genetic aetiology for their disease has been established. This group of disorders is discussed in Section 2 below.

1. Primary defects

Bone marrow stem cells differentiate into two distinct lymphocytic populations—the T- and B-lymphocytes. It is the B-lymphocytes which differentiate and mature into plasma cells capable of immunoglobulin synthesis. Thus, any limitation imposed on B-cell differentiation will lead to a state of hypoimmunoglobulinopathy. However, it is also now clear that the presence of T-lymphocytes is necessary for B-lymphocytes to be able to respond to most antigens and mature to make antibody. The mechanism of T- and B-lymphocyte cooperation is poorly understood and is known to involve macrophages. Thus a defective T-cell system will be manifested as (a) poor or absent cell-mediated immunity and (b) a poor humoral antibody response in those instances where T-lymphocyte cooperation is important. In the latter case, the individual will show a poor antibody response despite the fact that the B-cell system is functionally intact. It is clear that the complex cellular interactions required for immune competence could fail in many ways and this is reflected in the wide spectrum of immunodeficiency diseases.

Classification of primary hypoimmunoglobulinopathies. The primary hypoimmunoglobulinopathies may be defined as those syndromes resulting from defects in the expression of the B-cell system. An adequate classification of these deficiency diseases is not possible at this stage but certain diseases are clearly distinctive as regards aetiological basis, functional deficiency etc. and these are listed in Table 13. Nevertheless most cases of immunodeficiency defy classification and these have been termed "variable immunodeficiency" syndromes. This group includes cases previously classified as "congenital" autosomal or sporadic hypogammaglobulinaemia and primary "dysgammaglobulinaemia" of both childhood and adult life.

Description of syndromes

(a) Congenital panhypogammaglobulinaemia. In 1952 Bruton described a boy with an abnormally low level of "gammaglobulin" who had experienced frequent infections including pneumonia, pneumococcal septicaemia and meningitis. The recognition of a similar clinical presentation

Table 13

Classification of primary hypoimmunoglobulinopathies

	Defect	
	Antibody	Cell mediated
Congenital panhypogammaglobulinaemia	+	
Selective IgA deficiency	+	
Transient hypogammaglobulinaemia	+	
Immunodeficiency with ataxia telangiectasia	+	+
Immunodeficiency with thrombocytopenia and eczema (Wiskott–Aldrich syndrome)	+	+
Immunodeficiency with thymoma	+	+
Severe combined immunodeficiency	+	+
Variable immunodeficiency (common, largely unclassified)	+	+

in association with a low gammaglobulin in several other boys led to the suggestion that this was a disease with X-linked inheritance (X-linked agammaglobulinaemia). However, girls too can be affected and it is now more useful to consider as a separate sub-group patients having low levels of all immunoglobulins (panhypogammaglobulinaemia) in contrast to those having selective deficiencies of one or two classes.

X-linked agammaglobulinaemia usually presents between six months and the third year of life. Suppurative otitis media, repeated septicaemia and recurrent sinusitis are symptomatic of this deficiency syndrome. The pyogenic organisms *Staphylococcus aureus, Haemophilus influenzae, N. meningitidis* and *S. pneumoniae* present the greatest difficulty to the patient, and death will ensue if effective antibiotic treatment is not instituted. The resistance of most children to virus infections is normal.

The primary defect in X-linked agammaglobulinaemia is an almost complete failure to synthesize all five classes of immunoglobulin (see Fig. 14). With sensitive techniques some IgG (< 10 i.u./ml) can usually be detected and also traces of IgA and IgM. Metabolic studies have shown that the defect is one of synthesis rather than accelerated breakdown. As might be expected, the patients are unable to produce antibodies, even after repeated antigenic challenge, but their cellular immunity mechanisms are essentially normal (Rosen and Janeway, 1966). Tissue findings are characteristic in this disease. The lymph nodes have practically no primary follicles and on antigenic stimulation fail to form secondary germinal centres and plasma cells. Similarly, plasma cells are absent from the bone marrow, lymphoid organs and intestinal lamina propria. Almost all affected patients lack blood B-lymphocytes but have normal numbers of T-lymphocytes.

X-linked agammaglobulinaemia is transmitted as a sex-linked recessive characteristic (Janeway and Gitlin, 1957). The inheritance is compatible with a single gene defect leading to failure of synthesis of a critical enzyme. Since trace amounts of immunoglobulins are usually detectable it is unlikely that the immunoglobulin structural genes are involved. A possible linkage between the "agammaglobulinaemia" gene and the Xg blood group has been discussed by Schimke and Kirkpatrick (1970) but the genes are separated by at least 30 centimorgans (Rosen *et al.*, 1965).

NHS

AHS

S

Fig. 14   Immunoelectrophoresis of serum from a patient with infantile sex-linked agammaglobulinaemia (S) compared with a normal human serum (NHS), developed using a rabbit anti-human serum (AHS). Anode is to the right. Note the complete absence of the IgG arc in the cathodic region of the patient's serum. Reproduced by courtesy of Farbwerke Hoechst A. G. Frankfurt (M) from Laboratory Notes for Medical Diagnostics

The forms of hypogammaglobulinaemia which are not X-linked are usually characterized by the presence of blood B-lymphocytes. Thus rather than being a failure of B-cell development these patients may manifest a disease due to a failure in the maturation of B-lymphocytes to plasma cells. Serum immunoglobulins of one or more classes may be absent although IgM is usually normal or even high. The clinical presentation of these patients resembles that of boys with X-linked agammaglobulinaemia except that age of onset may be somewhat later.

In all patients with hypogammaglobulinaemia replacement injections of IgG (25 mg per kg per week given intramuscularly) are usually effective in reducing the frequency of infections.

(b) Selective IgA deficiency. Patients with selective IgA deficiency usually have very low levels of this immunoglobulin (< 10 i.u./ml) but normal values for IgG and IgM. This form of primary immunodeficiency is probably the most common and affects approximately one in 700 of the general population. Selected populations (e.g. atopics, various autoimmune disorders) frequently show a higher incidence of IgA deficiency. Patients with selective IgA deficiency have normal or increased numbers of blood lymphocytes with surface IgA (Grey et al., 1971) but lack IgA secreting plasma cells in their tissues (Crabbé and Heremans, 1968). There is evidence that a T-cell abnormality is responsible for the failure of IgA B-lymphocytes to mature into plasma cells.

Many individuals with IgA deficiency are apparently healthy but others experience recurrent upper respiratory tract infections or have various autoimmune diseases.

(c) Transient hypogammaglobulinaemia. All infants have physiologically low levels of IgG between the second and sixth month of life. A persistence of low levels until the twelfth month asssociated with recurrent infections was first described by Janeway and Gitlin (1957) and has been termed transient hypogammaglobulinaemia.

The clinical features of this syndrome are recurrent pyogenic infections similar to those associated with X-linked agammaglobulinaemia. In general the IgA and IgM levels are normal in these children but the diagnosis of the disease is usually only certain after all the immunoglobulin levels (IgG included) have returned to normal.

Unusually prolonged deficiencies of immunoglobulins have been reported in the sibs of some children with X-linked agammaglobulinaemia or severe combined immunodeficiency. It is possible therefore that the disorder represents the heterozygous expression of a defective gene (Kirkpatrick and Schimke, 1967; Soothill, 1968).

The treatment of these children aims at the prevention of serious infections until the immune response systems have matured. Immunoglobulin replacement therapy may be beneficial in certain cases.

(d) Immunodeficiency with ataxia telangiectasia. This disease is a complex multisystem disorder characterized by progressive cerebellar ataxia, oculocutaneous telangiectasia and severe sinopulmonary infections. The ataxia usually manifests itself when walking commences (Peterson et al., 1966) and the telangiectasia appears at about five years of age. The disease has been reviewed in depth by McFarlin et al. (1972).

The immunoglobulin abnormalities are variable but the commonest are a low IgE in about 80% of cases (Polmar et al. 1972) and a low IgA in about 70% of the patients. IgG is usually normal but IgM may be normal or apparently elevated. However, immunochemical studies have revealed that the IgM in a high proportion of patients exists in the monomeric (7S) form which diffuses rapidly in quantitative single radial diffusion to give falsely elevated values.

Pathologically the thymus remains embryonic, lacking both cortical and medullary organization. About one-third of the patients have a low blood lymphocyte count with either a normal or low T-lymphocyte population. Most patients have repeated infections with progressive lung damage.

The neurological symptoms also tend to be progressive and irreversible. Many patients develop malignancies of the reticuloendothelial and central nervous systems.

The disease is inherited as an autosomal recessive disorder. It appears that the genetic defect involves more than the immune system. Schimke and Kirkpatrick (1970) have suggested that the thymic dysplasia and other developmental abnormalities point towards an absent mesenchymal inducer substance (arising as a direct consequence of the mutant gene).

(e) Immunodeficiency with thrombocytopenia and eczema (Wiskott–Aldrich syndrome). In 1937 Wiskott described a familial syndrome characterized by thrombocytopenia, eczema, unexplained diarrhoea and recurrent infections (otitis, pneumonia and meningitis being the most common). Aldrich et al., 1954, delineated the disease as a male sex-linked recessive disorder.

The syndrome involves defects of both cellular immunity and antibody mediated immunity. Early in the course of the disease lymphoid tissues may appear to be normal but, with increasing age, a progressive deficiency of lymphocytes in the paracortical regions of the peripheral lymphoid tissue is noted (Cooper et al., 1968). Thymus architecture is usually normal with an adequate number of Hassall's corpuscles.

There is usually an immunoglobulin deficiency but the class involved and the degree of deficiency is variable. Under the age of one year most affected infants have a normal IgG level with normal or raised IgA and IgM but older children have low IgM levels with normal or low IgG and high IgA (West et al., 1962). Serum IgE is usually elevated in these patients (Berglund et al., 1968).

Blaese et al. (1971) have shown that patients with the Wiskott–Aldrich syndrome have an increased catabolism of both albumin and immuno-globulins but increased synthesis compensates for this in most cases. IgM is an exception and the serum level is therefore significantly lowered.

The children usually have absent or low titres of isoagglutinins and Forssman antibodies are usually lacking. The failure of these patients to respond to pneumococcal polysaccharides led to a suggestion that they have a general inability to recognize and respond to polysaccharide antigens (Cooper et al., 1968). Other studies have shown that patients with this disease make less antibody to particulate antigens such as Salmonella typhi, S. brucella and S. tularaemia vaccines than do normal individuals (Blaese, 1972).

The prognosis is poor and most boys die before the age of ten. Recent attempts at treatment include bone marrow grafts from HLA identical siblings (Bach et al., 1968; August et al., 1973) and transfer factor injections (Spitler et al., 1972).

(f) Immunodeficiency with thymoma. Twenty cases of benign thymoma associated with hypogammaglobulinaemia were reviewed by Jeunet and Good (1968). All patients were adults and the average age was 50 years (range 25 to 77 years). In each case the patient was in good health prior to the appearance, during adult life, of a clinical syndrome marked by repeated pulmonary infections. Of the 20 cases, 12 were females and eight were males. In seven of the cases the mediastinal tumour was known to have been present before the development of immunological deficiency.

In the patients described by Jeunet and Good (1968) a severe depression of the serum IgG level, associated with an absence of IgA and IgM, was noted in most of the cases studied. A patient with this syndrome deficient in IgA only has been described (Amman and Hong, 1973).

Biopsies of peripheral lymphoid tissues have usually revealed an abundance of lymphocytes but absence of germinal centres and a deficiency of plasma cells.

Humoral responses to typhoid and diphtheria antigens are reported to be grossly depressed in these patients and isohaemagglutinins are usually absent.

Thymectomy does not appear to influence the immunological deficiency or restore immunoglobulin levels but the administration of gammaglobulin has been found to improve the patients clinically.

(g) Severe combined immunodeficiency. The condition of severe combined immunodeficiency (SCID) was first described by Glanzmann and Rinicker (1950) in Switzerland and is frequently referred to as "Swiss type agammaglobulinaemia". The disease differs from X-linked agammaglobulinaemia in several respects. Infections are usually of earlier onset and of greater severity and the disease usually follows a rapidly fatal course.

There are variations both in the severity of the disease and the associated laboratory findings suggesting a range of genetic defects. Usually there is an absolute decrease in the number of circulating lymphocytes with few or no T-lymphocytes although B-lymphocytes with surface immunoglobulin may be present even in the absence of circulating immunoglobulins. Antibody formation is invariably impaired or absent and a marked decrease in total immunoglobulins is the general rule although selective deficiencies have also been observed.

Meuwissen (1974) showed that many patients with severe combined immunity deficiency lacked the enzyme adenosine deaminase (ADA) in their red cells. This appears to be a particular characteristic of those infants with the autosomal recessive form of the disease and genetic counselling may be useful for those heterozygous parents having half normal levels of ADA (Scott et al., 1974). Prenatal diagnosis of SCID also becomes possible with cultured cells from amniotic fluid (Hirschhorn *et al.*, 1975) since no ADA deficient Caucasian individuals have yet been identified other than infants with SCID. Since Polmar *et al.* (1975) have shown enhanced thymidine uptake by lymphocytes from an ADA deficient SCID patient when cultured with exogenous ADA *in vitro* it appears likely that the enzyme deficiency is the primary pathogenic lesion. When ADA was given to this patient, by transfusing 100 ml of frozen irradiated erythrocytes, a thymus shadow appeared on a chest X-ray, the blood lymphocyte count rose and these cells became responsive *in vitro* to mitogens.

Tishfield *et al.* (1974) have tentatively assigned the human ADA gene to chromosome 20 so that there is no linkage with the major histocompatibility (*HLA*) locus which is known to be on chromosome 6.

ADA is an aminohydrolase which plays an important role in purine reutilization by converting adenosine to inosine (see p. 541). The erythrocytes of ADA-deficient infants accumulate adenosine triphosphate and convert inosine monophosphate to inosine triphosphate at increased rates. Green and Chan (1973) have suggested that accumulations of adenosine in these infants may be lymphocytotoxic or interfere with pyrimidine biosynthesis by blocking the conversion of orotic acid to orotidine.

As mentioned above there are wide variations in the severity of the disease and some of these variants have attracted specific descriptions. De Vaal and Seynhaeve (1959) reported a profound immunodeficiency disease with a marked decrease in all white blood cells, lymphocytes and granulocytes. Infants with this defect, which is very rare, present during the first days of life with overwhelming infections and anaemia. The term reticular dysgenesis was suggested for this disease since defective stem cell maturation (after divergence of the red cell and platelet precursors) would account for the observed aberrations. The mode of inheritance is not clear.

Another rare variant of combined immunodeficiency is the association of the immune defect with short limbed dwarfism, first reported by McKusick and Cross (1966). The infants with this syndrome characteristically present at a few weeks of age with serious infections (staphylococcal septicaemia, otitis media) and foreshortening of the upper and lower extremities. There appears to be impairment of both humoral and cell mediated immunity in these children.

In some infants the disease may present at a later age and be associated with normal levels of one or more immunoglobulin classes (especially IgM) but with a severe depression of antibody response. This type of presentation has been described as "thymus dysplasia with immunoglobulin" and "Nezelof's syndrome".

The best treatment yet devised for these infants is a bone marrow graft from an HLA matched sibling. These patients are very susceptible to graft versus host (gvh) disease and grafts from donors mismatched at the *HLA-D* locus have usually been fatal. Similarly only irradiated blood or frozen erythrocytes should be transfused to such patients.

(h) Variable immunodeficiency. This group of immunodeficiency diseases was first proposed by Seligmann *et al.* (1968). It includes the various autosomal primary immunoglobulin deficiencies with variable onset and expression e.g. "sporadic hypogammaglobulinaemia", primary "dysgammaglobulinaemias" of both childhood and adult life, "acquired" agammaglobulinaemias and hypogammaglobulinaemias. Within these entities there is variability in the date of onset of disease and in the immunoglobulin patterns observed both from patient to patient and within families with multiple cases.

Usually the immunoglobulin deficiency in this group of patients is less severe than in the X-linked form of agammaglobulinaemia. The common clinical findings are sinusitis, pneumonia, malabsorption and pernicious anaemia. Patients with "acquired" agammaglobulinaemia may have low, normal or increased numbers of B-cells (Preud'homme *et al.*, 1973; Geha *et al.*, 1974). Similarly B-cell responses to mitogens are markedly variable in this disease.

A selective deficiency of one class of immunoglobulin has frequently been observed and it has become accepted practice to distinguish selective IgA deficiency as a separate entity (see Section III.D.1.b) but other selective deficiencies also occur and, for the present at least, are included within the "variable immunodeficiency" group. Primary isolated IgM deficiency (reviewed by Hobbs, 1975) is one such disease sub-group which was found to occur amongst 0·1% of British hospitalized patients. The majority of such patients are male and there is a high frequency of affected relatives. In general the serum level of IgM is more than two standard deviations below the population mean but is rarely completely absent. Sixty per cent of the patients reviewed by Hobbs (1975) had recurrent infections, principally involving the respiratory tract, intestinal tract and meninges.

In addition to selective deficiencies of immunoglobulin classes there are also well-documented deficiencies of IgG sub-classes. These are characterized by a failure to respond to certain antigens and an increased susceptibility to a limited range of microorganisms. Since IgG1 accounts for 70% of the IgG pool a deficiency of this sub-class is associated with the most severe clinical picture (Schur *et al.*, 1970; Yount *et al.*, 1970). More recently Yount (1975) has reported that almost 30% of patients with variable immunodeficiency and hypogammaglobulinaemia had detectable abnormalities in the concentration of various IgG sub-classes. Most commonly these patients were found to have a relative excess of IgG3. A high frequency of similar abnormalities in first degree relatives clearly indicates that the defect has a genetic basis although its nature remains obscure.

The true frequency of sub-class abnormalities is difficult to assess since the majority are never diagnosed. Oxelius (1974) described a family in which some individuals suffered repeated pyogenic infections typical of hypogammaglobulinaemia and yet their total IgG levels were normal. Only after further detailed investigations was it shown that these individuals were deficient in IgG2 and IgG4.

Another group of immunoglobulin disorders which come within the present category are characterized by normal levels of functionally ineffective immunoglobulin. This defect may be general and involve all classes of immunoglobulin or it may be restricted to one or two classes only. Blecher *et al.* (1968) described one such case in whom abnormal susceptibility to infections was suggested clinically by almost continuous infections by numerous different bacteria. Immunoglobulins (IgG, IgM and IgA) were present at normal concentrations for the age of the child but isohaemagglutinins were absent and there was failure to produce antibodies to several bacterial antigens (*Salmonella typhi* O, *S. paratyphi* B, O and H, and *S. pertussis*). Antibody production to other Salmonellae (O and H), staphylococci, diphtheria and tetanus toxoid was normal. Similarly, tests of cellular immunity were normal.

There is a wide range of specific immunodeficiencies and the heterogeneity of syndromes grouped under the heading "variable immunodeficiency" reflects the paucity of data in this field. More adequate evaluation of patients presenting with recurrent infections should assist the future definition of several distinct syndromes. Nevertheless, the complexity and inter-relationships between defects should warn against any optimism that a rational classification of all these diseases will soon be possible.

Potential primary defects? Kunkel *et al.* (1969) have described a family with one member whose serum was devoid of all the usual Gm genetic antigens. The serum lacked ordinary IgG1 and IgG3 proteins but contained instead hybrid molecules of the type IgG3–IgG1. Evidence was obtained to show that in the hybrid protein the N-terminal end was of IgG3 type and the C-terminal end was of IgG1 type. No defects in the light chains were detected. The concentration of the hybrid protein was far higher than the usual IgG3 level but below the normal mean for IgG1, suggesting that the rate of synthesis was influenced both by the gene coding for Fab–IgG3 and the gene for Fc–IgG1. The authors draw a parallel between this hybrid molecule and the $\alpha\beta$ chain hybrids established for Lepore-type haemoglobins. As shown in Fig. 15 an unequal homologous cross-over involving mispairing of heavy-chain cistrons would readily explain the deletion of Gm markers. This depicts diagrammatically the two major gene complexes found in Caucasians with IgG2, IgG3 and IgG1 cistrons placed in the order previously determined by Natvig *et al.* (1967). A mispairing of the IgG1 and IgG3 cistrons at meiosis followed by an intracistronic cross-over would result in a gamete with the genotype indicated at the bottom of Fig. 15. Kunkel *et al.* (1969) suggest that in view of the known homologies of IgG1 and IgG2 cistrons such mispairing might

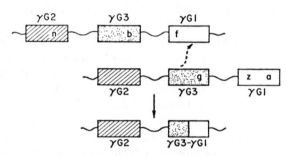

Fig. 15   Diagram of the consequence of mispairing of immunoglobulin genes at meiosis followed by intra-cistronic cross-over leading to an IgG3–IgG1 hybrid gene. The approximate position of the known genetic markers which would be deleted are indicated. Reproduced with permission from Kunkel *et al.* (1969)

well occur. The figure also shows how a single cross-over could result in a hybrid cistron lacking all known Gm genetic markers. The present case was probably homozygous for the hybrid gene since other family members (previously described by Steinberg *et al.*, 1968) appeared to have a similar "silent allele" in the heterozygous state.

Another hybrid protein has been reported by Natvig and Kunkel (1974). This was a myeloma protein present in the serum of a Negro patient. Laboratory investigations showed that this protein was a hybrid of IgG4 and IgG2 molecules with a probable cross-over point between the C_H2 and C_H3 domains i.e. the C_H1 and C_H2 regions were from an IgG4 molecule and the C_H3-region from an IgG2 molecule.

In the first patient described there was evidence that the hybrid molecules were still functional haemagglutinating antibodies but it is perhaps more likely that effector functions (e.g. complement fixation, macrophage binding etc.) would be modified by such a cross-over. Thus, homozygous individuals with this defect (for which no frequency data are available) might have normal levels of immunoglobulin and functional antibodies (in the sense of antigen binding) and yet immunity deficiency might become apparent if there was considerable interference with effector functions. Since IgM is presumably normal in such individuals and, moreover, extremely effective in complement activation, deficiencies of the IgG system might be concealed and special tests would be required to delineate such defects.

2. Secondary defects

Hypogammaglobulinaemia may occur as a secondary phenomenon subsequent to some other disorder. For example, in both the nephrotic syndrome and protein-losing enteropathy there is a loss of protein from the body but in neither case is this specific to the immunoglobulins.

Another group of secondary defects, more interesting to the student of genetic disorders, are those associated with disorders of catabolism.

Familial idiopathic hypercatabolic hypoproteinaemia. Waldmann *et al.* (1968) have described a disorder involving hypercatabolism of different classes of serum proteins, including IgG and albumin. Two siblings, a 34-year-old woman and a 17-year-old man (offspring of a first-cousin marriage) were shown to have a marked reduction of serum IgG (130 and 440 mg%) and of albumin (1900 and 2100 mg%). The concentrations of IgM and IgA were essentially normal. Laboratory investigations showed that, despite normal synthetic rates, the total body pools of IgG and albumin were significantly reduced. The fractional catabolic rate of IgG was found to be increased five-fold in both patients to 35% of the intravascular pool per day. There was no proteinuria and ^{57}Cr albumin clearance studies were normal indicating no abnormal gastrointestinal loss. The authors concluded that the patients were suffering from a disorder involving a familial defect in the endogenous catabolism of serum proteins.

Dystrophia myotonica. This is an inherited progressive muscular abnormality characterized by muscle wasting and weakness, myotonia, personality changes, lens opacities, skull radiographic changes, electrocardiographic changes and low serum levels of immunoglobulin G.

Wochner *et al.* (1966) studied 19 patients with myotonic dystrophy and found the serum concentrations and metabolism of albumin, IgM, IgA and IgD were all within normal limits. The serum concentration of IgG, however, was only 700 mg% (compared to 1200 mg% in healthy controls). Furthermore it was observed that the half-life of IgG in these patients was only 11·4 days (compared to 23 days in controls) and that the fractional catabolic rate for IgG was 14% of the intravascular pool per day (6·7 days in controls; see Table 6). Since the IgG synthetic rate was found to be normal it was concluded that the low serum IgG concentration was due to

680   M. W. TURNER

hypercatabolism. Further investigations were made into the nature of the
defect by isolating IgG from the serum of patients with the disease,
labelling it with a radioactive isotope and injecting it into both control
subjects and the original patients. The controls catabolized the IgG at
normal rates but the patients with myotonic dystrophy catabolized the
myotonic IgG at accelerated rates. It appears therefore that IgG is synthesized by these patients in normal amounts and is qualitatively indistinguishable from any other "normal" IgG.

In a study of 45 patients with myotonic dystrophy Bundey et al. (1970)
found that the serum concentrations of both IgG and IgM were low and, in
addition, there was an insignificant trend for IgA to be low also. This
suggests that the abnormally rapid catabolism of immunoglobulin may not
be specific for IgG. Bundey et al. (1970) proposed that the exceptionally
long half-life of IgG makes the detection of metabolic abnormalities in this
immunoglobulin much easier than for IgA and IgM with their short half-
lives (Table 6.). Measurement of serum concentrations of IgA and IgM may
be the most sensitive test available for the detection of abnormalities
involving these classes. Further studies in this field are required since the
results would have direct bearing on the ill-understood problem of control
mechanisms for immunoglobulin catabolism.

REFERENCES

Abele, D. C., Tobie, J. E., Gill, G. J., Contacos, P. G. and Evans, C. B. (1965).
   Am. J. Trop. Med. Hyg. 14, 191.
Aldrich, R. A., Steinberg, A. G. and Campbell, D. C. (1954). Pediatrics 13, 133.
Ammann, A. J. and Hong, R. (1973). Cellular Immunodeficiency Disorders. In
   "Immunologic Disorders in Infants and Children" (E. R. Stiehm and V. A.
   Fulginiti, Eds). Saunders, Philadelphia.
August, C. A., Hathaway, W. E. and Githens, J. G. (1973). J. Pediat. 82, 58.
Axelsson, U., Bachmann, R. and Hällen, J. (1966). Acta Med. Scand. 179, 235.
Bach, F. H., Albertini, R. J., Joo, P., Anderson, J. L. and Bortin, M. M. (1968).
   Lancet ii, 1364.
Bachman, R. (1965). Acta Med. Scand. 177, 593.
Benacerraf, B. and Katz, D. H. (1975). In "Immunogenetics and Immuno-
   deficiency" (B. Benacerraf, Ed.) MTP, Lancaster.
Benacerraf, B. and McDevitt, H. O. (1972). Science 175, 273.
Berg, T. and Johansson, S. G. O. (1969). Acta Paediat. Scand. 58, 513.
Berglund, G., Finnström, O., Johansson, S. G. O. and Möller, K. L. (1968).
   Acta Paediat. Scand. 57, 89.
Biozzi, G., Stiffel, C., Mouton, D. and Bouthillier, Y. (1975). In "Immunogenetics
   and Immunodeficiency" (B. Benecerraf, Ed.) MTP, Lancaster.

Blaese, R. M. (1972). *Ann. Int. Med.* **77**, 605.

Blaese, R. M., Strober, W., Levy, A. L. and Waldmann, T. A. (1971). *J. Clin. Invest.* **50**, 2331.

Blecher, T. E., Soothill, J. F., Voyce, M. A. and Walker, W. H. C. (1968). *Clin. Exp. Immunol.* **3**, 47.

Brown, A. K., Elves, M. W., Gunson, H. H. and Pell-Ilderton, R. (1967). *Acta Haemat.* **38**, 184.

Bruton, O. C. (1952). *Pediatrics* **9**, 722.

Bundey, S., Carter, C. O. and Soothill, J. F. (1970). *J. Neurol. Neurosurg. Psychiat.* **33**, 279.

Calvanico, B. R., Plaut, A. and Tomasi, T. B. (1972). *Fed. Proc. Abst.* 3124.

Capra, J. D. and Kindt, T. J. (1975). *Immunogenetics* **1**, 417.

Cohen, R. J., Bohannon, R. A. and Wallerstein, R. O. (1966). *Am. J. Med.* **41**, 274.

Cohen, S. and McGregor, I. A. (1963). *In* "Immunity to Protozoal Infections" (P. C. C. Garnham, Ed.) 123, Blackwell Scientific Publishers, Oxford.

Cohn, M. (1974). *In* "Progress in Immunology" (L. Brent and E. J. Holborow, Eds) Vol. 2, 261. North Holland and American Elsevier Publishing Companies, New York and Amsterdam.

Cooper, M. D., Chase, H. P., Lowman, J. T., Knvit, W. and Good, R. A. (1968). *Ann. J. Med.* **44**, 499.

Cooper, M. R., Dechatelet, L. R., La Vie, M. F., McCall, C. E., Spurr, C. L. and Bachner, R. L. (1972). *J. Clin. Invest.* **51**, 769.

Crabbé, P. A. and Heremans, J. F. (1968). *Am. J. Med.* **42**, 319.

Curtain, C. C., Kidson, C., Champness, D. L. and Gorman, J. G. (1964). *Nature, Lond.* **203**, 1366.

Dancis, J., Lind, J., Oratz, M., Imolens, J. and Vara, P. (1961). *Am. J. Obstet. Gynec.* **82**, 167.

De Coteau, W. E., Calvanico, N. J. and Tomasi, T. B. (1973). *Clin. Immunol. Immunopathol.* **1**, 192.

Despont, J. P. J. and Abel, C. A. (1974). *J. Immunol.* **112**, 1623.

Despont, J. P. J., Abel, C. A., Grey, H. M. and Penn, G. M. (1974). *J. Immunol.* **112**, 1517.

Deutsch, H. F. and Suzuki, T. (1971). *Ann. N. Y. Acad. Sci.* **190**, 472.

De Vaal, O. M. and Seynhaeve, V. (1959). *Lancet* **ii**, 1123.

Dorrington, K. J., Mihaesco, E. and Seligmann, M. (1970). *Biochim. Biophys. Acta* **221**, 647.

Dreyer, W. J. and Bennett, J. C. (1965). *Proc. Natn. Acad. Sci. USA* **54**, 864.

Edelman, G. M. (1970). *Biochemistry* **9**, 3197.

Edelman, G. M. and Gall, W. E. (1969). *Ann. Rev. Biochem.* **38**, 415.

Edelman, G. M. and Gally, J. A. (1962). *J. Exp. Med.* **116**, 207.

Edelman, G. M. and Poulik, M. D. (1961). *J. Exp. Med.* **113**, 861.

Engle, R. L. and Wallis, L. A. (1969). *In* "Immunoglobulinopathies: Immunoglobulins Immune Deficiency Syndromes, Multiple Myeloma and Related Disorders" Charles C. Thomas, Springfield, Illinois.

Faulkner, W. and Borella, L. (1970). *J. Immunol.* **105**, 786.

Fett, J. W., Deutsch, H. F. and Smithies, O. (1973). *Immunochemistry* **10**, 115.

Fleischman, J. B., Pain, R. and Porter, R. R. (1961). *Arch. Biochem. Biophys.* Suppl. **1**, 174.

Forte, F. A., Prelli, F., Yount, W. J., Jerry, L. M., Kochwa, S., Franklin, E. C. and Kunkel, H. G. (1970). *Blood* **36**, 137.

682   M. W. TURNER

Frangione, B. (1975). *In* "Immunogenetics and Immunodeficiency" (B. Benacerraf, Ed.) MTP, Lancaster.
Frangione, B. and Franklin, E. C. (1973). *Semin. Hematol.* **10**, 53.
Frangione, B. and Milstein, C. P. (1969). *Nature, Lond.* **224**, 597.
Frangione, B., Milstein, C. P. and Pink, J. R. L. (1969). *Nature, Lond.* **221**, 145.
Frangione, B., Lee, L. and Bloch, K. J. (1972). *Fed. Eur. Biochem. Soc.* **8**, 683.
Frangione, B., Lee, L., Haber, E. and Bloch, K. J. (1973). *Proc. Natn. Acad. Sci. USA* **70**, 1073.
Franklin, E. C. (1964). *J. Exp. Med.* **120**, 691.
Franklin, E. C. and Frangione, B. (1971). *Proc. Natn. Acad. Sci. USA* **68**, 187.
Franklin, E. C., Lowenstein, J., Bigelow, B. and Meltzer, M. (1964). *Am. J. Med.* **37**, 332.
Fudenberg, H. H. (1967). *In* "Proceedings of the Third International Congress of Human Genetics" (J. F. Crow and J. V. Neel, Eds) 233. Johns Hopkins Press, Baltimore.
Geha, R. S., Schreeberger, E., Merler, E. and Rosen, F. S. (1974). *New Eng. J. Med.* **291**, 1.
Gitlin, D. and Biasucci, A. (1969). *J. Clin. Invest.* **48**, 1433.
Gitlin, D., Kumate, J., Urrusti, J. and Morales, C. (1964). *J. Clin. Invest.* **43**, 1938.
Glanzmann, E. and Rinicker, P. (1950). *Ann. Paediat.* **175**, 1.
Good, R. A., Finstad, J., Gewurz, H., Cooper, M. D. and Pollara, B. (1967). *Am. J. Dis. Child.* **114**, 477.
Green, H. and Chan, T.-S. (1973). *Science* **182**, 836.
Grey, H. M., Abel, C. A., Yount, W. J. and Kunkel, H. G. (1968). *J. Exp. Med.* **128**, 1223.
Grey, H. M., Rabellino, E. and Pirofsky, B. (1971). *J. Clin. Invest.* **50**, 2368.
Heremans, J. F. (1959). *Clin. Chim. Acta* **4**, 639.
Hilschmann, N. and Craig, L. C. (1965). *Proc. Natn. Acad. Sci. USA* **53**, 1403.
Hirschhorn, R., Beratis, N., Rosen, F. S., Parkman, R., Stern, R. C. and Polmar, S. H. (1975). *Lancet* **i**, 73.
Hobart, M. J. and McConnell, I. (1975). "The Immune System: A Course on the Molecular and Cellular Basis of Immunity". Blackwell Scientific Publications, Oxford.
Hobbs, J. R. (1969). *Br. J. Haemat.* **16**, 599.
Hobbs, J. R. (1975). *In* "Immunodeficiency in Man and Animals". Birth Defects Original Article Series, Vol. 11. National Foundation, New York.
Hobbs, J. R. and Jacobs, A. (1969). *Clin. Exp. Immunol.* **5**, 199.
Houba, V. and Allison, A. C. (1966). *Lancet* **i**, 848.
Imhof, J. W., Ballieux, R. E., Mul, N. A. J. and Poen, H. (1966). *Acta Med. Scand.* **445**, 102.
Innes, J. and Newall, J. (1961). Lancet **i**, 239.
Isobe, T. and Osserman, E. F. (1974). *Blood* **43**, 505.
Janeway, C. A. and Gitlin, D. (1957). *Adv. Pediat.* **9**, 65.
Jeunet, F. S. and Good, R. A. (1968). *In* "Immunologic Deficiency Diseases in Man" (D. Bergsma, Ed.) Birth Defects Original Article Series Vol. 4 (1), 192. National Foundation, New York.
Johansson, S. G. O. and Bennich, H. (1967). *Immunology* **13**, 381.
Katz, D. H. and Benacerraf, B. (1975). *Transplant. Rev.* **22**, 175.
Kindt, T. J., Thunberg, A. L., Mudgett, M. and Klapper, D. G. (1974). *In* "The

Immune System: Genes, Receptors, Signals" (E. E. Sercarz, A. R. Williamson and C. F. Fox, Eds) 69. Academic Press, New York and London.

Kirkpatrick, C. H. and Schimke, R. N. (1967). *J. Am. Med. Assoc.* **200**, 105.

Kohler, H., Shimizu, A., Paul, C., Moore, V. and Putnam, F. W. (1970). *Nature, Lond.* **22**, 1318.

Kunkel, H. G., Smith, W. K., Joslin, F. G., Natvig, J. B. and Litwin, S. D. (1969). *Nature, Lond.* **223**, 1247.

Kyle, R. A. and Bayrd, E. D. (1966). *Am. J. Med.* **40**, 426.

Lam, C. W. K. and Stevenson, G. T. (1973). *Nature, Lond.* **246**, 419.

Lewis, A. F., Bergsagel, D. E., Bruce-Robertson, A., Schachter, R. K. and Connell, G. E. (1968). *Blood* **32**, 189.

Lewis, E. B. (1963). *Science* **142**, 1492.

Lumsden, W. H. R. (1965). *Wld. Hlth. Org. Tryp. Inf.* **1**.

Martin, N. H. (1970). *Br. J. Hosp. Med.* **3**, 662.

Mattern, P., Masseyeff, R., Michael, R. and Peretti, P. (1961). *Ann. Inst. Pasteur* **101**, 382.

Mellbye, O. J., Natvig, J. B. and Kvarstein, B. (1970). *In* "Protides of the Biological Fluids" (H. Peeters, Ed.). Pergamon Press, Oxford.

Meuwissen, H. J. (1974). *J. Pediat.* **84**, 315.

McDevitt, H. O. and Benacerraf, B. (1969). *Adv. Immunol.* **11**, 31.

McFarlin, D. E., Strober, W. and Waldmann, T. A. (1972). *Medicine* **51**, 281.

McKusick, V. A. and Cross, H. E. (1966). *J. Am. Med. Assoc.* **195**, 119.

Milstein, C. (1964). *J. Mol. Biol.* **9**, 836.

Milstein, C. (1966). *Proc. R. Soc. B.* **166**, 138.

Milstein, C. P., Steinberg, A. G., McLaughlin, C. L. and Solomon, A. (1974). *Nature, Lond.* **248**, 160.

Natvig, J. B. and Kunkel, H. G. (1974). *J. Immunol.* **112**, 1277.

Natvig, J. B., Kunkel, H. G. and Gedde-Dahl, T. (1967). *In* "Nobel Symposium 3 Gamma Globulins" (J. Killander, Ed.) 313. Almqvist and Wiksell, Stockholm.

Osserman, E. F. and Takatsuki, K. (1964). *Am. J. Med.* **37**, 351.

Oxelius, V. A. (1974). *Clin. Exp. Immunol.* **17**, 19.

Parkhouse, R. M. E. (1974). *In* "Progress in Immunology II" (L. Brent and E. J. Holborow, Eds) Vol. 1, 119. North Holland Publishing Co., Amsterdam.

Parr, D. M., Percy, M. E. and Connell, G. E. (1972). *Immunochemistry* **9**, 51.

Parsons, D. F., Darden, E. B. Jr., Lindsley, D. L. and Pratt, G. T. (1961). *J. Biophys. Biochem. Cytol.* **9**, 353.

Penn, G. M., Kunkel, H. G. and Grey, H. M. (1970). *Proc. Soc. Exp. Biol. Med.* **135**, 660.

Peterson, R. D. A., Cooper, M. D. and Good, R. A. (1966). *Am. J. Med.* **41**, 342.

Pink, J. R. L. and Milstein, C. (1967). *Nature, Lond.* **214**, 92.

Polmar, S. H., Waldmann, T. A. and Terry, W. D. (1972). *Am. J. Path.* **69**, 499.

Polmar, S. H., Wetzler, E., Stern, R. C. and Hirschhorn, R. (1975). *Lancet* **ii**, 743.

Porter, R. R. (1959). *Biochem. J.* **73**, 119.

Porter, R. R. (1962). *In* "Symposium on Basic Problems in Neoplastic Disease" (A. Gelhorn and E. Hirschberg, Eds) 177. Columbia University Press, New York.

Potter, M. and Boyce, C. R. (1962). *Nature, Lond.* **193**, 1086.

Potter, M. and Fahey, J. L. (1960). *J. Nat. Cancer Inst.* **24**, 1153.

Prahl, J. W. (1967). *Nature, Lond.* **215**, 1386.
Preud'Homme J.-L., Griscelli, C. and Seligmann, M. (1973). *Clin. Immunol. Immunopathol.* **1**, 241.
Putnam, F. W., Florent, G., Paul, C., Shinoda, T. and Shimizu, A. (1973). *Science* **182**, 287.
Putnam, F. W., Shimizu, A., Paul, C. and Shinoda, T. (1972). *Fed. Proc. Am. Soc. Exp. Biol.* **31**, 193.
Rambaud, J.-C., Bognel, C., Prost, A., Bernier, J. J., Quintrec, Y. L., Lamling, A., Danon, F., Hurez, D. and Seligmann, M. (1968). *Digestion* **1**, 321.
Roitt, I. M. (1974). "Essential Immunology" 2nd edn. Blackwell Scientific Publications, Oxford.
Rosen, F. S. and Janeway, C. A. (1966). *New Engl. J. Med.* **275**, 709.
Rosen, F. S., Hutchinson, G. and Allen, F. H. (1965). *Vox. Sang.* **10**, 729.
Rowe, D. S. and Fahey, J. L. (1965). *J. Exp. Med.* **121**, 185.
Rowe, D. S., McGregor, I. A., Smith, S. J., Hall, P. and Williams, K. (1968). *Clin. Exp. Immunol.* **3**, 63.
Sanders, J. H., Fahey, J. L., Finegold, I., Ein, D., Reisfeld, R. A. and Berard, C. (1969). *Am. J. Med.* **47**, 43.
Scheibel, I. F. (1943). *Acta Path. Microbiol. Scand.* **20**, 464.
Schimke, R. N. and Kirkpatrick, C. H. (1970). *In* "Modern Trends in Human Genetics" (A. E. H. Emery, Ed.) 68. Butterworths, London.
Schofield, F. D. (1957). *Trans. R. Soc. Trop. Med. Hyg.* **51**, 332.
Schur, P., Borel, H., Gelfand, E. W., Alper, C. A. and Rosen, F. S. (1970). *New Eng. J. Med.* **283**, 631.
Scott, C. D., Chen, S. H. and Giblett, E. R. (1974). *J. Clin. Invest.* **53**, 1194.
Seligmann, M., Danon, F., Mihaesco, C. and Fudenberg, H. H. (1967). *Am. J. Med.* **43**, 66.
Seligmann, M., Danon, F., Hurez, D., Mihaesco, E. and Preud'Homme, J.-L. (1968). *Science* **162**, 1396.
Smithies, O., Gibson, D. M., Fanning, E. M., Perry, M. E., Parr, D. M. and Connell, G. E. (1971). *Science* **172**, 574.
Solomon, A., Killander, J., Grey, H. M. and Kunkel, H. G. (1966). *Science* **151**, 1237.
Soothill, J. F. (1968). Lancet **i**, 1001.
Sorenson, G. D. (1961). *Exp. Cell. Res.* **25**, 219.
Spitler, L. E., Levin, A. S., Stites, D. P., Fudenberg, H. H., Pirofsky, B., August, C. A., Stiehm, E. R., Hitzig, W. H. and Gatti, R. A. (1972). *J. Clin. Invest.* **51**, 3216.
Steinberg, A. G., Muir, W. A. and McIntire, S. A. (1968). *Am. J. Hum. Genet.* **20**, 258.
Steinberg, A. G., Morell, A., Skvaril, F. and van Loghem, E. (1973). *J. Immunol.* **110**, 1642.
Stoop, J. W., Ballieux, R. E., Hijmans, W. and Zegers, B. J. M. (1971). *Clin. Exp. Immunol.* **9**, 625.
Terry, W. D. and Ein, D. (1971). *Ann. N. Y. Acad. Sci.* **190**, 467.
Tishfield, J. A., Creagan, R. P., Nichols, E. A. and Ruddle, F. H. (1974). *Hum. Hered.* **24**, 1.
Tobie, J. E. (1965). *In* "Progress in Protozoology" (J. Ludvik, J. Lau and J. Vavra, Eds) 164. Academic Press, New York and London.
Trevorrow, V. E. (1959). *Pediatrics* **24**, 746.

Virella, G., Nunes, M. A.-S. and Tamagnini, G. (1972). *Clin. Exp. Immunol.* **10**, 475.
Waldenström, J. (1944). *Acta Med. Scand.* **117**, 216.
Waldmann, T. A. and Strober, W. (1969). *Prog. Allergy* **13**, 1.
Waldmann, T. A., Miller, E. J. and Terry, W. D. (1968). *Clin. Res.* **16**, 45.
Walzer, P. D. and Kunkel, H. G. (1974). *J. Immunol.* **113**, 274.
Wang, A. C., Wang, I. T. F., McCormick, J. N. and Fudenberg, H. H. (1969). *Immunochemistry* **6**, 451.
Wang, A. C., Wilson, S. K., Hopper, J. E., Fudenberg, H. H. and Nisonoff, A. (1970). *Proc. Natn. Acad. Sci. USA* **66**, 337.
Wang, A. C., van Loghem, E. and Shuster, J. (1973). *Fed. Proc. Fed. Am. Soc. Exp. Biol.* **32**, 1003.
Wells, J. V., Fudenberg, H. H. and MacKay, I. R. (1971). *J. Immunol.* **107**, 1505.
Wells, J. V., Bleumers, J. F. and Fudenberg, H. H. (1973). *Proc. Natn. Acad. Sci. USA* **70**, 827.
West, C. D., Hong, G. R. and Holland, N. H. (1962). *J. Clin. Invest.* **41**, 2054.
Williams, R. C., Erikson, J. L., Polesky, H. F. and Swain, W. R. (1967). *Ann. Int. Med.* **67**, 309.
Williamson, A. R. (1977). Origin of antibody diversity. *In* "Immunochemistry" (L. E. Glynn and M. W. Steward, Eds) 141. Wiley and Sons Ltd., London.
Williamson, A. R. and Fitzmaurice, L. C. (1976). *In* "The Generation of Antibody Diversity" (A. J. Cunningham, Ed.) 183. Academic Press, London and New York.
Williamson, A. R. and McMichael, A. J. (1975). *In* "Molecular Approaches to Immunology" (E. E. Smith and D. W. Ribbons, Eds) 153. Academic Press, New York and London.
Wilson, I. D., Williams, R. C. and Tobian, L., Jr. (1967). *Am. J. Med.* **43**, 356.
Wiskott, A. (1937). *Mschr. Kinderheilk* **68**, 212.
Wochner, R. D., Drews, G., Strober, W. and Waldmann, T. A. (1966). *J. Clin. Invest.* **45**, 321.
Wolfenstein-Todel, C., Mihaesco, E. and Frangione, B. (1974). *Proc. Natn. Acad. Sci. USA* **71**, 974.
Yount, W. J. (1975). *In* "Immunodeficiency in Man and Animals" Birth Defects Original Article Series, Vol. 11. National Foundation, New York.
Yount, W. J., Kunkel, H. G. and Litwin, S. D. (1967). *J. Exp. Med.* **125**, 177.
Yount, W. J., Hong, R., Seligmann, M., Good, R. A. and Kunkel, H. G. (1970). *J. Clin. Invest.* **49**, 1957.

13 Coagulation Disorders

K. W. E. DENSON

Maternity Department, John Radcliffe Hospital, Oxford, England

I. THE MODERN CONCEPT OF BLOOD COAGULATION

In a chapter devoted to genetic variants of blood clotting factors it is necessary to consider briefly the modern concept of blood coagulation theory. It is likely that few other biological systems have received as much attention from research scientists in recent years.

For the first 40 years of this century the theory of Morawitz (1905) survived without modification. This theory envisaged only two clotting proteins and a mysterious substance termed thromboplastin

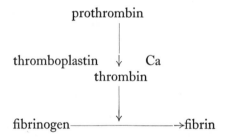

The theory did not of course explain all the facts and in the next two decades numerous groups of workers described additional new clotting factors and accelerators associated with syndromes and haemorrhagic diatheses. The growth of knowledge has been fragmentary and specialized with each group of workers inventing its own nomenclature and incomprehensible jargon. On the results of experiments devised with a stopwatch and a few clotting tubes, a single factor was known by as many as 10 or 12 different names and each group used the chosen name with the synonyms of other workers in parenthesis. The International Committee for the Nomenclature of Blood Clotting Factors was formed to examine this confusing terminology and contradictory evidence.

This Committee assigned a Roman numeral to each clotting factor, whose existence was proved beyond all reasonable doubt, and the acceptance of this nomenclature has done much to clarify the growth of knowledge and lay a foundation for common understanding and collaboration. The Roman numeral system, important synonyms and the properties of clotting factors are listed in Table 1.

Although normal variation in clotting factors is now well known (e.g. Table 9 on p. 453 or Lester et al., 1972), most of our knowledge of the factors has come from pathological variation, with which this chapter largely deals. Indeed, the existence of most of the clotting factors has been deduced

Table 1

The roman numeral system of nomenclature for blood clotting factors

Factor	Common important synonyms	Properties
I	fibrinogen	converted by thrombin to fibrin with the release of fibrinopeptides
II	prothrombin	precursor of thrombin adsorbed by $Al(OH)_3$, $BaSO_4$ and $Ca_3(PO_4)_2$; not present in serum
III	thromboplastin tissue factor	present in saline extracts of most tissues notable activity being present in lung and brain
IV	calcium	required for the interaction of all clotting factors except $XI + XII + IIa$
V	labile factor accelerator globulin	not adsorbed by $Al(OH)_3$, $BaSO_4$ and $Ca_3(PO_4)_2$; not present in serum; labile on storage; required by extrinsic and intrinsic clotting
VI		term not now used
VII	stable factor serum accelerator globulin autoprothrombin I proconvertin	adsorbed by $Al(OH)_3$, $BaSO_4$ and $Ca_3(PO_4)_2$ present in serum, stable on storage; not required by RVV[a]; not involved in intrinsic clotting
VIII	antihaemophilic factor A antihaemophilic globulin	not adsorbed by $Al(OH)_3$, $BaSO_4$, $Ca_3(PO_4)_2$ etc.; absent from serum
IX	autoprothrombin II antihaemophilic factor B Christmas factor plasma thromboplastin component	adsorbed by $Al(OH)_3$, $BaSO_4$, $Ca_3(PO_4)_2$ etc.; present in serum
X	Stuart-Prower factor autoprothrombin III	adsorbed by $Al(OH)_3$, $BaSO_4$, $Ca_3(PO_4)_2$; present in serum
XI	plasma thromboplastin antecedent (PTA) factor	adsorbed by glass, celite, sodium stearate etc.; not adsorbed by $Al(OH)_3$, $BaSO_4$; present in serum
XII	Hageman factor	adsorbed by glass, celite, sodium stearate etc.; not adsorbed by $Al(OH)_3$, $BaSO_4$; present in serum
XIII	fibrin stabilizing factor fibrinase	required for the formation of cross linkages in fibrin during the conversion of fibrinogen to fibrin by thrombin

[a]Russell's viper venom

from a study of abnormal physiology in patients with haemorrhagic diatheses, assumed to be the result of deficiency of various substances or
factors since fractions of blood plasma were shown to contain the corrective
substances. By the early sixties the existence of no less than 10 separate
clotting factors, thought to be proteins or glycoproteins, had been established, and the interaction of these substances together with the effect of
platelets or platelet substitute, calcium ions, autocatalytic mechanisms and
the presence of natural inhibitors to active intermediates, presented much
food for thought. The clotting factors identified at this time and the possible
sequence of their interaction are shown in Fig. 1. Two arbitrary systems

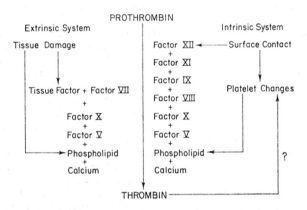

Fig. 1   Intrinsic and extrinsic sequence of clotting factor interactions

were envisaged although it was realized that these were in some way interrelated; the intrinsic system in which a prothrombin activator was formed
wholly from blood constituents and an extrinsic system in which prothrombin activator was formed by the intervention of tissue extracts. At this
time much work had been devoted to the study of the kinetics of the interaction of separated or "purified" clotting factors and it was becoming clear
that many of these behaved as enzymes or substrates in isolated reaction
sequences.

Macfarlane (1964) and Davie and Ratnoff (1964) crystallized the ideas of
the time by advancing their "enzyme cascade" and "waterfall sequence"
hypotheses to explain the interaction of blood clotting factors. The enzyme
cascade hypothesis is shown in Fig. 2. It is thought that the process of
clotting consists of a chain of enzyme substrate reactions forming a biochemical amplifier system, starting in shed blood with surface contact and
culminating in the formation of a fibrin clot. Such a scheme allows for an

explosive generation of large amounts of thrombin from smaller amounts of the precursor activators. The substrates or zymogens are present in an inactive form in plasma and become active enzymes during clotting. For example Hageman factor is activated by surface contact to form activated Hageman factor (factor XIIa) and this in turn converts the substrate PTA (factor XI) into the active enzyme factor XIa. The active enzyme factor

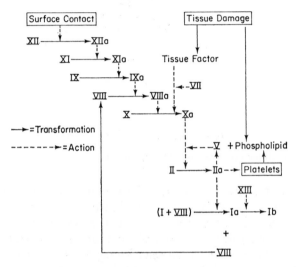

Fig. 2   The enzyme cascade hypothesis

XIa thus formed converts the substrate factor IX into activated factor IXa. Much experimental evidence has been accumulated which supports the enzyme cascade hypothesis and whilst there are minor objections it is reasonable to suppose that in broad outline an amplifier scheme is involved in which small amounts of Hageman factor and PTA are able to initiate the conversion of large amounts of fibrinogen to fibrin. An alternative scheme which is supported by more recent work (Hemker, 1969; Hemker et al., 1967a; Deggeller, 1968; Denson, 1967) is shown in Fig. 3 in which the intermediates XIa, VIIIa, VIIa and Va have been omitted. Autocatalytic processes also play a part; it is known for example that traces of thrombin activate both factor V and factor VIII. Platelet factor 3 liberated from platelets is also essential for the rapid reaction involving factors VIII, IXa and X and factors Xa, V and prothrombin. Calcium is necessary at all stages of clotting with the exception of the activation of Hageman factor although its precise mode of action remains ill-defined. The development of such powerful coagulants as factor IXa, factor Xa and thrombin would

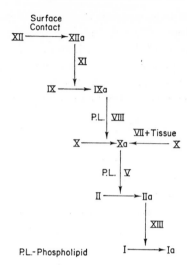

Fig. 3   The modified enzyme cascade hypothesis

ultimately result in massive intravascular thrombosis if provision were not made to remove them from the circulation, and naturally occurring inhibitors to these substances are present in plasma (Biggs *et al.*, 1970). Considering the clotting process as a whole it can be seen that the processes of enzyme amplifier systems, autocatalytic systems and natural inhibitor systems are in a state of delicate equilibrium and the process itself is a wonderful example of adaptation involving all the subtleties of nature.

A. The Isolation and Study of Clotting Factors

The isolation of the 10 known clotting factors has involved relatively crude methods of fractionation such as salt precipitation, differential adsorption on to inorganic precipitates, isoelectric precipitation, solvent extraction and differential stability to pH and temperature. More recent refinements in separation include electrophoresis, chromatography, ultracentrifugation and immunological methods. The clotting factors are all high molecular weight proteins or glycoproteins and the only end point in a study of the reactions has until recently been the fibrin clot. Synthetic tri- and tetra-peptides containing *p*-nitroaniline groups have now been developed which can be used to measure the proteolytic enzymes, factor Xa, thrombin and plasmin. After reaction with the appropriate enzyme, the liberated *p*-nitroaniline may be measured spectrophotometrically. (Blombäck *et al.*, 1974; Odegard *et al.*, 1976; Soria *et al.*, 1976). Purification has been assessed by functional activity in bioassays and biochemical criteria of homogeneity have not been applied. The bioassays have been developed empirically to measure the effects of replacement therapy in patients with a functional deficiency of one or more of the clotting factors. The structures of only fibrinogen (Blombäck *et al.*, 1968a; Blombäck and Blombäck, 1969); and prothrombin (Magnusson, 1969) have been determined with certainty, since these are the last two substrates concerned in the chain of reactions leading to the formation of fibrin, and of the clotting factors they are present at highest concentration.

The bioassay techniques used in the study of blood coagulation have been many and varied. Snake venoms have been used extensively. Russell's viper venom, for example, converts factor X to activated factor X, the venom from the Taipan snake converts prothrombin to thrombin in the absence of all known clotting factors, and *Bothrops jararaca* venom in common with many other snake venoms contains an enzyme which has a thrombin-like action and converts fibrinogen to fibrin (Denson, 1969). Staphylocoagulase converts prothrombin to thrombin coagulase, the concentration of which can be measured by its ability to clot fibrinogen. The use *in vivo* of the antibiotic ristocetin had been found to produce thrombocytopenia, and it was subsequently observed that this substance caused platelet aggregation when added to normal platelet-rich plasma, but not when added to platelet-rich plasma from the majority of patients with von Willebrand's disease. Many different thromboplastic substances, for example human placenta and mammalian brain and lung, have been used in the one-stage prothrombin time test, and this led to the chance observation that the plasma samples from a minority of patients with Christmas disease gave prolonged clotting times with ox brain thromboplastin.

II. ABNORMAL CLOTTING FACTORS

For many years it was accepted that the congenital coagulation defects were due to the absence in patients of specific coagulation factors and such defects were referred to as "factor-VIII deficiency, factor-IX deficiency" etc. Much evidence has accumulated recently which suggests that some patients possess abnormal forms of clotting factors which are nonfunctional in clotting factor assays. These abnormal clotting factors have been detected entirely by the use of immunological methods and their recognition suggests that even the patients who appear to lack recognizable immunological equivalents of the normal clotting factor may still have a structurally defective clotting factor rather than total deficiency of the normal clotting factor. The recognition only recently of the existence of genetic variants of clotting factors not unnaturally stems from the lack of knowledge concerning the normal clotting factors. It is conceivable that each of the congenital coagulation defects may represent a group of patients with a broader spectrum of structurally defective proteins similar to the broad spectrum of abnormal haemoglobins associated with the haemoglobinopathies. Table 2 shows a list of the clotting factors for which rare variants or evidence of heterogeneity have been found.

Table 2

Coagulopathies in which the presence of rare variants have or have not been demonstrated

Clotting factors	Presence of rare variants
Fibrinogen	
Prothrombin	
Factor VII	
Factor VIII	
Factor-VIII related antigen	yes
Factor IX	
Factor X	
Factor XIII	
Antithrombin III	
Factor V	
Factor XI	not found
Factor XII	

A. Rare Variants of Factor IX

Fantl et al. (1956) were the first workers to draw attention to the possible existence of antigenic variants of factor IX. They studied a patient with a factor-IX inhibitor and observed that the inhibitor was neutralized by a barium sulphate adsorbate from only one of three plasma samples from patients with Christmas disease (Haemophilia B; factor-IX deficiency). They suggested on the basis of these findings that there could be two forms of Christmas disease. Kidd et al., 1963, noted that the plasma of a child with Christmas disease had a prolonged one-stage prothrombin time when ox brain was used as thromboplastin. At the time this observation was difficult to explain but subsequent investigation of the inhibitory effect of some Christmas disease plasma samples to ox brain thromboplastin (Denson et al., 1968) revealed that there were at least two types of Christmas disease. In the larger group of patients the plasma samples had a normal one-stage prothrombin time with ox brain thromboplastin, the plasma samples were unable to neutralize factor-IX antibodies and did not give a precipitin line with factor-IX antibodies on immunodiffusion tests. It was concluded that this group lacked immunologically recognizable factor IX. In the smaller group of patients comprising about 1 in 10 of those studied, the plasma samples had an abnormal reaction to ox brain thromboplastin which became normal on the addition of factor-IX antibodies (Table 3),

Table 3

The clotting time of Christmas disease plasma with ox brain thromboplastin after the addition of a factor IX inhibitor

Plasma sample	Dilution of factor-IX inhibitor	One-stage clotting time with ox brain thromboplastin (seconds)
Haemophilia B+	—	160
	1 in 2	41
Haemophilia B+	—	126
	1 in 2	49
Haemophilia B+	—	107
	1 in 2	48
Haemophilia B−	—	45
	1 in 2	47
Normal	—	40
	1 in 2	45

and plasma eluates gave precipitin lines with the latter. On this evidence of immunological testing, Christmas disease was classified by these authors into two groups: Haemophilia B− for patients in whom no immunologically recognizable factor IX could be detected and Haemophilia B+ for patients in whom a functionally inactive but antigenically similar substance to factor IX was present.

Hougie and Twomey (1967) made similar observations on two siblings with Christmas disease. They obtained a prolonged one-stage prothrombin time on the plasma of the two boys with ox brain thromboplastin. The plasma also prolonged the ox brain thromboplastin time of normal plasma but when the patient's plasma was absorbed with $Al(OH)_3$ it ceased to lengthen appreciably the ox brain thromboplastin time of normal plasma. The defect was present to a lesser extent in the mother and maternal grandfather. Although they did not carry out immunological tests to demonstrate the presence of an antigenically similar form of factor IX in the plasma of the patient, their results suggested that the plasma contained an abnormal form of factor IX and they termed the defect in the family Haemophilia

B$_M$. Roberts *et al.* (1968) also demonstrated the presence of a genetic variant of Haemophilia B. About 10% of the Haemophilia B patients studied by them had material which cross-reacted (CRM) with an antibody to factor IX which developed in a patient with Christmas disease. The presence of CRM was not correlated with factor-IX level or clinical severity. Pfueller *et al.* (1969) prepared an antiserum to human factor IX and observed that the plasma samples from four out of 13 patients with Christmas disease were able to neutralize the anti-factor-IX component and deduced that a genetic variant of normal factor IX was present in these samples. It is interesting that the plasma samples which neutralized the antibody had a normal one-stage prothrombin time with ox brain and thus differed from the abnormal samples studied by Hougie and Twomey (1967) and Denson *et al.* (1968). Clearly then, there must be at least two genetic variants of factor IX.

Meyer *et al.* (1972a) have studied 22 patients with Christmas disease using both an inhibitor developing in a patient with Christmas disease (human antibody), and a rabbit antibody made to purified human factor IX (rabbit antibody). Antibody neutralization tests showed that 13 out of 22 samples failed to neutralize the human antibody, and nine samples had a normal or reduced capacity for neutralizing it. With the rabbit antibody, only one sample lacked immunologically detectable factor IX. On the basis of these results, Meyer *et al.*, 1972a, suggested that there could be four groups of Christmas disease: patients showing either the presence or absence of factor IX antigen with a normal or abnormal ox brain thromboplastin time. Lastly, Elodi and Puskas (1972) have tested a group of 14 patients with Christmas disease using an antibody neutralization test employing a rabbit antibody to human factor IX. Only two out of 14 plasma samples neutralized the antibody, whilst two of the remaining 12 plasma samples had slightly prolonged ox brain thromboplastin times. From all these results, it may be concluded that at least three, and possibly four variants of factor IX exist. It should be realized that our methods of investigation are, to say the least, crude, and as these become more sophisticated, it is likely that a broader spectrum of molecular variants will be recognized.

A further interesting variant of Christmas disease has been described by Veltkamp *et al.* (1970) and termed Haemophilia B Leyden, in which the patients' factor IX levels increase during life. Thirty-one male patients from three families were studied, and it was found that patients below the age of 15 years had factor IX levels of less than 2%, with severe clinical manifestations of Christmas disease. Adult patients had levels of factor IX between 20 and 60%, with the disappearance of bleeding symptoms during adult life. Two patients were studied over a period of time during which the factor IX levels rose from 1 to 20%, thus providing conclusive evidence for the increase in factor IX levels with advancing years. There is little doubt that this type of Christmas disease is a distinct entity and different from Haemophilia B m or Haemophilia B$^+$.

B. Hypoprothrombinaemia and Rare Variants of Prothrombin

1. Hypoprothrombinaemia

Josso et al. (1967) and Soulier and Prou Wartelle (1966) have investigated a patient with low levels of prothrombin by bioassay, and also by staphylocoagulase and immunological methods. They concluded that this patient had a true quantitative deficiency of prothrombin. Denson (unpublished observation) has tested the plasma of six patients from two families with mild hypoprothrombinaemia in whom the level of prothrombin ranged from 30 to 50% both by Taipan assay and antibody neutralization tests.

2. Genetic variants of prothrombin

Josso et al. (1971) have investigated a family in which four siblings had an unusual haemorrhagic syndrome. They showed that this was due to a defect in the prothrombin molecule transmitted as a genetic autosomal trait which they termed prothrombin Barcelona. The parents were shown to be heterozygous for the defect. The homozygous patients had 5% of prothrombin by one-stage clotting factor assay, 13% by two-stage clotting factor assay, and yet 100% by the staphylocoagulase assay, and 100% by the immunological assay of Laurell. The prothrombin in a euglobulin preparation from the plasma of one patient was converted to thrombin very slowly, but yielded the same amount of thrombin as a euglobulin preparation from normal plasma after six hours incubation. The abnormal prothrombin showed a different electrophoretic mobility from normal prothrombin, and prothrombin in the plasma of the heterozygous parents showed an intermediate electrophoretic mobility. Similar findings have been obtained by Shapiro et al. (1969) in several members of a family found to be heterozygous with respect to an abnormal prothrombin molecule. They termed this dysprothrombinaemia prothrombin Cardeza. The electrophoretic mobility of the prothrombin in their patients was normal, although in serum a band in addition to the normal fragment was present.

C. Rare Variants of Factor VIII

The possibility that haemophiliacs possess a functionally inactive form of factor VIII rather than total lack of factor-VIII substance has been

MAN—DD*

considered by many authors. To account for the very small incidence of antibodies to factor VIII developing in haemophiliacs following transfusion, it has been suggested that there might be two groups of patients, one group in which the plasma is completely deficient in factor VIII and a second group in which the plasma contains a functionally inactive but immunologically similar form of factor VIII to that found in normal persons. As a result of the total lack of normal antigen in the first group it was thought that this group might be more prone to develop antibodies to infused factor VIII (Dausset, 1959; Shanberge and Gore, 1957; Uszynski, 1966). The difficulties in testing the hypothesis of the existence of an inactive form of factor VIII are greater than those associated with the recognition of an abnormal form of factor IX because of the lack of suitable test systems for detecting factor VIII immunologically. It is doubtful whether true anti-factor-VIII precipitating antibodies have ever been demonstrated. Berglund (1962, 1963) prepared precipitating antibodies to factor-VIII preparations in rabbits but concluded that the resulting precipitin lines could be due to contaminating proteins. Uszynski (1967) obtained seven precipitin lines between an antiserum to Cohn fraction 1–0 and either normal or haemophilic plasma using immunoelectrophoresis. One of these lines was absent when a haemophilic plasma containing an antibody to factor VIII was tested and the same line was very faint when samples of plasma from patients with von Willebrand's disease were tested. Bidwell (1966) observed that haemophilic patients treated with animal factor VIII who developed antibodies produced precipitating antibodies only to contaminating proteins and Denson (1967) showed by immunodiffusion that rabbit antibodies to highly purified factor VIII produced several precipitin lines of non-identity with fibrinogen, any or all of which could have been due to contaminating proteins. Thus the most satisfactory technique which has been available to test for the presence of factor-VIII antigen in haemophilic plasma samples is to see whether the samples are able to neutralize factor-VIII antibodies using standard coagulation assays to measure residual inhibitor after first mixing the potential source of antigen and inhibitor. Using various modifications of this technique different authors have obtained variable results. Adelson et al. (1963) showed that Haemophilia A plasma was unable to neutralize anti-factor-VIII antibodies whilst Shanberge and Gore (1957) and Piper and Schreier (1964) concluded that haemophilic plasma had the same capacity as normal plasma to neutralize factor-VIII antibodies. The latter authors used the thromboplastin generation test, which does not measure factor VIII specifically, and antibodies prepared against crude anti-haemophilic globulin and thus it is possible that a different interpretation could be placed on their results. Berglund (1962) also obtained equivocal results using the thromboplastin generation test as a method of measuring factor-VIII inhibitor neutralization. Goudemand *et al.* (1963) obtained results using an acquired factor-VIII antibody from a haemophilic patient which indicated that normal plasma and serum neutralized the antibody, whilst haemophilic plasma only partially neutralized the antibody and von Willebrand plasma had no neutralizing capacity. Abildgaard *et al.* (1967) showed that normal plasma and serum both neutralized the acquired factor-VIII antibody from a haemophilic patient, but that aged normal plasma and plasma from 29 haemophilic patients had no neutralizing ability. Uszynski (1966) used an antiserum prepared in rabbits to fibrinogen-free factor VIII obtained by careful and selective heat precipitation of fibrinogen. He tested 20 plasma samples from severe haemophiliacs and obtained no neutralization of the factor-VIII antibody. McLester and Wagner (1965) studied haemophilic dogs and showed that canine plasma was unable to neutralize factor-VIII antibodies prepared against canine factor VIII and concluded that canine haemophilic plasma did not contain an immunological equivalent to factor VIII. These results are summarized in Table 4.

Table 4

The results of factor-VIII antibody neutralization tests by different authors

Authors and date	Type of antibody	Number of samples tested	Number of haemophilic samples able to neutralize antibody
Shanberge and Gore (1957)	rabbit	?	all
Berglund (1962)	rabbit	1	1
Adelson *et al.* (1963)	rabbit	?	none
Goudemand *et al.* (1963)	haemophilic	13	some (average results for all patients)
Piper and Schreier (1964)	rabbit	4	4
McLester and Wagner (1965)	rabbit (anti canine VIII)	6	none
Uszynski (1966)	rabbit	20	none
Abildgaard *et al.* (1967)	haemophilic	20	none
Denson *et al.* (1969)	haemophilic	48	4
Hoyer and Breckenridge (1968)	haemophilic	34	6
Feinstein *et al.* (1969)	acquired	54	2
Bennett and Huehns (1970)	rabbit	24	22
Zimmerman *et al.* (1971)	rabbit	14	14
Denson (1971)	rabbit	23	4
Gralnick *et al.* (1971)	goat	55	8

The results of all these investigations up to 1967 were thoroughly confusing and although the same basic technique of factor-VIII antibody neutralization had been used the technique itself was difficult to standardize and, with the many manipulations involved, required careful attention to detail. With the discovery of two forms of Haemophilia B it became apparent that perhaps there are two varieties of haemophiliacs, in one of which immunologically recognizable factor VIII is totally lacking and in the second an immunological equivalent is present. Certainly this hypothesis would explain some of the conflicting and confusing reports of earlier workers. More recently, Denson (1968), Hoyer and Breckenridge (1968), Denson *et al.* (1969) and Feinstein *et al.* (1969) have provided evidence to show that this is indeed the case. Denson *et al.* (1969) applied a similar terminology to that which had been applied to Haemophilia B; namely Haemophilia A$^+$ for patients whose plasma contained a functionally inactive but antigenic equivalent substance to factor VIII and Haemophilia A$^-$ for patients whose plasma was totally lacking in immunologically recognizable factor VIII. The principle of the technique used by Denson *et al.* (1969) for detecting the presence of factor-VIII antibody neutralizing material is described below.

1. Principle of the method of antibody neutralization

Stage 1. The haemophilic plasma sample and 100%, 50% and 25% dilutions of normal plasma were incubated with a carefully chosen dilution of an antibody to factor VIII. Incubation was continued to allow the neutralization of factor VIII to go to completion; it has been shown by Leitner *et al.* (1963) and Denson (1967) that the reaction between factor VIII and its antibody is complete in four to six hours. The dilution of antibody was chosen such that when normal plasma was used as antigen 0·1 to 0·5 units of antibody (or units of inhibitor) remained at this stage.

Stage 2 (determination of residual antibody). A human factor-VIII concentrate was then added to the mixture of antigen and antibody and incubation continued for a further two hours. The residual factor VIII was then determined by the two-stage factor-VIII assay (Denson, 1966) and the residual antibody calculated by reference to a curve relating residual factor VIII to antibody (inhibitor) units (Biggs and Bidwell, 1959). The amount of factor VIII-like protein in a haemophilic sample expressed as a percentage of that in normal plasma may also be obtained by reference to a curve relating percentage normal plasma added in the first stage to residual factor VIII in the second stage.

The results obtained when this technique of antibody neutralization was applied to 48 haemophilic plasma samples and 30 normal plasma samples are shown in Table 5. Four of the haemophilic samples (Haemo-

Table 5

Neutralization of a factor-VIII inhibitor by different plasma samples

Samples	Total number of tests	Units of inhibitor neutralized[a]		
		Range	Mean	S.D.
30 Normal	30	0·45–1·00	0·74	0·05
44 Haemophilia A⁻	72	−0·42–0·42	0·06	0·20
Haemophilia A⁺ (1)	3	0·62–1·04	0·79	0·22
(2)	3	0·48–0·82	0·65	0·17
(3)	3	0·54–0·70	0·64	0·09
(4)	4	0·74–0·86	0·81	0·05

[a] 1 unit of inhibitor is defined as the amount which will destroy 75 % of added factor VIII at a final concentration equivalent to the amount in average normal human plasma, after incubation for 2 hours at 37⁰

philia A⁺) had the same ability as normal plasma to neutralize the factor-VIII antibody while the remaining 44 samples (Haemophilia A⁻) neutralized only small amounts of antibody. Concentrates prepared by alcohol fractionation of plasma gave similar results to whole plasma; the concentrates from Haemophilia A⁻ plasma had little antibody neutralizing ability whilst those from Haemophilia A⁺ plasma had the same ability as those prepared from normal plasma to neutralize the antibody. Similar results were also obtained when five different factor-VIII antibodies were tested. In contrast to the results obtained by some earlier workers it was shown that normal serum had little antibody neutralizing ability. In six samples of plasma from patients with von Willebrand's disease who had very little factor-VIII activity no antibody neutralizing material was detected.

Hoyer and Breckenridge (1968) obtained similar results using a method for antibody neutralization employing the addition of normal plasma to the antigen–antibody mixture as a source of factor VIII and measuring the residual factor VIII by a one-stage method. They obtained neutralization of the factor-VIII antibody (presence of CRM positive material) in six out of 34 haemophilic plasma samples comprising two families out of a total of 27 families studied. These authors also related the percentage of factor VIII remaining after incubation of antibody, normal plasma and test plasma to the concentration of factor VIII in the test plasma by testing a series of dilutions of known factor-VIII levels by making mixtures of normal plasma and CRM negative haemophilic plasma.

It should be emphasized that this broad classification into Haemophilia A⁺ (or CRM positive) and Haemophilia A⁻ (or CRM negative) has been based entirely on the presence or absence of a substance in the plasma able to neutralize antibodies against antigenic determinants concerned in the clotting activity of factor VIII. The group showing the presence of such material may include several different molecular variants and the group unable to neutralize much antibody may still nevertheless contain an abnormal factor-VIII molecule which is antigenically unrecognizable by this particular immunological criterion. There seems to be no marked difference in clinical severity of the two groups. Of the four patients with Haemophilia A^+ described by Denson et al. (1969) three were severely affected and the fourth was only mildly affected.

It has been shown by Zimmerman et al. (1971) and Bennett and Huehns (1970) that an antibody to purified human factor VIII prepared in rabbits could be neutralized by all the plasma samples from a large group of haemophilic patients. This confirms the results of many earlier workers and refutes the conclusion of others. In order to resolve the apparent conflict in results it is necessary to have a clear understanding of the techniques used by different groups of workers.

It can be safely assumed that the factor-VIII molecule has a number of separate antigenic sites. If haemophilic patients do not lack factor VIII then the factor-VIII molecule which they possess is certainly defective in clotting activity and if they produce an antibody in response to infusion of normal factor VIII it is likely to be directed against the antigenic determinants (and quite likely a single antigenic determinant) associated with activity. Antibodies prepared in rabbits against normal factor VIII on the other hand are likely to be heterogeneous and directed against the multiple antigenic sites of the factor-VIII molecule. The mechanisms for the inactivation of factor-VIII activity by antibodies are probably of four types:

(a) direct combination of an antibody with the antigenic group associated with clotting activity;

(b) combination of antibodies and antigenic determinants adjacent to the part of the molecule responsible for the clotting activity resulting in steric hindrance and thus loss of activity;

(c) the combination of antibodies with multiple antigenic determinants resulting in lattice formation between the antigen complexes with ultimate precipitation and occlusion of the activity within the lattice or precipitate;

(d) it is also possible that the factor-VIII molecule is associated with a carrier protein or complexed with a von Willebrand factor, and antibodies to either the carrier protein or von Willebrand factor might be expected to neutralize the clotting activity by mechanism (c).

Thus antibodies developing in haemophilic patients will be neutralized by the antigenic determinant associated with clotting activity provided that the structural changes affecting the molecule in haemophilic plasma are not sufficiently great to render the antigen site unrecognizable (mechanisms (a) and (b)). The application of haemophilic antibodies in neutralization tests first pointed to the existence of molecular variants in Haemophilia A.

The heterogeneous antibodies developing in rabbits in response to injection of factor VIII will be active against many if not all of the antigenic determinants in factor VIII both from normal plasma and haemophilic plasma if the latter contains a structurally defective factor VIII, and may neutralize factor VIII by mechanism (c). The observations of Zimmermann *et al.* (1971) that all haemophilic plasma samples tested by them were able to neutralize the rabbit antibodies suggests that the majority of haemophilic samples contain a defective factor-VIII molecule.

Antibody neutralization tests using rabbit antibodies need to be interpreted with some caution. The rabbit serum may contain thrombin or active intermediates which may destroy factor VIII but which could be neutralized on addition of haemophilic plasma and thus give erroneous results. Rabbit antibodies also have extremely complex kinetics and it is doubtful whether these can be used to give sensible results in antibody neutralization tests. The rabbit antibodies are also directed against factor VIII-related protein (which is dealt with under the next section), and neutralization of factor VIII may be through mechanism (d) above.

D. Von Willebrand's Disease

The inheritance of von Willebrand's disease is autosomal and dominant and the disease is characterized by a variably low level of factor VIII, prolonged bleeding time and abnormal platelet aggregation with ristocetin. When haemophilic plasma is infused into a patient with von Willebrand's disease, a delayed and sustained rise in factor VIII occurs, which suggests that a factor is infused which is reduced or genetically abnormal in von Willebrand's disease, and that the infused factor determines the synthesis of factor VIII in the patient.

Zimmerman *et al.* (1971) using quantitative immunoelectrophoresis have obtained precipitin lines between a rabbit antibody to purified factor VIII and 3% alcohol concentrates from both normal and haemophilic plasma. The factor VIII-like antigen was present in normal amounts in the plasma of haemophilic patients. In the plasma of patients with von Willebrand's disease, the antigen was correlated with the level of biologically active factor VIII. Meyer *et al.* (1972b) have obtained similar results to those of Zimmerman *et al.* (1971) using a rabbit antibody to purified factor VIII. Bennett and Ratnoff (1972), Denson (1973) and Prentice *et al.* (1972) have found an increase in this "factor VIII-like" antigen during pregnancy and following physical exercise and the infusion of adrenaline which paralleled the increase in the biological activity of factor VIII. These workers have also found the antigen correlated with biologically active factor VIII in the plasma of patients with von Willebrand's disease.

In all these experiments using rabbit antibodies, there is some doubt whether the antigen which is being measured is factor VIII. It is possible that the increase in factor VIII which occurs during pregnancy, following exercise and the infusion of adrenaline, and post-operatively, is accompanied by a rise in the von Willebrand factor(s); in which case, the antigen being measured could be the von Willebrand factor or factor VIII complexed to the von Willebrand factor. The factor which is reduced or absent in von Willebrand's disease and causes abnormal retention of platelets in a glass bead column has been studied by Bouma et al. (1972). This factor was separated together with factor VIII by gel chromatography of normal plasma cryoprecipitate, and a rabbit antiserum was made to the fraction. The results of in vitro correction studies and immunological studies indicated that the factor VIII activity and von Willebrand's disease correcting activity are located either on the same molecule or on two separate molecules, of which the antigenic determinants could not be differentiated. Bennett et al. (1972) have obtained significant results by studying the factor VIII-like antigen and factor VIII biological activity in haemophilic and von Willebrand's disease patients following the infusion of factor VIII concentrates. They found that the characteristic delayed rise in biologically active factor VIII was not accompanied by a corresponding rise in the level of "factor VIII-like" antigen, and that infused antigen disappeared rapidly from the circulation. These results would suggest that the antigen being measured immunologically is the von Willebrand factor and not factor VIII. The highest protein purification of human factor VIII achieved by the above authors has been about 10 000. If these preparations consisted of pure factor VIII, then factor VIII would be present at a concentration of about 10 μg per ml in plasma, and this would be just sufficient to produce precipitin lines by the Laurell method using undiluted plasma. Michael and Tunnah (1966), however, reported a protein purification of over 500 000 for porcine factor VIII, and suggested a concentration of factor VIII in human plasma of about 0·10 μg per ml or less. If this degree of purification can be obtained for human factor VIII, then it would seem that the precipitin lines obtained by Zimmerman et al. (1971) and other workers could be due to a protein other than factor VIII. Recently, Stites et al. (1971) using haemagglutination inhibition have demonstrated the presence of "factor VIII" antigen in the plasma and serum of both haemophiliacs and normal persons. Their rabbit antibody was produced in response to the injection of a factor VIII preparation with a protein purification of 10 000, and the same antigen was coupled to the red cells by the chromic chloride technique. Clearly, if the protein purification of 500 000 for porcine factor VIII reported by Michael and Tunnah can be achieved with human factor VIII, it would seem likely that Stites *et al.* are detecting antigens other than factor VIII. By means of agarose gel filtration, Hoyer (1972) has studied the binding properties of human factor VIII using radio-labelled rabbit and human antibodies to factor VIII. He calculated that there could be 35 antigenic sites on the factor VIII molecule, and suggested that the human antibody to factor VIII had only weak binding properties, whereas the rabbit antibodies formed stable complexes. It seems equally likely that the antigen used in these experiments was not pure factor VIII. Hougie and Sargeant (1973) have made a careful study of the supernatant and precipitate from the 3% alcohol precipitate of normal plasma. On average, a quarter of the factor VIII biological activity remained in the supernatant, but this had no detectable antigenic activity when tested with the rabbit antibody to factor VIII-like antigen. They have suggested, on the basis of these results, that the antigen detectable by the rabbit antibody is the von Willebrand's disease factor, and is unlikely to be factor VIII.

Bloom *et al.* (1973) have studied normal and haemophilic plasma, and von Willebrand's disease plasma following the infusion of normal plasma cryoprecipitate. They used the technique of Sepharose 6B gel filtration, and in both normal plasma and treated haemophilic plasma they found a single high molecular weight peak of factor-VIII activity. In contrast, the plasma from a treated patient with von Willebrand's disease showed an additional peak of activity at a lower molecular weight. They concluded that there is a low molecular weight factor VIII which has the factor VIII biological activity and a high molecular weight fraction which is identical to the factor VIII-like antigen characterized by rabbit antisera. The latter has low molecular weight factor VIII complexed to it. Von Willebrand's disease could therefore be due to defective production of circulating high molecular weight component and haemophilia to abnormal synthesis of low molecular weight component. The important concept to emphasize is that factor VIII is normally complexed to the high molecular weight component (the von Willebrand's disease factor). In the plasma of treated patients with von Willebrand's disease, there is insufficient of the high molecular weight component to complex the low molecular weight factor VIII activity which has been produced *in vivo*.

By more detailed studies of factor VIII-related protein in von Willebrand's disease, it is now apparent that molecular variants of this protein exist. Holmberg and Nilsson (1972) have studied a group of 70 patients with von Willebrand's disease. Fifty-one of these patients fell into one group with low values for the factor VIII-like antigen, whilst 19 patients comprised a second group with normal values for factor VIII-like antigen. They suggested that patients with von Willebrand's disease fall into two separate genetical groups. Patients with von Willebrand's disease have been described with normal levels of factor-VIII activity and low levels of factor VIII-related antigen (Veltkamp and van Tilburg, 1973; Meyer *et al.*, 1973; Weiss *et al.*, 1973; Gralnick *et al.*, 1973). Other patients with von Willebrand's disease have been studied and their plasma shown to contain factor VIII-related protein with increased mobility on two-dimensional crossed immunoelectrophoresis (Kernoff *et al.*, 1974; Peake *et al.*, 1974) and other patients with similar qualitative defects in factor VIII-related protein have been described (Firkin *et al.*, 1973; Thompson *et al.*, 1974). Thus it is clear that patients with von Willebrand's disease can have a reduced concentration or a molecular variant of normal factor VIII-related antigen.

Although there has been much confusion in recent years about the relationship between factor-VIII activity and factor VIII-related antigen, Zimmerman and Edgington (1973) have shown that these activities reside on different molecules. These authors showed that heterologous antibodies covalently coupled to agarose beads were capable of binding and segregating factor-VIII activity from the antigen. Also homologous antibodies bound to agarose beads could bind only factor-VIII activity, but not the antigen. It also seems likely that factor VIII-related antigen is identical to the von Willebrand factor which is necessary for a normal bleeding time, normal retention of platelets to glass beads and ristocetin aggregation of platelets.

E. Rare Variants of Factor X

The evidence for the existence of antigenic variants of factor VIII rested entirely on the technique of antibody neutralization. The abnormal forms of factor IX were also detected by antibody neutralization and immuno-diffusion tests. The presence of normal factor-X antigen can also be detected by antibody neutralization and immunodiffusion, since it has been shown by Denson (1967) that antibodies to factor X neutralize the clotting activity and that the antibodies form precipitin lines with factor X. In addition to the immunological techniques for the recognition of factor-X antigen there are at least three different clotting factor activities by means of which the normal antigen can be characterized. The laboratory tests are centred around the activation of factor X and consist mainly of tests for intrinsic clotting, tests for extrinsic clotting using brain thromboplastin and clotting tests employing Russell's viper venom. Thus, apart from the presence or absence of factor-X protein, it is possible to study three different clotting activities compared with only one clotting activity in the case of factor VIII and factor IX. It is conceivable that abnormal forms of factor X may give abnormal results with one or more of these three different test systems. Denson *et al.* (1970) have recently studied samples from six patients with the factor-X defect (classical factor-X deficiency) employing three tests of clotting function: the kaolin cephalin clotting time as a measure of intrinsic clotting, the factor-X assay employing brain thromboplastin as a measure of extrinsic clotting and the factor-X assay employing

Table 6

Kaolin cephalin clotting times (KCCT) and the results of assays of factor X in the plasma of six patients from five unrelated families with "factor-X deficiency"

| | KCCT | | Factor X(%) | |
	Clotting time (seconds)	Ratio patient: normal	Brain extract assay	RVV assay
Prower	170	3·4	8	7
D.E.C.	310	5·8	2	96
R.E.D.	67	1·3	6	22
M.M.	114	1·9	1	<1
G.S.[a]	134	2·2	25	17
L.S.[a]	109	1·8	7	8
P.W. (factor-VII deficient)	53	1·0	100	100

[a] Lyophilized samples from two brothers

Russell's viper venom. Their results with these three test systems are shown in Table 6, from which it can be seen that these six patients gave widely differing results for the amount of factor X recorded by the different methods. The implication of these findings was that perhaps the patients had different abnormal factor-X molecules according to which method was used. They next tested the samples with the technique of factor-X antibody neutralization. The technique consisted of adding a dilution of the test plasma to a dilution of an antiserum to factor X for 12 minutes to allow the interaction of antibody and antigen to go to completion. A dilution of normal plasma was then added and incubation continued for a further 12 minutes for interaction of the residual antibody and the added normal factor X to go to completion. The factor-X assay components were next added (Denson, 1961) and the clotting times of the mixtures transformed to residual factor-X concentration by reference to a calibration curve relating clotting times to dilutions of 1 in 10, 1 in 100 and 1 in 1000 of normal plasma tested in the same way but substituting a dilution of normal rabbit serum for the antiserum and diluted citrate saline for the test plasma. Destruction of factor X by its antiserum is linear from 100% to 10% of residual factor X and the dilution of antiserum was chosen such that if the test plasma contained no factor-X antigen the antibody would neutralize 90% of factor X in the second addition of normal plasma. The results of applying this technique to the plasma samples from the six patients showed that they appeared to contain widely differing amounts of factor-X antigen. Serum eluates were prepared from the plasma samples and these were tested for the presence of factor-X antigen by the technique of immunodiffusion. An antiserum to factor X was placed in the centre well of an Ouchterlony plate and different serum eluates in the outer wells. The samples again displayed different characteristics. Some showed lines of identity with normal factor X whilst others showed lines of partial identity with characteristic "spur" formation. A summary of the results of applying these five techniques and the heterogeneity in the behaviour of factor X in the six plasma samples are shown in Table 7. The use of three tests of clot-

Table 7

Summary of results in the six patients with the factor-X defect

	Factor X activation			Factor-X antibody neutralization	Line of identity with normal factor X
	Intrinsic	Extrinsic	RVV		
Prower	Abn	Abn	Abn	+++	+
D.E.C.	Abn	Abn	Normal	+++	+
R.E.D.	Sl.Abn	Abn	Sl.Abn	+	−
M.M.	Abn	Abn	Abn	−	−
G.S.	Abn	Sl.Abn	Sl.Abn	−	−
L.S.	Abn	Abn	Abn	−	−

Abn: abnormal
Sl.Abn: slightly abnormal

ting coupled with the techniques of antibody neutralization and immunodiffusion have revealed that the haemorrhagic state defined as "factor-X deficiency" includes at least five distinct and separate abnormalities of factor X. This is an exceedingly rare defect and it is extraordinary that the first six patients investigated in this manner should show such differences. It seems likely from these results that different forms of factor X may result from a variety of genetical mutations involving alterations in the structure of the molecule as has been demonstrated for haemoglobin and many other proteins. Human factor X has not yet been purified nor has its structure been determined. Compared with haemoglobin, factor X is a trace protein in plasma and considerable technical difficulties accompany its isolation and purification and thus it will be some years before the factor-X molecule can be fingerprinted in the same way as the haemoglobin and fibrinogen molecules and these protein abnormalities clearly understood. The importance underlying the recognition of so many varieties of factor X is that when more sophisticated techniques can be applied to reveal the existence of abnormal forms of factor VIII, it is likely that Haemophilia A will also consist of a group showing a broad spectrum of genetic variants of factor VIII.

F. Rare Variants of Factor VII

Recently, Goodnight et al. (1971) have studied the plasma of two patients with a factor-VII defect by the technique of antibody neutralization. The plasma of one patient contained a protein which neutralized the factor-VII antibody, whilst plasma from the second patient failed to neutralize the antibody. They suggested on this evidence that there were at least two types of hereditary "factor-VII deficiency". Denson et al. (1972) have also studied nine patients from six different families with the factor-VII defect using the technique of antibody neutralization. Some of these patients recorded higher levels of factor VII by antibody neutralization than by assay and there was considerable variation within the group. There seems little doubt judging from these preliminary results that the factor-VII defect will, like the factor-X defect, consist of a spectrum of patients with different genetic variants of factor VII.

G. Rare Variants of Fibrinogen

Although hereditary deficiencies of fibrinogen have been known for many years (Macfarlane, 1938), attention has recently been focused on the recognition of molecularly abnormal fibrinogens. Fanconi in 1944 postulated the existence of dysfibrinogenaemia states due to the synthesis of molecularly abnormal fibrinogens and Biggs (1951) suggested that there was a difference between the fibrinogen of adults and newborn infants because plasma from the latter reacted more slowly with thrombin. Ingram (1955) studied two patients with prolonged thrombin clotting times and suggested that dysfibrinogenaemia might be a hereditary trait. At least eight families with dysfibrinogenaemia have now been described in which the affected patients had a normal concentration of fibrinogen but their plasma reacted more slowly than normal plasma with different concentrations of thrombin. Menaché (1964) described the first case of a patient with a clear-cut abnormally-reacting fibrinogen. The plasma of the patient did not produce a fibrin clot with intrinsic thomboplastin, extrinsic thomboplastin or normal amounts of thrombin. The addition of calcium to normal amounts of thrombin resulted in a translucid clot and higher concentrations of thrombin produced clotting of the plasma. The venom of *B. jararaca*, which has a thrombin-like action either with or without calcium, produced no clot and the patient's plasma inhibited the clotting of normal plasma by venom, thrombin or extrinsic thromboplastin. It was shown that the abnormality was confined to the fibrinogen and when this was separated by salt or alcohol precipitation and added to normal plasma, inhibition of the thrombin–fibrinogen reaction of the normal plasma occurred. At this time no chemical studies were undertaken to identify the amino acid constitution of the α (A) or β (B) chains of the fibrinogen, although it was clear that this constituted an aberrant molecular form of fibrinogen. It is interesting that this patient had no clinical history of abnormal bleeding. Beck *et al.* (1965) described a further patient with a mild haemorrhagic diathesis in which the abnormality could be attributed to the patient's fibrinogen and in which the findings were similar to those described by Menaché. The findings differed from those described by Menaché in that the patient's plasma did not interfere with the thrombin fibrinogen reaction of normal plasma and minor immunoelectrophoretic differences distinguished the fibrinogen from normal fibrinogen. Beck *et al.* (1965) recognized this as an abnormal variant of fibrinogen and adopting the nomenclature previously applied to abnormal haemoglobins, proposed the descriptive name fibrinogen Baltimore for this variant. Patients with plasma having a prolonged thrombin clotting time attributed to the presence of an abnormal fibrinogen molecule have also been described by von Felten *et al.* (1966), Mammen *et al.* (1968) (fibrinogen Detroit) and Forman *et al.* (1968) (fibrinogen Cleveland). The abnormal fibrinogens are inherited as autosomal dominant traits; some of the patients have a mild bleeding tendency whilst others are symptom free, and in two cases episodes of thromboembolism and post-operative thrombophlebitis have been recorded. It is not clear yet exactly what precise differences exist in the eight families so far studied. Forman *et al.* (1968) have compared their case with those of Beck *et al.* (1965), von Felten *et al.* (1966) and Mammen *et al.* (1968) and suggest that the abnormality is associated only with the aggregation of dissolved fibrin monomers or polymers. Menaché suggested that the defect in this case involved the stage of hydrolysis of the fibrinogen by thrombin. Fibrinogen Detroit (Mammen et al., 1968) has now been more extensively investigated than other fibrinogen variants. Ultracentrifuge studies gave the same molecular weight as normal fibrinogen and after sulphite cleavage the average molecular weights of the three chains were the same as for normal fibrinogen, but carbohydrate analysis showed significantly lower values for neutral sugars, sialic acid and hexosamines than for normal fibrinogen. Blombäck et al. (1968b) have shown differences in the amino acid sequence of the α (A) chain of fibrinogen Detroit. After reduction and alkylation, the N-terminal α (A) chain fragment was separated by fractionation on Sephadex G-100. This fraction was then digested with trypsin and the tryptic digest subjected to two-dimensional electrophoresis-chromatography. Comparison of the "fingerprint" of this fraction from normal fibrinogen and fibrinogen Detroit showed that one peptide was missing in the map from the abnormal fibrinogen. A second peptide spot was a different colour with the ninhydrin reagent, suggesting that this consisted of elements of the normal peptide in this position and elements of the missing peptide. The complete absence of one of the normal peptides indicated that the patient was homozygous for the fibrinogen Detroit trait. Further analysis of the differently coloured peptides indicated that the N-terminal part of the α (A) chain of the abnormal fibrinogen had arginine residue 19 of a normal α (A) chain replaced by a neutral amino acid, probably serine. Blombäck et al. (1968b) point out that the replacement of a strongly basic amino acid by a neutral hydroxy acid in the N-terminal portion of the α (A) chain may result in considerable conformational changes in the N-terminal disulphide knot of fibrinogen and these changes may directly or indirectly affect the active site for polymerization. More recently Blombäck and Blombäck (1969) have investigated the tryptic fingerprints of the α (A) chain fragments of fibrinogens Paris, Zürich, Baltimore and Louvain. These fibrinogens appear to contain only normal tryptic fragments. Hampton and Morton (1970) have also investigated in some detail the variant "Fibrinogen Oklahoma". Sulphite cleavage and acrylamide gel electrophoresis of this fibrinogen revealed a fast-moving chain in addition to the normal α (A) and β (B) chains. Two-dimensional chromatography on the sulphitolysed material showed that two basic peptides were missing but that an additional acidic peptide was present.

Fibrinogen is not difficult to fractionate and amongst the blood clotting proteins it is present in the greatest quantity and thus in view of the great amount of work already devoted to the structure of normal fibrinogen (for references see Blombäck et al. (1968a) and Blombäck (1967) amongst many others) it is likely that the abnormal fibrinogens will be the first of the group of abnormal clotting factors to be "fingerprinted" in detail.

H. Heterogeneity of Fibrinogen

Several reports of heterogeneity of human and bovine fibrinogens have been published. Finlayson and Mosesson (1970) have shown that single donor fibrinogen contains three types of fibrinogen which can be separated chromatographically. These authors separated two major peaks and subjected them to sulphitolysis and tryptic digestion. Peptide mapping of the α (A) and β (B) chains of both fibrinogens revealed no differences although the patterns for the γ-chains were difficult to interpret. Single donor bovine fibrinogen has been shown to contain two α (A) chains and two chromatographically distinct γ-chains (Takagi and Iwanga, 1969; Brummel and Montgomery, 1970; Gerbeck et al., 1969). Heterogeneity of the α (A) chain in which some chains contain a phosphorylated serine and some lack an N-terminal valine has been reported by Blombäck et al. (1963). Gaffney (1971) using the technique of isoelectric focusing on sulphitolysed fibrinogen has demonstrated fewer bands with single donor fibrinogen than pooled fibrinogens. These different reports in which various techniques have been used leave little doubt of the existence of heterogeneity in fibrinogen.

I. Factor XIII (Fibrin Stabilizing Factor) Deficiency

A functional deficiency in factor XIII of autosomal recessive inheritance is now well documented. Duckert (1971) tested samples from one large family using an antibody to purified factor XIII and found that all homozygotes without factor-XIII activity had about 50% of protein and heterozygotes about 60 to 80% of protein by immunodiffusion. He also demonstrated a line of immunological identity with normal factor XIII in the plasma samples of 13 patients from 10 families. McDonagh (1971) also observed that a commercial antiserum to factor XIII produced a precipitin line with factor XIII deficient plasma, but that the functional factor XIII neutralizing capacity of the antiserum was not affected. Thus it would seem that patients with a functional defect in factor XIII possess a genetic variant of normal factor XIII.

J. Factor V Deficiency

This disease is of autosomal recessive inheritance and occurs with about the same frequency as deficiencies of factor VII and factor X. Feinstein *et al.* (1970) used an acquired IgG antibody to factor V to test the plasma of patients with an hereditary deficiency of factor V, and could not demonstrate the presence of cross-reacting material. Thus to date, genetic variants of normal factor V have not been demonstrated.

K. Factor XI (PTA) Deficiency

This is believed to be transmitted as an incompletely recessive trait. The vast majority of cases recognized have been Jewish, and deficiency of factor XI gives rise to a mild bleeding tendency characterized by traumatic bleeding from injury, dental extraction or surgery. Antibodies to factor XI have been described in response to plasma infusion therapy (Nossel, 1976), but to date, genetic variants of normal factor XI have not been reported.

L. Antithrombin Deficiency

An inherited deficiency of antithrombin III (progressive antithrombin) has been described in several families, associated with a high incidence of thromboembolic disease (Egeberg, 1965; van der Meer *et al.*, 1973). Reduced levels of antithrombin in plasma samples from the patients were demonstrated by immunological methods and by clotting factor assays suggesting that these patients had a true quantitative deficiency of antithrombin III. Sas *et al.* (1974) described a family with the lowest reported levels of antithrombin III associated with a high incidence of venous thromboembolic disease. The members of this family had normal levels of antithrombin III by immunological methods, suggesting defective synthesis of the molecule, and they termed this defect antithrombin III Budapest.

M. Allelomorphic Forms of Factor XII

Hageman trait is transmitted as an autosomal recessive disorder (Ratnoff and Steinberg, 1962). Recently a state of moderate deficiency of factor XII has been described which was attributed to a second recessive mutant allele in the homozygous state (Bok *et al.*, 1965) and thus it would appear that there are two mutant alleles concerned in the Hageman defect. In addition to the abnormal genes there is evidence to suggest that two other alleles are present at the Hageman locus which determine the production of normal factor XII (Veltkamp *et al.*, 1965). The factor-XII levels were measured in 50 carriers of severe factor-XII deficiency and found to fall clearly into two groups. One group had levels of between 19 and 34% with a mean of 23% and the other group levels of between 41 and 83% with a mean of 61%. Interestingly, some sibships contained members of both groups suggesting that the normal parent had dissimilar alleles at the Hageman locus. Smink et al. (1967) have examined the plasma of 14 patients with Hageman trait using techniques of passive haemagglutination inhibition, immunodiffusion and an antibody neutralization test for clotting activity. They concluded that the patients lacked immunologically recognizable factor XII.

Deletion mapping (p. 445) has been used to establish the probable locus of a structural gene for the Hageman factor. A patient was described with 50% Hageman factor and deletion of the short arm of chromosome 6, which suggested that the structural gene for factor XII might be localized on 6p. Re-examination of the karyotype of this patient using the R and G banding techniques (p. 96) showed that the detected fragment was in fact translocated on to the distal end of 7q with loss of chromosome material from the latter. A further patient who was monosomic for the distal end of 7q was also shown to have 50% of Hageman factor. It was concluded that a structural gene for the Hageman factor is localized on the distal segment of 7q, probably the 7q 35 band (de Grouchy et al., 1974).

III. THE GENETICS OF HAEMOPHILIA AND VON WILLEBRAND'S DISEASE

Classical haemophilia (factor-VIII deficiency) is a sex-linked recessively inherited disease affecting males only and the patients have little or no factor-VIII activity. Von Willebrand's disease, inherited as an autosomal dominant, is characterized by a variably low factor-VIII level, prolonged bleeding time and/or defect in platelets and probably represents a heterogeneous group of defects. Von Willebrand's disease is differentiated from classical haemophilia by the characteristic response in the patient to the infusion of plasma fractions. Following the infusion of a factor-VIII concentrate the initial rise in factor VIII is proportional to the amount of factor VIII infused and unlike the response in haemophilic patients, the rise in the factor-VIII level is maintained for about a day. Increase in the factor-VIII level is also obtained following the infusion of serum, factor-VIII-free fibrinogen and haemophilic plasma, which suggests that the patients are able to synthesize factor VIII from a factor contained in the material infused. In contrast the infusion of plasma from patients with von Willebrand's disease into haemophilic patients produces no such rise in factor

VIII. Two models have been suggested to explain the genetics of factor-VIII synthesis in haemophilia and von Willebrand's disease by Graham *et al.* (1964). One of these, the regulatory hypothesis, is shown in Fig. 4 and is based on the operon theory of gene regulation (Jacob and Monod, 1961). The autosomal von Willebrand's genes produce the repressor which regulates the operon and structural genes for the production of an inducer.

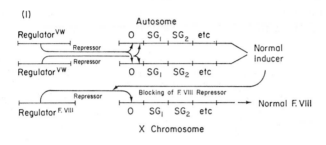

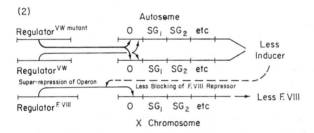

Fig. 4 Graham's regulatory hypothesis for the synthesis of factor VIII. (1) Normal factor VIII synthesis. (2) Synthesis of factor VIII in von Willebrand's disease. The heavy black lines indicate super repression of the autosomal operons and the dotted line less blocking of the X-chromosome repressor

The inducer acts on the regulator governing the operon and structural genes for the production of factor VIII on the X-chromosome. Haemophilia can be the result of mutations in the haemophilia operon or structural genes on the X-chromosome yielding products of different biological activity. In von Willebrand's disease a mutant regulator gene produces a super-repressor resulting in greater repression of the target operons and, since the repressor can act on either target operon, variable amounts of the inducer for the production of factor VIII on the X-chromosome are produced. When the inducer is absent or grossly deficient little or no factor VIII is produced by the X-chromosomes. The advantage of this hypothesis is that it explains satisfactorily the variable levels of factor VIII encountered in von Willebrand's disease. Also when haemophilic plasma containing a normal amount of the inducer produced autosomally is transfused to a patient with von Willebrand's disease the normal X chromosome is able to produce factor VIII.

A second and simpler model is the combining subunit hypothesis (Graham et al., 1964) which is shown in Fig. 5. The difficulty with this model is to explain the great variation in factor-VIII levels in von Willebrand's disease because the model predicts 50% or 0% factor VIII. It is possible that the defect in von Willebrand's disease produces a spectrum of genetic variants with A chains of variable biological activity.

	Normal Male.		Haemophilic Male.		von Willebrand's Disease. Heterozygous Male.		Homozygous Male.	
Loci.	Autosomal.	X.	Autosomal.	X.	Autosomal.	X.	Autosomal.	X.
Genes.	N N	N	N N	H	N VW	N	VW VW	N
Product.	A^N A^N	B^N	A^N A^N	B^H	A^N A^{VW}	B^N	A^{VW} A^{VW}	B^N
Combination of A and B Chains.	$A^N + B^N = 100\%$ F. VIII activity.		$A^N + B^H = 0\%$ F. VIII activity.		$\left.\begin{array}{l} 1/2A^N + 1/2B^N \\ 1/2A^{VW} + 1/2B^N \end{array}\right\} = 50\%$ F. VIII activity.		$A^{VW} + B^N = 0\%$ F. VIII activity.	
Free A Chains.	A^N in Excess.		A^N in Excess.		$\left.\begin{array}{l} 1/2A^N \\ 1/2A^{VW} \end{array}\right\}$ in Excess.		A^{VW} in Excess.	

Fig. 5 Graham's "Combining Subunit" hypothesis for the synthesis of factor VIII. N = normal gene; VW = mutant gene in von Willebrand's disease; H = mutant gene in haemophilia

At this stage of our knowledge, such models are tentative. Factor VIII is present in only trace concentration in plasma, and it has only been isolated in a relatively crude state. On the basis of recent separative procedures and biochemical studies, Graham and Barrow (1973) have suggested that it is the product of at least three genetic loci, and that there may be three components in the factor VIII/factor VIII-related antigen complex.

A. Observations on the Inheritance of Structurally Defective Clotting Factors

Denson et al. (1969) studied a family with Haemophilia A in which two siblings were found to be of different Haemophilia types. One sibling was found to be Haemophilia A$^+$ and had a plasma factor-VIII level of 13% whilst the other was Haemophilia A$^-$ and had a plasma factor-VIII level of 1%. In terms of clinical severity the brother classified as Haemophilia A$^+$ was only mildly affected whilst the Haemophilia A$^-$ brother was severely affected. The antibody neutralization tests were carried out at least seven times with different haemophilic inhibitors on samples from the brothers, because of this unusual finding. At this time it was thought that a simple sex-linked inheritance could not apply to these siblings but fresh plasma samples have recently been re-tested and both found to be Haemophilia A^+ with factor-VIII levels of 9·5% and 7·5%. A possible explanation for the discrepancy is that one brother had previously developed a weak antibody to factor VIII thereby reducing his own factor-VIII level from 7·5% to 1% and eliminating the greater part of his variant factor-VIII protein. Meyer and Larrieu (1970) have also observed that two first cousins were classified as Haemophilia A^+ and Haemophilia A^- when first tested but on subsequent testing they both proved to be Haemophilia A^+; a similar explanation could apply to these cousins.

The findings of Denson et al. (1970), in which patients were shown to have different structurally defective factor-X molecules showing different abnormalities by tests of intrinsic clotting or tests of extrinsic clotting employing Russell's viper venom or brain thromboplastin, appear now to shed some light on the work of Rabiner and Kretchmer (1961). These authors investigated a patient who appeared to have "factor-X deficiency" and whose plasma showed abnormal intrinsic clotting by the thromboplastin generation test and abnormal extrinsic clotting with either tissue extract or Russell's viper venom. The father had equally abnormal intrinsic clotting but almost normal extrinsic clotting and the mother had normal intrinsic clotting and almost normal extrinsic clotting. Both parents were clinically symptom free and since there appeared to be no consanguinity in the family it seems probable that the parents transmitted different genetic variants of factor X, the mother transmitting a relatively innocuous genetic variant which would go undetected by laboratory tests, the father transmitting a genetic variant demonstrable by laboratory tests but not resulting in clinical symptoms.

B. The Relationship Between the Development of Antibodies to Factor VIII or IX and the Type of Haemophilia A or Haemophilia B

It would be of great value in clinical practice to be able to say whether one of the two groups of patients Haemophilia A^+ or Haemophilia A^- was more prone to develop antibodies to factor VIII as the result of infusions of plasma or factor-VIII concentrates.

Six patients with Haemophilia A who had in the past developed antibodies to factor VIII but which subsequently disappeared have been tested by the technique of antibody neutralization (Denson, unpublished observation).

None of the six neutralized any appreciable amount of antibody and were classified as Haemophilia A⁻. Beck *et al.* (1969) have described two patients with mild haemophilia who developed antibodies to factor VIII during replacement therapy to cover major surgery. The antibodies inactivated not only transfused factor VIII but also the patient's own factor VIII and temporarily converted them to severe haemophiliacs. The antibodies subsequently disappeared and plasma samples from these two patients were tested by the technique of antibody neutralization. One neutralized antibody and was thus classified as Haemophilia A⁺ whilst the other failed to neutralize antibody and was classified Haemophilia A⁻ (Denson, unpublished observations). A similar finding has been reported by Crowell (1971).

C. The Dectection of Carriers of Haemophilia

The biological assay of factor VIII gives values below the normal range in about 50% of haemophilic carrier females. The remaining carriers give values which may, on the whole, be lower than average normal, but which overlap in the normal range and therefore cannot be classified by use of the factor-VIII assay. If it is accepted that all haemophiliacs possess a factor-VIII molecule which is functionally inactive, then carrier females will have a complement of normal biologically active factor VIII and a complement of biologically inactive factor VIII. The net results of this will be an excess of antigen over biological activity. Since only about one in 10 haemophiliacs have a factor-VIII molecule which is able to neutralize human antibody to factor VIII (Haemophilia A⁺), it would seem likely that the only carriers with antibody neutralizing material in excess of their biologically active factor VIII would be carriers of Haemophilia A⁺. Bouma *et al.* (1972) have recently studied a group of obligatory and potential female carriers of haemophilia, and concluded that the immunological detection of factor VIII using a haemophilic antibody does not offer any advantage over the factor-VIII assay alone in the detection of the carrier state, but may assist in the detection of the Haemophilia A⁺ variant.

Zimmerman *et al.* (1971), using rabbit antibody to purified factor VIII and the Laurell technique of immunoprecipitation, showed that in 23 out of 25 obligatory carriers of haemophilia, the relationship between biological factor-VIII activity and factor VIII-like antigen was significantly different from normal women at the 99% limit of confidence. This work is clearly a milestone in our progress in the recognition of female carriers of haemophilia. Meyer *et al.* (1972b) and Denson (1973) have confirmed these findings, and shown that in a large proportion of obligatory carriers, the factor

VIII-like antigen was two or more times greater than the biological activity of factor VIII. It does not matter if the antigen being measured proves to be the von Willebrand factor and not factor VIII, since its measurement provides a diagnostic tool whereby carriers of haemophilia may be classified with greater certainty.

D. Haemophilia in the female

Haemophilia A and B in the female has been described in more than 60 females and some of the possible causes are listed below (Graham, 1975):

   (a) Mis-diagnosis of von Willebrand's disease (vWd) or another autosomal form of haemophilia

   (b) Genetic homozygosity at haemophilia locus:
- (i) due to consanguinity in a haemophilia kindred
- (ii) due to incestuous illegitimacy
- (iii) due to mutation in the germ line of a person marrying into a haemophilia kindred
- (iv) due to post-zygotic somatic mutation

   (c) Genetic abnormality in a phenotypic female with one X-chromosome bearing the haemophilia gene
- (i) testicular feminization (46, XY)
- (ii) Turner's syndrome (45, XO)
- (iii) mosaicism (46, XX/XO) affecting factor-VIII synthesis
- (iv) isochromosome Y (46, XY qi) in a phenotypic female
- (v) isochromosome X (46, XXqi)

   (d) "Extreme lyonisation" in a heterozygote

   (e) Deletions, inversions, translocations etc. involving an X-chromosome

Haemophilia A in the female has been attributed to one or other of these causes. Recently two reports have appeared in which cytogenetic studies have revealed chromosome deletions. Samama *et al.* (1976) reported the case of a 10-year-old girl with a brother suffering from classical haemophilia. The factor-VIII level was less than 1% and the bleeding time was normal in both the girl and her brother, there was no evidence of consanguinity and paternity was not disproved by extensive blood grouping tests. Cytogenetic studies revealed a deletion of the long arm of an X-chromosome and the sex chromatin showed the presence of a phenotypic female.

A case of Haemophilia B has also been described in a 4-year-old girl and cytogenetic studies showed a deletion of the short arm of an X-chromosome (Spinelli *et al.*, 1976). In this case, there was no family history of haemophilia and coagulation studies on the relatives were normal.

IV. COMBINED "DEFICIENCY" OF CLOTTING FACTORS

Defects involving more than one clotting factor have been described. Deficiency of prothrombin, factor VII, factor IX and factor X as one defect of acquired or familial origin has been reported. This can also be induced by blocking the action of vitamin K on the synthesis of these factors by the use of oral anticoagulant drugs. Reports of combined deficiency of factors VIII and IX have also appeared in the literature, but the validity of such combined defects is open to some doubt. Patients with disseminated lupus erythematosus who develop an inhibitor, are sometimes found to have plasma levels of less than 20% of factor VIII and factor IX by one-stage methods due to the action of the inhibitor, although two-stage assays on the same plasma samples may record levels of 100% to 200%. To the unwary, such patients may appear to have a combined defect of both factor VIII and factor IX if only one-stage assay methods are used. Combined deficiency of factor-VII and factor IX has also been reported, but the apparent factor-VII deficiency in this defect could be the result of a genetic variant of factor IX interfering with the extrinsic assay of factor VII. It seems likely in the light of present day knowledge that many of these reported defects could be spurious. An exception is a defect involving both factor V and factor VIII which is now well-documented and poses interesting genetic questions. Several families of these patients have been studied (Oeri et al., 1954; Iversen and Bastrup Madsen, 1956; Seibert et al., 1958; Jones et al., 1962; Gobbi et al., 1967) and the levels of factors V and VIII and clinical severity have differed widely. The three patients described by Jones et al. (1962) had severe factor-V deficiency but only mild factor-VIII deficiency (factor-VIII levels of 11%, 19% and 20%) and were only moderately affected clinically. Two siblings investigated by Gobbi et al. (1967), had factor-V levels of 25% and 27% and factor-VIII levels of less than 2% and were severely affected clinically. Oeri et al. (1954), Seibert et al. (1958) and Jones et al. (1962) postulated that the combined defect depended on the recessive inheritance of a single mutant autosomal gene, which was responsible for the synthesis of a common precursor to both factors V and VIII. The infusion of normal plasma into these patients produces the expected rise in levels of factors V and VIII, which are not sustained as is the level of factor VIII when normal plasma is infused into a patient with von Willebrand's disease. Jones et al. (1962) postulated that infusion of plasma from a patient with haemophilia or pure factor-V deficiency might be expected to produce an increase of both factors in the recipient's plasma by providing the missing precursor. Gobbi et al. (1967) infused haemophilic plasma into one of their patients and obtained only a rise in the level of factor V, and so dismissed the hypothesis of the previous authors. To explain the low factor VIII in their female sibling they postulated instead three possible mechanisms on the basis that their patient was heterozygous for the factor-V defect and was a haemophilic carrier with one abnormal X chromosome. The three postulated mechanisms were:

(a) interaction between the autosomal factor-V gene and the heterozygous haemophilic gene;

(b) inversion of dominance, the factor-V gene acting as a "specific modifier" for the dominance of the haemophilic gene;

(c) inactivation of the normal X chromosome early in embryonic development.

The last hypothesis requires an additional rare phenomenon and thus a total of three extremely rare phenomena in one patient and although possible is very unlikely. It seems more likely that, despite the failure of the patient to synthesize factor VIII following the infusion of haemophilic plasma, a single mutant gene concerned in the synthesis of a common precursor for factor V and VIII is responsible for the defect. Further evidence for the hypothesis of a single mutant gene is provided by the natural groups into which several of the clotting factors fall. Prothrombin, factor VII, factor IX and factor X are physicochemically similar substances and are probably synthesized from a common precursor since the action of oral anticoagulants is to depress all four factors. Factor V, factor VIII and fibrinogen also exhibit close physicochemical similarity; factor V and factor VIII are absent in more primitive organisms and it is not unreasonable to suppose that factor V and factor VIII have evolved from a common precursor concerned in the synthesis of fibrinogen. Certainly if the combined factor-V and factor-VIII defect is the result of two separate mutant genes then one might expect a combined defect involving factor VII and factor VIII to occur with about the same frequency, since the gene frequencies for the factor-V defect and factor-VII defect are about equal. No such cases of combined factor-VII and factor-VIII defect have yet been reported.

V. RESISTANCE TO THE ORAL ANTICOAGULANT DRUGS

In Chapter 9, the concept was introduced that genetical variation might be revealed by differential response of patients to drug treatments. One of the most important areas of this new field of pharmacogenetics has been resistance to oral anticoagulants.

Oral anticoagulants, dicumarol, phenylindanedione, warfarin etc. act by lowering the plasma levels of factors II, VII, IX and X. It is thought by

Hemker *et al.* (1976b) that the drugs act by blocking the enzyme or co-enzyme responsible for the stage in synthesis from precursor protein to glycoprotein clotting factor. In Man, resistance to the anticoagulant drugs has been recorded in isolated cases (Blaustein *et al.*, 1953). Warfarin has been successfully used for the destruction of rats because of the high suscep-tibility of the animals to small doses (0·005 % incorporated into oatmeal) of the drug, which are not lethal to small domestic animals. Since 1960 appar-ently independent outbreaks of resistance to warfarin in *Rattus norvegicus* have occurred in Scotland, Denmark and Wales and this has now become a serious problem in the control of infestation. Greaves and Ayres (1967) crossed wild resistant rats with susceptible laboratory rats for five genera-tions and in extensive experiments produced the expected 50% of resistant progeny compatible with the hypothesis that warfarin resistance depends on a single autosomal gene with dominant effect. These workers have also demonstrated linkages between genes for coat colour and warfarin resist-ance and determined the linear order (Greaves and Ayres, 1969). Resistant rats have an abnormally high requirement for vitamin K and to explain the dominant effect of the mutant gene, Hermodson *et al.* (1969) have sug-gested that a protein involved in clotting factor synthesis which interacts with both warfarin and vitamin K is altered in resistant rats so that its affinity for both compounds is reduced, the affinity for warfarin being reduced more than that for vitamin K. Rats heterozygous for resistance have some of the altered protein and can thus synthesize clotting factors in the presence of relatively high levels of warfarin, which would account for resistance appearing as a dominant character. In the absence of warfarin, the normal protein would still be needed to synthesize clotting factors opti-mally, and rats homozygous for resistance would have only the abnormal protein and might have reduced synthesis of clotting factors associated with an apparent vitamin K deficiency. Monogenic recessive inheritance of a haemorrhagic trait in rats has been recorded. The gene involved is linked with the *p* locus and could be identical or allelic to the mutant gene respon-sible for warfarin resistance (Dunning and Curtis, 1939). As Greaves and Ayres (1969) state

> It is, however, of interest that a single biochemical mechanism has been pro-posed that connects the apparent dominance of Rw^2 (the mutant gene) in respect of resistance to warfarin with a recessive effect in respect of vitamin K requirements, because it would be expected on evolutionary grounds that a gene as rare as Rw^2 would normally be recessive and deleterious in its expres-sion.

REFERENCES

Abildgaard, C. F., Vanderheiden, J., Lindley, A. and Rickles, F. (1967). *Thromb. Diath. Haemorrh.* **18**, 354.

Adelson, E., Rheingold, J. J., Parker, O., Steiner, M. and Kirby, J. C. (1963). *J. Clin. Invest.* **42**, 1040.

Beck, E. A., Characke, P. and Jackson, D. F. (1965). *Nature, Lond.* **208**, 143.

Beck, P., Giddings, J. C. and Bloom, A. L. (1969). *Br. J. Haemat.* **17**, 283.

Bennett, B. and Ratnoff, O. D. (1972). *J. Lab. Clin. Med.* **80**, 256.

Bennett, B., Ratnoff, O. D. and Levin, J. (1972). *J. Clin. Invest.* **51**, 2597.

Bennett, E. and Huehns, E. R. (1970). *Lancet* **ii**, 956.

Berglund, G. (1962). *Br. J. Haemat.* **8**, 204.

Berglund, G. (1963). *Int. Arch. Allergy* **22**, 1.

Bidwell, E. (1966). *In* "Treatment of Haemophilia and Other Coagulation Disorders" (R. Biggs and R. G. Macfarlane, Eds) 98. Blackwell Scientific Publications, Oxford.

Bidwell, E. (1969). *In* "Annual Review of Medicine" **20**, 63 (A. C. de Graff and W. P. Creger, Eds). Annual Reviews Inc., Palo Alto, California.

Biggs, R. (1951). *In* "Prothrombin Deficiency" 55. Blackwell Scientific Publications, Oxford.

Biggs, R. and Bidwell, E. (1959). *Br. J. Haemat.* **5**, 379.

Biggs, R., Denson, K. W. E., Akman, N., Borrett, R. and Haddon, M. E. (1970). *Br. J. Haemat.* **19**, 283.

Blaustein, A., Schnayerson, N. and Wallach, R. (1953). *Am. J. Med.* **14**, 704.

Blombäck, B. (1967). "Fibrinogen to Fibrin Transformation. Blood Clotting Enzymology" (W. H. Seegers, Ed.) 143. Academic Press, New York and London.

Blombäck, B. and Blombäck, M. (1969). *In* "Human Blood Coagulation. Biochemistry, Clinical Investigations and Therapy" (H. C. Hemker, E. A. Loeliger and J. J. Veltkamp, Eds) 7. Leiden University Press.

Blombäck, B., Blombäck, M., Doolittle, R. F., Hessell, B. and Edman, P. (1963). *Biochem. Biophys. Acta* **78**, 563.

Blombäck, B., Blombäck, M., Henschen, A., Hessell, B., Iwanaga, S. and Woods, K. R. (1968a). *Nature, Lond.* **218**, 130.

Blombäck, M., Blombäck, B., Mammen, E. F. and Prasad, A. S. (1968b). *Nature, Lond.* **218**, 134.

Blombäck, M., Blombäck, B., Olsson, P., Srendson, L. (1974). *Thromb. Res.* **5**, 621.

Bloom, A. L., Giddings, J. C. and Peake, I. R. (1973). *Lancet* **i**, 661.

Bok, J., Veltkamp, J. J. and Loeliger, E. A. (1965). *Thromb. Diath. Haemorrh.* **13**, 8.

Bouma, B. N., Wiegerinck, Y., Sixma, J. J., van Mourick, J. A. and Mochtar, I. A. (1972). *Nature New Biol.* **236**, 104.

Brummel, M. C. and Montgomery, R. (1970). *Anal. Biochem.* **33**, 28.

Crowell, E. (1971). *Am. J. Med. Sci.* **260**, 261.

Dausset, J. (1959). *Vox Sanguinis* **30**, 428.

Davie, E. W. and Ratnoff, O. D. (1964). *Science* **145**, 1310.

Deggeller, K. (1968). "The Human Prothrombin Activating Enzyme". Thesis, Amsterdam.

Denson, K. W. E. (1961). *Acta Haemat.* **25**, 105.

Denson, K. W. E. (1966). *In* "Treatment of Haemophilia and Other Coagulation

Disorders" (R. Biggs and R. G. Macfarlane, Eds) 350. Blackwell Scientific Publications, Oxford.

Denson, K. W. E. (1967). "The Use of Antibodies in the Study of Blood Coagulation". Blackwell Scientific Publications, Oxford.

Denson, K. W. E. (1968). *Lancet* **ii**, 222.

Denson, K. W. E. (1969). *Toxicon* **7**, 5.

Denson, K. W. E. (1971). Proc. of the 7th Congress of the World Federation of Haemophilia, Teheran.

Denson, K. W. E. (1973). *Br. J. Haemat.* **24**, 451.

Denson, K. W. E. and Ingram, G. T. C. (1973). *Lancet* **i**, 157.

Denson, K. W. E., Biggs, R. and Mannucci, P. M. (1968). *J. Clin. Path.* **21**, 160.

Denson, K. W. E., Biggs, R., Haddon, M. E., Borrett, R. and Cobb, K. (1969). *Br. J. Haemat.* **17**, 163.

Denson, K. W. E., Lurie, A., De Cataldo, F. and Mannucci, P. M. (1970). *Br. J. Haemat.* **18**, 309.

Denson, K. W. E., Conard, J. and Samama, M. (1972). *Lancet* **i**, 1234.

Duckert, F. (1971). Transactions of the Conference of the International Society on Thrombosis and Haemostasis. Montreaux, Switzerland, 1970, 329.

Dunning, W. F. and Curtis, M. R. (1939). *Genetics* **24**, 70.

Edman, P. (1963). *Biochem. Biophys. Acta* **78**, 563.

Egeberg, O. (1965). *Thromb. Diath. Haemorrh.* **13**, 516.

Elodi, S. and Puskas, E. (1972). *Thromb. Diath. Haemorrh.* **28**, 489.

Fanconi, G. (1944). *Schweiz. Med. Wschr.* **22**, 255.

Fantl, P., Sawers, R. J. and Marr, A. G. (1956). *Aust. Ann. Med.* **5**, 163.

Feinstein, D. I., Chong, M. N. Y., Kasper, C. K. and Rapaport, S. I. (1969). *Science* **163**, 1071.

Feinstein, D. I., Rapaport, S. I., McGee, W. G. and Patch, M. J. (1970). *J. Clin. Invest.* **49**, 1578.

von Felten, A., Duckert, F. and Frick, P. G. (1966). *Br. J. Haemat.* **12**, 667.

Finlayson, J. S. and Mosesson, M. W. (1970). 13th International Congress of Haematology, Munich. J. F. Lehmanns Verlag, Munchen.

Firkin, B. G., Firkin, F. and Stott, L. (1973). *Austral. N.Z. J. Med.* **3**, 225.

Forman, W. B., Ratnoff, O. D. and Boyer, M. H. (1968). *J. Lab. Clin. Med.* **72**, 455.

Gaffney, P. J. (1971). *Nature, Lond.* **230**, 54.

Gerbeck, G. M., Yoshikawa, T. and Montgomery, R. (1969). *Arch. Biochem. Biophys.* **134**, 67.

Gobbi, F., Ascari, E. and Barbieri, M. (1967). *Thromb. Diath. Haemorrh.* **17**, 194.

Goodnight, S. H., Feinstein, D. I., Osterud, B. and Rapaport, S. I. (1971). *Blood* **38**, 1.

Goudemand, M., Foucaut, M. S., Hutin, A. and Parquet-Gerney, A. (1963). *Nouv. Rev. Fr. Hemat.* **3**, 703.

Graham, J. B. (1975). *In* "Handbook of Haemophilia" (K. M, Brinkhous and H. C. Hemker, Eds). Excerpta Medica, Amsterdam.

Graham, J. B. and Barrow, E. S. (1973). *Lancet* **ii**, 388.

Graham, J. B., McLester, W. D., Pons, K., Roberts, H. R. and Barrow, E. M. (1964). *In* "The Haemophilias" (K. M. Brinkhous, Ed.) 263. University of North Carolina Press, Chapel Hill.

Gralnick, H. R., Abrell, E. and Bagley, J. (1971). *Nature New Biol.* **230**, 16.

Gralnick, H. R., Coller, B. S. and Marchesi, S. L. (1973). *Nature New Biol.* **244**, 281.

Greaves, J. H. and Ayres, P. (1967). *Nature, Lond.* **215**, 877.

Greaves, J. H. and Ayres, P. (1969). *Nature, Lond.* **224**, 284.

de Grouchy, J., Turlea, C., Josso, F., Nedelee, G. and Nedelee J. (1974). *Human-genetik* **24**, 197.

Hampton, J. W. and Morton, R. O. (1970). 13th International Congress of Haematology, Munich. J. F. Lehmanns Verlag, Munchen.

Hemker, H. C. (1969). *In* "Human Blood Coagulation. Biochemistry Clinical Investigation and Therapy" (H. C. Hemker, E. A. Loeliger and J. J. Veltkamp, Eds) 54. Leiden University Press.

Hemker, H. C., Esnouf, M. P., Hemker, P. W., Swart, A. C. W. and Macfarlane, R. G. (1967a). *Nature, Lond.* **215**, 248.

Hemker, H. C., Veltkamp, J. J. and Loeliger, E. A. (1967b). *Thromb. Diath. Haemorrh.* **19**, 346.

Hermodson, M. A., Suttri, J. W. and Link, K. P. (1969). *Fed. Proc. Fed. Am. Soc. Exp. Biol.* **28**, 386.

Holmberg, L. and Nilsson, I. M. (1972). *Br. Med. J.* **3**, 317.

Hougie, C. and Sargeant, R. (1973). *Lancet* **i**, 616.

Hougie, C. and Twomey, J. J. (1967). *Lancet* **i**, 698.

Hoyer, L. W. (1972). *Blood* **39**, 481.

Hoyer, L. W. and Breckenridge, R. T. (1968). *Blood* **32**, 962.

Ingram, G. I. C. (1955). *J. Clin. Path.* **8**, 318.

Iversen, T. and Bastrup Madsen, P. (1956). *Br. J. Haemat.* **2**, 265.

Jacob, F. and Monod, J. (1961). *J. Mol. Biol.* **3**, 318.

Jones, J. H., Rizza, C. R., Hardisty, R. M., Dormandy, K. M. and Macpherson, J. C. (1962). *Br. J. Haemat.* **8**, 120.

Josso, F., Lavergne, J. M., Weiland, C. and Soulier, J. P. (1967). *Thromb. Diath. Haemorrh.* **18**, 311.

Josso, F., Monasteria De Sanchez, J., Lavergne, J. M., Menaché, D. and Soulier, J. P. (1971). *Blood* **38**, 9.

Kernoff, P. B. A., Gruson, R. and Rizza, C. R. (1974). *Br. J. Haemat.* **26**, 435.

Kidd, P., Denson, K. W. E. and Biggs, R. (1963). *Lancet* **ii**, 522.

Larrieu, M. J. and Meyer, D. (1970). 6th Congress of the World Federation of Haemophilia, Vienna, Austria.

Leitner, A., Bidwell, E. and Dike, G. W. R. (1963). *Br. J. Haemat.* **9**, 245.

Lester, R. H., Elston, R. C. and Graham, J. B. (1972). *Am. J. Hum. Genet.* **24**, 168.

Macfarlane, R. G. (1938). *Lancet* **i**, 309.

Macfarlane, R. G. (1964). *Nature, Lond.* **202**, 498.

Magnusson, S. (1969). *In* "Human Blood Coagulation. Biochemistry, Clinical Investigation and Therapy" (H. C. Hemker, E. A. Loeliger and J. J. Veltkamp, Eds) 18. Leiden University Press.

Mammen, E. F., Prasad, A. S. and An, C. C. (1968). Proceedings of the 16th Annual Symposium on Blood. College of Medicine, Wayne State University, 24.

McDonagh, J. (1971). Transactions of the Conference of the International Society on Thrombosis and Haemostasis, Montreux, Switzerland, 1970, 329.

McLester, W. D. and Wagner, R. H. (1965). *Am. J. Physiol.* **208**, 499.

Menaché, D. (1964). *Thromb. Diath. Haemorrh.* **13**, 173.

Meyer, D., Bidwell, E. and Larrieu, M. J. (1972a). *J. Clin. Path.* **25**, 433.

Meyer, D., Lavergne, J. M., Larrieu, M. J. and Josso, F. (1972b). *Thromb. Res.* **1**, 183.

Meyer, D., Dreyfus, M. D. and Larrieu, M. J. (1973). *Pathol. Biol.* **21**, 66.

Michael, S. E. and Tunnah, G. W. (1966). *Br. J. Haemat.* **12**, 115.

Morawitz, P. (1905). *Ergebn. Physiol.* **4**, 307.

Nossel, H. L. (1976). *In* "Human Blood Coagulation, Haemostasis and Thrombosis" (R. Biggs, Ed.). Blackwell Scientific Publication, Oxford.

Odegard, O. R., Lie, M. and Abidgaard, O. (1976) *Haemostasis* **5**, 265.

Oeri, J., Matter, M., Isenschmid, H., Hauser, F. and Koller, F. (1954). *Biblthca. Paediat.* **58**, 575.

Ouchterlony, O. (1958). *Prog. Allergy* **5**, 1.

Peake, I. R., Bloom, A. L. and Giddings, J. C. (1974). *New Engl. J. Med.* **291**, 113.

Pfueller, S., Somer, J. B. and Castaldi, P. A. (1969). *Coagulation* **2**, 213.

Piper, W. and Schreier, M. H. (1964). *Thromb. Diath. Haemorrh.* **11**, 423.

Prentice, C. R. M., Forbes, C. D. and Smith, S. M. (1972). *Thrombosis Res.* **1**, 493.

Rabiner, S. F. and Kretchmer, N. (1961). *Br. J. Haemat.* **7**, 99.

Ratnoff, O. D. (1970). Immunological Techniques and Coagulation Factors. 1st Conference of the International Society on Thrombosis and Haemostasis, Montreux, Switzerland.

Ratnoff, O. D. and Steinberg, A. G. (1962). *J. Lab. Clin. Med.* **59**, 980.

Roberts, H. R., Grizzle, J. E., McLester, W. D. and Penick, G. D. (1968). *J. Clin. Invest.* **47**, 360.

Samama, M., Perrotez, C., Houissa, R. and Hafsia, A. (1976). *Lancet* **i**, 151.

Sas, G., Blascó, G., Bánhegy, D., Jákó, J. and Pálos, L. Á. (1974). *Thromb. Diath. Haemorrh.* **32**, 105.

Seibert, R. H., Margolius, A., Jr. and Ratnoff, O. D. (1958). *J. Lab. Clin. Med.* **52**, 449.

Shanberge, J. N. and Gore, J. (1957). *J. Lab. Clin. Med.* **50**, 945.

Shapiro, S. S., Martinex, J. and Holborn, R. R. (1969). *J. Clin. Invest.* **48**, 2251.

Smink, M. M., Daniel, T. M., Ratnoff, O. D. and Stavitsky, A. B. (1967). *J. Lab. Clin. Med.* **69**, 819.

Soria, J., Soria, C. and Samama, M. (1976). *Pathol. Biol.* **24**, 725.

Soulier, J. P. and Prou Wartelle, O. (1966). *Nouv. Rev. Franc. Hêmat.* **6**, 623.

Spinelli, A., Schmidt, W. and Straub, P. W. (1976). *Br. J. Haemat.* **34**, 129.

Stites, D. P., Hershgold, E. J., Perlman, J. D. and Fudenberg, H. H. (1971). *Science* **171**, 196.

Takagi, T. and Iwanga, S. (1969). *Biochem. Biophys. Acta* **194**, 594.

Thompson, C., Forbes, C. D. and Prentice, C. R. M. (1974). *Lancet* **i**, 594.

Uszynski, L. (1966). *Thromb. Diath. Haemorrh.* **16**, 559.

Uszynski, L. (1967). *Thromb. Diath. Haemorrh.* **18**, 325.

Van der Meer, J., Stoepman van Dalen, E. A. and Jansen, J. M. S. (1973). 4th International Congress on Thrombosis and Haemostasis Abstracts, Vienna, 226.

Veltkamp, J. J. and van Tilburg, N. H. (1973). *New Engl. J. Med.* **289**, 882.

Veltkamp, J. J., Hemker, H. C. and Loeliger, E. A. (1965). *Thromb. Diath. Haemorrh.* **17**, 181.

Veltkamp, J. J., Meilof, J., Remmelts, H. G., van der Vleerk, D. and Loeliger, E. A. (1970). *Scand. J. Haemat.* **7**, 82.

Weiss, H. J., Hoyer, L. W., Pickles, F. R., Varma, A. and Rogers, J. (1973). *J. Clin. Invest.* **52**, 2708.

Zimmerman, T. S. and Edginton, T. S. (1973). *J. Exp. Med.* **138**, 1015.

Zimmerman, T. S., Ratnoff, O. D. and Powell, A. E. (1971). *J. Clin. Invest.* **50**, 244.

14 Unsolved Mendelian Diseases

G. R. FRASER

Department of Health and Welfare, Tunney's Pasture, Ottawa, Canada

I. INTRODUCTION

By a paradox common to all branches of science, the subject matter of this chapter covers a scope which is greater by several orders of magnitude than that of the remainder of the book. This is because the frontiers of ignorance encompass a far wider area than the frontiers of knowledge. Thus, the category of Mendelian disorders whose biochemical basis is unknown includes the vast majority of those already recognized at the clinical level as well as the far larger number not yet even identified. In this chapter an attempt, perforce largely speculative, will be made to define the extent of this terra incognita in very broad outline and to suggest some approaches whereby advances in the correlation of biochemical and genetical mechanisms with the clinical phenomena characterizing these Mendelian disorders may be made in the future.*

Since the pioneering work of Sir Archibald Garrod, it has been customary to regard autosomal recessive diseases, as a class, as being potential inborn errors of metabolism in which the lack or deviant action of an important enzyme or other protein molecule has led to a more or less typical constellation of clinical signs and symptoms. Certainly, other chapters in this book bear abundant evidence to the validity of this concept in the elucidation of the mechanisms of pathogenesis of, for example, galactosaemia, phenylketonuria, some types of haemolytic anaemia and the various disorders due to production of abnormal haemoglobins.

Nevertheless, it seems clear that the temptation should be resisted to seek such a holistic explanation for every autosomal recessive disorder in Man. For example, what is known about the pathogenesis of thalassaemia may serve as a model on which to base a discussion of several alternative hypothetical modes of gene action in the causation of such autosomal recessive disorders. Thus it seems that the primary genetical defect in thalassaemia lies in the synthesis, processing, transport or stability of the RNA (α-chain mRNA in α-thalassaemia, β-chain mRNA in β-thalassaemia). Total absence of mRNA in α^0- and β^0-thalassaemia suggests gene deletion, thereby focusing attention on the regulation of globin gene expression. In addition, it has been shown that some types of β-thalassaemia may arise from a process of unequal chromosomal crossing-over with the resultant formation of a gene product which represents the fusion of remnants of two distinct molecules, the β- and δ-chains of haemoglobin; such a mechanism could conceivably be subsumed under the concept of aberrant protein production even though it does not represent the classical type of single amino acid substitution. Hypotheses other than that of a regulatory gene mutation may readily be constructed, using this mechanism of unequal crossing-over, and applied to explain those types of thalassaemia in which the synthesis of two chains such as β and δ is affected (p. 617).

Another concept which may be relevant in the context of thalassaemia is that of a mutation controlling quantitative rather than qualitative aspects of protein production. This might possibly be effected by means of base-pair substitutions which alter the triplet concerned to another coding for the same amino acid but less efficacious in protein synthesis, but other mechanisms might also be involved. In this way, a whole series of "normal iso-alleles" may exist, coded to synthesize the β-chain of haemoglobin and the varying degrees of efficiency associated with them may account for part of the normal variation in haemoglobin synthesis. Under this hypothe-

* Excellent catalogues of such disorders may be found elsewhere (for example McKusick 1975)

sis, the possession of two alleles (which need not necessarily be the same which are from the lower extreme of the range of efficacy may result in thalassaemia. Part at least of the considerable variation in the clinical symptomatology of thalassaemia major ("homozygous" thalassaemia) may be due to the existence of many different combinations of such "normal isoalleles".

The set of four clinically distinct conditions related to α-thalassaemia which is discussed on p. 612 provides yet further examples of the dangers of attempting to force all defects into simple Mendelian frameworks. Kan et al. (1975) have now shown that these four α-thalassaemia states probably correspond to deletions of one to four of the α-chain structural genes, there being duplicate α-chain loci. However, it is by no means clear that the α-chain locus is duplicated in all populations, so that generalization from the work of Kan et al. is most hazardous.

Another important lesson which can be learned from studies of thalassaemia and the haemoglobinopathies is that the idea of an autosomal recessive disease being due to homozygosity for an abnormal allele can no longer be accepted in an unmodified form, even when classical amino acid substitutions are involved. Thus, a whole series of abnormal alleles of this type might exist, and different forms of such a disease may be due to homozygosity for distinct abnormal alleles at the same locus as well as to compound heterozygosity. An example is afforded by the diseases caused by the genotypes giving rise to SS, SC and CC patterns of haemoglobin production, and McKusick (1973) has reviewed the evidence for other examples. The segregation ratio of affected to total cases to be expected in such situations remains one-quarter, but these complications drastically affect the theory underlying quantitative relationships between frequencies of autosomal recessive conditions and the expected parental consanguinity rate (Dahlberg, 1948).

Anomalies in these relationships as well as minor inter-familial differences in clinical expression may be useful in the future as indicators of the possible existence of multiple pathological alleles determining several defective forms of some enzymes. This type of explanation may underlie the variability of the clinical and laboratory expressions of such inborn errors of metabolism as galactosaemia and phenylketonuria.

In the case of diseases defined thus far only at the clinical level (for example, childhood visual or auditory handicap), such failure of correspondence between parental consanguinity and population frequency has also been taken to indicate genetical heterogeneity and the delineation of very different as well as of apparently closely related clinical syndromes within these entities has suggested that the heterogeneity involves not only multiple alleles at the same loci but also entirely distinct loci. For example, Leber's tapeto–retinal degeneration and the Laurence–Moon–Bardet–Biedl syndrome are entirely distinct though both are autosomal recessive conditions which can give rise to serious visual handicap in childhood; they are presumably due to mutant alleles at entirely distinct loci. On the other hand, the condition described by Alström *et al.* (1959) is sufficiently similar to the Laurence–Moon–Bardet–Biedl syndrome to raise the suspicion that homozygosity or compound heterozygosity involving different mutant alleles at the same locus is involved.

Examples can be drawn from among solved Mendelian diseases of these phenomena. Thus, the methaemoglobinaemias are very similar clinical conditions determined at different loci by quite dissimilar biochemical mechanisms (p. 505), whereas the Lesch–Nyhan syndrome and some forms of mild gouty arthritis are distinct disorders but determined by allelic genes. The gradual unfolding of the molecular pathology of the classical mucopolysaccharidoses over the past five years has produced further examples of heterogeneity and affinity. Thus, the Hurler and Scheie syndromes, which clinicians readily distinguish, are now known to arise from homozygosity for distinct alleles at the same locus, and to involve deficiency of the same lysosomal enzyme, α-L-iduronidase. These disorders have been renamed mucopolysaccharidoses IH and IS, which allows a more rational description of the recently discovered Hurler-Scheie compound heterozygote as mucopolysaccharidosis I-HS. On the other hand enzyme studies on the clinically homogeneous entity known as Sanfilippo syndrome have demonstrated two distinct disorders, in one of which sulphaminidase and in the other α-N-acetyl-glucosaminidase are deficient. A third enzyme defect, involving α-glucosaminidase, has been tentatively identified. Regrettably, the new nomenclature for the Sanfilippo spectrum, mucopolysaccharidoses IIIA, IIIB and IIIC, fails to emphasize that the enzyme studies have implicated three separate gene loci (Dorfman and Matalon, 1976).

II. SCREENING TECHNIQUES

Limited screening techniques are available for the study of Mendelian disorders whose biochemical developmental bases are unknown. There are very few such conditions where karyotypic screening of one or more cases has not been performed and several gross chromosomal anomalies have been described in which excellent correlation between the clinical features and the chromosomal abnormality indicates a cause and effect relationship. Most of these conditions are potentially Mendelian in that, if reproduction of affected individuals were possible, transmission might be expected; such transmission has in fact been reported in the case of trisomy 21 (Down's syndrome). Furthermore, many syndromes involving translocations may be regarded as Mendelian, even though they do not give rise to patterns of inheritance characteristic of the classical types of transmission. Typically such a condition might occur in one or more sibs, but not in a parent. One of the parents, however, would be expected to be a balanced translocation carrier and so the condition might appear in his sibs and possibly his uncles and aunts; an increased incidence of abortions in the pregnancies of appropriate relatives has been regarded as a possible indication of the presence of this type of inheritance. It should be noted that this pattern of inheritance in general is not too dissimilar to that described as irregularly dominant or dominant with reduced penetrance of the mutant allele; it would not be unexpected, therefore, to find that some such irregularly dominant conditions are due to chromosomal rearrangements. In fact, a whole spectrum of heritable conditions might exist between the severe malformations associated with gross chromosomal anomalies and the "classical" Mendelian disorders presumably due to changes in the chromosomal DNA affecting a single base-pair.

Other screening procedures, which have been useful in the past in the detection and delineation of Mendelian disorders, have been examination of urines for excretion patterns of sugars, amino acids and mucopolysaccharides, supplemented by serum screening in appropriate cases, and such methods are likely to continue to be valuable in the future. Often discovery of a gross anomaly of excretion in this way leads to elucidation of the abnormality at the enzymatic level even though the unravelling of the full chain of cause and effect relationships between the enzymatic abnormality and the clinical manifestations remains elusive; nevertheless, such disorders may be regarded as representative of "solved" rather than "unsolved" Mendelian disease for the purpose of this discussion (see Chapter 10, Section III). There are ample reasons to suppose that extension of screening techniques to other types of laboratory determinations, both in urine and in blood, and to direct enzyme assays will lead to further transfers of Mendelian disease from the unsolved to the solved category. Steroids and other hormones, trace elements, lipids and immunoglobulins are examples of classes of compounds which may be included in such screening in the future. At the present time the main problem is that the methods available are too complex and expensive for screening applications, but this is likely to be remedied in the foreseeable future by the general introduction of automated machinery for analysis supplemented by computer interpretation of the results. Thus, it is likely that patients with a newly detected Mendelian disease, as well as cases of those which are already well recognized, will be subjected to a whole battery of screening tests as a routine and this approach is likely to bear further dividends. It should be noted that such a blanket approach, which may be likened to fishing with a net rather than in a more selective manner with a rod and line, is necessitated by our profound ignorance of the normal processes of intermediary metabolism in the human body. Were our knowledge more precise in this field, then the clues obtainable from the clinical symptomatology of Mendelian disease would give many more indications than they in fact do as to the exact type of metabolic error involved in aetiology.

In addition to blood and urine, such screening approaches are likely to be extended in a far more systematic manner to other tissues. The red blood cell, like haemoglobin, may be obtained in large quantities without harming the patient and it may be regarded as a laboratory in which a fair-sized sample of the organism's metabolic repertoire is performed. The lymphocyte is also useful in this respect and may be used to set up cell cultures in which normal and aberrant metabolic pathways may be studied at leisure; in addition, the skin is a relatively non-traumatic source for long-term cultures of this type. This somatic cellular approach suffers from a substantial disadvantage in that, at the present time, many metabolic errors can be detected only in tissues in which they exert their main effects *in vivo* such as the brain, liver, kidney, muscle or endocrine glands and, clearly, biopsy procedures to obtain specimens of such tissues are too traumatic to be used on a large scale for investigative purposes unless some immediate benefit to the patient is likely to accrue. There is no reason, however, why an extension of studies of this type to such organs, using tissues obtained *post mortem*, should not occur. In addition, leucocyte or skin fibroblast nuclei contain the genetical information used in other tissues and some way may be found to induce the activation of enzyme configurations appropriate to the clinical features of the disease under study. Investigations of this type, both at the organ level and the cell culture level, seem particularly appropriate in the growing class of Mendelian disorders which are associated with abnormal storage of cell products in various organs.

III. THE NATURE OF DOMINANCE

One of the major unsolved problems of Mendelian disease is the pathogenesis of abnormality in heterozygotes. It is, of course, a truism to say that a mutant allele is not dominant or recessive as the case might be; the qualification properly belongs to its effects. For example, the mutant allele determining the substitution in the β-chain responsible for haemoglobin S production may lead, in heterozygotes, to an increased predisposition to vascular thrombotic phenomena especially under conditions of reduced oxygen tension. This may be regarded as a dominant effect of relatively minor clinical importance (with the necessary qualifications of reduced penetrance and variable expressivity); the recessive effects of the same allele manifested as sickle-cell anaemia are well-known.

This is a clear-cut and simple situation and presents no major difficulties in interpretation. It is not easy, however, to understand the pathogenesis of the large class of diseases in which major clinical manifestations are common in the heterozygote and still less the mechanisms whereby many such heterozygotes escape most or all of the deleterious dominant effects of the mutant allele (reduced penetrance and variable expressivity).

In the improved understanding of these biological mechanisms lies one of the major avenues towards elucidation of unsolved Mendelian disease. Does the disease arise because the normal allele is inadequate quantitatively to ensure a vital metabolic function? Such a mechanism might be operative both in connection with a critical time-limited process as during the perinatal period of adaptation to extra-uterine existence or even earlier at equally critical conjunctures of embryonic existence. In the latter case, disturbances of switch-over from the production of one form of foetal protein to another one with a similar function may be involved; in general, complete ignorance prevails concerning the significance, for example, of the relationships governing such switch-overs in the production of various types of haemoglobin chains in foetal life. Such a quantitative inadequacy of a mutant allele might also be operative over a long-term period in post-natal existence, for example in the determination of Huntington's chorea and of dominant retinitis pigmentosa. On the other hand, the abnormality may be due to a qualitative effect, i.e. to the production by the mutant allele of a distinct product which either competes with that formed by the normal allele or which leads to the deposition of abnormal tissue (cf. p. 487). Such qualitative effects may be suspected in such diseases as spherocytosis and elliptocytosis, in which deposition of an abnormal constituent in the red blood cell membrane may be involved, and also in connective tissue disorders such as Marfan's syndrome. Diseases involving a break-down of an important process in adult life, such as retinitis pigmentosa and Huntington's chorea mentioned above in connection with possible quantitative defects, could of course equally well be due to qualitative defects whose effects make themselves felt gradually. An excellent example of this type of mechanism acting in a more acute manner, where the basic defect is known, is afforded by the clinical diseases due to unstable haemoglobins in which deposition of the abnormal product leads to destruction of the red blood cell.

A prototype for yet another class of mechanisms which may lead to dominant diseases is afforded by dominant polycythaemias caused by abnormal haemoglobins. In these conditions, a disturbance of physiological function (in this case oxygen affinity) of the abnormal protein product leads to over-activation of normal compensatory mechanisms which leads to pathological problems associated with production of an excess of red blood cells.

It is clear, in fact, that there is no unitary explanation of dominant disease just as there is no unitary explanation of recessive disease. However, just as the postulated one–to–one correspondence between a mutant allele and an aberrant protein product has been fruitful in the elucidation of many recessive inborn errors of metabolism and their relationships to specific enzyme defects, so greater knowledge of the nature of gene action in the causation of dominant conditions will throw new light on the pathogenesis of many types of unsolved Mendelian diseases. Nevertheless, as has already been discussed in Chapter 10, regulatory mechanisms in eukaryotes are for the most part completely unknown, and the early optimistic claims that many dominant disorders represent mutations at regulatory loci have virtually all been abandoned.

IV. UNSOLVED MENDELIAN DISEASE AND NEOPLASIA

It is clear that genetical factors play an important role in the causation of every type of neoplasm, both malignant and benign; it is equally clear that only a small minority of such diseases is primarily due to simple Mendelian mechanisms, i.e. single-gene inheritance. Ignorance prevails concerning the mode of operation of these mechanisms, but there are some clues available.

Certain neoplasms, for example uterine leiomyomas, appear to arise from one initial precipitating event in a single precursor cell, while others have a multicellular origin (p. 425). These neoplasms are not Mendelian traits; indeed, at least two of them have essentially viral aetiology. However, the same heterogeneity undoubtedly applies to those neoplasms which are simply inherited. Thus, some types of dominantly inherited neoplasms are multicentric in origin, and it has been postulated that they arise as a result of at least one somatic mutation occurring in each of these foci of origin (Nicholls, 1969). On the simplest hypothesis, such a somatic mutation is a replica of the germinal mutation already present in all cells of such heterozygotes, so that these cells are homozygous at the locus in question.

Such a hypothesis seems particularly applicable to tumours of embryonic and early postnatal life such as retinoblastoma. It may well be that other tumours of this type such as nephroblastoma (Wilms's tumour), hepatoblastoma and neuroblastoma also conform to this model. They may thus be similar to retinoblastoma in their pattern of inheritance, and, in fact, sporadic reports of affected offspring being born to survivors are appearing in the literature. Under this hypothesis, true phenocopies of these tumours would occur, in that occasionally identical somatic mutations of both alleles at the locus concerned would occur in a non-heterozygote; such tumours would be unicentric in origin, as opposed to the multicentricity of the germinally transmissible type. In the case of retinoblastoma, there are several supporting strands of evidence for this type of hypothesis. Thus, both transmissible and non-transmissible forms of this tumour occur and the transmissible types are usually bilateral while the non-transmissible are usually unilateral; this might indicate respective multi- and unicentric origin of these two classes of tumour. Such a hypothesis, of course, does not contribute to the elucidation of the role of these mutant alleles in the pathogenesis of the tumour, but one possible lead in this direction is afforded by instances of undoubted heterozygotes for the mutant allele causing retinoblastoma (i.e. unaffected persons through whom retino-blastoma is transmitted from grandparent to grandchild) who are appar-ently normal and also by patients both with retinoblastoma and with other embryonic cell tumours in whom spontaneous regression of the tumour has occurred. These findings may suggest a phenomenon of an immunological type, possibly a host reaction against the malignant cells. These types of causation may also be operative in other hereditary tumours not necessarily connected with embryonic cells such as trichoepithelioma or polyposis coli.

A second possible clue towards an eventual elucidation of these modes of aberrant gene action involved in the causation of simply inherited neoplasia is afforded by work on xeroderma pigmentosum, an autosomal recessive disorder characterized by extreme sensitivity to sunlight with the develop-ment of multiple carcinomatous changes in the skin at an early age. It has been shown that fibroblasts derived from patients with this disease are unable *in vitro* to perform adequate repair of damage to DNA bases. It is clear that precise definition of this difficulty which very probably involves deficient or totally absent function of an enzyme which is usually respon-sible for this stage of repair of DNA damage may have far-reaching implications in the field of carcinogenesis in general. It is of interest to note that xeroderma pigmentosum is one of the group of autosomal recessive diseases mentioned above where genetical heterogeneity involving homozy-gosity at different gene loci has been demonstrated (McKusick, 1973).

V. UNSOLVED MENDELIAN DISEASE AND COMPLEX INHERITANCE

Autosomal and sex-linked, recessive and dominant, single-gene inheritance does not satisfactorily explain all forms of inherited disease even though valiant efforts are often made to fit data to such simple models, using as powerful tools to this end the flexible and not too meaningful concepts of penetrance and expressivity. In the non-experimental situation in which human heredity is studied there is little opportunity to test more complex models involving interactions, epistatic and otherwise, between two or more genes but presumably such mechanisms do influence the determination of many unsolved Mendelian diseases. Again, in the haemoglobin field, a beginning has been made in this direction as, for example, in the study of the apparent interaction between alelles determining α-thalassaemia and the production of haemoglobin E (Tuchinda *et al.*, 1964) and between those determining α and β thalassaemia (Kan and Nathan, 1970). More complex interactions involving multiple genes are, of course, postulated *ex hypothesi* to account for the most important traits of the human species, including intelligence, height, visual refraction, personality and behaviour. Complete ignorance prevails concerning the physiological basis of these interactions and, in essence, only a bare beginning has been made in understanding these complex mechanisms.

In general, development, both pre- and post-natal, is determined under the influence of such interactions between multiple genetical and environmental factors and the majority of cases of congenital malformations can be explained in these terms. However, as in the case of disease appearing at any other stage in life, causation can be due primarily to a single factor, whether a gene in the instance of malformation syndromes or an environmental teratogen such as thalidomide or the rubella virus. The term primarily is used advisedly since genetically determined malformation syndromes are very variable in their symptomatology because of modification both by the environment and the residual genotype, and the effects of a teratogen are strongly influenced both by the maternal and foetal genotypes.

Phenomena of this same general class are probably also operative in the causation of many of the common diseases of adult life. Thus for example, diabetes mellitus is, apart from rare Mendelian forms, a disease resulting from the interaction of a complex genetical predisposition with elements of the life style of the individual such as diet, stress and lack of exercise. In contrast to a birth defect such as cleft palate where the environmental influences on the genotype are intra-uterine, act over a brief period and are, in the absence of knowledge of their specific nature, largely uncontrollable, in the case of diabetes mellitus, exogenous factors act over a long period and attempts at control of these effects may be made. Thus this is to a certain extent an avoidable disease especially if the complex genetically determined susceptibility could be defined. In this way preventive measures and advice could be specifically directed towards a small segment of the population at increased risk.

It is unfortunately not easy to identify such persons at risk in the case of diabetes but in the case of another common disease of adult life, athero-sclerosis, a substantial beginning has been made in this direction. Extensive family studies have revealed various types of hyperlipidaemias which in a proportion of cases, seem to be determined in a relatively simple genetical manner, even though complex inheritance is involved in the majority. In this complex determination it has been shown that the genotype at the *ABO* locus is of importance (see p. 167). These hyperlipidaemias are important predisposing factors in the pathogenesis of atherosclerosis and may be controlled by dietary measure or by drug treatment. Thus a sub-group of the population may be defined either by general screening or by careful investigation of relatives of patients to whom such preventive advice can be directed in the hope of avoiding premature morbidity and mortality due to cardiovascular disease in association with atherosclerosis. In the case of diabetes mellitus, even though systematic screening for a pre-diabetic state has not so far proved possible, the familial aggregation of the disease, determined by complex inheritance, suggests that avoidance of known exogenous factors connected with the life style is especially relevant in relatives of patients.

The situation in the case of atherosclerosis is an illustration of the fact that although multifactorial inheritance may involve interactions of many genes, some of these may play a major role in the determination of a particular disease. Although this does not imply that atherosclerosis is a Mendelian disease in the strict sense, definable Mendelian factors may in some cases be of substantial importance in causation.

Another aspect of this situation is the association of disease with poly-morphic marker systems discussed elsewhere in this volume (see p. 448), the extreme case being that of ankylosing spondylitis with the B27 antigen of the HLA system. In this case, either this particular antigenic marker or the product of another gene locus situated in the vicinity of that deter-mining HLA, and in linkage disequilibrium with it, plays a major role in the determination of this disease which is modified both by the residual genotype and by environmental circumstances. In general, the relation-ships of the HLA chromosomal region with disease is a topic which is being intensively studied at the present time (see p. 398).

Another type of inheritance which is presumably sometimes involved in the determination of hereditary disease is that mediated by non-nuclear or cytoplasmic factors which may include viruses. Such inheritance or inter-actions between nuclear and cytoplasmic factors has been invoked to explain, for instance, the bizarre pattern of transmission of Leber's optic atrophy, a disease which is believed to be determined at an X-chromosomal locus but which shows certain anomalous features such as failure to demonstrate transmission from an affected male to his descendants and a substantial excess over the 50% expected of carrier daughters born of carrier mothers. Cytoplasmic inheritance has also been invoked (Nance, 1969) as a possible explanation of types of malformations such as anencephaly and spina bifida, which are marked by familial aggregation but which do not show simple Mendelian segregation. On the other hand, Clarke *et al.* (1975) have suggested that residual pathological trophoblastic material either from a previous miscarriage or a co-twin may be responsible for lesions of the spina bifida and anencephaly group.

VI. IMMUNOLOGICAL VARIATION AND MENDELIAN DISEASE

Genetically determined variation in the structure of immunoglobulins plays a role in the determination of many diseases. In some cases, as mentioned in Chapter 12, the problem of causation may be regarded as being partially solved in that a reasonable correspondence seems to exist between a Mendelian pattern of transmission of a disease, its clinical features and a specific disturbance of immunoglobulin synthesis (for example, agammaglobulinaemia). In other cases, the immunoglobulin disturbance is only one of a whole complex of features whose interconnection is entirely unexplained, as in the case of the Louis Bar ataxia-telangiectasia syndrome.

It would appear that leukaemia and other malignancies of the reticulo-endothelial system are unusually common in persons with at least some of these simply inherited disorders of the immune mechanism and that this predisposition may, in some cases at least, be associated with a tendency towards chromosome breakage in cell cultures *in vitro*. This finding provides one of many pointers to an extremely large and potentially very important area of biological uncertainty—the question of interactions between gene mutations, both somatic and germinal (especially those which affect the production and structure of immunoglobulins), viruses, chromosomal aberrations, malignancy, ageing and auto-immune disease. Burch (1969) has summarized his own work and that of his collaborators arising from Burnet's "forbidden clone" theory of disturbed-tolerance autoimmunity in a thought-provoking synthesis bearing on this subject. Burch's (1976) extension of these ideas has already aroused substantial opposition (Cairns, 1976), as it runs counter to the generally held view that most neoplasms are largely of environmental causation. Less all-encompassing, but perhaps more easily investigable hypotheses, for example ones based on findings such as the possible relationship between arylhydrocarbon hydroxylase inducibility and lung cancer (p. 169; cf. also Harris *et al.*, 1976; Coomes *et al.*, 1976), may aid in a more concrete understanding of the inter-relationships mentioned above, and this will then one day help to bridge the gap between the role of mutant alleles in causing rare recessive Mendelian diseases independently of the residual genotype and their synergistic actions as contributory factors in heterozygotes to the pathogenesis of the common neoplastic and degenerative diseases associated with ageing.

VII. GENETICAL NOSOLOGY

While preceding sections have contained much speculation on the possible nature of mechanisms which may play an important role in the causation of genetically determined disease, the raw material which forms the basis and the essential prerequisite for the recognition, delineation and eventual elucidation of an unsolved Mendelian disease is the study of genetical heterogeneity at the clinical level, or genetical nosology. This subject remains as germinal to this problem as it was two and three centuries ago, despite all the epoch-making advances in molecular biology which have taken place in recent times.

The fact that we are virtually unable today to use the clues afforded by the association of pleiotropic effects of a particular mutant allele to pinpoint the site of the primary defect, let alone in the elucidation of the often lengthy chain of causation between this primary defect and the clinical expression of the disease, is of course largely due to our profound ignorance of the biochemical and physiological aspects of normal human metabolism. We may expect that these deficiencies will be slowly remedied and that an interchange of information will develop between investigators interested in normal and abnormal aspects of human biology. Thus, on the one hand, increased knowledge of the normal may help to provide solutions of Mendelian diseases while, on the other hand, such solutions may throw light on many important and currently obscure aspects of the nature and genetical control of normal bodily processes. In this latter respect, it has been suggested in the last section that recessive diseases involving defects of the immune mechanism may provide important clues for the elucidation of problems which go far beyond the range of these individually very rare conditions. This is but one of many potential applications of a rule enunciated by a very perspicacious physician, William Harvey, more than 300 years ago.

> Nature is nowhere accustomed more openly to display her secret mysteries than in cases where she shows traces of her workings apart from the beaten

path; nor is there any better way to advance the proper practice of medicine than to give our minds to the discovery of the usual law of Nature by careful investigation of cases of rarer forms of disease. For it has been found, in almost all things, that what they contain of useful or applicable nature is hardly perceived unless we are deprived of them, or they become deranged in some way.

This statement remains no less topical today, and the explosive growth of a masterly catalogue of Mendelian inheritance between its four successive editions (McKusick, 1966, 1968, 1971, 1975) attests to the continuing validity and vigour of this type of approach.

VIII. CONCLUSIONS

The field of unsolved Mendelian disease is vast indeed and, of necessity, no attempt has been made to cover it in any detail. Adequate introductions to a study of these Mendelian diseases which are unsolved at the biochemical level but at least partially defined clinically may be found elsewhere (for example McKusick (1975)); there exists, of course, an even larger group of diseases which at the present time are both unsolved and even undefined. Rather than trying to reproduce a catalogue of this type, attention has been paid to some specific principles illustrated by selected disorders which may be of use in pointing the way towards basic advances in the study of aberrant gene function involved in the causation of Mendelian disease.

The one gene–one enzyme hypothesis has long been superseded by one in which the correspondence emphasized is between a nucleotide triplet and an amino acid (cf. Chapter 2), but it is perhaps a valid speculation to question the exclusive nature of such a mechanism. As is now well-known (cf. Chapter 3), chromosomal DNA consists (by one method of classification) of three types: unique sequences, occurring precisely once in the haploid genome and constituting perhaps 50% of the nuclear DNA; repeated sequences, occurring a thousand times or more in the haploid genome; and highly reiterated sequences, repeated more than a millionfold in many cases. This third kind of DNA, which does not appear to be transcribed, usually represents only a few per cent of the nuclear DNA, and has been shown to lie near the centromeres. One special kind of DNA which codes for structural genes but which is reiterated is that which codes for rRNA. These genes, of which there may be more than 100, are generally clustered on chromosomes carrying nucleolar organizers.

Thus, there are in one sense four kinds of DNA, the roles of two of which are by no means clear. Britten and Davidson (1971) have suggested that highly repeated DNA may be non-functional but available for use in further evolution, or that it may provide recognition sites for transcription or replication. Walker (1971), on the other hand, has emphasized its possible importance for chromosomal structure or meiotic mechanics, on account of its centromeric location. Also, as the moderately reiterated DNA is interspersed with the unique sequence DNA, it seems possible that this intermediate form is involved in gene regulation. (See Lewin, 1974, for a general discussion.) However, there is as yet no real evidence confirming any of these hypotheses, and in addition they have no clear implications for the nature of deleterious effects on the phenotype which might arise from defects in the various reiterated sequences.

A final word of caution is required concerning the title of this chapter. Unsolved Mendelian disease implies the existence of the antithesis, solved Mendelian disease. Solved is used here in a very partial sense to connote a small step towards a definite solution, for example, the identification of a particular enzyme as being defective or of a particular amino acid substitution in the aberrant protein product of a mutant allele. Suggestions have been made (for example by Jerne (1969) in an article arrestingly entitled "The complete solution of immunology") that, in certain fields, there will one day be nothing left to discover. This seems to imply a fundamental misconception of the nature of knowledge. As the boundaries of ignorance recede slightly at one point, a compensatory advance occurs at another which may be many times greater than the retreat; a solution of any problem brings to light a host of new unsolved problems whose very existence was previously unknown. The other chapters of this book illustrate vast variation in the interpretation of a "solution" in this sense; in some fields a solution may represent only an infinitesimal advance beyond total ignorance. There is little danger that in the field of Mendelian disease, scientists will not find problems to tax their ingenuity to the utmost for many thousands of years to come. Knowledge, or conversely ignorance, is infinite and a definitive solution in the colloquial sense of the word will always be an illusory goal.

REFERENCES

Alström, C. H., Hallgren, B., Nilsson, L. B. and Asander, H. (1959). *Acta Psychiat. Scand.* **34** (Suppl.), 129.

Britten, R. J. and Davidson, E. H. (1971). *Quart. Rev. Biol.* **46**, 111.

Burch, P. R. J. (1969). "An Inquiry concerning Growth, Disease and Ageing". University of Toronto Press.

Burch, P. R. J. (1976). "The Biology of Cancer: A New Approach" M.T.P. Lancaster.

Cairns, J. (1976). *Nature, Lond.* **260**, 198.

Clarke, C. A., Hobson, D., McKendrick, O. M., Rogers, S. C. and Sheppard, P. M. (1975). *Br. Med. J.* **4**, 743.

Coomes, M. L., Mason, W. A., Muijsson, I. E., Cantrell, E. T., Anderson, D. E. and Busbee, D. L. (1976). *Biochem. Genet.* **14**, 671.

Dahlberg, G. (1948). "Mathematical Methods for Population Genetics" Karger, Basel.

Dorfman, A. and Matalon, R. (1976). *Proc. Natn. Acad. Sci. USA* **73**, 630.

Harris, C. C., Autrup, H., Connor, R., Barrett, L. A., McDowell, E. M. and Trump, B. F. (1976). *Science* **194**, 1067.

Jerne, N. K. (1969). *Aust. Ann. Med.* **4**, 345.

Kan, Y. W. and Nathan, D. G. (1970). *J. Clin. Invest.* **49**, 635.

Kan, Y. W., Dozy, A. M., Varmus, H. E., Taylor, J. M., Holland, J. P., Lie-Injo, L. E., Ganesan, J. and Todd, D. (1975). *Nature, Lond.* **255**, 255.

Lewin, B. (1974). "Gene Expression" Vol. 2 Eucaryotic Chromosomes John Wiley, New York.

McKusick, V. A. (1973). *Am. J. Hum. Genet.* **25**, 446.

McKusick, V. A. (1975). "Mendelian Inheritance in Man. Catalogs of Autosomal Dominant, Autosomal Recessive, and X-Linked Phenotypes" 4th ed. Johns Hopkins University Press, Baltimore.

Nance, W. E. (1969). *Nature, Lond.* **224**, 373.

Nicholls, E. M. (1969). *Hum. Hered.* **19**, 473.

Tuchinda, S., Rucknagel, D. L., Minnich, V., Boonyaprakob, U., Balankura, K. and Suvatee, V. (1964). *Am. J. Hum. Genet.* **16**, 311.

Walker, P. M. B. (1971). *Prog. Biophys. Mol. Biol.* **23**, 145.

Subject Index

A

AACE, *see* Electrophoresis

ABH secretion, 337–9
  association with disease, 166, 169
  H gene, 337–8
  linkage, 317–18, 437

ABO blood groups, 326–8, 335–7
  alleles, 335–7
  alkaline phosphatase, 168, 338
  antibodies, 332–3
  antigenic strength, 333–4
  association with disease, 166, 168, 179, 329, 450–1
  Ax and Am types, 335–6
  blood transfusion, 327–8
  Bombay phenotype, 329
  cholesterol level, 168, 443, 737
  chromosomal assignment, 115–16, 436
  "*cis-AB*" allele, 336, 346
  enzymic degradation, 340–2
  E. coli as B antigen, 326
  evolution, 361–2
  H gene, 337–8
  haptoglobin association, 163
  Lewis system, 338
  linkage, 115–16, 358, 437
  maternofoetal incompatibility, 328
  mortality, 170–2
  natural selection, 286
  naturally occurring antibodies, 326
  organ transplantation, 398
  paternity testing, 447–8
  primates, 361
  secretion, 337–9
  sex ratio, 170
  structure of antigens, 340–6
  synthesis, 337, 342–6
  tissue distribution, 337

Abortion
  chromosomal anomalies, 144
  therapeutic, 515–7

Absorption, *see* Histocompatibility antigens

Acatalasia (Takahara's disease), 542
  in Japan, 493
  pharmacogenetics, 455, 493
  tissues affected, 501

Accelerator globulin, *see* Factor V

Acetyl cholinesterase, 5, 13
  mouse, in cultured cells, 61
  polymorphism, 189–90
  alleles, 190

Acetylesterase, salivary, 121

Acetyl transferase, liver, 224
  antidepressant metabolism, 224
  isoniazid metabolism, 224, 455–6
  polymorphism, 224, 455–7
  alleles, 224
  gene frequencies, 224, 455

Achondroplasia, 478

Acid α-glucosidase, 190–1
  affinity electrophoresis, 191
  alleles, 191
  deficiency *see* Pompe's disease
  gene frequencies, 191

Acid lipase deficiency, *see* Lipase

Acid phosphatase, erythrocyte, 191–4
  ACP_1 191–2
    activity, 192–3
    alleles, 192
    chromosomal assignment, 108, 117–18, 436
    electrophoretic patterns, 192
    gene frequencies, 193–4
    heritability, 453
    linkage, 437
    thermostability, 192
  ACP_2, 194
  ACP_3, 194
    substrate specificity, 191

Acid phosphatase, lysosomal
  chromosomal assignment, 436
  deficiency, 511, 537
    antenatal diagnosis, 518

Clotting factors—*contd.*
  properties, 689
  synonyms, 689
Coagulation, *see* Blood coagulation
Cobalamin, *see* Vitamin B_{12}
Coeliac disease
  and HLA, 404–5
Collagen, 5
  in cultured cells, 66
  disorders classified, 542
  isoelectric point, 21
  Type I
    chromosomal assignment, 436
    defect, 479–80
  Type III
    defect, 480
  Type IV
    defect, 480
Collagenase inhibition, 292
Colour blindness
  X-linkage, 302, 356–7, 444
Combined deficiencies of clotting
    factors, 720–1
  Factors V and VIII, 720
    molecular mechanisms, 721
  Factors VII and IX, 720
  Factors IX and XIII, 720
  Prothrombin and Factors VII, IX
    and, X 720
Complement, 5, 294–6, 312–13
  component deficiencies, 542
  concentration, 311
  C1 component and hereditary angio-
    neurotic oedema, 484–5
    chromosomal assignment, 397,
      436
  C2 component, 310
    linkage, 310, 358
    mouse, 310
  C3 component, 294–5, 312–13
    function, 294–5
    immunoelectrophoresis, 279, 295
    linkage, 358, 437
    polymorphism, 295
    structure, 294–5
  C4 component, 309
    activity, 309
    chromosomal assignment, 436
    deficiency, 309, 542
    electrophoresis, 309
    linkage, 309

  mouse, 309
    variation, 309
  C6 component, 295, 312–13
    polymorphism, 295
  C8 component
    chromosomal assignment, 436
    linkage, 358, 397
    fixation, 24
    pathways, 294
    properdin factor B, 295–6, 312–13
      chromosomal assignment, 296,
        397–8, 436
      electrophoresis, 296
      function, 296
      linkage, 296, 338, 397–8, 437
Complementation, 42–5
  albinism, 43
  galactosaemia, 434–5
  mucopolysaccharidoses, 506–7
Complex inheritance, 735–8, *see also*
    Variance
Congenital adrenal hyperplasia, *see*
    Adrenal hyperplasias
Congenital hypothyroidism, *see* Hypo-
    thyroidism
Congenital non-spherocytic haemolytic
    anaemia (CNHA) and enzyme
    deficiencies, 482–3, 505
Congenital sucrose intolerance, 329,
    525–6
  multiple enzyme defect, 496–7
Congenital zonular pulverulent cataract
    chromosomal assignment, 436
    linkage, 437
Coombs test, *see* Blood grouping
    methods
Co-operativity, 569–70
Corticosteroids, mouse, in cultured
    cells, 66
Crigler-Najjar syndrome, 484, 538
Crohn's disease and sulphasalazine,
    455
Crossing-over, *see also* Linkage,
    Recombination
  unequal, 578–9
Cross-reacting antibodies, *see* Anti-
    bodies
Cross-reacting material (CRM), 24,
    31–2, 514
Crouzon's syndrome, 242
C3 proactivator, *see* Complement

DNA
amount per nucleus, 101–8
highly reiterated, 104–8
intermediate redundancy, 104–8
misreplication rate, 40
reassociation kinetics, 103–8
reiterated, 104–8, 740–1
repair, 121, 123–4
satellite, 105–8
unique, 104–8
DNA-RNA hybridization, 106–7, 477, 618–19
antenatal diagnosis, 519
Dominance, 477–84, 732–4
Dominant polycythaemias and abnormal haemoglobins, 478–9, 783–4
Dominantly inherited disorders, 477–84
Dopamine-β-hydroxylase activity, heritability, 453
Dosage compensation, 239
in *D. melanogaster*, 53, 124–5
Dosage effects, 112–23, 215, 509
autosomal aberration syndromes, 112–70
in *D. melanogaster*, 144
in Jimson weed (*Datura stramonium*), 125
trisomies, 112–14
Down's syndrome, see Trisomy, 21
Drosophila melanogaster
cross-veinless mutant, 39
dosage compensation, 124–5
homoeotic mutants, 54–5
xanthine dehydrogenase mutants, 52
Dubin-Johnson syndrome
gene frequency, 471
Duchenne muscular dystrophy, 486, 508
antenatal diagnosis, 444
gene frequency, 471
tissue necrosis, 496
X-inactivation, 132
Duffy blood groups
association with disease, 168
chromosomal assignment, 108–9, 358, 436
linkage, 437, 444
racial differences, 168, 172, 174
Duplication, see Gene duplication
Dwarfism on Krk, 174

Dysfibrinogenaemia, see Fibrinogen variants
Dystrophica myotonica, see Myotonic dystrophy
Dystrophy, see Duchenne muscular dystrophy; Myotonic dystrophy

E

Echo II sensitivity
chromosomal assignment, 436
EF_3 initiation factor, see Protein synthesis
Ehlers-Danlos syndrome, 479, 542
Elastase inhibition, 292
Elastin, 5
Electrophoresis, see also microcataphoresis
Electrophoresis, 187–8, 271–80, 475
agar immunoelectrophoresis, 271
ampholytes, 274
antibody antigen crossed, 278–9
cellulose acetate, 271, 274
gradient gel, 273–5
Gradipore, 273–5
immunodiffusion, 276–7
isoelectric focusing, 207, 274
ampholine, 274
polyacrylamide gel, 273
starch gel, 272–3
Elliptocytosis$_1$
chromosomal assignment, 436
linkage, 358, 437, 444
Ellis-van Creveld syndrome, 175
Embden-Meyerhof pathway, 20, 482–3
enzymes
activities, 482
chromosomal assignment, 444–5
rate-limiting, 482–3
Embryonic chains, see Haemoglobin
Embryonic haemoglobins, see Haemoglobin
Emphysema, 542
a_1-antitrypsin deficiency, 167, 293
Enolase$_1$, enolase$_2$
chromosomal assignment, 116, 436
inheritance, 206
linkage, 358
Enterokinase deficiency, 529
Enzyme cascade theory of clotting, 690–1

Gout, 470
Gouty arthritis, 539
  allelic to Lesch-Nyhan syndrome,
    540, 730
  phosphoribosyl   pyrophosphate
    synthetase, 32–3
Gower haemoglobins, see Haemoglobin
Gradient gel, see Electrophoresis
Gradipore, see Electrophoresis
Graft rejection, 373–4
Granulocyte-specific antigens, see Histo-
    compatibility antigens
Granulocyte transfusion, 400–1
Group specific component (Gc), 288–9,
    312–13
  AACE, 289
  alleles, 288–9
  association with disease, 167
  detection
    immunoelectrophoresis, 279, 288
  function, 289
  heterozygous advantage, 289
  linkage, 437
  tissue distribution, 288
Growth hormone
  rat, in cultured cells, 66
  regulation of biosynthesis, 55
Guanylate kinase
  chromosomal assignment, 436
Gunther's   disease,   see   Porphyria,
    Congenital erythropoietic

H

Haemagglutination   inhibition,   280,
    298–9
Haem: haem interaction, 569–70
Haem: oxygen reaction, 567–8, 586–8
  sigmoidal kinetics, 569–70
Haem pocket, see Haemoglobin structure
Haem stabilizing effect, 594–6
Haem structure, 564–5
Haem synthesis, 483–4, 498
  porphyria, 483–4, 498
Haemochromatosis, 488
Haemoglobin A, 16, 82, 520
  antigenic properties, 24–25
  electrophoretic pattern, 275
  function, 567–72
  goat, 601
  isoelectric point, 21

minor components, 576–7
  stability, 595–6
  structure of tetramer, 565–6
  thalassaemia, 610–12, 615–16, 618
Haemoglobin Abruzzo, 591
Haemoglobin Agenogi, 592
Haemoglobin Andrew Minneapolis, 591
Haemoglobin A2, 137, 573–6, 616
Haemoglobin A3, 576
Haemoglobin $a_2\epsilon_2$, see Haemoglobin
    Gower 2
Haemoglobin Bart's
  hydrops fetalis, 570
  oxygen affinity, 570
  trisomy 13 (D₁), 120
Haemoglobin Bethesda, 590
Haemoglobin Bibba (α-chain), unstable,
    479, 595
Haemoglobin Brighton, 590
Haemoglobin Bristol, 577
Haemoglobin C, 27, 610, 615–16
  antigenic properties, 25
  Harlem, 577, 584
  polymorphism, 581, 583–5
  sheep, 603
Haemoglobin chain α, 55–6, 563, 573,
    606
  gene, 82–3, 615
    chromosomal assignment, 436, 600
    duplication, 479, 602–3
    variants, 477, 577
Haemoglobin chain β, 55–6, 563, 573,
    605–6
  gene, 81–3, 615
    chromosomal assignment, 436, 600
    linkage, 81–3, 437
    rabbit, inserted sequence, 85–6
    variants, 477, 577
Haemoglobin chain γ, 563, 575
  gene, 81–3
    chromosomal assignment, 600
    duplication, 602
    linkage, 601–3
Haemoglobin chain δ, 563, 573, 605
  linkage, 437, 600–1
Haemoglobin chain ε (embryonic), 563,
    573
Haemoglobin chain ζ (embryonic), 563,
    573
Haemoglobin Chesapeake, 589–90, 592
Haemoglobin of Chironomus, 568

Immunoglobulins—*contd.*
  variable (V) region, 74–8, 638–40
    gene stitching, 78
    genetic control, 74–8
    sequence, 640
    somatic mutation, 79–80
    somatic recombination, 79–80
    variation, 310
Immunoproteins, *see* Immunoglobulins
Incompatibility, materno-foetal, *see* ABO blood groups, Rhesus blood groups
Inborn errors of metabolism
  classified, 521–43
  defined, 473
Inbreeding, 175, 729–30
Indophenol oxidase, *see* Superoxide dismutase
Initiation of protein synthesis, *see* Protein synthesis
Inosine triphosphatase
  chromosomal assignment, 436
Insulin, 9, 56, 59
  evolution, 179
  isoelectric point, 21
  species variation, 9
Interferon$_1$
  chromosomal assignment, 113, 436, 444
Interferon$_2$
  chromosomal assignment, 113, 436
Intermediary metabolism, 473
Interspecific cell hybridization, *see* Hybridization
Intraspecific cell hybridization, *see* Hybridization
Intrinsic clotting, *see* Clotting
Inv system, *see* Immunoglobulin polymorphism
Iodide transport defects, 490–1
Isochromosomes, *see* Chromosomes
Isocitrate dehydrogenase
  of *Drosophila melanogaster*, 125
  mitochondrial, 222
  soluble, 222
    chromosomal assignment, 436
    polymorphism, 222
    tissue distribution, 14
    in Trisomy 1 in mouse, 125
Isoenzyme, *see* Isozyme
Isolates, genetics of 172, 174–5

Isoleucine, 7
Isomaltase deficiency, *see* Congenital sucrose intolerance
Isovaleric acid sensitivity, 459–60
Isovaleric acidaemia (Isovaleryl-CoA dehydrogenase deficiency), 489, 527, 530
Isozymes (isoenzymes)
  defined, 13, 187
  detection, 187–8
  differentiation, 424
  genetically distinct, 14–17, 422–5
  nomenclature, 190
  origin, 14–19, 422–5
  structure, 14, 17–18, 53
  tissue distribution, 14, 424, 499–501

J

Jimson weed, *see* Dosage effects

K

Kangaroo tail tendon protein, 11
Kell blood groups
  blood transfusion, 328
  complexity, 359
  racial differences, 172, 174
Keratin, 5
  isoelectic point, 21
Ketoacidosis, infantile, 529
Ketone sensitivity, 459–60
Ketosteroids, mouse, in cultured cells, 66
Ketotic hyperglycinaemia, *see* Propionicacidaemia
Kidney graft, *see* Organ transplantation, Graft rejection
Klinefelter's syndrome, *see* X-chromosome
Km system, *see* Immunoglobulin polymorphism
Krabbe's disease (globoid leucodystrophy, galactocerebroside β-galactosidase deficiency), 534–5
  antenatal detection, 518
  metachromasia, 492
Kuru and Gc system, 167
Kynureninase deficiency, *see* Xanthurenic aciduria

Mutation—*contd.*
"frameshift", 42, 51, 579
frequency, 27
genetic code, 29–30
insertion, 42, 579
missense, 41
neutral, 158–9
nonsense, 41–2, 46
null, 51
point, 26–30, 40–5, 577–9
proportion detectable through electrophoresis, 27
protein structure, 51–2
rate, 580
regulation, 47
substitution, 26–7
structural gene, 474–7
termination 29–30, 579
trans-configuration, 47
transition, 26–7, 577
transversion, 26–7, 577
unequal crossing-over, 28
Myasthenia gravis and HLA, 404–5
Myeloma of mouse, 66
Myeloma proteins, 636, 657–9
Myelomatosis, 636
Myeloperoxidase deficiency, 500, 537
Myetiola, 60
Myocardial infarction
ABO blood groups, 168
Lp system, 167, 169
Myoglobin, 563–4, 569–70
isoelectric point, 21
Myosin, 5
isoelectric point, 21
Myotonic dystrophy
alcianophilia, 492
immunoglobulin defects, 679–80
linkage, 358
metachromasia, 492
thiopentane anaesthesia, 493

N

N-acetylhexosaminidase
deficiency of A component, *see* Tay-Sachs disease
deficiency of A and B components, *see* Sandhoff's disease
NADH oxidase deficiency, *see* Chronic granulomatous disease

Nail-patella syndrome
chromosomal assignment, 115, 436
linkage, 115, 358, 437
n-butyl mercaptan sensitivity, 459
Negroes
Duffy blood groups, 168, 172
cirrhosis of the liver, 168
V antigen, 172
Nephroblastoma (Wilms's tumour), 734
Nephrosis, congenital
antenatal diagnosis, 517–18
Neuroblastoma
of man, 66, 734
of mouse, 66
Neurofibroma, 426–7
Neutrality, evolutionary, 158–9, 178–9, 251
Niemann-Pick disease, 534–5
antenatal diagnosis, 518
Nine system, *see* Histocompatibility antigens
Nortriptyline metabolism, 457
Nuclear transplantation, 61
Nucleic acids, *see* DNA, RNA
Nucleolar organizers, 138–143, *see also* Chromosones, Trisomies
Xenopus laevis, 140
Nucleoproteins, *see also* Histones, 5
Nucleoside phosphorylase, 227, 424
chromosomal assignment, 119, 436
deficiency, 227–8
T-cell immune response, 227
electrophoretic patterns, 227
structure, 14
tissue distribution, 227
Numbers of genes in man, *see* Genes

O

Obesity, 470
Oculocutaneous albinism, *see* Albinism
Oestrogen, 58
One gene-one enzyme hypothesis, 39, 473
revision, 39–40
Operon theory, 44, 122, 126, 474–5, 485, 605–6, 715–16
Organ transplantation, 398–400, *see also* Histocompatibility antigens
ABO incompatibility, 398
HLA incompatibility, 398–400
survival time, 398–400

X

Xanthine dehydrogenase in *Drosophila melanogaster*
cis-regulation, 52
Xanthinuria, hereditary (xanthine oxidase deficiency), 540–1
hypouricaemia, 488
Xanthurenic aciduria, 494, 529
vitamin B_6 response, 495
Xavante Indians, 174–5
inbreeding, 174–5
population structure, 175
X-chromosome, 128–35, 425–8, 508–9
aneuploidy, 128–35
deletion, 431
of *Drosophila melanogaster*, 143
inactivation, 16–17, 61–2, 129–32, 215–16, 428–32
cellular mosaicism, 136
differentiation, 215, 431–2
dosage compensation, 129–135, 215
heterozygote detection, 511–12, 514
hybrid cells, 429
Lyon hypothesis, 428–32
origin of cells, 425
origin of tumours, 425–7
randomness, 132
reversibility, 429–30
Xg locus, 133
late replication, 97–8
recessive lethals of *Drosophila melanogaster*, 103
XO (Turner's syndrome), 136, 354–5
XX males, 134
XXX, 133, 136
XXY (Klinefelter's syndrome), 130, 135, 354–5
XXXY, 355
XYY, 135
Xeroderma pigmentosum, 543
antenatal diagnosis, 518

chromosomal assignment, 436
DNA repair, 124, 735
genetic heterogeneity, 435
Xg blood group, 354–5, *see also* X-linkage
antigens, 354–5
gene frequencies, 354, 430
materno-foetal incompatibility, 328
primates, 362
sex chromosome aneuploidy, 354
translocation, 430
X-linkage, 302, 345, 356–7
X-irradiation and DNA repair, 123
X-linkage, 109–110, 158, 160, 297–8, 302, 443
antenatal diagnosis, 444
Xm a_2-macroglobulin system, 302, 312–13
immunodiffusion, 302
polymorphism, 302
relationships to a_2-pregnancy associated globulin, 324
X-linkage, 302, 356–7
Xylitol dehydrogenase deficiency, *see* Pentosuria
Xylosidosis (β-xylosidase deficiency), 537

Y

Y-chromosome, 133
DNA, 107
extra, *see* X-chromosome
H–Y antigen, 133–4
satellite DNA, 145
sexual development, 133
variability, 99

Z

Zonular pulverulent cataract
chromosomal assignment, 436
zymogen activation, 40

Author Index

(Numbers in italics indicate the pages on which the references are listed in full.)

A

Aarskog, D., 293, *315*
Abbasi, K., *412*
Abbie, A. A., 170, *179*
Abbot, J., *89*
Abel, C. A., 668, *681, 682*
Abele, D. C., 655, *680*
Abeles, R. H., *553*
Abell, C. W., 204, *253*
Abelson, L. D., *414*
Abelson, N. M., 332, *369*
Aber, V. R., *258*
Abeyounis, C. J., *408*
Abildgaard, C. F., 699, *723*
Abildgaard, O., *726*
Abrahamson, S., 106, *146*
Abramson, R. K., 602, 609, *622*
Abrell, E., *724*
Adair, G. S., 570, *622*
Adam, A., *259, 265, 267, 321,* 356, *363, 463, 553, 557*
Adamek, R., *261*
Adams, H. R., *625*
Adams, W. S., *559*
Adamson, J., *629*
Adelson, E., 699, *723*
Adhya, S., *91*
Adinolfi, M., 332, 333, *363, 368*
Adornato, B., *554*
Aebi, H., *268*, 543, *544*
Afeltra, P., *263*
Affara N., *89, 625, 628*
Agar, J. A. M., *627*
Agliozzo, C. M., *320*
Agostini, R. M., *324*
Agostino, R., *180*
Aguilu, L. A., *365*
Ahern, E. J., 602, *622*
Ahmed, S. I., 81, *87*
Ahrons, S., 381, 391, 401, *408*

Ahuja, Y. R., *255*
Air, G. M., *87*
Aird, I., 165, *179*
Aisen, P., 288, *318*
Aitken, D. A., 108, *148*
Ajmar, F., 237, *253, 256*
Akedo, H., 195, *253, 263*
Akeroyd, J. H., 334, *365, 370*
Akira, H., *630*
Akman, N., *723*
Alberman, E. D., *147*
Akrivakis, A., *629*
Albers, J. J., 300, *315*
Albert, E. D., 381, 382, 398, 408, *415*
Albert, H., *322*
Albert, J., *550*
Albertini, R. J., *680*
Alberts, B. M., 58, *92,* 127, *154*
Albright, F., 493, *544*
Aldrich, R. A., 673, *680*
Aledort, L. M., *315*
Alepa, F. P., 280, *315*
Alfred, B. M., 236, *253*
Alhara, E., *464*
Allan, N. C., 585, *622*
Allan, T. M., *179*
Allard, D., *153*
Alldredyce, P. W., 117, *146, 464*
Allen, C. R., 273, 274, *315*
Allen, D. M., *553*
Allen, D. W., 576, 603, *622*
Allen, F. H., 296, *315, 317,* 326, 352, 359, *363, 364, 366,* 397, *408, 412, 684*
Allen, R. C., 291, *315*
Allfrey, V. G., 90
Allison, A. C., *151,* 161, 164, 165, *179,* 281, 298, *315,* 573, *624,* 655, *682*
Aloni, Y., 85, 87
Aloysia, M., 336, 337, *363*
Alper, C. A., 279, 295, 296, 297, *315, 316, 319, 321, 550, 684*

de Bruyn, C. H. M. M., 131, *147*
De Capoa, A., 142, *148*
De Cataldo, F., *724*
De Chatelet, L. R., *681*
De Coteau, W. E., 668, *681*
Defendi, V., 399, *412*
Degani-Bernard, O., *410*
De Garay, A. L., *255*
Deggeller, K., 691, *723*
Deggins, B. A., *90*
Degos, L., 381, *409, 411, 414*
de Greve, W. B., *462*
de Groot, W. P., *368, 371*
de Grouchy, J., *148, 149, 369,* 714, *725*
Dehay, C., *411*
Deicher, H., 282, *316*
Deininger, P. L., 106, *152*
Deisseroth, A., 520, *547*, 600, *624*
de Jong, W. W. W., 578, 596, *626*
de Josselin de Jong, J., *548*
de la Chapelle, A., 96, 135, *147, 148, 465*
de la Cruz, F. F., 109, *148*
de la Lande, I. A., 201, *255*
de Lange, G., *321*
Delarue, F., *369*
Dell'Acqua, G., *463, 548*
del Senno, L., *88, 623*
Delvin, E., *546, 556*
de Marchi, M., *415*
De Mars, R. I., 430, *463*, 476, 537, *552, 556*
DeMarsh, Q. B., *254, 544*
de Maugre, F., *546*
de Moor, P., 310, *320*
Denborough, M. A., 340, *364*
Denney, R. M., 109, *148, 462*
Denson, K. W. E., 691, 693, 694, 698, 699, 700, 701, 703, 706, 707, 709, 716, 717, 718, *723, 724, 725*
Dent, C. E., 488, *547*
Denton, R. L., 331, *364*
De Peretti, E., *556*
De Re, J., *463*
DerKaloustian, V. M., *464, 553*
Derom, F., *320*
de Romenf, J., *152*
Dern, R. J., *88*, 213, 238, *257*
De Robes-Sternberg, S., *323*
Dersjant, H., *408*
Desai, P. R., *418*

Descamps, B., 399, *408, 411*
Descos, F., *315*
Deshpande, C. K., *363*
Despont, J. P. J., 668, *681*
Detter, J. C., 196, 210, 211, 212, *257*, 510, *547*
de Turi, N., *152*
Deutsch, H. F., 203, *258, 262, 263, 464*, 662, *681*
Dev, V. G., *151*
de Vaan, G. A. M., *268*
Devadetta, S., 224, *257*
de Vaal, O. M., 675, *681*
Devictor, M., *153*
de Virgilis, S., *153*
Devivo, D., *554*
de Vries, A., *36*
de Weerd-Kastelein, E. A., 43, 44, *88*, 435, *462, 552*
de Wit-Verbeck, H. A., *463*
Deys, B. P., *261*, 429, *462*
Dhar, R., *87*
Dherte, P., 579, *624*
Diamond, L. K., *153, 259*, 331, 333, *364*
Diamond, R., *153*
Dick, H. M., *146, 409*
Dickerson, R. E., *35*, 564, *624, 626*
Dickler, H. B., *408*
Dickman, S. R., 194, *257*
Dierich, M. P., *316*
Dietz, A. A., 246, *257*
Di George, A. M., *553*
Diggs, L. W., 585, *626*
Dike, G. W. R., *725*
Di Mauro, S., 486, *544, 547*
Dintzis, H. M., 606, 607, *624, 627*
Dissing, J., 196, *257*, 449, *466*
Disthasongchan, P., *630*
Dittes, H., 139, *147, 148, 152*
Dittrich, A., *152*
Dixon, G. H., 283, 284, *316, 317, 320*, 323
Dobzhansky, Th., *481, 463*
Dodd, B. E., *268*
Dodd, M. C., 351, *363, 364*
Doenicke, A., 246, *257*
Doherty, P. C., 407, *411*
Dohmann, U., *318, 550*
Doi, R. H., 63, *88*
Dolivo, P., *419*
Domaniewska-Sobczak, K., *264*

Gandini, E., 425, 432, *463*, 512, *548*
Ganesan, J., *558, 626, 630, 742*
Gangaharam, P. R., *257*
Gannon, F., 58, 64, *89*
Ganschow, R., 57, *88, 92*
Gant, N., *149, 259*
Gardas, A., 346, *365*
Gardner, A. L., 101, *149*
Gardner, B., *366*
Gardner, F., *151*
Gardner, R. L., *91*, 431, *463*
Gardner-Medwin, D., 471, *548, 550*
Garel, J. P., 128, *149*
Garen, A., 607, *624*
Garnjobst, L., *461*
Garoff, H., *317, 318*
Garrick, M. D., *553*
Garrod, A. E., 39, *88*, 470, *548*
Garoff, H., 301, *318*
Garson, O. M., *261*
Garthwaite, E., *268*
Gartler, S. M., 129, 130, *149, 151, 153,*
    216, *258, 259, 262,* 425, 426, 427, 431,
    453, *461, 462, 463, 464,* 514, *548, 549*
Gartner, L. M., *544*
Garvin, A. J., *556*
Gatti, R. A., *419, 684*
Gaull, G. E., 528, *548*
Gautier, T., *550*
Gavendo, S., *558*
Gavin, J., *152,* 362, *365, 366, 368, 370,*
    371, 466
Gedde-Dahl, T. Jr., 296, *318, 416, 463,*
    465, 683
Geerdink, R. A., 234, *259, 324*
Geerthson, J. M. P., *149*
Geerts, S. J., 404, *412*
Geffen, L. B., 454, *465*
Geha, R. S., 676, *682*
Gehring, D., *259*
Gehring-Mueller, G., *623*
Geier, M. R., *151*
Geigerova, H., *182*
Gelb, A. G., *363, 368, 464*
Gelbart, W. M., 52, *87, 88*
Gelber, R., 456, *463*
Gelboin, H. V., *256, 461*
Gelehrter, T. D., 69, *92, 555*
Gelfand, E. W., *684*
Geller, L. N., *151*
Gelpi, A. P., 168, *180*

Gelsthorpe, K., 381, *411, 412*
Genest, K., 245, *261*
Gengozian, N., 361, *365*
Gentz, J., 527, *548*
George, D. L., *149*
George, P., 589, *624*
Georgiev, G. P., 64, *88*
Geraedts, J. P. M., 443, *463*
Gerald, P. S., 122, *147, 153,* 436, 444,
    461, 551, 580, 586, 588, 589, *624*
Gerbeck, G. M., 712, *724*
German, J. L., 108, *149,* 397, *412*
Gerner, E. M., *88*
Gerner, R. E., *553*
Gerok, W., *149, 559*
Gerritsen, T., 529, *548*
Gershowitz, H., 171, *182, 254,* 324
Gerson, B., *555*
Gerstley, B. J. S., *316*
Gertner, M., 492, 539, *548*
Geserick, G., 295, *322*
Gewurz, A., *415*
Gewurz, H., *415, 682*
Geyer, V. B., *364*
Gianelli, F., 115, *151, 465*
Gibbs, M. B., 334, 348, 349, *365, 370*
Giblett, E. R., 54, *88, 180,* 187, 188,
    190, 191, 192, 193, 196, 206, 213, 216,
    217, 227, 228, 235, 236, 237, 239, 240,
    249, 250, *256, 257, 259, 260, 266,* 272,
    280, 281, 285, 286, 287, 288, 289, 304,
    318, 323, 349, 351, 358, 362, *365, 366,*
    423, *461,* 540, *547, 548, 684*
Gibson, D. M., *684*
Gibson, Q. H., *35,* 505, *548,* 566, *624*
Giddings, J. C., *723, 726*
Giers, D., *149*
Gigliani, F., *263*
Giglioni, B., *91, 627, 630*
Gilbert, F., 477, *548*
Giles, C. M., 360, *365, 370,* 398, *414*
Giles, N. H., 81, *87*
Giliberti, P., *554*
Gill, G. J., *680*
Gillis, W. T., 460, *463*
Gillnas, T., *316*
Gilmour, R. S., 63, *91,* 599, *624*
Gilon, E., *557*
Ginsburg, V., *366, 370*
Githens, J. G., *680*
Gitlin, D., 651, 652, 671, 672, *682*

MAN—HH*

820 AUTHOR INDEX